Introduction to

Probability and Statistics for Science, Engineering, and Finance

Walter A. Rosenkrantz

Department of Mathematics and Statistics
University of Massachusetts at Amherst

CRC Press
Taylor & Francis Group
Boca Raton London New York

CRC Press is an imprint of the
Taylor & Francis Group, an **informa** business

A CHAPMAN & HALL BOOK

CRC Press
Taylor & Francis Group
6000 Broken Sound Parkway NW, Suite 300
Boca Raton, FL 33487-2742

First issued in paperback 2022

© 2009 by Taylor & Francis Group, LLC

No claim to original U.S. Government works
CRC Press is an imprint of Taylor & Francis Group, an Informa business

ISBN 13: 978-1-03-247778-7 (pbk)
ISBN 13: 978-1-58488-812-3 (hbk)
ISBN 13: 978-0-429-14204-8 (ebk)

DOI: 10.1201/9781584888130

Publisher's Note
The publisher has gone to great lengths to ensure the quality of this reprint but points out that some imperfections in the original copies may be apparent.

Library of Congress Cataloging-in-Publication Data

Rosenkrantz, Walter A.
   Introduction to probability and statistics for science, engineering, and finance / Walter A. Rosenkrantz.
      p. cm.
   Includes bibliographical references and index.
   ISBN 978-1-58488-812-3 (alk. paper)
   1. Probabilities. 2. Mathematical statistics. I. Title.

QA273.R765 2008
519.5--dc22   2008013044

Visit the Taylor & Francis Web site at
http://www.taylorandfrancis.com

and the CRC Press Web site at
http://www.crcpress.com

Preface

Student Audience and Prerequisites

This book is written for undergraduate students majoring in engineering, computer science, mathematics, economics, and finance who are required, or urged, to take a one- or two-semester course in probability and statistics to satisfy major or distributional requirements. The mathematical prerequisites are two semesters of single variable calculus. Although some multivariable calculus is used in Chapter 5, it can be safely omitted without destroying the continuity of the text. Indeed, the topics have been arranged so that the instructor can always adjust the mathematical sophistication to a level the students can feel comfortable with. Chapters and sections marked with an asterisk (*) are optional; they contain material of independent interest, but are not essential for a first course.

Objectives

My primary goal in writing this book is to integrate into the traditional one- or two-term statistics course some of the more interesting and widely used concepts in financial engineering. For example, the *volatility* of a stock is the standard deviation of its returns; *value at risk* (VaR) is essentially a confidence interval; a stock's β is the slope of the regression line obtained when one performs a linear regression of the stock's returns against the returns of the S&P500 index (the S&P500 index, itself, is used as a proxy for the *market portfolio*). The binomial distribution, it is worth noting, plays a fundamental role in the Cox-Ross-Rubinstein (CRR) model, also called the *binomial lattice model*, of stock price fluctuations. A passage to the limit via the central limit theorem yields the lognormal distribution for stock prices as well as the famous Black-Scholes option pricing formula.

Organization of the Book

Beginning with the first chapter on data analysis, I introduce the basic concepts a student needs in order to understand and create the tables and graphs produced by standard statistical software packages such as MINITAB, SAS, and JMP. The data sets themselves have been carefully selected to illustrate the role and scope of statistics in science, engineering, public health, and finance. The text then takes students through the traditional topics of a first course in statistics. Novel features include: (i) applications of traditional statistical concepts and methods to the analysis and interpretation of financial data; (ii) an introduction to modern portfolio theory; (iii) mean-standard deviation ($r - \sigma$) diagram of a collection of portfolios; and (iv) computing a stock's β via simple linear regression.

For the benefit of instructors using this text, I have included technical, even tedious details, needed to derive various theorems, including the famous Black-Scholes option pricing formula, because, in my opinion, one cannot explain this formula to students without a thorough understanding of the fundamental concepts, methods, and theorems used to derive them. These computational details, which can safely be omitted on a first reading, are contained in a section titled "Mathematical Details and Derivations," put at the end of most chapters.

Examples

The text introduces the student to the most important concepts by using suitably chosen examples of independent interest. Applications to engineering (queueing theory, reliability theory, acceptance sampling), computer performance analysis, public health, and finance are included as soon as the statistical concepts have been developed. Numerous examples, using both statistical software packages and scientific calculators, help to reinforce the student's mastery of the basic concepts.

Problems

The problems (there are 675 of them), which range from the routine to the challenging, help students master the basic concepts and give them a glimpse of the vast range of applications to a variety of disciplines.

Supplements

An Instructor's Solutions Manual containing carefully worked out solutions to all 675 problems is available to adopters of the textbook. All data sets, including those used in the worked out examples, are available in a CD-ROM to users of this textbook.

Contacting the Author

In spite of the copy editor's and my best efforts to eliminate all errors and typos it is an almost impossible task to eliminate them all. I, therefore, encourage all users of this text to send their comments and criticisms to me at *rkrantz@math.umass.edu*.

Acknowledgments

The publication of a statistics textbook, containing many tables and graphs, is not possible without the cooperation of a large number of highly talented individuals, so it is a great pleasure for me to have this opportunity of thanking them. First, I want to thank my editors at Chapman-Hall: Sunil Nair for initiating this project, and Theresa Delforn and Michele Dimont for guiding and prodding me to a successful conclusion. Shashi Kumar's technical advice with LaTeX is deeply appreciated and Theresa Gandolph, of the Instructional Technology Lab at George Washington University, gave me valuable assistance with the Freehand graphics software package. Professors Alan Durfee of Mount Holyoke College and Michael Sullivan of the University of Massachusetts (Amherst) provided me with some insights and ideas on financial engineering that were very useful to me in the writing of this book.

I am also grateful to the American Association for the Advancement of Science, the *American Journal of Clinical Nutrition*, the *American Journal of Epidemiology*, the American Statistical Association, the Biometrika Trustees, Cambridge University Press, Elsevier Science, Iowa State University Press, Richard D. Irwin, McGraw-Hill, Oxford University Press, Prentice-Hall, Routledge, Chapman & Hall, the Royal Society of Chemistry, *Journal of Chemical Education*, and John Wiley & Sons for permission to use copyrighted material. I have made every effort to secure permission from the original copyright holders for each data set, and would be grateful to my readers for calling my attention to any omissions so they can be corrected by the publisher.

Finally, I dedicate this book to my wife, Linda, for her patient support while I was almost always busy writing it.

CD materials can be found at www.routledge.com/9781584888123

Contents

1	**Data Analysis**	**1**
1.1	Orientation	1
1.2	The Role and Scope of Statistics in Science and Engineering	2
1.3	Types of Data: Examples from Engineering, Public Health, and Finance	5
	1.3.1 Univariate Data	5
	1.3.2 Multivariate Data	7
	1.3.3 Financial Data: Stock Market Prices and Their Time Series	9
	1.3.4 Stock Market Returns: Definition and Examples	13
1.4	The Frequency Distribution of a Variable Defined on a Population	17
	1.4.1 Organizing the Data	17
	1.4.2 Graphical Displays	18
	1.4.3 Histograms	22
1.5	Quantiles of a Distribution	26
	1.5.1 The Median	26
	1.5.2 Quantiles of the Empirical Distribution Function	27
1.6	Measures of Location (Central Value) and Variability	32
	1.6.1 The Sample Mean	32
	1.6.2 Sample Standard Deviation: A Measure of Risk	33
	1.6.3 Mean-Standard Deviation Diagram of a Portfolio	36
	1.6.4 Linear Transformations of Data	37
1.7	Covariance, Correlation, and Regression: Computing a Stock's Beta	38
	1.7.1 Fitting a Straight Line to Bivariate Data	40
1.8	Mathematical Details and Derivations	43
1.9	Chapter Summary	44
1.10	Problems	44
1.11	Large Data Sets	65
1.12	To Probe Further	70
2	**Probability Theory**	**71**
2.1	Orientation	71
2.2	Sample Space, Events, Axioms of Probability Theory	72
	2.2.1 Probability Measures	78
2.3	Mathematical Models of Random Sampling	84
	2.3.1 Multinomial Coefficients	93
2.4	Conditional Probability and Bayes' Theorem	94
	2.4.1 Conditional Probability	94
	2.4.2 Bayes' Theorem	97
	2.4.3 Independence	99
2.5	The Binomial Theorem	100
2.6	Chapter Summary	101
2.7	Problems	101
2.8	To Probe Further	111

3 Discrete Random Variables and Their Distribution Functions   **113**
  3.1 Orientation 113
  3.2 Discrete Random Variables 114
      3.2.1 Functions of a Random Variable 120
  3.3 Expected Value and Variance of a Random Variable 121
      3.3.1 Moments of a Random Variable 125
      3.3.2 Variance of a Random Variable 128
      3.3.3 Chebyshev's Inequality 130
  3.4 The Hypergeometric Distribution 130
  3.5 The Binomial Distribution 134
      3.5.1 A Coin Tossing Model for Stock Market Returns 140
  3.6 The Poisson Distribution 144
  3.7 Moment Generating Function: Discrete Random Variables 146
  3.8 Mathematical Details and Derivations 148
  3.9 Chapter Summary 150
  3.10 Problems 151
  3.11 To Probe Further 160

4 Continuous Random Variables and Their Distribution Functions   **161**
  4.1 Orientation 161
  4.2 Random Variables with Continuous Distribution Functions: Definition and
      Examples 162
  4.3 Expected Value, Moments, and Variance of a Continuous Random Variable 167
  4.4 Moment Generating Function: Continuous Random Variables 171
  4.5 The Normal Distribution: Definition and Basic Properties 172
  4.6 The Lognormal Distribution: A Model for the Distribution of Stock Prices 177
  4.7 The Normal Approximation to the Binomial Distribution 179
      4.7.1 Distribution of the Sample Proportion $\hat{p}$ 185
  4.8 Other Important Continuous Distributions 185
      4.8.1 The Gamma and Chi-Square Distributions 185
      4.8.2 The Weibull Distribution 188
      4.8.3 The Beta Distribution 188
  4.9 Functions of a Random Variable 189
  4.10 Mathematical Details and Derivations 191
  4.11 Chapter Summary 192
  4.12 Problems 192
  4.13 To Probe Further 202

5 Multivariate Probability Distributions   **205**
  5.1 Orientation 205
  5.2 The Joint Distribution Function: Discrete Random Variables 206
      5.2.1 Independent Random Variables 211
  5.3 The Multinomial Distribution 212
  5.4 Mean and Variance of a Sum of Random Variables 213
      5.4.1 The Law of Large Numbers for Sums of Independent and Identically
           Distributed (iid) Random Variables 220
      5.4.2 The Central Limit Theorem 222
  5.5 Why Stock Prices Have a Lognormal Distribution: An Application of the
      Central Limit Theorem 224
      5.5.1 The Binomial Lattice Model as an Approximation to a Continuous
           Time Model for Stock Market Prices 227

5.6	Modern Portfolio Theory	230
	5.6.1 Mean-Variance Analysis of a Portfolio	230
5.7	Risk Free and Risky Investing	232
	5.7.1 Present Value Analysis of Risk Free and Risky Returns	232
	5.7.2 Present Value Analysis of Deterministic and Random Cash Flows	235
5.8	Theory of Single and Multi-Period Binomial Options	237
	5.8.1 Black-Scholes Option Pricing Formula: Binomial Lattice Model	237
5.9	Black-Scholes Formula for Multi-Period Binomial Options	240
	5.9.1 Black-Scholes Pricing Formula for Stock Prices Governed by a Lognormal Distribution	242
5.10	The Poisson Process	243
	5.10.1 The Poisson Process and the Gamma Distribution	246
5.11	Applications of Bernoulli Random Variables to Reliability Theory	248
5.12	The Joint Distribution Function: Continuous Random Variables	251
	5.12.1 Functions of Random Vectors	254
	5.12.2 Conditional Distributions and Conditional Expectations: Continuous Case	256
	5.12.3 The Bivariate Normal Distribution	257
5.13	Mathematical Details and Derivations	258
5.14	Chapter Summary	263
5.15	Problems	263
5.16	To Probe Further	275

6 Sampling Distribution Theory — **277**

6.1	Orientation	277
6.2	Sampling from a Normal Distribution	277
6.3	The Distribution of the Sample Variance	282
	6.3.1 Student's t Distribution	284
	6.3.2 The F Distribution	285
6.4	Mathematical Details and Derivations	286
6.5	Chapter Summary	287
6.6	Problems	287
6.7	To Probe Further	290

7 Point and Interval Estimation — **291**

7.1	Orientation	291
7.2	Estimating Population Parameters: Methods and Examples	292
	7.2.1 Some Properties of Estimators: Bias, Variance, and Consistency	294
7.3	Confidence Intervals for the Mean and Variance	296
	7.3.1 Confidence Intervals for the Mean of a Normal Distribution: Variance Unknown	299
	7.3.2 Confidence Intervals for the Mean of an Arbitrary Distribution	300
	7.3.3 Confidence Intervals for the Variance of a Normal Distribution	302
	7.3.4 Value at Risk (VaR): An Application of Confidence Intervals to Risk Management	303
7.4	Point and Interval Estimation for the Difference of Two Means	304
	7.4.1 Paired Samples	305
7.5	Point and Interval Estimation for a Population Proportion	307
	7.5.1 Confidence Intervals for $p_1 - p_2$	309
7.6	Some Methods of Estimation	310
	7.6.1 Method of Moments	310

	7.6.2	Maximum Likelihood Estimators	312
7.7		Chapter Summary	316
7.8		Problems	316
7.9		To Probe Further	324

8 Hypothesis Testing — **325**

8.1		Orientation	325
8.2		Tests of Statistical Hypotheses: Basic Concepts and Examples	326
	8.2.1	Significance Testing	336
	8.2.2	Power Function and Sample Size	338
	8.2.3	Large Sample Tests Concerning the Mean of an Arbitrary Distribution	339
	8.2.4	Tests Concerning the Mean of a Distribution with Unknown Variance	340
8.3		Comparing Two Populations	344
	8.3.1	The Wilcoxon Rank Sum Test for Two Independent Samples	347
	8.3.2	A Test of the Equality of Two Variances	350
8.4		Normal Probability Plots	351
8.5		Tests Concerning the Parameter p of a Binomial Distribution	355
	8.5.1	Tests of Hypotheses Concerning Two Binomial Distributions: Large Sample Size	359
8.6		Chapter Summary	360
8.7		Problems	361
8.8		To Probe Further	372

9 Statistical Analysis of Categorical Data — **373**

9.1		Orientation	373
9.2		Chi-Square Tests	373
	9.2.1	Chi-Square Tests When the Cell Probabilities Are Not Completely Specified	376
9.3		Contingency Tables	377
9.4		Chapter Summary	383
9.5		Problems	383
9.6		To Probe Further	388

10 Linear Regression and Correlation — **389**

10.1		Orientation	389
10.2		Method of Least Squares	390
	10.2.1	Fitting a Straight Line via Ordinary Least Squares	392
10.3		The Simple Linear Regression Model	398
	10.3.1	The Sampling Distribution of $\hat{\beta}_1$, $\hat{\beta}_0$, SSE, and SSR	399
	10.3.2	Tests of Hypotheses Concerning the Regression Parameters	402
	10.3.3	Confidence Intervals and Prediction Intervals	403
	10.3.4	Displaying the Output of a Regression Analysis in an ANOVA Table	406
	10.3.5	Curvilinear Regression	408
10.4		Model Checking	411
10.5		Correlation Analysis	416
	10.5.1	Computing the Market Risk of a Stock	417
	10.5.2	The Shapiro–Wilk Test for Normality	421
10.6		Mathematical Details and Derivations	422
10.7		Chapter Summary	426
10.8		Problems	426
10.9		Large Data Sets	437

10.10 To Probe Further 439

11 Multiple Linear Regression   441

11.1 Orientation 441
11.2 The Matrix Approach to Simple Linear Regression 442
    11.2.1 Sampling Distribution of the Least Squares Estimators 447
    11.2.2 Geometric Interpretation of the Least Squares Solution 449
11.3 The Matrix Approach to Multiple Linear Regression 450
    11.3.1 Normal Equations, Fitted Values, and ANOVA Table for the Multiple Linear Regression Model 454
    11.3.2 Testing Hypotheses about the Regression Model 457
    11.3.3 Model Checking 460
    11.3.4 Confidence Intervals and Prediction Intervals in Multiple Linear Regression 462
11.4 Mathematical Details and Derivations 464
11.5 Chapter Summary 464
11.6 Problems 465
11.7 To Probe Further 468

12 Single Factor Experiments: Analysis of Variance   469

12.1 Orientation 469
12.2 The Single Factor ANOVA Model 469
    12.2.1 Estimating the ANOVA Model Parameters 473
    12.2.2 Testing Hypotheses about the Parameters 475
    12.2.3 Model Checking via Residual Plots 478
    12.2.4 Unequal Sample Sizes 480
12.3 Confidence Intervals for the Treatment Means; Contrasts 482
    12.3.1 Multiple Comparisons of Treatment Means 485
12.4 Random Effects Model 487
12.5 Mathematical Derivations and Details 489
12.6 Chapter Summary 490
12.7 Problems 490
12.8 To Probe Further 496

13 Design and Analysis of Multi-Factor Experiments   497

13.1 Orientation 497
13.2 Randomized Complete Block Designs 498
    13.2.1 Confidence Intervals and Multiple Comparison Procedures 507
    13.2.2 Model Checking via Residual Plots 508
13.3 Two Factor Experiments with $n > 1$ Observations per Cell 508
    13.3.1 Confidence Intervals and Multiple Comparisons 520
13.4 2^k Factorial Designs 522
13.5 Chapter Summary 540
13.6 Problems 540
13.7 To Probe Further 548

14 Statistical Quality Control   551

14.1 Orientation 551
14.2 $\overline{x}$ and R Control Charts 552
    14.2.1 Detecting a Shift in the Process Mean 557
14.3 p Charts and c Charts 559

14.4 Chapter Summary 562
14.5 Problems 562
14.6 To Probe Further 565

A Tables   **567**

A.1 Cumulative Binomial Distribution 568
A.2 Cumulative Poisson Distribution 570
A.3 Standard Normal Probabilities 572
A.4 Critical Values $t_\nu(\alpha)$ of the t Distribution 574
A.5 Quantiles $Q_\nu(p) = \chi_\nu^2(1-p)$ of the χ^2 Distribution 575
A.6 Critical Values of the $F_{\nu_1,\nu_2}(\alpha)$ Distribution 576
A.7 Critical Values of the Studentized Range $q(\alpha; n, \nu)$ 580
A.8 Factors for Estimating σ, $\overline{s}$, or $\overline{\sigma}_{RMS}$ and σ_R from $\overline{R}$ 584
A.9 Factors for Determining from $\overline{R}$ the Three-Sigma Control Limits for $\overline{X}$ and
R Charts 585
A.10 Factors for Determining from σ the Three-Sigma Control Limits for $\overline{X}$, R,
and s or $\overline{\sigma}_{RMS}$ Charts 586

Answers to Selected Odd-Numbered Problems   **589**

1 Data Analysis   **591**

2 Probability Theory   **597**

3 Discrete Random Variables and Their Distribution Functions   **601**

4 Continuous Random Variables and Their Distribution Functions   **609**

5 Multivariate Probability Distributions   **617**

6 Sampling Distribution Theory   **623**

7 Point and Interval Estimation   **625**

8 Hypothesis Testing   **631**

9 Statistical Analysis of Categorical Data   **637**

10 Linear Regression and Correlation   **641**

11 Multiple Linear Regression   **645**

12 Single Factor Experiments: Analysis of Variance   **647**

13 Design and Analysis of Multi-Factor Experiments   **653**

14 Statistical Quality Control   **659**

Index   **661**

Chapter 1

Data Analysis

Information, that is imperfectly acquired, is generally as imperfectly retained; and a man who has carefully investigated a printed table, finds, when done, that he has only a very faint and partial idea of what he has read; and that like a figure imprinted on sand, is soon totally erased and defaced.

William Playfair (1786), *Commercial and Political Atlas*

1.1 Orientation

Statistics is the science and art of collecting, displaying and interpreting data in order to test theories and make inferences concerning all kinds of phenomena. In brief, its main goal is to transform data into knowledge. Scientists, engineers, and economists use statistics to summarize and interpret the data before they can make inferences from it. Statistical software packages such as MINITAB[1] and SAS[2] are helpful because they produce graphical displays that are very effective for visualizing and interpreting the data. It is therefore of no small importance that we be able to understand and interpret the current methods of graphically displaying statistical data. However, it is worth reminding ourselves now and then, that the goal of a statistical analysis is primarily insight and understanding, not mindless, formal calculations using statistical software packages.

In this chapter we develop the concepts and techniques of data analysis in the context of suitable examples taken from science, engineering, and finance. In detail, the chapter is organized as follows:

Organization of Chapter

1. Section 1.2: The Role and Scope of Statistics in Science, and Engineering

2. Section 1.3: Types of Data: Examples from Engineering, Public Health, and Finance

3. Section 1.3.1: Univariate Data

4. Section 1.3.2: Multivariate Data

5. Section 1.3.3: Financial Data, Stock Market Prices, and Their Time Series

6. Section 1.3.4: Stock Market Returns: Definition, Examples

7. Section 1.4: The Frequency Distribution of a Variable Defined on a Population

8. Section 1.5: Quantiles of a Distribution: Medians, Quartiles, Box Plots.

---

[1] MINITAB is a registered trademark of Minitab, Inc.
[2] SAS is a trademark of the SAS Institute, Inc., Cary, North Carolina, USA.

9. Section 1.6: Measures of Location and Variability: Sample Mean and Sample Variance

10. Section 1.6.2: Standard Deviation: A Measure of Risk

11. Section 1.6.3: Mean-Standard Deviation Diagram of a Portfolio

12. Section 1.7: Covariance, Correlation, and Regression: Computing a Stock's Beta

13. Section 1.8: Mathematical Details and Derivations

14. Section 1.9: Chapter Summary

1.2   The Role and Scope of Statistics in Science and Engineering

A scientific theory, according to the philosopher of science Karl Popper, is characterized by "its falsifiability, or refutability, or testability" (Karl R. Popper, 1965), *Conjectures and Refutations: The Growth of Scientific Knowledge*, Harper Torchbooks). In practical terms this means a scientist tests a theory by performing an experiment in which he observes and records the values of one or more variables in which he is interested. The variable of interest for a chemist might be the atomic weight of an element; for an engineer it might be the lifetime of a battery for a laptop computer; for an investor it might be the weekly or annual rate of return of an investment; for the Bureau of the Census it is the population of each state so political power can be reapportioned in accordance with the U.S. Constitution. The next three examples illustrate how we use statistical models to validate scientific theories, interpret data, and grapple with the variability inherent in all experiments.

Example 1.1

Einstein's special relativity theory, published in 1905, asserts that the speed of light is the same constant value, independent of position, inertial frame of reference, direction, or time. The physicists A. A. Michelson and E. W. Morley (1881 and 1887) and D. Miller (1924) performed a series of experiments to determine the velocity of light. They obtained contradictory results. Michelson and Morley could not detect significant differences in the velocity of light, thus validating relativity theory, while Miller's experiments led him to the opposite conclusion. The most plausible explanation for the different results appears to be that the experimental apparatus had to detect differences in the velocity of light as small as one part in 100,000,000. Unfortunately, temperature changes in the laboratory as small as 1/100 of a degree could produce an effect three times as large as the effect Miller was trying to detect. In addition, Miller's interferometer was so sensitive that "A tiny movement of the hand, or a slight cough, made the interference fringes so unstable that no readings were possible"(R. Clark (1971), *Einstein: The Life and Times*, World Publishing, p. 329). In spite of their contradictory results, both scientists were rewarded for their work. Michelson won the Nobel Prize for Physics in 1907 and Miller received the *American Association for the Advancement of Science Award* in 1925. Reviewing this curious episode in the history of science, Collins and Pinch (1993) (*The Golem: What Everyone Should Know about Science*, Cambridge University Press, p. 40) write, "Thus, although the famous Michelson–Morley experiment of 1887 is regularly taken as the first, if inadvertent, proof of relativity, in 1925, a more refined and complete version of the experiment was widely hailed as, effectively, disproving relativity."

Although the debate between Michelson and Miller continued—they confronted each other at a scientific meeting in 1928 and agreed to differ—the physics community now accepts special relativity theory as correct. This demonstrates that *the validity of a scientific experiment strongly depends upon the theoretical framework within which the data are collected, analyzed and interpreted.* It is, perhaps, the British astronomer Sir Arthur Eddington (1882-1944) who put it best when he said: "...no experiment should be believed until confirmed by theory." From the statistician's perspective this also shows how experimental error can invalidate experimental results obtained by highly skilled scientists using the best available state of the art equipment; that is, *if the design of the experiment is faulty then no reliable conclusions can be drawn from the data.*

Example 1.2

Our second example is the famous breeding experiment of the geneticist Mendel who classified peas according to their shape *(round (r) or wrinkled `(w))* and color *(yellow (y) or green (g)).* Each seed was classified into one of four categories: $(ry) = (round, yellow)$, $(rg) = (round, green)$, $(wy) = (wrinkled, yellow)$, and $(wg) = (wrinkled, green)$. According to Mendelian genetics the frequency counts of seeds of each type produced from this experiment occur in the following ratios:

Mendel's predicted ratios: ry:rg:wy:wg $= 9 : 3 : 3 : 1$.

Thus, the ratio of the number of (ry) peas to (rg) peas should be equal to $9 : 3$, and similarly for the other ratios. An unusual feature of Mendel's model should be noted: it predicts a set of frequency counts (called a *frequency distribution*) instead of a single number. The actual and predicted counts obtained by Mendel for $n = 556$ peas appear in Table 1.1. The non integer values appearing in the third column come from dividing up the 556 peas into four categories in the ratios $9 : 3 : 3 : 1$. This means that the first category has $556 \times (9/16) = 312.75$ peas, the next two categories each have $556 \times (3/16) = 104.25$, and the last category has $556 \times (1/16) = 34.75$ of them.

Table 1.1 *Mendel's data*

Seed type	Observed frequency	Predicted frequency
ry	315	312.75
wy	101	104.25
rg	108	104.25
wg	32	34.75

Looking at the results in Table 1.1 we see that the observed counts are close to but not exactly equal to those predicted by Mendel. This leads us to one of the most fundamental scientific questions: How much agreement is there between the collected data and the scientific model that explains it? Are the discrepancies between the predicted counts and the observed counts small enough so that they can be attributable to chance or are they so large as to cast doubts on the theory itself? One problem with Mendel's data, first pointed out by Sir R. A. Fisher, *Annals of Science* (1936), pp. 115-137, is that every one of his data sets fit his predictions extremely well—too well in fact. Statisticians measure how well the experimental data fit a theoretical model by using the χ^2 (*chi-square* statistic, which we will study in Chapter 9). A large value of χ^2 provides strong evidence against the model. Fisher noted that the χ^2 values for Mendel's numerous experiments had more small values than would be expected from random sampling; that is, *Mendel's data were too good* to be true.[3] Fisher's method of analyzing Mendel's data is a good example of *inferential statistics,* which is the science of making inferences from the data based upon the theory of probability. This theory will be presented in Chapters 2 through 6.

The scope of modern statistics, however, extends far beyond the task of validating scientific theories, as important as this task might be. *It is also a search for mathematical models that explain the given data and that are useful for predicting new data.* In brief, it is a highly useful tool for inductive learning from experimental data. The next example, which is quite different from the preceding ones on relativity and genetics, illustrates how misinterpreting statistical data can seriously harm a corporation, its reputation, and its employees.

Example 1.3

The lackluster performance of the American economy in the last quarter of the twentieth century has profoundly affected its corporations, its workers, and their families. Massive layoffs, the household remedy of America's managers for falling sales and mounting losses, do not appear to have solved the basic problems of the modern corporation, which is to produce high quality products at competitive prices. Even the ultimate symbol of success, the CEO of a major American corporation, was no longer welcome at many college commencements– "too unpopular among students who cannot find entry level jobs."[4] W. E. Deming (1900–1993), originally trained as a physicist but employed as a statistician, became a highly respected, if not always welcomed, consultant to America's largest corporations. "The basic cause of sickness in American industry and resulting unemployment," he argued, "is failure of top management to manage." This was, and still is, a sharp departure from the prevalent custom of blaming workers first when poor product quality leads to decline in sales, profits, and dividends. To illustrate this point he devised a simple experiment that is really a parable because it is in essence a simple story with a simple moral lesson.

Deming's Parable of the Red Bead Experiment

(This example is adapted from W. Edwards Deming (1982), *Out of the Crisis,* MIT press, pp. 109-112.) Consider a bowl filled with 800 red beads and 3200 white beads. The beads in the bowl represent a shipment of parts from a supplier; the red beads represent the defective parts. Each worker stirs the beads and then, while blindfolded, inserts a special tool into the mixture with which he draws out exactly 50 beads. The aim of this process is to produce white beads; red beads, as previously noted, are defective and will not be accepted by the customers. The results for six workers are shown in Table 1.2. It is clear that the workers' skills in stirring the beads are irrelevant to the final results, the observed variation between them due solely to chance; or, as Deming writes:

> It would be difficult to construct physical circumstances so nearly equal for six people, yet to the eye, the people vary greatly in performance.

For example, looking at the data in Table 1.2 we are tempted to conclude that Jack, who produced only 4 red beads, is an outstanding employee and Terry, who produced nearly 4 times as many red beads as did Jack, is incompetent and should be immediately dismissed. To explain the data, Deming goes on to calculate what he calls *the limits of variation attributable to the system.* Omitting the technical details—we will explain his mathematical model, methods and results in Chapter 4 (Example 4.20)—Deming concludes that 99% of

---

[3]In spite of his fudged data, Mendelian genetics survived and is today regarded as one of the outstanding scientific discoveries of all time.

[4]*NY Times,* May 29, 1995.

the observed variation in the workers' performance is due to chance. He then continues: "The six employees obviously all fall within the calculated limits of variation that could arise from the system that they work in. There is no evidence in these data that Jack will in the future be a better performer than Terry. Everyone should accordingly receive the same raise. ...It would obviously be a waste of time to try and find out why Terry made 15 red beads, or why Jack made only 4."

Table 1.2 *Data from Deming's red bead experiment*

Name	Number of red beads produced
Mike	9
Peter	5
Terry	15
Jack	4
Louise	10
Gary	8
Total	51

In other words, the quality of the final product will not be improved by mass firings of poorly performing workers. The solution is to improve the system. Management should remove the workers' blindfolds or switch to a new supplier who will ship fewer red beads. This example illustrates one of the most important tasks of the engineer, which is *to identify, control, and reduce the sources of variation* in the manufacture of a commercial product.

1.3 Types of Data: Examples from Engineering, Public Health, and Finance

The correct statistical analysis of a data set depends on its type, which may consist of a single list of numbers (*univariate data*), data collected in chronological order (*time series*), *multivariate data* consisting of two or more columns of numbers, columns of *categorical data* (non-numerical data), such as ethnicity, gender, and socioeconomic status, as well as many other data structures of which there are too many to be listed here.

1.3.1 Univariate Data

The key to data analysis is to construct a mathematical model so that information can be efficiently organized and interpreted. Formally, we begin to construct the mathematical model of a univariate data set by stating the following definition.

Definition 1.1 *1. A population is a set, denoted $\mathcal{S}$, of well defined distinct objects, the elements of which are denoted by s.*

2. *A sample is a subset of $\mathcal{S}$, denoted by $A \subset \mathcal{S}$. We represent the sample by listing its elements as a sequence $A = \{s_1, \ldots, s_n\}$.*

To illustrate these concepts we now consider an example of a sample drawn from a population.

Example 1.4

The federal government requires manufacturers to monitor the amount of radiation emitted through the closed doors of a microwave oven. One manufacturer measured the radiation emitted by 42 microwave ovens and recorded these values in Table 1.3. The data in Table 1.3 is a *data set*. Statisticians call this *raw data*—a list of measurements whose values have not been manipulated in any way. The set of microwave ovens produced by this manufacturer is an example of a population; the subset of 42 microwave ovens whose emitted radiations were measured is a sample. We denote its elements by $\{s_1, \ldots, s_{42}\}$.

Defining a Variable on a Population We denote the radiation emitted by the microwave oven s by $X(s)$, where X is a function with domain $\mathcal{S}$. It is worth noting that the observed variation in the values of the emitted radiation come from the sampling procedure while the observed value of the radiation is a function of the particular microwave oven that is tested. From this point of view Table 1.3 lists the values of the function X evaluated on the sample $\{s_1, \ldots, s_{42}\}$. In particular, $X(s_1) = 0.15, X(s_2) = 0.09, \ldots, X(s_{42}) = 0.05$. Less formally, and more simply, one describes X as a *variable defined on a population*. Engineers are usually interested in several variables defined on the same population. For example, the manufacturer also measured the amount of radiation emitted through the open doors of the 42 microwave ovens. This data set is listed in Problem 1.1. In this text we will denote the ith item in the data set by the symbol x_i, where $X(s_i) = x_i$, and the data set of which it is a member by $X = \{x_1, \ldots, x_n\}$. Note that we use the same symbol X to denote both the data set and the function $X(s)$ that produces it. The sequential listing corresponds to reading from left to right across the successive rows of Table 1.3. In detail:

$$x_1 = 0.15, \; x_2 = 0.09, \ldots x_8 = 0.05, \ldots, x_{42} = 0.05.$$

We summarize these remarks in the following formal definition.

Definition 1.2 *1. The elements of a* data set *are the values of a real valued function* $X(s_i) = x_i$ *defined on a sample* $A = \{s_1, \ldots, s_n\}$.

2. The number of elements in the sample is called the sample size.

Table 1.3 *Raw data for the radiation emitted by 42 microwave ovens*

```
0.15 0.09 0.18 0.10 0.05 0.12 0.08
0.05 0.08 0.10 0.07 0.02 0.01 0.10
0.10 0.10 0.02 0.10 0.01 0.40 0.10
0.05 0.03 0.05 0.15 0.10 0.15 0.09
0.08 0.18 0.10 0.20 0.11 0.30 0.02
0.20 0.20 0.30 0.30 0.40 0.30 0.05
```

(Source: R. A. Johnson and D. W. Wichern (1992), *Applied Multivariate Statistical Analysis*, 3rd ed., Prentice Hall, p. 156. Used with permission.)

In many situations of practical interest we would like to compare two populations with respect to some numerical characteristic. For example, we might be interested in determining which of two gasoline blends yields more miles per gallon (mpg). Problems of this sort lead to data sets with a more complex structure than those considered in the previous section. We now look at an example where the data have been obtained from two populations.

Example 1.5 *Data analysis and Lord Rayleigh's discovery of argon*

Carefully comparing two samples can sometimes lead to a Nobel Prize. The discovery of the new element argon by Lord Rayleigh (1842-1919), Nobel Laureate (1904), provides an interesting example. He prepared volumes of pure nitrogen by two different chemical methods: (1) from an air sample, from which he removed all oxygen; (2) by chemical decomposition of nitrogen from nitrous oxide, nitric oxide, or ammonium nitrite. Table 1.4 lists the masses of nitrogen gas obtained by the two methods. Example 1.23 examines Rayleigh's data using the techniques of exploratory data analysis, which were not yet known in his time.

Table 1.4 *Mass of nitrogen gas (g) isolated by Lord Rayleigh*

From air (g)	From chemical decomposition (g)
2.31017	2.30143
2.30986	2.29890
2.31010	2.29816
2.31001	2.30182
2.31024	2.29869
2.31010	2.29940
2.31028	2.29849
.	2.29889

(Source: R.D. Larsen, Lessons Learned from Lord Rayleigh on the Importance of Data Analysis, J. Chem. Educ., vol. 67, no. 11, pp. 926-928, 1990. Used with permission.)

Example 1.6 *Comparing lifetimes of two light bulb filaments*

A light bulb manufacturer tested 10 light bulbs that contained filaments of type A and 10 light bulbs that contained filaments of type B. The values for the observed lifetimes of the 20 light bulbs are recorded in Table 1.5. The purpose of the experiment was to determine which filament had longer lifetimes. In this example we have two samples taken from two populations, each one consisting of $n = 10$ light bulbs. The lifetimes of light bulbs containing type A and type B filaments are denoted by the variables X, Y, where $X =$ lifetime of type A light bulb and $Y =$ lifetime of type B light bulb.

Table 1.5 *Lifetimes (in hours) of light bulbs with type A and type B filaments*

Type A filaments: 1293, 1380, 1614, 1497, 1340, 1643, 1466, 1627, 1383, 1711
Type B filaments: 1061, 1065, 1092, 1017, 1021, 1138, 1143, 1094, 1270, 1028

1.3.2 Multivariate Data

In the previous section we defined the concept of a single variable X defined on a population $\mathcal{S}$; multivariate data sets arise when one studies the values of one variable on two or more different populations, or two or more variables defined on the same population. Examples 1.7 and 1.8 below are typical of the kinds of multivariate data sets one encounters in public health studies.

Example 1.7 *Smoking during pregnancy and its effect on breast-milk volume*

The influence of cigarette smoking on daily breast-milk volume (g/day) and infant weight gain (grams) over 14 days was measured for 10 smoking and 10 non-smoking mothers. Smoking and non-smoking mothers are examples of two populations, and breast-milk volume is the variable defined on them. The first column (Group) represents the values of the Group variable, which identifies the mother as smoking (S) or non smoking (NS). It is example of a *categorical variable*; that is, its values are non-numerical. The two variables of interest for the infant population are birth weight and weight gain, respectively.

Table 1.6 *Biomedical characteristics of smoking/non-smoking mothers and their infants*

	Smoking (S)/Non Smoking (NS) Mother			Infant	
Group	Mother's height (cm)	Mother's weight (kg)	Milk volume(g/d)	Birth weight (kg)	Weight gain (kg)
S	154	55.3	621	3.03	0.27
S	155	53	793	3.18	0.15
S	159	60.5	593	3.4	0.35
S	157	65	545	3.25	0.15
S	160	53	753	2.9	0.44
S	155	56.5	655	3.44	0.4
S	153	52	895	3.22	0.56
S	152	53.1	767	3.52	0.62
S	157	56.5	714	3.49	0.33
S	160	70	598	2.95	0.11
NS	150	51	947	3.9	0.5
NS	146	54.2	945	3.5	0.35
NS	153	57.1	1086	3.5	0.51
NS	157	64	1202	3	0.56
NS	150	52	973	3.2	0.4
NS	155	55	981	2.9	0.78
NS	149	52.5	930	3.3	0.57
NS	149	49.5	745	3.3	0.54
NS	153	53	903	3.3	0.7
NS	148	53	899	3.06	0.58

(Source: F. Vio, G. Salazar, and C. Infante, Smoking During Pregnancy and Lactation and its Effects on Breast Milk Volume, Amer. J. of Clinical Nutrition, 1991, pp. 1011-1016. Used with permission.)

Example 1.8 *DNA damage from an occupational hazard*

The data in Table 1.7 come from a study of possible DNA damage caused by occupational exposure to the chemical 1,3-butadiene. The researchers compared the N-1-THB-ADE levels (a measure of DNA damage) among the exposed and control workers, labeled E and C respectively. Note that the data for this example are formatted in a single table consisting of two columns, labeled Group and N-1-THB-ADE. Each column represents the values of a variable defined on the worker population. The first column (Group) is another example of a categorical variable; that is, its values (E=Exposed, C=Control) are non-numerical. It is clearly a question of no small importance to determine whether or not cigarette smoking influences breast-milk volume, or exposure to the chemical 1,3-butadiene causes DNA damage. Statisticians have developed a variety of graphical and analytical methods to answer these and similar questions, as will be explained later in this chapter, beginning with Section 1.4.

Table 1.7 *DNA damage of exposed (E) and non-exposed (C) workers*

Group	N-1-THB-ADE
E	0.3
E	0.5
E	1
E	0.8
E	1
E	12.5
E	0.3
E	4.3
E	1.5
E	0.1
E	0.3
E	18
E	25
E	0.3
E	1.3
C	0.1
C	0.1
C	2.3
C	3.5
C	0.1
C	0.1
C	1.8
C	0.5
C	0.1
C	0.2
C	0.1

(Source: C. Zhao, et al., Human DNA adducts of 1,3 butadiene, an important environmental carcinogen, Carcinogenesis, vol. 21, no. 1, pp. 107-111, 2000. Used with permission.)

Example 1.9 *Parental socioeconomic data and their children's development*

The original StatLab population consists of 1296 member families of the Kaiser Foundation Health Plan living in the San Francisco Bay area during the years 1961–1972. These families participated in a Child Health and Development Studies project conducted under the supervision of the School of Public Health, University of California, Berkeley. The purpose of the study was "to investigate how certain biological and socioeconomic attributes of parents affect the development of their offspring." The 36 families listed in Table 1.16 in Section 1.11 at the end of this chapter, come from a much larger data set of 648 families to whom a baby girl had been born between 1 April 1961 and 15 April 1963. Ten years later, between 5 April 1971 and 4 April 1972, measurements were taken on 32 variables for each family, only 8 of which are listed here. **Key to the variables:** Reading from left to right the eight variables are: girl's height (in), girl's weight (lbs), girl's score on the Peabody Picture Vocabulary Test *(PEA)*, girl's score on the Raven Progressive Matrices test *(RAV)*, mother's height (in), mother's weight (lbs), father's height (in), father's weight (lbs).

1.3.3 Financial Data: Stock Market Prices and Their Time Series

Variables recorded in chronological order are called *time series*. A *time series plot* is a plot of the variable against time. Financial and economic data, for example, are obtained by recording—at equally spaced intervals of time, such as weeks, months, and years—the values of a variable, such as the stock price of the General Electric (GE) company, the S&P 500 index, the gross domestic product, the unemployment rate, etc. The S&P500 is a value weighted average of 500 large stocks; more precisely, the S&P500 consists of 500 stocks weighted according to their market capitalization (= number of shares × share price). For additional details concerning the makeup and computation of the S&P500 index, consult the references in the Section 1.12 at the end of this chapter. It is an example of a *portfolio,* which is is a collection of *securities,* such as stocks, bonds, and cash.

Example 1.10 *Weekly closing prices of the S&P500 index and other securities for the year 1999*

Table 1.17 in Section 1.11 gives the closing weekly prices for the S&P 500 index, the General Electric (GE) company, PIMCO total return bond mutual fund (PTTRX), and the Prudent Bear fund (BEARX) for the year 1999. These data were downloaded from the YAHOO website http://finance.yahoo.com/ . GE, PTTRX, and BEARX are the stock symbols that are used in order to view the price history of a security. Figure 1.1 is the time series plot of the closing averages of the S&P500 index and closing prices of GE stock for the year 1999 and Figure 1.2 displays the time series plot of the closing prices of the PTTRX and BEARX mutual funds for the year 1999. Formally, the price of the security is $X(t_i)$, where t_i denotes the ith week and X denotes its price. Its time series is the graph $(t_i, X(t_i))$, $(i = 1, 2, \ldots, n)$. In practice we ignore the underlying population on which the variable is defined and denote the sequence of closing prices by X_i, $i = 1, 2, \ldots, n$. The year 1999 is an example of what financial analysts call a *bull market*; it is characterized by increasing investor optimism concerning future economic performance and its favorable impact on the stock market. The opposite of a bull market is a *bear market*; which is characterized by increasing investor pessimism concerning future economic performance and its potentially unfavorable impact on the stock market. Clearly, investors in an S&P500 index fund were more pleased than those who invested in, say, the BEARX mutual fund. The year 2002 is a sobering example of a bear market (see Example 1.11). Financial data of the sort listed in Table 1.17 are used by investors to evaluate the returns (see Section 1.3.4 for the definition of return) and risks of an investment portfolio.

The Prudent Bear Fund's investment policy is most unconventional: it sells selected stocks when most investors are buying them and buys them when most investors are avoiding them. It "seeks capital appreciation," according to its prospectus, "primarily through short sales of equity securities when overall market valuations are high and through long positions in value-oriented equity securities when overall market valuations are low." It's an interesting example of a security that is negatively correlated with the stock market as will be explained in Section 1.7.

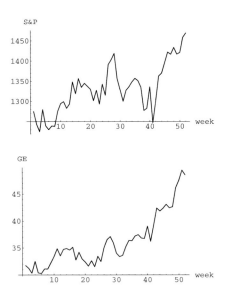

FIGURE 1.1:   Weekly closing prices of the S&P 500 and GE for the year 1999

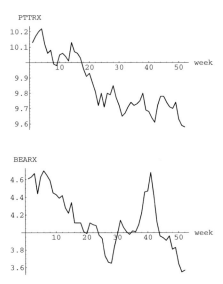

FIGURE 1.2:   Weekly closing prices of the PTTRX and BEARX mutual funds for the year 1999

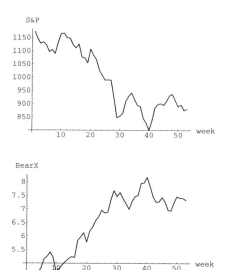

FIGURE 1.3:   Weekly closing prices of the S&P500 and BEARX mutual funds for the year 2002

Example 1.11 *The year 2002: An example of a bear market*

The time series plot of the closing prices for the S&P500 and BEARX mutual fund for the year 2002 are displayed in Figure 1.3 and Table 1.18 contains the data from which the time series was computed. For most, but not all, investors, the results were a grim contrast to the bull market of 1999, as Figure 1.3 clearly demonstrates. Indeed the return (see Equation 1.1 in Section 1.3.4) on the S&P index was a confidence shaking -25%.[5] Investors in the BEARX mutual fund, which suffered a -22.5% loss in 1999, received more cheerful news; their fund earned a 65% return. These data reveal not only the risks associated with investing in financial markets, but also suggest ways of reducing them. For example, instead of putting all your money into an S&P500 index fund, you could put 10% of your original capital into a stock or mutual fund that is negatively correlated with the S&P500. This certainly would have reduced your profits in 1999, but might have reduced your losses in 2002. We discuss this portfolio in some detail in Example 1.12 below. A more penetrating analysis of financial data requires the concept of *return*, which is discussed next (Section 1.3.4).

Example 1.12 *The effect of diversification on portfolio performance*

Looking at Figures 1.1 and 1.2 we see that the S&P500 performed well and BEARX poorly in 1999, while in 2002 the reverse was true (see Figure 1.3). This suggests that if we had invested a small portion of our capital in BEARX we would have reduced our gains in 1999 but also have reduced our losses in 2002. To fix our ideas, consider portfolio A consisting of a $9,000 investment in the S&P500 and a $1,000 investment in the BEARX mutual fund for the years 1999 and 2002; that is 90% of our capital is allocated to the S&P500 and 10% to the BEARX mutual fund. With the benefit of hindsight, which investment performed

---

[5]Note: The actual return—including dividends and reinvestments, but excluding commissions—was -22.1%.

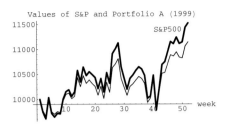

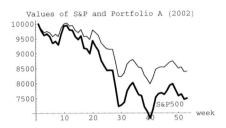

FIGURE 1.4: Weekly Value of a $10,000 investment in the S&P 500 index (thick curve) vs. Portfolio A, 1999 and 2002

better: Investing all $10,000 in the S&P500, or investing all of it in portfolio A? To answer this question look at Figure 1.4, which graphs the weekly price history of the two portfolios for the years 1999 and 2002, respectively. In 1999 the year-end value of an initial $10,000 investment in the S&P500 would have been worth $11,523, while Portfolio A would have been worth $11,145, which is $378 less. More precisely, the S&P earned 15.23% return (see Section 1.3.4 for definitions and more examples), while Portfolio A earned a still respectable 14.5% return. In 2002, although both portfolios lost money, the losses in the S&P500 were far more severe. In detail: the year-end value of the S&P portfolio would have been worth $7503 (a −25% return) while Portfolio A would have been worth $8,404 (a −16% return), a much smaller loss.

1.3.4 Stock Market Returns: Definition and Examples

The price history of a stock, although easy to understand from its time series graph, is not particularly useful for making prudent investment decisions. A $2 increase in the price of a $10 stock, which yields a 20% *positive return*, is much more significant than a $2 increase in the price of a $100 one, which yields only a 2% return. Of course, the price of the stock could fall, yielding a negative return.

Definition 1.3

The rate of return and total return are defined as

$$\text{rate of return} = \frac{\text{amount received} - \text{amount invested}}{\text{amount invested}}; \qquad (1.1)$$

$$\text{total return} = \frac{\text{amount received}}{\text{amount invested}}. \qquad (1.2)$$

We note for future reference that

$$\text{total return} = 1 + \text{ rate of return}. \qquad (1.3)$$

For example, the rate of return for an investor who buys a stock for $10 and sells it one year later for $12 is $0.20 = (12 - 10)/10$. When expressed in percentage terms we say the investment yielded a 20% return. Similarly, if the year end price falls to $9, the rate of return is $-0.10 = (9 - 10)/10$, equivalently, a -10% return or 10% loss. Suppose, instead of buying a stock, you deposit P_0 (amount invested) in a bank that pays simple interest at the rate r per year, with principal and interest guaranteed by the Federal Deposit Insurance Corporation (FDIC). At the end of one year you will have P_1 (amount received) in the bank, where $P_1 = P_0 + rP_0 = (1 + r)P_0$. The one year rate of return and total return in this case are $r = (P_1 - P_0)/P_0$ and $1 + r = P_1/P_0$, respectively. This is an example of a *risk free return*.

For investors, saving for their retirement, and confronted with several thousand mutual funds to invest in, the two most important variables are the investment returns and the level of risk they are willing to assume to achieve it. In practice, one compares the returns and risks of a stock, or of a portfolio of stocks, with a market portfolio, such as the S&P500 index. We then compute the returns of an investment in a publicly traded security, such as a stock or a mutual fund from their prices for $n + 1$ successive time periods, measured in days, weeks, months, etc., which are then used to calculate n returns. In detail, denote the sequence of stock prices by $\{S_0, S_1, \ldots, S_n\}$ and the corresponding sequence of returns by $\{r_1, r_2, \ldots, r_n\}$, where S_0 is the initial stock price and S_i is its price i time periods later. The time series of returns is given by

$$r_i = \frac{S_i - S_{i-1}}{S_{i-1}}, \ i = 1, 2, \ldots, n, \text{ thus} \tag{1.4}$$

$$1 + r_i = \frac{S_i}{S_{i-1}}. \tag{1.5}$$

Conversely, given an initial investment of W_0 that yields the sequence of returns $(r_1, \ldots, r_n)$ the corresponding sequence of values of this investment, denoted $(W_0, W_1, \ldots, W_n)$, is

$$W_{i+1} = W_i \times (1 + r_i), \ i = (0, 1, \ldots, n - 1). \tag{1.6}$$

$$W_{i+1} = W_0(1 + r_1) \times (1 + r_2) \times \cdots \times (1 + r_i), \ (i = 1.2 \ldots, n - 1). \tag{1.7}$$

Applying Equation 1.7 to the special case of a constant return $r = r_i, \ (i = 1.2 \ldots, n)$ (which corresponds to leaving the money in the bank earning a constant rate of interest r) yields the result

$$W_n = W_0 \times (1 + r)^n. \tag{1.8}$$

This result is useful for comparing the returns on a risky investment with the returns obtained by leaving one's money in the bank.

Example 1.13 *Computing the ten-year performance of a bond mutual fund*

The returns of a bond mutual fund for the ten years 1996-2005 are listed in column 2 of Table 1.8. At the beginning of 1996 an investor withdraws $10,000 from her savings account, invests it in the mutual fund, and leaves it there for the next ten years. (a) Find the values of her investment at the end of of each year from 1996-2005. **Solution** In this case

$W_0 = 10,000, \ W_1 = 10,000 \times (1 + 0.0238) = 10,238.$

$W_2 = 10,238 \times (1 + 0.0727) = 10,000 \times (1.0238) \times (1.0727) = 10,000 \times 1.09823$

$= 10,982.30$, etc.

The complete solution is displayed in column 4 of Table 1.8.

(b) Find the annualized rate of return for this ten-year period (1996-2005).

Solution The annualized rate of return for an initial investment of $\$W_0$ and terminal value W_n is the solution to the equation

$$(1 + r)^n = (1 + r_1) \times (1 + r_2) \times \cdots \times (1 + r_n). \tag{1.9}$$

It is the annual interest rate a bank must pay for n periods in order for the investor to have $\$W_n$ in the bank. Applying this result to the mutual fund data in Table 1.8 we see that the annualized rate of return for this investment is $r = 0.043$, or 4.3%, which is the solution to the equation

$$(1 + r)^{10} = \prod_{1 \leq i \leq 10} (1 + r_i) = 1.523604.$$

$$1 + r = \left(\prod_{1 \leq i \leq 10} (1 + r_i) \right)^{1/10} = (1.523604)^{1/10} = 1.043.$$

Consequently, $r = 0.043$. An alternative method is to take (natural) logarithms of both sides and obtain the equation

$$\ln(1 + r) = \frac{1}{10} \ln(1.523604) = 0.0421079.$$

Consequently, $r = \exp(\ln(1.523604)/10) - 1 = 0.043$.

Similar calculations (using logarithms) yield the following general formula for the annualized rate of return for an n year investment with initial and terminal values W_0, W_n, respectively:

$$r = \exp\left(\frac{1}{n} \ln(W_n/W_0) \right) - 1. \tag{1.10}$$

Table 1.8 *Ten-year performance of a bond mutual fund*

Year	Return (%)	$\prod_{1 \leq i \leq j}(1 + r_i)$	Value of $10,000 investment
1996	2.38	1.023800	10238.00
1997	7.27	1.098230	10982.30
1998	6.46	1.169176	11691.76
1999	-0.67	1.161342	11613.42
2000	8.34	1.258198	12581.98
2001	6.86	1.344511	13445.11
2002	8.18	1.454492	14544.92
2003	1.5	1.476309	14763.09
2004	2	1.505835	15058.35
2005	1.18	1.523604	15236.04

Example 1.14 *Weekly returns of the S&P500 and other securities, 1999*

The performance of an investment as noted earlier is measured by its return and its *risk*. We measure the risk of an investment by computing the standard deviation of its sequence of returns, concepts that will be explained in Section 1.6.2. Looking at Table 1.17 we see that the return on an investment in the S&P index for the week of 4-Jan-99 is $(1243.26 - 1275.09)/1275.09 = -0.025$; multiplying this quantity by 100 yields the percentage return of

-2.5%. Tables 1.19 and 1.20 give the 1999 and 2002 weekly returns for each of the securities whose weekly prices are listed in Tables 1.17 and 1.18. The closing prices of the S&P500 index on (Monday) Jan. 4, 1999, and (Monday) Dec. 27, 1999, were 1275.09 and 1469.25, respectively; consequently, the annual rate of return—excluding dividends, reinvestments, and commissions—for an investment in a mutual fund indexed to the S&P500 for the period (Jan. 4, 1999 − Dec. 27, 1999) is

$$\text{S\&P return } r = \frac{1469.25 - 1275.09}{1275.09} = 0.1523 = 15.23\% \text{ (Jan. 4, 1999 − Dec. 27, 1999)}$$

We gain a more profound understanding of the forces causing the fluctuations in stock market prices by viewing the time series of the returns for each of the securities listed in Table 1.19, which are displayed in Figures 1.5 and 1.6. We note the remarkable fact that, although the time series of the each of the securities prices displayed in these figures are quite different, the time series of their returns are quite similar. This suggests that the forces driving the returns of all heavily traded securities are also similar. A feature common to each of these time series is their unpredictability, that is, the returns are no more predictable than the sequence of heads and tails produced by tossing a coin. We return to this idea in Sections 3.5.1 and 5.5, where we construct a random walk model—called the binomial lattice model—of stock price fluctuations.

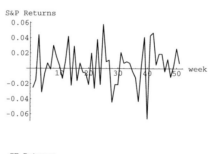

FIGURE 1.5: Time series of S&P500 and GE returns, 1999

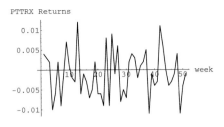

FIGURE 1.6: Time series of PTTRX and BEARX mutual fund returns, 1999

1.4 The Frequency Distribution of a Variable Defined on a Population

In this section we first examine the basic concept of the *frequency distribution of a variable defined on a population*. We then introduce a variety of tabular, graphical, and analytic methods that statisticians use to answer the question of how the data are distributed. Among the most widely used graphical displays are: *bar charts, dot plots, stem and leaf displays, histograms,* and *box plots*. These methods are elementary in the following sense: No knowledge of the calculus is required. Because our primary focus is on understanding and interpreting these graphs we will deemphasize the tedious details of their construction and instead produce these displays using statistical software packages.

1.4.1 Organizing the Data

We organize and display these data in order to extract information from them. Since the purpose of collecting the data in Table 1.3 was to determine the proportion of microwave ovens that emit radiation below an acceptable level, the easiest way to obtain this information is to first sort the data in *increasing* order of magnitude as in Table 1.9.[6]

Table 1.9 *Order statistics of the radiation data of Table 1.3*

```
0.01 0.01 0.02 0.02 0.02 0.03 0.05 0.05 0.05 0.05 0.05
0.07 0.08 0.08 0.08 0.09 0.09 0.10 0.10 0.10 0.10 0.10
0.10 0.10 0.10 0.10 0.11 0.12 0.15 0.15 0.15 0.18 0.18
0.20 0.20 0.20 0.30 0.30 0.30 0.30 0.40 0.40
```

---

[6]Both MINITAB and SAS have "commands" or "procedures" that sort the data in increasing order of magnitude.

The technique of *sorting,* or *ranking,* the observations of a data set in increasing order of magnitude is widely used in statistics and the following terminology is standard.

Definition 1.4 *When the observations* $\{x_1, x_2, \ldots, x_n\}$ *are sorted in increasing order*

$$\{x_{(1)} \leq x_{(2)} \leq \cdots \leq x_{(n)}\}$$

then the elements of this set are called the order statistics *and* $x_{(i)}$ *is called the ith* order statistic.

Looking at Table 1.9 we see that it is easier to extract useful information from sorted data. For example, the smallest observed value is 0.01, which appears as the 13th entry in the original data set (Table 1.3); thus $x_{(1)} = 0.01 = x_{13}$. The largest observed value is 0.40, which appears as the 20th entry, so $x_{(42)} = 0.40 = x_{20}$. This shows that in general $x_{(i)} \neq x_i$.

From the order statistics we see, for example, that the smallest observed value is 0.01 and the largest observed value is 0.40. Their difference, which equals $0.40 - 0.01 = 0.39$, is called the *sample range.* It is a crude measure of the amount of variability in the data. We can express the smallest and largest values in terms of the order statistics as follows:

$$x_{(1)} = \min\{x_1, \ldots, x_n\}.$$
$$x_{(n)} = \max\{x_1, \ldots, x_n\}.$$

The primary focus of a statistical study is the frequency distribution of the data set, and we now explain how we use order statistics to compute it.

Frequency distribution of a variable We compute the frequency distribution of a variable by counting the number of times that the value x appears. This number is denoted $f(x)$. Thus, the smallest observed value is 0.01 and occurs twice, so $f(0.01) = 2$; similarly $f(0.02) = 3$, $f(0.03) = 1$, etc. The value 0.10 has the highest observed frequency $f(0.10) = 9$; it is called the *mode.* In general, the mode is the value that occurs most often. A data set can have two or more modes.

Cumulative frequency distribution The cumulative frequency distribution counts the number of observed values *less than or equal to* x. It is denoted $F(x)$. Looking at Table 1.9 we see that there are six values less than or equal to 0.03; so, $F(0.03) = 6$. The cumulative frequency $F(x)$ is the sum of the frequencies of the values less than or equal to x. For example,

$$F(0.03) = f(0.01) + f(0.02) + f(0.03) = 2 + 3 + 1 = 6.$$

All major statistical software packages have programs that not only compute these distributions but display the results in attractive visual formats. We will study several of them in the next section.

1.4.2   Graphical Displays

The art of displaying data in an informative and visually attractive format is called *exploratory data analysis,* which is the name given by statisticians to a collection of methods for constructing graphical displays of data. One very informative display is the *horizontal bar chart.* It is a compact graphical display of the order statistics, frequency distribution, and cumulative frequency distribution. Figure 1.7 is the horizontal bar chart for the radiation data in Table 1.3. The information in the horizontal bar chart is laid out in columns as follows:

Frequency Distribution of Radiation Data (Table 1.3)

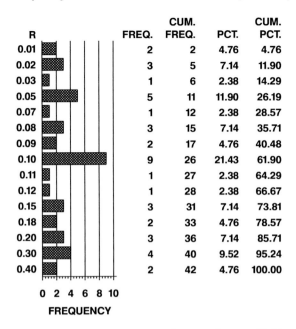

R		FREQ.	CUM. FREQ.	PCT.	CUM. PCT.
0.01		2	2	4.76	4.76
0.02		3	5	7.14	11.90
0.03		1	6	2.38	14.29
0.05		5	11	11.90	26.19
0.07		1	12	2.38	28.57
0.08		3	15	7.14	35.71
0.09		2	17	4.76	40.48
0.10		9	26	21.43	61.90
0.11		1	27	2.38	64.29
0.12		1	28	2.38	66.67
0.15		3	31	7.14	73.81
0.18		2	33	4.76	78.57
0.20		3	36	7.14	85.71
0.30		4	40	9.52	95.24
0.40		2	42	4.76	100.00

0 2 4 6 8 10

FREQUENCY

FIGURE 1.7: Horizontal bar chart for the radiation data in Table 1.3

- **Column 1:** The entries in the first column on the left, where R = emitted radiation, are the distinct observed values sorted in increasing order.

- **The bar chart:** Between the first and second columns is a visual display of the frequency of each observation, where the width of the bar to the right of the observed value x is proportional to the frequency $f(x)$.

- **Column 2 (Freq):** This is the frequency function $f(x)$. For example, to compute $f(0.09)$, scan down the first column until you locate 0.09 and then read across to 2 in the second column; that is $f(0.09) = 2$.

- **Column 3 (Cum. Freq):** This is the cumulative frequency distribution $F(x)$. To determine, for instance, how many microwave ovens emitted radiation in amounts less than or equal to 0.20 scan the first column down until you reach 0.20 and then read the entry 36 in column 3.

- **Columns 4 (Percent) and 5 (Cum. Percent):** Columns 4 and 5 are columns 2 and 3 restated in terms of percentages. For example, to determine the percentage of microwave ovens whose emitted radiation level equals 0.10, scan down the first column until you locate 0.10 and then read across to 21.43 in column 4; that is, 21.43% of the ovens emitted 0.10 amount of radiation. To determine the percentage of microwave ovens whose emitted radiation level is less than or equal to 0.10, read across to column 5 and read 61.90; that is, 61.90% of the ovens emitted radiation at levels less than or equal to 0.10.

Notation We use the symbol $\sharp(A)$ to denote the number of elements in the set A.
The Empirical Distribution Function The *empirical distribution function* $\hat{F}_n(x)$ is the proportion of the observed values less than or equal to x, so it equals $F(x)/n$, where $F(x)$ is the cumulative frequency distribution defined earlier. The formal definition follows.

Definition 1.5 *The empirical distribution function $\hat{F}_n$ of the data set $\{x_1, x_2, \ldots, x_n\}$ is the function defined by*

$$\hat{F}_n(x) = \frac{\sharp\{x_i : x_i \leq x\}}{n}.$$

That is, the empirical distribution function $\hat{F}_n(x)$ is the proportion of values that are less than or equal to x.

The function $100 \times \hat{F}_n(x)$ denotes the percentage of values that are less than or equal to x; it is this function that is displayed in the column labeled Cum. Percent in Figure 1.7.

The empirical distribution and the frequency distribution functions are related to one another by the equation:

$$\hat{F}_n(x) = \frac{1}{n} \sum_{y \leq x} f(y). \tag{1.11}$$

The empirical frequency distribution, denoted $\hat{f}(x)$, is the frequency distribution divided by the total number of observations; that is,

$$\hat{f}(x) = \frac{f(x)}{n}.$$

Note: Some authors call $\hat{F}_n(x)$ the *sample distribution function.*

Example 1.15

Suppose the goal of the manufacturer in Example 1.4 is to produce microwave ovens whose level of radiation emissions do not exceed 0.25. What proportion of the microwave ovens tested meet this standard?

Solution Looking at the order statistics in the horizontal bar chart (Table 1.7) we see that 36 microwave ovens emitted radiation less than or equal to 0.25; consequently, the proportion of ovens meeting the standard is $\hat{F}_n(0.025) = 36/42 = 0.857$. Notice that

$$\hat{F}_n(0.25) = \hat{f}(0.01) + \hat{f}(0.02) + \ldots + \hat{f}(0.18) + \hat{f}(0.20)$$
$$= 0.0476 + 0.0714 + \ldots + 0.0476 + 0.0714 = 0.857.$$

Example 1.16

Graph the empirical distribution function of the microwave oven radiation data (Table 1.3).

Solution The graph of $\hat{F}_n(x)$ is shown in Figure 1.8.

Note that this graph resembles a staircase with steps located at the order statistics. Looking at this staircase, we can see that the height of each step at x—that is, the size of the jump j—equals j/n where $j = f(x)$. In particular, the graph of the empirical distribution function starts with the first order statistic $x_{(1)} = 0.01$ whose frequency is $f(0.01) = 2$ (see Table 1.9) and plots the point $(0.01, 2/42) = (0.01, 0.0476)$. The next distinct order statistic is 0.02 and its frequency is $f(0.02) = 3$; so the point $(0.02, [f(0.01) + f(0.02)]/42) = (0.02, 5/42) = (0.02, 0.119)$ is plotted; this procedure continues, until we plot the last two points $(0.30, 40/42) = (0.30, 0.9524)$, and $(0.40, 42/42) = (0.40, 1)$. Note that $f(0.10) = 9$ so the jump at 0.10 equals $9/42 = 0.214$ while $f(0.25) = 0$ so the jump at 0.25 is zero; that is, $\hat{F}_n$ is continuous at $x = 0.25$.

The empirical distribution function is an example of a *piecewise defined function*; that is, the formula that produces the output $\hat{F}_n(x)$ depends on the interval where the input x is located.

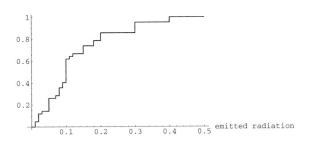

FIGURE 1.8: Empirical distribution function of the radiation data (Table 1.3)

Dot Plots Another useful way to display the radiation data in Table 1.3 is the *dot plot*, in which each observation x_i is displayed as a "dot" above the point x_i on the x axis, as shown in Figure 1.9. The frequency of x_i is the number of dots above x_i. The dot plot is more than an alternative visual display of the horizontal bar chart; it also provides a useful first glance at the *shape* and *variability* of the distribution. Looking at Figure 1.9 we see, for example, that the values are unevenly distributed along the x axis, with a single *peak* at $x = 0.10$ around which most of the data seem to be concentrated and a *tail* extending rightwards out to the value $x = 0.40$. A skewed distribution is one that has a single peak and a long tail in one direction. It is *skewed to the right,* or *positively skewed,* when the tail, as is the case here, goes to the right. Such a skewed distribution suggests that a few of the microwave ovens are emitting excessive amounts of radiation.

Stem and Leaf Plots One of the quickest ways of visualizing the shape of the distribution with a minimum of computational effort is the stem and leaf plot. Assume that each observation x_i consists of at least two digits as in the radiation data (Table 1.3). Then break each observation into a *stem* consisting of the leading digits and a *leaf* consisting of the remaining digits; decimal points are not used in a stem and leaf display—they are ignored. For the data in Table 1.3 we look only at the digits after the decimal point—we not only ignore the the decimal point but the leading digit 0 as well. The observation 0.15 has stem equal to 1 and a leaf equal to 5; 0.09 has stem 0 and leaf 9 and 0.10 has stem 1 and leaf 0. In detail:

data value	split	stem	leaf
.15 →	1 | 5	1	5
.09 →	0 | 9	0	9
.10 →	1 | 0	1	0

We then form two columns: the first column lists the stems and the second column contains the leaves corresponding to the given stem. The result is displayed in Table 1.10. Sorting the leaves in increasing order for each stem yields an *ordered* stem and leaf plot. An edited ordered stem and leaf plot produced by MINITAB is displayed in Table 1.11. The

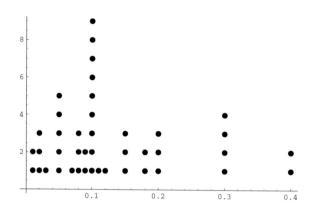

FIGURE 1.9:   Dot plot of emitted microwave radiation data (Table 1.3)

stem and leaf plot displays in condensed form all the information contained in the horizontal bar chart in Figure 1.7, including a rough look at the variability of the data.

Table 1.10 *Stem and leaf plot of the radiation data (Table 1.3)*

Stem	Leaf	Frequency
0	95858721215359825	17
1	5802000000505801	16
2	000	3
3	0000	4
4	00	2

Table 1.11 *Ordered stem and leaf plot of radiation data (Table 1.3)*

```
0  11222355555788899
1  0000000001255588
2  000
3  0000
4  00
```

1.4.3   Histograms

In this section we will demonstrate another way to visualize the distribution of a data set called the *histogram*; it is, in essence, a vertical bar chart constructed in a special way. Although it is possible to draw a histogram by hand, it is easier, as well as more reliable, to construct it via a suitable software package. Figure 1.10 is a computer generated histogram for the radiation data of Table 1.3.

Looking at this histogram we see that the x axis represents the values of the emitted radiation and the y axis represents the frequency values. The data have been grouped

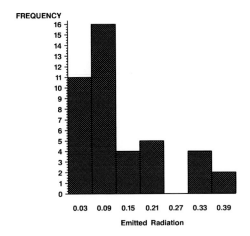

FIGURE 1.10: Computer generated histogram for the radiation data (Table 1.3)

into seven nonoverlapping intervals, called *class intervals*. Their midpoints, called the *class marks*, are 0.03, 0.09, 0.15, 0.21, 0.27, 0.33, 0.39, and they are clearly marked on the x axis.

To describe the class intervals in more detail we use the symbol $[a, b)$ to denote the set of points on the line that are less than b and greater than or equal to a; that is,

$$[a, b) = \{x : a \leq x < b\}.$$

The seven class intervals in Figure 1.10 are: $[0.0, 0.06), [0.06, 0.12), [0.12, 0.18), [0.18, 0.24), [0.24, 0.30), [0.30, 0.36), [0.36, 0.42]$. The lengths of the class intervals are called the *class widths*, which in this case equal 0.06, since the class intervals are of equal length.

Above each class interval is a rectangle whose base has a length equal to the class width and whose height is the class frequency; consequently, the areas of the rectangles are proportional to the class frequencies when the class widths are equal. Looking at the order statistics displayed in Table 1.9 we see that the frequency of the class interval $[0.0, 0.06)$ equals 11. The *relative (class) frequency* is the proportion of the values that are in the class interval, so the relative frequency of the class interval $[0.0, 0.06)$ equals $11/42 = 0.26$.

In the general case we construct a histogram by first dividing the sample range into a small number of nonoverlapping intervals, usually consisting of at least five but no more than twenty intervals, and then counting the number of values in each interval. In detail we proceed as follows:

1. Choose k nonoverlapping intervals that cover the sample range; they are denoted

$$[u_0, u_1), [u_1, u_2), \ldots [u_{k-1}, u_k].$$

The numbers u_i are called the *class boundaries.*

The midpoints of the class intervals are denoted by m_j; that is,

$$m_j = (u_j + u_{j-1})/2.$$

2. Count the number of observations in the jth class. This number is called the jth *class frequency* and is denoted by f_j, so that,

$$f_j = \sharp(x_i : u_{j-1} \leq x_i < u_j) \text{ and } f_n = \sharp(x_i : u_{n-1} \leq x_i \leq u_n).$$

The *relative frequency* of the jth class is the proportion of of observations in the jth class; it equals $\hat{f}_j = f_j/n$.

3. The computer draws the frequency histogram in the following manner: Above each class interval it constructs a rectangle with height proportional to the class frequency f_j.

4. The *relative frequency histogram* corresponds to choosing a rectangle whose area equals the relative frequency of the jth class.

Drawing a Histogram by Hand

Although we usually use computers to construct histograms, we can also construct them by hand—an exercise that can give some insight into their computation. There are no hard and fast rules about this kind of construction, but there are some steps and rules of thumb we can follow. We now give a method for determining the number of classes, the class marks, and the class frequencies. We then apply this method to show how to construct the histogram shown in Figure 1.10.

1. **Determining the number of classes:** The first problem one confronts in constructing a histogram is to determine the number of classes. There are no hard and fast rules about how to proceed, but there are some rules of thumb that are helpful to keep in mind. The first rule is that the number of classes should be at least five but no more than twenty; the exact number to be used depending on the sample size n and computational convenience. One rule[7] that has been suggested is this: Choose the number of classes k so that

$$2^{k-1} < n \le 2^k. \tag{1.12}$$

Thus for the radiation data (Table 1.3) we would choose $k = 6$ since $n = 42$ and

$$2^5 = 32 < 42 \le 64 = 2^6.$$

2. **Choosing the class marks:** The next rule is that the class intervals should be of equal length, although this is not always possible, and that the class marks should be easily computable. We shall now describe how these rules were applied to produce the frequency distribution in Table 1.12.

Looking at the order statistics in Figure 1.7 we see that the sample range equals $x_{(42)} - x_{(1)} = 0.40 - 0.01 = 0.39$. We next divide the sample range into an equal number of subintervals whose lengths are easily computable numbers. According to the rule stated in Equation 1.12 we should divide the data into 6 classes. This suggests that we divide the sample range into 6 equal parts, with the length of each interval being equal to $0.39/6 = 0.065$. Since this is not a particularly convenient number to work with we round it down to the more convenient number 0.06; this leads to the choice of 7 class intervals of total length 0.42 which covers the sample range. This leads to the class intervals

$$[0.0, 0.06), [0.06, 0.12), \ldots, [0.36, 0.42]$$

produced by the computer. In order to avoid any ambiguity in the classification procedure an observation, such as 0.12, that falls on a class boundary is assigned to the interval on the right, thus 0.12 is placed in the interval $[0.12, 0.18)$.

Table 1.12 _____

[7]Called Sturges' rule, see H. A. Sturges, *Journal of the American Statistical Association*, March 1926.

3. **Computing the class frequencies:** The *frequency table* is formed by counting the number of values that fall into each class interval and then recording the counts in the frequency column as illustrated in Table 1.12. The last column on the right records the cumulative frequency.

Frequency Table of Radiation Data		
class boundaries	frequency	cum.freq.
$0 \leq x < 0.06$	11	11
$0.06 \leq x < 0.12$	16	27
$0.12 \leq x < 0.18$	4	31
$0.18 \leq x < 0.24$	5	36
$0.24 \leq x < 0.30$	0	36
$0.30 \leq x \leq 0.36$	4	40
$0.36 \leq x < 0.42$	2	42

Example 1.17

Figure 1.11 is the frequency histogram of the weekly returns for the S&P500 index for the year 1999. The class marks are: -0.06, -0.04, -0.02, 0.00, 0.02, 0.04, 0.06, and the class width is 0.02. It is to be observed that the shape of the distribution is nearly symmetrical about the middle interval $[-0.01, 0.01]$ with the highest frequency, which equals 18.

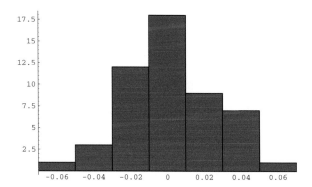

FIGURE 1.11: Histogram of S&P returns (1999)

1.5   Quantiles of a Distribution

In the previous sections we introduced a variety of graphical displays for studying the distribution of the data. In this section and the following one we examine several *numerical* measures that also describe important features of the data.

For example, for some engineers it may be enough to know that the median lifetime (defined in Section 1.5.1) of type A light bulbs (Table 1.5) is 1481.5 hours with a minimum lifetime of 1293 hours and a maximum lifetime of 1711 hours. We can summarize much of the information contained in a data set by computing suitable *percentiles* of the empirical distribution function. When we say, for instance, that a student's class rank is in the 90th percentile we mean that 90% of her classmates rank below her and 10% of her classmates rank above. Similarly, the $10th$ percentile of a data set plays an important role in reliability engineering since approximately 90% of the systems will operate this long without failure. This is sometimes called the *safe life*.

For many purposes, including mathematical convenience, it is easier to work with the *p-quantile* of a distribution, where $0 < p < 1$ is a fraction; it corresponds to the $100p\%$ percentile of the distribution. Quantiles are a useful tool for comparing two distributions via *Quantile–Quantile plots* or Q–Q plots, which some authors call *probability plots*. We will examine this important data analysis tool in Section 8.4 of this text. We begin our discussion by considering the important special case of the 50th percentile, which is also called the *median*.

1.5.1   The Median

A median of the data set $\{x_1, \ldots, x_n\}$ is a number denoted by $\tilde{x}$ that divides the data set in half, so that at least half the data values are greater than or equal to $\tilde{x}$ and at least half the data values are less than or equal to $\tilde{x}$.

We now describe a simple procedure for computing a median. In Example 1.18 we consider two cases: (1) n is even and (2) n is odd.

Example 1.18

(1) Given the lifetimes of type A filaments in Table 1.5 and (2) the failure times of Kevlar/epoxy pressure vessels (Table 1.21 in Section 1.11), find the median lifetime of the type A filaments and the median failure time of the pressure vessels.

1. **Solution** Looking at the lifetimes of type A filaments in Table 1.5 we see that $n = 10$, so the sample size is even. The order statistics for the light bulbs with type A filaments have two values (in **boldface**) that occupy the two middle positions.

   $1293 < 1340 < 1380 < 1383 < \mathbf{1466} < \mathbf{1497} < 1614 < 1627 < 1643 \ < \ 1711.$

   We define the median to be the average of the two middle terms, so that

$$\tilde{x} = \frac{1466 + 1497}{2} = 1481.5.$$

2. **Solution** Looking at the failure times of the pressure vessels in Table 1.21 we see that $n = 39$, so the sample size is odd. The data in this table are already arranged in increasing order; they are in fact the order statistics of the sample. The median is the 20th term in the ordered array, since there are 19 values less than $x_{(20)}$ and 19 values greater than $x_{(20)}$. Consequently, $\tilde{x} = x_{(20)} = 55.4$.

Equations 1.13 and 1.14 give a formula for the median in terms of the order statistics.

$$\tilde{x} = \frac{1}{2}\left(x_{(n/2)} + x_{(n/2+1)}\right), \text{ if n is even;} \tag{1.13}$$

$$\tilde{x} = x_{\left(\frac{n+1}{2}\right)}, \text{ if n is odd.} \tag{1.14}$$

Example 1.19

1. Compute the median lifetime of type A filaments in Table 1.5 via Equation 1.13.

   Solution Here $n = 10$ is even, so Equation 1.13 yields

   $$\tilde{x} = \frac{1}{2}\left(x_{(5)} + x_{(6)}\right) = \frac{1}{2}(1466 + 1497) = 1481.5,$$

   which is in agreement with the result obtained earlier.

2. Compute the median of the data in Table 1.21 via Equation 1.14.

   Here $n = 39$ is odd, so Equation 1.14 yields

   $$\frac{n+1}{2} = \frac{40}{2} = 20; \text{ so}$$
   $$\tilde{x} = x_{((n+1)/2)} = x_{(40/2)} = x_{(20)} = 55.4.$$

The median is an example of a *measure of location*, since it locates the *center* of a distribution. For this reason it is also called a measure of *central value*. (See Section 1.6.)

1.5.2 Quantiles of the Empirical Distribution Function

The inherent variability in a typical data set cannot be reduced to a single number such as the median. This is why, for example, economists use *quintiles* (the 20*th*, 40*th*, 60*th*, 80*th* percentiles) to more accurately describe the income distribution of a country. Similarly, many colleges and universities now report the SAT scores of students accepted for admission as a *range* of values instead of reporting, as they used to, the average SAT score. This minor shift in policy was deemed sufficiently newsworthy to be published in the *New York Times*.

> Syracuse and Colgate Universities have joined 50 other colleges and universities in pledging to release the SAT scores as a *range* rather than a single average score. Such scores are published by college guides and often requested by families of prospective students.
> In the standardized testing agreement the schools pledge not to provide an average test score. Instead, the pledge suggests presenting the range of scores of the middle 50% of students accepted.[8]

For example, the University of Florida reported in 1995 that the SAT scores of its entering freshman class had a mid 50% range of [1080, 1250]. This means that 50% of those accepted had SAT scores in the interval [1080, 1250]; in addition, 25% of those accepted had SAT scores below 1080 and 25% had SAT scores above 1250. The numbers 1080 and 1250 are examples of what are known as the lower and upper quartiles of a distribution.

---

[8]*New York Times,* Section 4A, 5 August 1990.

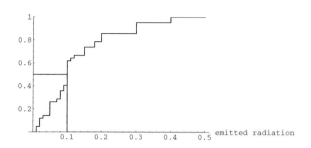

FIGURE 1.12:   Graphical determination of the median

As this example makes clear the quantiles of a distribution provide additional information concerning the distribution of the data.

A p quantile $(0 < p < 1)$ of the empirical distribution function is a number $Q(p)$ with the property that at least $100p\%$ of the data values are less than or equal to $Q(p)$ and at least $100(1-p)\%$ of the data values are greater than or equal to $Q(p)$. We first give a geometrical method for computing a p quantile. Afterwards, we give an analytic formula for computing it.

Example 1.20

Find the median of the radiation data (Table 1.3) from the empirical distribution function of the data.

Solution   Figure 1.12 is the graph of the empirical distribution function of the radiation data (Table 1.3), modified by connecting the jumps by vertical line segments. We obtain the median, which equals $Q(0.5)$, by drawing a horizontal line starting at the point $(0, 0.5)$ on the y axis and going to the right (or left) until it intersects the graph of the empirical distribution function. The x coordinate of this point is the median. Looking at Figure 1.12 we see that the point of intersection has coordinates $(Q(0.5), 0.5) = (0.10, 0.5)$, so $\tilde{x} = 0.10$. We obtain the quantile $Q(p)$ from the graph of the empirical distribution function in the same way, that is, we draw a horizontal line starting at the point $(0, p)$ on the y axis and going to the right (or left) until it intersects the graph of the empirical distribution function. The x coordinate of this point is the pth quantile $Q(p)$; the point of intersection has coordinates $(Q(p), p)$. When the empirical distribution function has a flat interval at the level p we define the pth quantile to be the midpoint of the flat interval. Since the height of the empirical distribution function at a point x gives the proportion of the data less than or equal to x, defining the pth quantile $Q(p)$ this way means that at least $100p\%$ of the data will be less than $Q(p)$ and at least $100(1-p)\%$ of the data will be greater than $Q(p)$.

The Upper and Lower Quartiles   The 25th and 75th percentiles are called the *lower* and *upper* quartiles and are denoted by $Q_1 = Q(0.25)$ and $Q_3 = Q(0.75)$, respectively.

Because drawing the graph of the empirical distribution function can be quite tedious we shall now give a formula for $Q(p)$ using the order statistics.

A formula for computing $Q(p)$

To calculate the quantile $Q(p)$ we proceed in a manner analogous to that used to calculate the median (Equations 1.13 and 1.14). We begin by computing np. There are two cases to consider, according to whether np is an integer or not.

1. **np is an integer:** In this case compute the npth and $(np+1)$th order statistics $x_{(np)}$ and $x_{(np+1)}$; $Q(p)$ is defined to be their average:

$$Q(p) = \frac{1}{2}\left(x_{(np)} + x_{(np+1)}\right). \tag{1.15}$$

2. **np is not an integer:** In this case there exists two consecutive integers r and $r+1$ so that np lies between them; that is, $r < np < r+1$. We define $Q(p)$ as

$$Q(p) = x_{(r+1)}. \tag{1.16}$$

Note: Formula 1.15 corresponds to computing the median when the sample size is even and Equation 1.16 corresponds to computing the median when the sample size n is odd.

Example 1.21

1. Compute $Q(0.10)$, or the safe life, of the lifetimes of the type A filaments (Table 1.5).

   Solution Here $n = 10$ and $p = 0.1$, so $np = 1$, an integer. Consequently,

   $$Q(0.10) = \frac{1}{2}(x_{(1)} + x_{(2)}) = 1,316.5.$$

2. Compute the lower and upper quartiles of the failure times of the pressure vessels in Table 1.21.

   Solution Here $n = 39$, $p = 0.25$; therefore, $9 < np = 9.75 < 10$. Thus,

   $$Q_1 = Q(0.25) = x_{(10)} = 9.1.$$

   Similarly, when $n = 39$, $p = 0.75$ we have $29 < np = 29.25 < 30$. Thus,

   $$Q_3 = Q(0.75) = x_{(30)} = 444.4.$$

Order statistics and quantiles It is worth pointing out that the ith order statistic $x_{(i)}$ is the $(i-0.5)/n$ quantile; that is,

$$x_{(i)} = Q\left(\frac{i-0.5}{n}\right). \tag{1.17}$$

We leave the derivation of Equation 1.17 to the reader as Problem 1.30.

The sample range and interquartile range Order statistics and percentiles are often used to measure the variability of a distribution. The *sample range* is the length of the smallest interval that contains all of the observed values; the *interquartile range* (IQR) is the length of the interval that contains the middle half of the data.

The sample range is defined by:

$$\text{sample range} = x_{(n)} - x_{(1)}; \qquad (1.18)$$

and the interquartile range is defined by

$$\text{interquartile range (IQR)} = Q_3 - Q_1. \qquad (1.19)$$

The interval $[Q_1, Q_3]$ is called *the middle* 50% range.

An *outlier* is any data value that lies outside the interval

$$(Q_1 - 1.5 \times IQR, \; Q_3 + 1.5 \times IQR).$$

Box plots   The *box plot* is a graphic display of the median, quartiles, and interquartile ranges of a data set.

Example 1.22

Figure 1.13 is the box plot for the radiation data (Table 1.3). The horizontal line locates the median and the lower and upper edges of the box locate the lower and upper quartiles; which, in this case are $Q_1 = 0.05$, $Q_3 = 0.18$, and the $IQR = 0.13$. The box represents the mid 50% range of the data. *Whiskers* are the lines that extend outward from the ends of the box to a distance of at most 1.5 units of IQR. More precisely, the whisker on the top starts at Q_3 and terminates at the number $\min(x_{(n)}, Q_3 + 1.5 \times IQR)$. In this case $x_{(42)} = 0.40$ and $Q_3 + 1.5 \times IQR = 0.18 + 0.195 = 0.375$, so

$$\min(x_{(42)}, Q_3 + 1.5 \times IQR) = \min(0.40, 0.375) = 0.375.$$

Similarly, the whisker on the bottom starts at Q_1 and terminates at the number $\max(x_{(1)}, Q_1 - 1.5 \times IQR)$. In this case $x_{(1)} = 0.01$ and $Q_1 - 1.5 \times IQR = -0.145$, so

$$\max(x_{(1)}, Q_1 - 1.5 \times IQR) = \max(0.01, -0.145) = 0.01.$$

Any value beyond these limits is marked with a star. Looking at Figure 1.13 we see a star at 0.40. Referring back to the original data set we see there are two outliers, both equal to 0.40.

Example 1.23 *Using side-by-side box plots to compare two populations*

Although the differences in the weights of nitrogen obtained by Lord Rayleigh from air (atmospheric nitrogen) and chemical decomposition (chemical nitrogen) appear to be too small to be of scientific significance (cf. Table 1.4), he understood that the differences were well outside his margin of error. However, as R.D. Larsen notes (op. cit.):

> Rayleigh's own data analysis was quite primitive. Despite his exceptional experimental skill, caution, and confidence in the validity of his results, he apparently merely compared the means of his various "chemical nitrogen samples" with the means of his "atmospheric nitrogen" samples. ... Surprisingly, however, (and even a bit disappointing) is that there is no evidence in his work of any plotting or graphing of any kind.

The box plot of Lord Rayleigh's data (Figure 1.14) confirms, 100 years too late, his observaion that there is a significant difference between the mass of "atmospheric nitrogen" and "chemical nitrogen." The difference was later accounted for by the discovery of the new element argon in air.

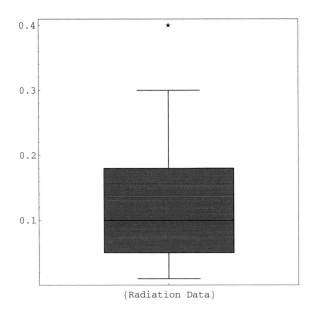

FIGURE 1.13: Box plot of the radiation data (Table 1.3)

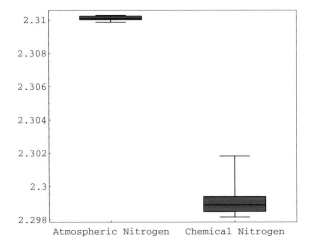

FIGURE 1.14: Side-by-side box plot of nitrogen obtained from air and nitrogen obtained from chemical decomposition (Table 1.4)

1.6   Measures of Location (Central Value) and Variability

In addition to such numerical measures such as the sample range, median, and lower and upper quartiles already discussed, we now introduce the *(arithmetic) mean, variance,* and *standard deviation.* The mean defines what physicists call the *center of mass* of the distribution. The standard deviation, like the interquartile range, is another measure of the variability of the distribution.

1.6.1   The Sample Mean

We use the term *average* in everyday speech to denote the arithmetic mean of a set of numbers. In statistics we call it the *sample mean* and it is, apart from the median, the most widely used numerical measure of the central value of a distribution.

Definition 1.6 *The sample mean $\overline{x}$ of the set of numerical data $\{x_1, \ldots, x_n\}$ is defined by*

$$\overline{x} = \frac{1}{n} \sum_{1 \leq i \leq n} x_i. \tag{1.20}$$

Example 1.24

1. The mean time before failure (MTBF) of a device, such as a light bulb, plays an important role in reliability engineering. Compute the MTBF of light bulbs with type A filaments (Table 1.5).

   Solution The sum of the 10 lifetimes of the type A filaments equals 14954.00. Consequently, the mean time before failure is given by

   $$\overline{x} = \frac{1293 + 1380 + \cdots + 1711}{10} = \frac{14954.00}{10} = 1495.4.$$

2. Compute the mean amount of radiation emitted by the microwave ovens (Table 1.3).

   Solution When the sample size is large, as it is in this example, then we use a statistical software package, since it is too tedious to compute the mean by hand. We obtain the value $\overline{x} = 0.128$.

Looking at the dot plot of the microwave radiation data in Figure 1.15 we notice that the mean is greater than the median. This occurs because the distribution is skewed to the right. In particular the two outliers (observations $x_{20} = x_{40} = 0.40$) increase the value of the mean but do not have any effect on the median. For instance, if the largest observed value were increased to 100 the median would remain unchanged, but the mean would be sharply increased.

It is helpful to think of the values displayed in the dot plot in Figure 1.15 as consisting of $n = 42$ point masses placed along a rod, each mass weighing $1/n = 1/42$ grams, say. Thus, we place $9/42$ grams at $x = 0.10$ since $\hat{f}(0.10) = 9/42$. The sample mean $\overline{x}$ can now be interpreted as the *center of mass* of this distribution.

   Mean, Median, or Mode?

We noted that in Example 1.24 the sample mean is larger than the sample median; this is fairly typical when the data, as is the case here, are skewed to the right. This raises the following question: Which measure of central value is appropriate? The answer is that

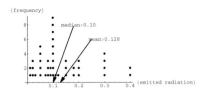

FIGURE 1.15: Dot plot of the emitted microwave radiation data of Table 1.3 displaying the sample mean $\bar{x} = 0.128$ and the median $\tilde{x} = 0.10$

it depends on the shape of the distribution and the variable type; that is whether or not the variable is numerical or categorical. For example, it makes no sense to talk about the average birth month, but it is reasonable to identify the mode, which is the month in which the most births occur. In other words:

1. The mode is a reasonable measure of central value for a categorical variable;

2. The median is an appropriate measure of central value when there are outliers skewing the distribution either to the right or the left. For instance, income data are skewed to the right and this explains why economists prefer to use the median income instead of the average income.

3. The sample mean is the most appropriate choice for the central value when the distribution appears symmetric about its mean.

1.6.2   Sample Standard Deviation: A Measure of Risk

The principal goal of collecting and analyzing financial data of the sort listed in Tables 1.17 and 1.19 is to make money, that is, to invest one's savings in a security, such as stocks, bonds, or real estate, in order to earn a positive return. The future returns of each these securities—excluding bank deposits guaranteed by the Federal Deposit Insurance Company (FDIC)—are subject to a variety of risks including stock price declines, bond defaults, and interest rate increases that are unknown at the time the investment is made. Some investments are inherently more risky than others; stocks, for example, are more risky than bonds. Which raises a fundamental question: How do we define and measure risk? Intuitively, *risk* is a measure of the uncertainty associated with the value of the return. Consider, for example, the performance of an investment in a mutual fund indexed to the S&P500 for the year 1999 (ignoring dividends and commissions). The risks, both upside and downside, are graphically illustrated in Figure 1.1 which plots the 1999 weekly (Monday) closing averages of the S&P500 and GE (if Monday is a holiday then Tuesday's price is listed).[9] On 12 July 1999 the value of the S&P500 index stood at 1418.78; just 3 weeks later its value had declined to 1300.29—a loss of 8.35%! This example shows that even though the average 1999 weekly return was positive, with $\bar{r} = 0.003$, it is still possible to

---

[9]Data obtained from http://finance.yahoo.com/.

suffer a major loss by buying and selling at the wrong time. We gain additional insight into the concept of risk by looking at the histogram of the 1999 weekly returns displayed in Figure 1.11; it reveals that the returns vary considerably from -0.07 to $+0.07$, with most of them clustering in the interval $[0.0, 0.02]$. The *sample standard deviation*, denoted s, is used to measure the spread of a distribution about its mean. It is a measure of the variability of the returns; that is, it is a measure of *how likely it is for the observed value to deviate from its mean*. Rational investors use returns and their standard deviations to guide them in their choice of investments. In particular, they prefer to invest in a security that offers them the highest rate of return for a given standard deviation.

Definition 1.7 *The* sample variance s_x^2 *and the* sample standard deviation $s_x = \sqrt{s_x^2}$ *of numerical data* $\{x_1, \ldots, x_n\}$ *are defined by*

$$s_x^2 = \frac{1}{n-1} \sum_{1 \le i \le n} (x_i - \overline{x})^2; \tag{1.21}$$

$$s_x = \sqrt{\frac{1}{n-1} \sum_{1 \le i \le n} (x_i - \overline{x})^2}. \tag{1.22}$$

Table 1.13 lists the mean return, sample range and standard deviation for each of the stocks listed in Table 1.19.

Table 1.13 *Weekly mean return, sample range, and volatility for securities in Table 1.19*

Stock	Mean Return	Sample Range	Standard Deviation
S&P	0.0031	$[-0.0663, 0.0577]$	$0.0255 = 2.55\%$
GE	0.0091	$[-0.0722, 0.0853]$	$0.0399 = 3.99\%$
PTTRX	-0.0011	$[-0.0113, 0.0120]$	$0.0058 = 0.58\%$
BEARX	-0.0046	$[-0.0679, 0.0544]$	$0.0290 = 2.90\%$

Note: When the data are univariate we drop the subscript and write s in place of s_x. To compute s^2 one proceeds in the following way: For each data value x_i compute $(x_i - \overline{x})^2$, then calculate the *sum of squares* $\sum_{1 \le i \le n} (x_i - \overline{x})^2$ and, finally, divide this sum by $n-1$. In practice one uses a statistical software package—such as Minitab, Excel, or JMP—to compute the sample mean and standard deviation for large data sets, and a pocket calculator for small ones.

Interpreting Table 1.13:

1. *Volatility* is another widely used term for the standard deviation of a stock's (or portfolio's) return. The volatility of a stock depends on the length of time period over which the returns are recorded. For example, the standard deviations listed in Table 1.13 are more accurately called weekly volatilities because they are standard deviations of the weekly returns. The values of monthly, quarterly, or annual volatilities are obtained by multiplying the weekly volatilities by $\sqrt{k}$, where k is the number of weeks in the month, quarter, or year. (A justification for this formula is given in Section 5.5.) To illustrate: The estimated monthly, quarterly, and annual volatility for GE stock are 0.0510, 0.0919, 0.1839, respectively. In practice, financial analysts estimate volatility using the most recent six or twelve month's data.

2. It is to be observed that GE's standard deviation is 56% greater than the S&P's. This illustrates a basic principle of modern portfolio theory, which is, that *diversification* reduces risk.

3. The mutual fund with the smallest standard deviation is the bond fund PTTRX, confirming that bonds are less risky than stocks.

4. Observe that the BEARX portfolio, for the 1999 year, had a negative return and higher risk than the S&P; this is an example of an *inefficient* portfolio. That is, given a choice between two portfolios with the same returns but different risks, an investor will choose the portfolio with the smaller risk. Similarly, given a choice between two portfolios with the same risks, but different returns, an investor will choose the portfolio with the higher return.

Computing the sample mean and sample variance from the frequency distribution The frequency distribution contains all the information in the sample; in particular, we can compute all numerical measures of location and variability from the frequency distribution itself. Equations 1.23 and 1.24, for example, give formulas for the sample mean and sample variance directly in terms of the frequency distribution.

$$\bar{x} = \frac{1}{n} \sum_x x f(x); \qquad (1.23)$$

$$s^2 = \frac{1}{n-1} \sum_x (x - \bar{x})^2 f(x) = \frac{1}{n-1} \left(\sum_x x^2 f(x) - n\bar{x}^2 \right). \qquad (1.24)$$

We derive Equations 1.23 and 1.24 by noting that

$$\sum_{1 \le i \le n} x_i = \sum_x x f(x) \text{ and}$$

$$\sum_{1 \le i \le n} (x_i - \bar{x})^2 = \sum_x (x - \bar{x})^2 f(x).$$

The right hand sides of the preceding equations are obtained by first grouping identical summands, x say, multiplying them by their frequency $f(x)$ and then performing the summation. We give the details of these computations in Example 1.25.

Example 1.25

Compute (1) the sample mean and (2) the sample variance for the radiation data (Table 1.3) using Equations 1.23 and 1.24.

1. **Solution** To compute the sample mean we apply Equation 1.23 to the frequency distribution shown in the "Freq" column of Figure 1.7. The computations yield

$$\bar{x} = \frac{1}{42}(0.01 \times 2 + 0.02 \times 3 + \cdots + 0.30 \times 4 + 0.40 \times 2)$$
$$= \frac{5.39}{42} = 0.128,$$

which agrees with the value obtained earlier (see Example 1.24).

2. **Solution** Similarly, to compute the sample variance we apply Equation 1.24 to the frequency distribution shown in the "Freq" column of Figure 1.7. The computations yield

$$s^2 = \frac{1}{41}((0.01)^2 \times 2 + (0.02)^2 \times 3 + \ldots + (0.40)^2 \times 2 - 42 \times (0.128)^2)$$
$$= \frac{1}{41}(1.1016 - 0.6881) = \frac{0.4135}{41} = 0.01.$$

Population Mean and Variance Sometimes a sample is not enough. The Constitution of the United States, for example, requires that every ten years a census be taken of the entire population. The basic purpose of the census is to reapportion the membership of the House of Representatives among the states. For us a *census* is a sample that consists of the entire population; that is, we measure the variable for *every* element of the population. The numerical measures obtained from a census, such as the median, mean, variance and standard deviation are called *population parameters*. The *population mean* and *population variance* are particularly important population parameters.

Definition 1.8 *Let* $\{x_1, \ldots, x_N\}$ *denote the data obtained from a census of a population consisting of* N *elements. We denote the population mean and population variance by the Greek letters* μ *(mu) and* σ^2 *(sigma squared). They are defined by*

$$\text{Population mean: } \mu = \frac{1}{N} \sum_{1 \leq i \leq N} x_i$$

$$\text{Population variance: } \sigma^2 = \frac{1}{N} \sum_{1 \leq i \leq N} (x_i - \mu)^2$$

$$\text{Population standard deviation: } \sigma = \sqrt{\frac{1}{N} \sum_{1 \leq i \leq N} (x_i - \mu)^2}$$

Remark 1.1 *Note that the divisor in the definition of the population variance is the population size* N *and not* $N - 1$.

When the size N of the population is large, however, the task of computing μ may be next to impossible. We therefore use the sample mean $\bar{x}$ and sample variance s^2, based upon a sample size $n < N$, to estimate μ and σ^2.

1.6.3   Mean-Standard Deviation Diagram of a Portfolio

The *mean-standard deviation diagram* is an important tool for investors comparing the risks and returns of various portfolios. It is the plots of the points $(s, \bar{r})$ for each portfolio under consideration.

Example 1.26

Table 1.14 lists six portfolios together with their risks s and mean returns $\bar{r}$. Figure 1.16 is the mean-standard deviation diagram for the this set of portfolios with risks and mean returns plotted on the x and y axes, respectively. Looking at Figure 1.16 we notice that portfolio A has a smaller mean return than portfolio B although both have the same risk. Portfolio A is an example of an *inefficient*. Portfolio F is also inefficient, but for a different reason: Its risk is higher than B's although both have the same return. Portfolio E has the smallest expected return, but zero risk. In this universe of portfolios B, C, D, and E are examples of *efficient portfolios* because no other portfolio has *simultaneously* a lower risk and higher return.

Table 1.14 *Risks and returns for six portfolios*

Portfolio	A	B	C	D	E	F
risk s	15	15	18	20	0	20
return $\bar{r}$	8	10	13	15	6	10

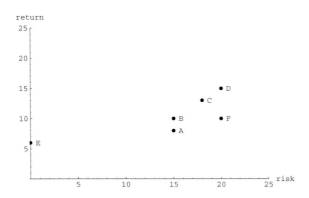

FIGURE 1.16: Mean standard deviation diagram of the six portfolios in Table 1.14

1.6.4 Linear Transformations of Data

It is sometimes necessary to rewrite the observations in a different system of units, for instance, transforming pounds (lb) to kilograms (kg), inches (in) to centimeters (cm) and degrees Fahrenheit ($^\circ F$) to degrees Celsius ($^\circ C$). We use the subscript notation s_x and s_y to distinguish between the sample standard deviation for the data sets X and Y. The transformed data set, denoted $Y = \{y_1, \ldots, y_n\}$, is obtained from the original one by computing $y_i = g(x_i)$, $(i = 1, \ldots, n)$ where $g(x)$ is a function. The most frequently used functions are the *linear functions* defined by

$$y = g(x) = ax + b.$$

We now give some examples.

1. To transform x (pounds) to y (kilograms) we set $a = 1/2.2$ and $b = 0$.

2. To transform x (inches) to y (centimeters) we set $a = 2.54$ and $b = 0$.

3. To transform x (degrees Celsius $^\circ C$) to y (degrees Fahrenheit $^\circ F$) we set $a = 9/5$ and $b = 32$.

We compute the mean, variance and standard deviation for data transformed by the linear function $g(x) = ax + b$ by applying Proposition 1.1. The proof is given in Section 1.8.

Proposition 1.1 *The sample mean, sample variance and sample standard deviation of the transformed data $y_i = ax_i + b, i = 1, 2, \ldots, n$ are given by*

$$\overline{y} = a\overline{x} + b;$$
$$s_y^2 = a^2 s_x^2;$$
$$s_y = |a| s_x.$$

Example 1.27 *Computing the sample mean and sample standard deviation of the weight increase of the infants of smoking mothers in pounds instead of kilograms*

Referring to the original data for smoking mothers (Table 1.6) a routine calculation yields the values $\overline{x} = 0.34(kgs)$, $s_x = 0.17$ for the mean and standard deviation of the infants' weight gain. The weight y in pounds is given by $y = 2.2 \times x(kg)$; therefore, the sample mean and sample standard deviation of the data in pounds is given by $\overline{y} = 2.2 \times 0.34 = 0.748(lbs)$ and $s_y = 0.34$.

1.7   Covariance, Correlation, and Regression: Computing a Stock's Beta

Horizontal bar charts, dot plots, stem and leaf plots, and histograms help us to understand the distribution of a single variable. *Scatter plots* are useful for studying how two variables, labeled X and Y, are related. For example, college admissions officials are interested in whether or not higher SAT scores (X) are associated with higher first year grade point averages (Y). The *correlation coefficient* between X and Y is a numerical measure of the closeness of the linear relation between them.

Example 1.28

College freshmen are given a math skills test at the beginning of the academic year and their grade point averages (GPA) are recorded at the end of the year. The results for 20 students are recorded in Table 1.15. The X column lists the student's score on the math test and the Y column lists the student's GPA. Figure 1.17 is the scatter plot of the bivariate data; it is a plot of the points $(89, 3.6)$, $(74, 2.9)$, $\dots$, $(78, 2.6)$. The scatter plot suggests that there is a *positive association* between the student's score on the math test and his GPA; that is, low math scores appear to be associated with low GPAs and high math scores with high GPAs. There are, however, a few points that do not appear to follow this pattern. For example, the student who scored an 80 on the math test had a GPA of 2.7, while another student who scored a 78 had a GPA of 3.0, very nearly equal to the 3.1 GPA of the student who had the highest math score of 93. To proceed further in our analysis of these data we bring in the concepts of covariance and correlation.

Definition 1.9 *Let $(x_1, y_1), \dots, (x_n, y_n)$ denote n observations on two variables (X, Y) with sample means $\overline{x}, \overline{y}$ and sample standard deviations s_x, s_y, respectively. The sample covariance s_{xy} and sample correlation r_{xy} are defined as*

$$\text{(sample covariance) } s_{xy} = \frac{1}{n-1} \left(\sum_{1 \leq i \leq n} (x_i - \overline{x})(y_i - \overline{y}) \right). \qquad (1.25)$$

$$\text{(sample correlation) } r_{xy} = \frac{s_{xy}}{s_x s_y}. \qquad (1.26)$$

From Equations 1.25 and 1.26 we see that the sign of $\sum_{1 \leq i \leq n} (x_i - \overline{x})(y_i - \overline{y})$ determines the sign of r_{xy}. In the context of Example 1.28 we would expect a student whose math score is above average to have an above average GPA, that is we would expect the product $(x - \overline{x})(y - \overline{y})$ to be positive because both factors $(x - \overline{x})$ and $(y - \overline{y})$ are. Similarly, students with math scores below average, so $x - \overline{x} < 0$, tend to have a below average GPA, so $y - \overline{y} < 0$; consequently, we would again expect their product $(x - \overline{x})(y - \overline{y})$ to be positive. In short, we say that the variables (X, Y) are *positively correlated* when increases in x tend

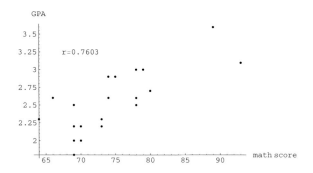

FIGURE 1.17: The scatter plot of GPA vs. math score shows a positive relation between the two variables

to be paired with increases in y. Conversely, when increases in x tend to be paired with decreases in y, we say that the variables are *negatively correlated*, as illustrated in Figure 1.18, and discussed further in Example 1.29 below. To simplify the notation we frequently drop the subscripts and write r in place of r_{xy}. For the bivariate data set in Table 1.15, the sample correlation $r = 0.781$. It can be shown (see the discussion following Equation 10.12) that $0 \le r^2 \le 1$ and therefore $-1 \le r \le 1$.

Table 1.15 *Math scores and GPAs of 20 students*

Math Score (X)	GPA(Y)	Math Score (X)	GPA(Y)
89	3.6	69	2.0
74	2.9	73	2.2
78	2.5	80	2.7
66	2.6	69	2.2
93	3.1	69	2.5
70	2.0	69	1.8
79	3.0	78	3.0
74	2.6	73	2.3
70	2.2	64	2.3
75	2.9	78	2.6

Example 1.29 *Comparing a stock to the S&P500 index*

A portfolio, as noted earlier, is a collection of securities, such as stocks, bonds, and cash. A portfolio can, of course, consist of just one stock, a strategy that has been described as, "the quickest way to get poor." In other words, an investor must avoid putting all her eggs in one basket. It is a question, then, of no small importance to understand how the risk of a portfolio depends on the risks of the individual stocks of which it is composed. A full treatment of this topic will be given in Chapter 5, Section 5.6. But some insight can be

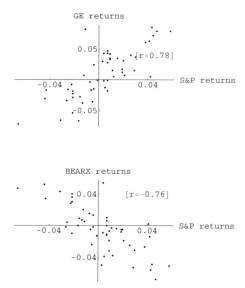

FIGURE 1.18:   The two scatter plots illustrate that GE returns are positively correlated, and BEARX mutual fund returns are negatively correlated, with S&P returns (1999)

gained by looking at the scatter plots of the returns of selected stocks vs the S&P500 listed in Table 1.19. The choice of the S&P500 as the X variable in both cases is not an arbitrary one; it is because the performance of mutual funds (and their managers) are measured against the performance of the *market portfolio*, which is often assumed to be the S&P500. Figure 1.18 displays the scatter plots of the returns of GE and BEARX vs. the returns of the S&P500 for the year 1999. Figure 1.18 shows that the returns on GE stock are positively correlated with the S&P500 (with $r = 0.7766$), while returns on the BEARX mutual fund are negatively correlated with it (with $r = -0.7566$). Including stocks or mutual funds in your portfolio that are negatively correlated with the market, while reducing your returns in bull markets, can reduce your losses in bear markets. This investment strategy is called diversification.

When are two variables strongly correlated? Weakly correlated? The consensus among statistical practitioners is that the correlation is weak when $|r| \leq 0.5$, moderate when $0.5 < |r| \leq 0.8$, and strong when $|r| > 0.8$. Computing r by hand is not recommended because it requires a series of tedious, error-prone computations that can be avoided through the use of either the statistical function keys on your calculator or a statistical software package in a computer laboratory.

1.7.1   Fitting a Straight Line to Bivariate Data

In Example 1.28 we studied the relationship between a student's score (x) on a math skills test and the student's GPA (y) at the end of the academic year. The scatter plot (Figure 1.17), as previously noted, suggests that higher GPAs are associated with higher math scores, which raises the following question: Can a student's math score be used to predict the student's GPA? We are particularly interested in a linear prediction formula of the form: $y(gpa) = b_0 + b_1 x(math\ score)$. From Figure 1.17 it appears that the GPA (y) varies *almost* linearly with respect to the math score (x). We say almost because it is

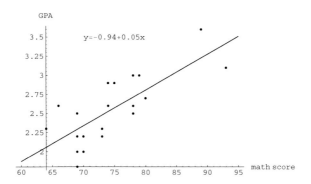

FIGURE 1.19: Regression line for the GPA vs. math score (Table 1.15)

clear that it is impossible to construct a line of the form $y = b_0 + b_1 x$ which passes through all the observations (x_i, y_i), $i = 1, \ldots n$; which leads us to the following problem: How do we choose b_0 and b_1 so that the line $y = b_0 + b_1 x$ "best fits" the observations (x_i, y_i)? In Section 10.2.1 we shall solve this problem via the method of *least squares*; the fitted line is also called the *regression line*. In the meantime we shall content ourselves with just giving the formulas for the slope b_1 and intercept b_0; we refer the reader to Sections 10.2.1 and 10.6 for additional discussion and mathematical details.

The slope–intercept formula for the line that best fits the data in the sense of the least squares criterion is

$$\text{(regression line) } y = b_0 + b_1 x, \text{ where} \tag{1.27}$$

$$\text{(slope of regression line) } b_1 = \frac{\text{sample covariance of (X,Y)}}{\text{sample variance of X}} = \frac{s_{xy}}{s_x^2} \tag{1.28}$$

$$(y \text{ intercept of regression line}) \ b_0 = \bar{y} - b_1 \bar{x} \tag{1.29}$$

$$\text{(fitted value) } \hat{y}_i = b_0 + b_1 x_i. \tag{1.30}$$

Figure 1.19 displays the regression (fitted) line obtained by the least squares method, with slope and intercept (rounded to two decimal place accuracy)

$$b_0 = -0.94, \ b_1 = 0.05.$$

Interpreting the Regression Line

1. Suppose a student's math score is $x = 80$; then the fitted value is

$$3.06 = -0.94 + 0.05 \times 80.$$

We interpret this value as follows: Consider the population of all students with a math score 80; then their mean GPA is 3.06.

2. We interpret the value of the slope $b_1 = 0.05$ as follows: Consider two groups of students with the math score of each student in the first group being 10 points higher than the second. Then the average GPA of the first group will be 0.05 higher than the second; that is a 10 point increase in the student's math score produces, on average, a 0.5 increase in the student's GPA, since $10 \times b_1 = 10 \times 0.05 = 0.5$.

3. The value $b_0 = -0.94$ states that the regression line intercepts the y axis at -0.94. This clearly is without any academic significance since a student's GPA can never be negative. This should serve as a warning against using the regression line to predict the response y when x lies far outside the range of the initial data, which in this case is the interval $[64, 93]$.

Example 1.30 *An application of regression to finance: Measuring a security's beta (β)*

The scatter plot of GE returns vs. the S&P returns, displayed in Figure 1.18, suggests a positive association between the returns of GE stock and the S&P500 index. To make this more precise, plot the regression line on the same coordinate axes used for the scatter plot, as shown in Figure 1.20. The equation of the regression line is $y = 0.005 + 1.28x$. In financial engineering the slope plays an important role in the *capital asset pricing model* (CAPM) and is denoted by the greek letter β. In this case we say that GE's beta is 1.28.

Finance interpretation of the slope of the regression line: There are various interpretations of a security's β, not all of them will be given here. One of them is that it is a measure of how sensitive its price movements are to future price changes of the *market portfolio*. In particular, a 2% increase (decrease) for the S&P500 index makes it highly probable that GE's price will increase (decrease) by $2.56\% = 1.28 \times 2\%$. It is useful to restate the formula for the slope of the regression line given in Equation 1.28 in the language of financial engineering. This leads to the following formal definition of a security's beta.

Definition 1.10 *Let $y = (y_1, \ldots, y_n)$ and $x = (x_1, \ldots, x_n)$ denote the returns for a security and market portfolio, respectively, then β defined as*

$$\beta = \frac{s_{xy}}{s_x^2} = \frac{Covariance(security\ returns,\ market\ returns)}{Variance(market\ returns)}$$

is called the beta *for the security.*

Remark: Because computing b_0, b_1 by hand requires a series of tedious, error prone computations, we recommend students use either the statistical function keys on a calculator or a statistical software package in the computer laboratory. Consult your instructor for the details on how to use a statistical software package. If these are not available then use Equations 1.28 and 1.29.

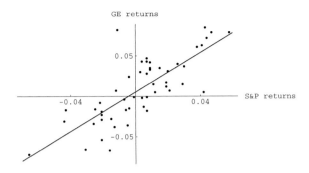

FIGURE 1.20:   GE returns vs. S&P500 returns (1999); $y = 0.005 + 1.28x$

1.8   Mathematical Details and Derivations

We prove Proposition 1.1 using the standard rules for manipulating sums; the details follow:

$$\sum_{1 \leq i \leq n} y_i = \sum_{1 \leq i \leq n} (ax_i + b)$$

$$= \sum_{1 \leq i \leq n} ax_i + \sum_{1 \leq i \leq n} b.$$

Therefore

$$\sum_{1 \leq i \leq n} y_i = a \left(\sum_{1 \leq i \leq n} x_i \right) + nb.$$

We complete the proof by dividing both sides by n. This yields

$$\overline{y} = a\overline{x} + b.$$

To prove $s_y^2 = a^2 s_x^2$ we proceed as follows:

$$s_y^2 = \sum_{1 \leq i \leq n} (y_i - \overline{y})^2$$

$$= \sum_{1 \leq i \leq n} ((ax_i + b) - (a\overline{x} + b))^2$$

$$= \sum_{1 \leq i \leq n} a^2 (x_i - \overline{x})^2$$

$$= a^2 \sum_{1 \leq i \leq n} (x_i - \overline{x})^2 = a^2 s_x^2.$$

1.9   Chapter Summary

Scientists, engineers, and financial consultants analyze their data by studying the frequency distribution of a variable defined on a population. They model data by postulating a theoretical distribution and test it by comparing the observed frequency distribution with the theoretical one. Consequently, the two most important concepts in this chapter are (i) the frequency distribution and (ii) the empirical distribution function of a variable defined on a population. Graphical methods, such as horizontal bar charts, dot plots, stem and leaf plots, histograms, and box plots are among the most effective methods for visualizing the shape of the distribution. Also useful are numerical measures such as quantiles, sample range, mean, variance, and standard deviation. Scatter plots are used to study how two variables are related. The correlation coefficient is a numerical measure of the closeness of the linear relation between them. In modern portfolio theory it is shown how the return and risk of a portfolio depends on the returns, risks, and correlations of the individual stocks of which it is composed.

1.10   Problems

Sections (1.2-1.4): Empirical distribution functions, frequency distributions, dot plots, stem and leaf plots, histograms, and applications to financial data

Problem 1.1 *The manufacturer of microwave ovens (see Example 1.4) also measured the radiation emitted through the open doors of the 42 microwave ovens and recorded the emitted radiation in the table below.*

```
0.30 0.09 0.30 0.10 0.10 0.12 0.09
0.10 0.09 0.10 0.07 0.05 0.01 0.45
0.12 0.20 0.04 0.10 0.01 0.60 0.12
0.10 0.05 0.05 0.15 0.30 0.15 0.09
0.09 0.28 0.10 0.10 0.10 0.30 0.12
0.25 0.20 0.40 0.33 0.32 0.12 0.12
```

(Source: R. A. Johnson and D. W. Wichern (1992), Applied Multivariate Statistical Analysis, *3rd ed., Prentice Hall, p. 169. Used with permission.)*
(a) Construct an ordered stem and leaf plot of this data set.
(b) Draw the histogram of this data set using 7 intervals with class width 0.1 and class marks $0.05, 0.15 \ldots, 0.65$. *Describe the shape of the distribution.*

Problem 1.2 *Statisticians prefer to work with distributions that are symmetrically distributed about a single well-defined peak. When the data are skewed, as is the case for the radiation data (see Figure 1.10), the statisticians Box and Cox have shown that transforming the data by means of the function* $y = g(x)$, *where*

$$g(x) = \frac{x^\lambda - 1}{\lambda} \ \ if \ \lambda \neq 0$$
$$g(x) = \ln y \ \ if \ \lambda = 0,$$

can produce a distribution that is more nearly symmetric for well chosen λ. With respect to the radiation data in Table 1.3 Johnson and Wichern have shown that $\lambda = 0.25$ is a good choice. The transformed data is shown in the table below.

```
-1.51  -1.81  -1.39  -1.75  -2.11  -1.65  -1.87
-2.11  -1.87  -1.75  -1.94  -2.50  -2.74  -1.75
-1.75  -1.75  -2.50  -1.75  -2.74  -0.82  -1.75
-2.11  -2.34  -2.11  -1.51  -1.75  -1.51  -1.81
-1.87  -1.39  -1.75  -1.33  -1.70  -1.04  -2.50
-1.33  -1.33  -1.04  -1.04  -0.82  -1.04  -2.11
```

(a) Make an ordered stem and leaf plot of the data.
(b) Draw the histogram of the data. Use the class width 0.30 and class marks
$\{-2.7, -2.4, -2.1, -1.8, -1.5, -1.2, -0.9\}$.
(c) Compare the histogram for the transformed data with the histogram for the original data set (Figure 1.10). Is the distribution of the transformed data symmetric or nearly so?

Problem 1.3 *Refer to the radiation data through the open door in Problem 1.1. To obtain a data set with a more symmetric distribution Johnson and Wichern used the same function as in Problem 1.2; that is, the transformed data is given by*

$$y_i = \frac{x_i^{0.25} - 1}{0.25}.$$

The transformed data set is shown in the following table.

```
-1.04  -1.81  -1.04  -1.75  -1.75  -1.65  -1.81
-1.75  -1.81  -1.75  -1.94  -2.11  -2.74  -0.72
-1.65  -1.33  -2.21  -1.75  -2.74  -0.48  -1.65
-1.75  -2.11  -2.11  -1.51  -1.04  -1.51  -1.81
-1.81  -1.09  -1.75  -1.75  -1.75  -1.04  -1.65
-1.17  -1.33  -0.82  -0.97  -0.99  -1.65  -1.65
```

Draw the histogram. Is its distribution more symmetric than the original data in Problem 1.1?

Problem 1.4 *Refer to the lifetimes of type B filaments in Table 1.5; compute:*
(a) the order statistics;
(b) the cumulative frequency distribution $F(x)$ for: $x = 1065, 1100, 1143$.
(c) Graph the empirical distribution function $\hat{F}_n(x)$.

Problem 1.5 *The following data are the measurements of the capacity (in ampere-hours) of 10 batteries.*
$\{157, 152, 146, 153, 151, 154, 149, 148, 153, 147\}$
(a) Compute the order statistics.
(b) Compute the empirical distribution function $\hat{F}_n(x)$ for $x = 145, 150, 150.5, 152.9, 153.1, 158$.
(c) Graph the empirical distribution function $\hat{F}_n(x)$.

Problem 1.6 *The following data record 14 measurements of the air concentration (measured in μ/gm^3) of Benzo(a)pyrene (BAP) in a house located near a metal pipe foundry.*

| 10 | 10 | 25 | 40 | 40 | 45 | 45 |
| 55 | 55 | 70 | 75 | 90 | 220 | 285 |

Let $\hat{F}(x)$ denote the empirical distribution function. Compute:

(a) $\hat{F}(5)$ (b) $\hat{F}(40)$ (c) $\hat{F}(55)$ (d) $\hat{F}(300)$

(e) Find all solutions to the equation $\hat{F}(x) = 0.5$.

Problem 1.7 *The following table lists* 17 *measurements of the burst strength of a rocket motor chamber.*

15.30	17.10	16.30	16.05	16.75	16.60
17.10	17.50	16.10	16.10	16.00	16.75
17.50	16.50	16.40	16.00	16.20	

(Source: S. Weerahandi & R. A. Johnson (1992), Testing Reliability in a Stress–Strength Model when *x* and *y* are Normally Distributed, *Technometrics, vol. 34, no. 1, pp. 83-91. Used with permission.)*

(a) Make a horizontal bar chart.

(b) Make the dot plot and describe the shape of the distribution as belonging to one of the following types: symmetric? skewed left? skewed right?

(c) Compute the order statistics.

(d) What percentage of the values are less than or equal to 16.10? greater than or equal to 16.10?

(e) What percentage of the values are less than or equal to 16.40? greater than or equal to 16.40?

(f) What percentage of the values are less than or equal to 16.75? greater than or equal to 16.75?

(g) Draw the graph of the empirical distribution function.

Problem 1.8 *The following table lists* 24 *measurements (psi) of the operating pressure of a rocket motor.*

7.7401	7.7749	7.7227	7.77925	7.96195	7.4472
8.0707	7.89525	8.0736	7.4965	7.5719	7.7981
7.8764	8.1925	8.01705	7.9431	7.71835	7.87785
7.2904	7.7575	7.3196	7.6357	8.06055	7.9112

(Source: I. Guttman et al. (1988), Confidence Limits for Stress–Strength Models with Explanatory Variables, *Technometrics, vol. 30, no. 2, pp. 161-168. Used with permission.)*

(a) Compute the order statistics.

(b) What percentage of the values are less than or equal to 7.78? greater than or equal to 7.78?

(c) What percentage of the values are less than or equal to 7.65? greater than or equal to 7.65?

(d) What percentage of the values are less than or equal to 7.95? greater than or equal to 7.95?

(e) Draw the frequency histogram.

(f) Comment on the shape of the distribution. Is it symmetric? skewed left? skewed right?

Problem 1.9 *An International Equity Fund reported the annual returns for the years 2001-2005 in the following table.*

year	2001	2002	2003	2004	2005
returns	-9.04%	-8.25%	26.13%	22.94%	11.55%

(a) At the beginning of 2001 Ms. Allen invests $10,000 in this mutual fund where it remains for the next five years. Find the values of her portfolio (to the nearest dollar) at the end of each year beginning at 2001 and enter your answers in the table below.

year	2001	2002	2003	2004	2005
value of portfolio					

(b) Compute Ms. Allen's annualized rate of return for the five year period 2001-2005.

(c) It is known that the annualized rate of return is always less than the average return. (It is a consequence of the arithmetic-geometric mean inequality.) Compute the investor's average rate of return for the same period and verify that it is larger than the annualized rate of return.

Problem 1.10 *An important performance measure of a stock market portfolio is to compute its annualized returns for the best three-year period, as well as the worst three-year period, over the past ten years. For the S&P500 the best three-year period (over the 10-year period 1995-2004) is 1997-1999, and the worst three-year period is 2000-20002. The table below lists the annual returns for the S&P500 for the 6 years (1997-2002).*

year	1997	1998	1999	2000	2001	2002
return	33.4%	28.6%	21.0%	-9.1%	-11.9%	-22.1%

(a) At the beginning of 1997 Mr. Fisher withdraws $10,000 from his savings account and invests it in the S&P500, where it remains for the next three years. Find the values of Mr. Fisher's portfolio (to the nearest dollar) at the end of 1997, 1998, and 1999 and enter your answers in the table below.

year	1997	1998	1999
value of portfolio			

(b) Find the annualized rate of return for the three-year period (1997-1999).

(c) At the beginning of 2000 Mr. Black, envious of Mr. Fisher's financial success, also withdraws $10,000 from his savings account and invests it in the S&P500, where it remains for the next three years. Find the values of Mr. Black's portfolio (to the nearest dollar) at the end of 2000, 2001, and 2002, and enter your answers in the table below.

year	2000	2001	2002
value of portfolio			

(d) Find the annualized rate of return for the three-year period (2000-2002).

Problem 1.11 *Refer to the weekly 1999 return data for GE listed in Table 1.19.*
(a) Draw the frequency histogram of the weekly returns and comment on its shape. Is it symmetric? skewed left? skewed right?
(b) Comment on the similarities and differences between the histogram obtained in part (a) and the histogram displayed in Figure 1.11.

Problem 1.12 *Refer to the weekly 1999 return data for PTTRX listed in Table 1.19.*
(a) Draw the frequency histogram of the weekly returns and comment on its shape. Is it symmetric? skewed left? skewed right?
(b) Comment on the similarities and differences between the histogram obtained in part (a) and the histogram displayed in Figure 1.11.

Problem 1.13 *Refer to the weekly 1999 return data for BEARX listed in Table 1.19.*
(a) Draw the frequency histogram of the weekly returns and comment on its shape. Is it symmetric? skewed left? skewed right?
(b) Comment on the similarities and differences between the histogram obtained in part (a) and the histogram displayed in Figure 1.11.

Problem 1.14 *Refer to the weekly 2002 price and return data for GE listed in Tables 1.18 and 1.20. Use a statistical software package to:*
(a) Draw the time series plot of the weekly returns of GE stock. Is there any pattern in this time series that an investor can take advantage of?
(b) Draw the frequency histogram of the weekly returns and comment on its shape. Is it symmetric? skewed left? skewed right?
(c) Comment on the similarities and differences between the graph obtained in part (b) and the graph displayed in Figure 1.11.
(d) Draw the time series plot of the weekly 2002 prices of GE stock and compare it to its performance in 1999 (Figure 1.1).

Problem 1.15 *Refer to the weekly 2002 price and return data for PTTRX listed in Tables 1.18 and 1.20. Use a statistical software package to:*
(a) Draw the time series plot of the weekly returns of PTTRX mutual fund. Is there any pattern in this time series that an investor can take advantage of?
(b) Draw the frequency histogram of the weekly returns and comment on its shape. Is it symmetric? skewed left? skewed right?
(c) Comment on the similarities and differences between the graph obtained in part (b) and the graph displayed in Figure 1.11.
(d) Draw the time series plot of the weekly 2002 closing prices of PTTRX mutual fund and compare it to its performance in 1999 (Figure 1.2).

Problem 1.16 *Refer to the weekly 2002 return data for BEARX listed in Table 1.20. Use a statistical software package to:*
(a) Draw the time series plot of the weekly returns of BEARX stock. Is there any pattern in this time series that an investor can take advantage of?
(b) Draw the frequency histogram of the weekly returns and comment on its shape. Is it symmetric? skewed left? skewed right?
(c) Comment on the similarities and differences between the graph obtained in part (b) and the graph displayed in Figure 1.11.

Problem 1.17 *Carbon dioxide (CO_2) emissions are believed to be associated with the phenomenon of global warming. The following data come from a sample of the twenty countries* *with the highest CO_2 emissions in 1989.*

Key to the variables: *The four variables are: X_1 = 1989 emissions, X_2 = 1989 emissions per capita, X_3 = 1987 GNP per capita, and X_4 = 1987 emissions per GNP. The units were not given.*

Country	1989 emissions X_1	1989 emissions per capita X_2	1987 GNP per capita X_3	1987 emissions per GNP X_4
USA	1328.3	5.37	18529	2.8
Soviet Union	1038.2	3.62	8375	4.3
China	651.9	0.59	294	19.0
Japan	284.0	2.31	15764	1.3
India	177.9	0.21	311	6.1
FRG	175.1	2.86	14399	2.1
UK	155.1	2.70	10419	2.6
Canada	124.3	4.73	15160	2.9
Poland	120.3	3.15	1926	17.6
Italy	106.4	1.86	10355	1.7
France	97.5	1.74	12789	1.4
GDR	88.1	5.40	11300	4.9
Mexico	87.3	1.01	1825	5.4
South Africa	76.0	2.20	1870	12.5
Australia	70.3	4.22	11103	3.6
Czechoslovakia	61.8	3.95	9280	4.5
ROK	60.3	1.42	2689	4.4
Rumania	57.9	2.50	6030	4.2
Brazil	56.5	0.38	2021	1.9
Spain	55.5	1.42	5972	2.1

(Source: UN Conference on Trade and Development, UNCTAD/RDP/DFP/1, United Nations, New York, 1992. Note: The data for USSR, FRG, GDR and Czechoslovakia were obtained prior to the breakup of the Soviet Union and Czechoslovakia and prior to the reunification of Germany. In this data set the countries are ranked (in decreasing order) according to their total 1989 CO_2 emissions data; this is simply a display of the order statistics corresponding to the variable X_1.)

(a) Looking at the data one sees that the USA leads the world in total CO_2 emissions. Now, rank the countries according to the variable X_2 (1987 emissions per capita). What is the rank of the USA? of China?

(b) Rank the countries according to the variable X_4 (1987 emissions per GNP). What is the rank of the USA? of China?

(c) How would you explain the differences in the rankings of the various countries according to the variables X_2 and X_4, respectively?

Problem 1.18 *The following table records the results of a study of lead absorption in children of employees who worked in a factory where lead is used to make batteries. In this study the authors matched 33 such children (exposed) from different families to 33 children (control) of the same age and from the neighborhood whose parents were employed in industries not using lead. The purpose of the study was to determine if children in the exposed group were at risk because lead was inadvertently brought home by their parents. The third column titled "Difference" records the differences of the lead levels between each child in the exposed group and control group.*

Exposed X	Control Y	Difference D = X − Y	Exposed X	Control Y	Difference D = X − Y
38	16	22	10	13	-3
23	18	5	45	9	36
41	18	23	39	14	25
18	24	-6	22	21	1
37	19	18	35	19	16
36	11	25	49	7	42
23	10	13	48	18	30
62	15	47	44	19	25
31	16	15	35	12	23
34	18	16	43	11	32
24	18	6	39	22	17
14	13	1	34	25	9
21	19	2	13	16	-3
17	10	7	73	13	60
16	16	0	25	11	14
20	16	4	27	13	14
15	24	-9			

(Source: D. Morton et al. (1982), Lead Absorption in Children of Employees in a Lead–Related Industry, *American Journal of Epidemiology, vol. 115, pp. 549-555. Used with permission.)*

Draw, for each of the three variables (Exposed, Control, and Difference), the frequency histogram and describe the shape of each distribution.

Problem 1.19 *Cadmium (Cd) is a bluish–white metal used as a protective coating for iron, steel and copper. Overexposure to cadmium dust can damage the lungs, kidney and liver. The federal standard for cadmium dust in the workplace is $200\mu g/m^3$. The following (simulated) data was obtained by measuring the levels of cadmium dust at 5-minute intervals over a 4-hour period.*

```
181   190   192   197   201   204   206   192
199   189   200   198   192   189   197   198
192   204   199   196   195   199   195   198
198   202   193   194   197   189   200   193
197   195   209   201   195   202   195   193
190   190   193   196   200   196   198   195
```

(a) Make an ordered stem and leaf plot of the data using an appropriate stem and leaf.
(b) Draw the frequency histogram. Do the data appear to be symmetrically distributed?
(c) Comparing this data with the federal standard would you work in this factory? Justify your answer.

Problem 1.20 *The following table records the measured capacitances of 40 capacitors rated at 0.5 micro farads ($\mu(F)$).*

```
0.5121   0.5059   0.4838   0.4981
0.4851   0.4964   0.4955   0.4990
0.5092   0.5075   0.4905   0.5026
```

0.5010	0.4863	0.5100	0.5042
0.4944	0.5084	0.5031	0.5072
0.5170	0.4957	0.5056	0.5038
0.5066	0.4935	0.5113	0.5186
0.4986	0.5040	0.4967	0.5001
0.4920	0.5001	0.4972	0.5061
0.5080	0.4842	0.4878	0.4948

(a) Make an ordered stem and leaf plot using the first two nonzero digits as the stem values (ignore the decimal point). Compute the sample range.
(b) Make a dot plot and describe the shape of the frequency distribution.
(c) Make the frequency histogram; use Sturges' rule, Equation 1.12, to guide you in choosing the number of classes. Comment on the shape of the distribution.

Section 1.5: Quantiles of a distribution; medians, quartiles, and boxplots

Problem 1.21 *The following data record ten measurements of the ozone concentration (in ppm) in downtown Los Angeles.* {4.0, 1.9, 7.1, 4.7, 4.9, 5.8, 6.1, 6.1, 7.3, 3.0}.
Compute (two decimal place accuracy):
(a) the order statistics.
(b) the sample median.
(c) the upper and lower quartiles and the IQR (inter quartile range).

Problem 1.22 *Let $\hat{F}(x)$ be the empirical distribution function for the ozone data of Problem 1.21. Compute the empirical distribution function $\hat{F}(x)$ for the following x values:*

(a) $\hat{F}(4.5)$ (b) $\hat{F}(5.2)$ (c) $\hat{F}(5.6)$ (d) $\hat{F}(1.1)$ (e) $\hat{F}(6.9)$

Problem 1.23 *Refer to the Benzo(a)pyrene levels of Problem 1.6. Compute:*
(a) the sample median.
(b) the upper and lower quartiles and the IQR (inter quartile range).
(c) Draw the boxplot and identify the outliers if there be any.

Problem 1.24 *The 1989 total charges (in dollars) for 33 female patients, age 30-49, admitted for circulatory disorders are recorded in the following table. The data were collected by an insurance company interested in the distribution of claims.*

2337 2179 2348 4765 2088 2872 1924 2294
2182 2138 1765 2467 3609 2141 1850 3191
3020 2473 1898 7787 6169 1802 2011 2270
3425 3558 2315 1642 5878 2101 2242 5746
3041

(Source: E. W. Frees (1994), Estimating Densities of Functions of Observations, *Journal of the American Statistical Association, vol. 89, no. 426, pp. 517–525. Used with permission.)*
(a) Draw the histogram and describe the shape of the distribution.
(b) Compute the median, the upper and lower quartiles and determine the outliers.
(c) Draw the box plot.

Problem 1.25 *Refer to the lifetimes of type B filaments in Table 1.5.*
(a) Compute the sample range and median of the lifetimes.
(b) Compute the lower and upper quartiles and the IQR. Are there any outliers?

Problem 1.26 *The annual income (in thousands of dollars) of 30 families in a metropolitan region are listed below. The data were collected to determine the profitability of locating a shopping mall in the region.*

19	51	43	43	51	23
23	23	31	43	67	31
23	99	67	51	91	55
31	27	35	31	35	35
39	27	47	71	111	63

(a) Compute the quintiles of the income distribution; that is, compute the 20th, 40th, 60th and 80th percentiles.
(b) Draw the box plot.

Problem 1.27 *The lifetimes (in hours) of 20 bearings are recorded in the following table.*

6278	3113	9350	5236	11584
12628	7725	8604	14266	6215
3212	9003	3523	12888	9460
13431	17809	2812	11825	2398

(Source: R. E. Schafer and J. E. Angus (1979), Estimation of Weibull Quantiles With Minimum Error in the Distribution Function, *Technometrics, vol. 21, no. 3, pp. 367-370. Used with permission.)*
The 10th percentile $Q(0.10)$ of the empirical distribution function is called, as noted earlier, the safe life.
(a) Compute $Q(0.10)$ and the median $\tilde{x} = Q(0.50)$.
(b) Compute the sample range, the upper and lower quartiles, and the IQR.
(c) Identify any outliers.   (d) Draw the box plot.

Problem 1.28 *The following data record the reverse–bias collector current (in micro-amperes) for a set of twenty transistors.*

Reverse–bias collector current (in micro-amperes) for a set of $n = 20$ transistors

0.20	0.16	0.20	0.48	0.92
0.33	0.20	0.53	0.42	0.50
0.19	0.22	0.18	0.17	1.20
0.14	0.09	0.13	0.26	0.66

(Source: M. G. Natrella (1963), Experimental Statistics, National Bureau of Standards, *vol. 91, U.S. Government Printing Office, Washington, D.C.)*
(a) Compute the sample range and median of the lifetimes.
(b) Compute the lower and upper quartiles and the IQR. Are there any outliers?
(c) Draw the box plot.

Problem 1.29 *In Example 1.20 we used the graph of the empirical distribution function (Figure 1.12) to compute the median of the radiation data (Table 1.3).*
(a) Use the same graphical method to compute the lower and upper quartiles.
(b) Compute the lower and upper quartiles using Equation 1.16. Your answers in both parts (a) and (b) should be the same.

Problem 1.30 *Show that the ith order statistic $x_{(i)}$ is the $(i - 0.5)/n$ quantile by deriving Equation 1.17.*

Problem 1.31 *Refer to the battery data of Problem 1.5.*
(a) Compute the sample range, median, Q_1, Q_3 and the IQR.
(b) Draw the box plot.

Problem 1.32 *Refer to the lifetimes of type A and type B filaments listed Table 1.5. Draw the side-by-side box plots of the two filament types. Using this information compare the lifetimes and variability of the type A and type B filaments.*

Problem 1.33 *Refer to the breast-milk volume data in Example 1.7 (Table 1.6).*
(a) Compute the median $\tilde{x}$, Q_1, Q_3, and IQR for the smoking mothers data. Identify all outliers, if any.
(b) Compute the median, Q_1, Q_3, and IQR for the non-smoking mothers data. Identify all outliers, if any.
(c) Draw the side-by-side box plots for the smoking and non-smoking mothers.
(d) Using the results of parts a, b, c above compare the two populations with respect to the effect, if any, that smoking has on milk volume.

Problem 1.34 *Refer to the infant weight gain data in Example 1.7 (Table 1.6).*
(a) Compute the median $\tilde{x}$, Q_1, Q_3, and IQR of the infant weight gain for the smoking mothers data. Identify all outliers, if any.
(b) Compute the median, Q_1, Q_3, and IQR of the infant weight gain for the non-smoking mothers data. Identify all outliers, if any.
(c) Draw the side-by-side box plots for the smoking and non-smoking mothers.
(d) Using the results of parts a, b, c above compare the two populations with respect to the effect, if any, that smoking has on infant weight gain.

Problem 1.35 *Refer to the DNA damage data given in Table 1.7.*
(a) Compute the median $\tilde{x}$, Q_1, Q_3, and IQR for the N-1-THB-ADE levels for the exposed workers. Identify all outliers, if any.
(b) Compute the median $\tilde{x}$, Q_1, Q_3, and IQR for the N-1-THB-ADE levels for the control group of workers. Identify all outliers, if any.
(c) Draw the side-by-side box plots for the exposed and control group of workers.
(d) Using the results of parts a, b, c above compare the two populations with respect to the effect, if any, that exposure to the chemical $1, 3$–butadiene has on the exposed group of workers.

Problem 1.36 *Refer to the failure times data of Table 1.21.*
(a) Compute the safe life $Q(0.10)$.
(b) Compute the sample range and the IQR.
(c) Identify all outliers.
(d) Draw the box plot.

Problem 1.37 *The times between successive failures of the air conditioning system of Boeing 720 jet airplanes is shown in the table below. It contains 97 values taken from 4 airplanes. This is an edited version of the original data set which contained the maintenance data on a fleet of 13 airplanes and 212 values. The purpose of the data collection was to obtain information on the distribution of failure times and to use this information for predicting reliability, scheduling maintenance, and providing spare parts. Notice that the* *number of observations is not the same for each airplane; the periods "." represent the missing values.*

Time intervals between successive failures of an air conditioning system. Missing values are displayed as "."

B7912	B7913	B7914	B8045	B7912	B7913	B7914	B8045
23	97	50	102	12	54	36	34
261	51	44	209	120	31	22	.
87	11	102	14	11	216	139	.
7	4	72	57	3	46	210	.
120	141	22	54	14	111	97	.
14	18	39	32	71	39	30	.
62	142	3	67	11	63	23	.
47	68	15	59	14	18	13	.
225	77	197	134	11	191	14	.
71	80	188	152	16	18	.	.
246	1	79	27	90	163	.	.
21	16	88	14	1	24	.	.
42	106	46	230	16	.	.	.
20	206	5	66	52	.	.	.
5	82	5	61	95	.	.	.

(Source: F. Proschan (1963), Theoretical Explanation of Observed Decreasing Failure Rate, *Technometrics, vol. 5, no. 3, pp. 375-383. Used with permission.)*
Compute the sample range, median, lower and upper quartiles, IQR , outliers, and draw the box plots for:
(a) Airplane B7912 (b) Airplane B7913   (c) Airplane B7914 (d) Airplane B8045

Problem 1.38 *Refer to the data on lead absorption in children of employees who worked in a factory where lead is used to make batteries (see Problem 1.18). Compare the exposed group with the control group by drawing the box plot for the difference variable $d_i = x_i - y_i$, ($i = 1, \ldots, 33$). What inferences can you make concerning the effect on children of the parents' exposure to lead in their workplace?*

Problem 1.39 *Refer to the capacitances data in Problem 1.20.*
(a) Compute the median, the lower and upper quartiles, and the IQR; are there any outliers?
(b) Draw the box plot.

Problem 1.40 *The following data gives the results of a study of 25 hospitalized schizophrenic patients who were treated with anti-psychotic medication, and after a period of time were classified as psychotic or non-psychotic by hospital staff. Samples of cerebrospinal fluid were taken from each patient and assayed for dopamine b-hydroxylase (DBH) activity. The units of measurement are omitted.*

```
Judged non-psychotic:
0.0104   0.0105   0.0112   0.0116   0.0130   0.0145   0.0154   0.0156
0.0170   0.0180   0.0200   0.0200   0.0210 0.0230   0.0252

Judged psychotic:
0.0150   0.0204   0.0208   0.0222   0.0226   0.0245 0.0270   0.0275
0.0306 0.0320
```

(Source: D.E. Sternberg, D.P. Van Kammen, P. Lerner and W.E. Bunney (1982), Schizophrenia: dopamine b-hydroxylase activity and treatment response, Science, vol. 216, pp. 1423-1425. Used with permission.)

(a) Compute the median, the lower and upper quartiles, and the IQR for the non-psychotic patients; are there any outliers?

(b) Compute the median, the lower and upper quartiles, and the IQR for the psychotic patients; are there any outliers?

(c) Draw the side-by-side box plots for the non-psychotics and psychotics.

Problem 1.41 *The manufacturer's cost and the selling price (in dollars) of light bulbs are denoted by C_1, C_2, respectively. The bulb is guaranteed for H hours; if the bulb fails before then the manufacturer guarantees a total refund. Let R denote the manufacturer's net revenue. Using the type A filaments data of Table 1.5 in Example 1.6 compute:*

(a) R when $C_1 = 1$, $C_2 = 2$ and $H = 1,400$.

(b) R when $C_1 = 1$, $C_2 = 2$ and $H = 1,500$.

Sections (1.6-1.6.3): Sample means, standard deviations, mean-standard deviation diagrams

Problem 1.42 *The analytical methods committee of the Royal Society of Chemistry reported the following results on the determination of tin in foodstuffs. The samples were boiled with hydrochloric acid under reflux for different times.*

Refluxing time (min)	Tin found (mg/kg)
30	55 57 59 56 56 59
75	57 55 58 59 59 59

(Source: The Determination of Tin in Organic Matter by Atomic Absorption Spectrometry, *The Analyst, vol. 108, pp. 109-115. Used with permission.)*

Compute the sample means, sample variances and standard deviations for the 30 and 75 minute refluxing times.

Problem 1.43 *Refer to Lord Rayleigh's data in Table 1.4. Compute:*

(a) The sample mean and standard deviation of the mass of nitrogen obtained from air.

(b) The sample mean and standard deviation of the mass of nitrogen obtained by chemical decomposition.

Problem 1.44 *Refer to the ozone level data of Problem 1.21. Compute the sample mean, sample variance and standard deviation of the ozone levels.*

Problem 1.45 *Refer to the Benzo(a)pyrene levels of Problem 1.6. Compute (two decimal place accuracy): the sample mean, sample variance, and sample standard deviation.*

Problem 1.46 *Eleven samples of the outflow from a sewage treatment plant were divided in two parts with one half going to a commercial laboratory (columns 1 and 2) and the other half sent to a state laboratory (columns 3 and 4). Each laboratory measured the biochemical oxygen demand (BOD) and the suspended solids (SS). The data appear in the following table.*

Commercial BOD	Lab SS	State BOD	Lab SS
6	27	25	15
6	23	28	13
18	64	36	22
8	44	35	29
11	30	15	31
34	75	44	64
28	26	42	30
71	124	54	64
43	54	34	56
33	30	29	20
20	14	39	21

(Source: R. A. Johnson and D. W. Wichern (1992), Applied Multivariate Statistical Analysis, 3rd ed., Prentice Hall, p. 223. Used with permission.)

(a) Compute the sample means, sample variances and standard deviations for biochemical oxygen demand (BOD) obtained by each laboratory. Compare the relative precision of their results. Are the mean biochemical oxygen demands obtained by the two laboratories in agreement with one another? Use the appropriate data analysis tools to guide your thinking.

(b) Compute the sample means, sample variances and standard deviations for the suspended solids (SS) for each laboratory. Compare the relative precision of their results. Are the mean amounts of the suspended solids obtained by the two laboratories in agreement with one another?

Problem 1.47 *Find the sample means, variances and standard deviations of the lifetimes of the type B filaments in Table 1.5.*

Problem 1.48 *Find the sample means, variances and standard deviations of the burst strengths of a rocket motor listed in Problem 1.7.*

Problem 1.49 *One measure of the effect of sewage effluent on a lake is to determine the concentration of nitrates in the water. Assume that the following data, which are simulated, were obtained by taking 12 samples of water, dividing each sample into two equal parts and measuring the nitrate concentration for each part by two different methods. We are interested in determining whether the two methods yield similar results.*

Two methods for measuring the amount of nitrates in water

Method 1 x_i	Method 2 y_i	Difference $d_i = x_i - y_i$
119	95	24
207	159	48
213	174	39
232	344	-112
208	256	-48
193	175	18
61	124	-63
376	360	16
314	350	-36
55	34	21
215	235	-20
101	198	-97

(a) Compute the sample means and sample standard deviations for each variable (method 1, method 2, difference) in the data set.
(b) Compute the box plots for each variable (method 1, method 2, difference) in the data set.
(c) Using the results in parts (a) and (b) answer, as best you can, the following question: Do the data indicate a difference between the two methods of measuring the nitrate concentration? Justify your answer.

Problem 1.50 *Refer to the DNA damage data given in Table 1.7.*
(a) Compute the sample mean, sample variance, and sample standard deviation for the N-1-THB-ADE levels for the exposed workers.
(b) Compute the sample mean, sample variance, and sample standard deviation for the N-1-THB-ADE levels for the control group of workers.
(c) Using the results of parts a and b above, compare the two populations with respect to the effect, if any, that exposure to the chemical 1,3-butadiene has on the exposed group of workers.

Problem 1.51 *The frequency distribution for the heights of 36 adult males is given in the table below. Key to the table: Height $= x$ and Freq $= f(x)$.*

HEIGHT	64.9	65	65.5	66	67	67.8	68	68.1	68.5	69	70
FREQ	1	2	1	1	1	1	1	1	1	3	3

HEIGHT	70.1	70.5	70.8	71	71.8	72	72.8	73	73.5	74	75
Freq	1	1	1	6	1	2	1	1	2	2	2

Compute the sample mean and sample standard deviation of the variable height. Hint: Use Equations 1.23 and 1.24.

Problem 1.52 *The frequency distribution for girls' scores on an achievement test is given in the table below. Key to the table: Score $= x$ and Freq $= f(x)$.*

SCORE	14	15	19	20	21	24	25	27	28	29	31	32
FREQ	1	1	1	1	1	1	3	1	1	3	1	2

SCORE	33	34	35	36	37	38	39	40	42	43	45
FREQ	1	3	3	1	2	3	1	2	1	1	1

Compute the sample mean and sample standard deviation of the girls' scores. Hint: Use Equations 1.23 and 1.24.

Problem 1.53 *Find the sample means, variances, and standard deviations of the bearing lifetimes of Problem 1.27.*

Problem 1.54 *Refer to the breast-milk volume data in Example 1.7 (Table 1.6).*
(a) Compute the sample mean, sample variance, and sample standard deviation of the milk

volume for smoking mothers.
(b) Compute the sample mean, sample variance, and sample standard deviation for non-smoking mothers.
(c) Using the results of parts a and above compare the two populations with respect to the effect, if any, that smoking has on milk volume.

Problem 1.55 *Refer to the infant weight gain data in Example 1.7 (Table 1.6).*
(a) Compute the sample mean, sample variance and sample standard deviation of the infant weight gain for smoking mothers.
(b) Compute the sample mean, sample variance and sample standard deviation of the infant weight gain for non-smoking mothers.
(c) Using the results of parts (a) and (b) above compare the two populations with respect to the effect, if any, that smoking has on infant weight gain.

Problem 1.56 *With reference to the failure times of the air conditioning systems in Problem 1.37 find the sample means, variances and standard deviations of:*
(a) Airplane B7912 (b) Airplane B7913   (c) Airplane B7914 (d) Airplane B8045
(e) How do you explain the fact that in each of the above cases the sample mean is greater than the sample median?

Problem 1.57 *Refer to the battery data of Problem 1.5. Compute the sample mean and sample standard deviation.*

Problem 1.58 *Refer to the reverse–bias collector current of Problem 1.28. Compute the sample mean and sample standard deviation.*

Problem 1.59 *Compute the sample mean, sample variance and sample standard deviation for the capacitances listed in Problem 1.20.*

Problem 1.60 *Refer to the data in Problem 1.18.*
(a) Compare the lead levels in the blood of the exposed and control groups of children by computing the sample mean, sample variance and sample standard deviation of the amount of lead in the blood of both groups.
(b) What inferences can you draw from this comparison?

Problem 1.61 *The data in Problem 1.18 come from an experiment where the goal is to control all the factors except the one we are testing. The factors of age and neighborhood were controlled by pairing the children of the same age and neighborhood; they differed with respect to their parents' exposure to lead. This is an example of paired data. One way of comparing the blood lead levels between the two populations is to compute the sample mean and sample standard deviation of D, the difference between the exposed and control group of children. Calculate the sample mean and sample standard deviation for the variable D and interpret your results.*

Problem 1.62 *The following table gives the concentration of thiol in lysate in the blood of two groups: a "normal" group and a group suffering from rheumatoid arthritis.*

Thiol Concentration
Normal	*Rheumatoid*
1.84	2.81
1.92	4.06
1.94	3.62
1.92	3.27
1.85	3.27
1.91	3.76
2.07	

(Source: J. C. Miller and J. N. Miller (1993), Statistics for Analytical Chemistry, 3rd ed., Prentice–Hall.)

(a) Compute the sample mean, sample variance and sample standard deviation of the thiol concentrate for the normal group.

(b) Compute the sample mean, sample variance and sample standard deviation of the thiol concentrate for the Rheumatoid group.

Problem 1.63 *Refer to the dopamine b-hydroxylase (DBH) activity data set of Problem 1.40.*

(a) Compute the sample mean and sample standard deviation of the DBH activity for the non-psychotic group.

(b) Compute the sample mean and sample standard deviation of the DBH activity for the psychotic group.

Problem 1.64 *The following data give the results of an experiment to study the effect of two different word processing programs on the length of time (measured in minutes) to type a text file. A group of 16 were randomly divided into 2 groups of 8 secretaries each, and each group was randomly assigned to one of the two word processing programs.*

Program	Typing Time (min)
A	13 12 9 18 13 17 15 20
B	18 17 13 18 15 22 16 22

(a) Compute the sample mean, sample variance and sample standard deviation of the time lengths for the group A secretaries.

(b) Compute the sample mean, sample variance and sample standard deviation of the time lengths for the group B secretaries.

Problem 1.65 *Compute the sample mean, sample variance and sample standard deviation for the cadmium levels displayed in Problem 1.19.*

Problem 1.66 *Refer to the weekly 2002 return data for GE listed in Table 1.20, Section 1.11. Use a statistical software package to compute:*

(a) the sample mean of the 2002 GE returns.

(b) the sample standard deviation of the 2002 GE returns.

(c) Compare these results to the mean and standard deviation of the 1999 GE returns given in Table 1.13.

Problem 1.67 *Refer to the weekly 2002 return data for PTTRX listed in Table 1.20, Section 1.11. Use a statistical software package to compute:*

(a) the sample mean of the 2002 PTTRX returns.

(b) the sample standard deviation of the 2002 PTTRX returns.

(c) Compare these results to the mean and standard deviation of the 1999 PTTRX returns given in Table 1.13.

Problem 1.68 *Refer to the weekly 2002 return data for BEARX listed in Table 1.20, Section 1.11. Use a statistical software package to compute:*
(a) the sample mean of the 2002 BEARX returns.
(b) the sample standard deviation of the 2002 BEARX returns.
(c) Compare these results to the mean and standard deviation of the 1999 BEARX returns given in Table 1.13.

Problem 1.69 *(a) Carefully plot the following risky portfolios on the graph below, with risk s plotted on the x axis and mean return $\bar{r}$ on the y axis.*

Portfolio	A	B	C	D	E	F	G	H
risk s	23	21	25	29	29	32	35	45
mean return $\bar{r}$	10	12.5	15	16	17	18	18	20

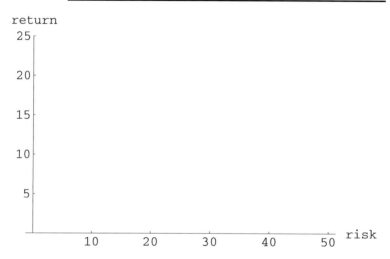

(b) Referring to the graph you constructed in part (a) you can see that five of these portfolios are efficient and three are not. Identify the inefficient ones and explain why they are inefficient.

Problem 1.70 *Suppose each observation x_i is bounded above and below by M, m, respectively; that is, assume that $m \leq x_i \leq M$, $i = 1, \ldots n$.*
(a) Show that $m \leq \bar{x} \leq M$.
(b) The writer Garrison Keillor has described the mythical town of Lake Wobegon as a place "where all the women are strong, the men are good looking and the children above average." Let x_i denote the grade point average (GPA) of the ith student and $\bar{x}$ denote the average GPA for all the students in Lake Wobegon high school. Explain why it is impossible for every student to have a GPA that is above the average. In other words, show that it is mathematically impossible to have $x_i > \bar{x}$ for every x_i. Hint: Use the result stated in part (a).

Problem 1.71 *The following data comes from an experiment that measures the weight loss for a group of 30 men who have been randomly assigned to three different weight loss programs labeled A, B, C, respectively.*

A	B	C	A	B	C
15.3	3.4	11.9	10	5.2	8.8
2.1	10.9	13.1	8.3	2.5	12.5
8.8	2.8	11.6	9.4	10.5	8.6
5.1	7.8	6.8	12.5	7.1	17.5
8.3	0.9	6.8	11.1	7.5	10.3

(a) As a first step toward comparing the three diets, compute the means and standard deviations of the weight losses for each of the three diets.
(b) Compute the side-by-side box plots of the weight losses for each of the three diets. Are there any outliers?
(c) Using the results of parts (a) and (b) comment on the similarities and differences between the weight loss data for each of the diets. Which diet is most effective? least effective?

Problem 1.72 *35 items were inspected and the number of defects per item were recorded:*
2 3 0 1 0 0 0
0 0 2 3 3 1 3
3 1 0 0 2 0 0
2 1 2 0 1 0 3
1 4 0 0 1 0 4
(a) Compute the mean and median number of defects per item.
(b) Compare the mean and median and explain why they differ.

Problem 1.73 *Suppose the data set $Y = \{y_1, \ldots, y_n\}$ is obtained from $X = \{x_1, \ldots, x_n\}$ by means of the linear transformation $y_i = ax_i + b$, where a, b are constants and $a > 0$.*
(a) Show that the order statistics for the data set Y are given by $y_{(i)} = ax_{(i)} + b$.
(b) Let $Q_x(p)$ and $Q_y(p)$ denote the quantiles of order p for the X and Y data sets, respectively. Show that $Q_y(p) = aQ_x(p) + b$.
(c) Let $\hat{F}_n$ and $\hat{G}_n$ be the empirical distribution functions of the data sets $\{x_1, \ldots, x_n\}$ and $\{y_1, \ldots, y_n\}$, respectively. Show that,

$$\hat{G}_n(y) = \hat{F}_n\left(\frac{y-b}{a}\right).$$

Section 1.7: Covariance, correlation, scatter plots, and regression lines

Problem 1.74 *The two columns of data in the table below are the living area X (in square feet) and the selling price $\$Y$ (in thousands of dollars) of 12 residential properties.*

Area(sq.feet)	Price(thousands of dollars)
1360	78.5
1940	175.7
1750	139.5
1550	129.8
1790	95.6

1750	110.3
2230	260.5
1600	105.2
1450	88.6
1870	165.7
2210	225.3
1480	68.8

(a) Draw the scatter plot of selling price vs. area.
(b) Compute the slope and intercept of the regression line.
(c) A person builds a 300 square foot addition to her home. What value is added to the home?

Problem 1.75 *The following data record the amount of water (x), in centimeters, and the yield of hay (y), in metric tons per hectare, on an experimental farm.*

water (x)	30	45	60	75	90	105	120
yield (y)	2.11	2.27	2.5	2.88	3.21	3.48	3.37

(a) Draw the scatter plot (x_i, y_i).
(b) Calculate the least squares estimates b_0 and b_1 and draw the graph of the regression line $y = b_0 + b_1 x$. Use the same axes on which you graphed the scatter plot.
(c) Compute the fitted value for $x = 45$.
(d) What is the effect on hay yield if the amount of water is increased by 10cm?

Problem 1.76 *The table below records the curbside weight (in units of 1000 lbs), fuel consumption in miles per gallon (mpg), engine displacement (disp), and fuel consumption in gallons per 100 miles (gpm) for 18 2004-05 model year automobiles. (Displacement is a measure of the size, in liters, of the automobile engine.)*
(a) Draw the scatter plot of mpg (y) vs. car weight (x).
(b) Calculate the least squares estimates b_0 and b_1 and draw the graph of the regression line $y = b_0 + b_1 x$. Use the same axes on which you graphed the scatter plot.
(c) Compute the fitted values for $x = 3.39, 4.07, 5.59$.
(d) What is the effect on an automobile's fuel efficiency (mpg) if its weight is increased by 500 lbs?

Automobile	Weight	mpg	Disp	gpm
Acura RL	3.86	25	3.5	4.000
Acura RSX	2.78	31	2	3.226
Acura TL	3.565	27	3.2	3.704
Audi A4	3.745	23	3	4.348
Audi A8	4.505	20	4.2	5.000
BMW3	3.39	26	3	3.846
BMW5	3.65	24	3	4.167
BMW7	4.505	21	4.4	4.762
Buick LeSabre	3.64	25	3.8	4.000
Buick Park Avenue	3.97	26	3.8	3.846
Cadillac CTS	3.62	24	3.2	4.167
Cadillac DeVille	4.07	23	4.6	4.348
Chevy Impala	3.655	24	3.8	4.167
Chevy Malibu	3.29	31	3.5	3.226
Chevy Suburban	5.59	17	5.3	5.882
Ford Crown Victoria	4.18	21	4.6	4.762
Infiniti FX	4.295	21	3.5	4.762
Lexus GX470	4.825	18	4.7	5.556

Problem 1.77 *Refer to the fuel efficiency data set of the previous Problem 1.76.*
(a) Draw the scatter plot of gpm (y) vs. car weight (x).
(b) Calculate the least squares estimates b_0 and b_1 and draw the graph of the regression line $y = b_0 + b_1 x$. Use the same axes on which you graphed the scatter plot.
(c) Compute the fitted values for $x = 3.39, 4.07, 5.59$.
(d) What is the effect on an automobile's fuel efficiency (gpm) if its weight is increased by 500 lbs?

Problem 1.78 *Refer to the fuel efficiency data set of Problem 1.76.*
(a) Draw the scatter plot of mpg (y) vs. engine displacement (x).
(b) Calculate the least squares estimates b_0 and b_1 and draw the graph of the regression line $y = b_0 + b_1 x$. Use the same axes on which you graphed the scatter plot.
(c) Compute the fitted values for $x = 2.0, 3.8, 5.3$.
(d) What is the effect on an automobile's fuel efficiency (mpg) if its displacement is increased by 0.5 liters?

Problem 1.79 *Refer to the fuel efficiency data set of Problem 1.76.*
(a) Draw the scatter plot of gpm (y) vs. engine displacement (x).
(b) Calculate the least squares estimates b_0 and b_1 and draw the graph of the regression line $y = b_0 + b_1 x$. Use the same axes on which you graphed the scatter plot.
(c) Compute the fitted values for $x = 2.0, 3.8, 5.3$.
(d) What is the effect on an automobile's fuel efficiency (gpm) if its weight is increased by 0.5 liters?

Problem 1.80 *Refer to the StatLab data set (Table 1.16).*
(a) Draw the scatter plot of mother's weight (y) vs. mother's height (x).
(b) Compute the sample correlation r.

Problem 1.81 *Sir Francis Galton (1822–1911), author of* Hereditary Genius *(1869) and a cousin of Charles Darwin (1809–1882), was interested in finding scientific laws governing the inheritance of physical traits. He was particularly interested in how the heights of parents were passed on to their children. Consider this question using the StatLab data set (Table 1.16).*
(a) Draw the scatter plot of the daughter's height (y) against the mother's height(x), and compute the sample correlation.
(b) Galton also considered the "midparent" height, obtained by taking the average of the heights of the mother and father. Define the new variable

$$midparent\ height = \frac{(mheight) + (fheight)}{2}$$

and draw the scatter plot of the daughter's height (y) against the midparent height (x), and compute the sample correlation.
(c) Comparing the scatter plots and sample correlations computed in parts (a) and (b) can you determine which variable—mother's height or midparent height—is most positively associated with the daughter's height?

Problem 1.82 *Refer to the weekly 2002 return data for GE listed in Table 1.20, Section 1.11. Use a statistical software package to:*
(a) Draw the scatter plot of GE returns (y) against the S&P returns (x) and compute the correlation coefficient. Compare this scatter plot to the one in Figure 1.18 and comment on the similarities and differences.
(b) Compute GE's beta for the year 2002.

Problem 1.83 *Refer to the weekly 2002 return data for PTTRX listed in Table 1.20, Section 1.11. Use a statistical software package to:*
(a) Draw the scatter plot of PTTRX returns (y) against the S&P returns (x), and compute the correlation coefficient.
(b) Compute PTTRX's beta for the year 2002.

Problem 1.84 *Refer to the weekly 2002 return data for BEARX listed in Table 1.20, Section 1.11. Use a statistical software package to:*
(a) Draw the scatter plot of BEARX returns (y) against the S&P returns (x), and compute the correlation coefficient. Compare this scatter plot to the one in Figure 1.18 and comment on the similarities and differences.
(b) Compute BEARX's beta for the year 2002.

1.11   Large Data Sets

Table 1.16 *StatLab data set consisting of 36 observations on 8 variables*

Girl's height	Girl's weight	PEA score	RAV score	Mother's height	Mother's weight	Father's height	Father's weight
55.7	85	85	34	66.0	130	70.1	171
48.9	59	74	34	62.8	159	65.0	130
54.9	70	64	25	66.1	138	70.0	175
53.6	88	87	43	61.8	123	71.8	196
53.4	68	87	40	62.8	146	68.0	163
59.9	93	83	37	63.4	116	74.0	180
53.1	72	81	33	65.4	220	68.1	173
52.2	84	74	37	62.3	120	72.0	150
56.8	68	72	21	65.4	141	71.0	150
53.8	76	64	31	66.4	184	75.0	235
49.0	51	72	29	64.8	107	70.0	145
51.9	59	87	38	66.9	114	72.8	197
55.7	78	72	19	65.6	151	74.0	204
53.4	73	78	27	63.6	123	71.0	220
50.3	60	77	35	65.4	143	65.0	150
55.8	70	75	29	63.3	150	69.0	180
55.2	78	71	25	63.7	149	69.0	170
58.8	106	60	20	64.3	205	72.0	235
53.6	76	79	36	58.3	121	75.0	190
53.3	64	63	32	63.3	147	70.0	190
49.9	50	80	42	65.3	134	68.5	160
53.8	56	66	28	66.4	117	65.5	130
53.9	68	65	38	65.6	151	69.0	160
53.1	68	69	15	63.8	150	73.5	154
56.1	64	63	24	62.0	110	70.8	183
51.7	56	70	14	59.6	138	64.9	169
48.0	64	51	40	65.1	128	67.0	146
51.3	58	80	45	62.0	132	71.0	165
50.5	60	75	29	65.4	144	71.0	190
57.3	79	96	38	68.1	143	73.5	185
52.8	64	94	25	66.3	147	71.0	180
51.3	56	82	34	61.0	113	71.0	149
50.1	56	82	39	64.6	116	66.0	145
48.7	50	74	35	62.8	132	67.8	166
57.7	89	90	35	66.8	134	70.5	200
54.8	65	64	32	67.0	121	73.0	175

(Source: J. L. Hodges Jr. et al. (1975), *Statlab*, McGraw-Hill. Used with permission.)

Table 1.17 *Weekly closing prices of the S&P500 index and other securities, 1999*

S&P	GE	PTTRX	BEARX	S&P	GE	PTTRX	BEARX
1275.09	31.74	10.13	4.61	1403.28	36.55	9.79	3.66
1243.26	31.23	10.17	4.63	1418.78	37.10	9.85	3.65
1225.19	30.34	10.20	4.67	1356.94	35.95	9.77	3.84
1279.64	32.51	10.22	4.44	1328.72	34.02	9.72	3.98
1239.40	30.38	10.12	4.63	1300.29	33.39	9.65	4.14
1230.13	30.26	10.06	4.70	1327.68	33.63	9.67	4.06
1239.22	31.12	10.08	4.65	1336.61	35.21	9.71	4.01
1238.33	31.10	9.99	4.59	1348.27	36.38	9.74	3.98
1275.47	32.26	9.98	4.45	1357.24	36.33	9.72	4.02
1294.59	33.40	10.05	4.43	1351.66	37.20	9.73	4.01
1299.29	34.85	10.06	4.39	1335.42	37.45	9.75	4.09
1282.80	33.54	10.04	4.42	1277.36	36.83	9.80	4.23
1293.72	34.73	10.01	4.28	1282.81	36.74	9.69	4.46
1348.35	34.9	10.13	4.22	1336.02	39.05	9.68	4.47
1319.00	34.61	10.07	4.34	1247.41	36.23	9.64	4.68
1356.85	35.12	10.06	4.11	1301.65	39.32	9.61	4.42
1335.18	32.78	10.03	4.11	1362.93	42.41	9.72	4.12
1345.00	34.20	9.96	4.11	1370.23	41.86	9.78	3.96
1337.80	32.96	9.91	4.01	1396.06	42.37	9.78	3.94
1330.29	32.39	9.93	3.99	1422.00	43.10	9.74	3.91
1301.84	31.64	9.87	4.11	1416.62	42.45	9.71	3.96
1327.75	32.63	9.81	4.09	1433.3	42.65	9.70	3.81
1293.64	31.53	9.72	4.08	1417.04	46.15	9.74	3.83
1342.84	33.46	9.80	3.97	1421.03	47.54	9.63	3.64
1315.31	32.49	9.71	3.93	1458.34	49.43	9.59	3.55
1391.22	35.06	9.80	3.73	1469.25	48.56	9.58	3.57

Table 1.18 *Weekly closing prices of the S&P500 index and other securities, 2002*

S&P	GE	PTTRX	BEARX	S&P	GE	PTTRX	BEARX
1172.51	39.61	10.10	4.44	921.39	27.96	10.36	7.34
1145.60	36.98	10.27	4.61	847.75	25.93	10.36	7.68
1127.58	37.41	10.25	4.69	852.84	27.18	10.30	7.46
1133.28	37.00	10.18	4.66	864.24	28.84	10.41	7.61
1122.20	35.64	10.23	4.83	908.64	31.68	10.38	7.39
1096.22	36.03	10.28	5.16	928.77	30.85	10.34	7.19
1104.18	35.89	10.29	5.25	940.86	31.53	10.41	6.99
1089.84	36.84	10.31	5.41	916.07	29.48	10.44	7.29
1131.78	38.33	10.26	5.22	893.92	27.67	10.46	7.46
1164.31	39.45	10.11	4.71	889.81	26.45	10.51	7.51
1166.16	39.05	10.10	4.80	845.39	26.16	10.53	7.95
1148.70	36.80	10.08	4.94	827.37	24.09	10.50	7.97
1147.39	36.34	10.08	5.06	800.58	23.64	10.50	8.17
1122.73	36.05	10.18	5.17	835.32	23.84	10.43	7.85
1111.01	32.60	10.21	5.24	884.39	26.24	10.35	7.46
1125.17	32.75	10.24	5.21	897.65	25.84	10.40	7.26
1076.32	30.61	10.28	5.84	900.96	25.60	10.47	7.27
1073.43	30.80	10.26	5.97	894.74	24.71	10.55	7.43
1054.99	29.78	10.25	6.12	909.83	23.49	10.52	7.27
1106.59	32.50	10.21	5.78	930.55	26.04	10.51	6.96
1083.82	31.68	10.26	6.18	936.31	26.70	10.47	6.93
1067.14	30.26	10.30	6.33	912.23	25.65	10.53	7.24
1027.53	29.35	10.29	6.57	889.48	25.11	10.57	7.46
1007.27	28.86	10.36	6.71	895.76	25.55	10.63	7.41
989.14	28.13	10.36	6.95	875.40	24.50	10.67	7.40
989.82	28.40	10.29	6.85	879.82	24.16	10.67	7.33
989.03	29.03	10.24	6.88				

Table 1.19 *Weekly returns of the S&P500 index and other securities, 1999*

S&P	GE	PTTRX	BEARX	S&P	GE	PTTRX	BEARX
-0.025	-0.016	0.004	0.004	0.011	0.015	0.006	-0.003
-0.015	-0.028	0.003	0.009	-0.044	-0.031	-0.008	0.052
0.044	0.072	0.002	-0.049	-0.021	-0.054	-0.005	0.036
-0.031	-0.066	-0.010	0.043	-0.021	-0.019	-0.007	0.040
-0.007	-0.004	-0.006	0.015	0.021	0.007	0.002	-0.019
0.007	0.028	0.002	-0.011	0.007	0.047	0.004	-0.012
-0.001	-0.001	-0.009	-0.013	0.009	0.033	0.003	-0.007
0.030	0.037	-0.001	-0.031	0.007	-0.001	-0.002	0.010
0.015	0.035	0.007	-0.004	-0.004	0.024	0.001	-0.002
0.004	0.043	0.001	-0.009	-0.012	0.007	0.002	0.020
-0.013	-0.038	-0.002	0.007	-0.043	-0.017	0.005	0.034
0.009	0.035	-0.003	-0.032	0.004	-0.002	-0.011	0.054
0.042	0.005	0.012	-0.014	0.041	0.063	-0.001	0.002
-0.022	-0.008	-0.006	0.028	-0.066	-0.072	-0.004	0.047
0.029	0.015	-0.001	-0.053	0.043	0.085	-0.003	-0.056
-0.016	-0.067	-0.003	0.000	0.047	0.079	0.011	-0.068
0.007	0.043	-0.007	0.000	0.005	-0.013	0.006	-0.039
-0.005	-0.036	-0.005	-0.024	0.019	0.012	0.000	-0.005
-0.006	-0.017	0.002	-0.005	0.019	0.017	-0.004	-0.008
-0.021	-0.023	-0.006	0.030	-0.004	-0.015	-0.003	0.013
0.020	0.031	-0.006	-0.005	0.012	0.005	-0.001	-0.038
-0.026	-0.034	-0.009	-0.002	-0.011	0.082	0.004	0.005
0.038	0.061	0.008	-0.027	0.003	0.030	-0.011	-0.050
-0.021	-0.029	-0.009	-0.010	0.026	0.040	-0.004	-0.025
0.058	0.079	0.009	-0.051	0.007	-0.018	-0.001	0.006
0.009	0.042	-0.001	-0.019				

Note: The returns listed in Table 1.19 do not include dividends, reinvestments, and commissions.

Table 1.20 *Weekly returns of the S&P500 index and other securities, 2002*

S&P	GE	PTTRX	BEARX	S&P	GE	PTTRX	BEARX
-0.023	-0.066	0.017	0.038	-0.068	-0.037	0.012	0.067
-0.016	0.012	-0.002	0.017	-0.080	-0.073	0.000	0.046
0.005	-0.011	-0.007	-0.006	0.006	0.048	-0.006	-0.029
-0.010	-0.037	0.005	0.036	0.013	0.061	0.011	0.020
-0.023	0.011	0.005	0.068	0.051	0.098	-0.003	-0.029
0.007	-0.004	0.001	0.017	0.022	-0.026	-0.004	-0.027
-0.013	0.026	0.002	0.030	0.013	0.022	0.007	-0.028
0.038	0.040	-0.005	-0.035	-0.026	-0.065	0.003	0.043
0.029	0.029	-0.015	-0.098	-0.024	-0.061	0.002	0.023
0.002	-0.010	-0.001	0.019	-0.005	-0.044	0.005	0.007
-0.015	-0.058	-0.002	0.029	-0.050	-0.011	0.002	0.059
-0.001	-0.012	0.000	0.024	-0.021	-0.079	-0.003	0.003
-0.021	-0.008	0.010	0.022	-0.032	-0.019	0.000	0.025
-0.010	-0.096	0.003	0.014	0.043	0.008	-0.007	-0.039
0.013	0.005	0.003	-0.006	0.059	0.101	-0.008	-0.050
-0.043	-0.065	0.004	0.121	0.015	-0.015	0.005	-0.027
-0.003	0.006	-0.002	0.022	0.004	-0.009	0.007	0.001
-0.017	-0.033	-0.001	0.025	-0.007	-0.035	0.008	0.022
0.049	0.091	-0.004	-0.056	0.017	-0.049	-0.003	-0.022
-0.021	-0.025	0.005	0.069	0.023	0.109	-0.001	-0.043
-0.015	-0.045	0.004	0.024	0.006	0.025	-0.004	-0.004
-0.037	-0.030	-0.001	0.038	-0.026	-0.039	0.006	0.045
-0.020	-0.017	0.007	0.021	-0.025	-0.021	0.004	0.030
-0.018	-0.025	0.000	0.036	0.007	0.018	0.006	-0.007
0.001	0.010	-0.007	-0.014	-0.023	-0.041	0.004	-0.001
-0.001	0.022	-0.005	0.004	0.005	-0.014	0.000	-0.009

Note: The returns listed in Table 1.20 do not include dividends, reinvestments, and commissions.

Table 1.21 *Ordered list of failure times of kevlar/epoxy pressure vessels at 86% stress level (4, 300 psi)*

2.2	4.0	4.0	4.6	6.1
6.7	7.9	8.3	8.5	9.1
10.2	12.5	13.3	14.0	14.6
15.0	18.7	22.1	45.9	**55.4**
61.2	87.2	98.2	101.0	111.4
144.0	158.7	243.9	254.1	444.4
590.4	638.2	755.2	952.2	1108.2
1148.5	1569.3	1750.6	1802.1	

(Source: R.E. Barlow, R.H. Toland and T. Freeman (1988), *A Bayesian Analysis of the Stress–Rupture Life of Kevlar/Epoxy Spherical Pressure Vessels*, in Accelerated Life Testing and Experts' Opinions in Reliability, eds. C. Clarotti and D. Lindley, Amsterdam, North Holland. Used with permission.)

1.12   To Probe Further

The role of William Playfair (1759-1826) in the development of graphical methods for visualizing complex data is recounted in Costigan–Eaves and Macdonald–Ross (1990), *William Playfair and Graphics*, Statistical Science, vol. 5, no. 3, pp. 318-326. The importance of identifying the sources of variation in the quality of a product or service is emphasized repeatedly in Deming (1982), *Out of the Crisis*, MIT Press.

The modern theory of portfolio analysis, including the trade-off between risk and return, and the capital asset pricing model is discussed in the following texts (among many others):

1. G.J. Alexander, W.F. Sharpe, and J.V. Bailey (2001), *Fundamentals of Investments*, 3rd ed. Prentice Hall.

2. R. Brealy and S. Myers (2003), *Principles of Corporate Finance*, 7th ed., McGraw-Hill .

3. B.G. Malkiel's (1996), *A Random Walk Down Wall Street*, 6th ed., W.W. Norton is a more popular and less technical account.

Comparing risks of alternative investment strategies is an important task for investors; expectations of high returns along with higher risks must be weighed against alternatives with lower returns and lower risks. A stock's β is a useful tool for comparing the riskiness of a stock with *market risk*, i.e., the risk of the *market portfolio*, defined here as the S&P500, which, as noted earlier, consists of 500 stocks value weighted according to their market capitalization (= number of shares × share price). For a discussion of weak, moderate, and strong correlations the student should consult G.W. Snedecor and W.G. Cochran, *Statistical Methods*, 7th ed., University of Iowa Press (1980).

Chapter 2

Probability Theory

Probabilities are numbers of the same nature as distances in geometry or masses in mechanics. The theory assumes they are given but need assume nothing about their numerical values or how they are measured in practice.
William Feller (1906-1970), American mathematician

2.1   Orientation

After analyzing the distribution of the data via the methods of exploratory data analysis the next step is *statistical inference*, which is a scientific method for making inferences about a population based upon data collected from a sample. The reliability of these inferences, and this is a point whose importance cannot be exaggerated, depends crucially on how one chooses the sample. The theoretical ideal, which is often quite difficult to implement, is that we select the sample via *random sampling*; that is, *every subset of the population of the same size has an equal chance (probability) of being selected.* If the sample is selected by a method which deviates significantly from the random sampling ideal, then the inferences will tend to be *biased* because, in some sense, the sample gives a distorted picture of the population. In their book *Statistical Methods* (7th ed., 1980, Iowa State University Press, Ames, Iowa, p. 6), Snedecor and Cochran describe an interesting class room experiment that illustrates how biased sampling can occur.

In a class exercise, a population of rocks of varying sizes was spread out on a large table so that all the rocks could be seen. After inspecting the rocks and lifting any that they wished, students were asked to select samples of five rocks such that their average weight would estimate as accurately as possible the average weight of the whole population of rocks. The average weights of the students' samples *consistently overestimated* the population mean.

The students' consistent overestimation of the population mean is an example of *biased sampling.* Snedecor and Cochran offer the following reasonable explanation for the observed bias: "A larger and heavier rock is more easily noticed than a smaller one."

Since different students will, in general, select different rocks for their sample, it follows that the value of the average weight of the rocks in the sample will also vary. This is an example of a *sampling experiment*; it is also an example of an experiment whose outcomes are subject to chance. In this chapter our primary goal is to construct a mathematical model of random sampling based upon the theory of probability. It is worth pointing out, however, that modern probability theory also plays an important role in financial engineering, where it is a basic tool for dealing with the uncertainties associated in managing and hedging a portfolio of stocks, bonds, and options, as discussed in Sections 5.6-5.9.1.

The basic concepts of probability theory originated in the analysis of games of chance by such renowned mathematicians as Blaise Pascal (1623-1662) and Pierre de Fermat (1601-

1665). Indeed, one of the first books ever published on probability theory was titled *The Doctrine of Chances: or, A Method of Calculating the Probability of Events in Play* (1718) by the 18th century mathematician Abraham de Moivre (1667-1754).[1] In the dedication of his book, de Moivre, anticipating the charge "that the Doctrine of Chances has a tendency to promote Play," refuted it by arguing, "that this Doctrine is so far from encouraging Play, that it is rather a guard against it, by setting in a clear light, the advantages and disadvantages of those games wherein chance is concerned . . . ." It is an interesting fact, and one that we shall take advantage of, that the concepts and techniques used by de Moivre and his contemporaries to analyze games of chance are precisely the tools we need to construct a mathematical model of random sampling. For these reasons we begin our introduction to probability theory with the analysis of some simple games of chance.

Organization of Chapter

1. Section 2.2: Sample Space, Events, Axioms of Probability

2. Section 2.3: Mathematical Models of Random Sampling

3. Section 2.4: Conditional Probability, Bayes' Theorem, and Independence

4. Section 2.5: The Binomial Theorem*

5. Section 2.6: Chapter Summary

2.2   Sample Space, Events, Axioms of Probability Theory

Before shipping a mass produced product to its customers a manufacturer inspects a small proportion of the output for defects. The number of defective items will vary from sample to sample. In Deming's red bead experiment (see Example 1.3) a worker selects 50 beads from a bowl filled with 800 red beads and 3200 white beads. The red beads represent the defective items. This is an important and well known example of a *sampling experiment*. The number of red beads produced by six workers were 9, 5, 15, 4, 10, 8 (see Table 1.2). The workers' performance, as previously noted, appears to vary greatly, with one worker producing 4 red beads and another worker producing 15, nearly four times as many! The problem for the quality control engineer is to account for this observed variation in the workers' performance: Is it to be found in the raw materials used or is it due to differences among the workers themselves? We will see later, as a result of the theory presented here and in the following chapters, that most of the observed variation in the workers' performance is due to random sampling and is not due to any intrinsic differences among them.

The key to understanding variations in experimental data is to construct a mathematical model of random sampling. *Probability theory* is that branch of mathematics that constructs, analyzes and compares probability models of random phenomena. We begin to construct our model by defining some important terms and basic concepts.

Selecting items for inspection is an example of *experiment* with unpredictable outcomes. A sample that contains 15 defective items is an example of an *event*. The set of logically possible outcomes of an experiment is called the *sample space* (corresponding to the experiment). The formal definition follows.

---

[1]The word *Play* for de Moivre is synonymous with a game of chance.

Definition 2.1 *A sample space $\mathcal{S}$ is the set of all logically possible outcomes resulting from some experiment. An element of the sample space is called a sample point, or more simply, an outcome, and is denoted by s.*

These terms are just convenient abstractions of some fairly simple ideas that are nicely illustrated in the intuitive context of gambling games such as: craps, roulette, poker and state lotteries. Indeed, P. S. Laplace (1749-1827), a great mathematician and one of the founders of the theory of probability, wrote, "It is remarkable that a science which commenced with the consideration of games of chance, should be elevated to the rank of the most important subjects of human knowledge." So we shall begin our introduction to probability models with a brief description of simple games of chance since the mathematical techniques used to analyze them are widely used in science, engineering, and finance.

Example 2.1 *Throwing two dice*

To describe the sample space corresponding to this experiment we assume the two dice can be distinguished from one another by color, the first one colored red, say, and the other colored white. We denote the result of throwing the two dice by the ordered pair (i, j), so $(1, 3)$ means the red die and white die showed a 1 and a 3, respectively. This experiment has a total of 36 outcomes listed in Table 2.1. This is an example of a *finite discrete sample space*, since it consists of a finite number of sample points.

Table 2.1 *Sample space for throwing two dice*

(1,1)	(1,2)	(1,3)	(1,4)	(1,5)	(1,6)
(2,1)	(2,2)	(2,3)	(2,4)	(2,5)	(2,6)
(3,1)	(3,2)	(3,3)	(3,4)	(3,5)	(3,6)
(4,1)	(4,2)	(4,3)	(4,4)	(4,5)	(4,6)
(5,1)	(5,2)	(5,3)	(5,4)	(5,5)	(5,6)
(6,1)	(6,2)	(6,3)	(6,4)	(6,5)	(6,6)

Events An *event A* is simply a subset of the sample space $\mathcal{S}$; in symbols $A \subset \mathcal{S}$. The event that contains no sample points at all is the *impossible event* and is denoted $\emptyset$.

Notation Unless otherwise stated, $\mathcal{S}$ will denote a fixed, but otherwise unspecified sample space. Subsets of the sample space are denoted by the capital letters A, B, C, etc.

Example 2.2 *Playing craps*

The game of craps is played when the player (also called the "shooter") rolls two dice; the outcome (i.e., the sample point) is the sum of the numbers shown on the dice. If, for example, the two dice show $(1, 3)$, then the outcome is 4. We begin with a partial list of events upon which the players can place a bet. **Notation:** The symbol A_i denotes the event that an i is thrown.

1. $A_2 = (1, 1)$

2. $A_{12} = \{(6, 6)\}$

3. $A_{11} = \{(5, 6), (6, 5)\}$

4. $A_7 = \{(3, 4), (4, 3), (2, 5), (5, 2), (1, 6), (6, 1)\}$

5. *Throwing a natural*: This means the player throws a 7 or an 11; there are 8 sample points in a natural:

$$\text{Natural} = \{(3,4), (4,3), (2,5), (5,2), (1,6), (6,1), (5,6), (6,5)\}.$$

6. *Throwing a point:* This means the player throws a $4, 5, 6, 8, 9$ or 10. This event contains 24 sample points, which are too many to be listed here; for our purposes the verbal description is entirely satisfactory.

7. Table 2.2 lists the 11 sample points for the game of craps.

Table 2.2 *Sample space for the game of craps displayed in the format of a horizontal bar chart*

A_i	Sample points in A_i
A_2	$(1,1)$
A_3	$(1,2), (2,1)$
A_4	$(1,3), (2,2), (3,1)$
A_5	$(1,4), (2,3), (3,2), (4,1)$
A_6	$(1,5), (2,4), (3,3), (4,2), (5,1)$
A_7	$(1,6), (2,5), (3,4), (4,3), (5,2), (6,1)$
A_8	$(2,6), (3,5), (4,4), (5,3), (6,2)$
A_9	$(3,6) (4,5), (5,4), (6,3)$
A_{10}	$(4,6) (5,5), (6,4)$
A_{11}	$(5,6), (6,5)$
A_{12}	$(6,6)$

Notation for discrete finite and infinite sample spaces

$$\mathcal{S} = \{s_1, \ldots, s_N\} \text{ (finite sample space)};$$
$$\mathcal{S} = \{s_1, \ldots, s_n, \ldots\} \text{ (discrete infinite sample space)}.$$

We say that the *event A occurred* if the outcome of the experiment is a sample point $s \in A$. Thus, if $(4,3)$ is thrown we say A_7 occurred.

Combinations of events Events can be combined in various ways to form more complex events and, conversely, complex events can be written as combinations of events that are simpler to analyze. An event, as defined earlier, is a subset of the sample space so it is quite natural to combine events using such basic concepts of set theory as: union, complement and intersection. A familiarity with the basic concepts of set theory is assumed.

The union of two events The event *A or B or both occurred* is called the *union* of the two events and is denoted by $A \cup B$. It consists of all sample points that are either in A or B or both.

Example 2.3

A *natural* occurs, as noted previously, when the dice show a 7 or 11. In the language of probability theory a natural is the union of the two events A_7 and A_{11} and is written $A_7 \cup A_{11}$. The operation of union can be extended to more than two events. Consider the event that a player throws a natural or a point. This event, denoted A, is a union of eight events and is written:

$$A = A_4 \cup A_5 \cup A_6 \cup A_7 \cup A_8 \cup A_9 \cup A_{10} \cup A_{11}.$$

The complement of an event The set of all sample points not in A is also an event; it is called the *complement of A* and is denoted by A'.

Example 2.4

Refer to the event A of the preceding Example 2.3. List the sample points in A'.

Solution The event A' occurs when the player fails to throw a natural or a point; we say that the player has thrown a *craps,* which means the player loses on the come-out throw. The event A' can be written in the following equivalent forms:

$$A' = A_2 \cup A_3 \cup A_{12} = \{(1, 1), (1, 2), (2, 1), (6, 6)\}.$$

The intersection of two events The event A *and B occurred simultaneously,* denoted by $A \cap B$, is called the *intersection* of the two events A and B. It consists of all those sample points common to both A and B; that is, a sample point $s \in A \cap B$ if and only if $s \in A$ and $s \in B$.

Example 2.5

Let $A = \{$an even number is thrown$\}$ and $B = \{$a number greater than 9 is thrown$\}$. Describe the events A, B, and $A \cap B$ in terms of the events $A_2, \ldots, A_{12}$.

Solution We express each of these events in terms of the A_i's as follows:

$$A = A_2 \cup A_4 \cup A_6 \cup A_8 \cup A_{10} \cup A_{12};$$
$$B = A_{10} \cup A_{11} \cup A_{12};$$
$$A \cap B = A_{10} \cup A_{12}.$$

Mutually disjoint events The events A and B are said to be *mutually disjoint* if $A \cap B = \emptyset$. More generally, the events $A_1, A_2, \ldots, A_n$ are mutually disjoint if

$$A_i \cap A_j = \emptyset, \ i \neq j.$$

Here are some examples of disjoint events.

1. Note that one throw of a pair of dice cannot simultaneously yield a 4 and a 9, so $A_4 \cap A_9 = \emptyset$.

2. Every sample point is either in A or in its complement A', but not in both; thus,

$$\mathcal{S} = A \cup A'; \ A \cap A' = \emptyset.$$

This is the simplest example of a *partition*, which is a decomposition of a sample space into a union of mutually disjoint sets.

Partition of a sample space The events $A_1, A_2, \ldots, A_n$ form a *partition* of the sample space $\mathcal{S}$ if:
(a) the events are mutually disjoint and (b) $\mathcal{S} = A_1 \cup A_2 \cup \ldots \cup A_n$. In other words, a partition divides the sample space into mutually disjoint events with the property that every point in the sample space belongs to exactly one of these events.

3. Referring to the sample space corresponding to one throw of a pair of dice (Table 2.1) it is easy to verify that the events $A_2, \ldots, A_{12}$ form a *partition* of the sample space because they are mutually disjoint, and they are exhaustive in the sense that every point in the sample space belongs to one, and only one, of these events. Using set theory notation we express this by writing

$$\mathcal{S} = A_2 \cup A_3 \cup \cdots \cup A_{12}.$$

Table 2.2 is a visual display of this partition.

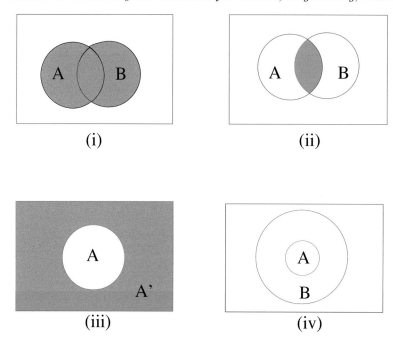

FIGURE 2.1:   Venn diagrams for: (i) $A \cup B$; (ii) $A \cap B$; (iii)A'; (iv) $A \subset B$

A **implies** B We say that the event A *implies* B when the occurrence of A implies the occurrence of B. This means that every sample point in A is also in B; that is, A is a *subset* of the event B, in symbols $A \subset B$.

Venn Diagrams

The set–theoretic operations and relations $A \cup B$, $A \cap B$, A', and $A \subset B$ can be represented graphically via *Venn diagrams* as in Figure 2.1. The sample space $\mathcal{S}$ is represented as a rectangle and the subsets A, B are represented as circles contained within the rectangle.

The Algebra of Sets   We next list, without proof, some facts about the set theoretic operations of union, intersection and complementation. We verify them by drawing the Venn diagram representations of the sets that occur on the left and right hand sides of the equal sign.

Proposition 2.1 *The operations of $\cup$, $\cap$ satisfy the following relations:*

$$A \cup B = B \cup A; \; A \cap B = B \cap A \; \textit{(Symmetry)} \tag{2.1}$$
$$A \cap (B \cup C) = (A \cap B) \cup (A \cap C) \; \textit{(Distributive law)} \tag{2.2}$$
$$(A \cup B)' = A' \cap B' \tag{2.3}$$
$$(A \cap B)' = A' \cup B' \tag{2.4}$$

Equations 2.3 and 2.4 are called *De Morgan's laws*.

Example 2.6

Consider the experiment of tossing a coin three times. Let A_i and B_i denote, respectively, the events *exactly i heads are thrown* and *at least i heads are thrown*, for $i = 0, 1, 2, 3$.

1. List the sample points in A_2 and B_2.

   Solution The sample space contains 8 points and are listed in Table 2.3. The symbols H and T stand for heads and tails, respectively.

Table 2.3 *Sample space for coin tossing experiment (n = 3)*

HHH	THH
HHT	THT
HTH	TTH
HTT	TTT

We now have all the information we need to list the sample points in A_2 and B_2.

$$A_2 = \{(HHT), (HTH), (THH)\}$$
$$B_2 = \{(HHT), (HTH), (THH), (HHH)\}$$

Note that $A_2 \subset B_2$, and more generally $A_i \subset B_i$; the latter assertion follows from the fact that if exactly i heads are thrown then at least i heads are thrown.

2. Express in terms of A_i, B_i the following events:

(a) One tail is thrown.

Solution If one tail is thrown then this means that two heads were thrown; consequently,
$$\text{One tail was thrown } = A_2.$$

(b) Only tails were thrown.

Solution This is equivalent to the assertion that no heads were thrown; consequently,
$$\text{Only tails were thrown } = A_0.$$

(c) At least two tails were thrown.

Solution This is equivalent to the statement that at most one head was thrown; thus,
$$\text{At least two tails were thrown } = B_2' = A_0 \cup A_1.$$

Example 2.7 *An application to reliability theory*

Let us now consider these concepts in the context of reliability theory. The *reliability* of a device is defined as the probability that it performs without failure for a specified period of time in a given environment. Consider a system—for example, a computer, monitor, and printer—consisting of three components each of which can be in only one of two states: (i) a functioning state and (ii) a non-functioning or failed state. The state of the system itself is denoted by W (functioning) and W' (non-functioning). Similarly, the state of the ith component is denoted by W_i (functioning) and by W_i' (non-functioning). Reliability engineers are interested in expressing the functioning state W of the system as a whole in terms of the functioning states of its components $W_i, (i = 1, 2, 3)$. We will consider two examples: (i) the components are linked in series and (ii) the components are linked in parallel.

1. **Components in series:** Linking the components in series means that the system functions if and only if each component is functioning; this can be expressed in set theoretic language as:

$$W = W_1 \cap W_2 \cap W_3.$$

2. **Components in parallel:** Linking the components in parallel means that the system functions if at least one of the components is functioning, i.e., the other two components serve as "backups." In this case the functioning state W can be expressed as:

$$W = W_1 \cup W_2 \cup W_3.$$

2.2.1   Probability Measures

At the beginning of every NFL football game an official flips a coin to decide who will receive the kickoff; this procedure is regarded by everyone as fair because everyone accepts without question that heads (H) and tails (T) are equally likely. The sample space corresponding to this experiment consists of two sample points $\mathcal{S} = \{H, T\}$. The assumption that the coin is "fair" means that each of the two sample points is assigned probability $1/2$. We use the symbol $P(s)$ to denote the probability assigned to the sample point s. Intuitively, $P(s)$ represents the chance that the event s occurs. With reference to the experiment of flipping a coin once, we write

$$P(H) = \frac{1}{2} \text{ and } P(T) = \frac{1}{2}.$$

Similarly, it is reasonable to assume (unless the dice are "loaded") that each of the 36 possible outcomes of throwing two dice (listed in Table 2.1) are equally likely. Thus $P(1,1) = \cdots = P(6,6) = 1/36$. Each of these examples is a special case of the equally likely probability measure which we define next.

Equally likely probability measure

Definition 2.2 *The equally likely probability measure P defined on a finite sample space $\mathcal{S} = \{s_1, s_2, \ldots, s_N\}$, assigns the same probability $P(s_i) = 1/N$ to each sample point.*

Example 2.8 *Equally likely probability measure for throwing two dice*

Assigning the equally likely probability measure to each sample point listed in Table 2.1 implies that $P((i,j)) = 1/36$, $i = 1, 2, \ldots, 6$, $j = 1, 2, \ldots, 6$.

Computing $P(A)$ when P is an equally likely probability measure

When P is the equally probability measure on a finite sample space containing N sample points and the set $A = \{s'_1, \ldots, s'_k\}$ contains k sample points, then $P(A)$ equals the k/N. This follows from Equation 2.7; the details follow.

$$P(A) = P(s'_1) + \cdots + P(s'_k) = \frac{1}{N} + \cdots + \frac{1}{N} \text{ (k times)}$$
$$= \frac{k}{N}.$$

Example 2.9 *Refer to Table 2.1. Compute $P(A_i)$, $i = 2, 3, \ldots 12$.*

Solution It is easy to check that $P(A_i) = (6 - |7-i|)/36$, $i = 2, 3, \ldots, 12$. Thus, $P(A_2) = (6 - |7-2|)/36 = (6-5)/36 = 1/36$, $P(A_2) = (6 - |7-3|)/36 = (6-4)/36 = 2/36$, etc.

Example 2.10   Playing the field *in craps means betting on the event $A = \{2, 3, 4, 9, 10, 11, 12\}$. It is also called a* field bet. *Compute $P(A)$.*

Solution Refer to Table 2.2 which displays the events A_i = (the number i is thrown) and their probabilities in the format of a horizontal bar chart. From this table we see that

$$A = A_2 \cup A_3 \cup A_4 \cup A_9 \cup A_{10} \cup A_{11} \cup A_{12}.$$

Since there are 16 sample points in A each one of which is assigned the probability $1/36$ it follows that $P(A) = 16/36$.

Notation We continue to use the "sharp" symbol $\sharp(A)$ to denote the number of sample points (outcomes) in the event A. It is also called the *number of outcomes favorable to A*. Thus,

$$P(A) = \frac{\sharp(A)}{\sharp(\mathcal{S})}.$$

Probability measure defined on a discrete sample space Equally likely probability measures are themselves a special case of a *probability measure* defined on a discrete sample space.

Definition 2.3 *A probability mass function (pmf) is a real valued function $P(s)$ defined on a discrete sample space $\mathcal{S} = \{s_1, \ldots, s_n, \ldots\}$ with the following properties:*

$$0 \le P(s_i) \le 1; \tag{2.5}$$

$$\sum_{1 \le i < \infty} P(s_i) = 1. \tag{2.6}$$

The probability of an event A

It is natural to define the probability of an event A to be equal to the sum of the probabilities of all the sample points in it; more precisely, the probability of a set A is given by

$$P(A) = \sum_{s_i \in A} P(s_i). \tag{2.7}$$

Example 2.11 *Playing craps: Rules of play*

Basic rules: Refer to Example 2.2. The player's first throw is called a "come-out" throw. Whether or not the player wins or loses depends not only on the come-out throw, but on subsequent throws as will now be explained. On a come-out throw:

1. If the player throws a natural (7 or 11) he wins.

2. If the player throws a "craps" (2, 3, or 12) he loses.

3. If the player throws a (4, 5, 6, 8, 9, or 10) he has thrown a "point." For the player to win he must "make the point," i.e., throw the same number again before he throws a 7—no other numbers matter. If the come-out throw, for example, is a 4, then the dice are thrown again and again until a 4 or 7 is thrown. If he throws a 4, the dice "pass" and the player wins. If, on the other hand, a 7 is thrown the dice miss out and the player loses.

4. Some examples:

    (a) Come-out throw: 2 (craps); player loses.

    (b) Come-out throw: 11 (natural); player wins.

(c) Come-out throw: 4 (point). Player then throws 11, 5, 9, 3, 4 (makes point); player wins.

(d) Come-out throw: 4 (point). Player then throws 3, 9, 11, 10, 5, 7 (sevens out); player loses.

What are the chances of winning? It can be shown that the probability the shooter wins, i.e., throws a natural or makes his point, is 0.4929, but the proof requires a detour that lies outside the scope of an introduction to basic probability theory. Readers interested in these questions should consult S. Ross, *A First Course in Probability Theory*, 6th ed., Prentice-Hall, 2002.

An infinite sample space: To determine whether or not the player "makes his point," e.g., "4 occurs before 7," is an event that cannot be realized in a finite sample space because it is possible, though not very likely, that even after 200 throws of the dice neither a 4 or 7 appeared. Consequently, our theory must include sample spaces corresponding to a possibly infinite sequence of dice throws. Since every sample point corresponds to an infinite sequence of integers from the set $\{2, 3, 4, 5, 6, 7, 8, 9, 10, 11, 12\}$ it is impossible to list even a single sample point, not to mention all the sample points in an event. In this case the sample point s represents the sequence $(x_1, x_2, \ldots, x_n, \ldots)$, $x_i \in \{2, 3, 4, 5, 6, 7, 8, 9, 10, 11, 12\}$.

The empirical probability function The frequency function $f(x)$ of a data set $X = \{x_1, x_2, \ldots, x_n\}$ generates a probability measure $\hat{f}(x)$ defined by the equation

$$\hat{f}(x) = \frac{f(x)}{n}. \tag{2.8}$$

We call $\hat{f}(x)$ the *empirical probability measure*. The empirical distribution function is obtained from the empirical probability function by noting that

$$\hat{F}_n(x) = \sum_{y \leq x} \hat{f}(y).$$

Example 2.12 *The empirical probability measure for the radiation data (Table 1.3)*

To compute the empirical probability measure we use Equation 2.8 and the second column of the horizontal bar chart (Figure 1.7) which lists the values of $f(x)$. Table 2.4 displays the values $\hat{f}(x)$.

Table 2.4 *Empirical probability measure for the radiation data set (Table 1.3)*

x	0.01	0.02	0.03	0.05	0.07	0.08	0.09	0.10
$\hat{f}(x)$	2/42	3/42	1/42	5/42	1/42	3/42	2/42	9/42

x	0.11	0.12	0.15	0.18	0.20	0.30	0.40
$\hat{f}(x)$	1/42	1/42	3/42	2/42	3/42	4/42	2/42

The empirical probability measure plays a fundamental role in the *bootstrap* method, see Example 2.19 (2) for additional details and references.

The frequency interpretation of probability: If a fair coin is tossed n times we would expect the relative frequency with which a head (H) appears to be nearly equal to its theoretical probability $P(H) = 1/2$; in other words we would expect the following relation to hold:

$$\frac{\text{the number of heads in } n \text{ tosses}}{n} \approx P(H) = \frac{1}{2} \text{ for } n \text{ large.}$$

Example 2.13

R. Wolf (1882) threw a die 20,000 times and recorded the number of times each of the six faces appeared. The results follow.

Face	1	2	3	4	5	6
Frequency	3407	3631	3176	2916	3448	3422

(Source: D. J. Hand et al. (1994), *Small Data Sets*, Chapman & Hall, London.)

When the dice are fair the expected frequency of occurrence for each face is $20,000 \times (1/6) = 3333.33$. We can now compare the observed frequencies with the expected frequencies to test whether the die is fair. We omit the details of the analysis since it involves an application of the chi-square test, which is discussed in Section 9.2.

In general, when an experiment is repeated n times, with n a large number, we would expect that the relative frequency with which the event A occurs to be approximately equal to its theoretical probability $P(A)$, that is

$$\lim_{n \to \infty} \frac{\text{the number of occurrences of the event } A \text{ in } n \text{ repetitions}}{n} = P(A). \tag{2.9}$$

Equation 2.9 is a mathematical formulation of this result called the *law of large numbers* and is discussed in Section 5.4.1 (Theorem 5.3).

Axioms for a Probability Measure

The function $P(A)$ defined by Equation 2.7 is an example of the more general concept of a *probability measure*, which is a function defined on the subsets A of $\mathcal{S}$ and satisfying the following axioms.

Definition 2.4 *Let $\mathcal{S}$ be a sample space corresponding to some experiment. A probability measure P is a function that assigns to each event $A \subset \mathcal{S}$ a number $P(A)$ satisfying the following axioms: For every event A*

$$0 \leq P(A) \leq 1; \tag{2.10}$$
$$P(\mathcal{S}) = 1. \tag{2.11}$$

For any two disjoint events A and B

$$P(A \cup B) = P(A) + P(B) \, (A \cap B = \emptyset). \tag{2.12}$$

More generally, for any finite sequence of mutually disjoint events $(A_i \cap A_j = \emptyset, i \neq j,)$ the probability of their union equals the sum of their probabilities:

$$P(A_1 \cup A_2 \cup \ldots \cup A_n) = \sum_{1 \leq i \leq n} P(A_i). \tag{2.13}$$

When the sample space contains an infinite number of outcomes and $A_1, A_2, \ldots, A_n, \ldots$, is an infinite sequence of mutually disjoint events, then we assume that

$$P(A_1 \cup A_2 \cup \ldots \cup A_i \cup \ldots) = \sum_{1 \leq i < \infty} P(A_n). \tag{2.14}$$

Axiom 2.14, called the *axiom of countable additivity*, plays an important role in more advanced treatises on probability theory.

Example 2.14 *An example of an infinite sequence of mutually disjoint events*

Consider the experiment of tossing a coin until the first time a head appears. Clearly, this is an experiment that can continue indefinitely. Denote by A_i the event that a head is thrown for the first time at the ith throw; this is a simple example of an infinite sequence of mutually disjoint events. The union $\cup_{1 \leq i < \infty} A_i$ is the event that a head ever appears.

Some Consequences of the Axioms

Proposition 2.2 *Probability of the complement: For every event A*

$$P(A') = 1 - P(A).$$

Proof It suffices to show that $P(A) + P(A') = 1$. Since $\mathcal{S} = A \cup A'$ and $A \cap A' = \emptyset$, it follows from Axioms 2.11 and 2.13 (here $n = 2$ and $A_1 = A$, $A_2 = A'$) that

$$1 = P(\mathcal{S}) = P(A) + P(A'). \tag{2.15}$$

In Example 2.15 we show that it is sometimes easier to compute $P(A)$ via the formula $P(A) = 1 - P(A')$ instead of computing it directly.

Example 2.15

Consider the experiment of tossing a coin $n = 3$ times (Example 2.6). Compute the probability of getting at least one head.

Solution Let A denote the event that at least one head is thrown. Its complement A' is the event that no heads were thrown. The event A' contains exactly one sample point, namely, $A' = (TTT)$, so $P(A') = 1/8$. Therefore, by Proposition 2.2

$$P(\text{ at least one head}) = 1 - P(\text{ no heads}) = 1 - \frac{1}{8} = \frac{7}{8}.$$

The computation of $P(B)$ is sometimes made easier by first partitioning B into two disjoint events and then using the following proposition, which generalizes Equation 2.15:

Proposition 2.3 *For every pair of events A, B*

$$P(B) = P(B \cap A) + P(B \cap A'). \tag{2.16}$$

Proof Using the partition $\mathcal{S} = A \cup A'$ and the distributive law (Equation 2.2) we can partition B into two disjoint events as follows:

$$B = B \cap \mathcal{S} = B \cap (A \cup A') = (B \cap A) \cup (B \cap A').$$

Thus,

$$P(B) = P((B \cap A) \cup (B \cap A'))$$
$$= P(B \cap A) + P(B \cap A') \text{ (by 2.12)}.$$

A Formula for $P(A \cup B)$

We next give a formula for computing $P(A \cup B)$ where the events A and B are not necessarily disjoint.

Proposition 2.4 *For arbitrary events A and B*

$$P(A \cup B) = P(A) + P(B) - P(A \cap B); \tag{2.17}$$
$$P(A \cup B) \leq P(A) + P(B). \tag{2.18}$$

Proof We begin with the partition
$A \cup B = A \cup (B \cap A')$, which can be verified by drawing Venn diagrams for $A \cup B$ and $A \cup (B \cap A')$.

$$A \cup B = A \cup (B \cap A'); \text{ therefore}$$
$$P(A \cup B) = P(A) + P(A' \cap B). \text{ Equation 2.16 implies}$$
$$P(A' \cap B) = P(B) - P(A \cap B). \text{ Therefore,}$$
$$P(A \cup B) = P(A) + P(B) - P(A \cap B).$$

The right hand side of Equation 2.17 is less than the right hand side of Equation 2.18 since it is obtained from the latter by subtracting the non-negative number $P(A \cap B)$.

Example 2.16

Let A, B denote events for which

$$P(A) = 0.3, \ P(B) = 0.8, \ P(A \cap B) = 0.2.$$

Using only this information and the basic properties of a probability measure compute $P(A \cup B)$, $P(A' \cap B')$, $P(A' \cap B)$.

Solutions

$$P(A \cup B) = P(A) + P(B) - P(A \cap B) \text{ (Proposition 2.4)}$$
$$= 0.3 + 0.8 - 0.2 = 0.9;$$
$$P(A' \cap B') = P([A \cup B]') \text{ (De Morgan's Laws)}$$
$$= 1 - P(A \cup B) = 1 - 0.9 = 0.1;$$
$$P(A' \cap B) = P(B) - P(A \cap B) \text{ (Proposition 2.3)}$$
$$= 0.8 - 0.2 = 0.6.$$

Proposition 2.4 will now be extended to the computation of the probability of the union of three events that are not assumed to be mutually disjoint.

Proposition 2.5

$$P(A_1 \cup A_2 \cup A_3) = P(A_1) + P(A_2) + P(A_3) - P(A_1 \cap A_2)$$
$$-P(A_1 \cap A_3) - P(A_2 \cap A_3) + P(A_1 \cap A_2 \cap A_3).$$

We omit the derivation since it is a special case of the inclusion-exclusion relation (see W. Feller, *An Introduction to Probability Theory and its Applications,* vol. 1, 3rd ed., (1968), John Wiley Press, p. 99).

Example 2.17

Given three events A_1, A_2 A_3 such that

$$P(A_1) = 0.3, \ P(A_2) = 0.6, \ P(A_3) = 0.4;$$
$$P(A_1 \cap A_2) = 0.2, \ P(A_1 \cap A_3) = 0.1, \ P(A_2 \cap A_3) = 0.3,$$
$$P(A_1 \cap A_2 \cap A_3) = 0.05.$$

Compute the following probabilities: (1) $P(A_1 \cup A_2 \cup A_3)$;
(2) $P(A_1' \cap A_2' \cap A_3')$.

Solutions

1. To compute $P(A_1 \cup A_2 \cup A_3)$ we apply Proposition 2.5.

$$P(A_1 \cup A_2 \cup A_3) = 0.3 + 0.6 + 0.4 - 0.2$$
$$-0.1 - 0.3 + 0.05 = 0.75.$$

2. To compute $P(A_1' \cap A_2' \cap A_3')$ we apply De Morgan's laws, or rather a straightforward extension of Equation 2.3 to computing the complement of the union of three events. Thus

$$A_1' \cap A_2' \cap A_3' = (A_1 \cup A_2 \cup A_3)'; \text{ consequently,}$$
$$P(A_1' \cap A_2' \cap A_3') = 1 - P(A_1 \cup A_2 \cup A_3) = 1 - 0.75 = 0.25.$$

2.3   Mathematical Models of Random Sampling

We return now to the main theme of this chapter, and that is to use the theory of probability to construct a mathematical model of the concept of a *random sample* of size k taken from a finite population consisting of n objects. We shall consider two basic methods for choosing a random sample, called *sampling without replacement*, and *sampling with replacement*, respectively.

As noted earlier, random sampling means that each subset of the same size has an equal probability of being the one that is drawn. This probability equals $1/N$ where N is the number of subsets of size k. Similarly, in order to compute the probability $P(A)$ of an event A it is necessary to compute $\sharp(A)/N$; that is, we must compute the number of ways that A occurs and then divide it by the total number of outcomes N. However, when either $\sharp(A)$ or N, or both, are very large numbers a complete listing of the sample points in A and in $\mathcal{S}$ is virtually impossible; nor would it provide much information even if it were possible to do so. The sample space corresponding to the experiment of tossing a coin ten times, for example, contains $N = 2^{10} = 1024$ sample points; the event *5 heads are thrown* has 252 sample points (these results are particular consequences of more general theorems to be proved shortly). Fortunately, in most cases the computation of N and $\sharp(A)$ can be simplified by reducing it to an equivalent problem in *combinatorial analysis*. For this reason a brief introduction to this subject is warranted.

The Multiplication Principle

The *multiplication principle* and its variations play a fundamental role throughout this section. Suppose experiment I has n_1 outcomes denoted $\{A_1, \ldots, A_{n_1}\}$ and experiment II has n_2 outcomes denoted $\{B_1, \ldots, B_{n_2}\}$, then performing the two experiments in succession has $n_1 \times n_2$ distinct outcomes. We denote an outcome of this experiment by the ordered pair (A_i, B_j), indicating that the first experiment produced the outcome A_i and the second experiment produced the outcome B_j. The set of all such outcomes can be displayed as an rectangular array consisting of n_1 rows and n_2 columns as in Figure 2.2. Looking at the array in Figure 2.2 we see that it has n_1 rows and n_2 columns, so it has $n_1 \times n_2$ elements.

It is worth noting that performing the two experiments in succession is itself an experiment called the *product of two experiments;* its sample space is denoted $\mathcal{S}_1 \times \mathcal{S}_2$, where $\mathcal{S}_1 = \{A_1, \ldots, A_{n_1}\}$ and $\mathcal{S}_2 = \{B_1, \ldots, B_{n_2}\}$. Because of its importance in probability and statistics some simple examples of this computation are worth studying in detail.

$$\begin{array}{llll}
(A_1, B_1) & (A_1, B_2) & \ldots & (A_1, B_{n_2}) \\
(A_2, B_1) & (A_2, B_2) & \ldots & (A_2, B_{n_2}) \\
\ldots & (A_i, B_j) & \ldots & \ldots \\
(A_{n_1}, B_1) & (A_{n_1}, B_2) & \ldots & (A_{n_1}, B_{n_2})
\end{array}$$

FIGURE 2.2: The sample space corresponding to the product of two experiments

Example 2.18

Describe the product experiment $\mathcal{S}_1 \times \mathcal{S}_2$, where:

1. $\mathcal{S}_1 = \mathcal{S}_2 = \{1, 2, 3, 4, 5, 6\}$. Here $n_1 = n_2 = 6$ and so there are $6 \times 6 = 36$ ordered pairs of the form (i, j). The set of ordered pairs is identical to the sample space corresponding to throwing two dice, or one die twice in succession (see Table 2.1).

2. $\mathcal{S}_1 = \mathcal{S}_2 = \{H, T\}$. Here $n_1 = n_2 = 2$ and so there are $2 \times 2 = 4$ ordered pairs of the form $\{(H, H), (H, T), (T, H), (T, T)\}$. This is the sample space corresponding to the experiment of throwing a coin twice.

3. Consider an experiment in which the yield of a chemical process depends on (i) the operating pressure and (ii) the operating temperature. To determine the optimum yield the experiment is run at $n_1 = 6$ different pressures and $n_2 = 4$ different temperatures. Here

$$\mathcal{S}_1 = \{P_1, \ldots, P_6\} \text{ and } \mathcal{S}_2 = \{T_1, \ldots, T_4\}.$$

A *complete trial* is an experiment that is run for all these possible combinations, which in this case equals $6 \times 4 = 24$.

In the preceding example the experiment is determined by making two choices, with 6 choices for the operating pressure and 4 choices for the operating temperature. The method of enumeration used above will now be extended to the case of three or more experiments. The result is intuitively obvious and so we shall omit the formal proof.

The generalized multiplication principle: Consider a sequence of k experiments where the ith experiment has n_i outcomes (or choices). Then the product experiment, which is the experiment corresponding to performing the k experiments in succession, has $n_1 \times n_2 \times \ldots \times n_k$ outcomes.

Example 2.19

The next three examples illustrate various applications of the multiplication principle.

1. *Tossing a coin k times:* Each experiment has only two outcomes H and T; thus $n_i = 2$, $i = 1, 2, \ldots, k$, and therefore the total number of outcomes in the sample space equals $2 \times \cdots \times 2 = 2^k$.

2. *The bootstrap:* A *bootstrap sample,* denoted by $\mathcal{X}^* = \{x_1^*, \ldots, x_n^*\}$, is a random sample of size n taken with replacement from the data set $X = \{x_1, x_2, \ldots, x_n\}$. Using the data generated by repeated resampling from the original data set is the starting point for a computer intensive method for studying the sampling distribution of certain statistics that are otherwise very difficult to compute (see B. Efron and R. J. Tibshirani (1993), *An Introduction to the Bootstrap,* Chapman & Hall, New York).

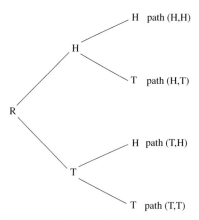

H path (H,H)

H

T path (H,T)

R

H path (T,H)

T

T path (T,T)

FIGURE 2.3: Tree diagram for coin tossing experiment ($k = 2$)

3. *License plates:* A license plate consisting of two letters followed by four digits is an ordered six-tuple of the form $(A, B, 1, 2, 3, 4)$, etc. The two letters can be chosen in 26^2 ways and the four digits can be chosen in any one of 10^4 ways. Consequently, the total number of such license plates is equal to $26^2 \times 10^4 =$6,760,000.

Tree Diagrams

When n_1 and n_2 are small then one can visualize the multiplication principle in the format of a *tree diagram*. The tree diagram corresponding to tossing a coin twice is displayed in Figure 2.3. We construct the tree diagram by listing in a column all the elements of S_1 and connecting each of these elements, called *nodes*, by a line segment, called a *branch* to the point labeled R called the *root*. Each of these nodes serves in turn as the root of another tree consisting of n_2 branches placed to the right of the node. Each branch corresponds to one of the n_2 elements of S_2. Each outcome of the experiment corresponds to a *path* through the tree, which can be identified by listing the nodes through which it passes. Thus HH denotes the event that two heads were thrown and also the path starting at R and connecting the two nodes labeled H.

Example 2.20 *A binomial tree model for stock price fluctuations*

A widely used model for stock price fluctuations assumes if the current price of a stock is S_0 then its price one time period later, denoted by S_1, is either $S_1 = S_0 u$ ($u > 1$) or $S_1 = S_0 d$ ($d < 1$). In other words, after one time period, the stock price either moves up by the factor u or down by the factor d. Similarly, after two periods its price is $S_2 = S_1 u$ or $S_2 = S_1 d$; consequently, after two time periods, its price is either $S_2 = S_0 u \cdot u = S_0 u^2$, $S_2 = S_0 ud$, or $S_2 = S_0 d \cdot d = S_0 d^2$. A more convenient way of recording the price history of a stock is the *binomial tree* as shown in Figure 2.4, where it is assumed that $S_0 = \$50$, $u = 1.02$, and $d = 0.99$. In detail: After one period $S_1 = 50 \times 1.02 = 51$ or $S_1 = 50 \times 0.99 = 49.50$. Similarly, after two up moves $S_2 = 50 \times (1.02)^2 = 52.02$. Note that the stock price after an up and down move is the same after a down and up move, that is $S_2 = S_0 ud = S_0 du = 50 \times 1.02 \times 0.99 = 50.49$. Finally, after two down moves $S_2 = 50 \times (0.99)^2 = 49.01$, where the price has been rounded to the nearest penny. Note that each path through the tree corresponds to a sequence of two elements from the finite set $\mathcal{S} = \{u, d\}$. This is an example of what computer scientists call a *string* of length two, or 2-string (also called a word of length 2). When the set $\mathcal{S}$ consists of two elements, which is the case here, it is called a *binary string*. Using our counting principle it is not too difficult

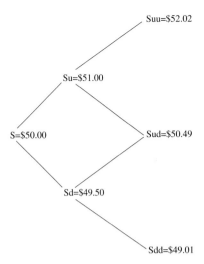

Suu=$52.02

Su=$51.00

S=$50.00

Sud=$50.49

Sd=$49.50

Sdd=$49.01

FIGURE 2.4: Binomial tree for stock price with two time periods

to see that there are 2^k binary strings of length k. For example, there are four paths through the binomial tree corresponding to the four 2-strings $\{(u, u), (u, d), (d, u), (d, d)\}$.

Example 2.21

An urn contains three objects labeled a, b, c. The experiment consists of successively drawing at random two objects, without replacement, from the urn. Draw the tree diagram and use it to list the elements of the sample space.

Solution The tree diagram corresponding to this experiment is shown in Figure 2.5. There are six outcomes in the sample space: $\mathcal{S} = \{ab, ac, ba, bc, ca, cb\}$.

Although tree diagrams are a useful visual display of the outcomes of a product of two experiments, they are of little practical value when either n_1 or n_2 are large. In these cases we will find it much easier to count the number of outcomes in an event, or in the sample space, by reducing it to an equivalent problem in *permutations* and *combinations;* concepts that play an important role in probability theory and in computer science (theory of algorithms).

Permutations

A *permutation of n objects taken k at a time* is an arrangement of k objects taken without replacement from n objects. There are six permutations of the three objects $\{a, b, c\}$ taken two at a time:

$$\{ab, ac, ba, bc, ca, cb\}.$$

Our goal here is to use the multiplication principle to count the number of different permutations of n objects taken k at a time.

Theorem 2.1 *The number of permutations of k objects taken from n distinct objects is denoted $P_{n,k}$ and equals*

$$P_{n,k} = n(n-1) \cdots (n - k + 1).$$

The special case $k = n$ is particularly important and is denoted by n! (pronounced n factorial).

$$n! = n(n-1) \cdots 3 \cdot 2 \cdot 1 = P_{n,n}.$$

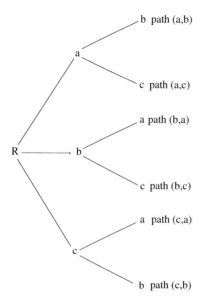

FIGURE 2.5:   Tree diagram for selecting two objects without replacement from the set $\{a, b, c\}$

Proof Denote the n distinct objects by the symbols $O_1, \ldots, O_n$. A permutation is an ordered k tuple of the form

$$(O_{x_1}, \ldots, O_{x_k}),$$

where we have n choices for O_{x_1}, $n - 1$ choices for $O_{x_2}, \ldots$, and, finally, we have $n - k + 1$ choices for O_{x_k}; consequently, by the general multiplication principle, the total number of such outcomes equals $n(n - 1) \cdots (n - k + 1)$.

Example 2.22

To illustrate, consider a lottery in which there are three winning tickets, corresponding to the first, second and third prizes, respectively. Suppose 100 lottery tickets are sold with first, second and third prizes to be determined in a random drawing. Compute the number of different ways of drawing the three winning tickets.

Solution Each drawing corresponds to a permutation of 3 objects taken from 100 objects. Thus, $k = 3$ and $n = 100$, so the number of such drawings equals
$P_{100,3} = 100 \times 99 \times 98 = 970{,}200$.

Example 2.23

Compute the following quantities: (1) $P_{8,3}$, (2) $P_{9,4}$, (3) $3!$, (4) $5!$.
  Solutions

1. $P_{8,3} = 8 \times 7 \times 6 = 336$

2. $P_{9,4} = 9 \times 8 \times 7 \times 6 = 3024$

3. $3! = 3 \times 2 \times 1 = 6$

4. $5! = 5 \times 4 \times 3 \times 2 \times 1 = 120$

Notation It is convenient to define $0! = 1$. With this convention we define $n!$ recursively as

$$n! = n \times (n-1)!, \; n = 1, 2, \ldots,.$$

Thus, $5! = 5 \times 4! = 5 \times 24 = 120$.

Example 2.24 *The birthday problem*

The birthday problem, first popularized by Feller (1968) (*An Introduction to Probability Theory and its Applications*, vol. 1, 3rd ed., John Wiley & Sons, Inc., New York) is an interesting application of the methods of random sampling with results that are at first quite surprising. The problem is this: There are k people in a room; what is the probability that at least two people in the room have a common birthday?
  Solution to the Birthday Problem:

1. **The sample space**: We begin by recording the birthday of the ith person as an integer x_i $(1 \le x_i \le 365)$; thus, if the ith person was born on January 1, then $x_i = 1$, and so on. Going around the room and recording the k birthdays produces a k string of the form $(x_1, \ldots, x_k)$. The sample space $\mathcal{S}$ of this experiment is just the set of all such k strings taken from the set $\mathcal{S} = \{1, 2, \ldots, 365\}$. It is not unreasonable to assume that each person has an equal chance of being born on any one of the 365 days of the year; in particular, we assume that each of these samples has an equal probability of being selected.

2. **An urn model**: This sample space can also be realized by means of the following sampling with replacement experiment: Consider an urn containing 365 balls numbered $1, 2, \ldots, 364, 365$. After carefully stirring the balls, pick one, record its value and then replace it in the urn. Repeat this procedure k times; the sample space corresponding to this experiment consists of the ordered k tuples of the form $(x_1, \ldots, x_k)$, where $x_i \in \{1, 2, \ldots, 365\}$. Applying the generalized multiplication principle we see that the total number of possible birthdays is equal to $365 \times \cdots \times 365 = 365^k$ (we ignore leap years).

3. **Computing the probability that at least two persons out of k have the same birthday:** We denote by A the event that at least two people in the room have the same birthday. It turns out to be much simpler to first compute the probability of the complement $A' = \{$ every one in the room has a different birthday $\}$, and then compute $P(A)$ via the formula $P(A) = 1 - P(A')$. Since the sample space is equipped with the equally likely probability measure we compute the probability of A' by first counting the number of outcomes favorable to A' and then dividing by the total number of outcomes. Now the number of ways k persons all have different birthdays is equal to the number of k tuples $(x_1, \ldots, x_k)$ in which all the $x_i's$ are different; but this is equivalent to drawing a sample of size k without replacement from $n = 365$ objects. Consequently, the number of sample points in A' equals $P_{365,k}$ and therefore the

$$\text{P}(k \text{ persons have different birthdays}) = \frac{P_{365,k}}{365^k}.$$

It is intuitively clear that as k increases the probability that everyone in the room has a different birthday decreases. More precisely, it is not too difficult to show that

$$\frac{P_{n,k}}{n^k} = \frac{n \cdot (n-1) \cdots (n-k+1)}{n \cdot n \cdots n}$$
$$= (1 - 1/n) \cdot (1 - 2/n) \cdots (1 - (k-1)/n), \; (k = 1, \ldots, n) \qquad (2.19)$$

and for k large enough, $P_{n,k}/n^k < 0.5$. When $n = 365$, $k = 23$, a calculation using a pocket calculator yields the result

$$\frac{P_{365,23}}{365^{23}} = 0.493 < 0.5 < \frac{P_{365,22}}{365^{22}} = 0.524.$$

Thus if one selects a group of 23 persons at random then the probability that at least 2 of them have a common birthday is equal to $1 - 0.493 = 0.507 > 1/2$, a result that Feller describes as astounding. What do you think?

Combinations

Consider a population consisting of n objects, labeled $O_1, \ldots, O_n$, a deck of cards, say, with $n = 52$. How many different 5-card poker hands are there? A 5-card poker hand is an example of a subset, or unordered sample, of size 5 taken from 52 objects. This problem is typical of many in probability and statistics for which it is important to count the number of subsets containing k elements taken from a finite population consisting of n objects, without regard to order. For instance, the number of subsets of size $k = 2$ taken from the $n = 3$ objects $\{a, b, c\}$ are the three subsets: $\{a, b\}, \{a, c\}, \{b, c\}$. On the other hand, the number of permutations of 2 objects taken from 3 objects are the $P_{3,2} = 6$ subsets: $(a, b), (b, a), (a, c), (c, a), (b, c), (c, a)$. The difference between a permutation and a combination is that the permutations (a, b) and (b, a) are different because they are ordered differently. We do not, however, distinguish between the subsets $\{a, b\}$ and $\{b, a\}$ because they contain the same elements. The number of subsets containing k elements taken from a finite population consisting of n objects is denoted by $C_{n,k}$ (read "n choose k"). It is called *the number of combinations of k objects taken from a set of n objects*; it is also called a *binomial coefficient*.

Another Interpretation of the Binomial Coefficients $C_{n,k}$

Suppose we have n objects $O_1, \ldots, O_n$ and two types of labels: k labels denoted $\boxed{\text{H}}$ and $n - k$ labels denoted $\boxed{\text{T}}$. We want to determine how many different ways these objects can be labeled using the label $\boxed{\text{H}}$ k times and the $\boxed{\text{T}}$ label $n - k$ times. If we think about it for a moment, each such labeling corresponds to first choosing the k objects to be labeled $\boxed{\text{H}}$, and then labeling the remaining objects with the label $\boxed{\text{T}}$. But this labeling process is equivalent to the number of different ways of choosing subsets of size k from n objects; consequently, there are $C_{n,k}$ ways of carrying out this labeling. In particular, there are $C_{n,k}$ ways of getting k heads (H) in n tosses of a coin. For each such outcome corresponds to labeling the integers $1, \ldots, n$ with k H's and $n - k$ T's. For instance the number of ways we can get 5 heads in 10 tosses of a coin is $C_{10,5}$. Theorem 2.2 gives we a formula for computing $C_{n,k}$.

Theorem 2.2 *A population S consisting of n distinct objects contains $C_{n,k}$ subsets of size $k : 0 \le k \le n$, where*

$$C_{n,k} = \frac{P_{n,k}}{k!} = \frac{n!}{k!(n-k)!}.$$

Notation The following alternate notation is also widely used:

$$C_{n,k} = \binom{n}{k}.$$

$C_{n,k}$ is also called a *binomial coefficient*. Before proceeding to the proof note that

$$\binom{n}{0} = \binom{n}{n} = 1.$$

This is a consequence of the convention that $0! = 1$.

Proof The assertion of the theorem is equivalent to the relation

$$C_{n,k} \times k! = P_{n,k},$$

which is a result of the multiplication principle as will now be explained. Every permutation can be obtained by first selecting a subset of k objects, which has $C_{n,k}$ outcomes, and then noting that each k element subset has $k!$ permutations. There are therefore $C_{n,k} \times k!$ ways of performing these two operations in succession. The proof is complete.

Example 2.25

Compute the following binomial coefficients: (1) $C_{3,2}$, (2) $C_{8,3}$, (3) $C_{10,5}$.

Solution

1. $C_{3,2} = \frac{3 \times 2}{2!} = 3$

2. $C_{8,3} = \frac{8 \times 7 \times 6}{3!} = 56$

3. $C_{10,5} = \frac{P_{10,5}}{5!} = \frac{30,240}{120} = 252$

Example 2.26 *Applications of binomial coefficients to games of chance*

The binomial coefficients are useful for computing the chances of winning in poker or winning a state lottery. The following examples illustrate some of the possible applications.

1. *Poker hands:* A poker hand consists of 5 cards taken from 52, i.e., a poker hand is a subset of size 5 taken from 52 cards; so $n = 52$, $k = 5$. The number of 5-card poker hands equals $C_{52,5} = 2,598,960$. Let us compute the probability of getting *four of a kind* in poker, for example, four kings, four aces, etc. There are 13 denominations and for each occurrence of four aces, say, there are 48 choices for the fifth card. Consequently, there are $13 \times 48 = 624$ ways of getting a hand with four of a kind. The probability of getting a four of a kind is then equal to $624 / \binom{52}{5}$.

2. *Bridge hands:* A bridge hand is a subset of size 13 taken from 52 cards; thus, $n = 52$, $k = 13$. The number of 13-card bridge hands equals $C_{52,13} = 635,013,559,600$.

3. *State lotteries:* Consider a lottery in which we have to pick 6 different digits out of $\{1, 2, \ldots, 40\}$ in order to win first prize. A winning ticket is a subset of size $k = 6$ taken from $n = 40$ objects. The sample is unordered because the order in which the winning numbers are drawn is not relevant. The probability of winning is $p = 1 / \binom{40}{6}$, where $\binom{40}{6} = 3,838,380$.

A recurrence relation for the binomial coefficients

A *recurrence relation* is a rule that defines each element of a sequence in terms of the preceding elements. Recurrence relations are particularly useful for computing the elements of the sequence. The binomial coefficients satisfy the following recurrence relation:

$$\binom{n}{k+1} = \frac{n-k}{k+1} \times \binom{n}{k}, \quad \binom{n}{0} = 1. \tag{2.20}$$

The derivation of Equation 2.20 is left to the reader.

Example 2.27

Use the recurrence relation to compute $\binom{5}{k}$, $(k = 1, 2, 3)$.

Solution

$$n = 5, \ k = 1 : \ \binom{5}{1} = \frac{5}{1} \times \binom{5}{0} = 5$$

$$n = 5, \ k = 2 : \ \binom{5}{2} = \frac{5-1}{1+1} \times \binom{5}{1} = 10$$

$$n = 5, \ k = 3 : \ \binom{5}{3} = \frac{5-2}{2+1} \times \binom{5}{2} = 10$$

The Binomial Theorem

In addition to their obvious importance in combinatorial analysis, the binomial coefficients have many interesting properties as well as many applications. They arise frequently, for example, in the analysis of *algorithms* in computer science (see for example Donald E. Knuth (1973), *The Art of Computer Programming: Fundamental Algorithms*, vol. 1, Addison-Wesley, Reading, MA.). The most useful application for us is the binomial theorem:

Theorem 2.3 *The binomial theorem*

$$(a+b)^n = \binom{n}{0} a^n + \cdots + \binom{n}{k} a^{n-k} b^k + \cdots + \binom{n}{n} b^n \tag{2.21}$$

Proof For a detailed proof see Section 2.5.

Example 2.28

Give the binomial expansion for $(a+b)^n$ for $n = 0, 1, 2, 3, 4$.

Solution

$$n = 0 : (a+b)^0 = 1$$

$$n = 1 : (a+b)^1 = a + b$$

$$n = 2 : (a+b)^2 = a^2 + 2ab + b^2$$

$$n = 3 : (a+b)^3 = a^3 + 3a^2b + 3ab^2 + b^3$$

$$n = 4 : (a+b)^4 = a^4 + 4a^3b + 6a^2b^2 + 4ab^3 + b^4$$

Looking at the binomial coefficients displayed above an interesting pattern emerges that is called the *Pascal triangle*. The vertex of the triangle, which contains the single entry 1, is called the zeroth row; it corresponds to the binomial expansion of $(a+b)^0 = 1$. The next two rows display the coefficients of the binomial expansion of $(a+b)^1 = a + b$, and $(a+b)^2 = a^2 + 2ab + b^2$, etc. The number $\binom{n+1}{k}$ in the $(n+1)$ st row is the sum of the two adjacent numbers in the row above; for instance, $4 = 1 + 3$, $6 = 3 + 3$, etc. In the general case we have the following *binomial identity*:

$$\binom{n}{k} + \binom{n}{k-1} = \binom{n+1}{k}. \tag{2.22}$$

The Pascal triangle is a visual display of the identity 2.22.

n=0				1				
n=1			1		1			
n=2		1		2		1		
n=3	1		3		3		1	
n=4	1	4		6		4		1

. . . etc.

FIGURE 2.6:   The Pascal triangle

Derivation of the binomial identity (Equation 2.22)

This identity will now be proved by a clever counting argument that avoids tedious algebraic computations. The binomial coefficient appearing on the right hand side of Equation 2.22 counts the number of subsets of size k taken from $(n + 1)$ objects; the left hand side also counts the number of k element subsets, but it does so in a slightly different way. To see this, pick one of the $n + 1$ elements and note that a k element subset either (i) does not contain this element, or (ii) it does. There are $\binom{n}{k}$ of the first kind and $\binom{n}{k-1}$ of the second; adding these two terms together yields Equation 2.22.

2.3.1   Multinomial Coefficients

We begin our discussion of multinomial coefficients by considering the following problem: How many distinguishable arrangements of the letters of the word ILLINOIS are there? This is an eight letter word consisting of three I's, two L's and one each of the letters N, O, and S. This is a generalization of the labeling problem considered earlier. We have eight labels divided into five subgroups denoted by the letters I, L, N, O, S. The problem is equivalent, then, to counting the number of different ways eight objects can be labeled using three labels of type I, two labels of type L, and one each of the labels N, O, and S. It is of course a special case of the following general problem:

Theorem 2.4 *Suppose we have k_1 labels of type 1, k_2 labels of type 2, $\ldots$, k_m labels of type m comprising a grand total of $n = k_1 + k_2 + \cdots + k_m$ labels. Then the total number of distinguishable arrangements is given by*

$$\binom{n}{k_1, \ldots, k_m} = \frac{n!}{k_1! \cdots k_m!}. \tag{2.23}$$

We give an example before sketching the proof.

Example 2.29

Compute the total number of distinguishable arrangements of the letters of the word ILLINOIS.

Solution ILLINOIS has $n = 8$ letters consisting of $k_1 = 3$ I's, $k_2 = 2$ L's and one each of the letters N, O, S, so $k_3 = k_4 = k_5 = 1$. Inserting these values into Equation 2.23 yields the result

$$\frac{8!}{3! \times 2! \times 1! \times 1! \times 1!} = 3360.$$

The term appearing in Equation 2.23 is called a *multinomial coefficient*.

Note that binomial coefficient is a special case of the multinomial coefficient with $m = 2$, $k = k_1$, $k_2 = n - k$.

Proof of Theorem 2.4   It suffices to give the proof for the case $m = 3$ since the proof in the general case is similar. There are $\binom{n}{k_1}$ ways of placing the k_1 labels of type 1 which leaves $n - k_1$ unlabeled positions. There are $\binom{n - k_1}{k_2}$ ways of placing k_2 labels of type 2 over the remaining $n - k_1$ positions. Once these two sets of labels have been placed the type 3 labels must go into the remaining positions. The total number of ways this can be done then is clearly the product $\binom{n}{k_1} \times \binom{n - k_1}{k_2}$. An easy calculation left to the reader shows that this product equals

$$\frac{n!}{k_1! k_2! k_3!}.$$

2.4   Conditional Probability and Bayes' Theorem

2.4.1   Conditional Probability

In one of the earliest studies (1936) establishing a "link" between smoking and lung cancer two British physicians reported that of 135 men with lung cancer, 122, or 90%, were heavy smokers. In non-technical language the inference is clear: If you have lung cancer, it is much more likely than not that you are a heavy smoker. Is the converse true? If you are heavy smoker does it necessarily follow that your chances of developing lung cancer are much higher than for a non-smoker? To answer questions like these we need to define the new concept of *conditional probability*, which is of fundamental importance in probability theory.

We shall assume that the data reported by the physicians came from a population $\mathcal{S}$ of N men, and we are interested in the two subpopulations:

1. $A = \{$men who are heavy smokers$\}$

2. $B = \{$men with lung cancer$\}$

Denote by $\sharp(B)$ the number of sample points in the set B. Thus $\sharp(B) = 135$ and $\sharp(B \cap A) = 122$. The probability that a person selected at random from the subpopulation B (men with lung cancer) is also in A (men who are heavy smokers) is clearly

$$\frac{\sharp(B \cap A)}{\sharp(B)} = \frac{122}{135} = 0.90. \tag{2.24}$$

In words we say: "The probability of the event A (the man is a heavy smoker) given the event B (the man has lung cancer) has occurred" is equal to 0.9. In symbols this is written

$$P(A|B) = \frac{\sharp(B \cap A)}{\sharp(B)} = 0.9.$$

Now the probability that a person selected at random from the population $\mathcal{S}$ is both a heavy smoker and has lung cancer is clearly

$$P(A \cap B) = \sharp(B \cap A)/N;$$

and, similarly, the probability that a person selected at random from the population $\mathcal{S}$ has lung cancer is clearly

$$P(B) = \sharp(B)/N.$$

Therefore,

$$P(A|B) = \frac{\sharp(B \cap A)}{\sharp(B)} = \frac{\sharp(B \cap A)/N}{\sharp(B)/N} = \frac{P(A \cap B)}{P(B)}.$$

This leads us to define the conditional probability of an event A given B as follows:

Definition 2.5 *Let A and B be two arbitrary events with $P(B) > 0$. The* **conditional probability** *of the event A given that B has occurred is defined to be*

$$P(A|B) = \frac{P(A \cap B)}{P(B)}. \tag{2.25}$$

We frequently use Definition 2.5 in the equivalent form:

$$P(A \cap B) = P(A|B) \times P(B). \tag{2.26}$$

Equation 2.26 is called the *product rule*. It is useful because it is often the case that one is given $P(A|B)$ and $P(B)$ and *not* $P(A \cap B)$ as in Example 2.30, where we use it to analyze the performance of a screening test to detect a disease.

Example 2.30

Consider a population partitioned into two groups depending on whether or not they are infected with a rare disease. To be specific, denote by D the event that a person selected at random has the disease and suppose that the *prevalence rate* of this disease is 1 person in 5,000; that is, we assume that $P(D) = 0.0002$. We are interested in studying the performance of a screening test for the disease. A screening test is not foolproof; it sometimes yields an incorrect result. These errors are of two types: *false positives* and *false negatives*. A false positive reading simply means that the person tested positive for the disease even though he is not infected. Similarly, a false negative means that even though he is infected the test failed to detect it. Let T^+ denote the event that the screening test is positive for a person selected at random. Suppose the manufacturer claims that its screening test is quite reliable in the sense that the rate of false positive readings is $2/100 = 0.02$ and the rate of false negative readings is $1/100 = 0.01$. Compute $P(T^+)$.

Solution It is interesting to note that in each case the the rate of false positive readings and the rate of false negative readings are conditional probabilities, although nowhere in the specifications is the term conditional probability actually used. To see this denote the event that a person taking the test yields a positive (negative) result by $T^+(T^-)$, and denote by $D(D')$ the event that the person has the disease (does not have the disease). A false positive means that the event T^+ occurred given that D' occurred. We express these specifications as conditional probabilities as follows:

$$P(T^+|D') = 0.02, \text{ and } P(T^-|D) = 0.01.$$

In order to evaluate the performance of the test it is necessary to calculate the probability that the test is positive, which is given by

$$P(T^+) = P(T^+ \cap D) + P(T^+ \cap D'). \tag{2.27}$$

Proof Equation 2.27 is a consequence of the fact that a positive response can occur for only one of two reasons: (i) the person being tested actually has the disease, or (ii) the person does not (false positive). This yields the partition

$$T^+ = (T^+ \cap D) \cup (T^+ \cap D').$$

Equation 2.27 is a straightforward application of Proposition 2.3 with $T^+ = B$ and $D = A$. Next, using Equation 2.27 and the product rule 2.26 we obtain the following value for $P(T^+)$:

$$P(T^+) = P(T^+ \cap D) + P(T^+ \cap D') \tag{2.28}$$
$$= P(T^+|D)P(D) + P(T^+|D')P(D'). \tag{2.29}$$

Inserting into Equation 2.29 the values $P(T^+|D') = 0.02$, $P(T^-|D) = 0.01$, $P(D) = 0.0002$, $P(D') = 0.9998$ yields the result that

$$P(T^+) = 0.99 \times 0.0002 + 0.02 \times 0.9998 = 0.02. \tag{2.30}$$

The reasoning used here is just a special case of a more general theorem called *the law of total probability.*

Theorem 2.5 *(The law of total probability): Let the events $A_1, A_2, \ldots, A_n$ be a partition of the sample space S and let B denote an arbitrary event. Then*

$$P(B) = \sum_{1 \le i \le n} P(B|A_i)P(A_i).$$

Before proving this result we give a typical application.

Example 2.31

In a factory that manufactures silicon wafers, machines M_1 and M_2 produce, respectively, 40% and 60% of the total output. Suppose that 2 out of every 100 wafers produced by machine M_1 and 3 out of every 200 wafers produced by machine M_2 are defective; what is the probability that a wafer selected at random from the total output is defective?
    Solution Let

$$D = \{\text{a wafer selected at random is defective}\};$$
$$M_1 = \{\text{a wafer selected at random is produced by machine 1}\};$$
$$M_2 = \{\text{a wafer selected at random is produced by machine 2}\}.$$

Figure 2.7 is the tree diagram corresponding to this experiment. The numbers appearing along the branches are the conditional probabilities.
    Clearly, the defective wafer is produced either by M_1 or M_2; and these two possibilities are represented in the tree diagram by the two paths labeled M_1D and M_2D respectively. This suggests that we compute $P(D)$ by "conditioning" on whether the wafer was produced by M_1 or M_2. More precisely we use Theorem 2.5 to show that

$$P(D) = P(D|M_1)P(M_1) + P(D|M_2)P(M_2).$$

The probabilities and conditional probabilities are derived from the information contained in the problem description. Specifically, it is easy to verify that

$$P(M_1) = 0.4, \ P(M_2) = 0.6;$$
$$P(D|M_1) = 0.02, \ P(D|M_2) = 0.015; \text{ thus,}$$
$$P(D) = 0.02 \times 0.4 + 0.015 \times 0.6 = 0.017.$$

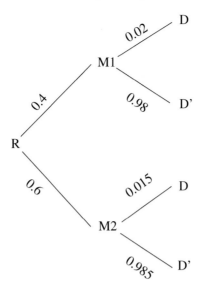

FIGURE 2.7: Tree diagram displaying conditional probabilities for Example 2.31

Proof of Theorem 2.5 The events $A_1, A_2, \ldots, A_n$ form a partition of $\mathcal{S}$; consequently

$$B = (B \cap A_1) \cup (B \cap A_2) \cup \ldots (B \cap A_n).$$

Therefore

$$P(B) = \sum_{1 \leq i \leq n} P(B \cap A_i) = \sum_{1 \leq i \leq n} P(B|A_i)P(A_i),$$

where we have used the product rule (Equation 2.26) in the last step.

Theorem 2.6 *The function $P(A|B)$, for fixed B, $P(B) > 0$, satisfies Axioms 2.10 through 2.14 of a probability measure, i.e.,*

$$0 \leq P(A|B) \leq 1, \tag{2.31}$$

$$P(\mathcal{S}|B) = 1. \tag{2.32}$$

$$\text{If } A_1 \cap A_2 = \emptyset, \text{ then } P(A_1 \cup A_2|B) = P(A_1|B) + P(A_2|B). \tag{2.33}$$

If $A_1, A_2, \ldots, A_n, \ldots$, is an infinite sequence of mutually disjoint events, then

$$P(A_1 \cup A_2 \cup \ldots \cup A_n \cup \ldots |B) = \sum_{n=1}^{\infty} P(A_n|B). \tag{2.34}$$

The proof is left to the reader, see Problem 2.15.

2.4.2 Bayes' Theorem

In Example 2.30 we computed the probability that the screening test for a rare disease is positive for a person selected at random. But this is not what the patient wants to know. What he wants to know is this: Does the positive test result really mean that he actually has the disease? Example 2.32 gives a partial and surprising answer.

Example 2.32

This is a continuation of Example 2.30; in particular we use the same notation. Suppose the screening test is positive: Compute the probability that the patient actually has the disease.

Solution We have to compute the conditional probability that the patient has the disease given that the test is positive; that is, we have to compute the following conditional probability:

$$P(D|T^+) = \frac{P(D \cap T^+)}{P(T^+)}.$$

We recall that the prevalence rate of the disease is 1 in 5,000, so $P(D) = 0.0002$. We showed earlier, see Equation 2.30, that $P(T^+) = 0.02$. Similarly,

$$P(T^+ \cap D) = P(T^+|D)P(D) = 0.99 \times 0.0002 = 0.000198 \approx 0.0002.$$

Therefore,

$$P(D|T^+) = \frac{P(D \cap T^+)}{P(T^+)} = \frac{0.0002}{0.02} = 0.01.$$

And so only 1% of those who test positive for the disease actually have the disease! What is the explanation for this result? The answer lies in the fact that the bulk of the positive readings come from the false positives. One should not conclude, however, that the test is without value. Prior to taking the test the chances of a person selected at random having the disease was 0.0002; after taking the test with a positive reading the chances have increased to 0.01. That is the probability he actually has the disease has increased by a factor of 50. Clearly, further testing of the patient is warranted. We may summarize these results in the following way: Before taking the test, the probabilities that a person selected at random has (or does not have) the disease is given by the *prior* probabilities:

$$P(D) = 0.0002, \ P(D') = 0.9998.$$

If the test yielded a positive result then the *posterior* probabilities are:

$$P(D|T^+) = 0.01, \ P(D'|T^+) = 0.99.$$

The reasoning used in Example 2.32 is a special case of a more general result called *Bayes' theorem*.

Theorem 2.7 *Let the events $A_1, A_2, \ldots, A_n$ be a partition of the sample space S and let B denote an arbitrary event satisfying the condition $P(B) > 0$. Then*

$$P(A_k|B) = \frac{P(B|A_k)P(A_k)}{\sum_{1 \le i \le n} P(B|A_i)P(A_i)}.$$

Proof $P(A_k|B) = P(A_k \cap B)/P(B)$(this is just Definition 2.5). The product rule 2.26 applied to the numerator yields

$$P(A_k \cap B) = P(B \cap A_k) = P(B|A_k)P(A_k).$$

Next apply Theorem 2.5 to the term $P(B)$ that appears in the denominator. Putting these results together yields

$$P(A_k|B) = \frac{P(A_k \cap B)}{P(B)} = \frac{P(B|A_k)P(A_k)}{\sum_{1 \le i \le n} P(B|A_i)P(A_i)}.$$

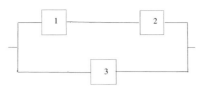

FIGURE 2.8: System of components for Example 2.33

2.4.3 Independence

Consider the following experiment: A card is drawn at random from a deck of 52 and its face value and suit are noted. The event that an ace was drawn is denoted by A and the event that a club was drawn is denoted by B. There are 4 aces so $P(A) = 4/52 = 1/13$, and there are 13 clubs, so $P(B) = 13/52 = 1/4$. $A \cap B$ denotes the event that the ace of clubs was drawn and since there is only one such card in the deck we see that

$$P(A \cap B) = 1/52 = (1/13) \times (1/4) = P(A) \times P(B).$$

Thus,

$$P(A|B) = \frac{P(A \cap B)}{P(B)} = \frac{1/52}{1/4} = \frac{1}{13} = P(A).$$

In other words, knowing that the card selected was a club did not change the probability that the card selected was an ace. We say that the event A is *independent* of the event B.

Definition 2.6 *The events A and B are said to be (probabilistically) independent if*

$$P(A \cap B) = P(A) \times P(B). \tag{2.35}$$

Clearly, if Equation 2.35 holds then

$$P(A|B) = P(A), \text{ and } P(B|A) = P(B).$$

Consequently, if A is independent of B then B is independent of A.

The definition of independence is easily extended to three or more events. To simplify the notation we shall only consider the case of three events.

Definition 2.7 *The events A, B, C are probabilistically independent if they are*

1. pairwise independent, i.e., the events $(A,B),(A,C),(B,C)$ are independent, and

2. the probability of the intersection of the three events equals the product of their probabilities:

$$P(A \cap B \cap C) = P(A) \times P(B) \times P(C).$$

Independence is often assumed, rather than proved as in Example 2.33.

Example 2.33

Consider the system consisting of three components linked together as indicated in the figure below. Let W_i be the event that component i is working and denote its probability by $P(W_i) = p$. Let W denote the event that the system is working. We define the reliability of the system to be $P(W)$. Compute $P(W)$.

Solution It is clear that this system works only if *both* W_1, W_2, and/or W_3 are working. Using set theory notation we can write

$$W = (W_1 \cap W_2) \cup W_3.$$

We assume that the failures of these components are mutually independent. Thus

$$
\begin{aligned}
P(W) &= P((W_1 \cap W_2) \cup W_3) \\
&= P(W_1 \cap W_2) + P(W_3) - P(W_1 \cap W_2 \cap W_3), \text{ by Proposition 2.4} \\
&= p^2 + p - p^3, \text{ by independence.}
\end{aligned}
$$

Thus, if $p = 0.9$ then $P(W) = 0.9^2 + 0.9 - 0.9^3 = 0.981$.

2.5   The Binomial Theorem*

Instead of proving Theorem 2.3 directly, we will prove the following special case first.

Theorem 2.8

$$(1+t)^n = \binom{n}{0} + \binom{n}{1} t + \cdots + \binom{n}{k} t^k + \cdots + \binom{n}{n} t^n. \qquad (2.36)$$

Proof It is obvious that $(1+t)^n$ is a polynomial of the nth degree, so it can be written as

$$f(t) = (1+t)^n = a_0 + a_1 t + a_2 t^2 + \cdots + a_k t^k + \cdots + a_n t^n.$$

We will prove the binomial theorem by showing that

$$a_k = \frac{P_{n,k}}{k!} = \binom{n}{k}.$$

In the ensuing sequence of calculations $f'(t)$ denotes the derivative of f and $f^{(k)}$ denotes the kth derivative.

$$
\begin{aligned}
f'(t) &= n(1+t)^{n-1} = a_1 + 2a_2 t + \cdots + na_n t^{n-1}; \\
f''(t) &= n(n-1)(1+t)^{n-2} = 2a_2 + 3 \cdot 2a_3 t + \cdots + n(n-1)a_n t^{n-2}, \text{ etc.}
\end{aligned}
$$

Since these equations are valid for all t we can set $t = 0$ in each of them to conclude

$$f(0) = 1 = a_0;$$

$$f'(0) = n = a_1;$$

$$f''(0) = n(n-1) = 2a_2;$$

and, in general,

$$f^{(k)}(0) = n(n-1) \cdots (n-k+1) = P_{n,k} = k!a_k.$$

Consequently,

$$a_k = \frac{P_{n,k}}{k!}.$$

Finally, we show that Theorem 2.8 implies Theorem 2.3. We begin with the algebraic identity

$$(a+b)^n = \left(a(1+\frac{b}{a})\right)^n = a^n \left(1+\frac{b}{a}\right)^n,$$

and then apply Theorem 2.8 to the term $\left(1+\frac{b}{a}\right)^n$ to conclude

$$\left(a(1+\frac{b}{a})\right)^n = a^n \sum_{k=0}^{n} \binom{n}{k} \left(\frac{b}{a}\right)^k.$$

Therefore,

$$a^n \left(1+\frac{b}{a}\right)^n = a^n \sum_{k=0}^{n} \binom{n}{k} a^{-k} b^k$$

$$= \sum_{k=0}^{n} a^n \binom{n}{k} a^{-k} b^k$$

$$= \sum_{k=0}^{n} \binom{n}{k} a^{n-k} b^k.$$

The proof of the binomial theorem is complete.

2.6 Chapter Summary

Statistical inference is a scientific method for making inferences about a population based upon data collected from a sample. To guard against the scientist's hidden biases the theoretical ideal is to collect the data via random sampling, which insures that *every subset of the population of the same size has an equal chance (probability) of being selected.* Probability theory is the mathematical framework within which we analyze the results of random sampling as well as other experiments whose outcomes are due to chance. Although it is not possible to predict in advance the result of one throw of a pair of dice it is possible to predict the distribution of the outcomes. It is by comparing the observed frequency distributions with those predicted by the mathematical theory that enables the scientist to make inferences about the population.

2.7 Problems

Section 2.2: Sample space, events, axioms of probability theory

Problem 2.1 *The sample space corresponding to the experiment of tossing a coin 4 times is listed below. Assume all 16 sample points have equal probabilities.*

HHHH	THHH
HHHT	THHT
HHTH	THTH

$$
\begin{array}{ll}
HHTT & THTT \\
HTHH & TTHH \\
HTHT & TTHT \\
HTTH & TTTH \\
HTTT & TTTT
\end{array}
$$

Let A_i and B_i denote, respectively, the events "exactly i heads are thrown" and "at least i heads are thrown," for $(i = 0, 1, 2, 3, 4)$. List the sample points in each of the following events and compute their probabilities:
(a) A_0 (b) A_1 (c) B_3 (d) B_4 (e) A_4

Problem 2.2 *This is a continuation of Problem 2.1. Which of the following statements are true? false?*
(a) $A_0' = B_1$ (b) $B_2 = A_2 \cup A_3 \cup A_4$ (c) $B_2 \subset A_2$ (d) $A_2 \subset B_2$

Problem 2.3 *This is a continuation of Problem 2.1. Express in terms of the sets A_i, B_i defined above the following events:*
(a) *No more than two tails were thrown.*   (b) *Exactly three tails were thrown.*
(c) *No tails were thrown.*   (d) *Only tails were thrown.*

Problem 2.4 *Refer to the sample space of Example 2.1 displayed in Table 2.1.*
(a) *List all sample points in the event $A = \{both\ faces\ are\ odd\}$.*
(b) *List all sample points in the event A'.*
(c) *Let $B = \{an\ even\ number\ is\ thrown\}$. Show that $A \subset B$.*
(d) *Describe, in plain English, the event $A' \cap B$.*
(e) *List all sample points in the event $C = \{an\ odd\ number\ is\ thrown\}$.*

Problem 2.5 *A quality control engineer observes that a printed circuit board with i defects $(i = 0, 1, 2, 3, 4)$ seems to occur with a frequency inversely proportional to $i + 1$; that is, the sample space corresponding to this experiment consists of the finite set of points $\mathcal{S} = \{s_0, s_1, s_2, s_3, s_4\}$, where s_i is the event that the circuit board has i defects and*

$$
P(s_i) = c/(i+1), \ i = 0, 1, \ldots 4.
$$

(a) *Compute the constant c. Hint: Use Equation 2.6 to derive an equation for c.*
(b) *What is the probability that a circuit board has no defects? at least one defect?*

Problem 2.6 *In an environmental study of the effect of radon gas on a population, blood samples were drawn from human volunteers, 100 cells were cultured and the number of cells that exhibited chromosome aberrations were counted. Denote by s_i the event that i chromosome aberrations occurred. Based on the observed frequency counts of these events the microbiologist feels that the following probability model is appropriate:*

$$
\mathcal{S} = \{s_0, s_1, s_2, s_3, s_4, s_5\}
$$
$$
P(s_i) = \frac{c(i+1)^2}{2^{i+1}}, \ (i = 0, 1, 2, 3, 4, 5).
$$

(a) *Determine the value of the constant c.*
(b) *What is the probability that a person selected at random from the population has exactly 3 chromosome aberrations? 3 or more?*

Problem 2.7 *In the paper* The Significant Digit Phenomenon, *American Mathematical Monthly, Vol. 102, No. 4, April 1995, the author proposes the following probability distribution for the first significant digit:*

$$P(s_d) = \log_{10}\left(1 + \frac{1}{d}\right), \; d = 1, 2, \ldots, 9.$$

Here, s_d is the event that the first significant digit is the integer d. Show, using only properties of the log *function, that*

$$\sum_{1 \le d \le 9} \log_{10}\left(1 + \frac{1}{d}\right) = 1.$$

Hint: It suffices to show that $\prod_{1 \le d \le 9}\left(1 + \frac{1}{d}\right) = 10$.

Problem 2.8 *Let W_i, $i = 1, 2, 3$ denote the event that component i is working. W_i' denotes the event that the ith component is not working. Express in set theoretic language the following events: Of the components W_1, W_2, W_3*

(a) only W_1 is working	*(b) all three components are working*
(c) none are working	*(d) at least one is working*
(e) at least two are working	*(f) exactly two are working*

Problem 2.9 *Let S denote the sample space for the game of craps (described in Table 2.1) and let A_i denote the events defined in Table 2.2. Express each of the following events in terms of the events A_i.*
(a) An odd number is thrown. (b) A number ≤ 6 is thrown (c) A number > 6 is thrown.

Problem 2.10 *Compute the probabilities of each of the events in the previous problem; assume all sample points have the same probability $1/36$.*

Problem 2.11 *A and B are events whose probabilities are given by $P(A) = 0.7, P(B) = 0.4, P(A \cap B) = 0.2$. Compute the following probabilities:*
(a) $P(A \cup B)$ (b) $P(A' \cap B')$ (c) $P(A' \cup B')$
(d) $P(A' \cap B)$ (e) $P(A \cap B')$ (f) $P(A' \cup B)$
(g) $P(A \cup B')$

Problem 2.12 *Given $P(A_1) = 0.5$, $P(A_2) = 0.4$, $P(A_3) = 0.4$, $P(A_1 \cap A_2) = .04$, $P(A_1 \cap A_3) = 0.1$, $P(A_2 \cap A_3) = 0.2$, $P(A_1 \cap A_2 \cap A_3) = 0.02$. Using this information, Venn diagrams, and Propositions 2.1 and 2.5 compute the following probabilities:*

(a) $P(A_1 \cup A_2 \cup A_3)$	*(b) $P(A_1' \cap A_2' \cap A_3')$*	*(c) $P((A_1 \cup A_2) \cap A_3)$*
(d) $P((A_1 \cup A_2) \cap A_3')$	*(e) $P(A_1' \cap A_2 \cap A_3)$*	*(f) $P(A_1' \cap A_2' \cap A_3)$*

Problem 2.13 *Which of the following statements are true? false?*
(a) $A \subset B$ then $P(B') \le P(A')$ (b) $P(A) = P(B)$ then $A \equiv B$
(c) $P(A \cap B) = 0$ then either $P(A) = 0$ or $P(B) = 0$

Problem 2.14 *Suppose $P(A) = 0.7$, $P(B) = 0.8$. Show that $P(A \cap B) \ge 0.5$.*

Problem 2.15 *Let B denote an event for which $P(B) > 0$. Define a set function P' via the recipe*

$$P'(A) = \frac{P(A \cap B)}{P(B)}.$$

Show that P' satisfies the axioms for a probability measure. $P'(A)$ is called "the conditional probability of A given that B has occurred."

Section 2.3: Discrete sample spaces, permutations, combinations, and binomial coefficients

Problem 2.16 *How many k-strings (words of length k) from the finite set $\mathcal{S} = \{1, 2, 3, 4, 5\}$ are there for:*
(a) $k = 1$?   (b) $k = 2$?   (c) $k = 3$?

Problem 2.17 *(a) Compute $P_{10,k}$ for $k = 1, 2, 3$.*
(b) Given that $P_{9,3} = 504$, compute $P_{9,k}$ for $k = 4, 5, 6$.

Problem 2.18 *Compute $\binom{n}{k}$ for:*

(a) $n = 10$, $k = 0, 1, 2, 3$	*(b) $n = 10$, $k = 7, 8, 9, 10$*
(c) $n = 12$, $k = 10, 11, 12$	*(d) $n = 12$, $k = 0, 1, 2$*

Problem 2.19 *Two digits are chosen without replacement from the set of integers $\{1, 2, 3, 4, 5\}$. (a) Describe the sample space and assign to each sample point the equally likely probability measure.*
(b) What is the probability that both digits are less than or equal to 3?
(c) What is the probability that the sum of the digits in the sample is greater than or equal to 4?

Problem 2.20 *Two digits are chosen without replacement from the set of integers $\{1, 2, 3, 4, 5, 6\}$. Let A_i, $i = 3, 4, \ldots, 11$ denote the event that the sum of the integers chosen equals i. Compute $P(A_i)$, $i = 3, 4, \ldots, 11$.*

Problem 2.21 *Each permutation of the integers $\{1, 2, 3, 4, 5\}$ determines a five digit number. If the numbers corresponding to all possible permutations are listed in increasing order of magnitude, i.e., $\{12345, 12354, \ldots 54321\}$, show that the 73rd number on the list is 41235. Hint: It is too tedious and unnecessary to list all the numbers in the list. A simple counting argument suffices.*

Problem 2.22 *Verify the binomial identity*

$$\binom{n}{k} + \binom{n}{k-1} = \binom{n+1}{k}$$

in the following cases:

(a) $n = 8$, $k = 3$	*(b) $n = 5$, $k = 2$*	*(c) $n = 10$, $k = 4$*
(d) $n = 9$, $k = 4$	*(e) $n = 12$, $k = 3$*	*(f) $n = 11$, $k = 7$*

Problem 2.23 *Using the recurrence Equation 2.20 compute:*

$$\binom{6}{k}, \ for \ k = 1, \ldots, 6.$$

Problem 2.24 *(a) Find the second term of the binomial expansion of $(2x + 3y)^4$.*
(b) Find the third term of the binomial expansion of $(2x + 3y)^4$.

Problem 2.25 *(a) Find the fifth term of the binomial expansion of $(1 + \sqrt{x})^9$.*
(b) Find the seventh term of the binomial expansion of $(1 + \sqrt{x})^9$.

Problem 2.26 *Find the value of the fourth term in the binomial expansion of $(2x + \frac{1}{x})^{10}$ when $x = 1/2$.*

Problem 2.27 *(a) In how many ways can two distinct numbers be chosen from the set $\{1, 2, \ldots, 100\}$ so that their sum is an even number?*
(b) In how many ways can three distinct numbers be chosen from the set $\{1, 2, \ldots, 100\}$ so that their sum is an even number?

Problem 2.28 *Consider the following variation on Feller's birthday problem (Example 2.24). The martian year is 669 days long.*
(a) There are k martians in a room; derive a formula for the probability that at least two martians in the room have a common birthday.
(b) Use your answer to part (a) to show that when $k \geq 31$ the probability that at least two martians in the room have a common birthday is 0.506.

Problem 2.29 *Consider the four-step binomial tree displayed next. Starting at the root node (point A) you can go one step up (u) or one step down (d) at each node. After four steps you arrive at one of the end points of the tree $\{0, 1, 2, 3, 4\}$.*
(a) How many different paths are there from A to 3? Hint: Each path through the tree corresponds to a sequence of u's and d's.
(b) How many different paths are there from A to 3 that pass through B?

4

3

A   B   2

1

0

Problem 2.30 *Poker dice is played by simultaneously rolling 5 dice.*
(a) Describe the sample space S; how many sample points are there?
(b) Show that P(no two dice show the same number) = 0.0926.
(c) What is the probability that at least two of the dice show the same number?

Problem 2.31 *The format of a Massachusetts license plate is 3 digits followed by 3 letters as in 736FSC. How many such license plates are there?*

Problem 2.32 *The format of a Virginia license plate is 3 letters followed by 4 numbers as in ABC1234. How many such license plates are there?*

Problem 2.33 *To win* megabucks *of the Massachusetts state lottery one must select 6 different digits out of $\{1,2,\ldots,36\}$ that match the winning combination. What is the probability of winning?*

Problem 2.34 *How many distinguishable arrangements of the letters of the word* MISSISSIPPI *are there?*

Problem 2.35 *Let S denote the sample space consisting of the 24 distinguishable permutations of the symbols 1,2,3,4 and assign to each permutation the probability $\frac{1}{24}$. Let A_i =event that the digit i appears at its natural place,e.g.,$\{2134\} \in A_4$.Compute:*
(a) $P(A_1 \cap A_2)$ *(b) $P(A_1)$*   *(c) $P(A_2)$*

Problem 2.36 *A group of 10 components contains 3 that are defective.*
(a) What is the probability that one component drawn at random is defective?
(b) Suppose a random sample of size 4 is drawn without replacement. What is the probability that exactly *one of these components is defective?*
(c) (Continuation of (b)): What is the probability that at least *one of these components is defective?*

Problem 2.37 *In how many ways can 12 people be divided into 3 groups of 4 persons (in each group) for an evening of bridge?*

Problem 2.38 *A bus starts with 5 people and makes 10 stops. Assume that passengers are equally likely to get off at any stop.*
(a) Describe the sample space and calculate the number of different outcomes.
(b) Compute the probability that no two passengers get off at the same stop.

Problem 2.39 *From a batch of 20 radios a sample of size 3 is randomly selected for inspection. If there are six defective radios in the batch what is the probability that the sample contains:*

(a) only defectives?	*(b) one defective and two non-defectives?*
(c) only non-defectives?	*(d) two defectives and one non-defective?*

Problem 2.40 *Consider the following two games of chance:*
(i) Person A throws 6 dice and wins if at least 1 ace appears.
(ii) Person B throws 12 dice and wins if at least 2 aces appears.
Who has the greater probability of winning?

Problem 2.41 *Consider a deck of four cards marked* 1, 2, 3, 4.
(a) List all the elements in the sample space S *coming from the experiment "two cards are drawn in succession and without replacement" from the deck. If* S *is assigned the equally likely probability measure compute the probability that:*
(b) the largest number drawn is a 4.
(c) the smallest number drawn is a 2.
(d) the sum of the numbers drawn is 5.

Problem 2.42 *An urn contains two nickels and three dimes.*
(a) List all elements of the sample space corresponding to the experiment "two coins are selected at random, and without replacement."
(b) Compute the probability that the value of the coins selected equals $0.10, $0.15, $0.20.

Problem 2.43 *A poker hand is a set of 5 cards drawn at random from a deck of 52 cards consisting of: (i) 4 suits (hearts, diamonds, spades and clubs) and (ii) each suit consists of 13 cards with face value (also called a denomination) denoted* {*ace,* 2,3,...,10, *jack, queen, king*}. *Find the probability of obtaining each of the following poker hands:*
(a) Royal flush ({10, *jack, queen, king, ace in single suit*}).
(b) Full house ({*one pair and one triple of the same denomination*}).

Problem 2.44 *A lot consists of 15 articles of which 8 are free of defects, 4 have minor defects, and 3 have major defects. Two articles are selected at random without replacement. Find the probability that:*

(a) both have major defects	*(b) both are good*
(c) neither is good	*(d) exactly one is good*
(e) at least one is good	*(f) at most one is good*
(g) both have minor defects	

Problem 2.45 *Twelve jurors are to be selected from a pool of 30, consisting of 20 men and 10 women.*
(a) What is the probability that all members of the jury are men?
(b) What is the probability all 10 women are on the jury?

Section 2.4: Conditional probability, Bayes' theorem, and independence

Problem 2.46 *A and B are events whose probabilities are given by*
$P(A) = 0.7, P(B) = 0.4, P(A \cap B) = 0.2.$ *Compute the following conditional probabilities:*

(a) $P(A|B)$	*(b)* $P(A'|B')$	*(c)* $P(B'|A')$
(d) $P(A'|B)$	*(e)* $P(A|B')$	*(f)* $P(B|A')$

Problem 2.47 *Suppose A and B are events in a sample space* S *and*
$P(A) = 0.4$, $P(B) = 0.3$, $P(A|B) = 0.3$. *Compute the following probabilities.*
(a) $P(A \cap B)$ *(b)* $P(A \cup B)$ *(c)* $P(A \cap B')$ *(d)* $P(B|A)$

Problem 2.48 *A and B are independent events whose probabilities are given by*
$P(A) = 0.7, P(B) = 0.4$. *Compute the following probabilities:*
(a) $P(A \cup B)$ *(b)* $P(A' \cap B')$ *(c)* $P(A' \cup B')$
(d) $P(A' \cap B)$ *(e)* $P(A \cap B')$ *(f)* $P(A' \cup B)$
(g) $P(A \cup B')$

Problem 2.49 *Two dice are thrown. If both of them show different numbers:*
(a) what is the probability that their sum is even?
(b) what is the probability that one of them is a six?

Problem 2.50 *Three dice are thrown. If all three of them show different numbers, what is the probability that one of them is a six?*

Problem 2.51 *The student body of a college is composed of 70% men and 30% women. It is known that 40% of the men and 60% of the women are engineering majors. What is the probability that an engineering student selected at random is a man?*

Problem 2.52 *Suppose the prevalence rate for a disease is 1/1,000. A diagnostic test has been developed which, when given to a person who has the disease, yields a positive result 99% of the time; while an individual without the disease shows a positive result only 2% of the time.*
(a) A person is selected at random and is given the test. What is the probability that the test is positive?
(b) Compute the probability that a person selected at random who tests positive actually has the disease.

Problem 2.53 *A new metal detector has been invented for detecting weapons in luggage. It is known that 1% of all luggage going through an airport contains a weapon. When a piece of luggage contains a weapon, an alarm (indicating a weapon is present) is activated 95% of the time. However, when the luggage does not contain a weapon the metal detector activates an alarm 10% of the time. Denote the event that a piece of luggage contains a weapon by W, the event that the alarm is activated by A^+, and the event that the alarm is not activated by A^-.*
(a) A piece of luggage is selected at random and tested; what is the probability that the alarm is activated?
(b) Suppose the alarm is activated. Compute the probability that the piece of luggage actually contains a weapon.

Problem 2.54 *In a bolt factory machines A, B, C manufacture, respectively, 20%, 30%, and 50% of the total. Of their output 3%, 2%, and 1% are defective.*
(a) A bolt is selected at random; find the probability that it is defective.
(b) A bolt is selected at random and is found to be defective. What is the probability it was produced by machine A?

Problem 2.55 *The CIA and FBI use polygraph machines—more popularly known as "lie detectors"—to screen job applicants and catch lawbreakers. A comprehensive review by a federal panel of distinguished scientists reported that "if polygraphs were administered to a group of 10,000 people that included 10 spies, 1,600 innocent people would fail the test and two of the spies would pass." (Source: Washington Post, May 1, 2006). Denote the event that a person taking the polygraph is a spy by S, the event that the polygraph is positive by T^+, and the event it is not by T^-.*
(a) A person is selected at random and tested; what is the probability that the person fails the polygraph test?
(b) Suppose a person fails the polygraph test. What is the probability that the person is a spy?

Problem 2.56 *An automobile insurance company classifies drivers into 3 classes: Class A (good risks), Class B (medium risks), Class C (poor risks). The percentage of drivers in each class is: Class A 20%; Class B 65%; Class C 15%. The probabilities that a driver in each of these classes will have an accident in 1 year are given by 0.01, 0.02, 0.03, respectively. After purchasing an insurance policy a driver has an accident within the first year. What is the probability that he is a class A risk? Class B risk? Class C risk?*

Problem 2.57 *Refer to Example 2.32. Suppose the results of the test were negative, i.e., the event T^- occurred. Compute the posterior probabilities in this case.*

Problem 2.58 *In the case (re: State vs. Pedro Soto et al.) the New Jersey Public Defender's office moved to suppress evidence against 17 black defendants (principally on charges of transporting illegal drugs) on grounds of selective enforcement, i.e., that blacks were more likely to be stopped by the police than were others, on the southern portion of the New Jersey Turnpike. The following study was undertaken by the defense: For 30 days and for one hour each day, cars were observed traveling between Junctions 1 and 3 of the NJ Turnpike, and the number of speeding cars was recorded. An interesting feature of the data is that 98.1% of the cars observed were driving faster than the posted speed limit of 60 mph, consequently the state police could legally stop nearly anyone they chose. The driver's racial identity (black or white) was recorded as well as the racial identities of the drivers who were stopped. The data collected are in the following table.*

	Driver		
	Black	*White*	*Total*
Stopped	127	148	275
Not Stopped	182	1605	1787
Total	309	1753	2062

(Source: Joseph B. Kadane & Norma Terrin, Missing Data in the Forensic Context, J. R. Statist. Soc. A (1997), vol. 160, part 2, pp. 351-357.)
(a) Compute the probability that a speeding driver is stopped.
(b) Suppose a speeding driver is stopped by the police; what is the probability that the driver is white?
(c) Suppose a speeding driver is white; what is the probability that the driver is stopped?
(d) Suppose a speeding driver is stopped by the police; what is the probability that the driver is black?
(e) Suppose a speeding driver is black; what is the probability that the driver is stopped?
(f) Do the data support, or refute, the defense claim of "racial profiling"? Justify your answer by referring to your answers to parts (a-e) above. Note: Justice Robert E. Francis of the Superior Court of New Jersey, in an unpublished opinion, found for the defense.

Problem 2.59 *A lot consists of 3 defective and 7 good transistors. Two transistors are selected at random. Given that the sample contains a defective transistor, what is the probability that both are defective?*

Problem 2.60 *(Continuation of previous problem) Consider the following procedure to find and remove all defective transistors: A transistor is randomly selected and tested until all 3 defective transistors are found. What is the probability that the third defective transistor will be found on:*
(a) the third test?   (b) the fourth test?   (c) the tenth test?

Problem 2.61 *A fair coin is tossed four times.*
(a) What is the probability that the number of heads is greater than or equal to three given that one of them is a head?
(b) What is the probability that the number of heads is greater than or equal to three given that the first throw is a head?

Problem 2.62 *A fair coin is tossed three times.*
(a) Let A denote the event that one head appeared in the first two tosses, and let B denote the event that one head occurred in the last two tosses. Are these events independent?
(b) Let A denote the event that one head appeared in the first toss, and let B denote the event that one head occurred in the last two tosses. Are these events independent?

Problem 2.63 *In the manufacture of a certain article, two types of defects are noted that occur with probabilities 0.02, 0.05, respectively. If these two defects occur independently of one another what is the probability that:*
(a) an article is free of both kinds of defects?
(b) an article has at least one defect of either type?

Problem 2.64 *Let W_i, $i = 1, 2, 3$ denote the event that component i is working. W_i' denotes the event that the ith component is not working. Assume the events W_i, $i = 1, 2, 3$ are mutually independent and $P(W_i) = 0.9$, $i = 1, 2, 3$. Compute the probabilities of the following events: Of the components W_1, W_2, W_3*

(a) only W_1 is working	*(b) all three components are working*
(c) none are working	*(d) at least one is working*
(e) at least two are working	*(f) exactly two are working*

Problem 2.65 *Consider the system consisting of three components linked together as indicated in the figure below. Let W_i be the event that component i is working and assume that $P(W_i) = 0.9$ and the events W_i, $i = 1, 2, 3$ are independent. Let W denote the event that the system is working. Compute $P(W)$. (It will be helpful to reread Example 2.33.)*

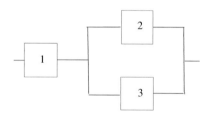

Problem 2.66 *Show that $P(A \cup B | B) = 1$.*

Problem 2.67 *(a) Show that S and A are independent for any A.*
(b) Show that the $\emptyset$ and A are independent for any A.

Problem 2.68 *Consider the sample space consisting of the 9 ordered triples:*
$(a, a, a), (b, b, b), (c, c, c), (a, b, c), (a, c, b), (b, a, c), (b, c, a), (c, a, b), (c, b, a)$ equipped with equally likely probability measure. Let A_k denote the event that the kth coordinate is occupied by a. Thus, $A_1 = \{(a, a, a), (a, b, c), (a, c, b)\}$, etc. Show that the events A_1, A_2, A_3 are:
(a) pairwise independent, i.e., $P(A_i \cap A_j) = P(A_i)P(A_j)$, but
(b) not mutually independent, i.e., $P(A_1 \cap A_2 \cap A_3) \neq P(A_1)P(A_2)P(A_3)$.

Problem 2.69 *Show that if A is independent of itself then $P(A) = 0$ or $P(A) = 1$.*

Problem 2.70 *Show that if $P(A) = 0$ or $P(A) = 1$ then A is independent of every event B.*

Problem 2.71 *Show that if A and B are independent then so are A and B'.*

Problem 2.72 *Suppose A, B, C are mutually independent events. Assume that $P(B \cap C) > 0$—this hypothesis is just to ensure that the conditional probabilities below are defined. (a) Show that $P(A|B \cap C) = P(A)$.   (b) Show that $P(A|B \cup C) = P(A)$.*

2.8   To Probe Further

There is still no better introduction to probability theory than Feller (1968), *An Introduction to Probability Theory and its Applications*, 3rd ed., vol. 1, John Wiley & Sons, Inc., New York. The birthday problem (Example 2.24) is discussed on p. 33. Problem 2.68 is adapted from Example (e), p. 127 of Feller, op. cit. The martian variation on the birthday problem (Problem 2.28) is due to Cormen, Leiserson, and Rivest (1990), *Introduction to Algorithms*, McGraw-Hill, New York. It is also a good reference for the applications of permutations, combinations, and recurrence relations to computer science, in particular to the analysis of algorithms. Problem 2.21 is an adaptation of a similar problem in Rohatgi (1976), *An Introduction to Probability and Mathematical Statistics*, John Wiley & Sons, Inc.

Chapter 3

Discrete Random Variables and Their Distribution Functions

It is the calculus of probabilities which alone can regulate justly the premiums to be paid for assurances; the reserve fund for the disbursements of pensions, annuities, discounts, etc. It is under its influence that lotteries and other shameful snares cunningly laid for avarice and ignorance have definitely disappeared.

Dominique Francois Jean Arago, French physicist (1786-1853)

3.1   Orientation

In this chapter we study the fundamental concept of the *distribution of a variable X defined on a sample space.* A variable defined on a sample space is called a *random variable.* It is analogous to the concept of a variable defined on a population S studied earlier in Chapter 1. There are other analogies as well: The analogue of the empirical distribution function is called the *distribution function* of X, and the analogues of the sample mean and sample variance are called the *expected value* and *variance* of X, respectively. Random variables are classified as *discrete* or *continuous*; the major difference between these two types is in the level and sophistication of the mathematical tools that one uses to study them. In this chapter we study only discrete random variables; the concept of a continuous random variable, which requires calculus in an essential way, will be treated in Chapter 4.

Organization of Chapter

1. Section 3.2: Discrete Random Variables and their Distribution Functions

2. Section 3.3: Expected Value of a Random Variable

3. Section 3.3.1: Moments of a Random Variable

4. Section 3.3.2: Variance of a Random Variable

5. Section 3.4: The Hypergeometric Distribution

6. Section 3.5: The Binomial Distribution

7. Section 3.5.1: A Coin Tossing Model For Stock Market Returns

8. Section 3.6: The Poisson Distribution

9. Section 3.7: Moment Generating Function: Discrete Random Variables*

10. Section 3.8: Mathematical Details and Derivations

11. Section 3.9: Chapter Summary

3.2   Discrete Random Variables

The number of heads that appear in n tosses of a coin, the payoff from a bet placed on a throw of two dice, the number of defective items in a sample of size n taken from lot of N manufactured items, and the payoff from a stock option on its expiration date are examples of functions whose values are determined by chance; such functions are called *random variables*. To illustrate, consider Table 3.1 below which lists: (i) the eight sample points corresponding to the experiment of tossing a coin three times, and (ii) the values of the random variable X that counts the number of heads for each outcome.

Table 3.1 *The random variable X that counts the number of heads in three tosses of a coin.*

$$X(TTT) = 0 \qquad X(HHT) = 2$$
$$X(TTH) = 1 \qquad X(HTH) = 2$$
$$X(THT) = 1 \qquad X(THH) = 2$$
$$X(HTT) = 1 \qquad X(HHH) = 3$$

Definition 3.1 *A random variable X is a real valued function defined on a sample space S. The value of the function at each sample point is denoted by $X(s)$. The set of values $\{X(s) : s \in S\}$ is called the* range *and is denoted R_X.*

The random variable X of Table 3.1 has range $R_X = \{0, 1, 2, 3\}$. It is an example of a *discrete random variable*; so called because its range is a discrete set. In the general case we say that the random variable X is discrete if its range is the discrete set $\{x_1, x_2, \ldots, x_n, \ldots\}$. Because it is convenient to do so we shall assume that the numbers in the range appear in increasing order, so $R_X = \{x_1 < x_2 < \ldots < x_n < \ldots\}$.

The event $\{X = x\}$

A random variable is often used to describe events; thus, with respect to the experiment of throwing a coin three times, the expression $X = 2$ is equivalent to saying "two heads are thrown." More generally, $X = x$ means that "x heads are thrown." In detail we write:

$$\{X = 0\} = (TTT)$$
$$\{X = 1\} = (HTT), (THT), (TTH)$$
$$\{X = 2\} = (HHT), (HTH), (THH)$$
$$\{X = 3\} = (HHH).$$

In the general case we use the shorthand notation $\{X = x\} = \{s : X(s) = x\}$.

The probability function (pf) of a discrete random variable X

The event $\{X = x\}$ has a probability denoted by $P(X = x)$. If x is not in the range $R_X = \{x_1, x_2, \ldots\}$ then the set $\{X = x\}$ is the empty set $\emptyset$; in this case $P(X = x) = 0$; consequently, it is only necessary to list the values $P(X = x), x \in R_X$. The function

$$f_X(x) = P(X = x),\ x \in R_X;\ f_X(x) = 0,\ x \text{ not in } R_X \qquad (3.1)$$

is called the *probability function (pf) of the random variable X*. Some authors also use the term *probability mass function* (pmf).

Distribution Function of a Random Variable X

For computational purposes it is often more convenient to work with the *distribution function* (df), (Definition 3.2 below), instead of the pf $f_X(x)$.

Definition 3.2 *The distribution function (df) $F_X(x)$ of the random variable X is the real valued function defined by the equation*

$$F_X(x) = P(X \leq x).$$

The distribution function of a discrete random value is given by

$$F_X(x) = \sum_{x_j \leq x} f_X(x_j).$$

The distribution function $F_X(x)$ determines the values $f_X(x_j)$ via the equation[1]

$$f_X(x_j) = F_X(x_j) - F_X(x_{j-1}). \tag{3.2}$$

Random variables are classified by their distribution functions with the the binomial, Poisson, normal, and chi-square among the most important of them.

Definition 3.3 *The symbol $X \overset{\mathcal{D}}{=} Y$ means that the random variables X and Y have the same distribution function.*

Notation To simplify the notation we sometimes drop the subscript X and simply write $f(x)$, $F(x)$.

Example 3.1 *Compute the probability function for the number of heads appearing in three tosses of a fair coin*

Solution The assumption that the coin is fair means that we assign the equally likely probability measure to the sample space. Since the sample space has 8 sample points each sample point has probability $1/8$ (see Table 3.2).

Table 3.2 *The probability and distribution functions for the number of heads in three tosses of a fair coin displayed in the format of a horizontal bar chart*

x	$f_X(x)$	$F_X(x)$	$\{s : X(s) = x\}$
0	1/8	1/8	TTT
1	3/8	4/8	HTT, THT, TTH
2	3/8	7/8	HHT, HTH, THH
3	1/8	1	HHH

Notice that the probability function in this example is nonnegative and sums to one; that is,

$$f_X(x) \geq 0 \text{ and that } \sum_{0 \leq x \leq 3} f_X(x) = 1.$$

This is just a special case of the following proposition:

Proposition 3.1 *Let X be a discrete random variable with probability function[1] $f_X(x)$. Then*

$$f_X(x) \geq 0 \text{ and } \sum_i f_X(x_i) = 1. \tag{3.3}$$

---

[1]We assume that $x_j < x_{j+1}$, $(j = 1, \ldots)$.

Derivation of Equation 3.3: This result follows from the fact that the events $A_i = \{X = x_i\}$ form a partition of the sample space $\mathcal{S}$; consequently, the axiom of countable additivity (Axiom 2.14) and the fact that $f_X(x_i) = P(A_i)$ together imply

$$1 = P(\mathcal{S}) = P(\cup_i A_i)$$
$$= \sum_i P(A_i)$$
$$= \sum_i f_X(x_i).$$

Any function satisfying the conditions listed in Equation 3.3 is called a *probability function* (pf).

Example 3.2

Compute the probability function of the random variable X that records the sums of the faces of two dice.

Solution The sample space is the set

$$\mathcal{S} = \{(i,j) : i = 1,\ldots,6; j = 1,\ldots,6\} \text{ (see Table 2.1)}.$$

The random variable X is the function $X(i,j) = i + j$. Thus, $X(1,1) = 2$, $X(1,2) = 3,\ldots,X(6,6) = 12$. The range of X is the set $R_X = \{2,3,\ldots,12\}$ and its probability function is displayed in Table 3.3 below.

Table 3.3 *The probability function for the random variable $X(i,j) = i+j$ displayed in the format of a horizontal bar chart*

x	$f_X(x)$	$F_X(x)$	$\{s : X(s) = x\}$
2	1/36	1/36	$(1,1)$
3	2/36	3/36	$(1,2), (2,1)$
4	3/36	6/36	$(1,3), (2,2), (3,1)$
5	4/36	10/36	$(1,4), (2,3), (3,2), (4,1)$
6	5/36	15/36	$(1,5), (2,4), (3,3), (4,2), (5,1)$
7	6/36	21/36	$(1,6), (2,5), (3,4), (4,3), (5,2), (6,1)$
8	5/36	26/36	$(2,6), (3,5), (4,4), (5,3), (6,2)$
9	4/36	30/36	$(3,6)\,(4,5), (5,4), (6,3)$
10	3/36	33/36	$(4,6)\,(5,5), (6,4)$
11	2/36	35/36	$(5,6), (6,5)$
12	1/36	1	$(6,6)$

Probability Histogram

Another way to visualize a pf is by constructing a *probability histogram*, as in Figure 3.1. Above each integer $x = 2,\ldots,12$ we construct a rectangle centered at x whose area equals $f_X(x)$.

Interpreting the probability histogram as a relative frequency histogram

If we throw a pair of fair dice n times, where n is a very large number, we would expect the relative frequency of the occurrences of the number $x(x = 2,\ldots,12)$ to be nearly equal to its probability $f_X(x)$; this is the frequency interpretation of probability discussed in the previous chapter. Of course in any actual experiment of tossing a pair of dice n times the relative frequency histogram would differ from the theoretical frequency displayed in Figure 3.1.

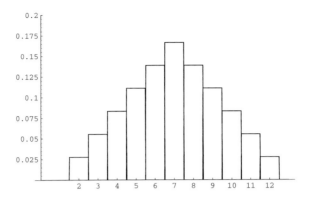

FIGURE 3.1:   Probability histogram of Example 3.2

Example 3.3

Compute and sketch the graph of the df $F_X(x)$ of the number X thrown with a pair of dice.
   Solution The values of the df $F_X(j)$, $(j = 2, \ldots, 12)$ are displayed in the third column of Table 3.3. We obtain the values of $F_X(x)$ for all x by noting that

$$F_X(x) = 0, \ x < 2;$$
$$F_X(x) = F_X(j) \text{ for } j \leq x < j + 1, \ (j = 2, \ldots, 12);$$
$$F_X(x) = 1, \ x \geq 12.$$

The graph is displayed in Figure 3.2.

Example 3.4 *Computing a probability function of a one-roll bet in craps*

The probability function of the payoff random variable W for a one-roll bet on a 12 with payoff of 30 to 1 is displayed in Table 3.4. W represents the gambler's *net* winnings. The gambler only wins \$30 when a 12 is thrown and loses \$1 in all other cases; consequently, $W(6, 6) = 30$ and $W(i, j) = -1$ for all other sample points. The range of W is the set consisting of the two points $\{-1, 30\}$.

Table 3.4 *The probability function for the payoff of a one-roll bet on the number 12*

x	-1	30
$f_W(x)$	$35/36$	$1/36$

Example 3.5

Compute the probability function of the number of hearts in a five-card poker hand.

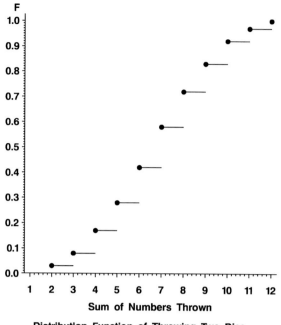

Distribution Function of Throwing Two Dice

FIGURE 3.2:   Graph of $F_X(x)$ (Example 3.3)

Solution The probability that a five-card poker hand contains x hearts is given by

$$f(x) = \frac{\binom{13}{x}\binom{39}{5-x}}{\binom{52}{5}}, \; x = 0, 1, 2, 3, 4, 5. \tag{3.4}$$

Derivation of Equation 3.4: The term in the denominator counts the total number of five-card poker hands each one of which is assigned the same probability; the term in the numerator counts the number of ways of getting x hearts and $5-x$ non-hearts in a five-card poker hand; their ratio is the probability of getting x hearts in a five-card poker hand. In detail: The number of ways of getting x hearts out of 13 is $\binom{13}{x}$ and the number of ways of getting $5-x$ non-hearts from the 39 non-hearts in the deck is $\binom{39}{5-x}$; consequently, the number of ways of getting x hearts and $5-x$ non-hearts in the poker hand is their product $\binom{13}{x}\binom{39}{5-x}$; this is the term in the numerator. There are $\binom{52}{5}$ sample points, and each one is assigned the same probability $1/\binom{52}{5}$; this is the term in the denominator.

Example 3.6 *For each of the following non-negative functions $f(x)$ determine the constant c so that it is a probability function.*

1. $f(x) = cx$; $x \in R_X = \{1, 2, 3, 4\}$; $f(x) = 0$ elsewhere.
   Solution The sum

$$\sum_{x \in R_X} cx = \sum_{1 \le x \le 4} cx = c \sum_{1 \le x \le 4} x = 10c = 1.$$

Consequently, $c = 1/10$.

2. $f(x) = c(5 - x^2)$; $x = -2, -1, 0, 1, 2$; $f(x) = 0$ elsewhere.

Solution The sum

$$\sum_{-2 \le x \le 2} c(5 - x^2) = c(1 + 4 + 5 + 4 + 1) = 15c = 1.$$

Consequently, $c = 1/15$.

Bernoulli Random Variables

A random variable X which assumes only the two values 0 and 1 with $P(X = 0) = 1 - p$ and $P(X = 1) = p$ is called *Bernoulli random variable*. These are the random variables of the sequence $X_1, \ldots, X_n$ used to model a Bernoulli sequence of trials. The Bernoulli probability function, denoted $f(x; p)$, is displayed in Table 3.5.

Table 3.5 *Bernoulli probability function*

x	0	1
$f(x; p)$	$1 - p$	p

The Geometric Distribution

The *geometric distribution* frequently arises in connection with so-called *waiting time problems*. For example, consider a sequence of coin tosses, with probability p for heads. Denote the number of trials until the first head is thrown by X. It is also called the *waiting time until the first success*. The probability of the first success at the kth trial equals

$$P(X = k) = q^{k-1}p, (k = 1, 2, \ldots), q = 1 - p,$$

since there are $k - 1$ failures before the first success occurs at the kth trial. In detail,

$$\begin{aligned} P(X = k) &= P(T, \ldots, T, H) \\ &= P(T) \cdots P(T)P(H) \\ &= q \times \cdots \times q \times p = q^{k-1}p, \ k = 1, 2, \ldots. \end{aligned}$$

Definition 3.4 *A random variable X is said to have a geometric distribution with parameter p if*

$$P(X = k) = pq^{k-1}, \ k = 1, \ldots. \tag{3.5}$$

To show that 3.5 defines a probability function we must show that it satisfies Equation 3.3. This is a consequence of the following well known property of the *geometric series*

$$\sum_{0 \le k < \infty} r^k = \frac{1}{1 - r}, \ |r| < 1. \tag{3.6}$$

Thus,

$$\sum_{1 \le k < \infty} pq^{k-1} = p \sum_{0 \le k < \infty} q^k$$

$$= p \times \frac{1}{1 - q} = p \times \frac{1}{p} = 1.$$

3.2.1   Functions of a Random Variable

In physics one defines the kinetic energy of a body of mass m with speed v to be equal to $(1/2)mv^2$. In statistical mechanics one assumes that the speed of a molecule in a gas is given by a random variable X with a df whose precise form need not concern us here. It follows that the kinetic energy of a molecule is given by the random variable $Y = (1/2)mX^2$, where m is the mass of the molecule. We write $Y = \phi(X)$, where $\phi(v) = (1/2)mv^2$. We say that Y is a *function of the random variable X*.

Computing the probability function of $\phi(X)$

Since the distribution of the kinetic energy can be related to such measurable quantities as the pressure of the gas, it is useful to have a method for computing the probability function of $\phi(X)$. Rather than give a general formula it is more instructive to work out specific examples.

Example 3.7

Suppose the random variable X has a pf given by the following table:

x	−2	−1	0	1	2
f(x)	0.1	0.2	0.3	0.3	0.1

1. Compute the pf of $Y = X^2$.

   Solution Here $\phi(x) = x^2$. Since $R_X = \{-2, -1, 0, 1, 2\}$ it follows that

   $$R_Y = \{-2^2, -1^2, 0^2, 1^2, 2^2\} = \{0, 1, 4\}.$$

   Moreover, you can easily verify that

   $$\{X^2 = 0\} = \{X = 0\};$$
   $$\{X^2 = 1\} = \{X = -1\} \cup \{X = 1\};$$
   $$\{X^2 = 4\} = \{X = -2\} \cup \{X = 2\}.$$

   Consequently,

   $$P(\{X^2 = 0\}) = P(\{X = 0\}) = 0.3;$$
   $$P(\{X^2 = 1\}) = P(\{X = -1\} \cup \{X = 1\}) = 0.5;$$
   $$P(\{X^2 = 4\}) = P(\{X = -2\} \cup \{X = 2\}) = 0.2.$$

   Therefore, the probability function of $Y = X^2$ is given by:

y	0	1	4
$f_Y(y)$	0.3	0.5	0.2

2. Compute the pf of $Y = 2^X$.

   Solution Here $\phi(x) = 2^x$, and the range of Y is the set

   $$R_Y = \{2^{-2}, 2^{-1}, 2^0, 2^1, 2^2\}$$
   $$= \left\{\frac{1}{4}, \frac{1}{2}, 1, 2, 4\right\}.$$

Proceeding as in the preceding case we see that

$$\{2^X = 2^x\} = \{X = x\};$$

consequently,

$$P(\{2^X = 2^x\}) = P(\{X = x\}) = f(x).$$

Therefore the probability function of $Y = 2^x$ is given by:

y	1/4	1/2	1	2	4
$f_Y(y)$	0.1	0.2	0.3	0.3	0.1

3.3 Expected Value and Variance of a Random Variable

In Chapter 1 we defined the sample mean $\overline{x}$ of a data set $\{x_1, \ldots, x_n\}$ to be the arithmetic mean

$$\overline{x} = \frac{1}{n} \sum_{1 \le i \le n} x_i.$$

We then derived an equivalent formula (Equation 1.23) for the sample mean which can be expressed as follows in terms of the empirical probability function (Equation 2.8):

$$\overline{x} = \sum_x x \hat{f}(x).$$

We define the *expected value* of a random variable in exactly the same way except that we use the probability function $f_X(x)$ in place of the empirical probability function $\hat{f}(x)$. The formal definition follows.

Definition 3.5 *The **expected value** of a random variable X with probability function $f_X(x)$ is denoted $E(X)$ and is defined to be the sum*

$$E(X) = \sum_x x\, P(X = x)$$

$$= \sum_x x\, f_X(x), \tag{3.7}$$

provided that the sum

$$\sum_x |x|\, f_X(x) < \infty \tag{3.8}$$

converges.

The expected value μ can be interpreted as the center of mass of a distribution of weights $f(x_i)$ located at the points x_i on the line.

Notation The expected value of X is also denoted by $\mu = E(X)$; to emphasize its dependence on X one also uses the notation μ_X.

An interpretation of $E(X)$ in terms of the average win per bet: Let X denote the payoff random variable for some gambling game. The gambler wins x_j $(j = 1, \ldots, n)$ dollars with probability $P(X = x_j)$. A negative value $x_j < 0$ is interpreted as a loss of x_j

dollars. According to the frequency interpretation of probability (Equation 2.9) we see that after N repetitions of the game the gambler has won x_j dollars approximately $NP(X = x_j)$ times. Consequently, after N plays W_N—his total winnings—is approximately equal to

$$W_N \approx \sum_{1 \le j \le n} N x_j P(X = x_j).$$

The average win per bet equals W_N/N; consequently,

$$\text{the average win per bet} \; = \frac{W_N}{N} \approx \sum_{1 \le j \le n} x_j P(X = x_j) = E(X).$$

Example 3.8 *Compute $E(X)$ for each of the following random variables:*

1. A Bernoulli random variable X with $P(X = 1) = p$.

   Solution We show that $E(X) = p$. The computation is simple but the result is quite useful:
   $$E(X) = 0 \times (1 - p) + 1 \times p = p. \tag{3.9}$$

2. A geometric random variable X with probability function

   $$f(x) = pq^{x-1}, \; x = 1, 2, \ldots.$$

   Solution We show that $E(X) = 1/p$. Recalling the interpretation of X as the waiting time until the first success we see that this result is intuitively plausible since a small probability of success implies a long waiting time for the first success.

   We use the recurrence relation

   $$f(x + 1) = qf(x), \; 1 \le x < \infty.$$

   Multiplying both sides of the recurrence relation by x and summing over x we obtain the equation

   $$\sum_{1 \le x < \infty} x f(x + 1) = \sum_{1 \le x < \infty} q x f(x) = q E(X).$$

   On the other hand,

   $$\sum_{1 \le x < \infty} x f(x + 1) = \sum_{1 \le x < \infty} (x - 1) f(x)$$
   $$= \sum_{1 \le x} x f(x) - \sum_{1 \le x} f(x)$$
   $$= E(X) - 1.$$

   Therefore,

   $$E(X) - 1 = q E(X).$$

   Solving for $E(X)$ yields the equation

   $$(1 - q)E(X) = 1.$$

   Thus, $E(X) = 1/(1 - q) = 1/p$.

3. Chuck-a-luck is a popular carnival game (so beware!) in which three fair dice are rolled. A bet placed on one of the numbers 1 through 6 has a payoff that depends on the number of times the number appears on the three dice. To fix our ideas suppose one bets \$1 on the number 4. The payoff random variable X equals $-\$1$ if none of the three dice shows a 4; otherwise $X = \$x$, where x is the number of dice that show the number 4. For instance, the payoff is \$3 if all three dice show the number 4. We now compute $E(X)$.

Solution The range $R_X = \{-1, 1, 2, 3\}$. The probability function is given by

$$P(X = -1) = \frac{5^3}{6^3}$$

$$P(X = 1) = \frac{3 \times 5^2}{6^3}$$

$$P(X = 2) = \frac{3 \times 5}{6^3}$$

$$P(X = 3) = \frac{1}{6^3}.$$

Consequently, $E(X) = -0.0789$.

Details of the computations: The sample space corresponding to throwing three dice has $6^3 = 216$ sample points each one of which has equal probability. The events $X = -1, 1, 2, 3$ contain 5^3, 3×5^2, 3×5, 1 sample points, respectively (why?).

4. W is the random variable giving the payoff of a one-roll bet on a 12.

Solution The payoff of a one-roll bet on a 12 is 30 to 1; its probability function $f(w)$ is given by

w	-1	30
$f(w)$	35/36	1/36

$$\text{Therefore, } E(W) = -1 \times \frac{35}{36} + 30 \times \frac{1}{36} = -\frac{5}{36} = -0.1389.$$

Definition 3.6 *Let W be the payoff random variable for a gambling game. The game is called:*

1. favorable, if $E(W) > 0$;

2. unfavorable, if $E(W) < 0$;

3. fair, if $E(W) = 0$.

The quantity $100 \times |E(W)|$ is called the *house percentage*.

Comment: A one roll bet on the number 12 with a 30 to 1 payoff is a sucker bet because the house percentage, which is nearly 14%, is so high.

Example 3.9

Let us now analyze the payoff random variable of a one-roll bet. A one roll bet pays you \$b if you win, otherwise you lose your stake of \$a. If you denote the probability of winning and losing by p and $1 - p$, respectively, then the probability function and the expected value of the payoff random variable W of a one-roll bet are given by:

$$P(W = -a) = 1 - p \text{ and } P(W = b) = p; \text{ so,}$$
$$E(W) = -a(1 - p) + bp = \mu. \tag{3.10}$$

To repeat: This means that if the event $B = \{W = b\}$ occurs, then you win b dollars; if B does not occur, then you lose your stake of a dollars. The ratio b/a ("b to a") is called the *house odds*. A *fair bet* is one for which the offered odds b'/a' are chosen so that $E(W) = 0$; in this case we call the ratio b'/a' the *true odds*.

Now the condition that $E(W) = -a'(1 - p) + b'p = 0$ implies that the true odds b'/a' must satisfy the condition

$$\textbf{True Odds: } \quad \frac{b'}{a'} = \frac{1 - p}{p}. \tag{3.11}$$

The true odds on a one-roll bet on a 12 are 35 to 1; the house odds are only 30 to 1. The true odds on a one-roll bet on an 11 are 34/2, or 17 to 1; the house odds are 15 to 1. It is this difference between the true odds and the house odds that guarantees the fantastic profits of the gambling casinos. More generally, if W is the payoff for some gambling game, we call $E(W) = \mu$ the *expected win per bet*.

Doubling the Bet: The Road to Financial Ruin

Over the weekend of 25-26 February 1995, the financial world was startled to learn that Baring Brothers, a highly respected merchant bank that had been in business for 233 years, collapsed into bankruptcy as a result of a series highly imprudent financial transactions by one of its junior bond traders. For the bank's senior management, its failure, coming as it did just before the annual bonuses were to be distributed, could not have occurred at a more embarrassing time. It was soon revealed that the trader was attempting to recoup his previous losses by committing ever increasing sums of the bank's capital to high risk bets on the future value of several indices based upon the prices of a complex hodgepodge of stocks and bonds. This, to use the currently fashionable terminology, is called *risk management.*[2] The bond trader's method, as we will now see, is a variant of an old and discredited gambling system called *doubling the bet*, which is a gambling strategy to cover your mounting losses by placing bigger and bigger bets.

Consider a gambler who has a probability p of winning on any one of a sequence of bets. The probability of winning for the first time at the xth bet is pq^{x-1}. To simplify the calculations we assume $p = q = 1/2$, so the probability of winning for the first time at the xth bet is $(1/2)^x$. We suppose the gambler's initial bet is 1 dollar and that he doubles his bet each time until he wins for the first time. Doubling the bet means that he bets 2^{x-1} dollars on the xth bet, where $x = 1, 2, \ldots$. How large must his initial capital be if he is sustain this betting system through the xth bet given that he lost his previous $x - 1$ bets? We solve this problem by computing the pf of X, the amount of capital the gambler needs to play this game until he wins for the first time.

To place his first bet he needs 1 dollar. If he loses he then bets 2 dollars, so he needs $1 + 2 = 2^2 - 1$ dollars to bet a second time. Repeating this reasoning we see that the gambler's cumulative losses from the preceding $(x - 1)$ unsuccessful bets followed by a bet of 2^{x-1} on the xth bet is given by the geometric series

$$X = 1 + 2 + \cdots + 2^{x-1} = 2^x - 1.$$

---

[2]From a mathematical perspective the major difference between gambling and risk management is that the latter activity does not take place in a gambling casino.

Therefore, the pf of X is given by

$$P(X = 2^x - 1) = 2^{-x}.$$

If he wins for the first time on the xth bet his net gain will be $2^x - (2^x - 1) = 1$ dollar. Consequently, when he wins he will be 1 dollar ahead. However, from the preceding remarks it is clear that he will need $2^x - 1$ dollars if he is to play the game through the xth play. Consequently, the random variable X has probability function given by

$$P(X = 2^x - 1) = \left(\frac{1}{2}\right)^x, \ x = 1, 2, \ldots.$$

The expected value of X is given by

$$\begin{aligned} E(X) &= \sum_{1 \le x < \infty} (2^x - 1) \left(\frac{1}{2}\right)^x \\ &= \sum_{1 \le x < \infty} (1 - 2^{-x}) \\ &= \infty. \end{aligned}$$

This series diverges to infinity since its xth term, which equals $(1 - 2^{-x})$, converges to 1 and not to zero. In other words, *no finite amount of money is sufficient to sustain this betting system.* In the Baring's bank failure the losses eventually reached a billion dollars.

3.3.1 Moments of a Random Variable

The kth moment μ_k, $k = 1, 2, \ldots$ of a random variable X is defined by the equation

$$\mu_k = E(X^k), \text{ where } k = 1, 2, \ldots.$$

It is clear from the definition that the first moment $\mu_1 = \mu$, the expected value. The moments of a random variable give useful information on the shape and spread of the distribution function of X. They are also used to construct estimators for population parameters via the so-called *method of moments* (see Chapter 7). For now we focus our attention on how to compute them. To compute the moments we note that $\mu_k = E(\phi(X))$ where $\phi(x) = x^k$. The next theorem gives a formula for computing the expected value of a function of a random variable X.

Theorem 3.1 *Let X be a discrete random variable with pf $f_X(x)$ and let $Y = \phi(X)$. Then*

$$E(\phi(X)) = \sum_x \phi(x) f_X(x), \text{ provided that} \tag{3.12}$$

$$\sum_x |\phi(x)| f_X(x) < \infty. \tag{3.13}$$

We omit the proof. The significance of the theorem is this: To compute $E(Y)$ it is not necessary to recompute the pf $f_Y(y)$ of $Y = \phi(X)$ and then compute $E(Y)$ using Equation 3.7. In terms of the probability function the kth moment is given by

$$\mu_k = \sum_x x^k f(x). \tag{3.14}$$

Example 3.10 *Compute the first and second moments for each of the following random variables.*

1. Let X denote the value of a number picked at random from the set of the first N integers $R_X = \{1, 2, \ldots, N\}$. Show that

$$\mu_1 = \frac{N+1}{2} \text{ and } \mu_2 = \frac{(N+1)(2N+1)}{6}. \tag{3.15}$$

Solution The probability function is the equally likely measure on the set R_X. To see why note that since each number is equally likely to be chosen the probability function is given by

$$f_X(x) = 1/N, \ (x = 1, \ldots, N); \ f_X(x) = 0, \text{ elsewhere.} \tag{3.16}$$

The probability function 3.16 is called the *discrete uniform distribution*. The formulas for the moments displayed in 3.15 are the consequences of the following well-known formulas for the sums of the powers of the first N integers:

$$\sum_{1 \le x \le N} x = \frac{N(N+1)}{2}; \ \sum_{1 \le x \le N} x^2 = \frac{N(N+1)(2N+1)}{6}. \tag{3.17}$$

Thus,

$$\mu_1 = \sum_{1 \le x \le N} x f_X(x) = \frac{\sum_{1 \le x \le N} x}{N} = \frac{N(N+1)}{2N} = \frac{(N+1)}{2}.$$

Similarly,

$$\mu_2 = \sum_{1 \le x \le N} x^2 f_X(x) = \frac{\sum_{1 \le x \le N} x^2}{N} = \frac{N(N+1)(2N+1)}{6N} = \frac{(N+1)(2N+1)}{6}.$$

2. Show that the first and second moments of the geometric distribution (see Equation 3.5) are given by

$$\mu_1 = \frac{1}{p} \text{ and } \mu_2 = \frac{2}{p^2} - \frac{1}{p}. \tag{3.18}$$

Solution Using the recurrence relation $f(x+1) = qf(x), \ 1 \le x < \infty$ we previously derived the result that $\mu_1 = 1/p$. Multiplying both sides of the recurrence relation by x^2 and summing over x we obtain the sums

$$\sum_{1 \le x < \infty} x^2 f(x+1) = \sum_{1 \le x < \infty} q x^2 f(x) = q E(X^2) = q\mu_2.$$

On the other hand, the left hand side of the preceding equation equals

$$\sum_{1 \le x < \infty} x^2 f(x+1) = \sum_{1 \le x < \infty} (x-1)^2 f(x)$$

$$= \sum_{1 \le x < \infty} x^2 f(x) - 2 \sum_{1 \le x < \infty} x f(x) + \sum_{1 \le x < \infty} f(x)$$

$$= \mu_2 - 2\mu_1 + 1.$$

Solving the equation $\mu_2 - 2\mu_1 + 1 = q\mu_2$ for μ_2 and using the result derived earlier that $\mu_1 = 1/p$ yields

$$(1-q)\mu_2 = p\mu_2 = \frac{2}{p} - 1.$$

Consequently,

$$\mu_2 = \frac{2}{p^2} - \frac{1}{p}.$$

3. The random variable X has the pf

x	−2	−1	0	1	2
f(x)	0.1	0.2	0.3	0.3	0.1

Compute (a) $E(X^2)$ and (b) $E(2^X)$.

Solutions

(a) The random variable $X^2 = \phi(X)$ where $\phi(x) = x^2$. Consequently,

$$E(X^2) = (-2)^2 \times 0.1 + (-1)^2 \times 0.2 + 1^2 \times 0.3 + 2^2 \times 0.1$$
$$= 1.3.$$

(b) The random variable $2^X = \phi(X)$ where $\phi(x) = 2^x$. Consequently,

$$E(2^X) = 2^{-2} \times 0.1 + 2^{-1} \times 0.2 + 2^0 \times 0.3 + 2^1 \times 0.3 + 2^2 \times 0.1$$
$$= 1.425.$$

The following corollary of Theorem 3.1 is particularly useful.

Corollary 3.1 *Let X be a random variable with expected value $E(X)$ and a, b arbitrary constants. Then*

$$E(aX + b) = aE(X) + b. \tag{3.19}$$

Before giving the derivation of 3.19 we note the following special cases:

1. Set $b = 0$ in Equation 3.19; then

$$E(aX) = aE(X).$$

2. Set $a = 1$ and $b = -\mu_X$ in Equation 3.19; then

$$E(X - \mu_X) = E(X) - \mu_X = 0.$$

Derivation of 3.19:
We choose $\phi(x) = ax + b$ in Equation 3.12 and obtain

$$E(aX + b) = \sum_x (ax + b) f_X(x)$$
$$= \sum_x ax f_X(x) + \sum_x b f_X(x)$$
$$= a \sum_x x f_X(x) + b \sum_x f_X(x)$$
$$= aE(X) + b.$$

Example 3.11

Given that $E(X) = 1.5$ compute the following expected values:
(1) $E(2X + 4)$, (2) $E(-3X - 0.5)$, (3) $E(0.1X)$.
   Solution

1. $E(2X + 4) = 2 \times 1.5 + 4 = 7$

2. $E(-3X - 0.5) = -3 \times 1.5 - 0.5 = -5$

3. $E(0.1X) = 0.1 \times 1.5 = 0.15$

3.3.2   Variance of a Random Variable

We defined the population variance σ^2 of a finite population by the equation

$$\sigma^2 = \frac{\sum_{1 \le i \le N} (x_i - \mu)^2}{N},$$

where μ is the population mean (see Definition 1.8). The population variance is a weighted average of the squared deviations of the values x_j from the population mean μ. The weight assigned to the squared deviation $(x_j - \mu)^2$ is $1/N$. We now extend this concept to random variables.

The *variance* of a random variable X is denoted by $V(X)$ and is defined as

$$V(X) = E((X - \mu_X)^2) = \sum_{1 \le j < \infty} (x_j - \mu_X)^2 f_X(x_j). \tag{3.20}$$

The variance of X is a weighted average of the squared deviations of the values x_j from the expected value μ_X; the weight assigned to $(x_j - \mu_X)^2$ is $f_X(x_j)$. Note also that the variance $V(X) = E(\phi(X))$ where $\phi(x) = (x - \mu_X)^2$.

Notation: The variance is frequently denoted by σ_X^2; sometimes we drop the subscript X and write σ^2. The *standard deviation* of X is the square root of the variance:

$$\text{standard deviation } \sigma_X = \sqrt{\sigma_X^2}.$$

A Shortcut Formula for $V(X)$

Computing the variance via Equation 3.20 requires tedious computations. The following formula for $V(X)$, which is algebraically equivalent to Equation 3.20, is called the *shortcut formula* for the variance; this is because it requires fewer arithmetic operations to compute.

$$V(X) = E(X^2) - \mu_X^2$$
$$= \sum_{1 \le i < \infty} x_i^2 f_X(x_i) - \mu_X^2. \tag{3.21}$$

We can also express $V(X)$ in terms of the first two moments by noting that

$$V(X) = \mu_2 - \mu_1^2.$$

Derivation of 3.21: Noting that

$$(X - \mu_X)^2 = X^2 - 2\mu_X X + \mu_X^2,$$

and taking expected values of both sides we conclude that

$$E((X - \mu_X)^2) = E(X^2) + E(-2\mu_X X) + E(\mu_X^2).$$

Since μ_X is a constant we have

$$E(-2\mu_X X) = -2\mu_X E(X) = -2\mu_X^2 \text{ and } E(\mu_X^2) = \mu_X^2.$$

Therefore,

$$E((X - \mu_X)^2) = E(X^2) - 2\mu_X^2 + \mu_X^2 = E(X^2) - \mu_X^2.$$

Example 3.12

1. Compute $V(X)$ where X is an Bernoulli random variable.

   Solution Recall that X takes on the two values 0 and 1 with probabilities $P(X = 0) = 1 - p$, $P(X = 1) = p$, respectively. We claim that the variance equals

   $$V(X) = p(1 - p). \tag{3.22}$$

   To see this note that $X^2 = X$ (why?), consequently $E(X^2) = E(X) = p$ and therefore

   $$V(X) = E(X^2) - E(X)^2 = p - p^2 = p(1 - p).$$

2. Compute the variance of the random variable with probability function

   $$f(x) = \frac{x}{10}; \ x \in R_X = \{1, 2, 3, 4\}; \ f(x) = 0 \text{ elsewhere}.$$

   Solution Using the shortcut formula we first compute

   $$E(X) = 3 \text{ and } E(X^2) = 10.$$

   Consequently,

   $$V(X) = 10 - 3^2 = 1.$$

A formula for $V(aX + b)$

We now derive a formula for computing the variance of the random variable $Y = aX + b$, which is obtained from X by means of the linear transformation $\phi(x) = ax + b$. For instance, if X records the temperature of some object in degrees Celsius then $Y = (9/5)X + 32$ is its temperature in degrees Fahrenheit. Here $a = 9/5$ and $b = 32$. The following formula for $V(aX + b)$ will prove to be quite useful.

Let X be a random variable with variance $V(X)$ and let a, b denote arbitrary constants; then

$$V(aX + b) = a^2 V(X). \tag{3.23}$$

Before giving the derivation of 3.23 we give an example.

Example 3.13

Suppose $Y = (9/5)X + 32$ and $V(X) = 5$. Compute $V(Y)$.
   Solution Apply Equation 3.23 with $a = 9/5$. Then

$$V(X) = \left(\frac{9}{5}\right)^2 \times 5 = \frac{81}{5} = 16.2.$$

Derivation of Equation 3.23:

We begin with the observation that

$$\mu_{aX+b} = E(aX + b) = aE(X) + b = a\mu_X + b \text{ and}$$
$$V(aX + b) = E([aX + b - \mu_{aX+b}]^2). \text{ Therefore,}$$
$$V(aX + b) = E([aX + b - (a\mu_X + b)]^2)$$
$$= a^2 E([X - \mu_X]^2)$$
$$= a^2 V(X).$$

3.3.3   Chebyshev's Inequality

The variance of a random variable controls the spread of its distribution about the expected value; in particular, a small variance implies that large deviations from the expected value are improbable. The precise version of this statement is called Chebyshev's inequality. This inequality is remarkable in that no knowledge of the pf $f(x)$ is required; it is only necessary to know μ and σ^2.

Theorem 3.2 *Let X be a random variable with expected value μ and variance σ^2. Then*

$$P(|X - \mu| \geq d) \leq \frac{\sigma^2}{d^2}; \tag{3.24}$$

$$P(|X - \mu| \leq d) \geq 1 - \frac{\sigma^2}{d^2}. \tag{3.25}$$

Proof Equations 3.24 and 3.25 are equivalent in the sense that either one can be derived from the other. The details of the proof appear in Section 3.8 at the end of this chapter.

Example 3.14

A random variable has expected value $\mu = 150$ and standard deviation $\sigma = 10$. We use Chebyshev's inequality to obtain an upper bound on the following probabilities:

- $P(|X - 150| \geq 15) \leq (10/15)^2 = 0.44$

- $P(|X - 150| \geq 20) \leq (10/20)^2 = 0.25$

- $P(|X - 150| \geq 35) \leq (10/35)^2 = 0.08$

Sometimes it is more illuminating to measure deviations from the mean μ in terms of units of standard deviation. More precisely, set $d = k\sigma$ in 3.24; the resulting inequality, which is also called Chebyshev's inequality, states that

$$P(|X - \mu| \geq k\sigma) \leq \frac{1}{k^2}. \tag{3.26}$$

For instance, the probability that the random variable X deviates from its mean μ by more than 3 units of standard deviation is less than $1/9$.

3.4   The Hypergeometric Distribution

The *hypergeometric distribution* arises when one takes a random sample of size n, without replacement, from a population of size N divided into two classes consisting of D elements of the first kind and $N - D$ elements of the second kind. Such a population is called *dichotomous*.

Example 3.15

We give two examples of dichotomous populations.

1. Consider a lot of 100 steel bolts, of which 5 are defective. Here, the population is split into *defective* and *non-defective* items, with $N = 100$, $D = 5$.

2. Consider a population of 1,000 school children 900 of whom have been vaccinated against the measles; here $N = 1,000$ and $D = 900$. In this case, the dichotomy is *vaccinated* and *non-vaccinated*.

The experiment of drawing a random sample of size n from a dichotomous population is best explained in the context of an *urn model*.

Example 3.16 *An urn model for the hypergeometric distribution*

We represent the dichotomous population from which the sample is being drawn as an urn filled with D beads colored red and $N - D$ beads colored white. We denote the number of red beads in a sample of size n by X. Compute the probability function of X.

Solution We begin by noting that the number of red beads x in the sample satisfies the following inequalities:

$$\max(0, n - (N - D)) \leq x \leq \min(n, D). \tag{3.27}$$

To see this we note that the number of red beads in the sample cannot exceed the sample size n nor the total number of red beads D; consequently, $x \leq \min(n, D)$. Similarly, $x \geq 0$ and $n - x \leq N - D$, since this inequality only expresses the fact that the number of white beads in the sample cannot exceed the total number of white beads. Solving this inequality for x yields $x \geq n - (N - D)$ and therefore $\max(0, n - (N - D)) \leq x$. Combining these two inequalities yields 3.27. The formula for the probability function is given in Equation 3.28.

Theorem 3.3 *Suppose an urn is filled with D beads colored red and $N - D$ beads colored white. A random sample of size n, without replacement, is drawn from the urn and the number of red beads in the sample is denoted by X. Then the pf $h(x) = P(X = x)$ is given by*

$$h(x) = \frac{\binom{D}{x}\binom{N - D}{n - x}}{\binom{N}{n}}, \tag{3.28}$$

where $\max(0, n - (N - D)) \leq x \leq \min(n, D)$.

We define $h(x) = 0$, elsewhere.

The pf $h(x)$ defined by Equation 3.28 is called the *hypergeometric distribution* with parameters n, N, D.

Derivation of Equation 3.28: We know from Theorem 2.2 that there are $\binom{N}{n}$ ways of selecting a sample of size n from a lot of size N. To each of these outcomes we assign the probability $\binom{N}{n}^{-1}$. There are

$\binom{D}{x} \times \binom{N-D}{n-x}$ ways of choosing x red beads from the D red ones and $n-x$ beads from the remaining $N - D$ white ones. The ratio of these two quantities is the pf $h(x)$ defined in Equation 3.28.

Applications of the hypergeometric distribution to acceptance sampling

Acceptance sampling is a set of methods for accepting or rejecting a lot of N items on the basis of inspecting a sample of size n items selected at random from the lot. In detail: Acceptance sampling requires that you specify three numbers:

- N = the number of items from which the sample is to be taken;

- n = the number of items to be sampled;

- c = maximum allowable number of defective items in a sample of size n. It is called the *acceptance number.*

The significance of the acceptance number c is this: the lot is rejected if the sample contains more than c defective items.

Example 3.17

Steel bolts are shipped in lots of 100. Ten bolts are selected at random and inspected. If one or more of these bolts is found to be defective the entire lot is rejected; if none of them are defective the lot is accepted. Suppose there are five defective bolts in the lot; what is the probability that the lot will be accepted?

Solution The defective bolts correspond here to the red beads and the non-defective bolts correspond to the white beads. The number of defective bolts in the sample, denoted X, has a hypergeometric distribution with parameters $N = 100, n = 10, D = 5$. The lot will be accepted only if there are no defectives in the sample; this means the acceptance number c equals 0. Consequently, the probability of acceptance equals $P(X = 0)$. Therefore,

$$P(acceptance) = P(X = 0) = \frac{\binom{95}{10}}{\binom{100}{10}} = 0.5838.$$

Is this a good acceptance sampling plan? The answer is that it depends on the percentage of defective items that is acceptable. Suppose the maximum allowable percentage of defective items in the lot is specified to be 1%. Since the percentage of defective items in this case equals $5\% > 1\%$, it follows that this lot should be rejected with high probability. However, the preceding calculation shows that the probability of accepting this lot equals 0.5838, which is uncomfortably high. One way to reduce the probability of accepting a defective lot is to double the sample size to $n = 20$, say. In this case the acceptance probability equals:

$$P(acceptance) = P(X = 0) = \frac{\binom{95}{20}}{\binom{100}{20}} = 0.3198.$$

This probability is still too high; at this point one should consider looking for an alternate supplier.

Computing the probability of acceptance

An important criterion for evaluating the performance of a sampling plan is to compute the probability of accepting the lot. Suppose we have a sampling plan defined by the two numbers (n, c) and the true number of defective items in the lot is denoted by D; what is the probability of acceptance? Since the number of defective items X in a sample of size n taken from a lot of size N containing D defective items has the hypergeometric distribution $h(x)$ it follows that probability of acceptance is given by

$$P(acceptance) = P(X \le c) = \sum_{0 \le x \le c} h(x). \tag{3.29}$$

The Mean and Variance of the Hypergeometric Distribution
Proposition 3.2 gives the formulas for the mean and variance of the hypergeometric distribution.

Proposition 3.2 *The random variable X has the hypergeometric distribution with parameters n, N, D; then its mean and variance are given by*

$$E(X) = n \times \frac{D}{N}; \; V(X) = \frac{N-n}{N-1} \times n \times \frac{D}{N}\left(1 - \frac{D}{N}\right). \tag{3.30}$$

The proof of 3.30 is omitted because it involves tedious algebraic manipulations of the binomial coefficients. Readers interested in the proof, however, can consult W. Feller, op. cit, pp. 232-233. Intuitively, the quantity D/N is the proportion of red beads in the sample which equals the probability that a bead drawn at random is red. The quantity $n(D/N)$ represents the expected number of red beads in a sample of size n. The quantity

$$n \times \frac{D}{N}\left(1 - \frac{D}{N}\right)$$

represents the variance of X if the sampling were done with replacement. In this case X has a binomial distribution (see Example 3.21 in Section 3.5). The quantity $(N - n)/(N - 1)$ is called the *finite population correction factor*.

Example 3.18

Suppose X has a hypergeometric distribution with parameters $n = 10, N = 10,000, D = 500$. Compute $E(X)$ and $V(X)$.
   Solution This is a straightforward application of Equation 3.30.

$$E(X) = 10 \times \frac{500}{10,000} = 0.5;$$

$$V(X) = \frac{10,000 - 10}{10,000 - 1} \times 10 \times 0.05 \times (1 - 0.05)$$

$$= 0.4703.$$

A Recurrence Relation for the Hypergeometric Probabilities
   A recurrence relation for the binomial coefficients was derived in Section 2.3 of Chapter 2 (2.20), so it is not surprising that we can derive a recurrence relation for the hypergeometric probabilities. The recurrence relation is given by

$$h(x + 1) = \frac{(n - x)(D - x)}{(x + 1)(N - D - n + x + 1)} \times h(x) \tag{3.31}$$

where $\max(0, n - (N - D)) \le x \le \min(n, D)$.

Derivation of 3.31: The recurrence relation is a consequence of the following formula for the ratio $R_h(x) = h(x+1)/h(x)$:

$$R_h(x) = \frac{h(x+1)}{h(x)}$$

$$= \frac{(n-x)(D-x)}{(x+1)(N-D-n+x+1)}, \tag{3.32}$$

where $\max(0, n-(N-D)) \leq x \leq \min(n, D)$.

We omit the derivation of Equation 3.32, since it is just a straightforward algebraic manipulation of the ratio of two binomial coefficients. Note that

$$R_h(x) = 0 \text{ for } x \geq \min(n, D),$$

which means that the recurrence relation 3.31 is valid for all integers $x \geq \min(n, D)$.

Example 3.19

Compute the probability function for the number of hearts in a five-card poker hand using the recurrence relation 3.31.

Solution In this case we have

$$n = 5, \ N = 52, \ D = 13, \ x = 0, 1, 2, 3, 4, 5; \text{ consequently,}$$

$$h(0) = \frac{\binom{13}{0}\binom{39}{5}}{\binom{52}{5}} = 0.2215, \text{ and}$$

$$R_h(x) = \frac{(5-x)(13-x)}{(x+1)(35+x)}, x = 0, 1, \ldots.$$

Therefore,

$$h(1) = \frac{5 \times 13}{35} \times 0.2215 = 0.4114,$$

$$h(2) = \frac{4 \times 12}{2 \times 36} \times 0.4114 = 0.2743,$$

$$h(3) = \frac{3 \times 11}{3 \times 37} \times 0.2743 = 0.0815,$$

$$h(4) = \frac{2 \times 10}{4 \times 38} \times 0.0815 = 0.0107,$$

$$h(5) = \frac{1 \times 9}{5 \times 39} \times 0.0107 = 0.0005.$$

3.5   The Binomial Distribution

A surprising variety of statistical experiments can be reduced to the analysis of the mathematical model of tossing a coin n times with probability p of tossing a head. The essential features of this model are that the results of the successive coin tosses are mutually independent and the probability of getting a head remains constant. It is the simplest example of n independent repetitions of an experiment under identical conditions. Our main interest is in obtaining the probability function for the number of heads in n tosses of a coin.

Proposition 3.3 *Let X denote the number of heads in n tosses of a coin with probability p for heads. Then the probability function of X is given by*

$$P(X = x) = \binom{n}{x} p^x (1-p)^{n-x}; \; x = 0, 1, \ldots, n. \tag{3.33}$$

Proof The sample space consists of 2^n sample points that can be represented as a sequence of H's and T's like so: $s = (T, H, H, T, \ldots, H, T)$. To each sample point s that consists of x $H's$ and of $n - x$ $T's$ we assign the probability $p^x(1-p)^{n-x}$. To justify this assignment of probabilities we reason as follows. We assume that: (i) the probability of tossing a head is the same for each toss and (ii) the outcomes of each of the coin tosses are mutually independent. Therefore, the probability of getting x $H's$ and of $n - x$ $T's$ equals:

$$\begin{aligned} P(T, H, \ldots, H, T) &= P(T)P(H) \cdots P(H)P(T) \\ &= (1-p)p \cdots p(1-p) \\ &= p^x(1-p)^{n-x}. \end{aligned}$$

The event $\{X = x\}$ contains all sample points s that consist of x $H's$ and $n - x$ $T's$, each one of which is assigned the same probability $p^x(1-p)^{n-x}$. In addition, there are $\binom{n}{x}$ such sample points, since each one of them corresponds to a labeling of the n trials using x H's and $n - x$ T's. Summing the probabilities of the sample points in the event $\{X = x\}$ yields Equation 3.33.

The probability function defined by Equation 3.33 is called the *binomial distribution;* it is one of the three most important discrete distributions; the other two being the hypergeometric and the Poisson distribution. To see why it is called the binomial distribution apply the binomial expansion (Equation 2.3) to the expression $(p + (1 - p))^n$, where $0 \le p \le 1$. We obtain the following expansion:

$$1 = 1^n = (p + (1-p))^n = \sum_{0 \le x \le n} \binom{n}{x} p^x (1-p)^{n-x}. \tag{3.34}$$

This shows that

$$f_X(x) = \binom{n}{x} p^x (1-p)^{n-x}; \; x = 0, 1, \ldots, n$$

defines a pf in the sense of Equation 3.3. Notice that the term that appears on the right hand side of Equation 3.33 also appears as the x th term (counting from 0) of the binomial expansion; which is the reason why the pf defined by Equation 3.33 is called the binomial distribution. The individual terms, denoted by $b(x; n, p)$ and defined by

$$b(x; n, p) = \binom{n}{x} p^x (1-p)^{n-x}; \; x = 0, 1, \ldots, n \tag{3.35}$$

are called *binomial probabilities.*

A Recurrence Relation for the Binomial Probabilities

We now derive a recurrence relation for the binomial probabilities that is useful for both computational and theoretical purposes:

$$b(x+1; n, p) = b(x; n, p) \times \frac{(n-x)}{(x+1)} \frac{p}{1-p};$$ (3.36)

$$b(0; n, p) = (1-p)^n.$$ (3.37)

The recurrence relation is a consequence of the following formula for the ratio $R_b(x) = b(x+1; n, p)/b(x; n, p)$:

$$R_b(x) = \frac{b(x+1; n, p)}{b(x; n, p)} = \frac{(n-x)}{(x+1)} \frac{p}{1-p} \text{ for } 0 \le x \le n.$$ (3.38)

We denote the distribution function of the binomially distributed random variable X by $B(x; n, p)$, and a formula for $B(x; n, p)$ is given in Equation 3.39 below.

$$B(x; n, p) = \sum_{0 \le j \le x} b(j; n, p).$$ (3.39)

Table A.1 in Appendix A lists the values of $B(x; n, p)$ for various values of the parameters n, p. The value of a binomial probabilitiy can be computed from the formula

$$b(x; n, p) = B(x; n, p) - B(x-1; n, p).$$ (3.40)

How to use the binomial tables

Example 3.20

Use Table A.1 in Appendix A to compute (1) $B(6; 10, 0.4)$ and (2) $b(6; 10, 0.4)$.
   Solution Table A.1 in Appendix A gives the values of $B(x; n, p)$ for $n = 5, 10, 15, 20, 25$ and $p = 0.05, 0.1, 0.15, 0.2, \ldots, 0.45, 0.50$.

1. To find $B(6; 10, 0.4)$ first look for the number 10 in the column headed by n and then look for the number 6 in the column headed by x. We then proceed to the right until we find the entry 0.945 in the column headed by 0.4. Thus $B(6; 10, 0.4) = 0.945$.

2. To compute $b(6; 10; 0.4)$ we use the formula

   $$b(6; 10; 0.4) = B(6; 10, 0.4) - B(5; 10, 0.4) = 0.945 - 0.834 = 0.111.$$

   To compute $B(x; n, p)$, $0.5 < p < 1$ we use the formula

$$B(x; n, p) = 1 - B(n - 1 - x; n, 1 - p).$$ (3.41)

The effect of the parameter p on the shape of the binomial distribution is understood by looking at their probability histograms shown in Figures 3.3, 3.4, and 3.5.
   The Mean and Variance of the Binomial Distribution Proposition 3.4 gives the formulas for the mean and variance of the binomial distribution. These formulas are derived in Example 3.27 using moment generating function techniques, a topic of independent interest (see Section 3.7). They can also be derived as corollaries of the addition formula for the mean and variance of sums of independent random variables (see Section 5.4).

Proposition 3.4 *Let X have a binomial distribution with parameters n, p; then*

$$E(X) = np;$$ (3.42)

$$V(X) = np(1-p).$$ (3.43)

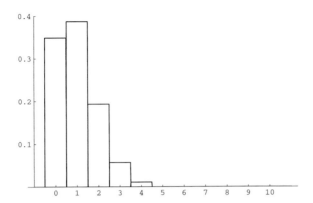

FIGURE 3.3: Probability histogram of binomial distribution ($n = 10$, $p = 0.1$)

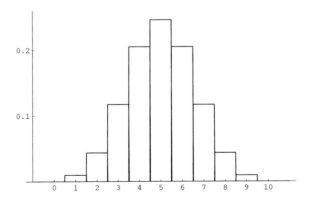

FIGURE 3.4: Probability histogram of binomial distribution ($n = 10$, $p = 0.5$)

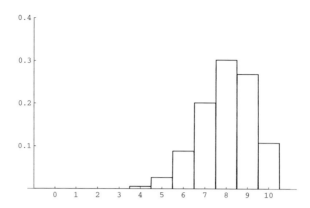

FIGURE 3.5:   Probability histogram of binomial distribution $(n = 10, p = 0.8)$

Example 3.21

Consider, once again, the urn model of the previous section: An urn contains D beads colored red and $N - D$ beads colored white. But this time, we take a random sample of size n, *with replacement*, from the urn and let X denote the number of red beads in the sample. Show that X has a binomial distribution with parameters n, $p = D/N$.

Solution The sample space consists of 2^n sequences of n letters, each letter being an R or a W. A sample point is simply a sequence of $R's$ and $W's$ denoted by $WWRW\ldots R$. Since we are sampling with replacement the number of red and white balls remains constant at each draw, and therefore the probability of drawing a red bead at each trial equals $p = D/N$; and, similarly, the probability of drawing a white bead equals $1 - p = (N - D)/N$. To each sample point we assign the probability $p^x(1 - p)^{n-x}$ where x and $n - x$ equal the number of $R's$ and $W's$ that appear in the sample. Reasoning just as we did in Proposition 3.3 we see that X has the binomial distribution with parameters n, p (see Equation 3.33).

The Binomial Approximation to the Hypergeometric Distribution

Suppose we have an urn filled with $10,000$ beads, 500 of which are red and the rest are white. If we take a sample (without replacement) of size 10 from the urn, then the number of red beads in the sample is a random variable X with hypergeometric distribution with parameters $n = 10$, $N = 10,000$, $D = 500$. Notice that each time a bead is drawn the proportion p' of red beads that remains satisfies the inequality $490/9990 = 0.049 \le p' \le 500/9990 = 0.0501$. The lower bound follows from the fact that the maximum number of red beads that can be drawn is 10, and the upper bound follows from the observation that the minimum number of red beads that can be drawn is 0. Consequently, the proportion of red beads that remain in the urn is close to $0.05 = 500/10,000 = D/N = p$. If we now take a sample of size 10 (with replacement) then the number of red beads in the sample has a binomial distribution with parameters $n = 10$ and $p = D/N = 500/10,000 = 0.05$. This suggests that there is very little difference in this case between sampling without replacement and sampling with replacement. More precisely, we have the following theorem, called the *binomial approximation to the hypergeometric*:

Theorem 3.4 *The random variable X has the hypergeometric distribution $h(x)$. Suppose N and D become large while their ratio remains fixed with $D/N = p$. Then*

$$\lim_{D,N\to\infty,\,D/N=p} h(x) = \binom{n}{x} p^x (1-p)^{n-x}; \; x = 0, 1, \ldots, n. \tag{3.44}$$

Proof We defer the proof to Section 3.8 at the end of this chapter.

A rule of thumb for approximating the hypergeometric distribution

It has been found that the binomial approximation is quite accurate when the sample size is less than 5% of the population size, that is whenever the ratio $n/N < 0.05$.

Example 3.22

Suppose $n = 10$, $N = 10,000$, $D = 500$; so $p = 500/10,000 = 0.05$. Compute the binomial approximations to: (1) $h(0)$; (2) $h(1)$; and compare them to the exact values.

Solution

1. The exact value of $h(0) = 0.5986$. The binomial approximation of $h(0)$ is $b(0; 10, 0.05)$, so

$$h(0) \approx (1 - 0.05)^{10} = 0.5987.$$

2. The exact value of $h(1) = 0.3153$. The binomial approximation of $h(1)$ is $b(1; 10, 0.05)$, so

$$h(1) \approx 10 \times 0.05 \times (1 - 0.05)^9 = 0.3151.$$

The exact hypergeometric probabilities and their binomial approximations are displayed in Table 3.6.

Example 3.23

Suppose a sampling plan uses a sample of size $n = 10$ taken from a lot of size $N = 10,000$, acceptance number $c = 1$, and $D = 500$. Compute the probability of acceptance using: (1) the hypergeometric distribution and (2) the binomial approximation to the hypergeometric distribution.

Solution

1. Using the hypergeometric distribution the probability of acceptance is given by

$$P(acceptance) = h(0) + h(1)$$
$$= 0.5986 + 0.3153 = 0.9139.$$

2. Using the binomial approximation to the hypergeometric distribution the probability of acceptance is given by

$$P(acceptance) \approx b(0; 10; 0.05) + b(1; 10; 0.05)$$
$$= 0.5987 + 0.3151 = 0.9138.$$

Table 3.6 *The binomial approximation to the hypergeometric distribution with $n = 10$, $N = 10000$, $D = 500$, $p = \frac{500}{10000} = 0.05$.*

x	b(x;10;0.05)	h(x)
0	0.5987	0.5986
1	0.3151	0.3153
2	0.0746	0.0746
3	0.0105	0.0104
4	0.0009	0.0007
5	0.0001	0.0000

3.5.1   A Coin Tossing Model for Stock Market Returns

Are Stock Market Returns Random? Some Empirical Evidence and a Theoretical Model

In Example 1.14 we showed that an investment in the S&P500 yielded a 15.23% annual return (excluding dividends). Looking again at Figure 1.1 we see that the path toward a 15.23% annual return on this investment was far from being a smooth one; indeed, short-term fluctuations in the price index appear to be totally unpredictable, a possibility that induces some discomfort among Wall Street analysts, who are paid enormous salaries to predict what the empirical data, and mathematical models of them, say is unpredictable. "Market analysts make forecasts," an eminent economist once observed, "not because they know, but because they are asked."[3] To see who is right, consider the weekly returns of the S&P500 for the period (Jan. 4, 1999–Dec. 27, 1999) listed in Table 1.19. Figure 3.6 is the time series plot of the returns, which is a visual display of their fluctuations. The reader will agree that it is difficult to discern any pattern in these fluctuations that an investor can profitably exploit. Indeed, the data suggest that the chance of earning a positive or negative return for each week does not depend on any previous pattern of returns. Moreover, Figure 3.6 suggests that the fluctuations of the S&P returns are truly random; that is, they are no more predictable than the results of flipping a coin. Are the empirical data consistent with the coin tossing model? Consider the Wall Street adage "The trend is your friend," suggesting that there is a tendency for periods with positive returns to be followed by positive ones and, similarly, negative returns tend to be followed by negative ones. If this were true, then the scatter plot of $\{(r_i, r_{i+1}), i = 1, 2, \ldots, n\}$ would contain more points in the northeast and southwest quadrants than in the northwest and southeast quadrants; that is, most of the points of the scatter plot would be of the form $(+, +)$, $(-, -)$. Consequently the correlation coefficient between the sequences $x = (r_1, \ldots, r_{n-1})$ and $y = (r_2, \ldots, r_n)$ would be positive. Looking at Figure 3.7 we see no such pattern: Indeed, the successive returns are weakly negatively correlated, with $r = -0.2560$. The economists' explanation of the unpredictability of stock market returns, which is called the *efficient market hypothesis*, is discussed in more detail in Section 5.5.

Of all the models proposed for the unpredictability of future returns of a stock the *binomial lattice model* is the simplest to understand, easiest to compute with, and yet it contains all the essential features of more sophisticated models, but free of all unnecessary technical details. Intuitively, the model asserts that next week's price is obtained by flipping a coin: If a head appears then the current price is increased by 2%, say, and if a tail appears then the current price is decreased by 1%. We will discuss this model in more detail in Section 4.6; in the meantime we present a simplified version containing all the essential ideas.

Example 3.24 *A coin tossing model for stock returns*

---

[3]John Kenneth Galbraith.

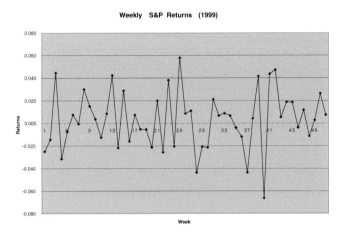

FIGURE 3.6: S&P returns (1999)

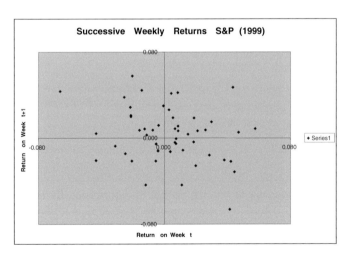

FIGURE 3.7: Each point represents the S&P returns on two successive weeks. The correlation between them is weakly negative, with $r = -0.2560$

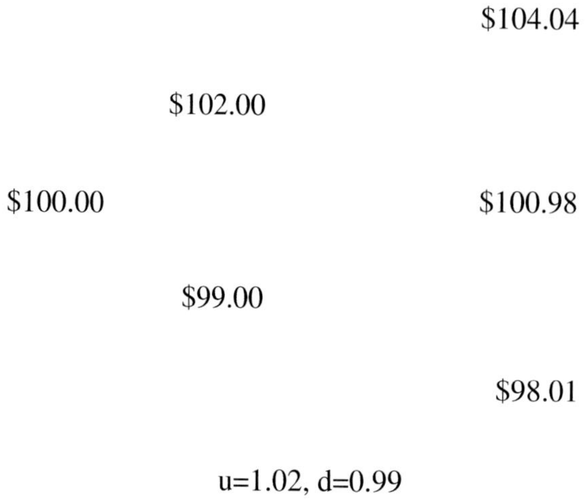

$104.04

$102.00

$100.00 $100.98

$99.00

$98.01

u=1.02, d=0.99

FIGURE 3.8:   Binomial lattice model for stock prices

Let S_i denote the stock price at the beginning of the ith time period $i = 0, 1, \ldots, i$. At the end of the period a coin, possibly biased, is tossed; if it comes up heads (H) then stock price moves up 2%; if it comes up tails (T) the price moves down 1%. Assume $P(heads) = 0.51$ and $P(tails) = 0.49$. Consequently,

$$P(S_i = 1.02 \times S_{i-1}) = 0.51 \text{ and } P(S_i = 0.99 \times S_{i-1}) = 0.49.$$

At the end of the second period, the coin is tossed again, and so on. Suppose the initial stock price $S_0 = \$100$. After one period the stock price is either $102 or $99; after two periods the stock price is either $\{98.01, 100.98, 104.04\}$ computed as follows:

$$98.01 = 100 \times (0.99)^2 \ (\text{TT})$$
$$100.98 = 100 \times (1.02 \times 0.99) \ (\text{HT or TH})$$
$$104.04 = 100 \times (1.02)^2 \ (\text{HH})$$

depending on whether no heads (TT), one head $\{(\text{HT}) \text{ or } (\text{TH})\}$, or two heads (HH) were thrown. Their probabilities are computed using the binomial distribution with parameters $n = 2$, $p = 0.51$; the detailed calculations follow:

$$P(S_2 = 98.01) = 0.2401; \ P(S_2 = 100.98) = 0.4998; \ P(S_2 = 104.04) = 0.2601.$$

Figure 3.8 represents the possible outcomes as a tree diagram. According to this model, then, the price at the beginning of the ith period is determined by the flip of a coin, with $S_i = 1.02 S_{i-1}$ if the coin comes up heads, and $S_i = 0.99 S_{i-1}$ if the coin comes up tails. Thus, the chance of earning a positive or negative return for each period does not depend on any previous pattern of heads or tails. This simple coin tossing model for stock market returns is easily generalized to the binomial lattice model or CRR (Cox-Ross-Rubinstein) model as will now be explained.

Example 3.25 *Binomial lattice model for stock prices*

The binomial lattice model assumes that if the stock price at the beginning of the ith period is S_i, then the stock price after one period assumes one of two possible values determined by the toss of a coin, with $P(heads) = p$ and $P(tails) = 1-p$. More precisely, $S_i = uS_{i-1}$, $u > 1$ (u for up) if the coin comes up heads, and $S_i = dS_{i-1}$, $d < 1$ (d for down) if the coin comes up tails; that is,

$$P\left(\frac{S_i}{S_{i-1}} = u\right) = p \text{ and } P\left(\frac{S_i}{S_{i-1}} = d\right) = 1 - p, \, d < 1 < u, \, (i = 1, 2, \ldots, n). \quad (3.45)$$

In other words, the successive values of the ratios $S_1/S_0, S_2/S_1, \ldots, S_n/S_{n-1}$ are determined by the sequence of heads and tails produced by tossing a coin. Using the algebraic identity

$$\frac{S_n}{S_0} = \frac{S_1}{S_0} \times \frac{S_2}{S_1} \cdots \times \frac{S_n}{S_{n-1}} \quad (3.46)$$

we see that

$$\frac{S_n}{S_0} = u^X d^{n-X} = d^n \left(\frac{u}{d}\right)^X, \quad (3.47)$$

where X and $n - X$ denote the number of up moves (which equals the number of heads), and the number of down moves, respectively. Therefore, X has a binomial distribution with parameters n, p; consequently, the probability distribution of the stock price after n periods is given by

$$P\left(S_n = S_0 u^x d^{n-x}\right) = \binom{n}{x} p^x (1 - p)^{n-x}; \, x = 0, 1, \ldots, n, \, S_0 = \text{initial price}. \quad (3.48)$$

Using moment generating function techniques described in Section 3.7 (cf. Theorem 3.7) one obtains the following equations for the expected price, expected return, and variance of a stock governed by the binomial lattice model. The mathematical details are given in Section 3.8.

$$E(S_n) = S_0(d(1 - p) + pu)^n \quad (3.49)$$
$$E(\text{return}) = (d(1 - p) + pu)^n - 1 \quad (3.50)$$
$$V(S_n) = S_0^2 \left[\left(d^2(1 - p) + pu^2\right)^n - (d(1 - p) + pu)^{2n}\right] \quad (3.51)$$

The mathematical analysis of returns is simplified by introducing the *logarithmic return*, denoted $\ln(S_i/S_{i-1})$, from which the return $(S_i - S_{i-1})/S_{i-1}$ itself is computed via the equation

$$\frac{S_i - S_{i-1}}{S_{i-1}} = \exp\left(\ln(S_i/S_{i-1})\right) - 1. \quad (3.52)$$

Taking logarithms of both sides of Equation 3.46, we see that the logarithmic return is *additive*, in the sense that

$$\ln\left(\frac{S_n}{S_0}\right) = \ln\left(\frac{S_1}{S_0}\right) + \ln\left(\frac{S_2}{S_1}\right) + \cdots + \ln\left(\frac{S_n}{S_{n-1}}\right). \quad (3.53)$$

It is this additive property of the logarithmic return that makes it more convenient to work with. Looking at Equation 3.45 we see that $\ln(S_i/S_{i-1})$ is a random variable whose distribution does not depend on the index i. Its mean μ and standard deviation σ—which represent the expected return and risk, respectively, of the stock—are of far more importance to investors than the parameters p, u, d. Given the values of p, u, d it is fairly straightforward to compute μ, σ. In practice, however, one is given the values μ and σ^2—usually obtained from empirical data—and from which one must compute the values p, u, d.

This leads to two equations in three unknowns, so there is no unique solution. It is not too difficult to show that if one chooses

$$p = 1/2, \; u = \exp(\mu + \sigma), \; d = \exp(\mu - \sigma) \tag{3.54}$$

in Equation 3.45, then $\ln(S_i/S_{i-1})$ has mean mean μ and standard deviation σ, that is

$$\mu = E\left(\ln\left(\frac{S_i}{S_{i-1}}\right)\right), \qquad \sigma^2 = V\left(\ln\left(\frac{S_i}{S_{i-1}}\right)\right). \tag{3.55}$$

Problem 3.60 asks the reader to supply the details.
Inserting these same values into Equation 3.47 yields the useful representation

$$\frac{S_n}{S_0} = \exp\left(2\sigma\left(X - \frac{n}{2}\right) + n\mu\right). \tag{3.56}$$

Taking logarithms of both sides of Equation 3.56 and using the fact that X is binomially distributed with parameters n, $p = 1/2$ (so $E(X) = n/2$ and $V(X) = n/4$) we find that

$$E\left(\ln\left(\frac{S_n}{S_0}\right)\right) = E\left(2\sigma\left(X - \frac{n}{2}\right) + n\mu\right) = n\mu, \tag{3.57}$$

$$V\left(\ln\left(\frac{S_n}{S_0}\right)\right) = V\left(2\sigma\left(X - \frac{n}{2}\right) + n\mu\right) = n\sigma^2. \tag{3.58}$$

These results are intuitively clear: the expected logarithmic return for n periods is just the sum of the expected logarithmic returns for each period.

3.6   The Poisson Distribution

Table I in Appendix A lists the values of $B(x; n, p)$ for selected values of n and p in the ranges $n = 2, 3, \ldots, 20$ and $p = 0.1, 0.2, \ldots 0.9$. There are many instances, however, where one must compute $B(x; n, p)$ for values of $n \gg 20$ and $p \ll 0.1$. (The symbol $n \gg 20$ means n is very much larger than 20; the symbol $p \ll 0.1$ means that p is very much smaller than 0.1.) As an example we consider the problem of determining whether or not the occurrence of 8 cases of Leukemia among 7076 children less than 15 years of age in a small city is so unexpectedly large as to warrant further study by public health officials. (The following account has been adapted from Larsen and Marx (1986), *An Introduction to Mathematical Statistics and its Applications*, Prentice-Hall, Englewood Cliffs, NJ). An examination of public health records in the neighboring towns showed that in the previous 5 years there were 286 diagnosed cases of leukemia out of 1,152,695 children. From this we estimate the probability that a child less than 15 years old will be stricken with leukemia to be $p = 286/1,152,695 = 0.000248$, a prevalence rate of 248 per 100,000. With this information we can now compute the probability of the occurrence of 8 or more cases of leukemia in a population of 7076 children. In a population of size n the number of occurrences of leukemia is a binomially distributed random variable X with parameters $n = 7076$, $p = 0.000248$. Now $P(X \geq 8) = 1 - P(X \leq 7)$, consequently,

$$P(X \geq 8) = 1 - \sum_{0 \leq j \leq 7} \binom{7076}{x} (0.000248)^x (0.999752)^{7076-x} \tag{3.59}$$

Although the summands occurring on the right hand side of the preceding equation are not to be found in any table, and are not easily computable, it is nevertheless possible to *approximate* the probabilities

$$\binom{7076}{x} (0.000248)^x (0.999752)^{7076-x}$$

by means of a technique due to the French mathematician S. Poisson, called the *Poisson approximation to the binomial distribution*. The Poisson approximation may be used whenever n is very large, p is very small and $np = E(X)$ is *moderate*. In the case $n = 7076$, $p = 0.000248$ we have $7076 \times 0.000248 = 1.75$. In other words, in the population of 7076 children less than 15 years old we would expect to see 1.75 cases of leukemia; instead, there were 8. Intuitively, this would seem to be a very rare event and we would therefore like to compute its probability.

Theorem 3.5 The Poisson approximation to the binomial. *Assume* $np_n = \lambda > 0$ *for* $0 < n < \infty$; *then*

$$\lim_{n \to \infty} b(x; n, p_n) = p(x; \lambda) = e^{-\lambda} \frac{\lambda^x}{x!}; \ x = 0, 1, \ldots, \lambda > 0. \tag{3.60}$$

We defer the proof of the Poisson approximation to the binomial to Section 3.8. In the meantime we give some applications.

Example 3.26

1. Use the Poisson approximation to estimate the probability of eight or more leukemia cases in a population of size $n = 7076$ when the probability of getting leukemia is $p = 0.000248$.

   Solution In this case $np = 7076 \times 0.000248 = 1.75 = \lambda$. Therefore,

   $$P(X \leq 7) \approx \sum_{0 \leq x \leq 7} e^{-1.75} \frac{1.75^x}{x!} = .999518; \text{ consequently}$$

   $$P(X \geq 8) \approx 1 - .999518 = 0.000482.$$

   What do we conclude from the data? Is it a statistical fluke, or are there environmental factors responsible for this unusual cluster of leukemia cases? Clearly, there is a cause for concern and further scientific studies will be necessary to resolve the mystery.

2. Suppose 1.5% percent of the ball bearings made by a machine are defective and the ball bearings are packed 200 to a box. Compute the pf of the number of defective ball bearings in the box and compare it to the Poisson approximation.

   Solution Denoting the number of defective ball bearings by X we see that X has a binomial distribution with parameters $n = 200$, $p = 0.015$, i.e.,

   $$P(X = x) = b(x; 200; 0.015).$$

   The Poisson approximation, with $\lambda = 200 \times 0.015 = 3$, yields

   $$P(X = x) \approx \frac{e^{-3} 3^x}{x!}, \ x = 0, 1, \ldots, 100.$$

   The following table gives an idea of the accuracy of the approximation.

x	$b(x; 200; 0.015)$	$p(x; 3)$
0	0.0487	0.0498
1	0.1482	0.1494
2	0.2246	0.2240
3	0.2257	0.2240

The Poisson Distribution

We now show that the quantities $p(x; \lambda) = e^{-\lambda}\lambda^x/x!$, $(x = 0, 1, \ldots,)$ define a probability function, a result that is not too surprising since $p(x; \lambda)$ is itself the limit of a sequence of probability functions.

Definition 3.7 *The random variable X is said to have a Poisson distribution with parameter $\lambda > 0$ if*

$$P(X = x) = e^{-\lambda}\frac{\lambda^x}{x!}; x = 0, 1, \ldots, . \tag{3.61}$$

That $p(x; \lambda)$ defines a pf follows from the calculation

$$\sum_{0 \le x < \infty} p(x; \lambda) = \sum_{0 \le x < \infty} e^{-\lambda}\frac{\lambda^x}{x!} = e^{-\lambda} \sum_{0 \le x < \infty} \left(\frac{\lambda^x}{x!}\right) = e^{-\lambda}e^{\lambda} = 1.$$

A Recurrence Relation for the Poisson Probabilities

We now derive a recurrence formula for the Poisson probabilities that has useful computational and theoretical applications. Let $R_p(x) = p(x + 1; \lambda)/p(x; \lambda)$, $x = 0, 1, \ldots$, then

$$R_p(x) = \frac{p(x + 1; \lambda)}{p(x; \lambda)} = \frac{\lambda}{x + 1}, x = 0, 1, \ldots. \tag{3.62}$$

We leave the derivation of 3.62 to the reader as Problem 3.65. Multiplying both sides of the Equation 3.62 by $(x + 1)$ we obtain the following recurrence relation for the Poisson probabilities:

$$(x + 1)p(x + 1; \lambda) = \lambda p(x; \lambda) \, (x = 0, 1, \ldots). \tag{3.63}$$

The Expected Value and Variance of the Poisson Distribution

The expected value and variance of the random variable X with Poisson distribution 3.60 are given by:

$$E(X) = \lambda; V(X) = \lambda. \tag{3.64}$$

The derivation of these formulas is a straight forward application of moment generating function techniques as discussed in Section 3.7. Problem 3.76 asks you to supply the details.

3.7 Moment Generating Function: Discrete Random Variables*

None of the distribution functions in this chapter are continuous, so they are not differentiable, which means the usual rules of the differential calculus cannot be applied to them. The *moment generating function* (mgf) is a useful way of representing a distribution function (in much the same way we use analytic geometry to represent points and curves in space) so that calculus methods can be applied. It is also, as its name indicates, a powerful tool for computing the moments of many important distributions, including the binomial and Poisson.

Definition 3.8 *The moment generating function (mgf) for a random variable X with probability function $f(x)$ is the function*

$$M_X(s) = E(e^{sX}) = \sum_{0 \leq j < \infty} e^{sx_j} f(x_j) \tag{3.65}$$

provided $E(e^{sX}) < \infty$ for $\{s : -\delta < s < \delta\}$, $\delta > 0$.

What makes the mgf particularly useful is the fact that when the mgf exists, it uniquely determines the distribution function (cf. Theorem 3.6). In particular, it is easy to see that if X and Y have the same probability function then their mgfs are equal. The converse, that $M_X(s) = M_Y(s)$ implies that $F_X(t) = F_Y(t)$ is also true, but much more difficult to prove; so, it is omitted.

Theorem 3.6 *Suppose $M_X(s) = M_Y(s)$, $\{s : -\delta < s < \delta\}$. Then $F_X(t) = F_Y(t)$. Conversely, if $F_X(t) = F_Y(t)$, and their mgfs exist, then $M_X(s) = M_Y(s)$.*

To illustrate, we compute the mgf for the binomial distribution.

Theorem 3.7 *Let X be a binomial random variable with, parameters n, p. Then*

$$M_X(s) = (pe^s + (1 - p))^n. \tag{3.66}$$

The derivation of Equation 3.66 is a straight forward application of Theorem 3.1 where $\phi(x) = e^{sx}$, and s is a fixed but unspecified real number.

$$M_X(s) = E(e^{sX}) = \sum_{0 \leq j \leq n} e^{sj} \binom{n}{j} p^j (1 - p)^{n-j}$$

$$= \sum_{0 \leq j \leq n} \binom{n}{j} (pe^s)^j (1 - p)^{n-j} = (pe^s + (1 - p))^n$$

Suppose, for example, that the mgf of a random variable X is $(0.4e^s + 0.6)^9$. It follows from Equation 3.66 that the distribution of X is binomial with parameters $n = 9$, $p = 0.4$.

Differentiating both sides of Equation 3.65 with respect to s, and interchanging the order of differentiation and summation on the right hand side, we see that

$$M'_X(s) = (\sum_{0 \leq j < \infty} e^{sx_j} f(x_j))' = \sum_{0 \leq j < \infty} (e^{sx_j})' f(x_j)$$

$$= \sum_{0 \leq j < \infty} x_j e^{sx_j} f(x_j) = E(Xe^{sX}).$$

Consequently,

$$M'_X(s) = E(Xe^{sX}); \text{ therefore } M'_X(0) = E(X). \tag{3.67}$$

Using similar reasoning one obtains formulas for the second and higher order derivatives and moments:

$$M_X''(s) = (\sum_{0 \le j < \infty} e^{sx_j} f(x_j))'' = \sum_{0 \le j < \infty} (e^{sx_j})'' f(x_j)$$

$$= \sum_{0 \le j < \infty} x_j^2 e^{sx_j} f(x_j) = E\left(X^2 e^{sX}\right)$$

and, more generally,

$$M_X^{(k)}(s) = E\left(X^k e^{sX}\right); \text{therefore } M_X^{(k)}(0) = E\left(X^k\right). \tag{3.68}$$

Here $M_X^{(k)}(s)$ denotes the kth derivative.

Example 3.27 *Computing the first two moments and variance of the binomial distribution*

In this case $M_X(s) = (pe^s + (1 - p))^n$; therefore

$$M_X'(s) = [(pe^s + (1 - p))^n]' = npe^s(pe^s + (1 - p))^{n-1}$$
$$E(X) = M_X'(0) = np$$
$$M_X''(s) = npe^s(pe^s + (1 - p))^{n-1} + n(n - 1)(pe^s)^2(pe^s + (1 - p))^{n-2}$$
$$E(X^2) = M_X''(0) = np + n(n - 1)p^2 = np(1 - p) + (np)^2$$
$$Var(X) = E(X^2) - (E(X))^2 = np(1 - p).$$

The mgfs for the Poisson and geometric distributions are:

$$M_X(s) = \exp(\lambda(e^s - 1)) \text{ (Poisson mgf)}. \tag{3.69}$$
$$M_X(s) = pe^s(1 - qe^s)^{-1} \text{ (geometric mgf)}. \tag{3.70}$$

Problems 3.76 and 3.77 ask you to derive Equations 3.69 and 3.70.

3.8   Mathematical Details and Derivations

Derivation of binomial approximation to the hypergeometric distribution (Theorem 3.4): Our proof uses the fact that the ratio $R_h(x)$ of the hypergeometric probabilities (Equation 3.32) converges to the the corresponding ratio $R_b(x)$ (Equation 3.38) for the binomial probabilities. In detail:

$$\lim_{D,N \to \infty, D/N=p} R_h(x) = \lim_{D,N \to \infty, D/N=p} \frac{(n - x)(D - x)}{(x + 1)(N - D - n + x + 1)}$$

$$= \frac{(n - x)}{(x + 1)} \frac{p}{1 - p} = R_b(x).$$

Additional details: The last step in the preceding sequence of calculations is justified by noting that

$$\lim_{D,N \to \infty, D/N=p} \frac{(D - x)}{(N - D - n + x + 1)} = \frac{p}{1 - p},$$

which follows by dividing numerator and denominator by N, using the fact that $p = D/N$, and noting that both x/N and n/N go to zero as N goes to infinity.

The proof will be completed by showing that

$$\lim_{D,N\to\infty,\, D/N=p} h(0) = (1-p)^n = b(0;n,p). \tag{3.71}$$

Proof of 3.71

$$\lim_{D,N\to\infty,\, D/N=p} h(0) =$$

$$= \lim_{D,N\to\infty,\, D/N=p} \frac{(N-D)(N-D-1)\ldots(N-D-n+1)}{N(N-1)\ldots(N-n+1)}$$

$$= \lim_{D,N\to\infty,\, D/N=p} \frac{(1-p)(1-p-1/N)\ldots(1-p-(n-1)/N)}{1(1-1/N)\ldots(1-(n-1)/N)}$$

$$= (1-p)^n. \tag{3.72}$$

Additional details: The next to the last line in the preceding sequence of equations is obtained by dividing the numerator and denominator by N; the limit follows from the fact that all terms of the form j/N go to zero as N tends to infinity. This completes the derivation of Equation 3.71. For instance, to show that

$$\lim_{D,N\to\infty,\, D/N=p} h(1) = b(1;n,p), \text{ we reason as follows:}$$

$$\lim_{D,N\to\infty,\, D/N=p} h(1) = \lim_{D,N\to\infty,\, D/N=p} R_h(0)h(0)$$

$$= R_b(0)b(0;n,p) = \frac{np}{1-p}(1-p)^n$$

$$= np(1-p)^{n-1} = b(1;n,p).$$

Derivation of Chebyshev's inequality: Chebyshev's inequality is a consequence of the observation that the sum of a set of non-negative numbers is greater than the sum taken over a subset. In detail,

$$\sigma^2 = \sum_x (x-\mu)^2 f(x) \geq \sum_{(x:|x-\mu|\geq d)} (x-\mu)^2 f(x) \geq \sum_{(x:|x-\mu|\geq d)} d^2 f(x)$$

$$= d^2 \sum_{(x:|x-\mu|\geq d)} f(x) = d^2 P(|X-\mu| \geq d).$$

We have thus shown that $d^2 P(|X-\mu| \geq d) \leq \sigma^2$. Dividing both sides of this inequality through by d^2 yields Chebyshev's inequality 3.24.

Derivation of the Expected Price and Variance of a Stock Governed by the Binomial Lattice Model: Equations 3.49, 3.50, and 3.51

Taking the expected value of both sides of Equation 3.47 we see that

$$E\left(\frac{S_n}{S_0}\right) = E\left(d^n \left(\frac{u}{d}\right)^X\right)$$

$$= d^n E(e^{sX}), \text{ where } s = \ln\left(\frac{u}{d}\right)$$

$$= d^n (p e^{\ln(u/d)} + 1 - p)^n = d^n (p\frac{u}{d} + 1 - p)^n$$

$$= (pu + d(1-p))^n.$$

To compute $E\left((S_n/S_0)^2\right)$ we note that it can be evaluated via the mgf method, in detail:

$$E\left((S_n/S_0)^2\right) = d^{2n} E\left(\left(\frac{u}{d}\right)^{2X}\right)$$

$$= d^{2n} E(e^{sX}), \text{ where } s = 2\ln\left(\frac{u}{d}\right)$$

$$= d^{2n}\left(p\exp\left(2\ln\left(\frac{u}{d}\right)\right) + (1-p)\right)^n$$

$$= \left(pu^2 + (1-p)^2\right)^n.$$

Derivation of the Poisson Approximation to the Binomial

The derivation is similar to that used to obtain the binomial approximation to the hypergeometric. That is, we begin with a recurrence relation for the Poisson probabilities and then show that it is the limit of the recurrence relation for the binomial. In detail: Now $p_n = \lambda/n$ implies that $\lim_{n\to\infty} p_n = 0$ and, therefore,

$$\lim_{n\to\infty} R_b(x) = \lim_{n\to\infty} \frac{(n-x)p_n}{(x+1)(1-p_n)}$$

$$= \lim_{n\to\infty} \frac{np_n - xp_n}{(x+1)(1-p_n)}$$

$$= \frac{\lambda}{x+1} = R_p(x), \ x = 0, 1, \ldots.$$

The next step is to use the following limit, which can be found in any standard calculus text:

$$\lim_{n\to\infty}\left(1 - \frac{\lambda}{n}\right)^n = e^{-\lambda}. \tag{3.73}$$

Setting $p_n = \lambda/n$ in Equation 3.73, and noting that $b(0; n, p_n) = (1-p_n)^n$ we obtain the Poisson approximation for $b(0; n, p_n)$:

$$\lim_{n\to\infty, np_n=\lambda} b(0; n, p_n) = e^{-\lambda} = p(0; \lambda).$$

To obtain the Poisson approximation for $x = 1$, say, we proceed as follows:

$$\lim_{n\to\infty, np_n=\lambda} b(1; n, p_n) = \lim_{n\to\infty, np_n=\lambda} R_b(0)b(0; n, p_n)$$

$$= R_p(0)p(0; \lambda) = \lambda e^{-\lambda} = p(1; \lambda).$$

A similar argument is used to derive the Poisson approximation for all integers x. This completes the proof of Equation 3.60.

3.9   Chapter Summary

In this chapter we studied the fundamental concept of the *distribution of a random variable X defined on a sample space*. It is analogous to the concept of a variable defined on a population $\mathcal{S}$ studied earlier in Chapter 1. Similarly, the *expected value* and *variance* of X are analogues of the sample mean and sample variance. Random variables are classified according their distribution functions and among the most important of them are the hypergeometric, binomial, and the Poisson. The binomial distribution plays a fundamental role in the widely used binomial lattice model for stock prices.

3.10   Problems

Section 3.2: Discrete random variables and their distribution functions

Problem 3.1 *The random variable X has the pf defined by*

$$f(x) = \frac{5 - x^2}{15}; \, x = -2, -1, 0, 1, 2; \, f(x) = 0 \text{ elsewhere.}$$

Compute:

(a) $P(X \leq 0)$	(b) $P(X < 0)$	(c) $P(X \leq 1)$
(d) $P(X \leq 1.5)$	(e) $P(|X| \leq 1)$	(f) $P(|X| < 1)$

Problem 3.2 *With reference to the random variable defined in Problem 3.1 compute:*
(a) The probability function of X^2
(b) The probability function of 2^X
(c) The probability function of $2X$

Problem 3.3 *The random variable X has the pf defined by*

$$f(x) = c(6 - x) \text{ for } x = -2, -1, 0, 1, 2; \, f(x) = 0 \text{ elsewhere.}$$

Compute:
(a) c (b) $F(x) = P(X \leq x)$ for $x = -2.5, -0.5, 0, 1.5, 1.7, 3$

Problem 3.4 *The df of a discrete random variable Y is given in the following table:*

$$F(y) = 0, y < -1$$
$$F(y) = 0.2, -1 \leq y < 0$$
$$F(y) = 0.5, 0 \leq y < 1$$
$$F(y) = 0.8, 1 \leq y < 3$$
$$F(y) = 1, y \geq 3$$

(a) Draw the graph of $F(y)$.
(b) Compute the pf of Y.
(c) Compute $P(0 \leq Y \leq 2)$.

Problem 3.5 *Consider the experiment of selecting one coin at random from an urn containing 4 pennies, 3 nickels, 2 dimes, and 1 quarter. Let X denote the monetary value of the coin that is selected. Compute the probability function of X.*

Problem 3.6 *The number of customers who make a reservation for a limousine service to an airport is a random variable X with probability function given by*

$$f(x) = \frac{1}{21}(5 - |x - 3|), \, (x = 1, 2, 3, 4, 5, 6).$$

The limousine has a 5 seat capacity and the cost is $10 per passenger. Consequently, the revenue R per trip is also a random variable. Compute the probability function of R.

Problem 3.7 *Consider a deck of 5 cards marked 1,2,3,4,5. Two of these cards are picked at random and without replacement; let $W = $ sum of the numbers picked. Compute the pf of W.*

Problem 3.8 *Suppose a lot of 100 items contains 10 defective ones, and a sample of 5 items is selected at random from the lot. Let X denote the number of defective items in the sample; compute $P(X = x)$ for $x = 0, 1, 2, 3, 4, 5$. Hint: This problem is similar to the computation of the probability function of the number of hearts in a five-card poker hand (Equation 3.4) with defective and non-defective playing the roles of heart and non-heart.*

Problem 3.9 *Suppose a lot of 50 items contains 4 defective ones, and a sample of 5 items is selected at random from the lot. Let X denote the number of defective items in the sample; compute the probability function of X.*

Problem 3.10 *Let X have the pf given by:*
$f(x) = cx$ for $x = 1, 2, 3, 4, 5, 6$; $f(x) = 0$ elsewhere.
Compute:

(a) c	*(b) $F(x)$ for $x = 0, 1, 5, 7$*	*(c) $P(X$ is odd $)$*
(d) $P(X$ is even $)$	*(e) $P(X < 4)$*	*(f) $P(2 \leq X \leq 4)$*

Problem 3.11 *Let X have the pf given by:*

$$f(x) = cx \text{ for } x = 1, 2, \ldots n; \ f(x) = 0 \text{ elsewhere.}$$

(a) Show that $c = \frac{2}{n(n+1)}$.

(b) Show that $F(x) = \frac{x(x+1)}{n(n+1)}$ for $x = 1, 2, \ldots n$.

$$\text{Hint: } \sum_{1 \leq i \leq x} i = \frac{x(x+1)}{2}.$$

Problem 3.12 *The random variable X has the pf defined by*

$$f(x) = \frac{1}{x(x+1)} \text{ for } x = 1, 2, \ldots, n, \ldots.$$

(a) Show that $\sum_{1 \leq x < \infty} \frac{1}{x(x+1)} = 1$.

$$\text{Hint: Use the algebraic identity: } \frac{1}{x(x+1)} = \frac{1}{x} - \frac{1}{x+1}.$$

(b) Show that

$$F(x) = P(X \leq x) = \frac{x}{x+1} \text{ for } x = 1, 2, \ldots, n, \ldots.$$

Problem 3.13 *The random variable Y has a geometric distribution*
$P(Y = j) = pq^{j-1}, \ j = 1, \ldots, ; \ 0 < p < 1$.

(a) Show that $P(Y > a) = q^a$.	*(b) Show that $P(Y > a + b \mid Y > b) = P(Y > a)$.*
(c) Show that $P(Y$ is odd $) = 1/(1 + q)$.	*(d) Compute the probability function of $(-1)^Y$.*

Hint for (c): Use Equation 3.6 with a suitable choice for r.

Problem 3.14 *With reference to Example (3.24) compute the distribution function of the stock price after four periods.*

Sections 3.3-3.3.2: Expected value (mean), variance, and moments of a random variable

Problem 3.15 *The number of defects on printed circuit board is a random variable X with pf given by*
$P(X = i) = c/(i + 1), i = 0, 1, \ldots 4.$
    (a) Compute the constant c.   *(b) Compute the mean and variance of X.*

Problem 3.16 *The number of cells (out of 100) that exhibit chromosome aberrations is a random variable X with pf given by*

$$P(X = i) = \frac{c(i + 1)^2}{2^{i+1}}, \ (i = 0, 1, 2, 3, 4, 5).$$

    (a) Determine the value of the constant c.   *(b) Compute the mean and variance of X.*

Problem 3.17 *The probability function for the number of heads X in three tosses of a fair coin is given in Table 3.2. Using this information compute:*
    (a) $E(X)$ *(b) $V(X)$. Use the shortcut formula.*

Problem 3.18 *Table 3.3 displays the probability function corresponding to the experiment of throwing two dice. Compute $E(X)$ and $V(X)$.*

Problem 3.19 *Compute $E(X)$ and $V(X)$ where X is the number of hearts in a five-card poker hand. Hint: Use Equation 3.4 to compute $f_X(x)$.*

Problem 3.20 *The payoff of a one-roll bet on the number 11 is 15 to 1.*
    (a) Compute the house percentage. Note: The definition of house percentage is given in Definition 3.6.
    (b) Compute the true odds. For the definition of true odds refer to Equation 3.11.

Problem 3.21 *The random variable X has pf defined by:*
$f(x) = \frac{1}{5}; x = 1, 2, 3, 4, 5; f(x) = 0$ *elsewhere. Compute:*

(a) $E(X)$	*(b) $E(X^2)$*	*(c) $V(X)$*
(d) $E(2^X)$	*(e) $E(\sqrt{X})$*	*(f) $E(2X + 5)$*
(g) $V(2X + 5)$		

Problem 3.22 *Suppose $E(X) = 2$, $V(X) = 9$. Compute:*
(a) $E(3X - 6)$
(b) $E(X^2)$
(c) $E(X + 1)^2$

Problem 3.23 *Suppose $E(X) = 2$, $E(X^2) = 6$, $E(X^3) = 22$. Compute:*
(a) $V(X)$
(b) $E((X - 1)^2)$
(c) $E((X - 1)^3)$

Problem 3.24 *The random variable X has pf given by the following table:*

x	-2	-1	0	1	2
$f(x)$	0.2	0.1	0.1	0.4	0.2

Compute:

(a) $E(X)$ (b) $E(X^2)$ (c) $V(X)$
(d) $E(2^X)$ (e) $E(2^{-X})$ (f) $E(\sin(\pi X))$

Problem 3.25 *The random variable X has the pf defined by*
$f(x) = \frac{5-x^2}{15}; x = -2, -1, 0, 1, 2; f(x) = 0$ *elsewhere.*
Compute:

(a) $E(X)$ (b) $E(X^2)$ (c) $V(X)$
(d) $E(2^X)$ (e) $E(2^{-X})$ (f) $E(\cos(\pi X))$.

Problem 3.26 *Let the random variable Y have the distribution function $F(y)$ defined by:*

$$F(y) = 0, y < -1$$
$$F(y) = 0.2, -1 \le y < 0$$
$$F(y) = 0.5, 0 \le y < 1$$
$$F(y) = 0.8, 1 \le y < 3$$
$$F(y) = 1, y \ge 3$$

Compute:
(a) $E(Y)$ (b) $E(Y^2)$ (c) $V(Y)$

Problem 3.27 *The random variable X has the pf defined by*
$f(x) = (6-x)/30$ *for* $x = -2, -1, 0, 1, 2; f(x) = 0$ *elsewhere.*
Compute:

(a) $E(X)$ (b) $E(X^2)$ (c) $V(X)$
(d) $E(2^X)$ (e) $E(2^{-X})$ (f) $E(\cos(\pi X))$

Problem 3.28 *Compute $E(W)$ and $V(W)$ for the random variable W of Problem 3.7.*

Problem 3.29 *Let X have the pf given by:*

$$f(x) = \frac{2x}{n(n+1)} \text{ for } x = 1, 2, \ldots n; f(x) = 0 \text{ elsewhere.}$$

Compute $E(X)$. Hint: Use Equation 3.17.

Problem 3.30 *The random variable X has the pf defined by*

$$f(x) = \frac{1}{x(x+1)} \text{ for } x = 1, 2, \ldots, n, \ldots.$$

Show that $E(X) = \infty$ by verifying that $\sum_{1 \le x < \infty} x f(x) = \infty$, and therefore it does not satisfy the finiteness condition 3.8.

Problem 3.31 *Let X be a Bernoulli random variable. Show that*

$$V(X) \le \frac{1}{4}.$$

Hint: The variance of a Bernoulli random variable is given by

$$V(X) = p(1 - p) \text{ (see Equation 3.22).}$$

Then show that $p(1 - p) \le 1/4$.

Problem 3.32 *The random variable X has the discrete uniform distribution (see Equation 3.16). Show that*

$$V(X) = \frac{(N - 1)(N + 1)}{12}.$$

Hint: Use the shortcut formula and Equation 3.15 for the moments of the discrete uniform distribution.

Sections 3.4-3.5: The hypergeometric and binomial distributions

Problem 3.33 *The random variable X has a hypergeometric distribution with parameters $n = 4$, $N = 20$, $D = 6$.*
(a) Compute $h(x), x = 0, 1, 2, 3, 4$ using Equation 3.28.
(b) Compute $h(x), x = 0, 1, 2, 3, 4$ using the recurrence Equation 3.31. Which method is the more efficient for computing the hypergeometric probabilities?

Problem 3.34 *A lot of 10 components contains 2 that are defective. A random sample of size 4 is drawn. Let X denote the number of defective components that are drawn.*

(a) Show that X has a hypergeometric distribution and identify the parameters N, D, n. Use the formula from part (a) to compute the following probabilities:

(b) $P(X = 0)$ (c) $P(X = 1)$ (d) $P(X > 1)$

Problem 3.35 *Suppose the sampling plan is to sample 10% of a lot of N items and the acceptance number $c = 0$. Compute the acceptance probability in each of the following cases:*
(a) $N = 50$, $D = 1$
(b) $N = 100$, $D = 2$
(c) $N = 200$, $D = 4$

Problem 3.36 *Suppose a sample of size 15 is taken from a lot of 50 items and the lot is accepted if the number X of defective items in the sample is less than or equal to one, i.e., we accept the lot if $X \le 1$. Compute the $P(acceptance)$ for the following values of D:*
(a) $D = 1$ (b) $D = 3$ (c) $D = 6$

Problem 3.37 *A group of 10 items contains 2 that are defective. Articles are taken one at a time from the lot and are tested. Let Y denote the number of articles tested in order to find the first defective item. Compute the pf of Y.*

Problem 3.38 *An urn contains 2 nickels and 3 dimes. We select two coins at random, without replacement. Let X= monetary value of the coins in the sample, e.g., if the sample consists of a nickel and dime then X=15 cents.*
(a) Compute the pf of X.   (b) Compute E(X).   (c) Compute V(X).

Problem 3.39 *An urn contains two white balls and four red balls. Balls are drawn one by one without replacement until two white balls have appeared. Denote by U the number of balls that must be drawn until both white balls have appeared. Compute the pf of U.*

Problem 3.40 *Compute each of the following binomial probabilities in two ways: (1) by using Equation 3.33, and (2) by using the Equation 3.40 together with the binomial tables in Appendix A.*
(a) $b(0; 20; 0.1)$, $b(1; 20; 0.1)$, $b(5; 20; 0.1)$
(b) $b(2; 10; 0.4)$, $b(4; 10; 0.4)$, $b(7; 10; 0.4)$

Problem 3.41 *The random variable X has the binomial distribution $B(x; n, p)$. Compute the following probabilities using the binomial tables.*
(a) $P(X \leq 4) = B(4; 10, p)$ for $p = 0.1, 0.4, 0.6, 0.8$
(b) $P(X > 9) = 1 - B(9; 20, p)$ for $p = 0.2, 0.5, 0.7, 0.9$

Problem 3.42 *For $n = 20$ and $p = 0.5$*
(a) Estimate $P(|X - 10| \geq 5)$ using Chebyshev's inequality.
(b) Compute $P(|X - 10| \geq 5)$ using the binomial tables.

Problem 3.43 *Show that $b(x; n, p) = b(n - x; n, 1 - p)$ and give an intuitive explanation for this identity.*

Problem 3.44 *(a) Compute $b(x; 15, 0.3)$ for $x = 0, 1, 2, 3$ directly using Equation 3.33.*
(b) Compute $b(x; 15, 0.3)$ for $x = 0, 1, 2, 3$ using the recurrence Equation 3.36.

Problem 3.45 *A multiple choice quiz has 10 questions each with four alternatives. A passing score is 6 or more correct. If a student attempts to guess the correct answer to each question what is the probability that he passes?*

Problem 3.46 *Let Y denote the number of times a 1 appears in 720 throws of a die. Compute $E(Y)$ and $V(Y)$.*

Problem 3.47 *A taxi company has a limousine with a seating capacity of N. The maintenance cost (to the taxi company) of each seat is $5, and the price per seat is $10. The number of reservations is a binomial random variable X with parameters $n = 10$, $p = 0.4$. Compute the expected net revenue when:*
(a) $N = 4$ (b) $N = 5$
(c) Which is more profitable to operate: a 4 or 5 seat van?

Problem 3.48 *The lifetime (measured in hours) of a randomly selected battery is a random variable denoted by T. Assume that $P(T \leq 3.5) = 0.10$. This means that 90% of the batteries have lifetimes that exceed 3.5 hours. A random sample of 20 batteries is selected for testing.*
(a) Let Y = the number of batteries in the sample whose lifetimes are less than or equal to 3.5 hours. Show that the pf of Y is binomial and identify the parameters n, p.
(b) Compute $E(Y)$ and $V(Y)$.
(c) Compute $P(Y > 4)$.
(d) Suppose 6 batteries (out of the 20 tested) had lifetimes that were less than or equal to 3.5 hours. If, in fact, the proportion of batteries with lifetimes T less than or equal to 3.5 hours is 0.10 what is the probability that 6 or more batteries (out of the 20 tested) would have lifetimes less than or equal to 3.5 hours? Is this result consistent with the assumption that $P(T \leq 3.5) = 0.10$? Justify your conclusions by computing $P(Y \geq 6)$ and then interpreting this probability.

Problem 3.49 *A sampling plan has $n = 20$ and the lot size $N = 1,000$.*
(a) Compute the acceptance probabilities for $D = 50$ acceptance number $c = 2$. Hint: Use the binomial approximation to the hypergeometric distribution.
(b) Compute the acceptance probabilities for $D = 50$ and acceptance number $c = 1$. (c) Compute the acceptance probabilities for $D = 50$ and acceptance number $c = 0$.

Problem 3.50 *It is known that 40% of patients naturally recover from a disease. A drug is claimed to cure 80% of all patients with the disease. An experimental group of 20 patients is given the drug. What is the probability that at least 12 recover if:*
(a) The drug is worthless and hence the true recovery rate remains at 40%?
(b) The drug is really 80% effective?

Problem 3.51 *(Continuation of Problem 3.50)*
 Suppose 200 patients are given the drug. Let R denote the number of patients who recover. Compute $E(R)$ and $V(R)$ for each of the following two cases:
(a) the drug is worthless.
(b) the drug is really 80% effective as claimed.

Problem 3.52 *Use Chebyshev's inequality to estimate the probability that the sample proportion obtained from a poll of voters with a sample size of $n = 2500$ differs from the true proportion by an amount less than or equal to 0.04.*

Problem 3.53 *Let X denote the number of heads in n tosses of a fair coin. Use Chebyshev's inequality to obtain a lower bound on the following probabilities:*
(a) $P(225 \leq X \leq 275)$ when $n = 500$.
(b) $P(80 \leq X \leq 120)$ when $n = 200$.
(c) $P(45 \leq X \leq 55)$ when $n = 100$.

Problem 3.54 *Suppose X is a random variable with $\mu_X = 20$, $\sigma_X = 1$. Using only this information show that $P(|X - 20| \geq 4) \leq 0.0625$.*

Problem 3.55 *Suppose X is a random variable with $\mu_X = 50$, $\sigma_X = 4$. Using only this information show that $P(42 \leq X \leq 58) \geq 0.75$.*

Problem 3.56 *Assume the price moves of a stock follows the binomial lattice model described in Example 3.25 with initial price $S_0 = \$50$ and parameters $u = 1.02$, $d = 0.99$, $p = 0.52$. Assume the time period is one week.*
(a) Give the formula for the probability distribution S_4, which is the stock price after one month.
(b) Compute the probability that after one month $S_4 \geq 51$.
(c) Compute the probability that after one month the price is up by at least 5%.
(d) Compute the probability that after one month the price is down by at least 3%.

Problem 3.57 *Compute the expected value and standard deviation (risk) of the return (after one month) on the stock described in Problem 3.56.*

Problem 3.58 *Assume the price moves of a stock follows the binomial lattice model described in Example 3.25 with initial price $S_0 = \$75$ and parameters $u = 1.015$, $d = 0.985$, $p = 0.52$. Assume the time period is one week.*
(a) Give the formula for the probability distribution S_4, which is the stock price after one month.

(b) Compute the probability that after one month $S_4 \geq 76$.
(c) Compute the probability that after one month the price is up by at least 5%.
(d) Compute the probability that after one month the price is down by at least 2%.

Problem 3.59 *Compute the expected value and standard deviation (risk) of the return (after one month) on the stock described in Problem 3.58.*

Problem 3.60 *Verify that choosing the parameters of the binomial lattice model as in Equation 3.54 yields the values for the mean and variance given in Equation 3.55.*

Problem 3.61 *R. Wolf (1882) threw a die $20,000$ times and recorded the number of times each of the six faces appeared. The results follow.*

Face	1	2	3	4	5	6
Frequency	3407	3631	3176	2916	3448	3422

(Source: D. J. Hand, et. al. (1994), Small Data Sets, Chapman& Hall, London.)
(a) Compute an upper bound on the probability that the number of times the number five appeared is greater than or equal to the observed value 3448. Assume that the dice are fair and use Chebyshev's inequality to obtain the upper bound.
(b) Compute an upper bound on the probability that the number of times the number four appeared is less than or equal to the observed value 2916.

Problem 3.62 *Can Chebyshev's inequality be improved? Consider the random variable X with the following probability function:*

$$P(X = \pm a) = p, \ P(X = 0) = 1 - 2p, \ (0 < p < 0.5).$$

Show that

$$P(|X| \geq a) = \frac{V(X)}{a^2}.$$

Problem 3.63 *The probability that a disk drive lasts 6000 or more hours equals 0.8. In a batch of $20,000$ disk drives how many would be expected to last 6000 or more hours?*

Section 3.6: The Poisson Distribution

Problem 3.64 *The random variable X has a binomial distribution with parameters $n = 20$ and $p = 0.01$. In each of the following cases compute: (i) the exact value of $P(X = x)$ and (ii) its Poisson approximation for:*
(a) $x = 0$; (b) $x = 1$; (c) $x = 2$.

Problem 3.65 *Verify Equation 3.62.*

Problem 3.66 *(a) Compute $p(x; 2)$, $x = 0, 1, 2, 3$ using Equation 3.60.*
(b) Compute $p(x; 2)$, $x = 0, 1, 2, 3$ using the recurrence relation 3.63.

Problem 3.67 *(a) Compute the Poisson probabilities $p(x; 1.5)$, $x = 0, 1, 2, 3$ using Equation 3.60.*
(b) Compute the Poisson probabilities $p(x; 1.5)$, $x = 0, 1, 2, 3$ using the formula

$$p(x; \lambda) = P(x; \lambda) - P(x - 1; \lambda),$$

where the values of the distribution function

$$P(x; \lambda) = \sum_{0 \le y \le x} p(y; \lambda)$$

are tabulated in Table A.2 in Appendix A.

Problem 3.68 *A lot of* 10,000 *articles has* 150 *defective items. A random sample of* 100 *articles is selected. Let* X *denote the number of defective items in the sample. Give the formula for:*
(a) the exact pf $f_X(x)$ *of* X.
(b) the binomial approximation of $f_X(x)$.
(c) the Poisson approximation of the binomial distribution of part b.

Problem 3.69 *If* 99% *of students entering a junior high school are vaccinated against the measles, what is the probability that a group of* 50 *students contain:*
(a) no unvaccinated students?
(b) exactly one unvaccinated student?
(c) two or more unvaccinated students?

Problem 3.70 *The prevalence rate for a rare disease is 3 per 10,000.*
(a) Let X *denote the number of cases (of the disease) in a small town of population* 8,000. *Describe the exact distribution of* X; *name it and identify the parameters.*
(b) What is the expected number of town residents with the disease?
(c) The Department of Public Health reports that six town residents have the disease. Is this to be expected or is it an unusual event? Discuss.

Problem 3.71 *The number of chocolate chips per cookie is assumed to have a Poisson distribution with* $\lambda = 4$. *Find the probability that:*
(a) a cookie has no chocolate chips.
(b) a cookie has 7, or more, chocolate chips.

Problem 3.72 *A batch of dough yields* 500 *cookies. How many chocolate chips should be mixed into the dough so that the probability of a cookie having no chocolate chips is* $\le .01$?

Problem 3.73 *A computer disk drive is warranted to last at least 4,000 hours. The probability that a disk drive lasts 4,000 or more hours equals* 0.9999.
(a) Let X *denote the number of disk drive failures during the warranty period. Describe the distribution of* X; *name it and identify the parameters.*
(b) In a batch of 20,000 disk drives how many would be expected to fail before the warranty expired?
(c) What is the probability that three or more disk drives fail during the warranty period?

Problem 3.74 *A taxi company has a limousine with a seating capacity of* N. *The cost of each seat is* $5 *and the price per seat is* $10. *The number of reservations* X *is Poisson distributed with parameter* $\lambda = 5$, *i.e.,* $P(X = x) = e^{-5}5^x/x!, x = 0, 1 \ldots$ *Compute the expected net revenue when:*
(a) $N = 4$ *(b) $N = 5$*
(c) Which is more profitable to operate: a 4 or 5 seat van?

Problem 3.75 *A sampling plan has $n = 100$, $c = 2$ and the lot size $N=10,000$. Compute the approximate acceptance probabilities for the following values of D:*
(a) $D = 50$ *(b) $D = 100$*   *(c) $D = 200$*
Hint: Use the Poisson approximation to the binomial.

Problem 3.76 *(a) Derive the formula for the mgf of the Poisson distribution displayed in Equation 3.69.*
(b) Use the mgf to compute the first two moments and the variance of the Poisson distribution.

Problem 3.77 *(a) Derive the formula for the mgf of the geometric distribution displayed Equation 3.70.*
(b) Use the mgf to compute the first two moments and the variance of the geometric distribution.

3.11   To Probe Further

The pioneering contributions to the theory of probability of the mathematicians J. Bernoulli (1654-1705), A. DeMoivre (1667-1754), P. S. Laplace (1749-1827), and S. D. Poisson (1781-1840) are described in Stigler (1986), *History of Statistics: The Measurement of Uncertainty before 1900*, The Belknap Press of Harvard University Press, Cambridge, Massachusetts. The debates among the mathematicians and philosophers of the 17th and 18th centuries, such as Pascal, Bernoulli, d'Alembert, and Laplace, on how a rational and prudent man should act when the future course of events is subject to chance is discussed in L. Daston (1988), *Classical Probabilty in the Enlightenment*, Princeton University Press. Among the many recent introductory text books on mathematics of finance we recommend the following:

1. J. Cvitanic and F. Zapatero (2004), *Introduction to the Economics and Mathematics of Financial Markets*, MIT Press. This book explains in more detail the intuitive ideas behind the concept of a risk-neutral probability measure.

2. John C. Hull (2003), *Options, Futures, and Other Derivatives*, 5th ed., Prentice-Hall.

3. D. G. Luenberger (1998), *Investment Science*, Oxford University Press.

Chapter 4

Continuous Random Variables and Their Distribution Functions

The principal element of the analysis of probabilities is an exponential integral that has presented itself in several very different mathematical theories... This same function is connected with general physics... We have discovered in recent years that it also represents the diffusion of heat in the interior of solid substances. Finally, it determines the probability of errors and mean results of numerous observations; it reappears in the questions of insurance and in all difficult applications of the science of probabilities.

J.B.J. Fourier (1768-1830), French mathematician

4.1   Orientation

In this chapter we study random variables whose range is not a discrete subset but is a subinterval of the real numbers. To compute probabilities, the distribution function, expected values, moments, etc., we use the concepts and methods of the calculus in an essential way. We then turn our attention to the study of the *normal distribution*, considered to be the most important probability distribution of all.

Organization of Chapter

1. Section 4.2: Random Variables with Continuous Distribution Functions: Definition and Examples

2. Section 4.3: Expected Value, Moments, and Variance of a Continuous Random Variable

3. Section 4.4: Moment Generating Function: Continuous Random Variables*

4. Section 4.5: The Normal Distribution: Definition and Basic Properties

5. Section 4.6: The Log Normal Distribution: A Model for the Distribution of Stock Prices

6. Section 4.7: Normal Approximation to the Binomial Distribution

7. Section 4.7.1: Approximate Distribution of the Sample Proportion $\hat{p}$

8. Section 4.8: Other Important Continuous Distributions

9. Section 4.8.1: The Gamma and Chi-Square Distributions

10. Section 4.8.2: The Weibull Distribution

11. Section 4.8.3: The Beta Distribution*

12. Section 4.9: Functions of a Random Variable

13. Section 4.10: Mathematical Details and Derivations

14. Section 4.11: Chapter Summary

4.2   Random Variables with Continuous Distribution Functions: Definition and Examples

Random variables whose range is an interval arise in a variety of applications. Consider, for example, the lifetime X of an electronic component. Theoretically the component can function forever or it can be defective and not function at all. The range of X in this case is the set of non-negative real numbers, denoted by $R_+ = [0, \infty)$. Informally, we say that a random variable X is *continuous* if its df $F_X(x)$ is a continuous function defined by a definite integral in the sense of Definition 4.1. This stands in sharp contrast to the discrete random variables studied in Chapter 3 whose dfs are discontinuous functions, as shown, for example, in Figure 3.2.

Definition 4.1 *A random variable X is said to be continuous if its df is a continuous function of the following type:*

$$F_X(x) = \int_{-\infty}^{x} f_X(t) \, dt, \ where \tag{4.1}$$

$$f_X(t) \geq 0 \ and \tag{4.2}$$

$$\int_{-\infty}^{\infty} f_X(t) \, dt = 1. \tag{4.3}$$

The function $f_X(x)$ is called the *probability density function* (pdf) of the random variable X. Figure 4.1 displays the graph of a pdf $f_X(x)$. Its distribution function has a simple geometric interpretation: It is the area under the curve $y = f_X(t)$ from $-\infty$ to x. Similarly, $P(a \leq X \leq b)$ is the area under the curve $y = f_X(x)$ from a to b. When there is little or no danger of confusion we shall drop the subscript X and simply write $F(x)$ and $f(x)$.

It follows from Definition 4.1 that $F_X(x)$ has the following properties:

$$0 \leq F_X(x) \leq 1 \tag{4.4}$$

$$a \leq b \text{ implies } F_X(a) \leq F_X(b) \tag{4.5}$$

$$\lim_{x \to -\infty} F_X(x) = 0 \quad \text{and} \quad \lim_{x \to +\infty} F_X(x) = 1. \tag{4.6}$$

Conversely, any continuous function $F(x)$ satisfying Equations 4.4, 4.5, and 4.6 is a distribution function of a random variable X.

We note that we can compute the probability $P(a \leq X \leq b)$ in terms of the distribution function as follows:

$$P(a \leq X \leq b) = F_X(b) - F_X(a).$$

One consequence of our definition is that

$$P(X = a) = \int_{a}^{a} f(x)dx = 0;$$

this is because the area under the curve from a to a is zero. From this it follows that when X is a continuous random variable the following probabilities are equal:

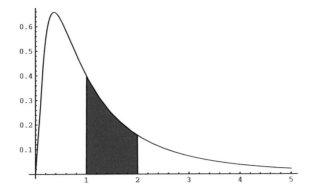

FIGURE 4.1: $P(1 \le X \le 2)$ equals the area under the curve between $x = 1$, $x = 2$, and the x axis

$$P(a < X \le b) = P(a < X < b) = P(a \le X < b) = P(a \le X \le b) = F_X(b) - F_X(a).$$

Using the fundamental theorem of the calculus one can derive the following important relation between the df F_X and the pdf f(x):

$$F'_X(x) = f(x) \text{ for all } x \text{ at which } f \text{ is continuous.} \qquad (4.7)$$

It is useful to restate the equation $F'_X(x) = f(x)$ in differential form like so:

$$P(x < X \le x + \triangle x) = F_X(x + \triangle x) - F_X(x)$$
$$\approx F'_X(x)\triangle x = f(x)\triangle x; \text{ more precisely,}$$
$$P(x < X \le x + \triangle x) = f(x)\triangle x + o(\triangle x), \triangle x \to 0. \qquad (4.8)$$

Here, $o(\triangle x)$ denotes a function with the property that $\lim_{\triangle x \to 0}(o(\triangle x)/\triangle x) = 0$.

The Exponential Distribution

The exponential distribution is widely used in the engineering sciences to model a variety of quantities, such as the length of time X ($\mu\,sec$) to run a program on an interactive computer system, the lifetime of a bearing, etc. Its pdf is given by

$$f(x) = \theta e^{-\theta x}, 0 < x < \infty, \theta > 0, \qquad (4.9)$$
$$f(x) = 0, \text{ elsewhere.} \qquad (4.10)$$

We will see later that the expected value of a random variable with an exponential distribution equals $1/\theta$. We obtain the distribution function by integrating the probability density function. Consequently,

$$F(x) = 0, x \le 0,$$
$$F(x) = \int_0^x \theta\, e^{-\theta t}\, dt = 1 - e^{-\theta x}, 0 \le x < \infty.$$

The graphs of the exponential pdf and its df, for $\theta = 2$, are displayed in Figure 4.2.

Example 4.1 *Let X have an exponential distribution with $\theta = 4$.*
Compute: (1) $P(0.1 \le X \le 0.3)$; (2) $P(X \ge 0.5)$.

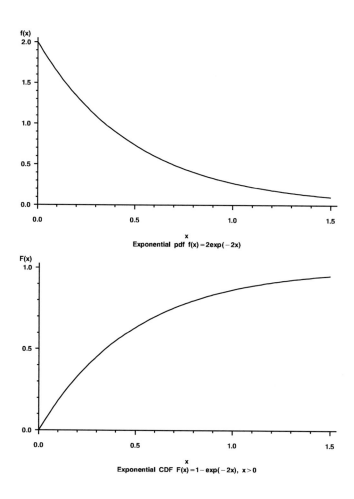

FIGURE 4.2:   Graphs of the pdf $f(x) = 2\exp(-2x)$, $x > 0$, $f(x) = 0$, $x \leq 0$ and its cdf $F(x) = 1 - 2\exp(-2x)$, $x \geq 0$, $F(x) = 0$, $x < 0$

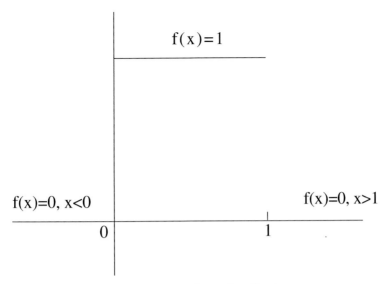

FIGURE 4.3: Graph of the pdf of the uniform distribution

Solution

$$(1)\ P(0.1 \leq X \leq 0.3) = e^{-4 \times 0.1} - e^{-4 \times 0.3} = 0.3691,\ \text{and}$$
$$(2)\ P(X \geq 0.5) = e^{-4 \times 0.5} = e^{-2} = 0.1353.$$

The Uniform Distribution

We say that the random variable U is *uniformly distributed* on the interval $[0, 1]$ if its pdf is given by

$$f(x) = 1,\ 0 < x < 1, \tag{4.11}$$
$$f(x) = 0,\ elsewhere. \tag{4.12}$$

The graph of $f(x)$ is displayed in Figure 4.3. Notice that $f(x)$ is another example of a function that is defined *piecewise*; that is, the formula that produces the output $f(x)$ depends on the interval where the input x is located. Consequently, its df $F(x)$ is also defined piecewise:

$$F(x) = 0,\ x \leq 0,$$
$$F(x) = \int_0^x 1\,dt = x,\ 0 \leq x \leq 1,$$
$$F(x) = 1,\ x \geq 1.$$

The graph of $F(x)$ is displayed in Figure 4.4.

The random variable U is the mathematical model of a *random number generator*, which is a computer program that produces a random number lying in the interval $[0, 1]$. In this instance randomness means that the number produced lies in the interval $[\alpha, \beta]$ with probability $\beta - \alpha$, where $0 \leq \alpha < \beta \leq 1$. The uniform distribution is widely used as a tool to simulate the dfs of other random variables (see Section 4.9 for additional details).

Example 4.2 *Let X be uniformly distributed on $[0, 1]$. Compute:*
(1) $P(0.25 \leq X \leq 0.6)$; (2) $P(X \geq 0.7)$.

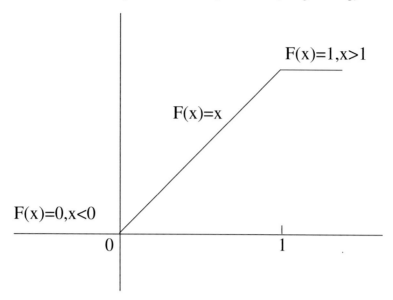

FIGURE 4.4:   Graph of the cdf of the uniform distribution

Solution

$$(1)\ P(0.25 \leq X \leq 0.6) = 0.6 - 0.25 = 0.35.$$
$$(2)\ P(X \geq 0.7) = 1 - 0.7 = 0.3.$$

The Uniform Distribution on the Interval $[a, b]$

A random variable X is uniformly distributed over the interval $[a, b]$ if the probability that X lies in a subinterval $[c, d] \subset [a, b]$ is proportional to the length of the subinterval. This means that its pdf is given by

$$f(x) = \frac{1}{b - a},\ a < x < b,$$
$$f(x) = 0,\ elsewhere.$$

The df of a random variable uniformly distributed over the interval $[a, b]$ is given by

$$F(x) = 0,\ x \leq a,$$
$$F(x) = \int_a^x \frac{1}{b - a}\, dt = \frac{x - a}{b - a},\ a \leq x \leq b,$$
$$F(x) = 1,\ x \geq b.$$

A random variable X that is uniformly distributed on $[0, L]$ is the mathematical model for the experiment *choosing a random point on an interval of length L.*

Choosing a random direction in the plane

The uniform distribution can also be used as a mathematical model for choosing a random direction in the plane, as will now be explained. A line through the origin is specified by the angle $\theta, 0 \leq \theta < 2\pi$ (measured in radians) that it makes with the positive x axis. Let Θ denote a random variable that is uniformly distributed over $[0, 2\pi)$. We say that $(\cos\Theta, \sin\Theta)$ is a "random point on the unit circle"; the line connecting $(0, 0)$ to $(\cos\Theta, \sin\Theta)$ defines a random direction, or random vector of unit length, in the plane.

Example 4.3 *An error to be avoided*

In this example we discuss a common error that occurs when students evaluate indefinite integrals instead of definite integrals. Consider the pdf defined by

$$f(x) = 2(1-x),\, 0 < x < 1, \tag{4.13}$$
$$f(x) = 0,\, elsewhere.$$

Using the formula $F'(x) = f(x)$ students compute $F(x)$ via the recipe

$$F(x) = \int 2(1-x)\, dx = 2x - x^2.$$

Unfortunately, this answer cannot be correct because $F(3) = -3$ and this is absurd since $0 \leq F(x) \leq 1$. The error, which is quite common, is to forget that $F_X(x)$ is obtained by computing a *definite integral*. In addition one should not forget that the function $f(x)$ is defined piecewise. The correct calculation is given below:

$$F(x) = 0 \,,\, x \leq 0,$$
$$F(x) = \int_0^x 2(1-t)\, dt = 2x - x^2,\, 0 \leq x \leq 1,$$
$$F(x) = 1,\, 1 \leq x < \infty.$$

Medians and Percentiles of a Continuous df

A median of a continuous df $F(x)$ is defined to be a solution to the equation $F(x) = 0.5$. Similarly, the $100p$th percentile of $F(x)$ is defined to be a solution to the equation $F(x) = p$. Although in all the examples considered in this text the solution will be unique, this will not be so in general. Little or no harm is done, however, by ignoring these possibilities. Since $F(x)$ is continuous and nondecreasing on every interval $[a, b]$ it follows from the *intermediate value theorem* of the calculus that given any p satisfying $0 < p < 1$ there exists at least one solution to the equation:

$$F(x) = p,\, 0 < p < 1; \text{ the solution is denoted by } q(p).$$

Example 4.4 *Compute the median of: (1) the exponential distribution and (2) the df defined by Equation 4.13.*

1. **Solution** The median of the exponential distribution satisfies the equation

$$1 - e^{-\theta q(0.5)} = 0.5 \text{ or, } q(0.5) = -\frac{\ln 0.5}{\theta} = \frac{\ln 2}{\theta}.$$

2. The median satisfies the quadratic equation

$$F(x) = 2x - x^2 = 0.5 \text{ which has two solutions: } x = 1 \pm \sqrt{0.5},$$

only one of which is the median. The solution $x = 1 + \sqrt{0.5} = 1.7071$ cannot possibly be the median since $F(1.7071) = 1$. Consequently, $q(0.5) = 1 - \sqrt{0.5} = 0.293$.

4.3 Expected Value, Moments, and Variance of a Continuous Random Variable

The definitions of $E(X)$, $E(\phi(X))$, $E(X^k)$ and variance for continuous random variables are similar to those already given for discrete random variables, except that sums are replaced by integrals.

Definition 4.2 *Let X be a continuous random variable with pdf $f(x)$. Then the $E(X)$ and $E(\phi(X))$ are defined by the equations:*

$$E(X) = \mu_X = \int_{-\infty}^{\infty} x f(x)\, dx, \text{ provided that} \tag{4.14}$$

$$\int_{-\infty}^{\infty} |x| f(x)\, dx < \infty. \tag{4.15}$$

$$E(\phi(X)) = \int_{-\infty}^{\infty} \phi(x) f(x)\, dx, \text{ provided that} \tag{4.16}$$

$$\int_{-\infty}^{\infty} |\phi(x)| f(x)\, dx < \infty. \tag{4.17}$$

The kth moment μ_k, $k = 1, 2, \ldots$ of a continuous random variable X is defined by the equation

$$\mu_k = E(X^k), \text{ where } k = 1, 2, \ldots.$$

The kth moment is computed by evaluating the integral

$$E(X^k) = \int_{-\infty}^{\infty} x^k f(x)\, dx. \tag{4.18}$$

The *variance* of a random variable X, is defined by

$$V(X) = E((X - \mu_X)^2) = \int_{-\infty}^{\infty} (x - \mu_X)^2 f(x)\, dx, \tag{4.19}$$

provided that

$$\int_{-\infty}^{\infty} (x - \mu_X)^2 f(x)\, dx < \infty.$$

Notation The standard deviation is denoted by $\sigma_X = \sqrt{V(X)}$. The shortcut formula (Equation 3.21) for computing $V(X)$ remains valid: $V(X) = E(X^2) - \mu_X^2$.

Remark 4.1 *The interpretation of the mean and variance as the center of mass and spread of the distribution remains valid in the continuous case.*

Example 4.5 *Compute the first two moments and the variance for the following random variables:*

1. The pdf of X is given by

$$f(x) = 2(1 - x),\ 0 < x < 1,$$
$$f(x) = 0,\ elsewhere.$$

Solution $E(X)$, $E(X^2)$ and $V(X)$ are computed by evaluating the following definite integrals and using the shortcut formula for computing the variance.

$$\mu_X = \int_{-\infty}^{\infty} x f(x)\, dx = \int_0^1 x 2(1 - x)\, dx = \frac{1}{3};$$

$$E(X^2) = \int_{-\infty}^{\infty} x^2 f(x)\, dx = \int_0^1 x^2 2(1 - x)\, dx = \frac{1}{6}, \text{ so}$$

$$V(X) = \frac{1}{6} - \left(\frac{1}{3}\right)^2 = \frac{1}{18}.$$

2. X has the uniform distribution. Then

$$E(X) = (a + b)/2; \; V(X) = \frac{(b-a)^2}{12}. \tag{4.20}$$

The derivation of Equation 4.20 is left to the reader (Problem 4.21).

3. X has the exponential distribution (Equation 4.9).

   Solution Using an appropriate integration by parts one can show that

$$\mu_X = \int_{-\infty}^{\infty} x f(x)\, dx = \int_0^\infty x\theta e^{-\theta x}\, dx = \frac{1}{\theta}; \tag{4.21}$$

$$E(X^2) = \int_{-\infty}^{\infty} x^2 f(x)\, dx = \int_0^\infty x^2 \theta e^{-\theta x}\, dx = \frac{2}{\theta^2}; \tag{4.22}$$

$$V(X) = \frac{1}{\theta^2}. \tag{4.23}$$

Example 4.6 *A mathematical model for internet traffic*

Many recent statistical analyses of internet traffic suggest that users send messages whose size, measured in bytes or packets, is a random variable T with finite mean and infinite variance, that is $V(T) = \infty$. This should not be a surprise when one realizes the huge sizes of music and video files, as well as security updates for computer operating systems that are transmitted across the internet. One widely used mathematical model for random variables with finite mean and infinite variance is the *Pareto distribution* defined here as

$$F(t) = 1 - \frac{1}{(1+t)^\alpha}, \; (t \geq 0); \; F(t) = 0, \; (t < 0), \; (1 < \alpha < 2). \tag{4.24}$$

It can be shown (Problem 4.27 asks you to supply the details) that

$$E(T) = \frac{1}{\alpha - 1}; \; E(T^2) = \infty, \; (1 < \alpha < 2) \tag{4.25}$$

The Pareto distribution is named after the Italian economist Vilfredo Pareto (1848-1923).

Example 4.7 *Computing a manufacturer's expected profit per component*

Suppose the lifetime T of an electronic component has pdf $f(t)$. The manufacturing costs and selling price of each component are C_1, C_2 dollars, respectively. The component is guaranteed for K hours; if $T \leq K$ then the manufacturer replaces the faulty component without charge. What is the manufacturer's expected profit per component?

   Solution The first step is to express the profit R as a function of the lifetime T:

$$R = C_2 - C_1, \text{ if } T > K, \tag{4.26}$$
$$R = C_2 - 2C_1, \text{ if } T \leq K. \tag{4.27}$$

Equation 4.26 asserts that if the component's lifetime exceeds the warranty period then the net profit is $C_2 - C_1$. The second equation asserts that if the component fails during the warranty then the manufacturer's profit is decreased by the cost of producing the defective component plus the cost of replacement, which equals $2C_1$. Thus R is a *discrete* random variable whose expectation is easily computed by the methods of Section 3.2; this yields the following formula for E(R):

$$E(R) = (C_2 - 2C_1) + C_1 P(T > K). \tag{4.28}$$

Example 4.8 *A variation of Example 4.7*

Suppose T is exponentially distributed with $\theta^{-1} = 1000$, which implies that $E(T) = 1,000$ hours. The manufacturing costs and selling price of each component are $6 and $10, respectively. If the item fails before 500 hours the manufacturer replaces the faulty component without charge. Compute the expected profit per customer.

 Solution The assumption that T has an exponential distribution means that

$$P(T > K) = e^{-K/1000}.$$

The length of the warranty period is 500 hours, so $K = 500$; the manufacturing costs and selling prices are $C_1 = \$6.00$, $C_2 = \$10.00$. Inserting these values into Equation 4.28 yields

$$E(R) = (10 - 12) + 6e^{-500/1000} = \$1.64.$$

Example 4.9 *A variation of Example 4.7*

Suppose T is exponentially distributed with $\theta^{-1} = 1000$, which implies that $E(T) = 1,000$ hours. The manufacturing costs and selling price of each component are $6 and $10, respectively. If the item fails before 500 hours the manufacturer replaces the faulty component without charge. Compute the expected profit per customer.

 Solution The assumption that T has an exponential distribution means that

$$P(T > K) = e^{-K/1000}.$$

The length of the warranty period is 500 hours, so $K = 500$; the manufacturing costs and selling prices are $C_1 = \$6.00$, $C_2 = \$10.00$. Inserting these values into Equation 4.28 yields

$$E(R) = (10 - 12) + 6e^{-500/1000} = \$1.64.$$

Chebyshev's Inequality
 Chebyshev's inequality (Equation 3.24), which we derived in the previous chapter, remains valid for continuous random variables:

Proposition 4.1 *Let X be a random variable with expected value μ and variance σ^2. Then*

$$P(|X - \mu| \geq d) \leq \frac{\sigma^2}{d^2}; \text{ equivalently,} \tag{4.29}$$

$$P(|X - \mu| \leq d) \geq 1 - \frac{\sigma^2}{d^2}. \tag{4.30}$$

We omit the proof, which is similar to the one already given, except that sums are replaced by integrals.

Example 4.10

Suppose X is uniformly distributed on the interval $[0, 4]$; consequently, $E(X) = 2$ and $V(X) = 4/3$. Using Chebyshev's inequality we see that

$$P(|X - 2| \geq 1.5) \leq \frac{16}{27} = 0.5926$$

whereas the exact answer is

$$P(|X - 2| \geq 1.5) = 0.25.$$

4.4   Moment Generating Function: Continuous Random Variables*

The definition of the moment generating function (mgf) for a continuous random variable is exactly the same as in the discrete case (Section 3.7) with sums replaced by definite integrals.

Definition 4.3 *The moment generating function of a continuous random variable X with pdf $f(x)$ is*

$$M_X(s) = E(e^{sX}) = \int_{-\infty}^{\infty} e^{sx} f(x) \, dx \qquad (4.31)$$

provided $E(e^{sX}) < \infty$ for all values s in an interval $(-\delta, \delta), \delta > 0$.

Differentiating both sides of Equation 4.31 with respect to s, and interchanging the order of differentiation and integration on the right-hand side, we see that

$$\frac{dM_X(s)}{ds} = \frac{d}{ds}\left(\int_{-\infty}^{\infty} e^{sx} f(x) \, dx\right) = \int_{-\infty}^{\infty} \left(\frac{de^{sx}}{ds}\right) f(x) \, dx = \int_{-\infty}^{\infty} x e^{sx} f(x) \, dx.$$

Evaluating the derivative at $s = 0$ yields a useful formula for the expected value, or first moment, of X:

$$M_X'(0) = \int_{-\infty}^{\infty} x f(x) \, dx = E(X) = \mu_1.$$

Differentiating both sides of Equation 4.31 k times with respect to s, and interchanging the order of differentiation and integration on the right-hand side we see that

$$M_X^{(k)}(s) = \frac{d^k}{ds^k}\left(\int_{-\infty}^{\infty} e^{sx} f(x) \, dx\right) = \int_{-\infty}^{\infty} \left(\frac{d^k e^{sx}}{ds^k}\right) f(x) \, dx = \int_{-\infty}^{\infty} x^k e^{sx} f(x) \, dx.$$

where $M_X^{(k)}(s), k = 1, 2, \ldots$, denotes the kth derivative. The next theorem, proof omitted, explains why $M_X(s)$ is called the moment generating function.

Theorem 4.1 *If the mgf exists for all $s \in (-\delta, \delta), \delta > 0$ then its kth derivative exists for all $s \in (-\delta, \delta)$, and*

$$M_X^{(k)}(0), = \int_{-\infty}^{\infty} x^k f(x) \, dx = E(X^k). \qquad (4.32)$$

To illustrate we compute the mgf for the exponential distribution.

Proposition 4.2 *Suppose X is an exponential random variable with parameter θ; then*

$$M_X(s) = \frac{\theta}{\theta - s}. \qquad (4.33)$$

Proof

$$M_X(s) = \int_0^{\infty} e^{sx} \theta e^{-\theta x} \, dx = \theta \int_0^{\infty} e^{sx - \theta x} \, dx$$

$$= \theta \int_0^{\infty} e^{(s-\theta)x} \, dx = \frac{\theta}{\theta - s}$$

4.5   The Normal Distribution: Definition and Basic Properties

It is a remarkable fact that the relative frequency histograms of many large data sets, such as the histogram of the weekly returns of the S&P500 index for the year 1999 (see Figure 1.11) and the histogram of a binomial distribution for large n, such as the one displayed in Figure 4.8, can be approximated by a bell shaped curve called a *normal probability density function*. The theoretical justification of this approximation is called the *central limit theorem* and is beyond the scope of this book.

Definition 4.4 *A random variable X with cumulative distribution function given by*

$$F(x; \mu, \sigma) = \int_{-\infty}^{x} \frac{1}{\sigma\sqrt{2\pi}} e^{-(t-\mu)^2/2\sigma^2} \, dt \tag{4.34}$$

is said to have a normal distribution *with mean μ and variance σ^2. It is convenient to use the symbol $X \overset{D}{=} N(\mu, \sigma^2)$ to say that X is a normally distributed random variable with mean μ and variance σ^2. (The fact that $E(X) = \mu$, $V(X) = \sigma^2$ will be demonstrated later in this section.)*

Its probability density function $f(x; \mu, \sigma) = F'(x; \mu, \sigma)$ is the integrand in Equation 4.34; that is

$$f(x; \mu, \sigma) = \frac{1}{\sigma\sqrt{2\pi}} e^{-(x-\mu)^2/2\sigma^2}; \; -\infty < x < \infty, \tag{4.35}$$

where

$$-\infty < \mu < \infty; 0 < \sigma < \infty.$$

Notation More generally, we use the symbol $X \overset{D}{=} Y$ to mean that X and Y have the same distribution.

Note: In the physics and engineering literature the normal distribution is also called the *Gaussian distribution*.

To prove that

$$\int_{-\infty}^{\infty} \frac{1}{\sigma\sqrt{2\pi}} e^{-(x-\mu)^2/2\sigma^2} \, dx = 1 \tag{4.36}$$

requires techniques from advanced calculus and so is omitted.

The graphs of two normal curves with the same mean ($\mu_1 = \mu_2 = 3$) but different variances ($\sigma_1 = 2$ and $\sigma_2 = 0.5$) are displayed in Figure 4.5.

The graphs of two normal curves with different means ($\mu_1 = 1.5 < 2.5 = \mu_2$) but the same variance ($\sigma_1 = \sigma_2 = 0.5$) are displayed in Figure 4.6.

Figures 4.5 and 4.6 give us some insight into the significance of the parameters μ and σ.

1. The maximum value of $f(x; \mu, \sigma)$ occurs at $x = \mu$ and its graph is bell shaped and *symmetric* about μ.

2. Looking at Figure 4.5 we see the *spread* of the distribution is determined by σ. A small value of σ produces a sharp peak at $x = \mu$; consequently, most of the area under this normal curve is close to μ. A large value of σ, on the other hand, produces a smaller, more rounded bulge at $x = \mu$. The area under this normal curve is less concentrated about μ.

3. Table 4.1 gives the value of the probability $P(\mu - k\sigma < X < \mu + k\sigma)$ for $k = 1, 2, 3$.

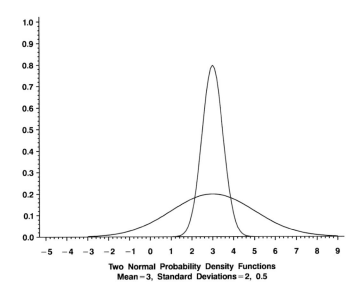

FIGURE 4.5: Two normal pdfs with the same mean ($\mu_1 = \mu_2 = 3$) but different variances ($\sigma_1 = 2$ and $\sigma_2 = 0.5$)

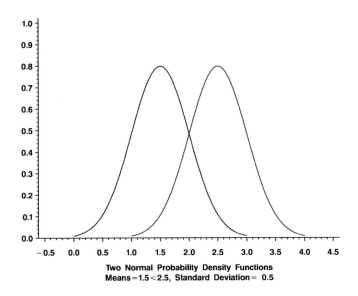

FIGURE 4.6: Two normal pdfs with different means ($\mu_1 = 1.5 < 2.5 = \mu_2$) but the same variance ($\sigma_1 = \sigma_2 = 0.5$)

Table 4.1 *Probabilities of intervals determined by the normal distribution*

k	$P(\mu - k\sigma < X < \mu + k\sigma)$
1	0.6826
2	0.9544
3	0.9974.

Note that most of the area under the normal curve is concentrated on the interval $[\mu - 3\sigma, \mu + 3\sigma]$; that is, 99% of the area under the normal curve lies within three units of standard deviation of the mean. Consequently, if σ is small then most of the area is concentrated in a small interval centered at μ; while if σ is large then the area is spread out over a larger interval.

Definition 4.5 *A normally distributed random variable with $\mu = 0$ and $\sigma = 1$ is said to have the* standard normal distribution; *it is denoted by the letter Z.*

The pdf of Z is given by:

$$f(x) = \frac{1}{\sqrt{2\pi}} e^{-x^2/2}; \quad -\infty < x < \infty. \tag{4.37}$$

Its cdf $\Phi(z)$ is the definite integral

$$\Phi(z) = \int_{-\infty}^{z} \frac{1}{\sqrt{2\pi}} e^{-x^2/2} \, dx. \tag{4.38}$$

We compute $P(Z \leq z) = \Phi(z)$ by using Table A.3 in the appendix.

Example 4.11

Use Table A.3 in the appendix to compute:
(1) $P(Z \leq 1.5)$; (2) $P(Z \leq -1.5)$; (3) $P(-1.5 < Z < 1.5)$.
   Solution

1. $P(Z \leq 1.5) = 0.9932$;

2. $P(Z \leq -1.5) = 0.0668$;

3. $P(-1.5 < Z < 1.5) = \Phi(1.5) - \Phi(-1.5) = 0.9932 - 0.0668 = 0.9264.$

   We now show that the mean $E(Z) = 0$ and variance $V(Z) = 1$; this is the content of the next proposition.

Proposition 4.3 *Let Z be a standard normal random variable; then,*

$$E(Z) = 0 \text{ and } V(Z) = 1.$$

Proof The proposition is proved by evaluating the following convergent improper integrals:

$$E(Z) = \int_{-\infty}^{\infty} \frac{1}{\sqrt{2\pi}} x e^{-x^2/2} \, dx = 0, \tag{4.39}$$

$$E(Z^2) = \int_{-\infty}^{\infty} \frac{1}{\sqrt{2\pi}} x^2 e^{-x^2/2} \, dx = \int_{-\infty}^{\infty} \frac{1}{\sqrt{2\pi}} x d(-e^{-x^2/2})$$

$$= \int_{-\infty}^{\infty} \frac{1}{\sqrt{2\pi}} e^{-x^2/2} \, dx = 1. \tag{4.40}$$

Some remarks on the calculations: Equation 4.39 follows at once from the fact that the integrand is an *odd* function; Equation 4.40 follows from a routine integration by parts, together with the fact that

$$\lim_{x \to \pm\infty} x e^{-x^2/2} = 0.$$

In many statistical applications it is necessary to compute $P(Z \geq z) = 1 - \Phi(z)$. We call this function the *tail* of the normal distribution. The *critical values* $z(\alpha)$ of the normal distribution are given by

$$P(Z > z(\alpha)) = \alpha. \tag{4.41}$$

Example 4.12

Compute the following critical values: $z(0.1)$, $z(0.05)$, $z(0.025)$.
   Solution Refer to Table A.3 in the appendix and look for the entry in the table that is closest to 0.1, which appears to be 0.8997. Consequently, $z(0.1) = 1.28$. Similarly, $z(0.05) = 1.645$ and $z(0.025) = 1.96$.

Standardizing a normal random variable

To compute the distribution function of a normal random variable X with mean μ and variance σ^2 we rescale it into a standard normal random variable via the transformation $Z = (X - \mu)/\sigma$. We can also use the transformation $X = \mu + \sigma Z$, $(\sigma > 0)$ to obtain a normal random variable with mean μ and variance σ^2 from the standard one. (The proof that $X = \mu + \sigma Z \stackrel{\mathcal{D}}{=} N(\mu; \sigma^2)$ is sketched in Proposition 4.4 below.) This yields the following formula for the distribution function of X.

$$\text{Let } X \stackrel{\mathcal{D}}{=} N(\mu; \sigma^2) \text{ then } P(X \leq x) = \Phi\left(\frac{x - \mu}{\sigma}\right), \tag{4.42}$$

where $\Phi(z)$ is the standard normal distribution whose values are listed Table A.3. Before deriving Equation 4.42 we illustrate its use in the next example.

Example 4.13 *Given that X is $N(25; 16)$ compute the following probabilities: (1) $P(X \leq 20)$, (2) $P(X > 33)$.*

   Solution

$$(1) \ P(X \leq 20) = \Phi\left(\frac{20 - 25}{4}\right) = \Phi(-1.25) = 0.1056$$

$$(2) \ P(X > 33) = 1 - P(X \leq 33) = 1 - \Phi\left(\frac{33 - 25}{4}\right) = 1 - \Phi(2.0) = 0.0228$$

Equation 4.42 for the distribution function of a normal random variable is a consequence of the next proposition, which is of independent interest.

Proposition 4.4 *The random variable $X = \mu + \sigma Z$, $(\sigma > 0)$, where $Z \stackrel{\mathcal{D}}{=} N(0,1)$, has a normal distribution with mean value μ and variance σ^2. Conversely, if $X \stackrel{\mathcal{D}}{=} N(\mu, \sigma^2)$ then the standardized random variable $(X - \mu)/\sigma$ has a standard normal distribution; that is*

$$\frac{X - \mu}{\sigma} \stackrel{\mathcal{D}}{=} N(0; 1). \tag{4.43}$$

Proof We begin with the observation that the event $(X \leq x)$ can be expressed in terms of an equivalent event involving Z; more precisely, we have the following relationship between X and Z:

$$(X \leq x) = (\mu + \sigma Z \leq x)$$
$$= \left(Z \leq \frac{x - \mu}{\sigma} \right).$$

Consequently,

$$P(X \leq x) = P(\mu + \sigma Z \leq x)$$
$$= P \left(Z \leq \frac{x - \mu}{\sigma} \right)$$
$$= \Phi \left(\frac{x - \mu}{\sigma} \right).$$

We leave it to the reader to derive Equation 4.43 and to show that

$$\frac{d}{dx} \Phi \left(\frac{x - \mu}{\sigma} \right) = \frac{1}{\sigma \sqrt{2\pi}} e^{-(x-\mu)^2/2\sigma^2}.$$

The representation $X = \mu + \sigma Z$ also implies that the parameters μ and σ^2 are the mean and variance of X. The details follow.

Corollary 4.1 *If X is $N(\mu, \sigma^2)$ distributed, then*

$$E(X) = \mu \ and \ V(X) = \sigma^2.$$

Proof Using the representation $X = \mu + \sigma Z$ we see that:

$$E(X) = E(\mu + \sigma Z) = E(\mu) + E(\sigma Z)$$
$$= \mu + \sigma E(Z) = \mu; \ \text{similarly,}$$
$$V(X) = V(\mu + \sigma Z)$$
$$= V(\sigma Z) = \sigma^2 V(Z) = \sigma^2.$$

Example 4.14

Graph the pdf of (i) $N(1; (1.5)^2)$ and (ii) the pdf of $(X - 1)/1.5$.
  Solution The pdf of $N(1; (1.5)^2)$ is displayed in Figure 4.7.
  The pdf of $(X - 1)/1.5$ is $N(0, 1)$ and is also displayed in Figure 4.7.
  The *critical value* $x(\alpha)$ of a normal random distribution with mean μ and variance σ^2 is defined by the condition that the area under the normal curve to the right of $x(\alpha)$ equals α. That is, $P(X \geq x(\alpha)) = \alpha$. We compute $x(\alpha)$ using the formula

$$x(\alpha) = \mu + \sigma z(\alpha), \ \text{where} \tag{4.44}$$

$z(\alpha)$ is the critical value of the standard normal distribution.
  The derivation of Equation 4.44 is left to the reader as an exercise (see Problem 4.38).

Moment generating function of a normal random variable

Theorem 4.2 *The moment generating function of a normal random variable X with mean μ and variance σ^2 is*

$$M_X(s) = e^{s\mu + s^2\sigma^2/2}. \tag{4.45}$$

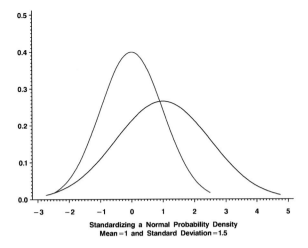

Standardizing a Normal Probability Density
Mean = 1 and Standard Deviation = 1.5

FIGURE 4.7: Standardizing a normal pdf: $\mu = 1$ and $\sigma = 1.5$

Equation 4.45, with $\mu = 0$, $\sigma = 1$ states that the mgf for the standard normal random variable is given by

$$M_Z(s) = e^{s^2/2}. \tag{4.46}$$

Conversely, Equation 4.46 implies Equation 4.45 as will now be shown. Using the fact that $X = \mu + \sigma Z$ it follows that

$$E(e^{sX}) = E(e^{s(\mu + \sigma Z)}).$$

Consequently,

$$E(e^{s(\mu+\sigma Z)}) = E(e^{s\mu}e^{s\sigma Z}) = e^{s\mu}E(e^{s\sigma Z})$$
$$= e^{s\mu}e^{(s\sigma)^2/2} = e^{s\mu + s^2\sigma^2/2}.$$

Therefore it suffices to derive Equation 4.46, which we now proceed to do:

$$E(e^{sZ}) = \int_{-\infty}^{\infty} e^{sx} \frac{1}{\sqrt{2\pi}} e^{-x^2/2} dx$$
$$= \int_{-\infty}^{\infty} \frac{1}{\sqrt{2\pi}} e^{-((x-s)^2 - s^2)/2} dx$$
$$= e^{s^2/2} \int_{-\infty}^{\infty} \frac{1}{\sqrt{2\pi}} e^{-(x-s)^2/2} dx = e^{s^2/2}.$$

4.6   The Lognormal Distribution: A Model for the Distribution of Stock Prices

Students interested in the fundamentals of financial engineering should familiarize themselves with the *lognormal model for stock prices*. An unusual feature of this model is that it is the logarithm of the stock price—not the stock price itself—that is normally distributed. We will show later that the lognormal model for stock prices is a consequence of the efficient market hypothesis discussed in Section 5.5 and the central limit theorem discussed in Section 5.4.2.

Definition 4.6 *A non-negative random variable W has a lognormal distribution with parameters μ, σ^2 if $\ln(W) \overset{\mathcal{D}}{=} N(\mu, \sigma^2)$.*

It can be shown, using our formulas for the moment generating function for a normally distributed random variable given in Equation 4.45, that the mean, second moment, and variance of W are

$$E(W) = e^{(\mu + \sigma^2/2)}; \; E(W^2) = e^{(2\mu + 2\sigma^2)}; \; V(W) = e^{(2\mu + \sigma^2)}(e^{\sigma^2} - 1). \qquad (4.47)$$

The application of the lognormal distribution to stock market prices is straightforward: Let S_0 denote the initial price of a stock and $S(t)$ its price at a future time t, where the unit of time is one year; then $\ln(S(t)/S_0) \overset{\mathcal{D}}{=} N(t\mu, t\sigma^2)$. That is, $S(t)/S_0$ has a lognormal distribution, in the sense of Definition 4.6, with parameters $t\mu$, $t\sigma^2$, respectively. It now follows from Proposition 4.4 that $\ln(S(t)/S_0) \overset{\mathcal{D}}{=} N(t\mu, t\sigma^2)$; consequently,

$$E(\ln((S(t)/S_0)) = t\mu, \qquad (4.48)$$
$$V(\ln((S(t)/S_0)) = t\sigma^2. \qquad (4.49)$$

(Note: In mathematical finance μ is the (expected) continuously compounded rate of return and σ is called the *volatility*; it is a measure of how likely the observed return deviates from the expected return.) Applying Equation 4.47 to the special case $W = S(t)/S_0$, and substituting $t\mu$ for μ and $t\sigma^2$ for σ^2, yields the following expressions for the mean and variance of $S(t)/S_0$:

$$E(S(t)/S_0) = e^{(t(\mu + \sigma^2/2))}; \; V(S(t)/S_0) = e^{t(2\mu + \sigma^2)}(e^{t\sigma^2} - 1). \qquad (4.50)$$

Before proceeding further it will be helpful for the reader to carefully study the following worked out numerical example.

Example 4.15

Suppose that a stock price $S(t)$, t measured in years, has a lognormal distribution with parameters $\mu = 0.10$ and $\sigma = 0.20$, with initial price $S_0 = \$50$.

- Evaluate the four quantities: $E(\ln(S(1)))$, $V(\ln(S(1)))$, $E((S(1)))$, $\sigma(S(1))$.
  Solution Use Equation 4.50, with $\mu = 0.10$, $\sigma = 0.20$, and $t = 1$; in detail,

  $$E(\ln(S(1)/50) = \mu = 0.10, \text{ therefore}$$
  $$E(\ln(S(1)) = \ln(50) + 0.10 = 4.0120$$

  $$V(\ln(S(1)/50) = (0.20)^2 = 0.04$$

  $$E((S(1)/50) = e^{(0.10 + (0.20)^2/2)} = e^{0.12} = 1.1275, \text{ so}$$
  $$E((S(1)) = 50 \times 1.1275 = 56.37$$

  $$\sigma((S(1)) = \sqrt{50^2 \times e^{(0.2 + (0.2)^2)}(e^{0.04} - 1)} = \sqrt{129.70} = 11.39$$

- Show that the probability that the stock's price will be up at least 30% after one year is 0.2090.

Solution An increase of at least 30% means that $S(1)/S_0 > 1.3$, equivalently, $\ln(S(1)/S_0) > \ln 1.3 = 0.2624$. By hypothesis $\ln(S(1)/S_0)$ has a normal distribution with parameters $\mu = 0.10$ and $\sigma = 0.20$, therefore

$$P\left(\ln(S(1)/S_0) > \ln(1.3)\right) = P\left(\ln(S(1)/S_0) > 0.2624\right)$$
$$= P\left(\frac{\ln(S(1)/S_0) - 0.10}{0.20} > \frac{0.2624 - 0.10}{0.20}\right)$$
$$= P(Z > 0.8120) = 0.2090.$$

- Show that the probability that the stock's price will be down at least 20% after one year is 0.0526.

Solution A decrease of at least 20% means that $S(1)/S_0 < 0.8$, equivalently, $\ln(S(1)/S_0) < \ln 0.8 = -0.2231$. Proceeding as in the previous case we see that

$$P\left(\ln(S(1)/S_0) < \ln(0.8)\right) = P\left(\ln(S(1)/S_0) < -0.2231\right)$$
$$= P\left(\frac{\ln(S(1)/S_0) - 0.10}{0.20} > \frac{-0.2231 - 0.10}{0.20}\right)$$
$$= P(Z < -1.62) = 0.0526.$$

4.7   The Normal Approximation to the Binomial Distribution

The problem of computing probabilities associated with the binomial distribution $b(x; n; p)$ becomes tedious, if not practically impossible, when n is large, $n > 25$ say. (Table A.1, for instance, lists only the binomial probabilities for $n = 5, 10, 15, 20, 25$.) How do we compute the distribution function of a binomial distribution when $n = 50$, or $n = 1000$? To compute these probabilities we turn to the most basic theorem in probability theory: The normal approximation to the binomial distribution is itself a special case of a more remarkable result known as the central limit theorem, which will be discussed elsewhere in this book (Section 5.4.2).

We denote by S_n the binomially distributed random variable with parameters n, p whose distribution function we wish to approximate. This is to emphasize the fact that its distribution depends on n. When n is large it is not an easy matter to compute accurately and quickly a probability of the form

$$P(\alpha \leq S_n \leq \beta) = \sum_{x=\alpha}^{\beta} b(x; n, p).$$

The normal approximation to the binomial states that for large n the probability histogram of the binomial is well approximated by the normal curve with $\mu = np$ and $\sigma^2 = np(1-p)$. Figure 4.8 is the graph of the normal pdf with $\mu = 5$, $\sigma^2 = 2.5$ superimposed on the probability histogram of the binomial distribution with parameters $n = 10$, $p = 0.5$ (so $np = 5$, $np(1-p) = 2.5$). The most striking feature of the histogram is its similarity to the graph of a normal pdf $\frac{1}{\sigma\sqrt{2\pi}} e^{-(x-\mu)^2/2\sigma^2}$.

Informal Version of the Normal Approximation to the Binomial

When the random variable S_n has a binomial distribution the Figure 4.8 suggests that the probability $P(\alpha \leq S_n \leq \beta)$ can be approximated by computing the $P(\alpha \leq X \leq \beta)$,

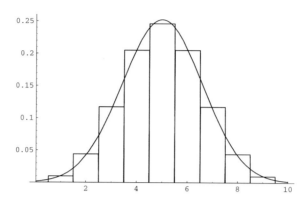

FIGURE 4.8:   Normal approximation to the binomial: $n = 10$, $p = 0.5$

where X is normally distributed with the same mean np and the same variance $np(1-p)$. More precisely, the normal approximation to the binomial asserts that for n large (one rule of thumb requires that $np \geq 5$ and $n(1-p) \geq 5$) the distribution of S_n is well approximated by a normal random variable with $\mu = np$ and $\sigma^2 = np(1-p)$.

This is the content of a famous theorem, due to the French mathematicians DeMoivre and Laplace, and called the *normal approximation to the binomial distribution*. The formula for the normal approximation is given by Equation 4.51:

$$P(\alpha \leq S_n \leq \beta) \approx P(\alpha \leq X \leq \beta)$$

$$= P\left(\frac{\alpha - np}{\sqrt{np(1-p)}} \leq \frac{X - np}{\sqrt{np(1-p)}} \leq \frac{\beta - np}{\sqrt{np(1-p)}}\right);$$

consequently,

$$P(\alpha \leq S_n \leq \beta) \approx \Phi\left(\frac{\beta - np}{\sqrt{np(1-p)}}\right) - \Phi\left(\frac{\alpha - np}{\sqrt{np(1-p)}}\right). \tag{4.51}$$

The significance of this theorem is that it enables us to reduce the tedious task of calculating binomial sums of the form

$$P(\alpha \leq S_n \leq \beta) = \sum_{x=\alpha}^{\beta} b(x; n, p)$$

to evaluating a corresponding area under the standard normal curve.

Example 4.16 *The random variable X has a binomial distribution with parameters $n = 10$, $p = 0.5$. Compute: (1) the exact value of $P(4 \leq X \leq 8)$ and (2) the normal approximation to $P(4 \leq X \leq 8)$.*

Solution

1. The exact answer is obtained from Table A.1 in the appendix:

$$P(4 \leq X \leq 8) = 0.989 - 0.172 = 0.817.$$

2. We obtain the normal approximation by computing the area under the approximating normal curve from $4 \leq x \leq 8$ (see Figure 4.8):

$$P(4 \leq X \leq 8) = P\left(\frac{4-5}{\sqrt{2.5}} \leq \frac{X-5}{\sqrt{2.5}} \leq \frac{8-5}{\sqrt{2.5}}\right)$$
$$\approx P(-0.63 \leq Z \leq 1.90)$$
$$= 0.9713 - 0.2643 = 0.7070 \text{ (normal approximation)}.$$

Example 4.17 *Let X denote the number of heads that appear when a fair coin is tossed 100 times. Use the normal approximation to estimate $P(X > 60)$ and $P(45 < X < 60)$.*

Solution X has a binomial distribution with parameters $n = 100$, $p = 0.5$. In particular $np = 50$, $\sqrt{np(1-p)} = 5$. Using the normal approximation we obtain

$$P(X > 60) = P(60 < X \leq 100) \approx \Phi\left(\frac{100-50}{5}\right) - \Phi\left(\frac{60-50}{5}\right)$$
$$= \Phi(5) - \Phi(2) = 1 - 0.9772 = 0.0228.$$

We estimate $P(45 < X < 60)$ in a similar way; thus,

$$P(45 \leq X \leq 60) \approx \Phi\left(\frac{60-50}{5}\right) - \Phi\left(\frac{45-50}{5}\right)$$
$$= \Phi(2) - \Phi(-1) = 0.9972 - 0.1587 = 0.8385.$$

Example 4.18 *Clinical trial of a new drug*

The natural recovery rate for a certain disease is 30%. A new drug is developed and the manufacturer claims that it boosts the recovery rate to 50%. Two hundred patients are given the drug. What is the probability that at least 80 patients recover if: (1) the drug is worthless? (2) It is really 50% effective as claimed?

Solution

1. If the drug is worthless then the the number of recoveries X is binomially distributed with parameters $n = 200$ and $p = 0.3$. Thus $np = 60$ and $\sqrt{np(1-p)} = \sqrt{60 \times 0.7} = \sqrt{42} = 6.48$. Using the normal approximation to the binomial we see that

$$P(80 \leq X \leq 200) \approx \Phi(21.6) - \Phi(3.09) = 1 - 0.999 = 0.001.$$

In other words, if we gave the drug to 200 patients and more than 80 recovered it would be most unlikely that the drug was completely ineffective.

2. If the drug is really as effective as claimed then X is binomially distributed with parameters $n = 200$ and $p = 0.5$. Thus, $np = 100$ and $\sqrt{np(1-p)} = \sqrt{50} = 7.07$. Using the normal approximation to the binomial once again we see that

$$P(80 \leq X \leq 200) \approx \Phi\left(\frac{100}{7.07}\right) - \Phi\left(\frac{-20}{7.07}\right) = 1 - 0.0023 = 0.9977.$$

Formal Statement of the Normal Approximation to the Binomial

Mathematical statisticians prefer to state the normal approximation to the binomial in the form of a limit theorem for the corresponding distributions as in Equation 4.52.

Theorem 4.3 *Let S_n denote the binomially distributed random variable with parameters n, p. Then*

$$\lim_{n \to \infty} P\left(\frac{S_n - np}{\sqrt{np(1-p)}} \le z \right) = \int_{-\infty}^{z} \frac{1}{\sqrt{2\pi}} e^{-x^2/2}\, dx. \qquad (4.52)$$

Although the proof of this important result is beyond the scope of this book its intuitive content and implications, as we have seen, are not.

The Continuity Correction The accuracy of the normal approximation is improved by using the *continuity correction:*

$$P(\alpha \le S_n \le \beta) \approx P(\alpha - 0.5 \le X \le \beta + 0.5)$$

$$= P\left(\frac{\alpha - np - 0.5}{\sqrt{np(1-p)}} \le \frac{X - np}{\sqrt{np(1-p)}} \le \frac{\beta - np + 0.5}{\sqrt{np(1-p)}} \right)$$

$$= \Phi\left(\frac{\beta - np + 0.5}{\sqrt{np(1-p)}} \right) - \Phi\left(\frac{\alpha - np - 0.5}{\sqrt{np(1-p)}} \right). \qquad (4.53)$$

Example 4.19 *The random variable Y has a binomial distribution with $n = 10$, $p = 0.5$, $np = 10$, $npq = 2.5$. Compute the value of $P(4 \le Y \le 8)$ using: (1) the binomial tables; (2) the normal approximation with the continuity correction; (3) the normal approximation without the continuity correction.*

Solution

(1) $P(4 \le Y \le 8) = 0.989 - 0.172 = 0.817$ (exact answer);

(2) $P(4 \le Y \le 8) \approx P\left(\frac{-1.5}{\sqrt{2.5}} \le Z \le \frac{3.5}{\sqrt{2.5}} \right)$

$$= P(-0.95 \le Z \le 2.21) = 0.9864 - 0.1711$$

$$= 0.8153 \text{ (with continuity correction)};$$

(3) $P(4 \le Y \le 8) \approx P(-0.63 \le Z \le 1.90)$

$$= 0.7070 \text{ (without continuity correction)}.$$

The continuity correction is graphically displayed in Figure 4.9. Note that for large n the effect of the correction term $0.5/\sqrt{n}$ becomes so small that it can be safely ignored; in the present case, however, the continuity correction significantly improves the accuracy of the normal approximation.

Example 4.20 *Analysis of Deming's red bead experiment*

We now apply the normal approximation to analyze and interpret the results of Deming's red bead experiment described earlier in Chapter 1 (Example 1.3).

Calculation of the limits of variation attributable to the system

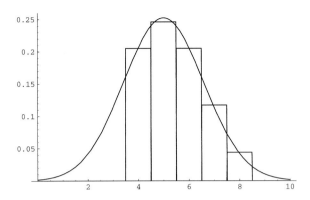

FIGURE 4.9: Graph illustrating the continuity correction of Example 4.19

Following Deming we now calculate how much of the observed variation is attributable to the system. Let X denote the number of red beads produced by a worker. The exact distribution of X is hypergeometric with

$$n = 50, \ N = 4000, \ D = 800, \ p = \frac{800}{4000} = 0.20.$$

Note that

$$\frac{n}{N} = \frac{50}{4000} = 0.0125 < 0.05,$$

consequently we can apply the binomial approximation to the hypergeometric and assume that X is $B(x; 50, p)$ distributed. Applying the normal approximation to the binomial, with $n = 50$, yields

$$P(50p - 3\sqrt{50p(1-p)} \le X \le 50p + 3\sqrt{50p(1-p)})$$
$$\approx \Phi(3) - \Phi(-3) = 0.9974.$$

Deming calls the quantities U, L defined by

$$L = 50p - 3\sqrt{50p(1-p)}, \ U = 50p + 3\sqrt{50p(1-p)}$$

the *limits of variation attributable to the system.* Since management does not know the true proportion of red beads in the mixture it uses the sample proportion denoted $\hat{p}$, which equals $51/300 = 0.17 = \hat{p}$. The estimated limits of variation attributed to the system are given by

$$L = 50\hat{p} - 3\sqrt{50\hat{p}(1-\hat{p})} = 8.5 - 7.97 = 0.53;$$
$$U = 50\hat{p} + 3\sqrt{50\hat{p}(1-\hat{p})} = 8.5 + 7.97 = 16.47.$$

Let us now apply these theoretical results to the actual data from Deming's red bead experiment. Looking at Table 1.2 we see that the observed variation in the workers' performance can be accounted for by chance alone; that is, with probability 0.99, the number of red beads produced by each worker lies within the interval $[0.53, 16.47]$.

Example 4.21

Overbooking occurs whenever the number of tickets sold by an airline exceeds the seating capacity of an aircraft. This is done to protect the airline from no shows. This practice is, unfortunately, not limited to airline reservations. Assume that a college can admit at most 820 freshmen. Assume that it sends out 1600 acceptances and that each student comes to the college with probability 0.5.

1. What is the probability that the school ends up with more students than it can accommodate?

   Solution Let T denote the total number of freshmen who decide to enroll. Clearly, T is binomially distributed with parameters $n = 1,600$ and $p = 0.5$. Consequently,

   $$E(T) = np = 800 \text{ and } \sqrt{V(T)} = \sqrt{np(1-p)} = \sqrt{400} = 20.$$

   The college ends up with more students than it can accommodate if $T \geq 821$; but, according to the normal approximation, $T \approx N(800, 400)$ and therefore

   $$P(821 \leq T < \infty) = P\left(\frac{821 - 800}{20} \leq \frac{T - E(T)}{\sqrt{V(T)}} < \infty\right)$$
   $$= P\left(\frac{21}{20} \leq Z < \infty\right) = 0.15.$$

2. Find the largest number n that they could accept if they wanted the probability of getting more students than they can accommodate to be less than 0.05.

   Solution The condition that $P(T \geq 821) = 0.05$ yields a quadratic equation for $\sqrt{n}$ that is derived by using the normal approximation. In detail:

   $$0.05 = P(T \geq 821) = P\left(\frac{T - (n/2)}{\sqrt{n}/2} \geq \frac{821 - (n/2)}{\sqrt{n}/2}\right).$$

   Consequently,

   $$\frac{821 - (n/2)}{\sqrt{n}/2} = 1.645.$$

   Setting $x = \sqrt{n}$ yields the following equation for x:

   $$\frac{821 - (x^2/2)}{x/2} = 1.645.$$

   Rearranging terms yields the quadratic equation:

   $$\frac{x^2}{2} + 0.8225x - 821 = 0.$$

   There are two solutions only one of which is meaningful, namely

   $$x = 39.7074 \text{ which implies that } n = x^2 = 1,576.$$

4.7.1 Distribution of the Sample Proportion $\hat{p}$

For the purposes of statistical inference it is more convenient to apply the normal approximation to the *sample proportion* $\hat{p} = S_n/n$. For example, a public opinion survey, such as the Gallup poll, usually reports the approval rating of a politician as a sample proportion. In this context, the normal approximation is a very useful tool for studying the fluctuations of $\hat{p}$ about its true value p; it is particularly useful for constructing a *confidence interval* for the unknown proportion p. A confidence interval is a measure of the accuracy, or reliability, of $\hat{p}$ as an estimator of p, and will be discussed in detail in Chapter 7.

To apply the normal approximation to $\hat{p} = S_n/n$ we begin by dividing the numerator and denominator of $\frac{S_n - np}{\sqrt{np(1-p)}}$ by n. This yields the normal approximation for the sample proportion $\hat{p}$:

$$\lim_{n\to\infty} \frac{S_n - np}{\sqrt{np(1-p)}} = \lim_{n\to\infty} \frac{S_n/n - p}{\sqrt{p(1-p)/n}}$$

$$= \lim_{n\to\infty} \frac{\hat{p} - p}{\sqrt{p(1-p)/n}} \overset{\mathcal{D}}{=} N(0,1). \tag{4.54}$$

Example 4.22 *Analysis of a public opinion poll*

Suppose the current approval rating of a politician is 52%. How likely is it that a poll based on a sample of size $n = 2{,}000$ would show that her approval rating would be less than 50%?

Solution We are interested in computing the probability $P(\hat{p} < 0.5)$ under the assumption that $p = 0.52$. Using the normal approximation as expressed in Equation 4.54 we proceed as follows:

$$P(\hat{p} < 0.5) = P\left(\frac{\hat{p} - 0.52}{\sqrt{0.52 \times 0.48/2{,}000}} < \frac{0.5 - 0.52}{0.0112} \right)$$

$$\approx P(Z < -1.79) = 0.0367.$$

4.8 Other Important Continuous Distributions

Although the most important distribution is the normal, there are several other distributions, such as the *gamma, Weibull*, and *beta* distributions, that are particularly useful in modeling a wide varety of phenomena in science and engineering.

4.8.1 The Gamma and Chi-Square Distributions

The *gamma distribution* is used to model all kinds of phenomena in statistics, engineering, and science. For example, the exponential, the chi-square, and Erlang distributions are themselves gamma distributions (we will treat these special cases in more detail later in Sections 6.3 and 5.10.1.

Definition 4.7 *A random variable X with pdf given by*

$$g(x; \alpha, \beta) = \left(\frac{1}{\Gamma(\alpha)\beta^{\alpha}} \right) x^{\alpha-1} e^{-x/\beta}; \ x > 0; \tag{4.55}$$

$$g(x; \alpha, \beta) = 0, \ x \le 0,$$

$$\text{where } 0 < \alpha < \infty, \ 0 < \beta < \infty,$$

is said to have a gamma distribution with parameters α and β.

The symbol $\Gamma(\alpha, \beta)$ denotes the gamma distribution with parameters α and β. The following additional points should be noted.

1. The constant $\Gamma(\alpha)\beta^{\alpha}$ is used to make $g(x; \alpha, \beta)$ a probability density function. The function $\Gamma(\alpha)$ is also of independent interest. It is called the *gamma function* and is defined as follows.

Definition 4.8 *The gamma function $\Gamma(\alpha)$ is defined by the convergent improper integral*

$$\Gamma(\alpha) = \int_0^{\infty} y^{\alpha-1} e^{-y} \, dy; \ \alpha > 0. \tag{4.56}$$

2. Properties of the gamma function.

$$\Gamma(1) = \int_0^{\infty} e^{-y} \, dy = 1; \tag{4.57}$$

$$\text{Recurrence relation: } \Gamma(\alpha) = (\alpha - 1)\Gamma(\alpha - 1). \tag{4.58}$$

The recurrence relation 4.58 is obtained by an integration by parts. The details are left to the reader as Problem 4.54. Except for special cases, computing $\Gamma(x)$ requires a computer; a noteworthy exception is the following explicit formulas for $\Gamma(n)$, n a positive integer greater than one, and $\Gamma(1/2)$:

$$\Gamma(2) = 1\Gamma(1) = 1;$$
$$\Gamma(3) = 2\Gamma(2) = 2 = 2!;$$
$$\Gamma(4) = 3\Gamma(3) = 3 \times 2 = 3!;$$

and, in general:

$$\Gamma(n) = (n-1)\Gamma(n-1) = (n-1)! \tag{4.59}$$
$$\Gamma(1/2) = \sqrt{\pi}. \tag{4.60}$$

The derivation of Equation 4.60 is left to the reader as Problem 4.53.

3. The gamma distribution is obtained by manipulating the integral defining the gamma function. Make the change of variable $y = x/\beta$, $\beta > 0$ in the integral 4.56, which is thereby transformed into

$$\Gamma(\alpha) = \int_0^{\infty} \left(\frac{x}{\beta} \right)^{\alpha-1} e^{-x/\beta} \left(\frac{1}{\beta} \right) dx.$$

Combining the the terms involving β in the denominator and dividing through by $\Gamma(\alpha)$ we obtain

$$\int_0^{\infty} \left(\frac{1}{\Gamma(\alpha)\beta^{\alpha}} \right) x^{\alpha-1} e^{-x/\beta} \, dx = 1.$$

It follows that the fucntion $g(x; \alpha, \beta)$ defined in Equation 4.55 is a probability density function.

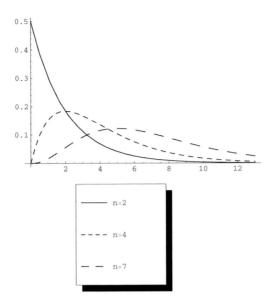

FIGURE 4.10:   The graphs of χ_n^2 for $n = 2, 4, 7$

4. Finally, note that a gamma random variable X is always non-negative, that is $P(X \geq 0) = 1$.

Except for certain special cases, such as $\alpha = 1$, we need a computer or a table to compute the values of the gamma distribution defined by

$$G(x; \alpha, \beta) = \int_0^x g(t; \alpha, \beta)\, dt.$$

Proposition 4.5 *The mean and variance of the gamma distribution with parameters α and β are given by*

$$E(X) = \alpha\beta \ and \ V(X) = \alpha\beta^2. \tag{4.61}$$

The proof is via a straightforward integration that we leave to the student as Problem 4.55.

By choosing special values for the parameters α, β one obtains many distributions that are widely used in statistics and probability theory; here are some examples:

1. *The exponential distribution* with parameter λ is obtained by setting $\alpha = 1$ and $\beta = \lambda^{-1}$.

2. *The chi-square distribution* with n degrees of freedom is a special case of the gamma distribution with $\alpha = n/2$, where n is a positive integer, and $\beta = 2$. A random variable with a chi-square distribution with n degrees of freedom is denoted by χ_n^2. The chi-square distribution plays an important role in statistical inference, particularly in K. Pearson's chi-square goodness of fit test.

   The graphs of χ_n^2 for $n = 2, 4, 7$ are displayed in Figure 4.10.

3. *The Erlang distribution* is a gamma distribution with parameters $\alpha = n$, $\beta = \lambda^{-1}$. It plays an important role in queuing theory (Section 5.10.1).

4.8.2   The Weibull Distribution

The *Weibull distribution* is widely used by engineers for modeling the lifetimes of a component. The non-negative random variable T with distribution function

$$F(t) = 1 - e^{-(t/\beta)^\alpha}, \ t \geq 0; \ F(t) = 0, \ t \leq 0, \tag{4.62}$$

is said to have a Weibull distribution with *shape parameter* $\alpha > 0$ and *scale parameter* $\beta > 0$. Observe that the exponential distribution is obtained as a special case by setting $\alpha = 1$, $\beta = 1/\theta$. The mean and variance of the Weibull distribution are conveniently expressed in terms of the gamma function:

$$E(T) = \beta\Gamma\left(1 + \frac{1}{\alpha}\right); \tag{4.63}$$

$$V(T) = \beta^2\left(\Gamma\left(1 + \frac{2}{\alpha}\right) - \Gamma\left(1 + \frac{1}{\alpha}\right)^2\right). \tag{4.64}$$

Example 4.23

The lifetime, measured in years, of a brand of TV picture tubes has a Weibull distribution with parameters $\alpha = 2$, $\beta = 13$. Compute the proportion of TV tubes that fail on a two-year warranty.

Solution To compute $P(T \leq 2)$ we use Equation 4.62, so

$$P(T \leq 2) = 1 - \exp\left(-(2/13)^2\right) = 0.0234.$$

4.8.3   The Beta Distribution*

The burden of computing the exact sum of binomial probabilities, which can be quite laborious for large n, is made much easier by representing the sum as a single definite integral. The integrand turns out to be a *beta distribution*, which has many other statistical applications, Bayesian estimation for example. The *tail probability* of the binomial distribution, with n fixed, is a function $\pi(p)$ of the parameter p and is defined as follows:

$$\pi(p) = \sum_{k \leq x \leq n} b(x; n, p). \tag{4.65}$$

The function $\pi(p)$ plays a useful role in testing hypotheses about a population proportion and in that context it is called a *power function*. We now derive a formula for the tail probability of the binomial distribution that reveals an unexpected connection between the binomial and the *beta distribution*.

Theorem 4.4 *The tail probability function of a binomial random variable X with parameters n, p is given by the following definite integral:*

$$\pi(p) = P(X \geq k|p) = \int_0^p n\binom{n-1}{k-1}t^{k-1}(1-t)^{n-k}\,dt. \tag{4.66}$$

Theorem 4.4 is an immediate consequence of the fact that

$$\pi'(t) = n\binom{n-1}{k-1}t^{k-1}(1-t)^{n-k}, \ \pi(0) = 0$$

and the fundamental theorem of the calculus. Section 4.10, at the end of the chapter, fills in all the details. The following consequence of Equation 4.66 is useful in hypothesis testing.

Proposition 4.6 *The function $\pi(p)$ is an* increasing *function of p; in particular*

$$\text{if } p_1 \le p_2 \text{ then } \pi(p_1) \le \pi(p_2).$$

Proof The integrand appearing in the right hand side of Equation 4.66 is a non-negative function; therefore, the definite integral, which is just the area under the integrand from 0 to p, increases as p increases.

The Beta Distribution

The integrand appearing on the right hand side of Equation 4.66 is a special case of the beta distribution which we now define. Bring in the *beta function* $B(\alpha, \beta)$ defined by the integral:

$$B(\alpha, \beta) = \int_0^1 x^{\alpha-1}(1-x)^{\beta-1}\, dx. \tag{4.67}$$

The beta function can be expressed in terms of the gamma function via the following formula:

$$B(\alpha, \beta) = \frac{\Gamma(\alpha)\Gamma(\beta)}{\Gamma(\alpha+\beta)}. \tag{4.68}$$

A random variable Y is said to have a *beta distribution* when its density function $f(x)$ has the form:

$$f(x) = \frac{1}{B(\alpha, \beta)} \times x^{\alpha-1}(1-x)^{\beta-1}\,(0 < x < 1,\ \alpha > 0,\ \beta > 0) \tag{4.69}$$

$$f(x) = 0, \text{ elsewhere.}$$

It is easy to see that the integrand in Equation 4.66 is of this form with $\alpha = k$ and $\beta = n - k + 1$. We also use the notation $Beta(\alpha, \beta)$ to denote a beta distribution with parameters α, β. Figure 4.11 displays the probability density functions of three beta distributions: $Beta(1.5; 2)$, $Beta(2, 1.5)$, $Beta(0.5, 2)$. Note that the density of $Beta(0.5, 2)$ defines a convergent improper integral at $x = 0$.

Proposition 4.7 *The mean and variance of the beta distribution with parameters α and β are given by:*

$$E(X) = \frac{\alpha}{\alpha + \beta}; V(X) = \frac{\alpha\beta}{(\alpha + \beta + 1)(\alpha + \beta)^2}. \tag{4.70}$$

We leave the derivation of Equation 4.70 to the reader (Problem 4.67).

4.9 Functions of a Random Variable

We can obtain a normal random variable with mean μ and variance σ^2 from the standard normal random variable via the formula $X = \mu + \sigma Z$. That is the content of Proposition 4.4. The relationship between Z and X can be expressed in the form $X = g(Z)$, where $g(t) = \mu + \sigma t$. We say that X is a *function of the random variable* Z. Conversely, if X is $N(\mu, \sigma^2)$ then the random variable defined by the function $Z = h(X)$ where $h(x) = (x-\mu)/\sigma$ has a standard normal distribution. It is this relationship between Z and X that we used to compute the df of X in terms of the df of Z. Now, it is a remarkable fact that most random variables, including many of those discussed in this text, can be represented as

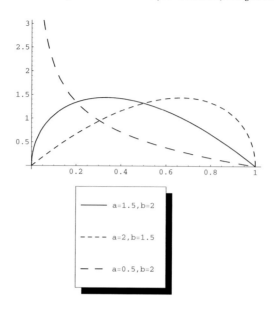

FIGURE 4.11:   The pdfs of three beta distributions

a function $g(U)$ of a uniformly distributed random variable U as will now be explained. This is very useful since most computers, including some pocket calculators, have a *random number generator* whose output u can be viewed as the value $U(s) = u$, where U denotes a random variable that is uniformly distributed over the interval $[0, 1]$.

Example 4.24 *Let U be uniformly distributed over the interval $[0, 1]$ and set $X = -\ln U$. Show that X is exponentially distributed with parameter $\theta = 1$.*

Solution In this case $X = g(U)$, where $g(u) = -\ln u$. Since $0 < U < 1$ it follows that $\ln 0 < \ln U < \ln 1$, so $-\infty < \ln U < 0$; therefore $0 < -\ln U < \infty$. Thus, the range of $-\ln U$ is $(0, \infty)$. Moreover,

$$(X \le x) = (-\ln U \le x) = (e^{-\ln U} \le e^x) = \left(\frac{1}{U} \le e^x\right) = (U \ge e^{-x}).$$

$$\text{Therefore, } P(X \le x) = P(U \ge e^{-x}) = 1 - e^{-x}.$$

Example 4.25 *Compute the distribution of $X = U^2$, where U is uniformly distributed on the interval $[0, 1]$.*

Solution The problem is to compute $F_X(x)$ in terms of $F_U(u)$; we proceed as follows: The event

$$(X \le x) \equiv (U^2 \le x) \equiv (-\sqrt{x} \le U \le \sqrt{x}).$$

Consequently,

$$F_X(x) = F_U(\sqrt{x}) - F_U(-\sqrt{x}) = F_U(\sqrt{x}),$$

since $F_U(u) = 0$, $u \le 0$.

Thus,

$$F_X(x) = 0,\ x \le 0,$$
$$F_X(x) = \sqrt{x},\ 0 \le x \le 1,$$
$$F_X(x) = 1,\ 1 \le x.$$

Example 4.26 *Compute the distribution of $X = Z^2$, where Z is $N(0,1)$.*

Solution Using exactly the same reasoning as in the preceding example we obtain

$$F_X(x) = \Phi(\sqrt{x}) - \Phi(-\sqrt{x}) = 2\Phi(\sqrt{x}) - 1,\ x \ge 0,\ \text{where}$$

$\Phi(z)$ is the standard normal distribution function. We obtain the density function of X by computing $F_X'(x)$ via the chain rule of the differential calculus:

$$F_X'(x) = 2\Phi'(\sqrt{x}) \frac{1}{2\sqrt{x}} = 2 \frac{1}{\sqrt{2\pi}} e^{-(\sqrt{x})^2/2} \frac{1}{2\sqrt{x}} = \frac{1}{\sqrt{2\pi}} x^{-1/2} e^{-x/2},\ x > 0. \quad (4.71)$$

This shows that Z^2 has a chi–square distribution with one degree of freedom. We have therefore proved the following proposition:

Proposition 4.8 *If Z is $N(0,1)$ then Z^2 has a chi-square distribution with one degree of freedom.*

Example 4.27

Compute the distribution function of $Y = U^{-1}$, where U is uniformly distributed on the interval $[0,1]$.

Solution The event

$$(Y \le y) \equiv (U^{-1} \le y) \equiv (U \ge y^{-1});\ \text{thus,}$$

$$F_Y(y) = P(U \ge y^{-1}) = 1 - F_U(y^{-1}).$$

Therefore,

$$F_Y(y) = 0,\ y \le 1;\ F_Y(y) = 1 - y^{-1},\ 1 \le y.$$

4.10 Mathematical Details and Derivations

Proof of Theorem 4.4:

Step 1: Differentiate both sides of Equation 4.65 with respect to p.

$$\frac{d}{dp}\pi(p) = \frac{d}{dp}\sum_{k \le x \le n} b(x; n,\ p) = \sum_{k \le x \le n} \frac{d}{dp} b(x; n, p)$$

Step 2: The derivative of $b(x; n, p)$ is given by

$$\frac{d}{dp} b(x; n, p) = n \binom{n-1}{x-1} p^{x-1}(1-p)^{n-x} - n \binom{n-1}{x} p^x (1-p)^{n-x-1}$$

where we set

$$\binom{n-1}{n} = 0.$$

Step 3: Inserting the formula for $\frac{d}{dp}b(x;n,p)$ obtained in step 2 into the right hand side of the expression for $\frac{d}{dp}\pi(p)$ in step 1 yields a "telescoping sum" with the result that

$$\frac{d}{dp}\pi(p) = n\binom{n-1}{k-1}p^{k-1}(1-p)^{n-k} > 0, \ 0 < p < 1. \tag{4.72}$$

The fact that the derivative $\pi'(p) > 0$ proves that $\pi(p)$ is an increasing function of p. We obtain the formula 4.66 for the tail of the binomial distribution by integrating both sides of Equation 4.72 and using the fundamental theorem of the calculus, which states that

$$\pi(p) - \pi(0) = \int_0^p \pi'(t)dt.$$

4.11   Chapter Summary

Engineers, both traditional and finacial, use random variables of the continuous type to model the lifetimes of manufactured components, the distribution of stock market returns, and prices, etc. The distribution function of a continuous random variable is given by

$$F_X(x) = \int_{-\infty}^x f_X(t)\,dt, \ \text{where} \ f_X(t) \ge 0, \ \text{and} \ \int_{-\infty}^\infty f_X(t)\,dt = 1.$$

The function $f_X(x)$ is called the *probability density function* (pdf) of the random variable X. The most important continuous distribution, however, is the *normal distribution,* which is used to model a wide variety of experimental data as well as to approximate the binomial distribution. The uniform, exponential, lognormal, gamma, Weibull, and beta distributions are widely used in engineering and finance.

4.12   Problems

Section 4.2

Problem 4.1 *The random variable Y has the exponential distribution*

$$F_Y(y) = 1 - \exp(-0.75y), 0 \le y < \infty \ \text{and} \ F_Y(y) = 0, \ y \le 0.$$

(a) Compute $P(1 < Y < 2)$.
(b) Compute $P(Y > 3)$.
(c) Compute the 95th percentile $q(0.95)$

Problem 4.2 *The random variable X has pdf given by*

$$f(x) = 2x, \ 0 < x < 1,$$
$$f(x) = 0, \ elsewhere.$$

(a) Sketch the graph of $f(x)$ for all values of x.
(b) Compute the df $F(x)$ for all values of x and sketch its graph.
(c) Shade the area corresponding to $P(0.25 < X < 0.5)$ and then compute its value via the formula $F(0.5) - F(0.25)$.
(d) Compute the median and 90th percentile of the distribution.

Problem 4.3 *Let Y denote a random variable with continuous df $F(y)$ defined below.*

$$F(y) = 0, y \leq 0,$$
$$F(y) = y^3/8, \, 0 < y \leq 2,$$
$$F(y) = 1, y > 2.$$

(a) Graph $F(y)$ for all y.
(b) Compute $P(0.2 < Y < 0.5)$, $P(Y > 0.6)$.
(c) Compute the pdf $f(y)$ for all y and sketch its graph.

Problem 4.4 *Consider the function $f(x) = c(1 + 2x)^{-3}, 0 \leq x < \infty$, and $f(x) = 0$ elsewhere.*
(a) Find the constant c so that $f(x)$ is a pdf.
(b) Compute the df $F_X(x) = P(X \leq x)$.
(c) Find the median and the lower and upper quartiles.

Problem 4.5 *The distribution of the lifetime in hours of a battery for a laptop computer is modeled as an exponential random variable with parameter $\theta = 1/3.5$. Compute the safe life of the battery, which is defined to be the 10th percentile of the distribution.*

Problem 4.6 *Consider the function $F(x) = 1/(1 + \exp(-x))$, $-\infty < x < \infty$.*
(a) Show that $F(x)$ satisfies Equations 4.4, 4.5, and 4.6 and is therefore a continuous distribution function. (This is an example of a logistic distribution.)
(b) Show that $F(-x) = 1 - F(x)$, $x \geq 0$. This means that $P(X \leq -x) = P(X \geq x)$, $x \geq 0$.
(c) Compute the pdf $f(x) = F'(x)$ and show that $f(-x) = f(x)$, that is, its graph is symmetric about the origin.

Problem 4.7 *Suppose $F(x)$ is the logistic distribution described in Problem 4.6. Compute:*
(a) the median.
(b) the 10th and 90th percentiles.
(c) the 5th and 95th percentiles.

Problem 4.8 *The percentage of alcohol in a certain compound is given by $100X$ where X is a random variable with pdf given by*

$$f(x) = k(1 - x)^2, \; 0 < x < 1; \; f(x) = 0, \; elsewhere.$$

(a) Find the value of k.
(b) Sketch the graph of $f(x)$ for all values of x.
(c) Compute the df $F(x)$ for all values of x.
(d) Sketch the graph of $F(x)$.
(e) What is the probability that the percentage of alcohol is less than 25%? More than 75%?
(f) Compute the median of the distribution.

Problem 4.9 *An archer shoots an arrow at a circular target. Suppose the distance between the center and impact point of the arrow is a continuous random variable X (meters) with pdf given by*

$$f(x) = 6x(1-x), 0 < x < 1; f(x) = 0, \text{ elsewhere.}$$

(a) Sketch the graph of $f(x)$.
(b) Compute the df $F(x)$ for all values of x and sketch its graph.
(c) The bull's eye of the target is a circle of radius 20 cm; what is the probability that the archer hits the bull's eye?
(d) Compute the 5th, 50th and 95th percentiles of the distribution.

Problem 4.10 *The random variable X has pdf given by*

$$f(x) = (1+x), -1 \le x \le 0, f(x) = (1-x), 0 \le x \le 1, f(x) = 0, elsewhere.$$

(a) Sketch the graph of $f(x)$.
(b) Compute the df $F(x)$ for all values of x and sketch its graph.
(c) Compute $P(-0.4 < X < 0.6)$.
(d) Compute the median of the distribution.
(e) Compute the 10th and 90th percentiles of the distribution.

Problem 4.11 *The lifetime T (in hours) of a component has a df given by*

$$F(t) = 0, t \le 1, F(t) = 1 - \frac{1}{t^2}, t \ge 1.$$

(a) Sketch the graph of $F(t)$.
(b) Compute the pdf $f(t)$ of T and sketch its graph.
(c) Find $P(T > 100)$.
(d) Compute $P(T > 150|T > 100)$, i.e., given that the component is still functioning after 100 hours, determine the probability that it will continue to function for another 50 hours.

Problem 4.12 *The thickness S (measured in millimeters) of a fiber thread has a pdf given by*

$$f(x) = kx^2(2-x), 0 < x < 2; f(x) = 0, \text{ elsewhere.}$$

(a) Find the value of k.
(b) Compute $P(0.5 < S < 1.5)$.
(c) Compute $P(S > 1)$.

Problem 4.13 *Suppose the lifetime of a fan on a diesel engine has an exponential distribution with $1/\theta = 28,700$ hours. Compute the probability of a fan failing on a:*
(a) 5,000 hour warranty.
(b) 8,000 hour warranty.

Problem 4.14 *Refer to the lifetime distribution in Problem 4.13. Compute the safe life of the diesel engine fan. (The safe life of a component is the 10th percentile of its lifetime distribution; it is the length of time that 90% of the components will operate without failure.)*

Problem 4.15 *Suppose the number of days to failure of a diesel locomotive has an exponential distribution with parameter $1/\theta = 43.3$ days. Compute the probability that the locomotive operates one entire day without failure.*

Problem 4.16 *Let X denote a random variable with continuous df $F(x)$ defined below.*

$$F(x) = 0, x \leq -1,$$
$$F(x) = (x^3/2) + 1/2, -1 \leq x \leq 1,$$
$$F(x) = 1, x \geq 1.$$

(a) Graph $F(x)$ for all x.
(b) Compute the pdf $f(x)$ for all x and sketch its graph.
(c) Compute the 10th and 90th percentiles of the distribution.
(d) Compute $P(X > 0.4)$.

Problem 4.17 *Which of the following functions $F(x)$ are continuous dfs of the form displayed in Definition 4.1? Calculate the corresponding pdf $f(x)$ where appropriate.*

$$(a)\ F(x) = 0, x \leq 0,$$
$$F(x) = \frac{x}{1+x}, x > 0;$$
$$(b)\ F(x) = 0, x \leq 2,$$
$$F(x) = 1 - \frac{4}{x^2}, x > 2;$$
$$(c)\ F(x) = 0, x < -\pi/2,$$
$$F(x) = \sin x, -\pi/2 < x < \pi/2,$$
$$F(x) = 1, x \geq \pi/2.$$

Problem 4.18 *Consider the random variable X with distribution function*

$$F(x) = 0, x \leq 0,$$
$$F(x) = 1 - \frac{1}{(1+x)^2}, x > 0.$$

(a) Graph $F(x)$ for all x.
(b) Compute the pdf $f(x)$ for all x and sketch its graph.
(c) Compute $P(1 < X < 2)$.
(d) Compute the median of the distribution.
(e) Verify that $q(p) = (1/\sqrt{1-p}) - 1$, $(0 < p < 1)$ is the $100p$th percentile of the distribution of F.
(f) Using the formula from part (e) find the values of the 5th and 95th percentiles.

Section 4.3

Problem 4.19 *Compute $E(X)$, $E(X^2)$, and $V(X)$ for the random variable X with pdf given by:*

(a) $f(x) = \dfrac{1}{2} - \dfrac{x}{4}, -1 \leq x \leq 1, f(x) = 0,\ elsewhere.$
(b) $f(x) = (1/2)\sin x, 0 \leq x \leq \pi,\ f(x) = 0,\ elsewhere.$
(c) $f(x) = 3(1-x)^2, \quad 0 < x < 1; \quad f(x) = 0,\ elsewhere.$
(d) $f(x) = x, 0 \leq x \leq 1, f(x) = 2 - x, 1 \leq x \leq 2, f(x) = 0,\ elsewhere.$
(e) $f(x) = (1+x), -1 \leq x \leq 0, f(x) = (1-x), 0 \leq x \leq 1, f(x) = 0,\ elsewhere.$
(f) $f(x) = (1/2)e^{-|x|}, -\infty < x < \infty.$

Problem 4.20 *Let S denote the random variable described in Problem 4.12. Compute:*
(a) $E(S)$ (b) $V(S)$

Problem 4.21 *If X is uniformly distributed over the interval $[a, b]$ show that:*
(a) $E(X) = (a + b)/2$.
(b) $V(X) = (b - a)^2/12$.

Problem 4.22 *Let Θ be uniformly distributed over the interval $[0, 2\pi]$. Compute:*
(a) $E(\cos\Theta)$, $V(\cos\Theta)$.
(b) $E(\sin\Theta)$, $V(\sin\Theta)$.
(c) $E(|\cos\Theta|)$.
(d) Does $E(\cos\Theta) = \cos(E(\Theta))$?

Problem 4.23 *Show that if U is uniformly distributed over $[0, 1]$ then $X = a + (b - a)U$ is uniformly distributed over $[a, b]$.*

Problem 4.24 *Verify Equations 4.21 and 4.22 by performing the integration by parts.*

Problem 4.25 *Let X denote a random variable with continuous df $F(x)$ defined below.*

$$F(x) = 0, x \le -1;$$
$$F(x) = (x^3/2) + 1/2, -1 \le x \le 1;$$
$$F(x) = 1, x \ge 1.$$

Compute $E(X)$, $E(X^2)$, and $V(X)$.

Problem 4.26 *Suppose the lifetime T of an electronic component is exponentially distributed with $\theta^{-1} = 1000$ and the warranty period is 700 hours. The manufacturing costs and selling price of each component are \$6.00 and \$10.00, respectively. If the item fails before K hours the manufacturer replaces the faulty component without charge.*
(a) Compute the expected profit per customer if the warranty period is: 700 hours; 900 hours.
(b) How long can the warranty period be extended before the expected profit turns into a loss? (Hint: Review Example 4.7.)

Problem 4.27 *Refer to Example 4.6. Derive the formulas for the mean and variance, given in Equation 4.24, of the random variable T with the Pareto distribution (Equation 4.24).*

Problem 4.28 *(a) Verify by direct calculation that $Q(p) = \exp(-\ln(1 - p)/\alpha) - 1$ is the p quantile of the Pareto distribution (Equation 4.24), that is show that $F(Q(p)) = p$.*
(b) Using part (a), compute the median of a Pareto distribution with $\alpha = 1.5$.

Section 4.5

Problem 4.29 *Let Z be a standard normal random variable. Use the table of the normal distribution to compute the following probabilities.*

(a) $P(Z < 1)$	(b) $P(Z < -1)$	(c) $P(|Z| < 1)$
(d) $P(Z < -1.64)$	(e) $P(Z > 1.64)$	(f) $P(|Z| < 1.64)$
(g) $P(Z > 2)$	(h) $P(|Z| > 2)$	(i) $P(|Z| < 1.96)$

Problem 4.30 *Estimate each of the following probabilities using Chebyshev's inequality and compare the estimates with the exact answers obtained from the tables of the normal distribution.*
(a) $P(|Z| > 1)$
(b) $P(|Z| > 1.5)$
(c) $P(|Z| > 2)$

Problem 4.31 *Find c such that*

(a) $P(Z < c) = .25$	*(b) $P(Z < c) = .75$*
(c) $P(|Z| < c) = 0.5$	*(d) $P(Z < c) = .85$*
(e) $P(Z > c) = 0.025$	*(f) $P(|Z| < c) = .95$*

Problem 4.32 *Miles per gallon (mpg) is a measure of the fuel efficiency of an automobile. The mpg of a compact automobile selected at random is normally distributed with $\mu = 29.5$ and $\sigma = 3$.*
(a) What proportion of the automobiles have a fuel efficiency that is greater than 24.5 mpg?
(b) What proportion of the automobiles have a fuel efficiency that is less than 34.0 mpg?

Problem 4.33 *The results of a statistics exam are assumed to be normally distributed with $\mu = 65$ and $\sigma = 15$. In order to get an A the student's rank must be in the 90th percentile or above. Similarly, in order to get at least a B the student's rank must be in the 80th percentile or above.*
(a) What is the minimum score required to get an A?
(b) What is the minimum score required to get at least a B?

Problem 4.34 *The distribution of the lifetime in hours of a battery for a laptop computer is modeled as a normal random variable with mean $\mu = 3.5$ and standard deviation $\sigma = 0.4$ Compute the safe life of the battery, which is defined to be the 10th percentile of the distribution.*

Problem 4.35 *Suppose that a manufacturing process produces resistors whose resistance (measured in ohms) is normally distributed with $\mu = 0.151$ and $\sigma = 0.003$. The specifications for the resistors require that the resistance equals 0.15 ± 0.005 ohms. What proportion of the resistors fail to meet the specifications?*

Problem 4.36 *The random variable $X \overset{D}{=} N(12, 36)$.*
(a) Compute $P(X < 0)$, $P(X > 20)$
(b) Find c such that: $P(|X - 12| < c) = 0.9, 0.95, 0.99$

Problem 4.37 *The random variable $X \overset{D}{=} N(60, 25)$. Compute:*
(a) $P(X < 50)$ (b) $P(X > 65)$
(c) The value of c such that $P(|X - 60| < c) = .95$
(d) The value of c such that $P(X < c) = 0.01$

Problem 4.38 *Derive Equation 4.44 for the critical value $x(\alpha)$ of the normal distribution $N(\mu, \sigma^2)$.*

Problem 4.39 *Assume that math SAT scores are normally distributed with $\mu = 500$ and $\sigma = 100$.*
(a) What proportion of the students have a math SAT score greater than 650?
(b) What proportion of the students have a math SAT score less than 550?
(c) Ten students are selected at random. What is the probability that at least one out of the ten scored at least 650 or better?
(d) (Part (c) continued) What is the probability that all ten scored at most 550?

Problem 4.40 *Suppose the combined SAT scores of students admitted to a college are normally distributed with $\mu = 1200$ and $\sigma = 150$. Compute the value of c for which $P(|X - 1200| < c) = 0.5$ The interval $[1200 - c,\ 1200 + c]$ is the interquartile range.*

Problem 4.41 *Refer to Example 4.15.*
(a) Compute: $E(\ln(S(0.5),\ \sigma(\ln(S(0.5)),\ E((S(0.5),\ \sigma(S(0.5))$.
(b) What is the probability that the stock's price will be up at least 30% after six months?
(c) What is the probability that the stock's price will be down at least 20% after six months?

Problem 4.42 *Suppose the distribution of a stock price $S(t)$ (t measured in years) is lognormal with parameters $\mu = 0.15$, volatility $\sigma = 0.25$, and initial price $S_0 = \$40$. Compute:*
(a) E(stock price after 6 months).
(b) σ(stock price after 6 months).
(c) Find the probability of a loss after 6 months.
(d) What is the probability that the stock's price will be up at least 20% after 6 months?
(e) What is the probability that the stock's price will be down at least 15% after 6 months?

Problem 4.43 *Suppose the distribution of a stock price $S(t)$ (t measured in years) is lognormal with parameters $\mu = 0.20$, volatility $\sigma = 0.30$, and initial price $S_0 = \$60$. Compute:*
(a) E(stock price after 3 months).
(b) σ(stock price after 3 months).
(c) Find the probability of a loss after 3 months.
(d) What is the probability that the stock's price will be up at least 15% after 3 months?
(e) What is the probability that the stock's price will be down at least 10% after 3 months?

Problem 4.44 *The random variable X has a binomial distribution with parameters $n = 25$ and $p = 0.5$.*
(a) Compute the exact value of $P(9 \leq X \leq 16)$ using the binomial tables in the text.
(b) Compute $P(9 \leq X \leq 16)$ using the normal approximation to the binomial.
(c) Compute the same probability in part (b) using the continuity correction. Does the continuity correction improve the accuracy of the approximation?

Problem 4.45 *Let X denote the number of heads that appear when a fair coin is tossed 300 times. Use the normal approximation to the binomial to estimate the following probabilities:*

(a) $P(X \geq 160)$	(b) $P(X \leq 140)$
(c) $P(|X - 150| \geq 20)$	(d) $P(135 < X < 165)$

Problem 4.46 *Let Y denote the number of times a 1 appears in 720 throws of a die. Using the normal approximation to the binomial approximate the following probabilities:*

(a) $P(Y \geq 130)$	(b) $P(Y \geq 140)$
(c) $P(|Y - 120| \geq 20)$	(d) $P(105 < Y < 120)$

Problem 4.47 *A certain type of seed has a probability of* 0.7 *of germinating. In a package of* 100 *seeds let Y denote the number of seeds that germinate.*
(a) Give the formula for the exact distribution of Y.
(b) Calculate the probability that at least 75% *germinate.*
(c) Calculate the probability that at least 80% *germinate.*
(d) Calculate the probability that no more than 65% *germinate.*

Problem 4.48 *The natural recovery rate for a certain disease is* 40%. *A new drug is developed and the manufacturer claims that it boosts the recovery rate to* 80%. *Two hundred patients are given the drug. What is the probability that at least* 100 *patients recover if:*
(a) the drug is worthless?
(b) it is really 80% *effective as claimed?*

Problem 4.49 *The number of chocolate chips per cookie is assumed to have a Poisson distribution with $\lambda = 4$. From a batch of 500 cookies, let Y denote the number of cookies without chocolate chips.*
(a) Describe the exact distribution of Y; name it and identify the parameters.
(b) Find $E(Y)$ and $V(Y)$.
(c) Use the central limit theorem to estimate the probability that 15, or more, cookies (out of 500) have no chocolate chips.

Problem 4.50 *Suppose that the velocity V of a particle of mass m is $N(0,1)$ distributed; denote its kinetic energy by $K = mv^2/2$.*
    Show that $E(K) = m/2$ and $V(K) = m^2/2$.

Problem 4.51 *Suppose the probability that a person cancels his hotel reservation is* 0.1 *and the hotel's capacity is* 1,000 *rooms. In order to minimize the number of empty rooms the hotel routinely overbooks. How many reservations above* 1,000 *should it accept if it wants the probability of overbooking to be less than* 0.01*?*

Section 4.8

Problem 4.52 *Consider the gamma pdf $f(x) = Cx^{1/2}e^{-x/2}$, $x > 0$; $f(x) = 0$, elsewhere. Identify the parameters α, β, and find the constant C. (Refer to Equation 4.55.)*

Problem 4.53 *(a) Show that $\Gamma(1/2) = \sqrt{\pi}$ by making the change of variables $x = u^2/2$ in the integral*

$$\Gamma(1/2) = \int_0^\infty x^{-1/2}e^{-x}\,dx.$$

(b) Using the result of part (a) compute $\Gamma(3/2)$, $\Gamma(5/2)$.

Problem 4.54 *Derive Equation 4.58 by performing a suitable integration by parts.*

Problem 4.55 *Derive the formulas for the mean and variance of the gamma distribution given in Equation 4.61.*

Problem 4.56 *(a) If X has a gamma distribution with $E(X) = \mu_1$ and $E(X^2) = \mu_2$ show that*

$$\alpha = \frac{\mu_1^2}{\mu_2 - \mu_1^2} \quad and \quad \beta = \frac{\mu_2 - \mu_1^2}{\mu_1}.$$

(b) Use the result of part (a) to find the parameters of the gamma distribution when $\mu_1 = 2$, $\mu_2 = 5$.

Problem 4.57 *The lifetime T of a tire, measured in units of $1,000$ miles, has a Weibull distribution with parameters $\alpha = 0.5$, $\beta = 15$. Compute:*
(a) the median mileage of the tire.   (b) the expected mileage $E(T)$.   (c) $P(T > 10)$.

Problem 4.58 *The lifetime, measured in years, of a TV picture tube has a Weibull distribution with parameters $\alpha = 2$, $\beta = 3$.*
(a) Compute the proportion of tubes that will fail on a: one year, two year, and three year warranty.
(b) Compute the expected profit per TV tube if the warranty is for one year and the cost of manufacture and selling price are $50 and $100, respectively. Hint: Use Equation 4.28.

Problem 4.59 *An engine fan life, measured in hours, has a Weibull distribution with $\alpha = 1.053$, $\beta = 26,710$. Compute:*
(a) the median life.
(b) the proportion that fail on an $8,000$ hour warranty.

Problem 4.60 *The lifetime T, measured in hours, of a disk drive has a Weibull distribution with parameters $\alpha = 0.25$, $\beta = 1000$.*
(a) Compute $E(T)$ and $V(T)$.
(b) What is the probability that the lifetime exceeds 6000 hours?
(c) Compute the probability of the disk drive failing on a $10,000$ hour warranty.

Problem 4.61 *Using Equations 4.68 and 4.58 compute:*
(a) $B(1/2, 1)$.   (b) $B(3, 2)$.   (c) $B(3/2, 2)$.

Problem 4.62 *Suppose that X has a beta distribution with parameters $\alpha = 3$, $\beta = 2$. Compute $E(X)$ and $V(X)$ using Equation 4.70.*

Problem 4.63 *Consider the beta pdf $f(x) = Cx^2(1 - x)^3$, $0 \le x \le 1$; $f(x) = 0$, elsewhere. Identify the parameters α, β, and find the constant C. (Refer to Equations 4.68 and 4.69.)*

Problem 4.64 *Suppose the proportion X of correct answers a student gets on a test has a beta distribution with parameters $\alpha = 4$, $\beta = 2$.*
(a) Compute the cdf $F_X(x)$.
(b) If a passing score is is 50%, what is the probability that a student passes?
(c) If 100 students take the test, what is the expected number and variance of the number of students who pass?
(d) If 100 students take the test, what is the probability that at least 85 of them pass the test?

Problem 4.65 *Graph the beta distribution for:*
(a) $\alpha = 2$, $\beta = 1$.   (b) $\alpha = 1$, $\beta = 2$.   (c) $\alpha = 1$, $\beta = 1$.   (d) $\alpha = 0.5$, $\beta = 1$.

Problem 4.66 *Suppose the proportion X of arsenic in copper has a beta distribution with $\alpha = 1.01$, $\beta = 100$. Compute $E(X)$ and $V(X)$.*

Problem 4.67 *Derive the formulas for the mean and variance of the beta distribution displayed in Equation 4.70. Hint: $E(X) = B(\alpha + 1, \beta)/B(\alpha, \beta)$ and $E(X^2) = B(\alpha + 2, \beta)/B(\alpha, \beta)$.*

Problem 4.68 *Show that the function $O(p)$ defined by*

$$O(p) = P(X \le c) = \sum_{x=0}^{c} b(x; n, p)$$

is a decreasing function of p.

Section 4.9

Problem 4.69 *Show that Z (the standard normal random variable) and $-Z$ have the same distribution.*

Problem 4.70 *Suppose the random variables X and $-X$ have the same distribution. Show that $E(X) = 0$.*

Problem 4.71 *Suppose the distribution of the random variable X is $Beta(\alpha, \beta)$. Show that the distribution of $1 - X$ is $Beta(\beta, \alpha)$.*

Problem 4.72 *Let U be a uniformly distributed random variable on the interval $[-1, 1]$. Let $X = U^2$.*
(a) Compute df $F(x) = P(X \le x)$ for all x.
(b) Compute the pdf $f(x)$ for all x.

Problem 4.73 *Let U be a uniformly distributed random variable on the interval $[-1, 1]$. Let $Y = 4 - U^2$.*
(a) Compute $G(y) = P(Y \le y)$ for all y.
(b) Compute the pdf $g(y)$ for all y.

Problem 4.74 *The random variable U is uniformly distributed on $[0, 1]$. Show that $Y = -\theta^{-1} \ln U$ has an exponential distribution with parameter θ, i.e., $P(Y < y) = 1 - e^{-\theta y}$.*

Problem 4.75 *Let $R = \sqrt{X}$ where X has an exponential distribution with parameter $\theta = 1/2$. Show that R has a Rayleigh distribution:*

$$F_R(r) = \int_0^r t e^{-t^2/2} \, dt, r > 0, F_R(r) = 0, \text{ elsewhere.}$$

Problem 4.76 *Let Y have pdf $g(y) = (1 + y)^{-2}, y > 0$ and $g(y) = 0$, elsewhere.*
(a) Compute the df $G(y)$ for all y.
(b) Let $X = 1/Y$. Verify that X and Y have the same df.

Problem 4.77 *Let $X = \tan U$, where U is uniformly distributed on the interval $[-\pi/2, \pi/2]$.*
(a) Show that

$$F_X(x) = \frac{1}{2} + \frac{1}{\pi} \arctan x.$$

(b) Show that

$$f_X(x) = \frac{1}{\pi(1 + x^2)}.$$

Problem 4.78 *The random variable T has df F given by*

$$F(t) = 0, t \leq 1, \ F(t) = 1 - \frac{1}{t^2}, t \geq 1.$$

Compute the df and pdf of $X = T^{-1}$.

Problem 4.79 *The random variable X has pdf f given by*

$$f(x) = (1+x), -1 \leq x \leq 0, \ f(x) = (1-x), 0 \leq x \leq 1, \ f(x) = 0, \ elsewhere.$$

Compute the df and pdf of $Y = X^2$.

Problem 4.80 *The random variable X has pdf F given by*

$$f(x) = 2(1-x), \ \ 0 < x < 1; \ \ f(x) = 0, \ elsewhere.$$

Compute the df and pdf of $Y = \sqrt{X}$.

Problem 4.81 *The first two moments and variance of the exponential distribution with parameter θ were derived in Example 4.5. These formulas can also be derived using the moment generating function technique. In detail:*
(a) Using Proposition 4.2 compute $M'_X(s)$ and verify that $M'_X(0) = \frac{1}{\theta}$.
(b) Using Proposition 4.2 compute $M''_X(s)$ and verify that $M''_X(0) = \frac{1}{\theta^2}$.

Problem 4.82 *Is the following statement true or false? Suppose the random variables X and Y have the same distribution function $F(x) = P(X \leq x) = P(Y \leq x)$. Then $X = Y$. If true, give a proof; if false, produce a counter example.*

Problem 4.83 *Which of the following statements are true. If false, give a counter example.*
(a) $E(X^2) = (E(X))^2$
(b) $E(1/X) = 1/E(X)$
(c) If $E(X) = 0$, then $X = 0$.

4.13   To Probe Further

The first statement of the normal approximation to the binomial in English appears in A. DeMoivre's (1738) *The Doctrine of Chances*. Motivating his research was the difficulty of computing the binomial probabilities for large values of n. "Although the solution of problems of chance," wrote DeMoivre, "often require that several terms of the binomial $(a + b)^n$ be added together, nevertheless in very high powers [that is when n is very large] the thing appears so laborious, and of so great a difficulty that few people have undertaken that task." Source: Stigler (1986), *History of Statistics: The Measurement of Uncertainty before 1900*, The Belknap Press of Harvard University Press, Cambridge, Massachusetts, p. 74.

The historical origins and scientific and philosophical implications of statistics are explored in Theodore M. Porter (1986), *The Rise of Statistical Thinking, 1820–1900,* Princeton University Press, Princeton, New Jersey. The quotation at the head of this chapter is also taken from the same source (p. 99). Statistical and theoretical analyses of internet data traffic are discussed in Walter Willenger et al., Long-Range Dependence and Data Network Traffic in Paul Doukhan et al., *Theory and Applications of Long-Range Dependence*, pp. 373-407, Birkhäuser, 2003. The lognormal model for the distribution of stock prices is discussed in John C. Hull (2003), *Options, Futures, and Other Derivatives*, 5th ed., Prentice-Hall. Example 4.21 is taken from J.G. Kemeny, H. Mirkil, J.L. Snell, and G.L. Thompson, (1959), *Finite Mathematical Structures*, Prentice–Hall, Englewood Cliffs, New Jersey.

Chapter 5

Multivariate Probability Distributions

Physicians have nothing to do with what is called the law of large numbers, a law which, according to a great mathematician's expression is always true in general and false in particular.
Claude Bernard, French physiologist (1813-1878)

5.1 Orientation

We frequently have the occasion to study the mutual interactions between two or more variables. How useful are a student's college grade point average (GPA) and score on the law school admissions test (LSAT) as predictor variables for a student's future performance in law school? How does diversification of an investment portfolio reduce risk, and how does it affect the returns? For example, when oil prices soar so do the stock prices of oil companies, while those of automobile manufacturers and airlines fall. Modern portfolio theory begins with an analysis of the joint variation of the different asset classes (such as stocks, bonds, and mutual funds) in the portfolio. In environmental studies the air quality of a metropolitan region is determined by simultaneously measuring several pollutants such as ozone, nitrogen oxide, hydrocarbons, etc. Before undertaking a plan to reduce the pollution levels the environmental scientist must study their interactions since reducing the level of just one pollutant could actually increase the combined harmful effects of the others. Statisticians use the *joint distribution function* to study the joint variations among the variables. We begin with the discrete case and follow with the continuous case, the only difference between them is that, in the latter case, we use some basic tools from multivariate calculus.

Organization of Chapter

1. Section 5.2: The Joint Distribution Function: Discrete Random Variables

2. Section 5.2.1: Independent Random Variables

3. Section 5.3: The Multinomial Distribution

4. Section 5.4: Mean and Variance of a Sum of Random Variables

5. Section 5.4.1: The Law of Large Numbers for Sums of Independent and Identically Distributed (iid) Random Variables

6. Section 5.4.2: The Central Limit Theorem

7. Section 5.5: Why Stock Prices have a Lognormal Distribution: An Application of the Central Limit Theorem

8. Section 5.5.1: The Binomial Lattice Model as an Approximation to a Continuous Time Model for Stock Market Prices*

9. Section 5.6: Modern Portfolio Theory; mean-standard deviation diagram of a portfolio; efficient portfolios; diversification*

10. Section 5.7: Risk Free and Risky Investing*

11. Section 5.7.1: Present Value Analysis of Risk Free and Risky Returns*

12. Section 5.8: Theory of Single and Multi-Period Binomial Options*

13. Section 5.9: Black-Scholes Formula for Multi-Period Binomial Options*

14. Section 5.9.1: Black-Scholes Formula for Stock Prices Governed by a Lognormal Distribution*

15. Section 5.10: The Poisson Process

16. Section 5.11: Applications of Bernoulli Random Variables to Reliability Theory*

17. Section 5.12: The Joint Distribution Function: Continuous Random Variables

18. Section 5.13: Mathematical Details and Derivations

19. Section 5.14: Chapter Summary

5.2   The Joint Distribution Function: Discrete Random Variables

The concept of a joint distribution function of X, Y is best understood when both random variables are discrete. Consider, for instance, the sample space corresponding to the experiment of rolling two dice (Example 3.2) which consists of the ordered pairs (i, j), $(i = 1, \ldots, 6;\ j = 1, \ldots, 6)$. Assume that the two dice have different colors, white and red, say. Denote by X the number appearing on the white die and Y the number appearing on the red die. Thus $X(i, j) = i$ and $Y(i, j) = j$. The ordered pair of random variables (X, Y) is called a *random vector*. The event (i, j) can be expressed as follows:

$$(i, j) = \{X = i\} \cap \{Y = j\}.$$

More generally, let X and Y be two discrete random variables, defined on the same sample space Ω with discrete ranges

$$R_X = \{x_1, x_2, \ldots, \}\ \text{and}\ R_Y = \{y_1, y_2, \ldots, \},\ \text{respectively.}$$

The intersection of the events $\{X = x\}$ and $\{Y = y\}$ is the event denoted by

$$\{X = x, Y = y\} = \{X = x\} \cap \{Y = y\}.$$

Definition 5.1 *The function $f(x, y)$ defined by*

$$f(x, y) = P(X = x, Y = y) \tag{5.1}$$

is called the joint probability mass function (jpmf) of the random vector (X, Y).

Example 5.1 *Compute the joint pmf for the experiment of throwing two dice,*

Solution Let X, Y denote the numbers on the faces of the two dice. Their joint pmf is

$$f(x,y) = \frac{1}{36}, \ x = 1, \ldots, 6; \ y = 1, \ldots, 6;$$
$$f(x,y) = 0, \ \text{elsewhere.}$$

The quantity $f(x,y) = P(X = x, Y = y)$ is the probability of an event and the events $\{X = x, Y = y\}$ form a partition of the sample space. Consequently, a joint pmf has the following two properties:

$$0 \leq f(x,y) \leq 1; \ \sum_x \sum_y f(x,y) = 1. \tag{5.2}$$

Example 5.2

A computer store sells 3 computer models priced at \$1600, \$2000, and \$2400, and two brands of monitors priced at \$400 and \$800, respectively. The joint pmf of X, the computer price, and Y, the monitor price, is given in Table 5.1. Interpretation: $P(X = 1600, Y = 400) = 0.30$ means that 30% of the computer systems sold consisted of the \$1600 computer and the \$400 monitor, etc.

Table 5.1

joint pmf of (X,Y)		
X\ Y	*400*	*800*
1600	0.30	0.25
2000	0.20	0.10
2400	0.10	0.05

Computing $P(X \in A, Y \in B)$

We use the joint pmf to compute $P(X \in A, Y \in B)$ via the formula

$$P(X \in A, Y \in B) = \sum_{x \in A} \sum_{y \in B} f(x,y). \tag{5.3}$$

The joint distribution function $F(x,y)$ is an important special case defined by

$$F(x,y) = P(X \leq x, Y \leq y). \tag{5.4}$$

For instance, with reference to the jpmf in Table 5.1 we see that

$$F(2000, 400) = P(X \leq 2000, Y \leq 400) = f(1600, 400) + f(2000, 400)$$
$$= 0.30 + 0.20 = 0.50.$$

Marginal Distribution Functions

The probability functions $f_X(x)$ and $f_Y(y)$ are obtained by summing the jpmf $f(x,y)$ over the variables x and y, respectively (see Equations 5.5 and 5.6).

$$f_X(x) = \sum_y f(x,y); \tag{5.5}$$

$$f_Y(y) = \sum_x f(x,y). \tag{5.6}$$

The probability functions f_X and f_Y are called the *marginal distributions*.

Derivation of Equations 5.5 and 5.6: The events $(Y = y)$, $y \in R_Y$ form a partition of the sample space, and therefore

$$(X = x) = \cup_y (X = x) \cap (Y = y); \text{ consequently}$$

$$f_X(x) = P(X = x) = P\left(\cup_y (X = x) \cap (Y = y)\right)$$
$$= \sum_y P(X = x, Y = y) = \sum_y f(x, y).$$

The derivation of Equation 5.6 is similar and is therefore omitted.

Example 5.3 *Refer to the joint pmf in Table 5.1. Compute the marginal probability functions.*

Solution To compute $f_X(1600)$, for instance, you sum the entries across the first row and obtain the value $f_X(1600) = 0.55$; to compute $f_Y(400)$ you sum the entries down the first column and obtain the value $f_Y(400) = 0.60$. The complete solution is displayed in the margins of Table 5.2 below.

Table 5.2

Joint & Marginal Probability Function of (X, Y)			
$X \backslash Y$	*400*	*800*	$f_X(x)$
1600	*0.30*	*0.25*	*0.55*
2000	*0.20*	*0.10*	*0.30*
2400	*0.10*	*0.05*	*0.15*
$f_Y(y)$	*0.60*	*0.40*	

Example 5.4 *This is a continuation of Example 5.2.*

The total cost of a computer system (computer and monitor) is given by the random variable $T = X + Y$. Compute the probability function $f_T(t)$ and $E(T)$.

Solution The range of T is the set of all possible values of $X + Y$. These values are listed in the first row of Table 5.3. We compute $f_T(2400)$ by noting that

$$(X + Y = 2400) = (X = 2000, Y = 400) \cup (X = 1600, Y = 800).$$

Therefore,

$$P(X + Y = 2400) = P(X = 2000, Y = 400) + P(X = 1600, Y = 800)$$
$$= 0.20 + 0.25 = 0.45.$$

We compute the remaining entries in the same way.

Table 5.3

t	*2000*	*2400*	*2800*	*3200*
$f_T(t)$	*0.30*	*0.45*	*0.20*	*0.05*

We compute $E(T)$ directly from Table 5.3.

$$E(T) = 2000 \times 0.30 + 2400 \times 0.45 + 2800 \times 0.2 + 3200 \times 0.05 = 2400.$$

Example 5.5 *Computing the jpmf from a sampling experiment*

In many cases the joint pmf is not given, but must be computed from the description of the sampling experiment. Consider, for example, a random sample of size two taken without replacement from an urn containing 2 nickles, 3 dimes, and 3 quarters. The number of nickles, dimes, and quarters in the sample are denoted by X, Y, Z, respectively. Compute the joint pmf of X and Y.

Solution The solution is displayed in Table 5.4.

Table 5.4

Joint pmf of X, Y				
$X \backslash Y$	0	1	2	$f_X(x)$
0	3/28	9/28	3/28	15/28
1	6/28	6/28	0	12/28
2	1/28	0	0	1/28
$f_Y(y)$	10/28	15/28	3/28	

Details of the computations: The denominator 28 in each of the non-zero terms is the total number of ways of choosing two coins from eight, which equals $C_{8,2} = 28$. The event $(X = 0, Y = 0)$ describes a sample consisting of two quarters only; the numerator is the number of ways of choosing two quarters from three, which equals $C_{3,2} = 3$. The other entries are computed in a similar way. For instance, the joint probabilities $f(0,1)$ and $f(1,0)$ are computed as follows:

$$f(0,1) = \frac{\binom{3}{1}\binom{3}{1}}{\binom{8}{2}} = \frac{9}{28}, \ f(1,0) = \frac{\binom{2}{1}\binom{3}{1}}{\binom{8}{2}} = \frac{6}{28}, \text{ etc.}$$

Note that $f(x,y) = 0$ whenever $x + y > 2$; this is because the total number of coins in the sample cannot exceed the sample size, which equals two.

Conditional Distribution of Y given $X = x$

The *conditional probability* of the event $Y = y$ given that $X = x$, which we denote by $P(Y = y|X = x)$, is defined in exactly the same way we defined the conditional probability $P(B|A) = P(A \cap B)/P(A)$; it is given by

$$P(Y = y|X = x) = \frac{P(Y = y, X = x)}{P(X = x)} = \frac{f(x,y)}{f_X(x)}, \ (f_X(x) > 0). \tag{5.7}$$

Example 5.6 *This is a continuation of Example 5.2.*

A customer buys a \$400 monitor. What is the probability that she also bought a \$1600 computer?

Solution

With reference to the joint pmf in Table 5.1 we see that

$$P(X = 1600|Y = 400) = \frac{f(1600, 400)}{f_Y(400)} = \frac{0.30}{0.60} = 0.50.$$

The *conditional distribution of Y given $X = x$* (where $f_X(x) > 0$) is the function $f(y|x)$ defined by:

$$f(y|x) = P(Y = y|X = x) = \frac{f(x,y)}{f_X(x)}, f_X(x) > 0. \tag{5.8}$$

The condition $f_X(x) > 0$ is intuitively clear, since the only x values that we observe are those for which $f_X(x) > 0$. We now verify that for each $x \in R_X$ the function $y \mapsto f(y|x)$ satisfies Equation 3.3 and is therefore a probability function. In detail:

$$\sum_y f(y|x) = \sum_y \frac{f(x,y)}{f_X(x)} = \frac{f_X(x)}{f_X(x)} = 1.$$

The *conditional expectation* of Y given $X = x$, denoted $E(Y|X = x)$, is the expected value of Y computed using the conditional distribution of Y given $X = x$. More precisely,

$$E(Y|X = x) = \sum_y y f(y|x). \tag{5.9}$$

Looking at Equation 5.9 we see that $E(Y|X = x)$ is a function of X, denoted $E(Y|X)$. Using Theorem 3.1 to compute its expected value we obtain the useful result that $E(E(Y|X)) = E(Y)$. In detail:

$$E(E(Y|X)) = \sum_x E(Y|X = x)f(x) = \sum_x (\sum_y y f(y|x))f(x)$$

$$= \sum_x (\sum_y y f(x,y)) = \sum_y y(\sum_x f(x,y)) = \sum_y y f_Y(y) = E(Y).$$

The product rule for computing the joint pmf: There is a product rule for computing joint probabilities that is similar to the product rule for events (Equation 2.26).

$$P(X = x, Y = y) = P(Y = y|X = x) \times P(X = x). \tag{5.10}$$

In the general case the joint pmf is formatted as in Table 5.5.

Table 5.5

	\multicolumn{5}{c}{*Joint pmf of (X, Y)*}
$X \backslash Y$	y_1	$\cdots$	y_j	$\cdots$	$f_X(x)$
x_1	$f(x_1, y_1)$	$\cdots$	$f(x_1, y_j)$	$\cdots$	$\sum_y f(x_1, y)$
$\vdots$	$\vdots$	$\vdots$	$\vdots$	$\vdots$	$\vdots$
x_i	$f(x_i, y_1)$	$\cdots$	$f(x_i, y_j)$	$\cdots$	$\sum_y f(x_i, y)$
$\vdots$	$\vdots$	$\vdots$	$\vdots$	$\vdots$	$\vdots$
$f_Y(y)$	$\sum_x f(x, y_1)$	$\cdots$	$\sum_x f(x, y_j)$	$\cdots$	

The marginal distribution functions f_X and f_Y are obtained by simply summing the entries in Table 5.5 across the rows and columns with the row and columns sums placed in the corresponding rightmost column and last row, respectively.

The joint pmf of n Random Variables

The concept of a joint pmf is easily extended to three or more discrete random variables.

Definition 5.2 *The joint pmf of n discrete random variables $X_1, \ldots, X_n$ is defined to be*

$$f(x_1, \ldots, x_n) = P(X_1 = x_1, \ldots, X_n = x_n). \tag{5.11}$$

5.2.1 Independent Random Variables

The concept of independence for events developed in Section 2.4.3 will now be extended to two or more random variables. We recall that the events A and B are said to be independent if

$$P(A \cap B) = P(A)P(B).$$

Definition 5.3 *We say that the random variables X and Y are mutually independent if*

$$P(X = x, Y = y) = P(X = x)P(Y = y) \tag{5.12}$$

for every x and y. We say that the random variables $X_1, \ldots, X_n$ are mutually independent if

$$P(X_1 = x_1, \ldots, X_n = x_n) = P(X_1 = x_1) \cdots P(X_n = x_n) \tag{5.13}$$

for every x_i, $i = 1, \ldots, n$.

We can express condition 5.12 for independence in terms of the marginal distribution functions f_X and f_Y as follows:

Proposition 5.1 *The random variables X and Y are independent if and only if their joint probability function is the product of their marginal distribution functions, i.e.,*

$$f(x, y) = f_X(x)f_Y(y), \text{ for all } x, y. \tag{5.14}$$

More generally, the random variables $X_1, \ldots, X_n$ are mutually independent if and only if their joint pmf is the product of their marginal distribution functions, i.e.,

$$f(x_1, \ldots, x_n) = f_{X_1}(x_1) \cdots f_{X_n}(x_n). \tag{5.15}$$

Bernoulli trials process: A *Bernoulli trials process* is a sequence of independent and identically distributed Bernoulli random variables $X_1, \ldots, X_n$. It is the mathematical model of n repetitions of an experiment under identical conditions, each experiment producing only two outcomes called *success/failure* or *heads/tails*, etc.

Example 5.7 *A mathematical model for coin tossing*

The simplest mathematical model of a Bernoulli sequence of trials is tossing a coin n times with probability p for heads. Let $X_i = 1$ if the ith toss comes up heads and set $X_i = 0$ if the ith toss comes up tails. The sample space consists of 2^n sequences of n digits, each digit being a 1 or a 0, so each sample point can be represented as a sequence of $1's$ and $0's$ in the following format:

$$s = (0, 1, 1, 0, \ldots, 1, 0) = (Tail, Head, Head, Tail, \ldots, Head, Tail).$$

The random variables X_i are identically distributed with $P(X_i = 0) = 1 - p$, $P(X_i = 1) = p$ since the probability of tossing a head is the same for each toss. The coin is said to be fair when $p = 0.5$.

It is reasonable to assume that the outcomes of the ith and jth tosses are independent events (the coin has no memory) and therefore X_i and X_j are mutually independent random variables; indeed, the same reasoning shows that the random variables $X_1, \ldots, X_n$ are mutually independent, and therefore, by Equation 5.13, each sample point consisting of x $1's$ and $n - x$ $0's$, is assigned the the probability $p^x q^{n-x}$. In detail,

$$P(0, 1, \ldots, 1, 0) = P(X_1 = 0)P(X_2 = 1) \cdots P(X_{n-1} = 1)P(X_n = 0)$$
$$= qp \cdots pq$$
$$= p^x q^{n-x}.$$

The Bernoulli trials process is widely used to model a variety of random phenomena; here are additional examples.

1. **Binomial Lattice Model for Stock Prices:** See Example 3.25.

2. **Quality Control:** As items come off a production line they are inspected for defects. When the ith item inspected is defective we record $X_i = 1$ and set $X_i = 0$, otherwise.

3. **Clinical Trials:** Patients with a disease are given a medication; if the ith patient recovers we set $X_i = 1$ and set $X_i = 0$, otherwise.

Definition 5.4 *Bernoulli trials process*

A *Bernoulli trials process* is a sequence of independent and identically distributed (iid) random variables $X_1, \ldots, X_n$, where each random variable takes on only one of two values 0, 1. The number $p = P(X_i = 1)$ is called the probability of success and $1 - p = P(X_i = 0)$ is called the probability of failure. It is easy to verify that $E(X_i) = p$. The sum

$$X = \sum_{1 \le i \le n} X_i, \text{ where } P(X_i = 1) = p, \ P(X_i = 0) = 1 - p, \tag{5.16}$$

counts the number of heads in n tosses of a coin with $P(heads) = p$; so it has a binomial distribution with parameters n, p (see Proposition 3.3); which is why a Bernoulli trials process is also called a *binomial experiment*.

5.3   The Multinomial Distribution

We now generalize the Bernoulli trials process to the case where each experiment results in one of $k \ge 2$ mutually exclusive and exhaustive outcomes denoted by $C_1, C_2, \ldots, C_k$; these outcomes are sometimes called *cells* and also *categories*. We denote their probabilities by $P(C_i) = p_i$, where $\sum_{1 \le i \le k} p_i = 1$. The values $p_i = P(C_i)$ are also called *cell probabilities*.

Example 5.8

Rolling a six-sided die n times is a simple example of a multinomial experiment. Here, of course, $k = 6$ and A_i denotes the event that an i is thrown. The assertion that the die is fair is equivalent to asserting that $P(A_i) = 1/6$. We shall use the term *multinomial experiment* to describe this model.

Let $Y_1, Y_2, \ldots, Y_k$ denote the frequencies of the outcomes $C_1, C_2, \ldots, C_k$, respectively, and note that $\sum_{1 \le i \le k} Y_i = n$, so these random variables are **not** independent. A formula for the joint distribution of these variables is given in Equation 5.17; it is called the *multinomial distribution.*

$$P(Y_1 = y_1, \ldots, Y_k = y_k) = \binom{n}{y_1, \ldots, y_k} p_1^{y_1} \cdots p_k^{y_k}$$

$$= \frac{n!}{y_1! \cdots y_k!} p_1^{y_1} \cdots p_k^{y_k}. \tag{5.17}$$

We shall not derive this formula in detail, except to point out that it is similar to the counting argument used to derive the expression for the multinomial coefficients in 2.23. Observe that each Y_i has a binomial distribution with parameters n, p_i (why?); consequently, $E(Y_i) = np_i$. The observed value y_i is called the *observed frequency* and the value np_i is called the *expected frequency* or *theoretical frequency.*

Example 5.9 *Consider the experiment of tossing 3 fair coins and repeating this experiment 8 times. Let C_i, $(i = 0, 1, 2, 3)$ denote the event that i heads were thrown. Let $Y_i =$ the number of times i heads occurred. Compute: (1) $E(Y_i)$ $(i = 0, 1, 2, 3)$ and (2) $P(Y_0 = 1, Y_1 = 3, Y_2 = 3, Y_3 = 1)$.*

Solution The random vector (Y_0, Y_1, Y_2, Y_3) has a multinomial distribution with probabilities p_i given by

$$p_i = P(C_i) = \binom{3}{i} 2^{-3}, \ (i = 0, 1, 2, 3),$$

and $n = 8$ repetitions of the experiment.

1. Each Y_i has the binomial distribution with parameters $n = 8$ and p_i as just defined. Consequently,

$$E(Y_i) = 8p_i = \binom{3}{i} \ (i = 0, 1, 2, 3).$$

2. We compute $P(Y_0 = 1, Y_1 = 3, Y_2 = 3, Y_3 = 1)$ by inserting the values $y_0 = 1, y_1 = 3, y_2 = 3, y_3 = 1, n = 8$ (and the values for p_i just computed) into Equation 5.17. Thus,

$$P(Y_0 = 1, Y_1 = 3, Y_2 = 3, Y_3 = 1) = \frac{8!}{1!3!3!1!} \frac{1}{8} \left(\frac{3}{8}\right)^3 \left(\frac{3}{8}\right)^3 \frac{1}{8}$$

$$= 0.0487.$$

5.4 Mean and Variance of a Sum of Random Variables

The main results of this section are some useful formulas for the mean and variance of sums of random variables expressed in terms of the means and variances of the individual summands. We begin with the simplest case of two summands. All our results are consequences of the following basic theorem whose proof is omitted.

Theorem 5.1 *Let (X, Y) denote a random vector with joint pmf $f(x, y)$ and let $\phi(x, y)$ denote a function of two real variables x, y. Then $\phi(X, Y)$ is a random variable, and*

$$E\left(\phi(X, Y)\right) = \sum_x \sum_y \phi(x, y) f(x, y) \ \textit{(discrete case)}. \tag{5.18}$$

Examples and applications

1. For random variables X and Y and arbitrary constants a and b we have

$$E(aX + bY) = aE(X) + bE(Y). \tag{5.19}$$

Derivation of 5.19: Choosing $\phi(x, y) = ax + by$ in Equation 5.18 and computing the summations yields

$$
\begin{aligned}
E(aX + bY) &= \sum_x \sum_y (ax + by) f(x, y) \\
&= \sum_x \sum_y ax f(x, y) + \sum_y \sum_x by f(x, y) \\
&= \sum_x ax \sum_y f(x, y) + \sum_y by \sum_x f(x, y) \\
&= a \sum_x x f_X(x) + b \sum_y y f_Y(y) \\
&= aE(X) + bE(Y).
\end{aligned}
$$

2. We now extend Equation 5.19 to the expectation of a finite sum of random variables.

   Let $X_1, \ldots, X_n$ be a sequence of random variables and $a_1, \ldots, a_n$ be a sequence of arbitrary constants; then

$$E\left(\sum_{1 \le i \le n} a_i X_i \right) = \sum_{1 \le i \le n} a_i E(X_i). \tag{5.20}$$

Equation 5.20 is called the *addition rule* for computing the expectation of a sum of random variables; its straightforward proof is omitted.

Some applications of the addition rule:

1. Consider an iid sequence $X_1, X_2, \ldots, X_n$ with $E(X_i) = \mu$. Choosing $a_i = 1/n$, $i = 1, 2, \ldots, n$ we obtain the sample mean

$$\overline{X} = \frac{\sum_{1 \le i \le n} X_i}{n}, \text{ consequently}$$

$$E(\overline{X}) = \sum_{1 \le i \le n} \frac{E(X_i)}{n} = \sum_{1 \le i \le n} \frac{\mu}{n} = \mu. \tag{5.21}$$

2. Show that the expected value of a binomial random variable equals np (cf. Equation 3.42). We compute its variance in Example 5.13.

   Solution We use the fact that a binomial random variable with parameters n, p can be represented as a sum of n iid Bernoulli random variables $X = \sum_{1 \le i \le n} X_i$ with $E(X_i) = p$, $i = 1, 2, \ldots, n$ (see Definition 5.4 and Equation 5.16). Choosing $a_i = 1$, $i = 1, 2, \ldots, n$ in Equation 5.20 we see that

$$E(X) = E\left(\sum_{1 \le i \le n} X_i \right) = \sum_{1 \le i \le n} E(X_i) = p + \cdots + p = np.$$

The same argument shows that if $X = \sum_{1 \le i \le n} X_i$ and $E(X_i) = \mu$, $i = 1, 2, \ldots n$ then $E(X) = E\left(\sum_{1 \le i \le n} X_i \right) = n\mu.$

The Covariance $Cov(X,Y)$

Let X and Y be two random variables with means μ_X, μ_Y and variances σ_X^2, σ_Y^2. We seek a formula for the variance of the sum, denoted by $V(X+Y)$. The computation of $V(X+Y)$ requires the new concept of *covariance* of X and Y, denoted $Cov(X,Y)$, and defined by

$$Cov(X,Y) = E\left((X-\mu_X)(Y-\mu_Y)\right). \tag{5.22}$$

It corresponds to choosing $\phi(x,y) = (x-\mu_X)(y-\mu_Y)$ in Equation 5.18. The covariance measures how the two variables vary together. Suppose, for example, that (X,Y) represents the heights of a mother and her son. We would expect that a mother taller than average (so $X > \mu_X$) is more likely to have a son taller than average (so $Y > \mu_Y$) with the inequalities reversed when the mother is shorter than average. This suggests that the product $(X-\mu_X)(Y-\mu_Y)$ is more likely to be positive since both factors are more likely to have the same sign. In other words, we would expect that $Cov(X,Y) = E\left((X-\mu_X)(Y-\mu_Y)\right) \geq 0$. Intuitively, a positive value of the covariance indicates that X and Y tend to increase together while a negative value indicates that an increase in X is accompanied by a decrease in Y. A classic example of the latter is the drop in stock market averages that accompanies an increase in interest rates.

Shortcut formula for the covariance of X and Y

As in the case of the variance there is also a shortcut formula for the computation of the covariance:

$$Cov(X,Y) = E(XY) - \mu_X\mu_Y. \tag{5.23}$$

Derivation of Equation 5.23:

Expand the product

$$(X-\mu_X)(Y-\mu_Y) = XY - \mu_X Y - \mu_Y X + \mu_X\mu_Y.$$

Then take expected values of both sides like so:

$$\begin{aligned}
E((X-\mu_X)(Y-\mu_Y)) &= E(XY - \mu_X Y - \mu_Y X + \mu_X\mu_Y) \\
&= E(XY) - \mu_X E(Y) - \mu_Y E(X) + \mu_X\mu_Y \\
&= E(XY) - 2\mu_X\mu_Y + \mu_X\mu_Y \\
&= E(XY) - \mu_X\mu_Y.
\end{aligned}$$

Note that $Cov(X,X) = Var(X)$.

Example 5.10 *Compute $Cov(X,Y)$ for the random vector with joint pmf given in Table 5.4.*

Solution This problem illustrates the computational efficiency of the shortcut formula. For instance, there is only one non-zero summand in the computation of $E(XY) = 1 \times 1 \times (6/28)$. A routine computation using the marginal distribution functions yields $\mu_X = 0.5$, $\mu_Y = 0.75$. Consequently,

$$Cov(X,Y) = 0.214 - 0.5 \times 0.75 = -0.1607.$$

Proposition 5.2 tells us that the covariance of independent random variables equals zero.

Proposition 5.2 *Let X and Y be independent random variables. Then*

$$E(XY) = E(X)E(Y), \text{ and therefore } Cov(X,Y) = E(XY) - E(X)E(Y) = 0. \tag{5.24}$$

Proof of Proposition 5.2 The hypothesis of independence means that $f(x,y) = f_X(x)f_Y(y)$; consequently,

$$E(XY) = \sum_x \sum_y xyf(x,y) = \sum_x \sum_y xyf_X(x)f_Y(y)$$

$$= \sum_x xf_X(x) \sum_y yf_Y(y) = E(X)E(Y).$$

Moment generating function of a sum of independent random variables A useful consequence of Proposition 5.2 is that the moment generating function of a sum of independent random variables is the product of their generating functions.

Theorem 5.2 *Suppose X and Y are independent random variables with moment generating functions $M_X(s)$, $M_Y(s)$, respectively. Then*

$$M_{(X+Y)}(s) = M_X(s) \times M_Y(s). \tag{5.25}$$

Theorem 5.2 is an immediate consequence of the fact that the expected value of the product of independent random variables is the product of their expected values (Proposition 5.2). In detail:

$$M_{(X+Y)}(s) = E\left(e^{s(X+Y)}\right) = E\left(e^{sX}e^{sY}\right)$$

$$= E\left(e^{sX}\right) \times E\left(e^{sY}\right) = M_X(s) \times M_Y(s).$$

More generally, if $(X_1, X_2, \ldots, X_n)$ is a sequence of independent random variables, then

$$M_{(X_1+\cdots X_n)}(s) = M_{X_1}(s) \times \cdots \times M_{X_n}(s). \tag{5.26}$$

Example 5.11 *Moment generating function of a binomial random variable*

The moment generating function of a binomial random variable was computed previously (Theorem 3.7). We now compute it using Theorem 5.2 and the fact that a binomial random variable (with parameters n, p) can be represented as the sum $X = X_1 + \cdots + X_n$, where each X_i, $(i = 1, \ldots, n)$ is a Bernoulli random variable with $P(X_i = 1) = p$ and $P(X_i = 0) = 1-p$ (see Definition 5.4 and Equation 5.16). It is easy to see that $M_{X_i}(s) = pe^s + 1 - p$, $(i = 1, 2, \ldots, n)$; consequently

$$M_X(s) = (pe^s + 1 - p) \times \cdots \times (pe^s + 1 - p) = (pe^s + 1 - p)^n.$$

The variance of a sum of random variables The following addition rule is useful for computing the variance of a sum of random variables in terms of the variances and covariances of the summands:

$$V(X + Y) = V(X) + V(Y) + 2Cov(X, Y); \tag{5.27}$$

$$V\left(\sum_{1 \le i \le n} X_i\right) = \sum_{1 \le i \le n} V(X_i) + 2\sum_{i < j} Cov(X_i, X_j). \tag{5.28}$$

We give an application of Equation 5.27 before deriving it.

Example 5.12 *An application of Equation 5.27*

Suppose X and Y are random variables with
$E(X) = 3$, $E(Y) = 2$; $E(X^2) = 26$, $E(Y^2) = 45$, $E(XY) = -14$. Using only this information, compute:
(a) $V(X)$ and $V(Y)$
(b) $Cov(X,Y)$ and $\rho(X,Y)$
(c) $V(X+Y)$
(d) Are X and Y independent?

Solution
(a) $V(X) = 26 - 9 = 17$ and $V(Y) = 45 - 4 = 41$
(b) $Cov(X,Y) = -14 - 6 = -20$ and $\rho(X,Y) = -20/(\sqrt{17}\sqrt{41}) = -0.7576$
(c) $V(X+Y) = 17 + 41 + 2 \times (-20) = 58 - 40 = 18$
(d) X and Y are not independent because $Cov(X,Y) = -20 \neq 0$.

Derivation of 5.27: We proceed in the usual way by first writing down the definition of $V(X+Y)$ and then expanding and rearranging the terms as follows:

$$V(X+Y) = E\left([(X+Y) - \mu_{X+Y}]^2\right) = E\left([(X - \mu_X) + (Y - \mu_Y)]^2\right)$$
$$= E\left((X - \mu_X)^2\right) + E\left(Y - \mu_Y)^2\right) + 2E\left((X - \mu_X)(Y - \mu_Y)\right)$$
$$= V(X) + V(Y) + 2Cov(X,Y).$$

The derivation of 5.28 is similar and is omitted.

The addition rule for computing the variance of a sum of independent random variables: When X and Y are independent (so $Cov(X,Y) = 0$) Equation 5.28 implies that the variance of the sum of two independent random variables equals the sum of the variances:

$$V(X+Y) = V(X) + V(Y). \tag{5.29}$$

Similarly, Equation 5.28 implies that the variance of a sum of mutually independent random variables $X_1, \ldots, X_n$ (so $Cov(X_i, X_j) = 0$, $i \neq j$,) equals the sum of the variances:

$$V(T = V(X_1 + \cdots + X_n) = V(X_1) + \cdots + V(X_n). \tag{5.30}$$

When the independent random variables have the same variance $\sigma^2 = V(X_i)$, $i = 1, \ldots, n$ Equation 5.30 yields the following useful result.

Proposition 5.3 *Let $X_1, \ldots, X_n$ be a sequence of independent random variables with the same variance $\sigma^2 = V(X_i)$, $i = 1, \ldots, n$. Then*

$$V\left(\sum_{1 \leq i \leq n} X_i\right) = \sum_{1 \leq i \leq n} V(X_i) = n\sigma^2. \tag{5.31}$$

Example 5.13 *Show that the variance of a binomial random variable equals $np(1-p)$ using Equation 5.31.*

Solution We first represent the binomial random variable as a sum of n independent Bernoulli random variables: $X = \sum_{1 \leq i \leq n} X_i$. (The reader might find it helpful to review Equation 3.43). A Bernoulli random variable X_i has variance $V(X_i) = p(1-p)$. The sum of the variances appearing on the right hand side of Equation 5.31 equals $np(1-p)$. Consequently, the variance of a binomial random variable with parameters n, p equals $np(1-p)$.

More generally the variance of the linear combination

$$a_1 X_1 + \cdots + a_n X_n,$$

where $X_1, \ldots, X_n$ is a sequence of independent random variables and $a_1, \ldots, a_n$ is a sequence of constants is given by

$$V(a_1 X_1 + \cdots + a_n X_n) = V(a_1 X_1) + \cdots + V(a_n X_n)$$
$$= a_1^2 V(X_1) + \cdots + a_n^2 V(X_n). \tag{5.32}$$

The variance of the sample mean $\overline{X}$

The variance of the sample mean $\overline{X}$ of a sequence of independent and identically distributed random variables $X_1, \ldots, X_n$, is given by

$$V(\overline{X}) = \frac{\sigma^2}{n}. \tag{5.33}$$

Derivation of 5.33: This formula for $V(\overline{X})$ is a special case of the addition rule (Equation 5.32) for variances where $a_i = 1/n$ and $V(X_i) = \sigma^2$, $(i = 1, \ldots n)$.

Computing the covariance under a change of scale

We recall that changing the system of units in which X is measured (from Fahrenheit to Celsius, for example) means replacing X by $aX + b$; this is called a change of scale. Unfortunately, the usefulness of the covariance as a measure of association is marred by the fact that its value depends on the systems of units used. The following proposition tells us how $Cov(X, Y)$ behaves under a simultaneous change of scale for X and Y.

Proposition 5.4 *Let the random variables $aX + b$, $cY + d$ be obtained from X, Y via a linear change of scale; then*

$$Cov(aX + b, cY + d) = acCov(X, Y). \tag{5.34}$$

The derivation of Equation 5.34 is a straightforward exercise (Problem 5.24).

The correlation coefficient

We now introduce the concept of the *correlation coefficient* $\rho(X, Y)$, defined by

$$\rho(X, Y) = \frac{Cov(X, Y)}{\sigma_X \sigma_Y}. \tag{5.35}$$

We say that the random variables are *uncorrelated* when $\rho = 0$. Independent random variables are always uncorrelated because their covariance $Cov(X, Y) = 0$, which implies that $\rho(X, Y) = 0$ as well. The converse is not true; that is, it is easy to construct a pair of uncorrelated random variables that are dependent (see Problem 5.65).

The correlation coefficient measures the strength of the linear relationship between the random variables X and Y. For instance it can be shown that $\rho = \pm 1$ implies that there exists constants a, b such that $Y = aX + b$; if $\rho = 1$ then $a > 0$, while $\rho = -1$ implies that $a < 0$. The correlation coefficient plays an important role in regression analysis and so we will defer a more detailed discussion of it until then.

The correlation coefficient has the following properties:

$$-1 \le \rho(X, Y) \le 1; \tag{5.36}$$
$$\rho(aX + b, cY + d) \quad = \quad \rho(X, Y) \text{ if } ac > 0; \tag{5.37}$$
$$\rho(aX + b, cY + d) \quad = \quad -\rho(X, Y) \text{ if } ac < 0. \tag{5.38}$$

The derivations of Equations 5.37 and 5.38 are left to you (Problem 5.25). The derivation of 5.36 requires Schwarz's inequality and is therefore omitted.

Notation: When there is no chance of confusion we drop the explicit dependence on X, Y and we write

$$\rho = \rho(X, Y).$$

Example 5.14

Refer to the joint pmf of Table 5.4. Compute the correlation coefficient.

Solution The essential steps in the computation of ρ follow:

$$E(X) = 0.5; \ E(Y) = 0.75;$$
$$V(X) = 0.32; \ V(Y) = 0.40;$$
$$E(XY) = \frac{6}{28}, \ \text{and therefore}$$
$$Cov(X, Y) = 0.214 - 0.5 \times 0.75 = -0.1607;$$
$$\rho(X, Y) = \frac{-0.1607}{\sqrt{0.32 \times 0.40}} = -0.449.$$

Since $Cov(X, Y) \neq 0$ it follows that X and Y are not independent.

Example 5.15

Suppose X and Y are random variables with

$$E(X) = 4 \ , E(Y) = -1 \ ; E(X^2) = 41 \ , E(Y^2) = 10 \ , E(XY) = 6.$$

Using only this information compute:

1. $V(X + Y)$.

   Solution The basic idea of the computation is to use Equation 5.27, which means we first compute $V(X)$, $V(Y)$, $cov(X, Y)$. Now

$$V(X) = E(X^2) - E(X)^2 = 41 - 16 = 25;$$
$$V(Y) = E(Y^2) - E(Y)^2 = 10 - 1 = 9;$$
$$Cov(X, Y) = E(XY) - E(X)E(Y) = 6 - (4 \times (-1)) = 10;$$
$$V(X + Y) = V(X) + V(Y) + 2Cov(X, Y) = 25 + 9 + 2 \times 10 = 56.$$

2. $\rho(X, Y)$.
   Solution

$$\rho(X, Y) = \frac{Cov(X, Y)}{\sigma_X \sigma_Y} = \frac{10}{5 \times 3} = \frac{10}{15} = 0.667.$$

3. Are X and Y independent?

   Solution No; because $\rho(X, Y) = 0.6667 \neq 0$.

5.4.1   The Law of Large Numbers for Sums of Independent and Identically Distributed (iid) Random Variables

Many problems in probability theory and statistical inference can be reduced to the study of the distribution of the *sample total* $T = \sum X_i$ and *sample mean* $\overline{X} = \sum \overline{X}_i / n$ of a sequence of independent and identically distributed random variables $X_1, \ldots, X_n$. The mathematical model of a *random sample* taken from a distribution is a sequence of independent and identically distributed (iid) random variables. The formal definition follows.

Definition 5.5 *A random sample of size n taken from a distribution F is a sequence of n mutually independent random variables $X_1, \ldots, X_n$ with a common distribution function $F(x) = P(X_i \leq x)$, $i = 1, 2, \ldots, n$.*

The distribution $F(X)$ is sometimes called the *parent distribution*. The mean and variance of $\overline{X}$ are given by

$$E(\overline{X}) = \mu \text{ and } V(\overline{X}) = \frac{\sigma^2}{n}.$$

These results follow from the addition rules for the mean and variance of a sum of independent random variables (Equations 5.21 and 5.33).

The *law of large numbers* is a precise statement of the proposition that the long run relative frequency of an event approaches a limit that equals its probability. In its most general form (cf. Theorem 5.3 at the end of this section) it is an assertion about the limiting behavior of the sample mean of an iid sequence of random variables.

The Law of Large Numbers for the Sample Proportion

The proportion of successes in a Bernoulli sequence of trials (cf. Example 5.7) is called the *sample proportion* and is denoted by $\hat{p}$, where

$$\hat{p} = \frac{\sum_{1 \leq i \leq n} X_i}{n} = \overline{X}, \text{ so } E(\hat{p}) = p \text{ and } V(\hat{p}) = \frac{p(1-p)}{n} \leq \frac{1}{4n},$$

where the inequality on the right is derived in Problem 3.31. Using Chebyshev's inequality we obtain the following upper bound on the probability that the sample proportion deviates from the true proportion by an amount greater than or equal to the quantity d:

$$P\left(|\hat{p} - p| \geq d\right) \leq \frac{p(1-p)}{nd^2} \leq \frac{1}{4nd^2}. \tag{5.39}$$

Consequently,

$$\lim_{n \to \infty} P\left(|\hat{p} - p| \geq d\right) \leq \lim_{n \to \infty} \frac{1}{4nd^2} = 0. \tag{5.40}$$

In other words, as the sample size increases the distribution of $\hat{p}$ becomes more and more concentrated around the true proportion p. For example, the Gallup poll, using a sample size of $n = 3,500$, predicted that Ronald Reagan would win the 1980 election with 55.3% of the popular vote. In this case, the sample proportion $\hat{p}$ equals 0.553, which is just the proportion of voters in the sample who said they would vote for Reagan. On election day, the final results were somewhat different, as Reagan's percentage of the vote was actually 51.6%; we note that the margin of error (predicted $-$ actual) equals 3.7%.

Example 5.16

Use Chebyshev's inequality to estimate the probability that the sample proportion obtained from a poll of voters based on a sample size of $n = 3,500$ differs from the true proportion by an amount less than or equal to 0.03.

Solution We have to compute the

$$P(|\hat{p} - p| < 0.03) = 1 - P(|\hat{p} - p| \geq 0.03).$$

We use inequality 5.39 with $n = 3500$ we obtain the upper bound

$$P(|\hat{p} - p| \geq 0.03) \leq \frac{1}{4 \times 3500 \times (0.03)^2} = 0.0794.$$

Therefore

$$P(|\hat{p} - p| < 0.03) = 1 - P(|\hat{p} - p| \geq 0.03) \geq 1 - 0.0794 = 0.9206.$$

$$\begin{aligned} P(|\hat{p} - p| < 0.03) &= 1 - P(|\hat{p} - p| \geq 0.03) \\ &\geq 1 - 0.0794 = 0.9206. \end{aligned}$$

The next example illustrates an important link between the binomial and empirical distribution functions.

Example 5.17 *Let $X_1, \ldots, X_n$ denote a sequence of mutually independent random variables with the same distribution $F(x)$; we call such a sequence a* random sample of size n *from the df $F(x)$. Show that the random variable $n\hat{F}_n(x)$ has a binomial distribution with parameters n, $p = F(x)$.*

Solution This result plays an important role linking probability theory to statistical inference, so we state it separately as Proposition 5.5.

Proposition 5.5 *Let $\hat{F}_n(x)$ be the empirical distribution of a random sample $X_1, \ldots, X_n$ taken from the df $F(x)$. Then $n\hat{F}_n(x)$ has a binomial distribution with parameters n, $p = F(x)$.*

Proof of Proposition 5.5

Define the related sequence Y_i as follows: $Y_i = 1$, if $X_i \leq x$; $Y_i = 0$, if $X_i > x$. In other words, $Y_i = 1$ if the ith element of the sample $X_i \leq x$, and is equal to zero elsewhere. Thus, the sum $\sum_{1 \leq i \leq n} Y_i$ counts the number of observations in the sample that are less than or equal to x. Consequently,

$$\sharp\{X_i \leq x\} = \sum_{1 \leq i \leq n} Y_i, \text{ and, therefore}$$

$$\hat{F}_n(x) = \frac{\sharp\{X_i \leq x\}}{n} = \frac{\sum_{1 \leq i \leq n} Y_i}{n}.$$

Note that the Y_i's are mutually independent, since the X_i's are; in addition, they only take on the two values 0 and 1; therefore, the sequence $Y_1, \ldots, Y_n$ forms a Bernoulli sequence of trials with

$$P(Y_i = 1) = P(X_i \leq x) = F(x).$$

We have thus shown that

$$n\hat{F}_n(x) = \sum_{1 \leq i \leq n} Y_i; \tag{5.41}$$

consequently, $n\hat{F}_n(x)$ has a binomial distribution with parameters n, $p = F(x)$ as claimed.

An application to the fundamental theorem of mathematical statistics

Let $\hat{F}_n(x)$ denote the empirical distribution function corresponding to the iid sequence $X_1, \ldots, X_n$ with the distribution $F(x)$. The *fundamental theorem of mathematical statistics* in its simplest form, states that the empirical distribution function $\hat{F}_n(x)$ is approximately equal to the distribution function $F(x)$ from which the random sample is drawn. In detail: According to Proposition 5.5 $n\hat{F}_n(x)$ has a binomial distribution with parameters n, $p = F(x)$. Applying Theorem 5.3 in this context yields the following version of the fundamental theorem of mathematical statistics:

$$\text{For every } d > 0, \ \lim_{n\to\infty} P\left(|\hat{F}_n(x) - F(x)| > d\right) = 0. \tag{5.42}$$

In other words, the empirical distribution $\hat{F}_n(x)$ function converges to the (unknown) df $F(x)$; that is,

$$\lim_{n\to\infty} \hat{F}_n(x) = F(x). \tag{5.43}$$

The law of large numbers for the sample proportion is, as noted earlier, just a special case of the law of large numbers for an iid sequence of random variables with a finite mean and variance.

Theorem 5.3 *(The Law of Large Numbers) Let $X_1, \ldots, X_n$ denote a sequence of independent random variables with $E(X_i) = \mu$ and $V(X_i) = \sigma^2$; then for every $d > 0$*

$$\lim_{n\to\infty} P\left(|\overline{X} - \mu| > d\right) = 0. \tag{5.44}$$

Proof Applying Chebyshev's inequality to $\overline{X}$ we see that

$$P\left(|\overline{X} - \mu| > d\right) \leq \frac{\sigma^2}{nd^2}; \text{ consequently, } \lim_{n\to\infty} P\left(|\overline{X} - \mu| > d\right) \leq \lim_{n\to\infty} \frac{\sigma^2}{nd^2} = 0.$$

5.4.2   The Central Limit Theorem

The most important result in probability theory is the central limit theorem, a far reaching and extremely useful generalization of the normal approximation to the binomial distribution discussed previously in Section (4.7). It states that the distribution of the *sample total* T_n, where $T_n = \sum_{1 \leq i \leq n} X_i$, and X_i, $i = 1, \ldots, n$ is an iid sequence, is approximately normal with mean $E(T_n)$ and $V(T_n)$. More precisely,

Theorem 5.4 *The central limit theorem (CLT) Let $T_n = \sum_{1 \leq i \leq n} X_i$, where $X_1, \ldots, X_n$ be an iid sequence of random variables with mean $E(X_i) = \mu$ and variance $V(X_i) = \sigma^2$ (so $E(T_n) = n\mu$ and $\sigma(T_n) = \sigma\sqrt{n}$). Then,*

$$\lim_{n\to\infty} P\left(\frac{T_n - n\mu}{\sigma\sqrt{n}} \leq x\right) = P(Z \leq x) = \Phi(x) \tag{5.45}$$

where Z is a standard normal random variable.

It is noteworthy that the normal approximation to the binomial is an immediate consequence of the CLT, as will now be shown. Suppose the iid sequence in Theorem 5.4 is Bernoulli with parameter $p = P(X_i = 1)$. In this case $E(X_i) = p$ and $V(X_i) = p(1 - p)$, so the hypotheses of the CLT are satisfied with $\mu = p$ and $\sigma = \sqrt{p(1 - p)}$. Substituting these values for μ and σ into Equation 5.45 yields Equation 4.51. The CLT is particularly useful because it yields a computable approximation to the distribution of the sample total and sample mean in terms of the normal distribution. It is, as noted earlier, a far reaching and

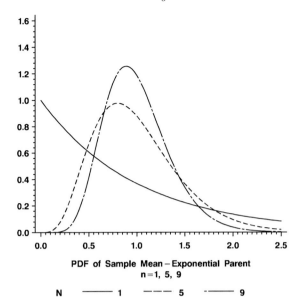

FIGURE 5.1: Probability density function of sample mean: exponential parent $(n = 1, 5, 9)$

remarkable generalization of the normal approximation to the binomial. The only conditions imposed are: (i) the random variables $X_1, \ldots, X_i, \ldots$ are iid and (ii) that their means and variances are finite. A sequence of iid Bernoulli random variables is just an important special case. The proof of this important result requires some techniques from advanced calculus, so it is omitted.

There is a version of the CLT for the *sample mean* $\overline{X} = \left(\sum_{1 \leq i \leq n} X_i \right) / n = T_n / n$ that is of particular importance in statistical applications, such as estimation and hypothesis testing.

Theorem 5.5 *The central limit theorem for the sample mean. Let $X_1, \ldots, X_n$ be an iid sequence of random variables with mean $E(X_i) = \mu$ and variance $V(X_i) = \sigma^2$. Then, for n sufficiently large, the distribution of the sample mean is approximately normal with mean and standard deviation $E(\overline{X}) = \mu$ and $\sigma(\overline{X}) = \sigma/\sqrt{n}$. More precisely,*

$$\lim_{n \to \infty} P\left(a \leq \frac{\overline{X} - \mu}{\sigma/\sqrt{n}} \leq b \right) = \Phi(b) - \Phi(a). \tag{5.46}$$

Equation 5.46 is algebraically equivalent to Equation 5.45 (just divide numerator and denominator by n). The elementary algebraic computations are left to the reader.

Figure 5.1 displays the pdfs of the sample mean $\overline{X}$ when the parent distribution is the exponential distribution with $\theta = 1$ and $n = 1, 5, 9$.

The most noteworthy feature is the fact that as the sample size increases the shape of the distribution of $\overline{X}$ appears to be very close to a normal distribution.

Example 5.18 *Suppose a machine set for filling 1 lb boxes of sugar yields a weight W with $E(W) = 16.0$ oz and $\sigma(W) = 0.2$. A carton of sugar contains 48 boxes of sugar. (1) Describe the distribution of the weight T of the carton. (2) Compute the probability that the total weight of the carton exceeds 48.2 lbs.*

Solution

1. The total weight of the carton $T = \sum_{i=1}^{48} W_i$, with $E(W_i) = 16.0$ oz and $\sigma(W_i) = 0.2$. It follows from the central limit theorem that T is approximately normally distributed with $\mu = 48 \times 16 = 768$ oz and $\sigma(T) = 0.2 \times \sqrt{48} = 1.39$.

2. We first convert 48.2 lbs to ounces. Thus,

$$P(T > 48.2 \text{ lbs}) = P(T > 771.2 \text{ oz.})$$
$$= P\left(\frac{T - 768}{1.39} > \frac{771.2 - 768}{1.39}\right)$$
$$= P(Z > 2.30) = 0.0107.$$

5.5   Why Stock Prices Have a Lognormal Distribution: An Application of the Central Limit Theorem

Empirical studies of the stock market suggest that the sequence of returns from a stock, stock index, or mutual fund are no more predictable than the sequence of heads and tails produced by tossing a coin (cf. Figure 3.6). Economists explain this unpredictability in terms of the *efficient market hypothesis*, which states that the current price of a stock always reflects all available information, and will change only with the arrival of new information. Since new information cannot be predicted, neither can a future change in price; consequently, the value of the return in one period yields no useful information about its value in any other period. The binomial lattice model (cf. Example 3.25) explicitly assumes that stock price fluctuations are determined by a sequence of coin tosses; that is, after each period the stock price increases by a factor $u > 1$ with probability p if the coin comes up heads, and decreases by a factor $d < 1$ with probability $1 - p$ if the coin comes up tails. A mathematical characterization of this unpredictability, called the *random walk model*, states that the sequence of returns $(S_i - S_{i-1})/S_{i-1}$, $i = 1, 2, \ldots$, is an iid sequence of random variables. It follows at once that the sequence of total returns, $S_i/S_{i-1} = 1 + (S_i - S_{i-1})/S_{i-1}$, $i = 1, 2, \ldots$ is also an iid sequence of random variables. In the real world, however, stocks are traded continuously throughout the day and their price ratios do not go up and down by fixed amounts, which is why we now extend the theory beyond the simple coin tossing model discussed earlier in Section 3.5.1.

The Continuous Time Random Walk Model for Stock Prices: We denote the stock price at time t by $S(t)$ and its initial price by $S_0 = S(0)$. A widely used model for stock price fluctuations assumes that $S(t)$ has a lognormal distribution. It is a remarkable fact that this model is a consequence of the efficient market hypothesis and the central limit theorem as will now be shown.

- **Efficient market hypothesis:** The assumptions for the continuous time model for stock prices are similar to the ones for the binomial lattice model just described. Divide the time axis into intervals of equal length $1/n$ with end points $t_i = i/n$, $i = 0, 1, \ldots,$ and assume that

$$\frac{S(t_i)}{S(t_{i-1})}, \quad i = 1, 2, \ldots; t_i = i/n, \text{ is an iid sequence.} \tag{5.47}$$

Consequently,

$$\ln\left(\frac{S(t_i)}{S(t_{i-1})}\right), \quad i = 1, 2, \ldots; t_i = i/n, \text{ is also an iid sequence.} \tag{5.48}$$

- **Continuity assumptions:** We make the reasonable assumptions that the stock price, its expected value and variance vary continuously with t. More precisely, we assume that

$$\lim_{\tau \to t} S(\tau) = S(t), \tag{5.49}$$

$$\lim_{\tau \to t} E\left(\ln\left(\frac{S(\tau)}{S(s)}\right)\right) = E\left(\ln\left(\frac{S(t)}{S(s)}\right)\right), \tag{5.50}$$

$$\lim_{\tau \to t} V\left(\ln\left(\frac{S(\tau)}{S(s)}\right)\right) = V\left(\ln\left(\frac{S(t)}{S(s)}\right)\right). \tag{5.51}$$

We are now going to show, using the central limit theorem (Proposition 5.4) and the efficient market hypothesis, that $\ln(S(t)/S_0)$ is normally distributed, i.e., $S(t)/S_0$ has a lognormal distribution in the sense of Definition 4.6.

Proposition 5.6 *Assume the price $S(t)$ of a stock satisfies the efficient market hypothesis and continuity assumptions (Equations 5.49, 5.50, and 5.51). Then $S(t)/S_0$ has a lognormal distribution with parameters $t\mu$ and $t\sigma^2$, where*

$$\mu = E\left(\ln\left(\frac{S(1)}{S_0}\right)\right) \ and \ \sigma^2 = V\left(\ln\left(\frac{S(1)}{S_0}\right)\right). \tag{5.52}$$

More generally, one can show that for $t_0 < t_1$

$$\ln\left(\frac{S(t_1)}{S(t_0)}\right) \stackrel{\mathcal{D}}{=} N((t_1 - t_0)\mu, (t_1 - t_0)\sigma^2). \tag{5.53}$$

The basic idea of the proof is to represent $\ln(S(t)/S_0)$ as a sum of independent, identically distributed random variables so that the central limit theorem can be brought to bear. We begin with the special case $t = 1$; the proof in the general case is given in Section 5.13. Taking logarithms of both sides of the algebraic identity

$$\frac{S(1)}{S_0} = \frac{S(1/n)}{S_0} \times \frac{S(2/n)}{S(1/n)} \times \cdots \times \frac{S((n-1)/n)}{S((n-2)/n)} \times \frac{S(n/n)}{S((n-1)/n)} \tag{5.54}$$

yields

$$\ln\left(\frac{S(1)}{S_0}\right) = \sum_{1 \le j \le n} \ln\left(\frac{S(j/n)}{S((j-1)/n)}\right). \tag{5.55}$$

Computing the expected values and variances of the left and right hand sides of Equation 5.55, we obtain

$$\mu = E\left(\ln\left(\frac{S(1)}{S_0}\right)\right) = \sum_{1 \le j \le n} E\left(\ln\left(\frac{S(j/n)}{S((j-1)/n)}\right)\right),$$

$$\sigma^2 = V\left(\ln\left(\frac{S(1)}{S_0}\right)\right) = \sum_{1 \le j \le n} V\left(\ln\left(\frac{S(j/n)}{S((j-1)/n)}\right)\right).$$

Now, the random variables $\ln(S(j/n)/S((j-1)/n)$, in addition to being mutually independent, have the same mean and variance, since we are assuming they also have the same distribution. Denoting their common mean and variance by μ_n and σ_n^2, respectively, we see that $n\mu_n = \mu$ and $n\sigma_n^2 = \sigma^2$. Therefore,

$$E\left(\ln\left(\frac{S(j/n)}{S((j-1)/n)}\right)\right) = \mu/n,$$

$$V\left(\ln\left(\frac{S(j/n)}{S((j-1)/n)}\right)\right) = \sigma^2/n.$$

Since $\ln(S(1)/S_0$ is a sum of n iid random variables with mean μ/n and variance σ^2/n, the CLT can be applied to yield

$$\lim_{n\to\infty} P\left(\frac{\ln(S(1)/S_0) - n(\mu/n)}{\sqrt{n(\sigma^2/n)}} \le z\right) = P\left(\frac{\ln(S(1)/S_0) - \mu}{\sigma} \le z\right)$$

$$= \Phi(z).$$

This shows that

$$\left(\frac{\ln(S(1)/S_0) - \mu}{\sigma}\right) \stackrel{\mathcal{D}}{=} N(0,1);$$

which is equivalent to the assertion that the distribution of $S(1)/S_0$ is lognormal with parameters μ, σ^2. This completes the proof of Proposition 5.6 for the special case $t = 1$. The proof in the general case uses the central limit theorem in the same way, but involves some technical details that are dealt with in Section 5.13.

The Stochastic Process $S(t)$**:** The time series $S(t)$ of Proposition 5.6 is an example of a *stochastic process* depending on the time parameter t; that is, for each t, $S(t)$ is a random variable. It follows that $W(t)$, defined as

$$W(t) = \frac{\ln(S(t)/S_0) - t\mu}{\sigma} \stackrel{\mathcal{D}}{=} N(0,t) \tag{5.56}$$

is also a stochastic process, one that is of fundamental importance in the analysis of stock price dynamics. This process is called *Brownian motion*. An immediate consequence of Equation 5.56 is a representation of $S(t)$ itself as the exponential of a Brownian motion process. More precisely,

$$S(t) = S_0 \exp(t\mu + \sigma W(t)). \tag{5.57}$$

It follows from Proposition 5.6 that the stochastic process $W(t)$ has the following properties:

1. $W(0) = 0$ and $W(t) \stackrel{\mathcal{D}}{=} N(0,t)$.

2. $W(t) - W(s)$ is normally distributed with

$$E(W(t) - W(s)) = 0 \text{ and } V(W(t) - W(s)) = E((W(t) - W(s))^2) = t - s.$$

3. For any sequence of times $0 \le t_1 < t_2 < \ldots < t_n$ the random variables

$$W(t_2) - W(t_1), W(t_3) - W(t_2), \ldots, W(t_n) - W(t_{n-1})$$

are mutually independent.

The last assertion follows from the assumption that the random variables $\ln\left(S(t_i)/S(t_{i-1})\right)$, $i = 1, 2, \ldots n$ are mutually independent, and the representation

$$\ln\left(S(t_i)/S(t_{i-1}, \right) = (t_i - t_{i-1})\mu + \sigma(W(t_i) - W(t_{i-1}),$$

which implies that the random variables $(W(t_i) - W(t_{i-1}))$, $i = 1, 2, \ldots n$ are also mutually independent.

Note: The stochastic process $\exp\left(t\mu + \sigma W(t)\right)$ on the right hand side of Equation 5.57 is called *geometric Brownian motion* with *drift parameter* μ and *volatility* σ.

Example 5.19 *The price $S(t)$ (t measured in years) of a stock is governed by a geometric Brownian motion with parameters $\mu = 0.09$, volatility $\sigma = 0.18$, and initial price $S_0 = \$50$.*

1. Show that the probability that the stock price increases by at least 50% in one year is 0.0401.
   Solution The assumptions $\mu = 0.09$ and $\sigma = 0.18$ together imply that

$$\ln(S(t)/S_0) \stackrel{\mathcal{D}}{=} N(0.09 \times t, (0.18)^2 \times t).$$

An increase of at least 50% means that $(S(1)/S_0) \geq 1.5$. Therefore

$$P((S(1)/S_0) \geq 1.5) = P(\ln(S(1)/S_0) \geq \ln(1.5))$$
$$= P\left(\frac{\ln(S(1)/S_0) - 0.09}{0.18} \geq \frac{0.4055 - 0.09}{0.18}\right)$$
$$= P(Z \geq 1.75) = 0.0401.$$

2. Show that the probability that the stock price loses at least 25% of its value in one year is 0.0179.
   Solution A decrease of at least 25% means that $(S(1)/S_0) \leq 0.75$. Therefore

$$P((S(1)/S_0) \leq 0.75) = P(\ln(S(1)/S_0) \leq \ln(0.75))$$
$$= P\left(\frac{\ln(S(1)/S_0) - 0.09}{0.18} \leq \frac{-0.2877 - 0.09}{0.18}\right)$$
$$= P(Z \leq -2.1) = 0.0179.$$

5.5.1 The Binomial Lattice Model as an Approximation to a Continuous Time Model for Stock Market Prices*

The essential feature of the binomial lattice model as discussed in Section 3.5.1 (cf. Example 3.25) is that the values of the stock price ratios $S_1/S_0, S_2/S_1, \ldots, S_n/S_{n-1}$ are determined by the sequence of heads and tails observed by tossing a coin. A slight modification of the model yields, as we now demonstrate, a very good approximation to continuous time models of the sort discussed in the preceding section. The approximation itself is easy to describe: Divide the time axis into small time steps of length $1/n$ and denote the corresponding sequence of stock prices by $S_n(t)$, $t = k/n$, $k = 1, 2, \ldots$. We define $S_n(t)$ for all t by setting $S_n(t) = S_n(k/n)$ for $k/n < t < (k+1)/n$. This is reasonable because the stock price (in the binomial lattice model) changes only at the discrete times $t = k/n$. Now every non-negative real number t lies in exactly one of the disjoint intervals $k/n \leq t < (k+1)/n$. The number k is the unique integer satisfying the inequality $k \leq nt < k+1$; we write $k = [nt]$, where

$$[x] = \text{ the greatest integer less than or equal to } x. \tag{5.58}$$

We call $[x]$ is the *greatest integer function*; $[nt]$ is the number of intervals of length $1/n$ contained in the interval $[0, t]$.

The following examples are easily verified: $[0.25] = 0$, $[1.99] = 1$. Note that $[nt] \leq nt < [nt] + 1$ and therefore $[nt]/n \leq t < ([nt] + 1)/n$. Consequently,

$$\lim_{n \to \infty} \frac{[nt]}{n} = t. \tag{5.59}$$

Using the greatest integer function notation we write the stock price as $S_n(t) = S_n([nt]/n)$. Recall that the binomial lattice is determined by three parameters that, in this particular case, depend on n and denoted (u_n, d_n, p_n).

Theorem 5.6 *Consider the binomial lattice model corresponding to the three parameters (u_n, d_n, p_n) defined as*

$$p_n = 1/2, \; u_n = \exp\left(\frac{\mu}{n} + \frac{\sigma}{\sqrt{n}}\right), \; d_n = \exp\left(\frac{\mu}{n} - \frac{\sigma}{\sqrt{n}}\right). \tag{5.60}$$

Then

$$\lim_{n \to \infty} S_n(t) = S(t) = S_0 \exp\left(t\mu + \sigma W(t)\right).$$

Before proving the theorem we give an example showing how to use it.

Example 5.20 *(Continuation of Example 5.19).*

The price $S(t)$ (t measured in years) of a stock is governed by a geometric Brownian motion with parameters $\mu = 0.09$, volatility $\sigma = 0.18$, and initial price $S_0 = \$50$. Find suitable parameters of the binomial lattice model that approximates the stock price with a basic time period of 3 months. Use Equation 5.60.

Solution For this case $n = 4$, $\mu = 0.09$, $\sigma = 0.18$. Consequently,

$$u_4 = \exp\left(\frac{0.09}{4} + \frac{0.18}{2}\right) = \exp(0.1125) = 1.1191,$$

$$d_4 = \exp\left((\frac{0.09}{4} - \frac{0.18}{2}\right) = \exp(-0.0675) = 0.9347.$$

Proof of Theorem 5.6 It follows from Equation 3.47 that the stock price at time t is given by

$$S_n(t) = S_0 u_n^{X_n(t)} d_n^{[nt] - X_n(t)}, \tag{5.61}$$

where $X_n(t)$ and $[nt] - X_n(t)$ denote, respectively, the number of up moves and down moves of the stock price in the time interval $[0, t]$. Now the number of up moves is the number of heads in $[nt]$ tosses of a coin with $P(\text{heads}) = p_n$; consequently, $X_n(t)$ has a binomial distribution with parameters $[nt]$, p_n. We choose these parameters so that the means and variances of the binomial lattice equal those of the continuous time model one is trying to approximate. This yields the following two equations:

$$E\left(\ln\left(\frac{S_n(j/n)}{S_n((j-1)/n)}\right)\right) = \mu/n, \tag{5.62}$$

$$V\left(\ln\left(\frac{S_n(j/n)}{S_n((j-1)/n)}\right)\right) = \sigma^2/n. \tag{5.63}$$

Since we have only two equations and three unknowns (u_n, d_n, p_n), we are free to fix the value of one of these parameters and then solve for the other two. Setting $p_n = 1/2$ and solving for u_n, d_n we obtain the values displayed in Equation 5.60. Problem 5.33 asks you to carry out the calculations to verify this assertion.

Inserting the values for u_n, d_n given in Equation 5.60 into Equation 5.61 yields the following more useful formula for the stock price:

$$S_n(t) = S_0 \exp\left(X_n(t)\left(\frac{\mu}{n} + \frac{\sigma}{\sqrt{n}}\right) + ([nt] - X_n(t))\left(\frac{\mu}{n} - \frac{\sigma}{\sqrt{n}}\right)\right). \qquad (5.64)$$

After some additional algebraic manipulations of the right hand side of Equation 5.64 (so that the central limit theorem can be brought to bear) we obtain the stock price formula:

$$S_n(t) = S_0 \exp\left(\frac{[nt]}{n}\mu + \sigma\left(\frac{X_n(t) - [nt]/2}{\sqrt{n}/2}\right)\right) = S_0 \exp\left(\frac{[nt]}{n}\mu + \sigma W_n(t)\right), \qquad (5.65)$$

where $W_n(t)$, $n = 1, 2, \ldots$ is the sequence of stochastic processes defined as

$$W_n(t) = \frac{X_n(t) - [nt]/2}{\sqrt{n}/2},$$

and $X_n(t)$, as previously noted, is binomial with parameters $[nt]$ and $p = 1/2$; therefore $E(X_n(t)) = [nt]/2$ and $V(X_n(t)) = [nt]/4$. When n is large, the normal approximation to the binomial implies that

$$\lim_{n\to\infty} \frac{X_n(t) - [nt]/2}{\sqrt{n}/2} \overset{\mathcal{D}}{=} N(0, t).$$

It is noteworthy that the sequence of stochastic processes $W_n(t)$, $n = 1, 2, \ldots$ share the following properties with Brownian motion defined earlier (cf. Equation 5.56):

- $E(W_n(t)) = 0$ and $V(W_n(t)) = t$

- $\lim_{n\to\infty} W_n(t) \overset{\mathcal{D}}{=} N(0, t)$

Indeed, it can be shown, by methods beyond the scope of this course, that $W_n(t)$ itself converges to Brownian motion, that is

$$\lim_{n\to\infty} W_n(t) = W(t). \qquad (5.66)$$

Remark The limit theorem stated in Equation 5.66 is a special case of the following, slightly more general result.

Theorem 5.7 *Suppose $X_n(t)$ is binomial with parameters $[nt]$, p, so $E(X_n(t)) = [nt]p$ and $\sqrt{V(X_n(t))} = \sqrt{[nt]p(1-p)}$. Let*

$$W_n(t) = \frac{X_n(t) - [nt]p}{\sqrt{np(1-p)}}.$$

Then

$$\lim_{n\to\infty} W_n(t) = W(t). \qquad (5.67)$$

Remark Setting $p = 1/2$ in Equation 5.67 yields Equation 5.66.

Proof Equation 5.66 and the fact that $\lim_{n\to\infty}[nt]/n = t$ together imply

$$\lim_{n\to\infty} S_n(t) = \lim_{n\to\infty} S_0 \exp\left(\frac{[nt]}{n}\mu + \sigma W_n(t)\right) = S_0 \exp\left(t\mu + \sigma W(t)\right) = S(t). \qquad (5.68)$$

5.6 Modern Portfolio Theory*

A *portfolio* is a collection of assets, such as stocks, bonds, and cash. A portfolio, of course, can consist of just one stock, a strategy that has been described as, "the quickest way to get poor." In other words, an investor must avoid putting all his eggs in one basket. The key insight of modern portfolio theory is that it is possible to create a portfolio of several risky assets that reduces an investor's overall risk while still earning a reasonable return. This is called *diversification*.

5.6.1 Mean-Variance Analysis of a Portfolio*

The benefits of diversification are best understood by means of a simple example that we have adapted from B.G. Malkiel's *A Random Walk Down Wall Street*, (1996), 6th ed., p. 236, W.W. Norton.

Example 5.21

Consider a resort island stock market with only two stocks: A, a fashionable beach resort and B, an umbrella manufacturer; denote the returns for each stock by R_A and R_B, respectively. The returns, of course, depend on the weather, of which there are only two types: a sunny season and a rainy one. We suppose that, on the average, half the seasons are sunny and half are rainy; consequently, $P(\text{sunny season}) = P(\text{rainy season}) = 0.5$. A sunny season results in a 50% return for the resort and a 25% loss for the umbrella manufacturer. The fortunes of the two companies are reversed when the season is a rainy one: The resort suffers a loss of 25% and the umbrella manufacturer enjoys a return of 50%. In other words, "an investor who bought stock in the umbrella manufacturer would find that half the time he earned a 50% return and half the time he lost 25% of his investment. Similarly, an investment in the resort would produce the same results." Thus R_A and R_B have the joint distribution function given by

$R_A \backslash R_B$	−25%	50%
−25%	0	0.5
50%	0.5	0

A routine calculation of the means, standard deviations, and correlation of the returns yields

$$\overline{R}_A = \overline{R}_B = 12.5\% \text{ and } \sigma_A = \sigma_B = 37.5\%, \rho(A, B) = -1.$$

Consequently, investing only in A (or only in B) produces a 12.5% return with a standard deviation of 37.5%. Consider, now, the return R_C of a portfolio weighted equally between the two stocks; that is, $R_C = (R_A + R_B)/2$. It is easy to see that $R_A + R_B = 25\%$, which implies that $R_C = 12.5\%$, so $\overline{R}_C = 12.5\%$, and $\sigma_C = 0$; in other words, the diversified portfolio C earns a risk free 12.5% return!

Table 5.6

Portfolio	A	B	C
risk σ	37.5%	37.5%	0
return $\overline{R}$	12.5%	12.5%	12.5%

Looking at Table 5.6 we see that portfolios A, B, and C all have the same 12.5% expected return, but they do not all have the same risks. Given a choice between two portfolios with the same expected returns, an investor will always choose the less risky one, i.e., the one with the smaller standard deviation. Using this criterion an investor will choose C over A because it has the same expected return, but is less risky. For this reason, we say that portfolio A is *inefficient* relative to portfolio C, and for the same reason, portfolio B is inefficient relative to portfolio C.

Calculating Expected Return and Standard Deviation of a Portfolio Consisting of Two or More Assets

We now consider the more general case where an investor has a sum of money to allocate, in possibly unequal proportions, among n assets (e.g., growth stocks, dividend paying stocks, US Treasury, and corporate bonds) for a particular length of time called the *holding period*. At the end of of the holding period one computes the rate of return $R = (X_1 - X_0)/X_0$, where X_0, X_1 denote the initial and final values of the investment, respectively. Denote the returns, variances, and covariances of the n assets by R_i, σ_i, $\sigma_{ij} = cov(R_i, R_j)$, respectively. Denote the proportion of the portfolio invested in the ith asset by w_i, $i = 1, 2, \ldots, n$, so

$$\sum_{1 \leq i \leq n} w_i = 1, \ w_i \geq 0.$$

The return R on the portfolio is then a weighted average of the returns on the n assets; more precisely, its return, expected return, and variance are

$$R = \sum_{1 \leq i \leq n} w_i R_i, \tag{5.69}$$

$$\overline{R} = \sum_{1 \leq i \leq n} w_i \overline{R}_i, \tag{5.70}$$

$$V(R) = \sum_{1 \leq i \leq n} w_i^2 \sigma_i^2 + \sum_{i \neq j} \sum w_i w_j \sigma_{ij}. \tag{5.71}$$

Referring back to Example 5.21 we see that Portfolio C consists of two equally weighted assets, so $n = 2$ and $w_1 = w_2 = 0.5$.

From Equation 5.69 it is clear that we can form an infinite number of portfolios from the n assets, not all of them being of equal interest to the investor. The set of all portfolios obtained from all possible linear combinations of the two assets is called the *feasible set*. The *efficient set theorem* (G.J. Alexander, W.F. Sharpe, and J.V. Bailey, (2001),*Fundamentals of Investments,* 3rd ed., Prentice Hall, p. 147) tells us that an investor will choose from the subset of portfolios that

1. offers maximum expected return for varying levels of risk, and

2. offers minimum risk for varying levels of expected return.

The set of portfolios satisfying these two conditions is known as the *efficient set*, or as the *efficient frontier*. Determining the efficient frontier is not an easy matter in general, so we simplify the analysis by considering the case of two assets in some detail.

- We say that asset 1 is *less risky* than asset 2 when $\overline{R}_1 = \overline{R}_2$ but $\sigma_1 < \sigma_2$.

- A *risk free* asset R_f, such as a treasury bill, is modeled as a constant random variable, so $\sigma_f = 0$. It is easy to see that the covariance of a risk free asset with a risky one is 0, that is $cov(R_f, R) = 0$.

- Consider now the case where the investor's portfolio is a combination of just two assets, so $n = 2$ in Equation 5.69, i.e., $R = w_1 R_1 + w_2 R_2$, $w_1 + w_2 = 1$, where $0 \leq w_i \leq 1$ is the proportion of the initial sum invested in asset i. Since $w_1 + w_2 = 1$ we can set $w_1 = 1 - t$ and $w_2 = t$; consequently, the rate of return for the portfolio R is a function of t; we denote this portfolio and its return by $R(t)$ and $\overline{R}(t)$, respectively:

$$R(t) = (1 - t)R_1 + tR_2, \ (0 \leq t \leq 1), \tag{5.72}$$
$$\overline{R}(t) = (1 - t)\overline{R}_1 + t\overline{R}_2, \ (0 \leq t \leq 1). \tag{5.73}$$

Note that $t = 0$ means that your portfolio consists only of asset 1, and similarly $t = 1$ means that your portfolio consists only of asset 2. Denoting the standard deviation of the portfolio $R(t)$ by $\sigma(t, \rho)$ and using the fact that $\sigma_{12} = \rho\sigma_1\sigma_2$, it is not too difficult to show that

$$\sigma^2(t, \rho) = (1 - t)^2\sigma_1^2 + 2t(1 - t)\sigma_{12} + t^2\sigma_2^2 \tag{5.74}$$
$$= (1 - t)^2\sigma_1^2 + 2t(1 - t)\rho\sigma_1\sigma_2 + t^2\sigma_2^2. \tag{5.75}$$

From Equation 5.75 it follows that $\rho < \rho'$ implies that $\sigma^2(t, \rho) \leq \sigma^2(t, \rho')$; consequently,

$$\sigma^2(t, -1) \leq \sigma^2(t, \rho) \leq \sigma^2(t, \rho') \leq \sigma^2(t, 1), \ -1 < \rho < \rho' < 1. \tag{5.76}$$

When $\rho = \pm 1$ the formulas for $\sigma(t, \pm 1)$ are given by

$$\sigma^2(t, 1) = ((1 - t)\sigma_1 + t\sigma_2)^2, \tag{5.77}$$
$$\sigma(t, 1) = (1 - t)\sigma_1 + t\sigma_2, \tag{5.78}$$
$$\sigma^2(t, -1) = ((1 - t)\sigma_1 - t\sigma_2)^2, \tag{5.79}$$
$$\sigma(t, -1) = |(1 - t)\sigma_1 - t\sigma_2|. \tag{5.80}$$

The set of efficient portfolios is obtained, as in Example 1.26 and Figure 1.16, by first constructing the mean-standard deviation diagram of the feasible set; that is, one plots the the points $(\sigma, \overline{R})$ for each portfolio in the feasible set.

5.7   Risk Free and Risky Investing*

The safest investment, and sometimes the most prudent one, is cash; that is, an investor deposits an amount A, called the *principal*, in a bank that pays interest of $r\%$ per annum. The *future value* of this investment at time t, where t is measured in years, is denoted $V(t)$; clearly $V(0) = A$. The future value depends on whether the interest is simple interest or compound interest. We note that depositing money into a bank is an example of an investment with a *risk free return* r, since, unlike a stock, its future value at time t is known. The comparison of risky returns, e.g., from stocks, with those from risk free investments plays an important role in financial analysis. *Present value analysis* is a criterion for choosing, among various investments, the one that is most preferable.

5.7.1   Present Value Analysis of Risk Free and Risky Returns*

Simple interest is the product of the principal, the interest rate, and the length of time it is on deposit; that is, the amount of interest earned is proportional to the length of time the principal remains on deposit. In particular, the interest earned after one year is Ar, after six months $Ar/2$, and after t years it is Art. The future value of the investment at time t is the principal plus the interest earned for the period, consequently

$$V(t) = \text{principal} \ + \ \text{interest} = A + Art = A(1 + rt). \tag{5.81}$$

Consider, by way of contrast, the return on an investment in a stock for one year, say, with initial and terminal values denoted by S_0, S_1, respectively. Suppose further that the value of the stock at year's end is a random variable given by the binomial lattice model, i.e.,

$$P\left(\frac{S_1}{S_0} = u\right) = p \text{ and } P\left(\frac{S_1}{S_0} = d\right) = 1 - p, \ d < 1 < u.$$

The expected total return, as one can verify by direct calculation, is

$$E\left(\frac{S_1}{S_0}\right) = pu + (1 - p)d.$$

Investors are classified according to their *risk tolerance*. An investor is considered *risk-averse*, *risk-neutral*, or *risk-seeking* accordingly as the expected return is greater than, equal to, or less than the risk free return. For a prudent investor the expected return must exceed the risk free return in order to compensate him for the increased risk; which is why such an investor is characterized as *risk-averse*. There are situations, such as *power ball lotteries*, where the the payoff is so huge—50 million dollars or more in some instances—while the cost of the lottery ticket is only a dollar, that many people, including the author of this book, wait in long lines to purchase a ticket, even though the expected return is negative. Table 5.7 summarizes the three types of risk tolerance.

Table 5.7 *Classification of investors according to their risk tolerance*

$$E\left(\frac{S_1}{S_0}\right) > 1 + r \ \textit{(risk-averse)}$$

$$E\left(\frac{S_1}{S_0}\right) = 1 + r \ \textit{(risk-neutral)}$$

$$E\left(\frac{S_1}{S_0}\right) < 1 + r \ \textit{(risk-seeking)}$$

When the investment is risk-neutral, that is, when $E(S_1/S_0) = 1 + r$ we obtain Equation 5.82 for p that is of independent interest:

$$E\left(\frac{S_1}{S_0}\right) = pu + (1 - p)d = 1 + r \text{ (risk-free total return).} \tag{5.82}$$

The solution to this equation, denoted by q, is given by

$$q = \frac{R - d}{u - d}, \ R = 1 + r. \tag{5.83}$$

The probability measure q, $1 - q$ (called the *risk-neutral probability measure*) plays an important role in the Black-Scholes formula for computing the fair price of an option. This is discussed in Section 5.8.1, where it is shown, in the paragraph just after Equation 5.104, that $d < R < u$, and therefore $0 < q < 1$.

Compound interest: Equation 5.81 must be modified when the interest is compounded periodically, say semi-annually. In this case the bank credits the account with the interest earned in the preceding six months, which is $A(r/2)$ and then pays interest on the new principal $V(1/2) = A(1 + (r/2))$ for the next six months. Consequently,

$$V(1) = V(1/2) \left(1 + \frac{r}{2}\right) = A\left(1 + \frac{r}{2}\right)^2.$$

To illustrate, consider a bank that pays depositors the interest rate $r = 0.06$, compounded semi-annually; consequently,

$$V(1) = A\left(1 + \frac{0.06}{2}\right)^2 = A(1 + 0.0609).$$

If the interest is compounded quarterly, then

$$V(1) = A\left(1 + \frac{0.06}{4}\right)^4 = A(1 + 0.061364).$$

Notice that 6% compounded semi-annually is equivalent to an *effective annual interest rate* of 6.09%, and when compounded quarterly the effective annual interest rate is 6.1364%. In general, the future value $V(1)$ depends not only on the interest rate, but on how frequently it is is compounded. In particular, the future value $V(1)$ of a deposit A earning interest at $r\%$ per annum, and compounded m times per year, is given by

$$V(1) = A\left(1 + \frac{r}{m}\right)^m. \tag{5.84}$$

The effective annual interest rate, denoted r_{eff}, is defined by the equation

$$1 + r_{eff} = \left(1 + \frac{r}{m}\right)^m. \tag{5.85}$$

The formula for the future value after t years is

$$V(t) = A\left(1 + \frac{r}{m}\right)^{[mt]}. \tag{5.86}$$

Continuous compound interest: Letting $m \to \infty$ in Equation 5.86 we obtain the formula for computing continuous compound interest:

$$V(t) = \lim_{m\to\infty} A\left(1 + \frac{r}{m}\right)^{[mt]} = \lim_{m\to\infty} A\left(\left(1 + \frac{r}{m}\right)^m\right)^{[mt]/m} = Ae^{rt}. \tag{5.87}$$

It follows from Equation 5.87 that $t^{-1}\ln(V(t)/V(0)) = r$ is the *continuously compounded rate of return* for the time period t.

Table 5.8 lists the future values $V(1)$ and the effective annual interest rates for various compounding frequencies.

Table 5.8 *The future value and effective annual interest rate of a $10,000 deposit after one year when the annual interest rate of 6% is compounded m times per year*

Compounding frequency m	Future value $V(1)$	Effective annual interest rate
$m = 1$ *(yearly)*	10,600.00	6%
$m = 2$ *(semiannually)*	10,609.00	6.09%
$m = 4$ *(quarterly)*	10,613.64	6.1364%
$m = 12$ *(monthly)*	10,616.78	6.1678%
$m = 52$ *(weekly)*	10,617.00	6.17%
$m = 365$ *(daily)*	10,618.31	6.1831%
$m = \infty$ *(continuous compounding)*	10,618.37	6.1837%

Example 5.22 *Bank One pays depositors 6.1% per annum, while Bank Two pays 6%, compounded quarterly. Which bank offers the better deal?*

Solution The correct answer is Bank Two because its effective annual interest rate $r_{eff} = 6.136\%$ is greater than Bank One's 6.1%.

Present value analysis: How much money should one put in the bank today in order to pay for a future obligation such as a tuition payment or retirement annuity? The correct answer depends on the compounding frequency m and the interest rate r. To simplify matters, assume the future payment is due one year from now, and the current interest rate r is compounded m times, so, the interest rate per period is r/m. Denoting the amount due by $V(1)$ and solving Equation 5.84 for A yields

$$A = \frac{V(1)}{\left(1 + \frac{r}{m}\right)^m}. \tag{5.88}$$

The amount A is called the *present value* of the future payment $V(1)$ and the term in the denominator— $(1 + (r/m))^{-m}$—is called the *discount factor*. Similar reasoning, but using Equations 5.86 and 5.87 shows that the present value of a future payment C at time t is

$$A = \frac{C}{\left(1 + \frac{r}{m}\right)^{[mt]}} \text{ (compounding frequency m)}, \tag{5.89}$$

$$A = Ce^{-rt} \text{ (continuous compound interest)}. \tag{5.90}$$

Example 5.23 *Find the amount A to be deposited in a bank account paying an interest rate of 6% per annum, compounded four times per year, in order to make a tuition payment of* $5000 *due in: (i) one year, (ii) 18 months.*

Solution (i) The amount A is the present value of $V(1) = \$5000$ with $m = 4$, $r/m = 0.06/4 = 0.015$. Substituting these values into Equation 5.88 we obtain the value

$$A = 4710.92 = \frac{5000}{(1 + 0.015)^4} = \frac{5000}{1 + 0.061364}.$$

This example illustrates the fact that *a dollar today is worth more than a dollar tomorrow* and why an interest rate is a measure of the *time value of money.*

(ii) In this case the interest rate and compounding period remain the same as in (i), only the time period changes, which is 1.5 years. Thus, A is the present value of $V(1.5) = \$5000$. Inserting these values into Equation 5.89 yields

$$A = 4572.71 = \frac{5000}{(1 + 0.015)^{[4 \times 3/2]}} = \frac{5000}{(1 + 0.015)^6} = \frac{5000}{1.0934}.$$

5.7.2 Present Value Analysis of Deterministic and Random Cash Flows*

The simplest problem in mathematical finance is determining the monthly payments required to pay off an installment loan for a car or a home mortgage. The monthly payments, denoted A, generate, for the lender, a *cash flow* $\{x_1, \ldots, x_n\}$, where x_i is the amount paid by the borrower at the beginning of the ith period. In this case $\{x_1 = A, \ldots, x_n = A\}$. The process of paying off a loan with n equal payments is called *amortization*. The method for computing the payment is simple: The present value of the cash flow sequence $(A, A, \ldots, A)$ must equal the amount of the loan, where the present value of the cash flow is defined as the sum of the present values of the payoffs x_i at the ith period.

Definition 5.6 *The present value (PV) of the cash flow $\{x_1, \ldots, x_n\}$* *(negative values of x_i are allowed) at the current interest rate of r per period is*

$$PV = \sum_{1 \le i \le n} \frac{x_i}{(1+r)^i}. \tag{5.91}$$

Remark We say that two cash flows are *equivalent* if their present values are the same.

The formula for the payment A for amortizing a loan of $\$P_0$ is given in Equation 5.92 below. We postpone its derivation to give some numerical examples.

$$\text{Amortizing a loan: } A = \frac{P_0 r}{1 - (1+r)^{-n}} \tag{5.92}$$

Applications and Examples:

1. A person borrows $\$100,000 = P_0$ for 15 years at 5.5% interest, to be repaid in equal monthly installments. Compute the monthly payment A.
   Solution The number of payments is $n = 12 \times 15 = 180$, and the monthly interest rate is $r = 0.055/12 = 0.0046$. Therefore,

$$A = \frac{P_0 r}{1 - (1+r)^{-n}} = \frac{100,000 \times (0.055/12)}{1 - (1 + (0.055/12))^{-180}} = 817.08.$$

2. Suppose the person in the previous problem borrows $\$100,000$ for 15 years at 6.25% interest. Compute the monthly payment A.
   Solution The number of payments is still $n = 180$, but the monthly interest rate is $r = 0.0625/12 = 0.0052$. Therefore,

$$A = \frac{P_0 r}{1 - (1+r)^{-n}} = \frac{100,000 \times (0.0625/12)}{1 - (1 + (0.0625/12))^{-180}} = 857.42.$$

Derivation of Equation 5.92:   To avoid disrupting the continuity of the derivation we need the following result for the sum of a finite geometric series:

$$\sum_{1 \le k \le n} t^k = \frac{t - t^{n+1}}{1 - t}. \tag{5.93}$$

To complete the argument we reason as follows: The present value of the cash flow $\{A, \ldots, A\}$ must equal the amount of the loan P_0; consequently,

$$P_0 = \sum_{1 \le i \le n} \frac{A}{(1+r)^i} = A\left(\frac{1 - (1+r)^{-n}}{r}\right). \tag{5.94}$$

The expression on the right hand side of Equation 5.94 is derived by applying Equation 5.93 with $t = (1+r)^{-1}$ to show that

$$\sum_{1 \le i \le n} \frac{1}{(1+r)^i} = \frac{(1+r)^{-1} - (1+r)^{-(n+1)}}{1 - (1+r)^{-1}}$$

$$= \frac{1 - (1+r)^{-n}}{(1+r) - 1} = \frac{1 - (1+r)^{-n}}{r}.$$

Clearly, Equation 5.94 is algebraically equivalent to Equation 5.92.

The preceding theory is easily extended to the case when the ith payoff is a random variable X_i; its present value is, therefore, also a random variable defined as

$$PV = \sum_{1 \le i \le n} \frac{X_i}{(1+r)^i}. \tag{5.95}$$

The expected present value of the random cash flow is defined as

$$E(PV) = \sum_{1 \le i \le n} \frac{E(X_i)}{(1+r)^i}. \tag{5.96}$$

Example 5.24 *For the situation considered in Example 5.19 compute the expected present value of the return on the stock after six months if the current rate of interest is 5%.*

Solution The stock price follows a geometric Brownian motion with parameters $\mu = 0.09$, $\sigma = 0.18$ and $t = 0.5$. Thus

$$E\left(\frac{S(0.5)}{S_0}\right) = \exp(0.5(0.09 + (0.18)^2/2)) = 1.0545.$$

The discount factor is $1.025 = (1 + 0.05/2)$; so the expected present value of the total return is $1.0288 (= 1.0545/1.025)$ and the expected present value of the return is 0.0288, or 2.88%.

5.8 Theory of Single and Multi-Period Binomial Options*

What is a fair price for a financial asset or an option to buy it—such as a stock or bond—with an uncertain future return? And, if the price be fair (however we define it), is the decision to buy it a reasonable one? These kinds of questions are not new; they go back, as Lorraine Daston has shown,[1] to the 17th and 18th centuries when mathematicians and philosophers such as B. Pascal (1623-1662), J. Bernoulli (1654-1705), J. d'Alembert (1717-1783), and P. S. Laplace (1749-1827) grappled with them. We are particularly interested here in determining the fair price of an *option*, which is the right to buy (*call option*) or sell (*put option*) an asset at a fixed price K called the *strike price* by a certain date known as the *expiration date* or *maturity*.

5.8.1 Black-Scholes Option Pricing Formula: Binomial Lattice Model*

A *call option* with a *strike price* K and an expiration date T is the right, but not the obligation, to buy a stock at the price K on the expiration date. Similarly, a *put option* with *strike price* K and expiration date T is the right, but not the obligation, to sell a stock at the price K on the expiration date. To simplify the analysis we assume the stock price is governed by the binomial lattice model and we consider only *European options*, which allows the option to be exercised only on the expiration date. An *American option* may be exercised at any time up to and including the expiration date.

Consider now the option price C on a stock whose price at maturity is $S(T)$: If $S(T) < K$ then the option is worthless because you could buy the stock at the market price $S(T)$, which

---

[1] L. Daston, *Classical Probabilty in the Enlightenment*, Princeton University Press, 1988.

is less than the strike price K. On the other hand, if $S(T) > K$ then exercising the option and selling the stock yields an immediate profit of $S(T) - K$. The notation x^+, defined in Equation 5.97, is very useful for expressing the option payoff as a function of the stock price and strike price.

$$x^+ = \max(x, 0) = x \text{ when } x \geq 0, \text{ and } x^+ = 0 \text{ when } x < 0 \qquad (5.97)$$

The payoff of a call option, using this notation, is $(S(T) - K)^+ = \max(0, S(T) - K)$. Suppose, for example, the strike price $K = 40$ and $S(T) = 42$; then the call option payoff is $\max(0, 42 - 40) = \max(0, 2) = 2$. On the other hand, if $S(T) = 39.50$ then the call option payoff is $\max(0, 39.50 - 40) = \max(0, -0.50) = 0$. A put option is analyzed in the same way: If $S(T) < K$ then the put option is worth $K - S(T)$ because you could buy the stock at the market price $S(T)$ and sell it for the strike price K, yielding an immediate profit of $K - S(T)$. On the other hand, if $S(T) > K$ then the put option is worthless. (Why would you sell your stock for K, when you could sell it at the market price $S(T) > K$?) Therefore, the payoff of a put option is $(K - S(T))^+ = \max(0, K - S(T))$.

Notation: Puts and calls are just two of the great variety of options available to investors, so it is convenient to have a notation for option payoffs no matter how they are defined. In particular, we suppose the *option payoff* is given by $g(S(T))$, where g is a non-negative function called the *payoff function*. The payoff function for a call is $g(x) = (x - K)^+$, and for a put it is $g(x) = (K - x)^+$.

Single period options: Binomial lattice model According to this model, the stock price after one period is given by

$$S_1 = S_0 u^x d^{1-x}, \ x = 0, 1,$$

where $x = 0$ if the stock price moves down, and $x = 1$ if the stock price moves up. Denote the payoff function for a single period option by

$$f(1, x) = g(S_0 u^x d^{1-x}), \ x = 0, 1.$$

Thus $f(1, 0)$ is the option payoff if the stock price moves down and $f(1, 1)$ is the option payoff if the stock price moves up. It follows from these remarks that when the stock price follows the binomial lattice model the value of the call option at the expiration date is either $f(1, 1)$, $f(1, 0)$, where

$$f(1, 1) = (S_0 u - K)^+ \text{ (if the stock price moves up)},$$
$$f(1, 0) = (S_0 d - K)^+ \text{ (if the stock price moves down)}.$$

The Replicating Portfolio

We now determine the fair price C of an option by means of an argument of independent interest. Let $R = 1 + r$, where r is the risk free interest rate, e.g., the interest rate of a US Treasury bill for the option period. Create a portfolio by purchasing x dollars worth of stock and y dollars worth of US Treasury bonds. Note that at the end of the period the value of the bond portion of the portfolio is Ry. At the expiration date the portfolio value is either $W(1, 1)$, $W(1, 0)$, where

$$W(1, 1) = ux + Ry \text{ (if the stock price moves up)},$$
$$W(1, 0) = dx + Ry \text{ (if the stock price moves down)}.$$

We now choose x, y so that the value of the portfolio matches exactly the value of the option, that is

$$W(1, 1) = ux + Ry = f(1, 1) \text{ (if the stock price moves up)}, \qquad (5.98)$$
$$W(1, 0) = dx + Ry = f(1, 0) \text{ (if the stock price moves down)}. \qquad (5.99)$$

Solving Equations 5.98 and 5.99 for x, y and $x + y$ (the cost of setting up the portfolio) yields

$$x = \frac{f(1,1) - f(1,0)}{u - d}, \quad y = \frac{uf(1,0) - df(1,1)}{R(u - d)}, \tag{5.100}$$

$$\begin{aligned} x + y &= \left(\frac{f(1,1) - f(1,0)}{u - d} + \frac{uf(1,0) - df(1,1)}{R(u - d)} \right) \\ &= \frac{1}{R} \left(\frac{R - d}{u - d} f(1,1) + \frac{u - R}{u - d} f(1,0) \right). \end{aligned} \tag{5.101}$$

This portfolio has been constructed so that it and the option have the same payoff, which is why it is called the *replicating portfolio*. It follows that the cost of the option must equal the cost of setting up the replicating portfolio. This reasoning is a special case of the *one price axiom*[2]:

Axiom 5.1   *Consider two investments costing C_1 and C_2, respectively. If the payoff from the first investment is always identical to that of the second one, then $C_1 = C_2$.*

Consequently, the option price C is given by

$$C = \frac{1}{R} \left(\frac{R - d}{u - d} f(1,1) + \frac{u - R}{u - d} f(1,0) \right), \quad R = 1 + r. \tag{5.102}$$

Risk-Neutral Probability Measure Looking at Equation 5.102 we see that the coefficients of $f(1,1)$, $f(1,0)$ add to one, that is

$$\frac{R - d}{u - d} + \frac{u - R}{u - d} = 1.$$

Thus, Equation 5.102 can be written as

$$C = \frac{1}{R}(qf(1,1) + (1 - q)f(1,0)), \tag{5.103}$$

where

$$q = \frac{R - d}{u - d}, \ (d < R < u). \tag{5.104}$$

It is, of course, the same q obtained earlier in Equation 5.83. The assertion $d < R < u$ follows from the fact if $R > u$ then no one would buy a stock whose return is less than the risk free return. Similarly, if $R < d$, then no one would buy the bond because the stock would always have a higher return. Consequently, $0 < q < 1$, so $(q, 1 - q)$ is a probability measure, called the *risk-neutral probability measure*. Equation 5.103 can be interpreted as stating the value of the option is its expected future value, computed using the risk-neutral probability measure, discounted at the risk-free rate.

Example 5.25

A stock price is currently \$40. It is known that at the end of one month it will be either \$42 or \$38. The risk free interest rate is 8% per year. What is the value of a one-month European call option with a strike price of \$39?

---

[2]S.M. Ross (2003), *An Elementary Introduction to Mathematical Finance*, 2nd ed., Cambridge University Press, p. 65.

Solution When $S = 42$ the value of the call option $f(1,1) = 42 - 39 = 3$, and when $S = 38$ the option is worthless, so $f(1,0) = 0$. In addition the monthly interest rate $r = 0.08/12 = 0.0067$; consequently,

$$R = 1.0067, \ u = 1.05, d = 0.95.$$

Inserting these values into Equation 5.102 yields the value $f = \$1.69$ for the call.

5.9   Black-Scholes Formula for Multi-Period Binomial Options*

Since options, as well as stocks, are traded continuously we need a formula that gives us the option price at any time prior to maturity. (Although the option can only be exercised on the expiration date it can be sold at any time before then.) Consider the case of an option with an expiration time T, $\{T = 1, 2, \ldots\}$ and payoff function $g(S(T))$; the single period option case, discussed in the preceding section, corresponds to the case $T = 1$. We assume the interest rate for each period is a constant r until maturity. We are interested in the price of the option at any time prior to maturity. We denote by $f(t, x)$ the option price at time t after x up moves and $t - x$ down moves of the stock price. In particular, $f(T, x)$ is the option payoff on the expiration date given that the stock price at maturity $S(T) = S_0 u^x d^{T-x}$. Consequently,

$$f(T, x) = g(S_0 u^x d^{T-x}), \ \{x = 0, 1, \ldots, T\}. \tag{5.105}$$

It is to be observed that the option price $f(t - 1, x)$ can be regarded as a one-period option expiring at time t with payoffs $f(t, x + 1)$ and $f(t, x)$ with probabilities $q, 1 - q$, respectively. This leads to the following useful modification of Equation 5.103:

$$f(t - 1, x) = \frac{1}{R}(qf(t, x + 1) + (1 - q)f(t, x)), \ t \le T. \tag{5.106}$$

To obtain the option price at any time $t < T$ we first compute $f(T - 1, x)$, $\{x = 0, 1, \ldots, T - 1\}$ using Equation 5.106 and then use it again to compute $f(T - 2, x)$, $\{x = 0, 1, \ldots, T - 2\}$, and so on. At the Tth stage we obtain $f(0,0) = C$. To illustrate the basic idea consider the following two-period example.

Example 5.26 *This is a variation of Example 5.25. The current price of a stock is $40; at the end of each month the price increases or decreases by 5%, and the current interest rate is 8% per annum. What is the value of a European call option with a strike price of $K = 41$ and a two-month expiration date?*

Solution The monthly interest rate r is 0.0067, $R = 1.0067$, with $u = 1.05$, $d = 0.95$. The risk-neutral probability $q = 0.567 = (1.0067 - 0.95)/(1.05 - 0.95)$. The stock price and option payoff as a function of the number of up moves are displayed in the next table.

number of up moves	$x = 0$	$x = 1$	$x = 2$
stock price	36.10	39.90	44.10
option payoff	$f(2,0) = 0$	$f(2,1) = 0$	$f(2,2) = 3.10$

To compute the values $f(1,1)$, $f(1,0)$ we use Equation 5.106, with $t = 2$, which yields

$$f(1,1) = \frac{1}{R}(qf(2,2) + (1-q)f(2,1)) = \frac{0.567 \times 3.10}{1.0067} = \$1.75,$$

$$f(1,0) = \frac{1}{R}(qf(2,1) + (1-q)f(2,0)) = 0.$$

Substituting the values for $f(1,1)$, $f(1,0)$ just obtained into Equation 5.106 with $t = 1$ we obtain the formula

$$C = f(0,0) = \frac{1}{R^2}(q^2 f(2,2) + 2q(1-q)f(2,1) + (1-q)^2 f(2,0)) \qquad (5.107)$$

$$= \frac{1}{(1.0067)^2} \times (0.567)^2 \times 3.10 = \$0.9834 = \$0.98.$$

The reader will recognize the right hand side of Equation 5.107 as the expected value of $R^{-2}f(2, X(2))$ where the distribution of $X(2)$ is binomial with parameters 2 and risk-neutral probability q. More generally, suppose the distribution of $X(T)$ is binomial with parameters T and risk-neutral probability q. An expectation with respect to the risk-neutral probability measure is denoted $E_Q(h(X(T)))$ and is defined by

$$E_Q(h(X(T))) = \sum_{0 \le x \le T} h(x) \binom{T}{x} q^x (1-q)^{T-x}. \qquad (5.108)$$

Using this notation, Equation 5.107 can be written as $C = E_Q \left(R^{-2}f(2, X(2))\right)$ (with $h(x) = R^{-2}f(2,x)$). The variable $R^{-2}f(2, X(2))$ is the present value of a two-period option payoff, discounted at the risk free interest rate, which suggests the following generalization:

Theorem 5.8 *(Black-Scholes formula for multi-period binomial options) The price of an option for a stock governed by a binomial lattice model is the expected value of its discounted payoff, computed with respect to the risk-neutral probability measure. More precisely, the price of an option for a stock following the binomial lattice model with expiration time T, payoff function $g(S(T))$, and interest rate r per period, is*

$$E_Q \left(R^{-T}g(S_0 u^{X(T)} d^{T-X(T)})\right) = R^{-T} \sum_{0 \le x \le T} g\left(S_0 u^x d^{T-x}\right) \binom{T}{x} q^x (1-q)^{T-x}.$$

Remark We emphasize that the distribution of $X(T)$ is binomial with parameters T, q.

Proof The simplest way to understand Equation 5.109 is to realize that the option price—just like the stock price—is governed by a random walk defined as follows: Starting at $(0,0)$ the random walk moves to the site $(1,0)$ with probability $1-q$ and to the site $(1,1)$ with probability q. If at time $t-1$ the random walk is at the site $(t-1, x)$ then it moves to (t, x) with probability $1-q$ and to $(t, x+1)$ probability q. It is clear, then, that at time T the position of the random walk at time T is a binomial random variable, denoted $X(T)$, with parameters T, q. Consequently, the probability of receiving the option payoff $g\left(S_0 u^x d^{T-x}\right)$ is given by

$$P\left(g(S_0 u^{X(T)} d^{T-X(T)}) = g\left(S_0 u^x d^{T-x}\right)\right) = \binom{T}{x} q^x (1-q)^{T-x}. \qquad (5.109)$$

This completes the proof of Theorem 5.8.

Example 5.27 *This is a variation of Example 5.26. The current price of a stock is $40; at the end of each month the price increases or decreases by 5%, and the current interest rate is 8% per annum. What is the value of a European put option with a strike price of $K = 39$ and a two-month expiration date?*

Solution The parameters R, u, d and the risk neutral probability q are the same as in Example 5.26; the payoff function is, however, different and is displayed in the next table.

number of up moves	$x = 0$	$x = 1$	$x = 2$
stock price	36.10	39.90	44.10
option payoff	$f(2,0) = 2.90$	$f(2,1) = 0$	$f(2,2) = 0$

$$f(1,1) = \frac{1}{R}(qf(2,2) + (1-q)f(2,1)) = 0$$

$$f(1,0) = \frac{1}{R}(qf(2,1) + (1-q)f(2,0))\frac{0.443 \times 2.90}{1.0067} = \$1.2473$$

Substituting the values for $f(1,1)$, $f(1,0)$ just obtained into Equation 5.107 we obtain the formula

$$f(0,0) = \frac{1}{R^2}(q^2 f(2,2) + 2q(1-q)f(2,1) + (1-q)^2 f(2,0))$$

$$= \frac{1}{(1.0067)^2} \times (0.443)^2 \times 1.2473 = \$0.2415 = \$0.28.$$

5.9.1 Black-Scholes Pricing Formula for Stock Prices Governed by a Lognormal Distribution*

The original Black-Scholes option pricing formula (see Equation 5.110 below) is based on the lognormal model (or, more accurately, the geometric Brownian motion model) for stock prices discussed in Sections 4.6 and 5.5 and is derived here via the binomial lattice approximation given in Section 5.5.1 (cf. Equation 5.68).

Theorem 5.9 *The option price for a stock governed by a geometric brownian motion $S(t) = S_0 \exp(t\mu + \sigma W(t))$, with maturity $T(years)$, payoff function $g(S(T))$, and annual interest rate r is*

$$C(T) = E\left(e^{-rT} g\left(S_0 \exp((r - \sigma^2/2)T + \sigma W(T))\right)\right). \tag{5.110}$$

Because the proof is lengthy and tedious it is given in Section 5.13 at the end of this chapter. Instead, we apply the option pricing formula of Equation 5.110 to two important special cases: (i) a European call option $(g(x) = (x - K)^+)$, and (ii) a European put option $(g(x) = (K - x)^+)$. Inserting these particular choices for g into Equation 5.110, a series of straightforward but tedious calculations yields explicit formulas for the option price in terms of the standard normal distribution function $\Phi(z)$.

$$C(T) = S_0 \Phi(d) - K \exp(-rT)\Phi(d - \sigma\sqrt{T}) \text{ (price of European call)} \tag{5.111}$$

$$C(T) = K \exp(-rT)\Phi(\sigma\sqrt{T} - d) - S_0 \Phi(-d) \text{ (price of European put)} \tag{5.112}$$

$$\text{where } d = \frac{\ln(S_0/K) + (r + \sigma^2/2)T}{\sigma\sqrt{T}} \tag{5.113}$$

Example 5.28 *Suppose the current price of a stock is $40, its volatility $\sigma = 0.30$, and the current interest rate is 5.5%, compounded continuously. What is the price of a call option with strike price $44 and an expiration date of: (a) three months? (b) six months?*

Solution (a) In this case $S_0 = 40$, $K = 44$, $r = 0.055$, $\sigma = 0.30$, $T = 0.25$. Inserting these values into Equations 5.111 and 5.113 we obtain the result that $d = -0.468735$ and $C(T) = \$1.15$. (b) In this case $S_0 = 40$, $K = 44$, $r = 0.055$, $\sigma = 0.30$, $T = 0.50$. Inserting these values into Equations 5.111 and 5.113 we obtain the result that $d = -0.213594$ and $C(T) = \$2.27$.

5.10 The Poisson Process

The Poisson distribution is more than just an approximation to the binomial; it, together with the binomial and the normal, is now regarded, and rightly so, as one of the three fundamental distributions in probability theory. It plays a important role in operations research, particularly in the analysis of *queueing systems.*

Example 5.29 *Queueing systems*

A *queueing system* denotes any service facility to which *customers* or *jobs* arrive, receive service, and then depart. We use the word customer in a general sense to refer to: (i) a telephone call arriving at a telephone exchange, (ii) an order for an component stocked in a warehouse, (iii) a broken component brought to a repair shop, (iv) a packet of digital data arriving at some node in a complex computer network. The *service time S* denotes the amount of time required to service the customer. The length of a telephone call and the time to repair a broken component are examples of service times. In the most typical applications both the customer arrivals and their service times are random variables. Queueing systems are classified according to

- the *input process*, which denotes the probability distribution of the customer arrivals,

- the *service distribution*, which denotes the probability distribution of the service time, and

- the *queueing discipline*, which refers to the order of service, e.g., First Come First Served (FCFS), etc.

Example 5.30

We now show that, under intuitively reasonable hypotheses, the input process to a queueing system has a Poisson distribution. To fix our ideas let $X(t)$ denote the number of phone calls, say, arriving at an exchange during the time interval $[0, t]$. In the general case we speak of an "event" occurring in the time interval $[0, t]$. Notice that number of telephone calls received during the time interval $[t_1, t_2]$ equals $X(t_2) - X(t_1)$. We are interested in computing the probability function $P(X(t) = k)$. Under certain reasonable conditions as explained below, we will show that $X(t)$ has a Poisson distribution with parameter λt, i.e.,

$$P(X(t) = k) = e^{-\lambda t} \frac{(\lambda t)^k}{k!}; \; k = 0, 1, \ldots \tag{5.114}$$

Equation 5.114 has some interesting consequences that are worth noting before we proceed to its derivation. Since $X(t)$ has a Poisson distribution with parameter λt it follows at once that $E(X(t)) = \lambda t$. Consequently, the expected number of phone calls in an interval of time of length t is proportional to the length of the interval. Notice that the constant $\lambda = E(X(t))/t$, so λ is the expected (or average) number of phone calls per unit time; and this is why λ is called the *intensity* of the Poisson process.

An intuitive derivation of 5.114: Partition the interval $[0, t]$ into n subintervals of equal length t/n. With each subinterval we associate the Bernoulli random variable X_i defined as follows:

$$X_i = 1, \text{ if a phone call was received during the } i\text{th time interval;}$$
$$X_i = 0 \text{ if no phone call was received during the } i\text{th time interval.}$$

We observe that this model is mathematically equivalent to tossing a coin n times, with the role of heads and tails now played by the arrival or non-arrival of a phone call during the ith time interval. To complete the analogy we make the reasonable assumptions that:

- the probability p_n that a phone call arrives is the same for each subinterval, and the number of arrivals in disjoint time intervals are mutually independent;

- the probability p_n is approximately proportional to the length of the interval, that is $p_n \approx \lambda\,(t/n)$.

Consequently, $S_n = \sum_{1 \le i \le n} X_i$ has a binomial distribution with parameters n, p_n and it is reasonable to hope that S_n is a good approximation to $X(t)$ as $n \to \infty$. Since $p_n \approx \lambda t/n$ it follows at once that $\lim_{n\to\infty} np_n = \lambda t$. Therefore, by the Poisson approximation to the binomial, we get

$$P(X(t) = k) = \lim_{n\to\infty} P(S_n = k) = e^{-\lambda t}\frac{(\lambda t)^k}{k!}; \; k = 0, 1, \dots. \tag{5.115}$$

Example 5.31

Consider a warehouse in which auto parts are stocked in order to satisfy consumer demands. Suppose orders for parts arrive at the warehouse according to a Poisson process at the rate $\lambda = 1.5$ per hour. The warehouse is open 8 hours per day. The warehouse begins the day with an initial inventory of 15 parts. What is the probability that the warehouse exhausts its inventory after:

(a) 4 hours of operation
(b) 8 hours of operation
(c) What is the probability that at the end of 8 hours the warehouse has exactly 5 parts in inventory? At least 5 parts in inventory?

Solution Let $X(t)$ denote the number of orders for the electronic component during the time period $[0, t]$ where t is measured in hours. $X(t)$ has a Poisson distribution with parameter $1.5\,t$; consequently,

$$P(X(t) = k) = p(k; 1.5\,t), \; k = 0, 1, \dots$$

(a) Here $t = 4$ and the problem is reduced to computing $P(X(4) \ge 15)$. But $X(4)$ has a Poisson distribution with parameter $1.5 \times 4 = 6$. Using Table A.2 we see that $P(X(4) \ge 15) = 1 - 0.9995 = 0.0005$.

(b) Here $t = 8$, and therefore $X(8)$ has a Poisson distribution with parameter $1.5 \times 8 = 12$. Proceeding exactly as in part (a) above we obtain $P(X(8) \ge 15) = 1 - 0.772 = 0.228$.

(c) The warehouse will have 5 components in inventory provided $X(8) = 10$; therefore, the answer is given by $P(X(8) = 10) = p(10; 12) = 0.105$. The warehouse will have at least 5 components in inventory provided $X(8) \leq 10$; the problem is solved by computing $P(X(8) \leq 10) = 0.347$.

The input process $X(t)$ has several additional properties that are worth considering in some detail. For example, the same reasoning that led us to conclude that the number of phone calls arriving during the time interval of length t is Poisson distributed with parameter λt can also be used to show that the number of phone calls arriving during any subinterval $[s, s + t]$ of length t, is Poisson distributed with parameter λt. Moreover, it is intuitively clear that the number of phone calls in non-overlapping time intervals are mutually independent random variables in the sense of Definition 5.13. We summarize this discussion in the following definition.

Definition 5.7 *We say that* $X(t), t \geq 0$ *is a Poisson process with intensity* $\lambda > 0$ *if*

1. *for* $s \geq 0$ *and* $t > 0$ *the random variable* $X(s + t) - X(s)$ *has the Poisson distribution with parameter* λt, *i.e.,*

$$P(X(t + s) - X(s) = k) = e^{-\lambda t} \frac{(\lambda t)^k}{k!}; \; k = 0, 1, \ldots, \; and$$

2. *for any time points* $0 = t_0 < t_1 < \ldots < t_n$, *the random variables*

$$X(t_1) - X(t_0), X(t_2) - X(t_1), \ldots, X(t_n) - X(t_{n-1})$$

are mutually independent.

The Poisson process is another example of a *stochastic process,* a collection of random variables indexed by the time parameter t.

Example 5.32

A simple model for the random movement of a particle on the line (cf E. Parzen (1962), *Stochastic Processes,* Holden-Day, p. 80) assumes that its velocity $v(t)$ is the stochastic process

$$v(t) = v(0)(-1)^{N(t)}, \; \text{where} \tag{5.116}$$
$$N(t) \text{ is a Poisson process with parameter } \lambda, \text{ and}$$
$$P(v(0) = \pm v) = \frac{1}{2}, \; v(0) \text{ is independent of } N(t). \tag{5.117}$$

Intuitively, $N(t)$ is the cumulative number of collisions with other particles that occurred up to time t, and, after each collision, the particle reverses its velocity from $\pm v$ to $\mp v$. $v(t)$ is called the *random telegraph signal.* The statistical analysis of a stochastic process begins with the computation of its mean value $E(v(t))$ and covariance $Cov(v(s), v(t))$, which in this case are:

$$E(v(t)) = 0, \; Cov(v(t) \cdot v(s)) = v^2 \exp(-2\lambda|t - s|). \tag{5.118}$$

It is easy to see that $E(v(t)) = 0$, since $E(v(0)) = 0$ and $v(0)$ is independent of $N(t)$. Problem 5.54 asks you to compute the covariance.

5.10.1   The Poisson Process and the Gamma Distribution

The gamma distribution, and the exponential distribution in particular, are associated in a natural way with the Poisson process, as will now be shown. Let messages arrive at a computer node according to a Poisson process $X(t)$ with intensity $\lambda > 0$ (the reader might find it helpful to review Example 5.29). Denote by W_n the waiting time until the *nth* message arrival; we define $W_0 = 0$. Thus, W_1 is the waiting time to the arrival of the first message, W_2 is the waiting time to the arrival of the second message, etc. We now prove that the distribution of the waiting time until the arrival of the *nth* message has an Erlang distribution.

Theorem 5.10 *Suppose messages arrive to a computer node according to a Poisson process with arrival rate λ; then the distribution of W_n, the waiting time until the nth message arrival, is a gamma distribution with parameters $\alpha = n$, $\beta = \lambda^{-1}$; that is*

$$P(W_n \le t) = \int_0^t \frac{\lambda^n x^{n-1} e^{-\lambda x}}{(n-1)!}\, dx \tag{5.119}$$
$$= 0,\ t < 0.$$

Proof   We begin with the observation that the event $(W_n \le t) = (X(t) \ge n)$. This is because the right hand side is the event that at least n messages arrived during the time interval $[0, t]$, which is equivalent to the assertion that the waiting time until the *nth* message arrival is less than or equal to t. Now,

$$P(W_n \le t) = P(X(t) \ge n) = \sum_{k=n}^{\infty} e^{-\lambda t}\frac{(\lambda t)^k}{k!} = 1 - \sum_{k=0}^{n-1} e^{-\lambda t}\frac{(\lambda t)^k}{k!}.$$

We have thus shown that

$$F_{W_n}(t) = P(W_n \le t) = 1 - \sum_{k=0}^{n-1} e^{-\lambda t}\frac{(\lambda t)^k}{k!}, \tag{5.120}$$

$$F'_{W_n}(t) = \frac{\lambda^n t^{n-1} e^{-\lambda t}}{(n-1)!}. \tag{5.121}$$

The derivation of Equation 5.121 is left to the reader (Problem 5.51).

The random variable $T_i = W_i - W_{i-1}$ is called the *ith inter-arrival time*. Notice that W_n can be written as the "telescoping sum"

$$W_n = W_1 + (W_2 - W_1) + \cdots + (W_n - W_{n-1}) = T_1 + \cdots + T_n. \tag{5.122}$$

Theorem 5.11 *The random variables $T_1, \ldots, T_n$ are mutually independent and exponentially distributed with parameter λ.*

Proof   We consider only the case $n = 1$ (the general case requires a more complex and lengthy argument that we omit). It is easy to see that $T_1 = W_1$ so it has an Erlang distribution with $n = 1$. Substituting $n = 1$ into the right hand side of Equation 5.120 we see that $P(W_1 \le t) = 1 - e^{-\lambda t}$, which is an exponential distribution with parameter λ. This yields the following alternative version of Theorem 5.10, which is frequently useful.

Theorem 5.12 *Let T_i, $i = 1, \ldots, n$ denote n exponential random variables, which are independent and identically distributed (with parameter λ); then, $\sum_{1 \le i \le n} T_i$ has a gamma distribution with parameters $\alpha = n, \beta = \lambda^{-1}$.*

Proof There is not much to prove; the theorem is a consequence of the representation $W_n = \sum_{1 \leq i \leq n} T_i$ (Equation 5.122), Theorem 5.11, and Theorem 5.10 (with $\alpha_i = 1, \beta = 1/\lambda$).

Another Interpretation of $W_n = \sum_{1 \leq i \leq n} T_i$

Suppose we have n electronic components with iid exponentially distributed lifetimes denoted $T_1, \ldots, T_n$. If the remaining $n - 1$ components are to be used as spares then W_n represents the total lifetime of all the components. The following example is a typical application of the previous theorems.

Example 5.33 *How many spare parts for the space shuttle?*

Suppose a critical component of an experiment to be performed on the space shuttle has an expected lifetime $E(T) = 10$ days and is exponentially distributed with parameter $\lambda = 0.1$. As soon as a component fails it is replaced by a new one whose lifetime T has the same exponential distribution. If the mission lasts ten days and two spare components are on board, what is the probability that these three components last long enough for the space shuttle to accomplish its mission? More generally, suppose the space shuttle is equipped with n such components, including $n - 1$ spare; denote the probability that these n components last long enough to complete their mission by $r(n)$. Our problem is to find a formula for $r(n)$.

Solution Denote the lifetime of the *ith* component by T_i and note that the total lifetime has the same distribution as W_n since

$$W_n = \sum_{1 \leq i \leq n} T_i, \text{ and therefore}$$

$$r(n) = P(W_n > 10) = \sum_{k=0}^{n-1} e^{-\lambda t} \frac{(\lambda t)^k}{k!} \text{ where } \lambda = 0.1, \, t = 10. \tag{5.123}$$

Since $\lambda t = 1$, $r(n)$ is easily computed via the formula:

$$r(n) = e^{-1}\left(1 + 1 + \frac{1}{2!} + \cdots + \frac{1}{(n-1)!}\right).$$

The values of $r(n)$ for $n = 2, 3, 4, 5$ are displayed below:

$$r(2) = e^{-1}(1 + 1) = 0.74;$$
$$r(3) = e^{-1}(1 + 1 + 1/2) = 0.92;$$
$$r(4) = e^{-1}(1 + 1 + 1/2 + 1/6) = 0.981;$$
$$r(5) = e^{-1}(1 + 1 + 1/2 + 1/6 + 1/24) = 0.9963.$$

If we use two spares, so the total number of components is three, then the probability that the space shuttle completes its mission is given by $r(3) = 0.92$. Since each launch costs tens of millions of dollars it is clear that the probability of failure is uncomfortably high. How many components are needed to increase the probability of success to 0.99? Since $r(4) = 0.981 < 0.99 < 0.9963 = r(5)$, it is now clear that one needs at least five components.

5.11 Applications of Bernoulli Random Variables to Reliability Theory*

We recall that a Bernoulli random variable X assumes only two values: 0, 1. In spite of their deceptively simple structure Bernoulli random variables are exceedingly useful in a wide variety of problems, some of which have already been discussed. Some additional applications will now be given.

Example 5.34 *Computing the reliability of a system of n components*

Consider a complex system such as a space telescope, nuclear reactor, etc., consisting of n components each of which has only two states: (i) a functioning state and (ii) a failed state. The state of the *ith* component can be represented as an Bernoulli random variable X_i where $X_i = 1$ means the *ith* component is functioning and $X_i = 0$ means the *ith* component has failed. The *reliability* p_i of the *ith* component is defined by $p_i = P(X_i = 1) = E(X_i)$. Similarly, the state of the system as a whole is also an Bernoulli random variable denoted by X, where $X = 1$ means the system is functioning and $X = 0$ means that the system has failed. The reliability r of the system is defined to be

$$r = P(X = 1) = E(X). \tag{5.124}$$

The random variable X is called the *system function*. It is a function, possibly a complicated one, of the states of its components. Reliability engineers classify an n component system into three classes: *series* system, *parallel* system, and *k-out-of-n* system as explained in more detail below.

1. **Series system**   A system of components connected in series functions if and only if each of its components functions (Figure 5.2 (i)). Consequently, the system function X is given by

   $$X = X_1 \times X_2 \times \cdots \times X_n. \tag{5.125}$$

   Equation 5.125 states that $X = 1$ if and only if each factor on the right equals 1, i.e., the system functions if and only if each of its components functions.

2. **Parallel system**   A parallel system functions if and only if *at least* one of its components functions (Figure 5.2 (ii)); consequently, the system function is given by

   $$X = 1 - (1 - X_1)(1 - X_2) \cdots (1 - X_n). \tag{5.126}$$

   Suppose, for example, that $X_2 = 1$, so $1 - X_2 = 0$. This implies that the product $(1 - X_1)(1 - X_2) \cdots (1 - X_n) = 0$, thus $X = 1$. In other words, if the second component is functioning then so is the system. Clearly, the same argument shows that $X = 1$ if $X_i = 1$ for at least one i.

3. **k-out-of-n system**   A k-out-of-n system functions if and only if at least k of the n components function. An airplane that functions if and only *at least* two of its three engines functions is an example of a 2–out–of–3 system; its system function is given by

   $$X = X_1 X_2 X_3 + X_1 X_2 (1 - X_3) + X_1 (1 - X_2) X_3 + (1 - X_1) X_2 X_3. \tag{5.127}$$

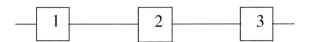

(i) Components linked in series

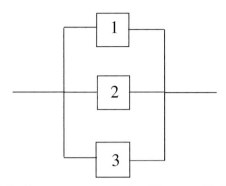

(ii) Components linked in parallel

FIGURE 5.2:   Components linked in series (i) and in parallel (ii)

Computing the Reliability of a System

Avionic systems are designed so that the failure of one component does not lead to a chain reaction of failures of the other components. In particular, the three jet engines are each installed in such a way that that their systems are *independent* of one another, so the failure of a fuel pump or oil filter in one engine does not affect the operation of similar components in the other two engines. This notion of independence is expressed mathematically as follows:

Definition 5.8 *We say that two components are independent if and only if their corresponding system functions X_1 and X_2 are mutually independent random variables.*

More generally, we say that n components are mutually independent if their corresponding system functions $X_1, \ldots, X_n$ are mutually independent random variables:

$$P(X_1 = x_1, \ldots, X_n = x_n) = P(X_1 = x_1) \cdots P(X_n = x_n). \qquad (5.128)$$

If we make the assumption that the components are independent then the reliability of the system is easily computed directly from Equations 5.125, 5.126, and 5.127.

1. **Reliability of a Series System:** According to Equation 5.125 the system function $X = X_1 \times \cdots X_n$ and the reliability of the system is given by

$$r = E(X) = E(X_1 \times \cdots X_n) = E(X_1) \times \cdots E(X_n) = p_1 \cdots p_n.$$

2. **Reliability of a Parallel System:** In this case the system function $X = 1 - (1 - X_1)(1 - X_2) \cdots (1 - X_n)$ and the reliability of the parallel system

$$r = E(X) = E\left(1 - (1 - X_1) \cdots (1 - X_n)\right)$$
$$= 1 - E(1 - X_1) \cdots E(1 - X_n) = 1 - (1 - p_1) \cdots (1 - p_n).$$

3. **Reliability of a 2-out-of-3 System:** We recall that the system function of a 2-out-of-3 system is given by

$$X = X_1 X_2 X_3 + X_1 X_2 (1 - X_3) + X_1 (1 - X_2) X_3 + (1 - X_1) X_2 X_3.$$

Therefore, the reliability of this system is given by

$$r = E(X) = p_1 p_2 p_3 + p_1 p_2 (1 - p_3) + p_1 (1 - p_2) p_3 + (1 - p_1) p_2 p_3.$$

If, in addition to being independent, the components have the same reliability $p = P(X_i = 1)$ then the previous formulas simplify to the following:

$$r = p^n \text{ (Series system)} \tag{5.129}$$
$$r = 1 - (1 - p)^n \text{ (Parallel system)} \tag{5.130}$$
$$r = p^3 + 3p^2(1 - p) \text{ (2-out-of-3 system)} \tag{5.131}$$

Example 5.35 *Some applications*

1. Suppose $p = 0.95$. How many components must be linked in parallel so that the system reliability $r = 0.99$?
   Solution It is clear that n satisfies the equation:

$$1 - (1 - 0.95)^n = 0.99, \text{ equivalently}$$
$$0.05^n = 0.01; \text{ taking logarithms of both sides yields}$$
$$n = \frac{\ln 0.01}{\ln 0.05} = 1.5372.$$

Since n must be the *smallest* integer satisfying the inequality $n \geq 1.5372$, we must choose $n = 2$.

2. Suppose the system consists of $n = 3$ components linked in parallel. What must the value of p be so that $r = 0.99$? In this case we have to solve the following equation for p:

$$1 - (1 - p)^3 = 0.99 \text{ which yields}$$

$$p = 1 - 0.01^{1/3} = 0.7846.$$

3. Calculating the reliability of a *k-out-of-n* system. Assume the system consists of n independent components each of which has the same reliability p. The system functions if and only if at least k-out-of-n components function. It is easy to see that the total number of functioning components X has a binomial distribution with parameters n, p. Consequently, the reliability of the system is given by

$$r = P(X \geq k) = \sum_{k \leq x \leq n} \binom{n}{x} p^x (1 - p)^{n-x}.$$

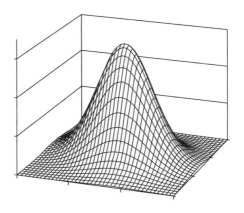

Graph of a Joint Density Function

FIGURE 5.3:   Graph of a joint density function

5.12   The Joint Distribution Function: Continuous Random Variables

Intuitively a density function of a continuous random variable X is a non-negative function $f(x)$ that assigns to each subinterval of the line $[a, b]$ a probability (or a mass) defined by the definite integral

$$P(a \leq X \leq b) = \int_a^b f(x)dx.$$

Thus, $P(a \leq X \leq b)$ can be represented as the area under the curve $y = f(x)$; $a \leq x \leq b$.

In a similar way the joint density function of a continuous random vector (X, Y) is a non-negative function of two variables $f(x, y)$ that assigns to each rectangle in the plane

$$[a, b] \times [c, d] = \{(x, y) : a \leq x \leq b, c \leq y \leq d\}$$

a probability (or mass) defined by the double integral:

$$P(a \leq X \leq b, c \leq Y \leq d) = \int_a^b \int_c^d f(x, y)\, dydx. \tag{5.132}$$

The joint density function of a continuous random vector is sketched in Figure 5.3.

Thus $P(a \leq X \leq b, c \leq Y \leq d)$ can be represented as the volume underneath the surface $z = f(x, y)$ with rectangular base $[a, b] \times [c, d]$.

Definition 5.9 *We say that $f(x, y)$ is a joint density function (jdf) if it satisfies the following two conditions:*

$$f(x, y) \geq 0, \tag{5.133}$$

$$\int_{-\infty}^{\infty} \int_{-\infty}^{\infty} f(x, y)\, dydx = 1. \tag{5.134}$$

Example 5.36 *Show that*

$$f(x,y) = \frac{1}{3}(x+y), 0 < x < 1, 0 < y < 2; \ f(x,y) = 0, \ elsewhere,$$

is a joint density function.

To verify that $f(x,y)$ is a joint density function we must show that

$$\int_{-\infty}^{\infty} \int_{-\infty}^{\infty} f(x,y) \, dy dx = \frac{1}{3} \int_0^1 \int_0^2 (x+y) \, dy dx = 1.$$

It suffices to show that

$$\int_0^1 \int_0^2 (x+y) \, dy dx = 3.$$

We evaluate this double integral in the usual way by first evaluating

$$\int_0^2 (x+y) \, dy = xy + \frac{y^2}{2} \Big|_{y=0}^{y=2} = 2x + 2,$$

and then evaluating

$$\int_0^1 (2x+2) \, dx = x^2 + 2x \big|_0^1 = 3.$$

This proves that $f(x,y)$ is a joint density function.

   The marginal density functions The probability density functions of the random variables X and Y are called the *marginal probability density functions*. Proposition 5.7 gives a formula for computing them.

Proposition 5.7 *If (X,Y) have the joint density function $f(x,y)$ then the marginal probability density functions $f_X(x)$ and $f_Y(y)$ are given by*

$$f_X(x) = \int_{-\infty}^{\infty} f(x,y) \, dy, \ and \ f_Y(y) = \int_{-\infty}^{\infty} f(x,y) \, dx.$$

Example 5.37 *Compute the marginal pdfs and expected values for the random variables X, Y whose joint distribution function is given in Example 5.36.*

   Solution The marginal pdfs are obtained by evaluating the following integrals:

$$f_X(x) = \int_{-\infty}^{\infty} f(x,y) \, dy$$

$$= \int_0^2 \frac{1}{3}(x+y) \, dy = \frac{1}{3}(2x+2), \ 0 < x < 1,$$

$$f_X(x) = 0, \ elsewhere.$$

Notice that the answer $f_X(x) = \frac{1}{3}(2x+2)$, $-\infty < x < \infty$ cannot possibly be correct, since $f_X(-3) = -4/3 < 0$, and a marginal density function can never assume a negative value. Similarly,

$$f_Y(y) = \int_{-\infty}^{\infty} f(x,y) \, dx$$

$$= \int_0^1 \frac{1}{3}(x+y) \, dx = \frac{1}{3}(\frac{1}{2} + y), \ 0 < y < 2,$$

$$f_Y(y) = 0, \ elsewhere.$$

We compute $E(X)$ and $E(Y)$ by evaluating the following integrals:

$$E(X) = \int_0^1 x \frac{1}{3}(2x+2)\,dx = \frac{5}{9};$$

$$E(Y) = \int_0^2 y \frac{1}{3}(\frac{1}{2}+y)\,dy = \frac{11}{9}.$$

The concept of a joint density is easily extended to three or more random variables.

Definition 5.10 *We say that* $f(x_1, x_2, \ldots, x_n)$ *is the joint density of n continuous random variables* $X_1, X_2, \ldots, X_n$ *if*

$$P(a_1 \leq X_1 \leq b_1, \ldots, a_n \leq X_n \leq b_n) = \int_{a_1}^{b_1} \cdots \int_{a_n}^{b_n} f(x_1, \ldots, x_n)\,dx_n \ldots dx_1.$$

Example 5.38 *The concept of a uniformly distributed random vector*

We give an informal definition first. We say that (X, Y) is uniformly distributed over a region $\mathcal{R}$ of finite area $A(\mathcal{R})$ if the probability that the random vector lies in $\mathcal{C} \subset \mathcal{R}$ is proportional to the area of $\mathcal{C}$. The formal definition follows.

Definition 5.11 *Let* $\mathcal{R}$ *denote a region of the plane with finite area* $A(\mathcal{R})$ *and let* $\mathcal{C} \subset \mathcal{R}$. *We say that the random vector is uniformly distributed over the region* $\mathcal{R}$ *if*

$$P((X, Y) \in \mathcal{C}) = \frac{A(\mathcal{C})}{A(\mathcal{R})}.$$

Let us now derive the formula for the joint density function $f(x, y)$. We recall that the area $A(\mathcal{R})$, can be expressed as the double integral of the function $f(x, y) \equiv 1$ computed over the region $\mathcal{R}$:

$$A(\mathcal{R}) = \int \int_{\mathcal{R}} 1\,dy\,dx.$$

The joint density function in this case is the (piecewise) defined function

$$f(x, y) = \frac{1}{A(\mathcal{R})} \text{ for } (x, y) \in \mathcal{R} \text{ and } f(x, y) = 0, \text{ elsewhere.}$$

Example 5.39 *Uniform distribution over the unit square*

Let (X, Y) have the joint density given by

$$f(x, y) = 1, \, 0 < x < 1, \, 0 < y < 1; \, f(x, y) = 0, \text{ elsewhere.}$$

Here $\mathcal{R}$ is the unit square, so $A(\mathcal{R}) = 1$.

1. To compute $P(X < 0.4, Y < 0.6) = P(0 < X < 0.4, 0 < Y < 0.6)$ we only have to compute the area of the rectangle $\mathcal{C} = \{(x, y) : 0 < x < 0.4, \, 0 < y < 0.6\}$, which equals $0.4 \times 0.6 = 0.24$.

2. To compute $P(X < Y)$ we first sketch the region $\mathcal{C} = \{(x, y) : x < y \text{ and } 0 \leq x \leq 1, 0 \leq y \leq 1\}$ and then compute its area. It is easy to verify that $\mathcal{C}$ is the triangle with vertices $(0, 0)$, $(0, 1)$, $(1, 1)$ so its area equals $1/2$; consequently, $P(X < Y) = 1/2$.

Independent random variables The definition of independent random variables in the continuous case is similar to the discrete case (Definition 5.3).

Proposition 5.8 *The continuous random variables X and Y are independent if and only if their joint density function is the product of their marginal density functions, i.e.,*

$$f(x, y) = f_X(x)f_Y(y), \text{ for all } x, y. \tag{5.135}$$

More generally, the random variables $X_1, \ldots, X_n$ are mutually independent if and only if their joint density function is the product of their marginal densities, i.e.,

$$f(x_1, \ldots, x_n) = f_{X_1}(x_1) \times \cdots \times f_{X_n}(x_n). \tag{5.136}$$

5.12.1   Functions of Random Vectors

The concepts of independence, covariance, and correlation are easily extended to continuous random vectors, the only difference being that sums are replaced by integrals. In particular, all the results concerning expected value and variance of a sum of discrete random variables obtained in Section 5.4 carry over without change. For this reason we shall content ourselves with a brief outline of the basic results.

A Formula for $E(\phi(X, Y))$

To compute $E(\phi(X, Y))$ in the continuous case we use the following theorem, which is the analogue of Equation 5.18.

Theorem 5.13 *Let (X, Y) denote a continuous random vector with joint distribution function $f(x, y)$ and let $\phi(x, y)$ denote a function of two variables. Then $\phi(X, Y)$ is a random variable and*

$$E(\phi(X, Y)) = \int_{-\infty}^{\infty} \int_{-\infty}^{\infty} \phi(x, y)f(x, y)\, dx dy. \tag{5.137}$$

Examples and Applications

1. **The addition rule for computing $E(aX + bY)$**

   For random variables X and Y and arbitrary constants a and b we have

   $$E(aX + bY) = aE(X) + bE(Y).$$

   Proof Let $\phi(x, y) = ax + by$ in Equation 5.137; then use the fact that the double integral of a sum of two functions is the sum of the double integrals. Thus,

$$
\begin{aligned}
E(aX + bY) &= \int_{-\infty}^{\infty} \int_{-\infty}^{\infty} (ax + by)f(x, y)dxdy \\
&= \int_{-\infty}^{\infty} \int_{-\infty}^{\infty} axf(x, y)dxdy + \int_{-\infty}^{\infty} \int_{-\infty}^{\infty} byf(x, y)dxdy \\
&= \int_{-\infty}^{\infty} ax\left(\int_{-\infty}^{\infty} f(x, y)dy\right)dx + \int_{-\infty}^{\infty} by\left(\int_{-\infty}^{\infty} f(x, y)dx\right)dy \\
&= a\int_{-\infty}^{\infty} xf_X(x)dx + b\int_{-\infty}^{\infty} yf_Y(y)dy \\
&= aE(X) + bE(Y).
\end{aligned}
$$

2. Using this result we can extend the addition rule (Equation 5.20) to sums of continuous random variables:

$$E\left(\sum_{1 \le i \le n} a_i X_i\right) = \sum_{1 \le i \le n} a_i E(X_i).$$

3. Computing $E(XY)$: Let $\phi(x,y) = xy$ in Equation 5.137; then evaluate the double integral

$$E(XY) = \int_{-\infty}^{\infty} \int_{-\infty}^{\infty} xy f(x,y)\, dx dy.$$

Example 5.40 *Compute $E(XY)$ for the random vector of Example 5.36.*

Solution We compute the double integral

$$\int_0^1 \int_0^2 xy \frac{1}{3}(x+y)\, dy dx = E(XY)$$

in the usual way by integrating first with respect to y

$$\int_0^2 xy \frac{1}{3}(x+y)\, dy = \frac{1}{3}\left(\frac{x^2 y^2}{2} + \frac{xy^3}{3}\right)\Big|_{y=0}^{y=2} = \frac{2}{3}x^2 + \frac{8}{9}x,$$

and then integrating with respect to x

$$\int_0^1 \left(\frac{2}{3}x^2 + \frac{8}{9}x\right) dx = \frac{2}{3}.$$

Thus, $E(XY) = 2/3$.

4. **The expected value of the product of independent random variables**

   Proposition 5.2 states that the expected value of the product of two independent (discrete) random variables equals the product of their expected values. In the next proposition we extend this result to the case where the random variables are continuous.

   Proposition 5.9 *Let X and Y be independent random variables. Then*

$$E(XY) = E(X)E(Y). \tag{5.138}$$

   Proof If X and Y are independent then their joint density function $f(x,y) = f_X(x)f_Y(y)$; therefore,

$$E(XY) = \int_{-\infty}^{\infty} \int_{-\infty}^{\infty} xy f_X(x) f_Y(y)\, dx dy = \int_{-\infty}^{\infty} x f_X(x)\, dx \cdot \int_{-\infty}^{\infty} y f_Y(y)\, dy$$
$$= E(X)E(Y).$$

5. **A formula for $Cov(X,Y)$**

   We define the covariance as in the discrete case and compute it by evaluating the double integral

$$Cov(X,Y) = E\left((X - \mu_X)(Y - \mu_Y)\right) = \int_{-\infty}^{\infty} \int_{-\infty}^{\infty} (x - \mu_X)(y - \mu_Y) f(x,y)\, dy dx.$$

6. The shortcut formula for the covariance of X and Y still holds:
$Cov(X, Y) = E(XY) - \mu_X \mu_Y$.

Proposition 5.10 *Let X and Y be independent random variables. Then*

$$Cov(X, Y) = 0. \tag{5.139}$$

Proof We omit the proof since it the same as the one given in the discrete case (cf. Proposition 5.2).

Example 5.41 *Show that the random variables of Example 5.36 are not independent.*

Solution In our previous discussion we showed that $E(X) = 5/9$, $E(Y) = 11/9$ and $E(XY) = 2/3$; consequently,

$$Cov(X, Y) = \frac{2}{3} - \frac{55}{81} = -0.0123 \neq 0.$$

Since $Cov(X, Y) \neq 0$ it follows that X and Y are not independent.

7. **The correlation coefficient:** The correlation coefficient is defined by $\rho(X, Y) = \frac{Cov(X,Y)}{\sigma_X \sigma_Y}$; which is the same formula as in the discrete case (see Equation 5.35).

Example 5.42 *Compute the correlation coefficient for the random variables X and Y of Example 5.36.*

Solution In the previous worked out problem we showed that $Cov(X, Y) = -0.0123$. A straightforward computation, the details of which are left to the reader, yields the result that

$$V(X) = 0.0802 \text{ and } \sigma(X) = 0.2832;$$
$$V(Y) = 0.2840 \text{ and } \sigma(Y) = 0.5329; \text{ thus,}$$
$$\rho(X, Y) = \frac{-0.123}{0.2832 \times 0.5329} = -0.0818.$$

8. **The variance of a sum of random variables**

The following addition rule for computing the variance of a sum of random variables in terms of the variances and covariances of the summands is an extension of formulas previously derived for discrete random variables. We omit the derivations.

Addition rule for computing $V(X + Y)$

$$V(X + Y) = V(X) + V(Y) + 2Cov(X, Y);$$
$$V\left(\sum_{1 \leq i \leq n} X_i\right) = \sum_{1 \leq i \leq n} V(X_i) + 2\sum_{i<j} Cov(X_i, X_j).$$

5.12.2   Conditional Distributions and Conditional Expectations: Continuous Case

The concepts of conditional distribution and conditional expectation will now be extended to the case where the random vector (X, Y) is of the continuous type. We emphasize that the underlying concepts are the same in both the discrete and continuous cases; the only difference is that we replace sums by integrals in some of the definitions and computations.

Definition 5.12 *Let* (X, Y) *be a continuous random vector with joint density function* $f(x, y)$. *The conditional density* $Y = y$ *given* $X = x$, *denoted* $f(y|x)$, *is defined to be*

$$f(y|x) = \frac{f(x, y)}{f_X(x)}, \ f_X(x) > 0, \ and \ 0 \ elsewhere. \tag{5.140}$$

We leave it to the reader to verify that $f(y|x)$ defined by Equation 5.140 is a pdf for each x such that $f_X(x) > 0$.

The conditional density $y \mapsto f(y|x)$ has an expectation that is given by

$$E(Y|X = x) = \int_{-\infty}^{\infty} y f(y|x) dy. \tag{5.141}$$

We call $E(Y|X = x)$ the *conditional expectation of* Y *given* $X = x$. Note that $E(Y|X = x)$ is a function of the random variable X that equals $E(Y|X = x)$ when $X = x$; this random variable is denoted $E(Y|X)$ and is also called the conditional expectation of Y given X. It is an interesting and useful result that

$$E(E(Y|X)) = E(Y). \tag{5.142}$$

The proof of Equation 5.142 is given in Section 5.13. The function $E(Y|X = x)$ is also called the *regression of* Y *on* X and is denoted $\mu_{Y|x}$.

Example 5.43 *A continuation of Example 5.36.*

Compute $f(y|x)$ and $E(Y|X = x)$.

    Solution

$$f(y|x) = \frac{(x + y)/3}{(2x + 2)/3} = \frac{x + y}{2x + 2}, \ 0 < x < 1; \ 0 < y < 2, \ so$$

$$E(Y|X = x) = \int_0^2 y f(y|x) dy = \frac{1}{2x + 2} \int_0^2 y(x + y) \, dy = \frac{6x + 8}{6x + 6}, \ 0 < x < 1.$$

5.12.3   The Bivariate Normal Distribution

We have seen that under certain reasonable hypotheses a random variable X has an approximate normal distribution; this is the content of the normal approximation to the binomial and its generalization, the *central limit theorem*, which was discussed in Section 5.4.2. It is also true, but not easy to prove, that under certain reasonable conditions, but too technical to state here, that sums of mutually independent random vectors have an approximate *bivariate normal distribution* as defined below.

Definition 5.13 *We say that the random vector* (X, Y) *has a bivariate normal distribution with mean vector* (μ_X, μ_Y), *variances* (σ_X^2, σ_Y^2), *and correlation coefficient* ρ *if*

$$f(x, y) = \frac{1}{2\pi \sigma_X \sigma_Y \sqrt{1 - \rho^2}} e^{-\frac{1}{2(1-\rho^2)} \left[\left(\frac{x - \mu_X}{\sigma_X} \right)^2 - 2\rho \frac{(x - \mu_X)(y - \mu_Y)}{\sigma_X \sigma_Y} + \left(\frac{y - \mu_Y}{\sigma_Y} \right)^2 \right]}. \tag{5.143}$$

It can be shown by a straightforward, but tedious calculation, that if (X, Y) has the bivariate normal density function 5.143 then

$$E(X) = \mu_X, \ E(Y) = \mu_Y,$$
$$V(X) = \sigma_X^2, \ V(Y) = \sigma_Y^2, \ \rho(X, Y) = \rho.$$

In addition, the marginal and conditional distributions are also normally distributed. More precisely, it can be shown that

$$f_X(x) = N(\mu_X, \sigma_X^2),$$
(5.144)

$$f_Y(y) = N(\mu_Y, \sigma_Y^2),$$
(5.145)

$$f(y|x) = N\left(\mu_Y + \rho\frac{\sigma_Y}{\sigma_X}(x - \mu_X); \sigma_Y^2(1 - \rho^2)\right),$$
(5.146)

$$f(x|y) = N\left(\mu_X + \rho\frac{\sigma_X}{\sigma_Y}(y - \mu_Y); \sigma_X^2(1 - \rho^2)\right).$$
(5.147)

Consequently,

$$\mu_{Y|x} = \mu_Y + \rho\frac{\sigma_Y}{\sigma_X}(x - \mu_X); \mu_{X|y} = \mu_X + \rho\frac{\sigma_X}{\sigma_Y}(y - \mu_Y).$$
(5.148)

Example 5.44

Suppose the heights of sons (Y) and mothers (X) (in inches) is a bivariate normal distribution with

$$\mu_X = 62.48, \mu_Y = 68.65, \sigma_X = 5.712, \sigma_Y = 7.344, \rho = 0.5.$$

Then

$$\mu_{Y|X=x} = 68.65 + 0.5 \times \frac{7.344}{5.712} \times (x - 62.48).$$

If the mother's height were 65 inches then this formula "predicts" that the son's height would be

$$\mu_{Y|X=65} = 68.65 + 0.5 \times \frac{7.344}{5.712} \times (65 - 62.48) = 70.27 \text{ inches}.$$

5.13   Mathematical Details and Derivations

Proof of Proposition 5.6 for t**:** We begin with the simpler task of showing that the mean and variance of $\ln(S(t)/S_0)$, are given by

$$E\left(\ln\left(\frac{S(t)}{S_0}\right)\right) = t\mu,$$
(5.149)

$$V\left(\ln\left(\frac{S(t)}{S_0}\right)\right) = t\sigma^2.$$
(5.150)

It follows from Equation 5.149 that

$$\mu = \frac{1}{t}E\left(\ln\left(\frac{S(t)}{S_0}\right)\right)$$

is the continuously compounded expected rate of return of the stock. (Continuous compound interest is discussed in Section 5.7.1.) The parameter σ is called the *volatility*; it is a measure of the uncertainty of the stock's returns. Empirical data indicate that the volatilities of most stocks are between 20% and 50%, i.e., $0.20 \leq \sigma \leq 0.50$. Just as in the case $t = 1$, we divide the time axis into intervals of equal length $1/n$, so their endpoints $t_i = i/n$ (cf. Equation 5.47).

Suppose now that $t = k/n$. In this case the algebraic identity of Equation 5.54 becomes

$$\frac{S(t)}{S_0} = \frac{S(k/n)}{S_0} = \frac{S(1/n)}{S_0} \times \frac{S(2/n)}{S(1/n)} \times \cdots \frac{S((k-1)/n)}{S((k-2)/n)} \times \frac{S(k/n)}{S((k-1)/n)}.$$

Applying the same reasoning as in the case $t = 1$ we see that

$$E\left(\ln\left(\frac{S(t)}{S_0}\right)\right) = \sum_{1 \le j \le k} E\left(\ln\left(\frac{S(j/n)}{S((j-1)/n)}\right)\right) = k \times \frac{\mu}{n} = t\mu,$$

$$V\left(\ln\left(\frac{S(t)}{S_0}\right)\right) = \sum_{1 \le j \le k} V\left(\ln\left(\frac{S(j/n)}{S((j-1)/n)}\right)\right) = k \times \frac{\sigma^2}{n} = t\sigma^2.$$

The proof in the general case is a consequence of the the fact that $\lim_{n\to\infty}[nt]/n = t$ and the continuity assumptions (Equations 5.49, 5.50, and 5.51) and Equation 5.59, which imply that

$$\lim_{n\to\infty} S([nt]/n) = S(t), \tag{5.151}$$

$$\lim_{n\to\infty} E\left(\ln\left(\frac{S([nt]/n)}{S(s)}\right)\right) = E\left(\ln\left(\frac{S(t)}{S(s)}\right)\right), \tag{5.152}$$

$$\lim_{n\to\infty} V\left(\ln\left(\frac{S([nt]/n)}{S(s)}\right)\right) = V\left(\ln\left(\frac{S(t)}{S(s)}\right)\right). \tag{5.153}$$

Consequently,

$$E\left(\ln\left(\frac{S(t)}{S_0}\right)\right) = \lim_{n\to\infty} E\left(\ln\left(\frac{S([nt]/n)}{S_0}\right)\right) = \lim_{n\to\infty} [nt] \times \frac{\mu}{n} = t\mu,$$

$$V\left(\ln\left(\frac{S(t)}{S_0}\right)\right) = \lim_{n\to\infty} V\left(\ln\left(\frac{S([nt])/n}{S_0}\right)\right) = \lim_{n\to\infty} [nt] \times \frac{\sigma^2}{n} = t\sigma^2.$$

Since $\ln(S([nt]/n)/S_0$ is a sum of $[nt]$ iid random variables with mean μ/n and variance σ^2/n, the CLT can be applied to yield

$$\lim_{n\to\infty} P\left(\frac{\ln(S([nt]/n)/S_0) - [nt](\mu/n)}{\sqrt{[nt]\sigma^2/n}} \le z\right) = P\left(\frac{\ln(S(t)/S_0) - t\mu}{\sqrt{t}\sigma} \le z\right)$$

$$= \Phi(z).$$

This shows that

$$\left(\frac{\ln(S(t)/S_0) - t\mu}{\sqrt{t}\sigma}\right) \overset{\mathcal{D}}{=} N(0,1);$$

which is equivalent to the assertion that the distribution of $S(t)/S_0$ is lognormal with parameters $t\mu$, $t\sigma^2$.

Derivation of Equation 5.53:   Define the stochastic process $\tilde{S}(t) = S(t_0 + t)$, and note that $\tilde{S}(t)$ satisfies the same hypotheses of Proposition 5.6; it is simply the time series of the same stock price, but with new initial price $\tilde{S}(0) = S(t_0)$ and terminal price $\tilde{S}(t_1 - t_0) = S(t_1)$. Substituting $\tilde{S}(t)$ for $S(t)$ in Equation 5.52 yields

$$\ln\left(\frac{S(t_1)}{S(t_0)}\right) = \ln\left(\frac{\tilde{S}(t_1 - t_0)}{\tilde{S}_0}\right) \overset{\mathcal{D}}{=} N((t_1 - t_0)\mu, (t_1 - t_0)\sigma^2),$$

which is Equation 5.53.

Proof of Black-Scholes Formula: Theorem 5.9 The proof is quite similar to the method used in Section 5.5.1 to obtain geometric Brownian motion as the limit of a sequence of simpler binomial lattice models. The first step is to approximate the continuous model by a sequence of discrete binomial lattice models with time periods of length $1/n$, and parameters p_n, u_n, d_n as previously defined in Equation 5.60:

$$p_n = 1/2, \ u_n = \exp\left(\frac{\mu}{n} + \frac{\sigma}{\sqrt{n}}\right), \ d_n = \exp\left(\frac{\mu}{n} - \frac{\sigma}{\sqrt{n}}\right).$$

We then compute the option price for the approximating binomial lattice model, with maturity T using Theorem 5.8. More precisely, fix n, assume $T = k/n = [nT]/n$, and let $C_n(T)$ denote the option price for this model. It follows from Theorem 5.8 that

$$C_n(T) = E_{Q_n}\left(R_n^{-[nT]}g(S_0 u_n^{X_n(T)} d_n^{[nT]-X_n(T)})\right), \ R_n = 1 + \frac{r}{n}, \tag{5.154}$$

where the distribution of $X_n(T)$ is binomial with parameters $[nT]$, q_n. E_{Q_n} signifies an expectation with respect to the risk-neutral probability measure $(q_n, 1 - q_n)$ defined as

$$q_n = \frac{R_n - d_n}{u_n - d_n} = \frac{(1 + r/n) - \exp\left(\mu/n - \sigma/\sqrt{n}\right)}{\exp\left(\mu/n + \sigma/\sqrt{n}\right) - \exp\left(\mu/n - \sigma/\sqrt{n}\right)}. \tag{5.155}$$

Although the formula for the risk neutral probability in Equation 5.155 is a complicated one, its asymptotic behavior as $n \to \infty$ is easier to describe:

$$\lim_{n \to \infty} q_n = \frac{1}{2}, \tag{5.156}$$

$$\lim_{n \to \infty} \sqrt{n}\left(q_n - \frac{1}{2}\right) = \frac{r - \mu + \sigma^2/2}{2\sigma}. \tag{5.157}$$

For the derivation of Equations 5.156 and 5.157 see the discussion just before Equation 5.165.

The Black-Scholes formula (Equation 5.109) is obtained by by letting $n \to \infty$ in Equation 5.109 and showing that $\lim_{n \to \infty} C_n(T) = C(T)$; that is

$$\lim_{n \to \infty} C_n(T) = \lim_{n \to \infty} E_{Q_n}\left(R_n^{-[nT]}g(S_0 u_n^{X_n(T)} d_n^{[nT]-X_n(T)})\right) \tag{5.158}$$

$$= E\left(e^{-rT}g\left(S_0 \exp((r - \sigma^2/2)T + \sigma W(T))\right)\right). \tag{5.159}$$

Reasoning as in Equation 5.87 we see that

$$\lim_{n \to \infty} R_n^{-[nT]} = \lim_{n \to \infty} \left(1 + \frac{r}{n}\right)^{-[nT]} = e^{-rT}. \tag{5.160}$$

So it suffices to show that

$$\lim_{n \to \infty} S_0 u_n^{X_n(T)} d_n^{[nT]-X_n(T)} = S_0 \exp((r - \sigma^2/2)T + \sigma W(T)). \tag{5.161}$$

The passage to the limit is eased by transforming the right hand side of Equation 5.158 into a form to which Theorem 5.7 can be applied:

$$S_0 u_n^{X_n(T)} d_n^{[nT]-X_n(T)} =$$

$$= S_0 \exp\left(X_n(T) \left(\frac{\mu}{n} + \frac{\sigma}{\sqrt{n}} \right) + ([nT] - X_n(T)) \left(\frac{\mu}{n} - \frac{\sigma}{\sqrt{n}} \right) \right)$$

$$= S_0 \exp\left(2\sigma \left(\frac{X_n(T) - [nT]/2}{\sqrt{n}} \right) + \frac{[nT]\mu}{n} \right). \tag{5.162}$$

It suffices to study $\lim_{n\to\infty} 2\sigma \left(\frac{X_n(T)-[nT]/2}{\sqrt{n}} \right)$ since we already know that $\lim_{n\to\infty} [nT]\mu/n = T\mu$:

$$2\sigma \left(\frac{X_n(T) - [nT]/2}{\sqrt{n}} \right) = \sigma \left(\frac{X_n(T) - [nT]q_n}{\sqrt{n}/2} \right) + 2\sigma \left(\frac{[nT]q_n - [nT]/2}{\sqrt{n}} \right).$$

It follows from Equation 5.156 that $\lim_{n\to\infty} \sqrt{q_n(1-q_n)} = 1/2$ and

$$\lim_{n\to\infty} 2\sigma \left(\frac{[nT]q_n - [nT]/2}{\sqrt{n}} \right) = \lim_{n\to\infty} \frac{[ntT]}{n} \sqrt{n}(q_n - 1/2)$$

$$= (r - \mu - \sigma^2/2)T$$

and from Theorem 5.7 that

$$\lim_{n\to\infty} \left(\frac{X_n(T) - [nT]q_n}{\sqrt{n}/2} \right) = \lim_{n\to\infty} \left(\frac{X_n(T) - [nT]q_n}{\sqrt{nq_n(1-q_n)}} \right)$$

$$= W(T).$$

Combining these results yields

$$\lim_{n\to\infty} 2\sigma \left(\frac{X_n(T) - [nT]/2}{\sqrt{n}} \right) + \frac{[nT]\mu}{n} = \sigma W(T) + (r - \sigma^2/2)T.$$

This completes the proof of Equation 5.161.

Derivation of Equations 5.156, 5.157:

The derivation requires Taylor's approximation (with remainder term) of the exponential function $\exp(x)$ given in Equation 5.165:

$$\exp(x) = 1 + x + \frac{x^2}{2} + R(x), \text{ where } |R(x)| \le M|x^3| \text{ as } x \to 0. \tag{5.163}$$

Setting $x = \frac{\mu}{n} \pm \frac{\sigma}{\sqrt{n}}$ in Equation 5.163 we obtain the following estimate on the size of the remainder term:

$$R\left(\frac{\mu}{n} \pm \frac{\sigma}{\sqrt{n}} \right) \le M' n^{-3/2}. \tag{5.164}$$

Equation 5.164 is a consequence of the estimate for the remainder term given in Equation 5.165 and a straightforward sequence of algebraic calculations sketched below:

$$R\left(\frac{\mu}{n} \pm \frac{\sigma}{\sqrt{n}} \right) \le M \left(\frac{\mu}{n} \pm \frac{\sigma}{\sqrt{n}} \right)^3$$

$$\le M \left(\frac{1}{\sqrt{n}} \left(\frac{\mu}{\sqrt{n}} \pm \sigma \right) \right)^3$$

$$\le M' n^{-3/2}, \, n \to \infty.$$

We assume the reader is familiar with the "big oh" notation $x_n = O(g(n))$, which means that there is a positive constant M such that $|x_n| \le M|g(n)|$. For example, $x_n = O(n^{-3/2})$ means that $|x_n| \le Mn^{-3/2}$; consequently, $\lim_{n\to\infty} x_n = 0$. It is to be observed that $x_n = O\left(n^{-3/2}\right)$ implies that $cx_n = O\left(n^{-3/2}\right)$, c an arbitrary constant, and $n^{1/2}x_n = O\left(n^{-1}\right)$. We use these facts repeatedly in the calculations below. Using similar methods and this notation, it is not difficult to show that

$$\left(\frac{\mu}{n} \pm \frac{\sigma}{\sqrt{n}}\right)^2 = \frac{\sigma^2}{n} + O(n^{-3/2}).$$

Applying the Taylor approximation to each of the expressions on the left hand side of the equations below yields the approximations on the right:

$$\exp\left(\frac{\mu}{n} \pm \frac{\sigma}{\sqrt{n}}\right) = 1 + \frac{\mu}{n} \pm \frac{\sigma}{\sqrt{n}} + \frac{\sigma^2}{2n} + O\left(n^{-3/2}\right), \tag{5.165}$$

$$\left(1 + \frac{r}{n}\right) - \exp\left(\frac{\mu}{n} - \frac{\sigma}{\sqrt{n}}\right) = \frac{\sigma}{\sqrt{n}} + \left(r - \mu - \frac{\sigma^2}{2}\right)\frac{1}{n} + O\left(n^{-3/2}\right), \tag{5.166}$$

$$\exp\left(\frac{\mu}{n} + \frac{\sigma}{\sqrt{n}}\right) - \exp\left(\frac{\mu}{n} - \frac{\sigma}{\sqrt{n}}\right) = \frac{2\sigma}{\sqrt{n}} + O\left(n^{-3/2}\right). \tag{5.167}$$

From the definition of q_n given in Equations 5.155, 5.166, and 5.167 we obtain the following asymptotic formulas for the risk-neutral probability.

$$
\begin{aligned}
q_n &= \frac{(1 + r/n) - \exp\left(\mu/n - \sigma/\sqrt{n}\right)}{\exp\left(\mu/n + \sigma/\sqrt{n}\right) - \exp\left(\mu/n - \sigma/\sqrt{n}\right)} \\
&= \frac{\frac{\sigma}{\sqrt{n}} + \left(r - \mu - \frac{\sigma^2}{2}\right)\frac{1}{n} + O\left(n^{-3/2}\right)}{\frac{2\sigma}{\sqrt{n}} + O\left(n^{-3/2}\right)} \\
&= \frac{\sigma + \left(r - \mu - \sigma^2/2\right)\frac{1}{\sqrt{n}} + O\left(n^{-1}\right)}{2\sigma + O\left(n^{-1}\right)}.
\end{aligned}
\tag{5.168}
$$

Looking at Equation 5.168 it is easy to see that $\lim_{n\to\infty} q_n = 1/2$. The evaluation of $\lim_{n\to\infty} \sqrt{n}(q_n - 1/2)$ is somewhat more complicated; in detail,

$$
\begin{aligned}
q_n - 1/2 &= \frac{2\left(\left(\sigma + \left(r - \mu - \sigma^2/2\right)n^{-1/2} + O\left(n^{-1}\right)\right) - \left(2\sigma + O\left(n^{-1}\right)\right)\right)}{4\sigma + O\left(n^{-1}\right)} \\
&= \frac{2\left(r - \mu - \sigma^2/2\right)n^{-1/2} + O\left(n^{-1}\right)}{4\sigma + O\left(n^{-1}\right)}.
\end{aligned}
$$

Consequently,

$$\lim_{n\to\infty} \sqrt{n}(q_n - 1/2) = \lim_{n\to\infty} \frac{2\left(r - \mu - \sigma^2/2\right) + O\left(n^{-1/2}\right)}{4\sigma + O\left(n^{-1}\right)} = \frac{\left(r - \mu - \sigma^2/2\right)}{2\sigma}.$$

This completes the proof of Equation 5.157.

Proof of Equation 5.142:

$$
\begin{aligned}
E(E(Y|X)) &= \int_{-\infty}^{\infty} E(Y|X=x) f_X(x) dx \\
&= \int_{-\infty}^{\infty} \left(\int_{-\infty}^{\infty} y f(y|x) dy \right) f_X(x) dx \\
&= \int_{-\infty}^{\infty} y \left(\int_{-\infty}^{\infty} f(y|x) f_X(x) dx \right) dy \\
&= \int_{-\infty}^{\infty} y \left(\int_{-\infty}^{\infty} f(x,y) dx \right) dy \\
&= \int_{-\infty}^{\infty} y f_Y(y) dy = E(Y).
\end{aligned}
$$

5.14 Chapter Summary

The joint distribution function tells us how one variable influences another. It is also a useful tool for deriving formulas for the mean and variance of a sum of random variables. Statisticians measure the interactions between variables by computing conditional distributions, correlations, and conditional expectations. Financial advisors use these concepts to analyze the risks and returns of an investor's portfolio. Stochastic processes—such as the Brownian motion, geometric Brownian motion, and the Poisson process—are characterized by their joint distribution functions.

5.15 Problems

Sections 5.2-5.4.2

Problem 5.1 *Let (X,Y) have the joint pmf specified in the table below:*

$X \backslash Y$	2	3	4	5
0	1/24	3/24	1/24	1/24
1	1/12	1/12	3/12	1/12
2	1/12	1/24	1/12	1/24

(a) Compute the marginal probability functions.
(b) Are X and Y independent?
(c) Compute μ_X and σ_X.
(d) Compute μ_Y and σ_Y.

Problem 5.2 *Let (X,Y) have the joint pmf displayed in Problem 5.1. Compute the following probabilities:*
(a) $P(X \leq 1, Y \leq 3)$.
(b) $P(X > 1, Y > 3)$.

(c) $P(X = 2, Y > 2)$.
(d) $P(Y = y|X = 0) : y = 2, 3, 4, 5$.
(e) $P(Y = y|X = 1) : y = 2, 3, 4, 5$.
(f) $P(Y = y|X = 2) : y = 2, 3, 4, 5$.

Problem 5.3 *Let X and Y have the joint pmf given in Table 5.4. Compute:*
(a) $P(Y \leq 1|X = x)$ *for $x = 0, 1, 2$.*
(b) $P(Y = y|X = 0)$ *for $y = 0, 1, 2$.*

Problem 5.4 *The (discrete) joint pmf of X, Y is $f(x, y) = c(2x+y)$, $x = 1, 2, 3; y = 1, 2, 3$.*
(a) *Show that $c = 1/54$.*
(b) *Display the joint pmf and marginal pmfs in the format of Table 5.2.*
(c) *Using the joint pmf computed in part (b) compute the pmf of $X + Y$.*
(d) *Compute $P(X \leq Y)$.*
(e) *Compute $P(X - Y > 0)$.*

Problem 5.5 *Let (X, Y) have the joint pmf given in Problem 5.4. Compute:*
(a) $E(X), E(X^2), V(X)$.
(b) $E(Y), E(Y^2), V(Y)$.
(c) $E(XY)$.
(d) $Cov(X, Y)$ *and $\rho(X, Y)$.*

Problem 5.6 *The (discrete) joint pmf of X, Y is $f(x, y) = cxy$, $x = 1, 2, 3; y = 1, 2$.*
(a) *Show that $c = 1/18$.*
(b) *Display the joint pmf and marginal pmfs in the format of Table 5.2.*
(c) *Compute $P(Y = 1|X = x)$, $x = 1, 2, 3$.*
(d) *Are X, Y independent random variables? Justify your answer.*
(e) *Using the joint pmf computed in part (b) compute the pmf of $X + Y$.*
(f) *Compute $P(Y < X)$.*
(g) *Compute $P(X - Y \geq 0)$.*

Problem 5.7 *Let (X, Y) have the joint pmf given in Problem 5.6. Compute:*
(a) $E(X), E(X^2), V(X)$.
(b) $E(Y), E(Y^2), V(Y)$.
(c) $E(XY)$.
(d) $Cov(X, Y)$ *and $\rho(X, Y)$.*

Problem 5.8 *Two numbers are selected at random and without replacement from the set $\{1, 1, 1, 1, 0, 0, 0, 0, 0, 0\}$. Let X_i, $(i = 1, 2)$ denote the ith number that is drawn.*
(a) *Compute the joint pmf of (X_1, X_2). Are (X_1, X_2) mutually independent?*
(b) *Compute the pmf of $X_1 + X_2$.*

Problem 5.9 *Refer to Example 5.5.*
(a) *Compute the joint and marginal probability functions of (Y, Z).*
(b) *The monetary value (in cents) of the coins in the sample is given by $W = 5X + 10Y + 25Z$. Compute $E(W)$.*

Problem 5.10 *Consider the experiment of drawing in succession and without replacement two coins from an urn containing two nickles and three dimes. Let X_1 and X_2 denote the monetary values (in cents) of the first and second coins drawn respectively. Compute:*
(a) *The joint pmf of the random variables X_1, X_2.*

(b) Compute the probability function of the random variable $T = X_1 + X_2$, which represents the total monetary value of the coins drawn.

Problem 5.11 *A bridge hand is a set of 13 cards selected at random from a deck of 52. Compute the probability of:*
(a) getting x hearts;
(b) getting y clubs;
(c) getting x hearts and y clubs.

Problem 5.12 *Non-transitive dice: Consider 3 six-sided dice whose faces are labeled as follows:*

$$Die\ 1 = \{5, 7, 8, 9, 10, 18\};$$
$$Die\ 2 = \{2, 3, 4, 15, 16, 17\};$$
$$Die\ 3 = \{1, 6, 11, 12, 13, 14\}.$$

Let Y_i denote the number that turns up when the ith die is rolled. Assume that the three random variables Y_1, Y_2, Y_3 are mutually independent. Show that $P(Y_1 > Y_2) = P(Y_2 > Y_3) = P(Y_3 > Y_1) = 21/36$.

Problem 5.13 *Weird dice: Consider 2 six-sided dice whose faces are labeled as follows:*

$$Die\ 1 = \{1, 2, 2, 3, 3, 4\},$$
$$Die\ 2 = \{1, 3, 4, 5, 6, 8\}.$$

Let Y_i denote the number that turns up when the ith die is rolled. Assume that the two random variables Y_1, Y_2 are independent. Compute the probability function of $Y = Y_1 + Y_2$ and display your results in the format of a horizontal bar chart as in Table 3.3.

Problem 5.14 *A fair die is thrown n times and the successive outcomes are denoted by $X_1, \ldots, X_n$. Let*

$$M_n = \max(X_1, \ldots, X_n).$$

Compute the conditional probabilities p_{ij} defined by

$$p_{ij} = P(M_n = j \mid M_{n-1} = i),\ (i = 1, \ldots, 6;\ j = 1, \ldots, 6).$$

Display your answers in the format of a 6×6 matrix, with first row given by

$$p_{11}, p_{12}, \ldots p_{16},$$

and ith row

$$p_{i1}, p_{i2}, \ldots p_{i6}.$$

Problem 5.15 *Compute the probability function of $X + Y$, where X and Y are mutually independent random variables with probability functions*

x	0	1	2
$f(x)$	0.25	0.5	0.25

y	0	1
$g(y)$	0.5	0.5

Problem 5.16 *Let $(X_1, \ldots, X_n)$ be mutually independent random variables having the discrete uniform distribution (see Equation 3.16).*
(a) Compute the probability function for the random variable M_n defined by

$$M_n = \max(X_1, \ldots, X_n).$$

Hint: Compute $P(M_n \leq x)$.
(b) Compute the probability function for the random variable L_n defined by

$$L_n = \min(X_1, \ldots, X_n).$$

Hint: Compute $P(L_n \geq x)$.

Problem 5.17 *A die is thrown 6 times. Let $Y_i =$ the number of times that i appears.*
(a) Compute $P(Y_1 = 1, \ldots, Y_6 = 1)$.
(b) Compute $P(Y_1 = 2, Y_3 = 2, Y_5 = 2)$.
(c) Compute $E(Y_i)$, $i = 1, \ldots, 6$.

Problem 5.18 *The output of steel plate manufacturing plant is classified into one of three categories: no defects, minor defects, and major defects. Suppose that $P(\text{no defects}) = 0.75$, $P(\text{minor defects}) = 0.20$, and $P(\text{major defects}) = 0.05$. A random sample of 20 steel plates is inspected. Let Y_1, Y_2, Y_3 denote the number of steel plates with no defects, minor defects, and major defects, respectively.*
(a) Compute $E(Y_i)$, $i = 1, 2, 3$.
(b) Compute $P(Y_1 = 15, Y_2 = 4, Y_3 = 1)$.

Problem 5.19 *The "placebo effect" refers to treatment where a substantial proportion of the patients report a significant improvement even though the treatment consists of nothing more than giving the patient a sugar pill or some other harmless inert substance. Ten patients are given a new treatment for symptoms of a cold. The patients' reactions fall into one of three categories: helped, harmed, and no effect. From previous experience with placebos it is known that $P(\text{helped}) = 0.50$, $P(\text{harmed}) = 0.30$, and $P(\text{no effect}) = 0.20$. Let Y_1, Y_2, Y_3 denote the number of patients in each of the three categories (helped, harmed, no effect). Assuming that the new treatment is no more effective than a placebo compute:*
(a) $E(Y_i)$, $i = 1, 2, 3$.
(b) $P(Y_1 = 5, Y_2 = 3, Y_3 = 2)$.

Problem 5.20 *Suppose that math SATs are $N(500, 100^2)$ distributed. Six students are selected at random. What is the probability that of their math SAT scores: 2 are less than or equal to 450, 2 are between 450 and 600, and 2 are greater than or equal to 600?*

Problem 5.21 *The random variables X and Y are independent with $E(X) = 2$, $V(X) = 9$, $E(Y) = -3$, $V(Y) = 16$. Compute:*
(a) $E(3X - 2Y)$
(b) $V(3X)$
(c) $V(-2Y)$
(d) $V(3X - 2Y)$

Problem 5.22 *Suppose X and Y are random variables with*

$$E(X) = 9, E(Y) = 6; E(X^2) = 405, E(Y^2) = 232, E(XY) = -126.$$

Using only this information, compute:
(a) $V(X)$ and $V(Y)$
(b) $Cov(X,Y)$ and $\rho(X,Y)$
(c) $V(X+Y)$
(d) Are X and Y independent?

Problem 5.23 *The random variables $X_1, \ldots, X_{40}$ are independent with the same probability function given by*

x	-1	0	1
$f(x)$	0.2	0.2	0.6

(a) Compute the mean and variance of the sample total $T = X_1 + \ldots + X_{40}$.
(b) What is the largest value that T can assume? The smallest?
(c) What does the central limit theorem tell you about the distribution of T? Use it to compute $P(T > 25)$ and $P(T < 0)$.

Problem 5.24 *Let X, Y, W be random variables with finite means and variances.*
(a) Show that $Cov(aX, cY) = ac\,Cov(X,Y)$.
(b) Show that $Cov(X + Y, W) = Cov(X, W) + Cov(Y, W)$.
(c) Give the details of the derivation of Equation 5.34.

Problem 5.25 *Derive the properties 5.37 and 5.38 of the correlation coefficient.*

Problem 5.26 *Suppose the sequence $X_1, \ldots, X_{12}$ is a random sample taken from the uniform distribution on the interval $[0,1]$ (Equation 4.11).*
(a) Compute the mean and variance of the sample total $T = X_1 + \ldots + X_{12}$.
(b) What is the largest value that T can assume? The smallest?
(c) Use the central limit theorem to compute approximate values for $P(T > 8)$ and $P(T < 5)$.

Problem 5.27 *Let $\overline{X}$ denote the mean of a random sample of size 16 taken from an exponential distribution with parameter $\theta = 1/4$.*
(a) Compute the mean and variance of $\overline{X}$.
(b) Use the central limit theorem to compute approximate values for $P(2.4 < \overline{X} < 5.6)$.

Problem 5.28 *A mathematical model for a random sine wave $X(t)$ assumes that $X(t) = A\cos(t + \Theta)$ where A (the amplitude) and Θ (the phase) are mutually independent random variables, Θ is uniformly distributed over $[0, 2\pi]$, and $E(A) = \mu_1$ and $E(A^2) = \mu_2$ are both finite.*
(a) Show that $E(X(t)) = 0$ for every t.
(b) Show that $Cov(X(t_1), X(t_2)) = (\mu_2/2)\cos(t_1 - t_2)$.
Hint: $\cos(\phi_1)\cos(\phi_2) = (1/2)(\cos(\phi_1 + \phi_2) + \cos(\phi_1 - \phi_2))$.

Problem 5.29 *Suppose X_1, X_2 are independent Poisson random variables with parameters λ_1, λ_2. Show that $X_1 + X_2$ is a Poisson random variable with parameter $\lambda = \lambda_1 + \lambda_2$. Hint: Compute the moment generating function $M_{X_1 + X_2}(s)$ using Theorem 5.2 and Equation 3.69.*

Sections 5.5–5.5.1

Problem 5.30 *The price $S(t)$ (t measured in years) of a stock is governed by a geometric Brownian motion with parameters $\mu = 0.12$, volatility $\sigma = 0.20$, and initial price $S_0 = \$75$. Compute:*
(a) E(stock price after 6 months).
(b) σ(stock price after 6 months).
(c) Show that the probability of a loss after six months is 0.3357.
(d) What is the probability that the stock's price will be up at least 20% after six months?
(e) What is the probability that the stock's price will be down at least 15% after six months?

Problem 5.31 *Suppose the distribution of a stock price $S(t)$ (t measured in years) is governed by a geometric Brownian motion with parameters $\mu = 0.15$, volatility $\sigma = 0.25$, and initial price $S_0 = \$40$. Compute:*
(a) E(stock price after 1 year).
(b) σ(stock price after 1 year).
(c) Find the probability of a loss after 1 year.
(d) What is the probability that the stock's price will be up at least 25% after 1 year?
(e) What is the probability that the stock's price will be down at least 15% after 1 year?

Problem 5.32 *Suppose the distribution of a stock price $S(t)$ (t measured in years) is governed by a geometric Brownian motion with parameters $\mu = 0.20$, volatility $\sigma = 0.30$, and initial price $S_0 = \$60$. Compute:*
(a) E(stock price after 6 months).
(b) σ(stock price after 6 months).
(c) Find the probability of a loss after 6 months.
(d) What is the probability that the stock's price will be up at least 15% after 6 months?
(e) What is the probability that the stock's price will be down at least 10% after 6 months?

Problem 5.33 *Verify that choosing the parameters of the binomial lattice model as in Equation 5.60 yields Equations 5.62 and 5.63 for the mean and variance.*

Sections 5.6–5.9.1

Problem 5.34 *Refer to the resort island stock market (Example 5.21) with the modified joint distribution function given below.*
(a) Compute the expected returns, standard deviations, of R_A, R_B and the correlation between them.
(b) Compute the expected return $\overline{R}_C$ and standard deviation of a portfolio C equally weighted equally between the two stocks; that is, $R_C = (R_A + R_B)/2$. Is portfolio C more efficient than either A or B? Justify your answer.

$R_A \backslash R_B$	-25%	50%
-25%	0.1	0.4
50%	0.4	0.1

Problem 5.35 *Given the correlation between securities A and B is $\rho(A, B) = 0.30$, with expected returns and standard deviations listed below, compute the expected return and standard deviation of a portfolio composed of 40% of A and 60% of B.*

Security	Return	σ	Proportion
A	0.10	0.20	0.40
B	0.15	0.28	0.60

Problem 5.36 *Given the correlation between securities A and B is $\rho(A, B) = 0.20$, with expected returns and standard deviations listed below, compute the expected return and standard deviation for each of the following portfolios.*
(a) Portfolio 1: 50% of stock A and 50% of stock B.
(b) Portfolio 2: 25% of stock A, 75% of stock B.
(c) Portfolio 3: 75% of stock A, 25% of stock B.

Security	Return	σ
A	12%	0.08
B	8%	0.05

Problem 5.37 *The estimated standard deviations and correlations for three stocks are given in the table below.*

Stock	σ	A	B	C
A	0.30	1	−0.63	0.52
B	0.20	−0.63	1	−0.60
C	0.36	0.52	−0.60	1

(a) Compute the standard deviation of a portfolio composed of 50% of B and 50% of C.
(b) Compute the standard deviation of a portfolio composed of 40% of A, 20% of B, and 40% of C.

Section 5.7

Problem 5.38 *A lottery pays the winner $10 million. The prize money is paid out at the annual rate of $500,000 each year, for twenty years, with the first payment at the **beginning** of the first year. What is the present value of this prize at:*
(a) 4% interest?
(b) 8% interest?

Problem 5.39 *An International Equity Fund reported the annual returns for the years 2001-2005 in the Table 5.9.*

Table 5.9

year	2001	2002	2003	2004	2005
returns	-9.04%	-8.25%	26.13%	22.94%	11.55%

(a) At the beginning of 2001 Ms. Allen invests $10,000 in this mutual fund where it remains for the next five years. Find the values of her portfolio (to the nearest dollar) at the end of each year beginning at 2001 and enter your answers in the table below.

year	2001	2002	2003	2004	2005
value of portfolio					

(b) Compute the investor's annualized rate of return for the five year period 2001-2005.
(c) It is known that the annualized rate of return is always less than the average this return. (This is a consequence of the arithmetic-geometric mean inequality.) Compute the investor's average rate of return for the same period and verify that it is larger than the annualized rate of return.

Problem 5.40 *This is a continuation of Problem 1.10. An important performance measure of a stock market portfolio is to compute its annualized returns for the best three year period, as well as the worst three year period, over the past ten years. For the S&P500 the best three year period (over the ten year period 1995-2004) is 1997-1999, and the worst three year period is 2000-20002. The table below lists the annual returns (excluding commissions, but including dividends and their reinvestments) for the S&P500 for the six years (1997-2002).*

year	1997	1998	1999	2000	2001	2002
return	33.4%	28.6%	21.0%	-9.1%	-11.9%	-22.1%

(a) At the beginning of 1997 Mr. Fisher takes $10,000 in savings out of the bank and invests it in the S&P500, where it remains for the next six years. Find the values of Mr. Fisher's portfolio (to the nearest dollar) at the end of 2000, 2001, and 2002 and enter your answers in the table below.

year	2000	2001	2002
value of portfolio			

(b) Find the annualized rate of return for the six year period (1997-2002).

Problem 5.41 *Dealer A offers 0% financing for a $10,000 car. The customer pays $1,000 down and $300 per month for the next 30 months. Dealer B does not give 0% financing but takes $1,000 off the price. Which deal is better (for the customer) if the annual interest rate r is:*
(a) $r = 10\%$?Hint: Compute and compare the present values for both deals.
(b) $r = 5\%$?

Sections 5.8-5.9.1

Problem 5.42 *The price $S(t)$ (t measured in years) of a stock is governed a geometric Brownian motion with expected return $\mu = 0.16$, volatility $\sigma = 0.35$, and current price $S_0 = \$38$.*
(a) An option on the stock has a maturity date of 6 months. Compute the expected price of the stock on the maturity date.
(b) What is the probability that a European call option on the stock with strike price $K = \$40$ and maturity date of 6 months will be exercised?
(c) What is the probability that a European put option on the stock with strike price $K = \$40$ and maturity date of 6 months will be exercised?

Problem 5.43 *For the situation considered in Problem 5.42:*
(a) Find suitable parameters of the binomial lattice model that approximates the stock price (not the option price) with a basic time period of 3 months.
(b) Draw the binomial tree for the binomial lattice model of part (a), (similar to Figure 3.8), and enter the prices of the stock after one year (at the appropriate nodes of the lattice).
(c) Compute the probability distribution of the year end stock prices and display them in the following format.

s					
$P(S = s)$					

Problem 5.44 *The price $S(t)$ (t measured in years) of a stock is governed by a geometric Brownian motion with parameters $\mu = 0.09$, volatility $\sigma = 0.18$, and initial price $S_0 = \$50$.*
(a) An option on the stock has a maturity date of 3 months. Compute the expected price of the stock on the maturity date.
(b) What is the probability that a European call option on the stock with strike price $K = \$52$ and maturity date of 3 months will be exercised?
(c) What is the probability that a European put option on the stock with strike price $K = \$52$ and maturity date of 3 months will be exercised?

Problem 5.45 *A stock price is currently $100. Over each of the next two six-month periods it is expected to go up by 10% or down by 10%. The risk free interest rate is 8% per year.*
(a) What is the value of a one-year European call option with a strike price of $100?
(b) What is the value of a one-year European put option with a strike price of $100?

Problem 5.46 *The current price of a stock is $94, and the price of one three-month European call option with a strike price $K = 95$ is $4.70(per share). You are offered two investment strategies:*

- *Strategy 1: Buy 100 shares @ $94 per share: total investment = $9400, or*

- *Strategy 2: Buy 2000 options @ $4.70 per option: total investment = $2000 \times 4.70 = \$9400$. Consequently, the initial investment is the same for both strategies. It is known that at the expiration date the stock price will be either up by 10% with probability 0.6 or down by 10% with probability 0.4.*

(a) Compute the expected rate of return $\bar{r}_1$ and standard deviation σ_1 for strategy 1.
(b) Compute the expected rate of return $\bar{r}_2$ and standard deviation σ_2 for strategy 2.
(c) Using the results from parts (a) and (b), which strategy is the more risky one? Justify your answer with a brief, intuitive, explanation.

Problem 5.47 *Refer to Example 5.28. Suppose the current price of a stock is $40, its volatility $\sigma = 0.30$, and the current interest rate is 5.5%, compounded continuously. What is the price of a put option with strike price $36 and an expiration date of*
(a) three months?   *(b) six months?*

Sections 5.10-5.11

Problem 5.48 *Telephone calls arrive according to a Poisson process $X(t)$ with intensity $\lambda = 2$ per minute. Compute:*
(a) $E(X(1))$, $E(X(5))$.
(b) $E(X(1)^2)$, $E(X(5)^2)$.
(c) What is the probability that no phone calls arrive during the interval 9:10 to 9:12?

Problem 5.49 *$X(t)$ is a Poisson process with intensity $\lambda = 1.5$. Compute:*
(a) $P(X(2) \leq 3)$.
(b) $P(X(2) = 2, X(4) = 6)$. Hint: $X(2)$ and $X(4) - X(2)$ are independent random variables.

Problem 5.50 *An inventory control model. Consider a warehouse containing an inventory of 10 high voltage transformers. If at the end of the week the inventory level falls below 2 (≤ 2), an order is placed to return the inventory level to 10. Suppose orders per week arrive according to a Poisson process with $\lambda = 6$.*

(a) Find the probability that at the end of the week the inventory falls below 2.
(b) Find the probability that at the end of the week the demand for the transformers exceeds the initial inventory, i.e., the number of transformers ordered is ≥ 11.

Problem 5.51 *Let W_n have an Erlang distribution as described in Equation 5.120. Verify the formula for the pdf displayed in Equation 5.121.*

Problem 5.52 *A critical component of a system needs to function continuously without failure for 5 days. The distribution of the component's lifetime is exponential with expected lifetime $E(T) = 2.5$. As soon as he component fails it is immediately replaced by a spare component with the same lifetime distribution. Suppose the system begins with n components, including $n - 1$ spares. Find the probability that the system functions continuously for the five day period for:*
(a) $n = 3$;   *(b) $n = 4$;*   *(c) $n = 5$.*

Problem 5.53 *Let $X(t)$ be a Poissont process with intensity λ.*
(a) Show that

$$\lim_{t \to \infty} E\left(\left|\frac{X(t)}{t} - \lambda\right|^2\right) = 0.$$

(b) Use Chebyshev's inequality and part (a) to show that for every $d > 0$

$$\lim_{t \to \infty} P\left(\left|\frac{X(t)}{t} - \lambda\right| > d\right) = 0.$$

Problem 5.54 *Refer to Example 5.32.*
Derive the formula for $Cov(v(t), v(s))$ given in Equation 5.118.

Problem 5.55 *By considering the various possibilities verify that the 2-out-of-3 system function is given by Equation 5.127.*

Problem 5.56 *For the system displayed below compute:*
(a) the system function X in terms of $X_1, \ldots, X_n$ and
(b) the reliability of the system, assuming $P(X_i = 1) = p$ and the components are independent.

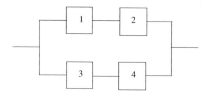

Problem 5.57 *For the system displayed below compute:*
(a) the system function X in terms of $X_1, \ldots, X_n$ and
(b) the reliability of the system, assuming $P(X_i = 1) = p$ and the components are independent.

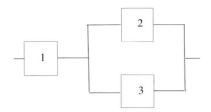

Problem 5.58 *High priority freight service (Source: Wayne Nelson (1982),* Applied Life Data Analysis, *John Wiley and Sons, Inc., New York). A high priority freight train required three locomotives for a one–day run. If any one of these locomotives failed the train was delayed and the railroad had to pay a large penalty. Analysis of statistical data indicated that the locomotives failed independently and their failure times could be approximated by an exponential distribution with expected failure times $\lambda^{-1} = 43.3$ days. Thus such a system is a series system.*
(a) Compute the reliability of the system.
(b) To reduce the chance of delay the railroad added a fourth locomotive; thus the train was delayed only if two or more locomotives failed. This is an example of a 3-out-of-4 system. Determine the reliability provided by the addition of a fourth locomotive.

Section 5.12

Problem 5.59 *Let (X, Y) have the joint density given by*

$$f(x, y) = 1/4, \ -1 < x < 1, \ -1 < y < 1; \ f(x, y) = 0 \ elsewhere.$$

Compute:

(a) $P(X < 0.4, Y < 0.6)$	*(b)* $P(X < 0.4)$	*(c)* $P(Y < 0.6)$
(d) $P(X < Y)$	*(e)* $P(Y > 2X)$	*(f)* $P(Y = X)$
(g) $P(X + Y < 0.5)$	*(h)* $P(X + Y < 1)$	*(i)* $P(X + Y < 1.5)$

Hint: No integrations are required. Just sketch the regions and compute the areas using elementary geometry.

Problem 5.60 *Given*

$$f(x, y) = c(x + 2y), \ 0 < x < 1, \ 0 < y < 1; \ f(x, y) = 0, \ elsewhere, \ find:$$

(a) c. (b) $f_X(x)$ and $f_Y(y)$.
(c) μ_X, μ_Y. (d) σ_X, σ_Y.

Problem 5.61 *Let the random vector (X, Y) have the joint density function*

$$f(x, y) = xe^{-xy-x}, \ for \ x > 0, \ y > 0$$
$$f(x, y) = 0, \ elsewhere.$$

Compute:
(a) $f_X(x), \mu_X, \sigma_X$.   *(b) $f_Y(y), \mu_Y, \sigma_Y$.*
(c) Are X and Y independent?

Problem 5.62 *Let (X, Y) be uniformly distributed over the unit circle $\{(x, y) : x^2 + y^2 \leq 1\}$. Its joint distribution function is given by*

$$f(x, y) = \frac{1}{\pi}, \text{ for } x^2 + y^2 \leq 1$$
$$f(x, y) = 0, \text{ elsewhere.}$$

Compute:
(a) $P(X^2 + Y^2 \leq 1/4)$. (b) $P(X > Y)$. (c) $P(X = Y)$. (d) $P(Y < 2X)$.
(e) Let $R = X^2 + Y^2$. Compute $F_R(r) = P(R \leq r)$.
Hint: No calculations involving double integrals are necessary. Just sketch the region of integration and evaluate the probability by first computing the corresponding area.

Problem 5.63 *Let X, Y have the joint pdf*
$f(x, y) = c(x + xy), 0 \leq x \leq 1, 0 \leq y \leq 1, f(x, y) = 0, elsewhere.$
(a) Compute the value of c.
(b) Compute the marginal pdfs $f_X(x)$ and $f_Y(y)$.
(c) Are the random variables X and Y independent?
(d) Compute $P(X + Y \leq 1)$ by first writing it as a double integral of the density $f(x, y)$ over an appropriate region of the plane. Then evaluate the double integral.

Problem 5.64 *Let X, Y have the joint pdf*
$f(x, y) = \frac{6}{7}(x + y)^2, 0 \leq x \leq 1, 0 \leq y \leq 1, f(x, y) = 0, elsewhere.$ By integrating over an appropriate region compute the following probabilities:
(a) $P(X + Y \leq 1)$.   (b) $P(Y > X)$.
(c) Compute the marginal pdfs: $f_X(x)$ and $f_Y(y)$.

Problem 5.65 *Let (X, Y) be uniformly distributed over the unit circle $\{(x, y) : x^2 + y^2 \leq 1\}$. Its joint distribution function is given in Problem 5.62. Find:*
(a) the marginal distributions.
(b) Show that $Cov(X, Y) = 0$, but that X and Y are not independent.

Problem 5.66 *Let X, Y have joint density function given by*

$$f(x, y) = \frac{3}{5}\left(xy + y^2\right), \ 0 \leq x \leq 2, \ 0 \leq y \leq 1$$
$$f(x, y) = 0, \text{ elsewhere.}$$

Compute:
(a) $P(Y < 1/2|X < 1/2)$.   (b) $f(y|x)$.   (c) $E(Y|X = x)$.

Problem 5.67 *Let the random vector (X, Y) have the joint density function*

$$f(x, y) = xe^{-xy-x}, \text{ for } x > 0, \ y > 0)$$
$$f(x, y) = 0, \text{ elsewhere.}$$

Compute: (a) $f(y|x)$.   (b) $\mu_{Y|x}$.

Problem 5.68 *Let X and Y have a bivariate normal distribution with parameters $\mu_X = 2, \mu_Y = 1, \sigma_X^2 = 9, \sigma_Y^2 = 16, \rho = 3/4$. Compute:*

(a) $P(Y < 1)$	(b) $P(Y < 1|X = 0)$
(c) $E(Y|X = 0)$	(d) $V(X + Y)$

Problem 5.69 *Let X and Y have a bivariate normal distribution with parameters* $\mu_X = 2$, $\mu_Y = 1$, $\sigma_X^2 = 9$, $\sigma_Y^2 = 16$, $\rho = -3/4$. *Compute:*

(a) $P(Y < 3)$	(b) $P(Y < 3 \mid X = 2)$
(c) $E(Y \mid X = 2)$	(d) $V(X + Y)$

Problem 5.70 *Let X and Y have a bivariate normal distribution with parameters* $\mu_X = 2$, $\mu_Y = 1$, $\sigma_X^2 = 9$, $\sigma_Y^2 = 16$, $\rho = 3/4$. *Compute:*

(a) $P(Y < 1)$	(b) $P(Y < 1 \mid X = 0)$
(c) $E(Y \mid X = 0)$	(d) $V(X + Y)$

Problem 5.71 *Let X and Y have a bivariate normal distribution with parameters* $\mu_X = 2$, $\mu_Y = 1$, $\sigma_X^2 = 9$, $\sigma_Y^2 = 16$, $\rho = -3/4$. *Compute:*

(a) $P(Y < 3)$	(b) $P(Y < 3 \mid X = 2)$
(c) $E(Y \mid X = 2)$	(d) $V(X + Y)$

Problem 5.72 *Let X and Y have a bivariate normal distribution with parameters* $\mu_X = 2$, $\mu_Y = 1$, $\sigma_X^2 = 9$, $\sigma_Y^2 = 16$, $\rho = -2/3$. *Compute:*

(a) $P(X < 4)$	(b) $P(X < 4 \mid Y = 1)$
(c) $E(X \mid Y = 1)$	(d) $V(X + Y)$

5.16   To Probe Further

There are several texts that offer the student an elementary introduction to stochastic processes and reliability theory. Among the better ones are:

1. H.M. Taylor and S. Karlin (1984), *An Introduction to Stochastic Modeling*, Academic Press, Orlando, FL.

2. S.M. Ross (1989), *Introduction to Probability Models*, 4th ed., Academic Press, San Diego, CA.

3. R. Barlow and F. Proschan (1975), *Theory of Reliability and Life Testing*, Holt, Rinehart and Winston, New York.

The story of how Galton discovered the bivariate normal distribution is presented in Stigler's (1986) *History of Statistics: The Measurement of Uncertainty before 1900*, The Belknap Press of Harvard University Press, Cambridge, Massachusetts.

The quotation at the head of this chapter is quoted in Porter (1986), *The Rise of Statistical Thinking*, Princeton University Press Princeton, New Jersey, p. 161.

Problem 5.12 appears in C. Wang (1993), *Sense and Nonsense of Statistical Inference*, Marcel Dekker, New York, and Problem 5.13 appears in J. Gallian (1995), *Math Horizons*, Feb., pp. 30-31.

To prudently manage the huge amounts of money currently flowing into stocks, bonds, and mutual funds, investors and money managers have turned to recent important developments in financial engineering, including the mathematical analysis of options, futures contracts, etc. (discussed in this chapter), the *capital asset pricing model* (CAPM), which are discussed in more detail and more depth in one or more of the books listed below. These fundamental results in financial engineering are due primarily to F. Black, M. Scholes, R. Merton, and W.F. Sharpe, who were awarded the Nobel Memorial Prize in economics, except for F. Black, who died before the prize was awarded.

1. M. Capinski and T. Zastawniak (2003), *Mathematics for Finance*, Springer.

2. J. Cvitanic and F. Zapatero (2004), *Introduction to the Economics and Mathematics of Financial Markets*, MIT Press.

3. A. Etheridge (2002), *A Course in Financial Calculus*, Cambridge University Press.

4. J. C. Hull (2003), *Options, Futures, and Other Derivatives,* 5th ed., Prentice-Hall.

5. S. M. Ross (2003), *An Elementary Introduction to Mathematical Finance,* Cambridge University Press.

Chapter 6

Sampling Distribution Theory

Everybody believes in the normal approximation, the experimenters because they think it is a mathematical theorem, the mathematicians because they think it is an experimental fact.

G. Lippman, French physicist (1845-1921)

6.1 Orientation

In this chapter we study the distribution of the sample mean of a random sample taken from an arbitrary distribution (cf. Definition 5.5). The exact distribution of the sample mean is, in general, not an easy matter to determine, except for some important special cases. For example, it can be shown that the sample mean taken from a normal distribution is again normally distributed with mean μ and standard deviation $\sigma/\sqrt{n}$. The *central limit theorem* asserts that for large values of the sample size n, $n > 30$, say, the distribution of the sample mean is *approximately normal* with mean μ and standard deviation $\sigma/\sqrt{n}$. We then introduce other distributions, such as the *chi-square* (denoted χ^2), t, and F distributions related to the normal distribution as well as to others. For example, the chi-square distribution is a special case of the gamma distribution, and it can be shown that t^2 has an F distribution (Section 6.3.2).

Organization of Chapter

1. Section 6.2: Sampling from a Normal Distribution

2. Section 6.3: The Distribution of the Sample Variance

3. Section 6.3.1: Student's t Distribution

4. Section 6.3.2: The F Distribution

5. Section 6.4: Mathematical Details and Derivations

6.2 Sampling from a Normal Distribution

The term *sampling from a normal population* refers to the important special case when the parent df is a normal distribution. Snedecor and Cochran (1980), *Statistical Methods*, 7th ed., Iowa State University Press, Ames, Iowa, list several situations in which it is reasonable to assume that the sample comes from a normal population.

1. The distributions of many variables such as heights, weights, and SAT scores are approximately normal.

2. Even if the distribution of X_i is not normal, the distribution of the sample mean $\overline{X}$ is approximately normal, provided the sample size is sufficiently large.

3. Sometimes a suitable transformation of the data such as $Y_i = g(X_i), i = 1, 2, \ldots, n$ produces a data set that is approximately normal.

4. To these examples we can add one more: stock market returns (cf. Proposition 5.6).

Since the random variables X_i are identically distributed, they have a common mean $E(X_i) = \mu$ and common variance $V(X_i) = \sigma^2$. We call μ and σ^2 the *population mean* and *population variance*, respectively. We then speak of a random sample of size n from a distribution (or population) with mean μ and variance σ^2.

Computing the distribution function of the sample total T and sample mean $\overline{X}$ from a normal distribution

The calculation of the df of the sample total T is not an easy matter except for certain special cases such as when the $X_i's$ have a Bernoulli distribution; in this case, as discussed earlier, T has a binomial distribution with parameters n, p. We can also compute the exact distribution of T and $\overline{X}$ when the parent distribution is normal. These results follow from Theorem 6.1, which tells us that sums of mutually independent, normally distributed random variables are again normally distributed.

Theorem 6.1 *Let the n random variables X_i, $i = 1, 2, \ldots, n$ be mutually independent and normally distributed, with $X_i \overset{D}{=} N(\mu_i, \sigma_i^2)$. Then,*

$$X = \sum_{1 \le i \le n} a_i X_i \overset{D}{=} N(\mu, \sigma^2),$$

where

$$\mu = \sum_{1 \le i \le n} a_i \mu_i, \text{ and } \sigma^2 = \sum_{1 \le i \le n} a_i^2 \sigma_i^2.$$

Proof The proof that X is again normally distributed is omitted since it requires sophisticated mathematical techniques that are beyond the scope of this course. The assertion that $E(X) = \sum_{1 \le i \le n} a_i \mu_i$ is a consequence of the addition rule for the expected value of a sum:

$$E\left(\sum_{1 \le i \le n} a_i X_i \right) = \sum_{1 \le i \le n} a_i \mu_i.$$

Similarly, the assertion that

$$V\left(\sum_{1 \le i \le n} a_i X_i \right) = \sum_{1 \le i \le n} a_i^2 \sigma_i^2$$

is an consequence of the rule for calculating the variance of a sum of independent random variables. The significance of the result is this: The distribution of $\sum_{1 \le i \le n} a_i X_i$ is again normally distributed. The following corollary is an consequence of Theorem 6.1.

Corollary 6.1 *Let $X_1, \ldots, X_n$ be a random sample from a normal population with mean μ and variance σ^2. Then the distribution of the sample total and sample mean are both normal with parameters given by*

$$T \overset{D}{=} N(n\mu, n\sigma^2) , \text{ and}$$

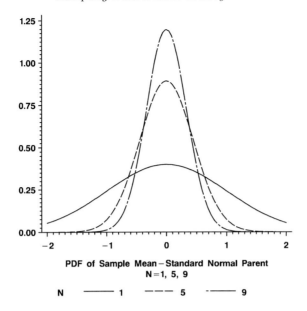

PDF of Sample Mean – Standard Normal Parent
N = 1, 5, 9

N ——— 1   - - - 5   ·——— 9

FIGURE 6.1: Probability density function of sample mean, standard normal parent $(N = 1, 5, 9)$

$$\overline{X} \stackrel{\mathcal{D}}{=} N\left(\mu, \frac{\sigma^2}{n}\right).$$

We emphasize the following points: $\overline{X}$ is again normally distributed with the same mean as each individual X_i but with variance reduced by a factor of $1/n$. To show that the distribution of T is normal we apply Theorem 6.1 with $a_i = 1$. Similarly, the distribution of $\overline{X}$ is also normal because $\overline{X} = T/n$. Figure 6.1 shows how the shape of the sampling distribution changes as the sample size n increases. Here, the parent distribution is $N(0, 1)$. Looking at Figure 6.1 you will notice that the variability of the sample mean $\overline{X}$ decreases as the sample size increases. This is a consequence of the fact that its variance $V(\overline{X}) = \sigma^2/n$ goes to zero as $n \to \infty$. Chebyshev's inequality tells us that the sample standard deviation $\sigma(\overline{X}) = \sigma/\sqrt{n}$ is a useful measure of how much the sample mean $\overline{X}$ deviates from the population mean μ. For this reason the quantity $\sigma(\overline{X})$ is called the *standard error of the mean*. The following examples (and the problems) illustrate a variety of applications of Theorem 6.1.

Example 6.1 *Examples and applications of Theorem 6.1*

The time for a worker to assemble a component is normally distributed with mean $\mu = 15$ minutes and standard deviation $\sigma = 2$. Denote the mean assembly times of 16 day shift workers and 9 night shift workers by $\overline{X}$ and $\overline{Y}$, respectively. Assume that the assembly times of the workers are mutually independent. (1) Describe the distribution of $\overline{X} - \overline{Y}$. (2) Suppose the observed mean assembly times of the day shift and night shift workers are $\overline{x} = 14.5$ and $\overline{y} = 16$ minutes, respectively. (Thus $\overline{x} - \overline{y} = -1.5$.) Compute $P(\overline{X} - \overline{Y} < -1.5)$.
   Solution

1. $\overline{X}$ is normally distributed with $\mu = 15$ and $\sigma^2(\overline{X}) = 4/16$ and $\overline{Y}$ is normally distributed with $\mu = 15$ and $\sigma^2(\overline{Y}) = 4/9$. Consequently, $\overline{X} - \overline{Y}$ is normally distributed with mean

0 and standard deviation

$$\sigma(\overline{X} - \overline{Y}) = \sqrt{\sigma^2(\overline{X}) + \sigma^2(\overline{Y})} = \frac{5}{6}.$$

2.

$$P(\overline{X} - \overline{Y} < -1.5) = P\left(\frac{\overline{X} - \overline{Y}}{5/6}\right) = P(Z < -1.8) = 0.0359.$$

Example 6.2 *Determining the maximum number of passengers on a cable car*

A cable car has a load capacity of 5,000 lbs. Assume that the weight of a randomly selected person is given by a normally distributed random variable W with mean 175 lbs and standard deviation 20. Determine the maximum number of passengers allowed on the cable car so that their total weight exceeds the 5,000 lb weight limit with a probability smaller than 0.05.

   Solution Let n denote the number of passengers allowed on the cable car. The first step is to describe the distribution of $T = \sum_{1 \le i \le n} W_i$, where W_i is the weight of the ith person. Each $W_i \overset{D}{=} N(175, 400)$; consequently, T is $N(n \times 175, n \times 400)$, by Corollary 6.1. Therefore,

$$\frac{T - n \times 175}{20\sqrt{n}} \overset{D}{=} N(0, 1).$$

An overload occurs if $T > 5,000$, i.e., if

$$\frac{T - n \times 175}{20\sqrt{n}} > \frac{5,000 - n \times 175}{20\sqrt{n}}.$$

This event will have probability 0.05 provided

$$\frac{5,000 - n \times 175}{20\sqrt{n}} = 1.645,$$

which, after some algebra, is transformed into the equation

$$175n + 32.9\sqrt{n} - 5,000 = 0.$$

Substituting $x = \sqrt{n}$ one obtains the quadratic equation

$$175x^2 + 32.9x - 5,000 = 0,$$

which has only one non-negative solution $x = 5.25$; and therefore $n = x^2 = 27.58$. Now the solution has to be an integer and consequently, $n = 27$ is the maximum number of passengers allowed on the cable car.

   Some Applications of the χ^2 Distribution

   The distribution of a sum of squares of n independent standard normal random variables has a χ^2_n distribution. The normalized sample variance of a random sample taken from a normal distribution also has a chi-square distribution, but with $n - 1$ degrees of freedom. A precise statement of this result is given in Theorem 6.4.

Theorem 6.2 *Let $Z_1, \ldots, Z_n$ be iid standard normal random variables, then the sum of squares*

$$\chi^2 = \sum_{1 \le i \le n} Z_i^2$$

has a chi-square distribution with n degrees of freedom.

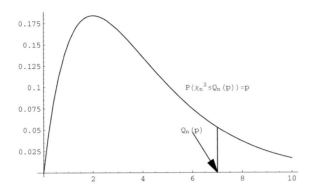

FIGURE 6.2:   Percentile of the chi-square distribution

Proof   We shall give the proof only for the special case $n = 1$ (the case $n \geq 2$ requires more advanced techniques and is therefore omitted). In Example 4.26 of Chapter 4 we showed that the pdf $f(x)$ of Z^2 is given by the following formula (Equation 4.71):

$$f(x) = \frac{1}{\sqrt{2\pi}} x^{-1/2} e^{-x/2}, \ x > 0; \ f(x) = 0, \ x \leq 0.$$

This is a gamma distribution with parameters $\alpha = 1/2$, $\beta = 2$.

Percentiles of the Chi-Square Distribution: Table A.5 lists the values of the $100p$th percentile $Q_n(p)$, where

$$P(\chi^2 \leq Q_n(p)) = p. \tag{6.1}$$

Figure 6.2 displays the relation between p and $Q_n(p)$.

The critical value of the chi-square distribution is the point $\chi_n^2(\alpha)$ on the x axis such that

$$P(\chi^2 > \chi_n^2(\alpha)) = \alpha. \tag{6.2}$$

It follows that the critical value is given by

$$\chi_n^2(\alpha) = Q_n(1 - \alpha),$$

since

$$P(\chi^2 \leq \chi_n^2(\alpha)) = 1 - P(\chi^2 > \chi_n^2(\alpha)) = 1 - \alpha.$$

Example 6.3 *Use Table A.5 to compute the following critical values of the chi-square distribution: (1) $\chi_{10}^2(0.95)$; (2) $\chi_{10}^2(0.05)$.*

Solution

1. For $\alpha = 0.95$, $n = 10$, the critical value $\chi_{10}^2(0.95) = Q_{10}(0.05) = 3.940$.

2. For $\alpha = 0.05$, $n = 10$, the critical value $\chi_{10}^2(0.05) = Q_{10}(0.95) = 18.307$.

Warning: The α that appears in Equation 6.2 is not to be confused with the α that appears in the definition of the gamma distribution (Equation 4.55).

It is an interesting and useful result that sums of mutually independent χ^2 random variables are also χ^2 distributed. A precise statement of this result follows; the proof uses advanced methods and is therefore omitted.

Theorem 6.3 *Let $Y_1, \ldots, Y_k$ independent chi-square random variables with $n_1, \ldots, n_k$ degrees of freedom, respectively, then their sum $\sum_{i=1}^{k} Y_i$ has a chi-square distribution with $n = n_1 + \ldots + n_k$ degrees of freedom.*

The χ^2 distribution has other useful applications as we illustrate in the next example.

Example 6.4 *The distribution of the error R of a range finder device*

Suppose we have a device that records the position of an object located at coordinates (a, b) in the plane as

$$(X, Y) = (a + X_1, b + X_2),$$

where the random variables $X_i, i = 1, 2$ are iid and $X_i \stackrel{D}{=} N(0, \sigma^2)$. In other words the X, Y coordinates recorded by the device are subject to normally distributed random errors as specified above. The random variable R defined by

$$R = \sqrt{(X - a)^2 + (Y - b)^2} = \sqrt{X_1^2 + X_2^2}$$

represents the distance between the true position of the object and the recorded position. Show that

$$P(R \le r) = P\left(\chi_2^2 \le \left(\frac{r}{\sigma}\right)^2\right).$$

Solution   We begin with the observation that $X_i/\sigma \stackrel{D}{=} N(0, 1)$, $i = 1, 2$ and therefore Theorem 6.2 implies

$$\frac{R^2}{\sigma^2} = \frac{X_1^2}{\sigma^2} + \frac{X_2^2}{\sigma^2} \stackrel{D}{=} \chi_2^2.$$

Finally,

$$P(R \le r) = P\left(\frac{R^2}{\sigma^2} \le \left(\frac{r}{\sigma}\right)^2\right)$$

$$= P\left(\chi_2^2 \le \left(\frac{r}{\sigma}\right)^2\right).$$

6.3   The Distribution of the Sample Variance

In Sections 6.2 and 5.4.2 we derived some results concerning the df of the sample mean $\overline{X}$ taken from: (i) a normal distribution and (ii) an arbitrary distribution. In this section we study the distribution of the *sample variance* s^2, which is defined by the equation

$$\text{(Sample Variance) } s^2 = \frac{1}{n-1} \sum_{1 \le i \le n} (X_i - \overline{X})^2. \tag{6.3}$$

We remark that the *sample standard deviation* s is just the square root of s^2.

The following proposition will be used in Chapter 7 to show that s^2 is an *unbiased estimator* of the population variance σ^2.

Proposition 6.1 *Let $X_1, \ldots, X_n$ denote a random sample of size n taken from an arbitrary distribution function F, with mean μ and variance σ^2; then*

$$E(s^2) = \sigma^2. \tag{6.4}$$

Proof To prove 6.4 we use the algebraic identity

$$\sum_{1 \le i \le n} (x_i - c)^2 = \sum_{1 \le i \le n} (x_i - \overline{x})^2 + n(\overline{x} - c)^2. \tag{6.5}$$

We give the derivation of 6.5 in Section 6.4 at the end of this chapter.

We return to the proof of Proposition 6.1. We first substitute X_i and $\overline{X}$ for x_i and $\overline{x}$ in Equation 6.5 and get

$$\sum_{1 \le i \le n} (X_i - c)^2 = \sum_{1 \le i \le n} (X_i - \overline{X})^2 + n(\overline{X} - c)^2.$$

Put $c = \mu$ and take the expectation of both sides to deduce

$$E\left(\sum_{1 \le i \le n} (X_i - \mu)^2 \right) = E\left(\sum_{1 \le i \le n} (X_i - \overline{X})^2 \right) + nE(\overline{X} - \mu)^2.$$

Next, observe that

$$E(X_i - \mu)^2 = V(X_i) = \sigma^2, \text{ and } nE(\overline{X} - \mu)^2 = nV(\overline{X}) = \sigma^2.$$

Consequently, $E(\sum_{1 \le i \le n} (X_i - \mu)^2) = n\sigma^2$, and therefore

$$n\sigma^2 = E\left(\sum_{1 \le i \le n} (X_i - \overline{X})^2 \right) + \sigma^2.$$

Solving this equation for σ^2 yields Equation 6.4.

If in Proposition 6.1 we assume that the random sample comes from a normal distribution then it is possible to compute the exact distribution of the sample variance s^2; this is the content of the next theorem whose proof lies outside the scope of this text.

Theorem 6.4 *Let s^2 be the sample variance of a random sample taken from a normal distribution. Then*

$$\frac{(n-1)s^2}{\sigma^2} \text{ is } \chi^2_{n-1} \text{ distributed.}$$

Example 6.5

Suppose a random sample of size $n = 19$ is taken from a normal distribution with $\sigma^2 = 9$. Compute the probability that the sample standard deviation s lies between 2 and 4.

Solution Our problem is to compute the probability $P(a \le s \le b)$ where $a = 2$, $b = 4$. We sketch the general procedure first, and then apply it to the particular case:

$$P(a \le s \le b) = P\left(\frac{(n-1)a^2}{\sigma^2} \le \frac{(n-1)s^2}{\sigma^2} \le \frac{(n-1)b^2}{\sigma^2} \right)$$

$$= P\left(\frac{(n-1)a^2}{\sigma^2} \le \chi^2_{n-1} \le \frac{(n-1)b^2}{\sigma^2} \right).$$

Substituting

$$n - 1 = 18, \sigma^2 = 9, a^2 = 4, b^2 = 16 \text{ into}$$

the last equation and using the chi-square tables we compute the following approximation:

$$P(2 \le s \le 4) = P(8 \le \chi^2_{18} \le 32) \approx 0.98 - 0.02 = 0.96.$$

6.3.1   Student's t Distribution

The sample mean taken from a normal distribution with mean μ and variance σ^2 is again normally distributed with mean μ and variance σ^2/n. That is the content of Corollary 6.1; consequently,

$$\frac{\sqrt{n}(\overline{X} - \mu)}{\sigma} \overset{\mathcal{D}}{=} N(0, 1).$$

In practice the population variance σ^2 will not be known and it is therefore reasonable to study the distribution of $\sqrt{n}(\overline{X} - \mu)/s$, where s is the sample standard deviation.

Theorem 6.5 *Let s be the sample standard deviation of a random sample of size n taken from a normal distribution. Then*

1. $\overline{X}$ and s are mutually independent random variables and

2.

$$\frac{\sqrt{n}(\overline{X} - \mu)}{s} \quad \text{is } t_{n-1} \text{ distributed.} \tag{6.6}$$

Here t_ν denotes student's t distribution with ν degrees of freedom as described in Definition 6.1 below.

Definition 6.1 *We say that the random variable t_ν, where $\nu = 1, 2, \ldots$, has student's t distribution with ν degrees of freedom, if its pdf is given by:*

$$f(t|\nu) = c(\nu)\left(1 + \frac{t^2}{\nu}\right)^{-(\nu+1)/2} , \quad -\infty < t < \infty;$$

$$c(\nu) = \frac{\Gamma([\nu + 1]/2)}{\sqrt{\pi\nu}\,\Gamma(\nu/2)}.$$

Remark 6.1 *It can be shown that*

$$\lim_{\nu \to \infty} f(t|\nu) = \frac{1}{\sqrt{2\pi}} e^{-t^2/2}.$$

In particular, $f(t|\nu)$ resembles a normal curve, and the standard normal curve itself serves as an approximation for large values of ν, e.g., for $\nu > 30$.

The graphs of student's t distribution for $\nu = 1, 4, 16$ are displayed in Figure 6.3.

The proof of Theorem 6.5, which is too advanced to be included here, is based on the following result whose proof is also outside the scope of this text.

Theorem 6.6 *Let $Z \overset{\mathcal{D}}{=} N(0, 1)$ and $V \overset{\mathcal{D}}{=} \chi_n^2$ be mutually independent; then*

$$\frac{Z}{\sqrt{V/n}} \overset{\mathcal{D}}{=} t_n. \tag{6.7}$$

The t distribution plays an important role in em estimation and *hypothesis testing*, as will be explained in more detail elsewhere.

The critical values $t_\nu(\alpha)$ are defined in the usual way by the condition

$$P(t > t_\nu(\alpha)) = \alpha,$$

and are tabulated in Table A.4 of Appendix A.

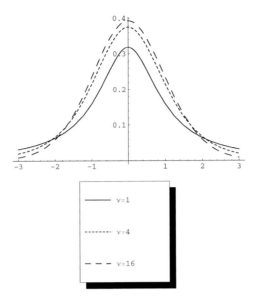

FIGURE 6.3: Graphs of student's t_ν for $\nu = 1, 4, 16$

Example 6.6

Compute: (1) $t_{12}(0.05)$; (2) $t_{12}(0.01)$; (3) $t_{15}(0.025)$; (4) c, where $P(t_6 \geq c) = 0.05$ (5) c, where $P(|t_6| \geq c) = 0.05$.
   Solution

(1) $t_{12}(0.05) = 1.782$;

(2) $t_{12}(0.01) = 2.681$;

(3) $t_{15}(0.025) = 2.131$;

(4) If $P(t_6 \geq c) = 0.05$ then $c = 1.943$;

(5) If $P(|t_6| \geq c) = 0.05$ then $c = 2.447$.

6.3.2   The F Distribution

The F distribution is widely used in statistical inference, notably in the analysis of variance (ANOVA). For this reason we shall postpone discussing its applications until then; in the meantime, we shall content ourselves with giving its definition.

Theorem 6.7 *Let U_1 and U_2 be independent χ^2 random variables, with ν_1 and ν_2 degrees of freedom, respectively. Then, the random variable F defined by*

$$F = \frac{U_1/\nu_1}{U_2/\nu_2}$$

has the F distribution with ν_1 and ν_2 degrees of freedom.

The graph of a typical F density function is displayed in Figure 6.4.
   The critical value of the F distribution is the point $F_{\nu_1,\nu_2}(\alpha)$ on the x axis such that

$$P(F > F_{\nu_1,\nu_2}(\alpha)) = \alpha \text{ (see Figure 6.4)}.$$

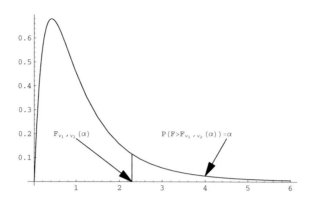

FIGURE 6.4:   Graph of pdf of an F distribution

Table A.6 in the appendix displays the critical values for $\alpha = 0.05$, 0.01 and for selected values of ν_1 and ν_2.

Example 6.7

Compute: (1) $F_{2,27}(0.05)$ and (2) $F_{4,40}(0.01)$.
  Solution

$$F_{2,27}(0.05) = 3.35;$$
$$F_{4,40}(0.01) = 3.83.$$

Corollary 6.2 *If t_n has a t distribution with n degrees of freedom then t_n^2 has an F distribution with $\nu_1 = 1, \nu_2 = n$ degrees of freedom, respectively.*

  Proof Using Equation 6.7 we see that

$$t_n^2 = \left(\frac{Z}{\sqrt{V/n}} \right)^2 = \frac{Z^2}{V/n}.$$

Observe that the term Z^2 in the numerator is χ_1^2 (why?) and the term in the denominator is χ_n^2. The corollary is then a special case of Theorem 6.7.

6.4   Mathematical Details and Derivations

We now derive the algebraic identity 6.5 restated here for the reader's convenience:

$$\sum_{1 \le i \le n} (x_i - c)^2 = \sum_{1 \le i \le n} (x_i - \overline{x})^2 + n(\overline{x} - c)^2.$$

We begin with the algebraic identity

$$(x_i - c)^2 = ((x_i - \overline{x}) + (\overline{x} - c))^2$$
$$= (x_i - \overline{x})^2 + 2(x_i - \overline{x})(\overline{x} - c) + (\overline{x} - c)^2.$$

Therefore,

$$\sum_{1 \le i \le n} (x_i - c)^2 =$$

$$= \sum_{1 \le i \le n} (x_i - \overline{x})^2 + 2(x_i - \overline{x})(\overline{x} - c) + (\overline{x} - c)^2$$

$$= \sum_{1 \le i \le n} (x_i - \overline{x})^2 + \sum_{1 \le i \le n} 2(x_i - \overline{x})(\overline{x} - c) + \sum_{1 \le i \le n} (\overline{x} - c)^2.$$

The middle summand equals zero because $\sum_{1 \le i \le n} (x_i - \overline{x}) = 0$ implies

$$\sum_{1 \le i \le n} 2(x_i - \overline{x})(\overline{x} - c) = 2(\overline{x} - c) \sum_{1 \le i \le n} (x_i - \overline{x}) = 0.$$

The last summand is the constant term $(\overline{x} - c)^2$ added n times, so

$$\sum_{1 \le i \le n} (\overline{x} - c)^2 = n(\overline{x} - c)^2.$$

Consequently,

$$\sum_{1 \le i \le n} (x_i - c)^2 = \sum_{1 \le i \le n} (x_i - \overline{x})^2 + n(\overline{x} - c)^2.$$

6.5 Chapter Summary

Many problems in probability theory and statistical inference can be reduced to the study of the distribution of the *sample mean* $\overline{X}$ of a sequence of iid random variables $X_1, \ldots, X_n$. The exact distribution of $\overline{X}$ is not easy to compute except in certain special cases such as sampling from the Bernoulli or normal distributions. The central limit theorem (CLT) tells us that for large values of the sample size n the distribution of the sample mean is approximately normal with mean μ and standard deviation $\sigma/\sqrt{n}$. It is a remarkable fact that the CLT remains valid assuming only that the parent distribution has a finite mean and finite variance. We then introduced several new families of distributions including the *gamma*, the *chi-square*, the t, and F distributions. These distributions are related to one another as well as to the normal distribution itself. For instance, the chi-square distribution is a special case of the gamma distribution, and t^2 has an F distribution.

6.6 Problems

Section 6.2

Problem 6.1 *Let X_1, X_2, X_3 be three independent, identically distributed normal random variables with $\mu = 50$, $\sigma^2 = 20$. Let $X = X_1 - 2X_2 + 2X_3$. Compute:*
(a) $E(X)$. (b) $V(X)$. (c) $P(\mid X - 50 \mid \leq 25)$.
(d) The 90th percentile of the distribution of X.

Problem 6.2 *A toy consists of three parts whose respective weights (measured in grams) are denoted by X_1, X_2, X_3. Assume that the these weights are independent and normally distributed with*

$$X_1 \overset{D}{=} N(150, 36), \ X_2 \overset{D}{=} N(100, 25), \ X_3 \overset{D}{=} N(250, 60).$$

(a) Give a formula for the distribution of total weight of the toy $T = X_1 + X_2 + X_3$; name it and identify the parameters.
(b) What proportion of the toys have weights greater than 520 grams?
(c) Suppose the toys are shipped 24 to a box. The box weighs 200 grams. Calculate the expected value and variance of the total weight of the box.
(d) Describe the distribution of the total weight W of the box; name it and identify the parameters. What is the probability that the total weight of the box is greater than 12,300 grams?

Problem 6.3 *Suppose the time (measured in minutes) to repair a component is normally distributed with $\mu = 65$, $\sigma = 10$.*
(a) What is the proportion of components that are repaired in less than one hour?
(b) Give a formula for the distribution of the total time required to repair 8 components and use it to find the probability that 8 components are repaired in an eight hour working day.

Problem 6.4 *Let $X_1, \ldots, X_n$ be iid $N(2, 4)$ distributed random variables.*
(a) If $n = 100$ compute $P(1.9 < \overline{X} < 2.1)$.
(b) How large must n be so that $P(1.9 < \overline{X} < 2.1) = 0.9$?

Problem 6.5 *Let $X_1, \ldots, X_9$ be iid $N(2, 4)$ distributed and $Y_1, \ldots, Y_4$ be iid $N(1, 1)$ distributed. The Y_i's are assumed to be independent of the X_i's.*
(a) Describe the distribution of $\overline{X} - \overline{Y}$.
(b) Compute $P(\overline{X} > \overline{Y})$.

Problem 6.6 *Let $\overline{X}$ denote the sample mean of a random sample of size $n_1 = 16$ taken from a normal distribution $N(\mu, 36)$ and let $\overline{Y}$ denote the sample mean of a random sample of size $n_2 = 25$ taken from a different normal distribution $N(\mu, 9)$. Assume $\overline{X}$ and $\overline{Y}$ are independent.*
(a) Describe the distribution of $\overline{X} - \overline{Y}$. Name it and identify the parameters.
(b) Compute $P(\overline{X} - \overline{Y} > 5)$. (c) Compute $P(|\overline{X} - \overline{Y}| > 5)$.

Problem 6.7 *To determine the difference, if any, between two brands of radial tires, 12 tires of each brand are tested. Assume that the lifetimes of both brands of tires come from the same normal distribution with $\sigma = 3300$.*
(a) Describe the distribution of the difference of the sample means $\overline{X} - \overline{Y}$. Name it and identify the parameters.
(b) The engineers recorded the following values for the mean lifetimes for each brand:

$$\overline{x} = 38500, \overline{y} = 41000 \ so \ \overline{x} - \overline{y} = -2500.$$

If, in fact, the lifetimes of both brands have the same distribution what is the probability that the difference between the sample means is at least as large as the observed difference 2500? That is, compute $P(|\overline{X} - \overline{Y}| > 2500)$.
(c) (Part (b) continued): Compute $P(\overline{X} - \overline{Y} < -2500)$.

Problem 6.8 *Assume that the miles per gallon (mpg) of two brands of gasoline come from the same normal distribution with $\sigma = 2.0$. Four cars are driven with brand A gasoline and six cars are driven with brand B gasoline. Denote the average mpg obtained with brand A and brand B gasolines by $\overline{X}$ and $\overline{Y}$, respectively.*
(a) Describe the distribution of the difference of the sample means $\overline{X} - \overline{Y}$. Name it and identify the parameters.
(b) The engineers recorded the following mpgs for each brand:

$$\overline{x} = 32.5 \ and \ \overline{y} = 31, \ so \ \overline{x} - \overline{y} = 1.5.$$

If, in fact, the mpg of both brands have the same distribution what is the probability that the difference between the sample means is at least as large as the observed difference 1.5? That is, compute $P(|\overline{X} - \overline{Y}| > 1.5)$.
(c) (Part (b) continued): Compute $P(\overline{X} - \overline{Y} > 1.5)$.

Problem 6.9 *An elevator has a weight capacity of 3000 lbs. Assume that the weight X of a randomly selected person is $N(175, 400)$ distributed, i.e., $\mu = 175, \sigma = 20$. Let W_n denote the total weight of n passengers.*
(a) Give the formulas for: $E(W_n)$ and $V(W_n)$.
(b) Describe the distribution of W_n.
(c) If 18 people get on the elevator what is the probability that the total weight of the passengers exceeds the weight capacity of the elevator?
(d) What is the maximum number N of persons allowed on the elevator so that the probability of their combined weight exceeding the $3,000$ lb load limit is less than 0.05?

Problem 6.10 *Boxes of cereal are filled by machine and the net weight of the filled box is a random variable with mean $\mu = 10$ oz and variance $\sigma^2 = 0.5$ oz^2. A carton of cereal contains 48 boxes. Let $T =$ total weight of carton.*
(a) Give a formula for the approximate distribution of T using the central limit theorem.
(b) Use the central limit theorem to estimate $P(T > 31$ lbs).

Section 6.3

Problem 6.11 *Use Table A.5 in Appendix A to compute $\chi_n^2(\alpha)$ for:*
(a) $n = 10$, $\alpha = 0.1$ (b) $n = 10$, $\alpha = 0.05$
(c) $n = 15$, $\alpha = 0.1$ (d) $n = 20$, $\alpha = 0.9$

Problem 6.12 *Use Table A.5 in Appendix A to compute numbers $a < b$ such that*

$$P(a < \chi_n^2 < b) = p \ and \ satisfying \ the \ conditions$$
$$P(\chi_n^2 < a) = P(\chi_n^2 > b).$$

(a) $n = 10$, $p = 0.9$
(b) $n = 15$, $p = 0.9$
(c) $n = 10$, $p = 0.95$
(d) $n = 20$, $p = 0.95$

Problem 6.13 *An artillery shell is fired at a target. The distance (in meters) from the point of impact to the target is modeled by the random variable $7.22\chi_2^2$. If the shell hits within 10 meters of its target it is destroyed.*
(a) Find the probability that the target is destroyed by the shell.
(b) If shells are fired repeatedly, how many rounds are required to ensure a 90% probability of destroying the target? Note: The decision of how many shells to fire must be made ahead of time, since we cannot see whether we have hit the target.

Problem 6.14 *The speed V of a molecule in a gas at equilibrium has pdf*

$$f(v) = a\, v^2\, e^{-b\, v^2}, \; v \geq 0; \; f(v) = 0, \; otherwise.$$

Here $b = m/2kT$, where m is the mass of the molecule, T is the absolute temperature, and k is Boltzmann's constant.
(a) Compute the constant a in terms of b.
(b) Let $Y = (m/2) \times V^2$ denote the kinetic energy of the molecule. Compute the df $G(y) = P(Y \leq y)$ of Y.
(c) Compute the pdf $g(y) = G'(y)$ of Y.

Problem 6.15 *Suppose a random sample of size $n = 16$ is taken from a normal distribution with $\sigma^2 = 5$. Compute the probability that the sample standard deviation s lies between 1.5 and 2.9.*

Problem 6.16 *The volatility of a stock is the standard deviation of its returns. Assume the weekly returns of a stock are normally distributed with $\sigma = 0.042$. Let s denote the sample standard deviation of the returns based on a sample of 13 weekly returns. (This corresponds to recording the returns for one quarter of the year.) Find:*
(a) $P(s < 0.0522)$;   (b) $P(s > 0.0556)$.

Problem 6.17 *Use Table A.4 in Appendix A to compute $t_\nu(\alpha)$ for:*
(a) $t_9(0.05)$;   (b) $t_9(0.01)$;   (c) $t_{18}(0.025)$;   (d) $t_{18}(0.01)$.

Problem 6.18 *Find the value c such that:*
(a) $P(t_{14} \geq c) = 0.05$; (b) $P(|t_{14}| \geq c) = 0.05$; (c) $P(t_{14} \geq c) = 0.01$; (d) $P(|t_{14}| \geq c) = 0.01$.

Problem 6.19 *A random sample of size 9 is taken from a normal distribution $\mu = 15$ and σ unknown. Suppose $s = 1.613$. Find:*
(a) $P(14 < \overline{X} < 16)$;   (b) $P(13.76 < \overline{X} < 16.24)$;   (c) $P(13.20 < \overline{X} < 16.80)$.

Problem 6.20 *Use Table A.6 in Appendix A to compute $F_{\nu_1, \nu_2}(\alpha)$ for:*
(a) $F_{3,20}(0.05)$;   (b) $F_{3,20}(0.01)$;   (c) $F_{4,30}(0.05)$;   (d) $F_{4,30}(0.01)$.

6.7   To Probe Further

The long road to the central limit theorem is traced in Stigler's (1986) *History of Statistics: The Measurement of Uncertainty before 1900*, The Belknap Press of Harvard University Press, Cambridge, Massachusetts.

Chapter 7

Point and Interval Estimation

In astronomy, no experiment should be believed until confirmed by theory.

Sir Arthur Eddington, British astronomer (1882-1944)

7.1 Orientation

"You don't have to eat a whole ox to know the meat is tough," so wrote Samuel Johnson (1709-1784), English critic and lexicographer. It is an amusing reminder that a primary goal of a statistical analysis is to make inferences about a population based on a sample drawn from that population. A widely used model in financial engineering, for example, assumes that the stock prices have a lognormal distribution; more precisely, it is assumed that $\ln(S(t)/S_0)$ is normally distributed with mean $t\mu$ and variance $t\sigma^2$ (Proposition 5.6). In financial engineering σ is called the *volatility* and plays an important role in estimating the *value at risk (VaR)*, to be discussed in Section 7.3.4, as well as in the Black-Scholes option pricing formula discussed earlier (Theorem 5.9). Using an incorrect value for the volatility can lead to unacceptably large losses in a portfolio. Consequently, it is a question of no small importance to determine the accuracy and reliability of a parameter estimate. In practice, these parameters can only be estimated from historical stock market data of the sort listed in Tables 1.19 and 1.20. It sometimes happens, however, that a population parameter, such as the population mean, can be estimated by two (or even more!) different methods. For example, one can use either the sample mean $\overline{X}$ or sample median $\widetilde{X}$ to estimate the mean of a normal population. Which one is better? And what criteria do we use to compare them? These are some of the questions we will discuss in this chapter.

Organization of Chapter

1. Section 7.2: Estimating Population Parameters: Methods and Examples

2. Section 7.3: Confidence Intervals for the Mean and Variance

3. Section 7.3.4: Value at Risk (VaR): An Application of Confidence Intervals to Risk Management

4. Section 7.4: Point and Interval Estimation for the Difference of Two Means

5. Section 7.5: Point and Interval Estimation for a Population Proportion

6. Section 7.6: Some Methods of Estimation, including the method of moments and the maximum likelihood method*

7. Section 7.7: Chapter Summary

7.2   Estimating Population Parameters: Methods and Examples

The basic principles and problems of estimation are best understood in the context of specific examples based on real data.

Example 7.1 *Estimating the lifetime of a bearing*

Consider the lifetimes (in hours) of 20 bearings listed in Problem 1.27 and reproduced in Table 7.1.

Table 7.1

6278	3113	9350	5236	11584
12628	7725	8604	14266	6215
3212	9003	3523	12888	9460
13431	17809	2812	11825	2398

(Source: R. E. Schafer and J. E. Angus (1979), *Estimation of Weibull Quantiles With Minimum Error in the Distribution Function*, Technometrics, vol. 21, no. 3, pp. 367-370. Used with permission.)

Among the population parameters of interest are:

1. the expected lifetime μ, or mean time before failure (MTBF).

2. the standard deviation σ of the lifetimes.

3. the safe life, defined as the 10th percentile $Q(0.10)$ of the lifetime distribution.

Consider the following three methods for estimating the expected lifetime of the bearings:

- the sample mean: $\bar{x} = (x_1 + \cdots + x_{20})/20 = (6278 + \ldots + 2398)/20 = 8568$.

- the sample median: $\tilde{x} = (x_{(10)} + x_{(11)})/2 = (8604 + 9003)/2 = 8803.5$.

- the 5% trimmed mean: $\bar{x}_{0.05} = (x_{(2)} + \ldots + x_{(19)})/2 = 8397.39$. A trimmed mean discards the k largest and k smallest observations and computes the mean of the remaining $n - 2k$ observations; it is denoted $\overline{X}_\alpha$, where $\alpha = k/n$. In this case $k = 1$, $n = 20$, and $\alpha = 0.05$.

Of these three estimators which one is best? To answer this question we need some criteria to distinguish a "good" estimator from a poor one. We will see that for most applications, but not all, the sample mean is a good estimator because it is *unbiased* and *consistent*, statistical terms that will soon be explained.

We begin with the observation that the sample mean, sample median, and trimmed mean are functions of the random sample called *(point) estimators* of the mean, and their particular values $(8568, 8803.5, 8397.39$, respectively) are (point) estimates of the expected lifetime. To simplify the terminology we call them estimators and estimates, respectively. More generally, an estimator is a *statistic*, which is defined to be a function of the random sample. In particular, a statistic is itself a random variable; consequently, its values will vary from sample to sample. Statisticians use the generic symbol θ to denote the unknown parameter and the statistic used to estimate it is denoted $\hat{\theta} = \hat{\theta}(X_1, \ldots, X_n)$. The sample mean, for example, is the statistic

$$\overline{X} = \frac{\sum_{1 \leq i \leq n} X_i}{n} = \hat{\theta}(X_1, \ldots, X_n).$$

The set of allowable values θ is called the *parameter space* and is denoted by Θ (capital theta). Suppose, for example, we are sampling from a Bernoulli population with $P(X_i = 1) = p$, $P(X_i = 0) = 1 - p$. The parameter p of a Bernoulli distribution is called a *population proportion*, and the parameter space in this case is the set $\Theta = \{p : 0 < p < 1\}$. Among the most important examples of estimators are:

1. The *sample mean $\overline{X}$*, an estimator of the population mean μ (the notation $\hat{\mu}$ is also used), is defined by

$$\overline{X} = \frac{\sum_{1 \leq i \leq n} X_i}{n} = \hat{\theta}(X_1, \ldots, X_n), \Theta = \{\mu : -\infty < \mu < \infty\}.$$

Refer to Example 7.1, where it is shown that $\hat{\mu} = 8568$.

2. The *sample variance s^2*, an estimator of the population variance σ^2, is defined by

$$s^2 = \frac{1}{n-1} \sum_{1 \leq i \leq n} (X_i - \overline{X})^2 = \hat{\theta}(X_1, \ldots, X_n), \Theta = \{\sigma^2 : \sigma^2 > 0\}.$$

Refer to Example 7.1, where it is shown that $s^2 = 19,771,045.66$.

3. The *sample standard deviation $s = \sqrt{s^2}$*, an estimator of the standard deviation σ (the notation $\hat{\sigma}$ is also used), is defined by

$$s = \sqrt{s^2} = \sqrt{\frac{1}{n-1} \sum_{1 \leq i \leq n} (X_i - \overline{X})^2}, \Theta = \{\sigma : \sigma > 0\}.$$

Refer to Example 7.1: In this case $s = \sqrt{19,771,045.66} = 4446.46$.

Generalizing from these three examples we see that a point estimator $\hat{\theta}$ consists of the following two elements:

1. a random sample $X_1, \ldots, X_n$ whose common distribution (also called the parent distribution) depends on the unknown values of one or more parameters, and

2. a *statistic* $\hat{\theta} = \hat{\theta}(X_1, \ldots, X_n)$, which is a function of the random sample that *does not depend on any of the unknown parameters*. To repeat: A statistic is a random variable, since it is a function of the random sample. Given a statistical model for the data, an *estimator* is a statistic that provides a plausible value for the the unknown parameter θ. More precisely, an estimator of θ is a function of the random sample, and $\hat{\theta}(x_1, \ldots, x_n)$ is the value of the estimator when $X_1 = x_1, \ldots, X_n = x_n$; this observed value of the statistic is called an *estimate*. Depending on the context we use the symbol $\hat{\theta}$ to denote either the statistic $\hat{\theta}(X_1, \ldots, X_n)$ or its value $\hat{\theta}(x_1, \ldots, x_n)$.

7.2.1   Some Properties of Estimators: Bias, Variance, and Consistency

How accurate an estimator of the population mean is the sample mean? To begin with, we know that $E(\overline{X}) = \mu$ (Equation 5.21); that is, the expected value of the sample mean equals the population mean. Consequently, the sample mean is an *unbiased estimator* of the population mean. More generally, we say that an estimator $\hat{\theta}$ of θ is *unbiased* of θ when

$$\text{Unbiased estimator } E(\hat{\theta}) = \theta. \tag{7.1}$$

Otherwise it is is said to be *biased,* and the quantity $b(\hat{\theta}) = E(\hat{\theta}) - \theta$ is called the *bias.* The sample variance is another example of an unbiased estimator (of the population variance) because, as we have previously shown $E(s^2) = \sigma^2$ (Proposition 6.1). Note that $s = \sqrt{s^2}$ is a biased estimator of σ, because $E(s) = E\left(\sqrt{s^2}\right) \leq \sqrt{E(s^2)} = \sigma$. (This is a consequence of Jensen's inequality and the fact that $g(x) = \sqrt{x}$ is a concave function.)

To determine the accuracy of the sample mean we need to know its standard deviation so Chebyshev's inequality (Section 3.3.3) can be brought to bear. It follows from Proposition 5.3 that the variance, and standard deviation of $\overline{X}$ are

$$V(\overline{X}) = \frac{\sigma^2}{n} \text{ and } \sigma(\overline{X}) = \frac{\sigma}{\sqrt{n}}. \tag{7.2}$$

It is to be observed that $\sigma(\overline{X}) = \sigma/\sqrt{n}$ is *not* a statistic since its value depends on the unknown parameter σ. In general, statisticians determine the accuracy of an unbiased estimator $\hat{\theta}$ of θ by computing its standard deviation $\sqrt{V(\hat{\theta})}$, and for this reason it is called the *standard error of the estimator.*

Definition 7.1 *The* standard error of an estimator $\hat{\theta}$, denoted $\sigma(\hat{\theta})$, is

$$\text{Standard error of the estimator: } \sigma(\hat{\theta}) = \sqrt{V(\hat{\theta})}.$$

The reader will recall that the spread of the distribution of $\overline{X}$ about μ is controlled by the magnitude of its standard deviation $\sigma(\overline{X})$. When $\sigma(\overline{X})$ is large the more likely it is for $\overline{X}$ to deviate significantly from the mean; and when it is small a large deviation from the mean is highly unlikely. In other words, the standard deviation is a reasonable measure of the accuracy of the estimator. We call $\sigma(\overline{X})$ the *standard error of the sample mean.* Of course, when σ is unknown, which is the typical situation, $\sigma(\overline{X})$ is not computable. We therefore replace σ by its estimate s (the sample standard deviation) and obtain the *estimated standard error,* denoted $s(\overline{X})$, and defined by the equation:

$$\text{(Estimated standard error)} s(\overline{X}) = \frac{s}{\sqrt{n}}. \tag{7.3}$$

For example, the estimated standard error for the lifetimes in Table 7.1 is $994.26 = 4446.46/\sqrt{20}$.

When the expression for $\sigma(\hat{\theta})$ contains unknown population parameters that can be estimated from the sample data, then we compute the *estimated standard error!s$(\hat{\theta})$* by substituting these estimates into the formula for $\sigma(\hat{\theta})$. This is exactly how we derived Equation 7.3 for the estimated standard error of the sample mean. The estimated standard error $s(\hat{\theta})$ plays important role in constructing confidence intervals for population parameters as will be explained in Section 7.3.

Estimating a function $g(\theta)$ of the parameter θ   To estimate the standard deviation σ we note that $\sigma = \sqrt{\sigma^2}$; consequently, it seems reasonable to use the estimator $s = \sqrt{s^2}$ to estimate σ. This example suggests the following useful principle for estimating $g(\theta)$: If $\hat{\theta}$ is a (good) point estimator of θ, then $g(\hat{\theta})$ is also (good) point estimator $g(\theta)$. For instance, in Example 7.1 we obtained the estimate $s^2 = 19,771,045.66$ for the variance σ^2 of the lifetimes in Table 7.1; consequently, we use $\sqrt{19,771,045.66} = 4446.46$ to estimate the standard deviation σ. For a precise version of this principle see Proposition 7.2 in Section 7.6.2 at the end of this chapter. In particular, this principle is valid for the class of *maximum likelihood estimators* (MLE); see Definition 7.7 for the details.

Comparing two unbiased estimators Suppose $\hat{\theta}_1$ and $\hat{\theta}_2$ are both unbiased estimators; how do we decide which one is better? The answer is to choose the estimator with the smaller variance. To see why this is a reasonable choice let us suppose that $V(\hat{\theta}_1) < V(\hat{\theta}_2)$. From Chebyshev's inequality (Theorem 3.2), we see that

$$P(|\hat{\theta}_i - \theta| \geq \delta) \leq \frac{V(\hat{\theta}_i)}{\delta^2}. \tag{7.4}$$

It follows from Equation 7.4 and the condition $V(\hat{\theta}_1) < V(\hat{\theta}_2)$ that $\hat{\theta}_1$ is more likely to be closer to θ than $\hat{\theta}_2$; so it is a more accurate estimator of θ. This leads to the principle of *minimum variance unbiased estimation:*

> When comparing two unbiased estimators choose the one with the smaller variance. Among all unbiased estimators choose the one with minimum variance. This estimator is called the minimum variance unbiased estimator (MVUE).

Consistent Estimators

The sample mean is a *consistent estimator* of the population mean; that is, when the sample size n is large then $\overline{X}$ is close to μ with high probability. More precisely, the law of large numbers (Theorem 5.3) applied to the sample mean yields the following result:

$$\lim_{n\to\infty} P\left(|\overline{X} - \mu| > \delta\right) = 0, \text{ for every } \delta > 0. \tag{7.5}$$

In addition to being a consistent estimator of μ the sample mean $\overline{X}$ is asymptotically normal; indeed, it follows from Theorem 5.5 that

$$\lim_{n\to\infty} \sqrt{n}(\overline{X} - \mu) \overset{\mathcal{D}}{=} N(0, \sigma^2), \text{ where}$$

σ^2 is the population variance. It is noteworthy that, as the sample size $n \to \infty$, the median $\widetilde{X}$ and trimmed mean $\overline{X}_\alpha$ are also asymptotically normal. In particular, suppose the parent distribution is symmetric about the mean (so the mean and the median are the same), then it can be shown that (for the precise mathematical details see the books by Lehman and Bickel and Doksum listed at the end of this chapter)

$$\lim_{n\to\infty} \sqrt{n}(\widetilde{X} - \mu) \overset{\mathcal{D}}{=} N(0, \widetilde{\sigma}^2) \text{ and } \lim_{n\to\infty} \sqrt{n}(\overline{X}_\alpha - \mu) \overset{\mathcal{D}}{=} N(0, \sigma_\alpha^2).$$

The consistency of $\widetilde{X}$ and $\overline{X}_\alpha$ follow at once. For example,

$$\lim_{n\to\infty} P\left(|\widetilde{X} - \mu)| > \epsilon\right) = \lim_{n\to\infty} P\left(\frac{\sqrt{n}|(\widetilde{X} - \mu)|}{\widetilde{\sigma}} > \frac{\sqrt{n}\epsilon}{\widetilde{\sigma}}\right)$$

$$= \lim_{n\to\infty} 2 \times \left(1 - \Phi\left(\frac{\sqrt{n}\epsilon}{\widetilde{\sigma}}\right)\right) = 0.$$

When the parent distribution is normal it can be shown that $\sigma^2/\tilde{\sigma}^2 < 1$ and $\sigma^2/\sigma_\alpha^2 < 1$; consequently, $\overline{X}$ is the better estimator in this case. In particular, if the lifetime data in Table 7.1 comes from a normal distribution then $\overline{X}$ is the best estimator. There are, however, parent distributions for which $\tilde{X}$ and $\overline{X}_\alpha$ perform better than $\overline{X}$ because their asymptotic variances are smaller. Readers interested in exploring these and related questions should consult the books by Lehman, and Bickel and Doksum op. cit.

From these examples we arrive at the standard definition of consistency.

Definition 7.2 *We say that an estimator $\hat{\theta}_n$ of θ based on a random sample of size n is a consistent estimator of θ if*

$$\lim_{n\to\infty} P\left(|\hat{\theta}_n - \theta| > \delta\right) = 0, \ \textit{for every } \delta > 0. \tag{7.6}$$

Additional examples of consistent estimators:

1. The sample variance is a consistent estimator of the population variance.

2. The sample standard deviation is a consistent estimator of the population standard deviation.

3. A sample quantile is a consistent estimator of a population quantile. For the proofs we refer the reader to the books by Lehman, and Bickel and Doksum op.cit.

7.3   Confidence Intervals for the Mean and Variance

Suppose $\overline{X}$ is the sample mean of a random sample drawn from a normal distribution, then $\overline{X} \stackrel{\mathcal{D}}{=} N(\mu, \sigma^2/n)$ (Corollary 6.1). Therefore $P(\overline{X} > \mu) = 0.5 = P(\overline{X} < \mu)$; thus, the value of $\overline{X}$ fluctuates above and below the mean μ; and since $P(\overline{X} = \mu) = 0$ (why?) it follows that $\overline{X} \neq \mu$ with probability one. Therefore, simply reporting the value of $\overline{X}$ tells us nothing about the magnitude of the discrepancy $|\overline{X} - \mu|$. For this reason, instead of simply providing the estimate, it is more informative to report an interval of values that contains the unknown parameter with high probability; such an interval is called a *confidence interval*. Proposition 7.1 illustrates how this is done in the special case when the data are assumed to be normally distributed with known variance; the case when the variance is unknown will be discussed in the next section.

Proposition 7.1 *Suppose $X_1, \ldots, X_n$ is a random sample from a normal distribution $N(\mu, \sigma^2)$, so $\sqrt{n}(\overline{X} - \mu)/\sigma \stackrel{\mathcal{D}}{=} N(0,1)$. Then*

$$P\left(\overline{X} - z(\alpha)\frac{\sigma}{\sqrt{n}} \leq \mu\right) = 1 - \alpha; \tag{7.7}$$

$$P\left(\mu \leq \overline{X} + z(\alpha)\frac{\sigma}{\sqrt{n}}\right) = 1 - \alpha; \tag{7.8}$$

$$P(\overline{X} - z(\alpha/2) \times \sigma/\sqrt{n} \leq \mu \leq \overline{X} + z(\alpha/2) \times \sigma/\sqrt{n}) = 1 - \alpha. \tag{7.9}$$

Here, $z(\alpha)$ denotes the critical value of the standard normal distribution defined in Equation 4.41; of course $z(\alpha)$ is also the $100(1-\alpha)\%$ percentile of the standard normal distribution, that is $P(Z \leq z(\alpha)) = 1 - \alpha$. Equations 7.7, 7.8, and 7.9 are easily derived from the fact that $\sqrt{n}(\overline{X} - \mu)/\sigma \stackrel{\mathcal{D}}{=} N(0,1)$; consequently,

$$P\left(\frac{\sqrt{n}(\overline{X} - \mu)}{\sigma} \leq z(\alpha)\right) = 1 - \alpha. \tag{7.10}$$

We obtain Equation 7.7 by solving the inequality inside the parentheses of the left hand side of Equation 7.10 for μ:

$$\left(\frac{\sqrt{n}(\overline{X} - \mu)}{\sigma} \leq z(\alpha)\right) = \left(\overline{X} - z(\alpha)\frac{\sigma}{\sqrt{n}} \leq \mu\right).$$

Therefore

$$P\left(\frac{\sqrt{n}(\overline{X} - \mu)}{\sigma} \leq z(\alpha)\right) = P\left(\overline{X} - z(\alpha)\frac{\sigma}{\sqrt{n}} \leq \mu\right) = 1 - \alpha.$$

Equations 7.8 and 7.9 are derived in a similar fashion; we omit the details.

It follows from Equation 7.9 that the (random) interval $[L, U]$, where

$$\text{(lower limit) } L = \overline{X} - z(\alpha/2)\sigma/\sqrt{n}, \text{ and} \tag{7.11}$$
$$\text{(upper limit) } U = \overline{X} + z(\alpha/2)\sigma/\sqrt{n}, \tag{7.12}$$

contains μ with probability $1 - \alpha$; which is why it is called a $100(1-\alpha)\%$ confidence interval for μ. It is sometimes more convenient to express the confidence interval in the equivalent form:

$$\overline{X} \pm z(\alpha/2)\frac{\sigma}{\sqrt{n}}.$$

Example 7.2 *Constructing a 95% confidence interval for μ*

Suppose we are given a random sample of $n = 10$ observations taken from a normal distribution $N(\mu, 4)$ with $\overline{x} = 15.1$. Construct a 95% confidence interval for μ.

Solution Set $\alpha = 0.05$, then compute: $z(\alpha/2) = z(0.025) = 1.96$; $z(\alpha/2)\frac{\sigma}{\sqrt{n}} = 1.96 \times 2/\sqrt{10} = 1.24$. The confidence interval $[L, U]$ is then given by

$$L = \overline{X} - z(\alpha/2)\frac{\sigma}{\sqrt{n}} = 15.1 - 1.24 = 13.86; \ U = \overline{X} + z(\alpha/2)\frac{\sigma}{\sqrt{n}} = 15.1 + 1.24 = 16.34.$$

It is worth repeating that the confidence interval $[L, U]$ is a (random) interval whose end points are the random variables displayed in Equations 7.11 and 7.12, respectively. We summarize this discussion by giving a formal definition of a confidence interval.

Definition 7.3 *A random interval of the form $[L, U]$, where L and U are statistics satisfying the condition $L \leq U$, is said to be a $100(1 - \alpha)\%$ confidence interval for the parameter θ if*

$$P(L \leq \theta \leq U) \geq 1 - \alpha.$$

One Sided Confidence Intervals

There are many situations in which an engineer is interested in obtaining a *lower bound* on a population parameter: an industrial engineer is never concerned about the system reliability being too high, but he is worried about it possibly being too low. We obtain a lower bound by constructing a $100(1 - \alpha)\%$ lower one sided confidence interval as will now be explained.

Definition 7.4 *A random interval of the form* $[L, \infty)$ *is a* $100(1 - \alpha)\%$ *lower one sided confidence interval for the parameter* θ *if*

$$P(L \leq \theta) \geq 1 - \alpha.$$

Similarly, a random interval of the form $(-\infty, U]$ *is a* $100(1-\alpha)\%$ *upper one sided confidence interval for the parameter* θ *if*

$$P(\theta \leq U) \geq 1 - \alpha.$$

To illustrate the concept we construct a $100(1 - \alpha)\%$ lower one sided confidence interval for the mean of a normal distribution based on a sample of size n drawn from a normal population $N(\mu, \sigma^2)$. It follows from Equation 7.7 that the random interval $[L, \infty)$ where

$$L = \overline{X} - z(\alpha)\frac{\sigma}{\sqrt{n}} \qquad (7.13)$$

contains μ with probability $1 - \alpha$.

Example 7.3 *A lower confidence interval for the compressive strengths of bricks*

The data set 7.14 records nine determinations of the compressive strength of bricks measured in units of 100 pounds per square inch. Assume the data are normally distributed with $\sigma = 5$. Obtain a lower 95% confidence interval for the mean compressive strength μ.

$$45.2, \ 50.4, \ 65.9, \ 52.3, \ 48.3, \ 49.6, \ 51.3, \ 62.2, \ 53.8 \qquad (7.14)$$

Solution We use Equation 7.13 with $\overline{x} = 53.22$, $n = 9$ and $z(0.05) = 1.645$; consequently, the lower bound L is given by

$$L = 53.22 - 1.645 \times \frac{5}{3} = 50.48,$$

so the 95% lower confidence interval is $[50.48, \infty)$.

An Interpretation of the Confidence Interval: Each random sample $X_1, \ldots, X_n$ produces a random interval $[\overline{X} - z(\alpha/2) \times \sigma/\sqrt{n}, \ \overline{X} + z(\alpha/2) \times \sigma/\sqrt{n}]$ some of which will contain μ and some will not. All we can say, prior to doing the experiment, is that m repetitions of this experiment will produce m confidence intervals, of which approximately $100(1 - \alpha)\%$ of them contain the parameter μ. After doing the experiment the confidence interval $[L, U]$ either contains μ or it does not.

Choosing the Sample Size

A frequently asked question is this: How large must the sample size be in order for $\overline{X}$ to be within d units of the unknown parameter μ? The $100(1 - \alpha)\%$ confidence interval 7.9 states that with probability $(1 - \alpha)$ $\overline{X}$ is within d units of the unknown parameter μ, where

$$d = z(\alpha/2) \times \frac{\sigma}{\sqrt{n}}.$$

We note in passing that the width of the confidence interval equals $2d$. Thus, the accuracy of the estimator $\overline{X}$ is determined by d, the half width of the confidence interval (also called the *margin of error*). It follows that given d, σ and α we can solve for n and obtain the following equation for the sample size:

$$n = \left(\frac{z(\alpha/2)\sigma}{d}\right)^2. \qquad (7.15)$$

When σ is unknown, as is frequently the case, then we can estimate it using data obtained from previous studies.

Example 7.4 *Finding the minimum sample size using historical data*

Refer to the data in Problem 1.5. How large a sample size is needed if our estimate of the battery capacity is to be within 1.5 units of μ?

Solution On the basis of the data already collected we have the estimate $s = 3.4641$. To play it safe we round it up and assume that $\sigma = 3.5$. Setting $d = 1.5$ and $\sigma = 3.5$ in Equation 7.15 yields

$$n = \left(\frac{1.96 \times 3.5}{1.5} \right)^2 = 20.92; \text{ since } n$$

must be an integer, we round it up to $n = 21$.

Looking at Equation 7.15 we see that the margin of error depends on α. In particular, as $\alpha \to 0$ we have $z(\alpha/2) \to \infty$. Thus, for a fixed sample size n there is a trade-off between the accuracy of the estimator as measured by the half width of the confidence interval and the confidence level as measured by $1 - \alpha$.

7.3.1 Confidence Intervals for the Mean of a Normal Distribution: Variance Unknown

The results derived in the previous section are not very practical since it is unrealistic to assume that the variance is known but the mean is not. We now turn our attention to the problem of constructing confidence intervals for the mean of a normal distribution $N(\mu, \sigma^2)$ where σ is unknown.

In this case we use Theorem 6.5, which states that

$$\frac{\sqrt{n}(\overline{X} - \mu)}{s} \overset{\mathcal{D}}{=} t_{n-1},$$

where s is the sample standard deviation and t_{n-1} is student's t distribution with $n-1$ degrees of freedom. Therefore,

$$P\left(\frac{\sqrt{n}(\overline{X} - \mu)}{s} \le t_{n-1}(\alpha) \right) = 1 - \alpha. \tag{7.16}$$

From the symmetry of the t distribution we obtain

$$P\left(-t_{n-1}(\alpha/2) \le \frac{\sqrt{n}(\overline{X} - \mu)}{s} \le t_{n-1}(\alpha/2) \right) = 1 - \alpha. \tag{7.17}$$

Solving the inequalities for μ in Equation 7.17 yields the $100(1 - \alpha)\%$ confidence interval $[L, U]$, where

$$L = \overline{X} - t_{n-1}(\alpha/2)s/\sqrt{n}, \text{ and} \tag{7.18}$$
$$U = \overline{X} + t_{n-1}(\alpha/2)s/\sqrt{n}. \tag{7.19}$$

The confidence interval is also expressed in the form:

$$\overline{X} \pm t_{n-1}(\alpha/2)\frac{S}{\sqrt{n}}.$$

Example 7.5 *Constructing a 95% confidence interval: Variance unknown*

Refer to the data of Problem 1.7. Construct a 95% confidence interval for the chamber burst strength.

Solution The sample mean and standard deviation are given by: $\bar{x} = 16.49$ and $s = 0.584$. (The reader should use a calculator or statistical software package to verify these values.) To compute a 95% confidence interval we set $n = 17$, $\alpha = 0.05$ and then compute

$$t_{n-1}(\alpha/2) = t_{16}(0.025) = 2.12; \quad t_{16}(0.025)\frac{0.584}{\sqrt{17}} = 0.30.$$

Inserting these values into Equations 7.18 and 7.19 yields

$$L = \bar{X} - t_{16}(0.025)\frac{0.584}{\sqrt{17}} = 16.49 - 0.30 = 16.19;$$

$$U = \bar{X} + t_{16}(0.025)\frac{0.584}{\sqrt{17}} = 16.49 + 0.30 = 16.79.$$

The confidence interval is [16.19, 16.79].

One Sided Confidence Intervals, Unknown Variance

We conclude this section by giving the recipe for constructing one sided confidence intervals when the standard deviation is unknown.

Solving the inequality in Equation 7.16 for μ we obtain

$$P(\bar{X} - t_{n-1}(\alpha)\frac{s}{\sqrt{n}} \le \mu) = 1 - \alpha. \tag{7.20}$$

We interpret this result in the usual way: The $100(1 - \alpha)\%$ lower confidence interval is given by

$$[\bar{X} - t_{n-1}(\alpha)\frac{s}{\sqrt{n}}, \infty). \tag{7.21}$$

Example 7.6 *Problem 1.5 lists the capacities (measured in ampere hours) of ten batteries; an easy calculation shows that $\bar{x} = 151$ and $s = 3.46$. From these data construct a 95% lower confidence interval for the battery capacity.*

Solution To construct the lower confidence interval set $n = 10$, $s = 3.46$ in Equation 7.21. In detail:

$$\frac{s}{\sqrt{n}} = \frac{3.46}{\sqrt{10}} = 1.0941, \text{ and therefore}$$

$$t_9(0.05)\frac{s}{\sqrt{n}} = 1.833 \times 1.0941 = 2.01;$$

$$\bar{X} - t_9(0.05)\frac{s}{\sqrt{n}} = 151 - 2.01 = 148.99.$$

The corresponding 95% one sided confidence interval is given by [148.99, ∞).

We note for future reference that the $100(1 - \alpha)\%$ upper confidence interval is given by

$$P\left(\mu \le \bar{X} + t_{n-1}(\alpha)\frac{s}{\sqrt{n}}\right) = 1 - \alpha. \tag{7.22}$$

7.3.2   Confidence Intervals for the Mean of an Arbitrary Distribution

In this section we drop the assumption that the random sample comes from a normal distribution; instead, we assume it comes from an arbitrary distribution with population mean μ and population variance σ^2. We make the additional assumption that the sample size n is large enough ($n \geq 30$, say) so that the distribution of $\sqrt{n}(\overline{X} - \mu)/\sigma$ is close to the distribution of Z, a standard normal random variable. Thus,

$$P\left(-z(\alpha/2) \leq \frac{\sqrt{n}(\overline{X} - \mu)}{\sigma} \leq z(\alpha/2)\right) \approx P(-z(\alpha/2) \leq Z \leq z(\alpha/2)). \qquad (7.23)$$

The left hand side of Equation 7.23 is therefore approximately equal to $1 - \alpha$. Consequently,

$$P\left(-z(\alpha/2) \leq \frac{\sqrt{n}(\overline{X} - \mu)}{\sigma} \leq z(\alpha/2)\right) \approx 1 - \alpha.$$

The pair of inequalities inside the parentheses may be solved for μ, which gives

$$P(\overline{X} - z(\alpha/2) \times \sigma/\sqrt{n} \leq \mu \leq \overline{X} + z(\alpha/2) \times \sigma/\sqrt{n}) \approx 1 - \alpha. \qquad (7.24)$$

It follows from Equation 7.24 that the interval $[L, U]$, where

$$L = \overline{X} - z(\alpha/2)\frac{\sigma}{\sqrt{n}} = \overline{X} - z(\alpha/2)\sigma(\overline{X}),$$

$$U = \overline{X} + z(\alpha/2)\frac{\sigma}{\sqrt{n}} = \overline{X} + z(\alpha/2)\sigma(\overline{X}),$$

contains μ with a probability that is approximately equal to $1 - \alpha$, so it is called an *approximate* $100(1 - \alpha)\%$ confidence interval for μ.

When σ is unknown, which is usually the case, then so is $\sigma(\overline{X})$ and one replaces it with the estimated standard error $s(\overline{X})$; the approximate $100(1 - \alpha)\%$ confidence interval $[L, U]$ now takes the form:

$$L = \overline{X} - z(\alpha/2)\frac{s}{\sqrt{n}}; U = \overline{X} + z(\alpha/2)\frac{s}{\sqrt{n}}. \qquad (7.25)$$

Example 7.7 *Constructing an approximate 95% confidence interval*

Refer to the 36 measurements of the heights of ten-year-old girls obtained from the Statlab data set (Table 1.16). Construct an approximate 95% confidence interval for the mean of the girls' heights.

Solution We assume that these measurements constitute a random sample of size $n = 36$ from a distribution with unknown mean and unknown variance. Using a computer we obtained the following values for the sample mean, sample standard deviation, and estimated standard error.

$$\bar{x} = 53.34,\ s = 2.91,\ s(\overline{X}) = \frac{2.91}{\sqrt{36}} = 0.49.$$

An approximate 95% confidence interval for the mean height of ten-year-old girls is given by

$$53.34 \pm 1.96 \times 0.49 = 53.34 \pm 0.95.$$

The confidence interval is $[52.39, 54.29]$.

One sided confidence intervals

In a similar way, one constructs approximate $100(1 - \alpha)\%$ upper and lower one sided confidence intervals $(-\infty, U], [L, \infty)$, where

$$L = \overline{X} - z(\alpha)\frac{s}{\sqrt{n}}; U = \overline{X} + z(\alpha)\frac{s}{\sqrt{n}}. \qquad (7.26)$$

7.3.3   Confidence Intervals for the Variance of a Normal Distribution

To construct a confidence interval for the variance σ^2 we use Theorem 6.4, which asserts that

$$\frac{(n-1)s^2}{\sigma^2} \text{ is } \chi^2_{n-1} \text{ distributed.}$$

Let $\chi^2_{n-1}(\alpha)$ denote the critical value of the chi-square distribution defined by

$$P(\chi^2 \geq \chi^2_{n-1}(\alpha)) = \alpha.$$

$\chi^2_n(\alpha)$ is, of course, the $100(1-\alpha)\%$ percentile of the chi-square distribution, so

$$P(\chi^2 \leq \chi^2_{n-1}(\alpha)) = 1 - \alpha. \tag{7.27}$$

Therefore,

$$P\left(\chi^2_{n-1}(1-\alpha/2) \leq \frac{(n-1)s^2}{\sigma^2} \leq \chi^2_{n-1}(\alpha/2)\right) = 1 - \alpha. \tag{7.28}$$

Solving these inequalities for σ^2 in the usual way we arrive at the following $100(1-\alpha)\%$ confidence interval for σ^2 and σ:

$$P\left(\frac{(n-1)s^2}{\chi^2_{n-1}(\alpha/2)} \leq \sigma^2 \leq \frac{(n-1)s^2}{\chi^2_{n-1}(1-\alpha/2)}\right) = 1 - \alpha, \tag{7.29}$$

$$P\left(\sqrt{\frac{(n-1)s^2}{\chi^2_{n-1}(\alpha/2)}} \leq \sigma \leq \sqrt{\frac{(n-1)s^2}{\chi^2_{n-1}(1-\alpha/2)}}\right) = 1 - \alpha. \tag{7.30}$$

Estimating the volatility of a stock from its stock price history

The annual volatility σ of a stock is a parameter that plays an important role in financial engineering. It appears in the formula for the distribution of the stock price at a future time t (Equation 5.52 in Section 5.5) and in the Black-Scholes option pricing formula (Equation 5.154 in Section 5.9.1). The volatility is estimated by recording the prices $(S_0, S_1, \ldots, S_n)$ at $n+1$ time intervals of fixed length T and computing the sample standard deviation of the n returns $(r_1, r_2, \ldots, r_n)$, where $r_i = (S_i - S_{i-1})/S_{i-1}$, $i = 1, 2, \ldots n$. Recall that T is measured in years, so $T = 1/52$, $1/13$ for weekly returns and quarterly returns, respectively. According to the lognormal model for stock prices the standard deviation of the returns for a time period of length T is $\sigma(T) = \sigma\sqrt{T}$, where σ is the annual volatility of the stock's returns. Equivalently, $\sigma(T)/\sqrt{T} = \sigma$. Since $\widehat{\sigma(T)}$ is an estimate of $\sigma(T)$ it follows that $\widehat{\sigma(T)}/\sqrt{T}$ is an estimate of $\sigma/\sqrt{T}$. Consequently, σ itself can be estimated by $\hat{\sigma}$, where

$$\hat{\sigma} = \frac{\widehat{\sigma(T)}}{\sqrt{T}}. \tag{7.31}$$

Example 7.8 *Estimating the annual volatility of GE's stock price (1999)*

Refer to Example 1.14 and Table 1.13, which gives the mean weekly return and the weekly volatility (standard deviation of the weekly returns) for each of the stocks listed in Table 1.19. Compute a 95% confidence interval for the annual volatility of GE's stock price.

Solution We first obtain a 95% confidence interval for $\sigma(T)$, $T = 1/52$ using Equation 7.30 and then use Equation 7.31 to find a 95% confidence interval for σ. In detail, inserting the values

$$\alpha = 0.025, \ n = 51, \ s = 0.0399, \ \chi^2_{50}(0.025) = 32.357, \chi^2_{50}(0.975) = 71.420, \ T = \frac{1}{52}$$

into Equation 7.30 we obtain the 95% confidence interval for $\sigma(T)$:

$$\left[\sqrt{\frac{50 \times (0.0399)^2}{71.4720}} \le \sigma(T) \le \sqrt{\frac{50 \times (0.0399)^2}{32.357}}\right] = [0.0334, 0.0496].$$

The corresponding confidence interval for $\sigma(T)/\sqrt{T}$, with $T = 1/52$ is obtained by multiplying the endpoints of the interval $[0.0334, 0.0496]$ by $\sqrt{52}$. Therefore, a 95% confidence interval for GE's annual volatility is $[\sqrt{52} \times 0.0334, \sqrt{52} \times 0.0496] = [0.2409, 0.3577]$.

7.3.4 Value at Risk (VaR): An Application of Confidence Intervals to Risk Management

An important factor in managing a portfolio of stocks and bonds is estimating the probability of a large loss for a specific period, e.g., daily, weekly, monthly, etc. Specifically, denote the initial and end-of-period values of the portfolio by W_0, W_1, respectively, and assume the portfolio's rate of return is a normally distributed random variable $R \overset{D}{=} N(\mu, \sigma^2)$. The loss, which is the negative of the gain, is given by $L = -(W_1 - W_0) = -W_0 R$, which is an immediate consequence of the fact that $R = (W_1 - W_0)/W_0$. Consequently, L is also a normally distributed random variable. In particular, the probability that the loss exceeds $x is given by

$$P(L \ge x) = \Phi\left(\frac{-(\mu + (x/W_0))}{\sigma}\right). \tag{7.32}$$

The derivation of Equation 7.32 proceeds along a familiar path:

$$P(L \ge x) = P(-W_0 R \ge x) = P\left(R \le -\frac{W_0}{x}\right)$$

$$= P\left(\frac{R - \mu}{\sigma} \le \frac{-(\mu + (x/W_0))}{\sigma}\right)$$

$$= P\left(Z \le \frac{-(\mu + (x/W_0))}{\sigma}\right) = \Phi\left(\frac{-(\mu + (x/W_0))}{\sigma}\right).$$

In practice, the probability $\alpha = P(L \ge x)$ is fixed—$\alpha = 0.01$ is the usual choice—and then one uses Equation 7.32 to find the the value x, interpreted here as the amount such that the probability of a loss exceeding it is α. Equivalently, the probability of a loss not exceeding x is $1 - \alpha$. The value x is called the *value at risk (VaR)* at the $100(1 - \alpha)\%$ confidence level. It is easy to verify that the solution to the equation

$$\Phi\left(\frac{-(\mu + (x/W_0))}{\sigma}\right) = \alpha \text{ is } x = -W_0(\mu + \sigma z(\alpha)). \tag{7.33}$$

Problem 7.31 asks you to supply the details.

Example 7.9 *Finding the 99% value at risk*

A portfolio has a weekly return of $\mu = 0.002$ and weekly volatility $\sigma = 0.025$. The current value of the portfolio is \$1 million. Find the 99% VaR for the one week period.

Solution Inserting $W_0 = 1,000,000$, $\mu = 0.002$, $\sigma = 0.025$, $z(0.01) = 2.33$ into Equation 7.33 yields the result:

$$x = 1,000,000(0.002 + 0.025 \times 2.33) = \$60,250.$$

Interpretation: The probability is 0.99 that we will lose no more than \$60,250 in the next week; equivalently, the probability of losing more than \$60,250 is less than 0.01.

7.4   Point and Interval Estimation for the Difference of Two Means

The methods of the previous sections will now be applied to a comparative study of two populations. For example, we might be interested in the difference in weight loss obtained by following two different diets, or the difference in tensile strength between two types of steel cable. To study these and similar questions we take random samples of size n_1 and n_2 from each of the two populations, which we assume to be normally distributed with possibly different means, denoted by μ_1 and μ_2, but with a common variance σ^2. We make the additional assumption that the random samples taken from each population are mutually independent. The sample means from each population will be denoted by $\overline{X}_1$ and $\overline{X}_2$, and the sample variances by s_1^2 and s_2^2, respectively. Notice that $\overline{X}_1 - \overline{X}_2$ is an unbiased estimator of the difference between the means, denoted by $\Delta = \mu_1 - \mu_2$. When the variances of the two populations are different the problem of comparing the two means becomes somewhat more complicated and will not be discussed here.

As noted earlier, $\overline{X}_1 - \overline{X}_2$ is an unbiased estimator of Δ and it is therefore natural to use it for the purpose of constructing confidence intervals. Before doing so, however, we need an estimator for the common variance. Of course, s_1^2 and s_2^2 are both unbiased estimators of σ^2, but it is known that the best combined estimator is the *pooled* estimator s_p^2 defined by

$$s_p^2 = \frac{(n_1 - 1)s_1^2 + (n_2 - 1)s_2^2}{n_1 + n_2 - 2}. \tag{7.34}$$

Because $\overline{X}_1$ and $\overline{X}_2$ are independent, it follows that

$$V(\overline{X}_1 - \overline{X}_2) = V(\overline{X}_1) + V(\overline{X}_2)$$
$$= \sigma^2/n_1 + \sigma^2/n_2 = \sigma^2(1/n_1 + 1/n_2).$$

The estimated standard error, therefore, is given by

$$s(\overline{X}_1 - \overline{X}_2) = s_p\sqrt{\frac{1}{n_1} + \frac{1}{n_2}}.$$

The confidence intervals are computed using the statistic t defined in the next theorem (we omit the proof).

Theorem 7.1 *The statistic t defined by*

$$t = \frac{(\overline{X}_1 - \overline{X}_2) - (\mu_1 - \mu_2)}{s_p\sqrt{\frac{1}{n_1} + \frac{1}{n_2}}} \tag{7.35}$$

has a student t distribution with $n_1 + n_2 - 2$ degrees of freedom.

Consequently,

$$P\left(\frac{(\overline{X}_1 - \overline{X}_2) - (\mu_1 - \mu_2)}{s_p\sqrt{\frac{1}{n_1} + \frac{1}{n_2}}} \leq t_{n_1+n_2-2}(\alpha)\right) = 1 - \alpha. \tag{7.36}$$

Confidence Intervals for the Difference of Two Means

We can now construct a $100(1 - \alpha)\%$ confidence interval for $\mu_1 - \mu_2$ by proceeding in exactly the same way as in Section 7.3.1; omitting the details the final result is

$$L = \overline{X}_1 - \overline{X}_2 - t_{n_1+n_2-2}(\alpha/2)s_p\sqrt{\frac{1}{n_1} + \frac{1}{n_2}}; \tag{7.37}$$

$$U = \overline{X}_1 - \overline{X}_2 + t_{n_1+n_2-2}(\alpha/2)s_p\sqrt{\frac{1}{n_1} + \frac{1}{n_2}}. \tag{7.38}$$

One sided confidence intervals are similarly obtained. The lower and upper $100(1 - \alpha)\%$ one sided confidence intervals are:

$$\left[\overline{X}_1 - \overline{X}_2 - t_{n_1+n_2-2}(\alpha)s_p\sqrt{\frac{1}{n_1} + \frac{1}{n_2}}, \infty\right); \tag{7.39}$$

$$\left(-\infty, \overline{X}_1 - \overline{X}_2 + t_{n_1+n_2-2}(\alpha)s_p\sqrt{\frac{1}{n_1} + \frac{1}{n_2}}\right]. \tag{7.40}$$

Example 7.10 *Lord Rayleigh's discovery of argon (see Examples 1.5 and 1.23)*

It is a useful exercise to analyze Lord Rayleigh's experimental data (Table 1.4), which led him to the discovery of argon and a Nobel Prize. Notice that the sample mean of nitrogen obtained from the air sample ($\overline{x}_1 = 2.31011$) is 0.46% greater than the sample mean of nitrogen obtained from chemical decomposition ($\overline{x}_2 = 2.29947$). One way to determine whether the nitrogen obtained from the air sample is heavier than nitrogen obtained from chemical decomposition is to calculate a 95% confidence interval for the difference in their sample means. A straightforward calculation, using Equations 7.37 and 7.38, yields the 95% confidence interval $[0.00962, 0.01166]$. In detail:

$$\overline{x}_1 = 2.31011, \overline{x}_2 = 2.29947; \overline{x}_1 - \overline{x}_2 = 0.01064;$$

$$s_1 = 0.000143, s_2 = 0.00138;$$

$$s_p = \sqrt{\frac{6 \times (0.000143)^2 + 7 \times (0.00138)^2}{13}} = 0.00102;$$

$$t_{13}(0.025)s_p\sqrt{\frac{1}{6} + \frac{1}{7}} = 0.00102;$$

$$L = 0.00962 = 0.01064 - 0.00102; U = 0.01166 = 0.01064 + 0.00102.$$

The 95% confidence interval $[0.0095, 0.01178]$ lies on the positive real axis, so it does not contain 0; consequently, the experimental data supports Rayleigh's claim that the nitrogen gas obtained from the air sample contains a small amount of a heavier gas, which later turned out to be argon.

7.4.1 Paired Samples

Paired samples arise when we study the response of an experimental unit to two different treatments. Suppose, for example, we are conducting a clinical trial of a blood pressure reducing drug. To measure its effectiveness (or ineffectiveness), we give the drug to a random sample of n patients. The data set, then, consists of the n pairs $(x_1, y_1), \ldots, (x_n, y_n)$, where x_i denotes the ith patient's blood pressure before, and y_i denotes the ith patient's blood pressure after the treatment. It is customary to call the two data sets $x_1, \ldots, x_n$ and $y_1, \ldots, y_n$ as treatment 1 and treatment 2, respectively. The variable $d_i = x_i - y_i$ represents the difference in blood pressure produced by the drug on the ith patient; a positive value for d_i indicates that the drug reduced the blood pressure of the ith patient. This is an example of what is called *self pairing*, in which a single experimental unit is given two treatments. There are, however, many other ways in which such pairs are selected. For example, identical twins are frequently used in psychology tests and mice of the same sex and genetic strain are used in biology experiments. After the pair has been selected one of the two units is randomly selected for treatment 1 and the other for treatment 2. The advantages of pairing are intuitively clear: it reduces the variation in the data that is due to causes other than the treatment itself.

From these considerations we are led naturally to the following model for the paired sample.

Definition 7.5 *Mathematical model of a paired sample: The data come from a sequence of paired random variables* $(X_i, Y_i), i = 1, \ldots, n$ *that denote the response of the ith experimental unit to the first and second treatments, respectively; their difference, which is the variable we are particularly interested in, will be denoted by* $D_i = X_i - Y_i$. *In addition, we assume that the random variables* D_i, $(i = 1, \ldots, n)$ *are mutually independent, normally distributed random variables with a common mean and variance denoted by* $E(D_i) = \Delta$, $V(D_i) = \sigma_D^2$. *The sample variance is denoted by* s_D^2 *where*

$$s_D^2 = \frac{1}{n-1} \sum_{1 \leq i \leq n} (D_i - \overline{D})^2.$$

In general, the pair of random variables X_i, Y_i will not be independent, since the patient's final blood pressure, say, will be highly correlated with his initial blood pressure. To estimate Δ and construct confidence intervals we begin with the fact that $D_1, \ldots, D_n$ is, by hypothesis, a random sample from the normal distribution $N(\Delta, \sigma_D^2)$. Therefore

$$\overline{D} \overset{\mathcal{D}}{=} N(\Delta, \frac{\sigma^2}{n}); \text{ consequently,}$$

$$\frac{\sqrt{n}(\overline{D} - \Delta)}{s_D} \overset{\mathcal{D}}{=} t_{n-1}, \text{ and therefore}$$

$$P\left(-t_{n-1}(\alpha/2) \leq \frac{\sqrt{n}(\overline{D} - \Delta)}{s_D} \leq t_{n-1}(\alpha/2)\right) = 1 - \alpha.$$

Solving for Δ in the usual way produces the $100(1-\alpha)\%$ confidence interval $[L, U]$, where:

$$L = \overline{D} - t_{n-1}(\alpha/2) \times s_D/\sqrt{n}; \ U = \overline{D} + t_{n-1}(\alpha/2) \times s_D/\sqrt{n}. \quad (7.41)$$

Example 7.11 *Comparing execution times on two different processor configurations*

In a study to compare two different configurations of a processor, the times to execute six different workloads (W1,..., W6) was measured for each of the two configurations denoted by *one cache* and *no cache*, respectively. The results, obtained from computer simulations, are displayed in Table 7.2. The difference in the execution times for each of the workloads is displayed in the rightmost column. Compute a 95% confidence interval for the difference in execution times between the one cache and no cache configurations.

Table 7.2	*Workload*	*One Cache*	*No Cache*	$d_i = x_i - y_i$
	W1	*69*	*75*	*-6*
	W2	*53*	*83*	*-30*

W3	61	89	-28
W4	54	77	-23
W5	49	83	-34
W6	50	83	-33

Solution   This is a paired sample with workload as the experimental unit and one cache and no cache memories denoting the two treatments. We obtain a 95% confidence interval for Δ by first computing

$$\overline{D} = -25.67, \; s_D = 10.41, \; \text{and } t_5(0.025) = 2.571.$$

The left and right endpoints of the confidence interval are given by

$$L = -25.67 - t_5(0.025) \times \frac{10.41}{\sqrt{6}} = -36.60;$$

$$U = -25.67 + t_5(0.025) \times \frac{10.41}{\sqrt{6}} = -14.74.$$

We can also write the confidence interval $[-36.60, -14.74]$ in the following equivalent form:

$$-25.67 \pm t_5(0.025) \frac{10.41}{\sqrt{6}} = -25.67 \pm 10.93.$$

The data indicate that the execution times for a one cache memory are less than the times for a no cache memory.

7.5   Point and Interval Estimation for a Population Proportion

Estimating the proportion of defectives in a lot of manufactured items or the proportion of voters who plan to vote for a politician are examples of estimating a population proportion. The Bernoulli trials process, described in Section 5.2.1, is a mathematical model of this sampling procedure. We repeat, briefly, the essential details: Let $X_1, \ldots, X_n$ denote a random sample from a Bernoulli distribution with $P(X_i = 0) = 1 - p$, $P(X_i = 1) = p$, so

$$f(x;p) = p^x(1-p)^{1-x}, \; x = 0, 1.$$

The mean, variance, standard deviation, and estimated standard error of the sample proportion $\hat{p}$ are given by:

$$E(\hat{p}) = p; \; V(\hat{p}) = \frac{p(1-p)}{n}; \tag{7.42}$$

$$\sigma(\hat{p}) = \sqrt{\frac{p(1-p)}{n}}; \; s(\hat{p}) = \sqrt{\frac{\hat{p}(1-\hat{p})}{n}}. \tag{7.43}$$

We now assume that the sample size is large enough so that the normal approximation to the binomial can be brought to bear; one rule of thumb is that $np > 5$ and $n(1-p) > 5$. Of course, when p is unknown then we assume $n\hat{p} > 5$ and $n(1-\hat{p}) > 5$. In any event, assuming the validity of the normal approximation to the binomial yields

$$\frac{\hat{p} - p}{s(\hat{p})} \approx N(0, 1).$$

Therefore

$$P\left(-z(\alpha/2) \le \frac{\hat{p} - p}{s(\hat{p})} \le z(\alpha/2)\right) \approx 1 - \alpha.$$

The inequalities inside the parentheses can be solved for p in exactly the same way we solved a similar pair of inequalities for μ in Section 7.3 (see Equation 7.24). We have thus shown that

$$P(\hat{p} - z(\alpha/2) \times s(\hat{p}) \le p \le \hat{p} + z(\alpha/2) \times s(\hat{p})) \approx 1 - \alpha. \qquad (7.44)$$

Putting these results together in the usual way yields an approximate $100(1-\alpha)\%$ confidence interval for p of the form $[L, U]$ where

$$L = \hat{p} - z(\alpha/2)\sqrt{\frac{\hat{p}(1 - \hat{p})}{n}}; \qquad (7.45)$$

$$U = \hat{p} + z(\alpha/2)\sqrt{\frac{\hat{p}(1 - \hat{p})}{n}}. \qquad (7.46)$$

One Sided Confidence Intervals for p

We now say a few words about an upper one sided confidence interval for p; such confidence intervals are of importance to an engineer who is interested in obtaining an upper bound on the proportion of defective items in a lot of manufactured items. We shall content ourselves with merely stating the result. Since

$$P\left(p \le \hat{p} + z(\alpha)\sqrt{\frac{\hat{p}(1 - \hat{p})}{n}}\right) \approx (1 - \alpha), \text{ it follows that}$$

$[0, U]$, is a $100(1 - \alpha)\%$ upper one sided confidence interval, where

$$U = \hat{p} + z(\alpha)\sqrt{\frac{\hat{p}(1 - \hat{p})}{n}}. \qquad (7.47)$$

Example 7.12 *Confidence intervals for the proportion defective*

The proportion of defective memory chips produced by a factory is p, $0 < p < 1$. Suppose 400 chips are tested and 10 of them are found to be defective. Compute a two sided and one sided upper 95% confidence interval for the proportion of defective chips.

Solution First compute the sample proportion and its estimated standard error:

$$\hat{p} = 10/400 = 0.025 \text{ and } \sqrt{\frac{\hat{p}(1 - \hat{p})}{400}} = 0.0078.$$

The choice $z_{0.025} = 1.96$ leads to an (approximate) 95% two sided confidence interval $[L, U]$ where

$$L = 0.025 - 1.96 \times 0.0078 = 0.0097;$$
$$U = 0.025 + 1.96 \times 0.0078 = 0.0403.$$

An approximate 95% one sided upper confidence interval is given by $[0, U]$, where

$$U = 0.025 + 1.645 \times 0.0078 = 0.0378.$$

Choosing the sample size

How large must the sample size be in order for our point estimate $\hat{p}$ to be within d units of the unknown proportion p? The $100(1 - \alpha)\%$ confidence interval defined by Equations

7.45 and 7.46 states that with (approximate) probability $(1 - \alpha)$ our point estimate $\hat{p}$ is within d units of the unknown parameter p, where

$$d = \frac{z(\alpha/2)}{2\sqrt{n}} \geq z(\alpha/2) \times \sqrt{\frac{\hat{p}(1 - \hat{p})}{n}}. \tag{7.48}$$

To prove 7.48 observe that $p(1 - p) \leq 1/4, 0 \leq p \leq 1$.
Therefore, $p(1 - p)/n \leq 1/4n$. Consequently,

$$z(\alpha/2) \times \sqrt{\frac{\hat{p}(1 - \hat{p})}{n}} \leq \frac{z(\alpha/2)}{2\sqrt{n}}, \text{ no matter what the value of } p \text{ is.}$$

We now use this result to determine how large a sample size is needed in order to estimate p with a given accuracy. Solving the inequality 7.48 for n yields the result that

$$n \geq \left(\frac{z(\alpha/2)}{2d}\right)^2. \tag{7.49}$$

Example 7.13 *Finding the minimal sample size to accurately estimate a proportion*

What is the minimum sample size required to estimate the proportion of voters in favor of a political party with a margin of error $d = 0.03$ and a confidence level of 90%?
Solution In this case $d = 0.03$, $\alpha/2 = 0.05$, $z(0.95) = 1.645$ and therefore n satisfies the equation

$$0.03 = \frac{1.645}{2\sqrt{n}}, \text{ so } n = 752.$$

If one is willing to live with a somewhat larger sampling error, $d = 0.05$ say, then n satisfies the equation

$$0.05 = \frac{1.645}{2\sqrt{n}}, \text{ so } n = 270.$$

7.5.1 Confidence Intervals for $p_1 - p_2$

Confidence intervals for the difference between two proportions arise in a wide variety of applications. For example, we might be interested in comparing the proportion of defective items produced by two machines M_1 and M_2, respectively. Suppose machine M_i produces n_i items and X_i of them are defective. We denote the proportion of defective items produced by M_i by $\hat{p}_i$. The estimated standard error of the difference between the two proportions is given by

$$s(\hat{p}_1 - \hat{p}_2) = \sqrt{\frac{\hat{p}_1(1 - \hat{p}_1)}{n_1} + \frac{\hat{p}_2(1 - \hat{p}_2)}{n_2}}. \tag{7.50}$$

Assuming, as in the previous section, that both sample sizes are large enough so that the normal approximation to the binomial is valid, it can be shown that the difference between the sample proportions is also approximately normally distributed with mean $p_1 - p_2$ and standard deviation $s(\hat{p}_1 - \hat{p}_2)$. Using this result, we obtain the following approximate $100(1 - \alpha)\%$ two sided confidence interval $[L, U]$ for $p_1 - p_2$:

$$L = \hat{p}_1 - \hat{p}_2 - z(\alpha/2)s(\hat{p}_1 - \hat{p}_2); \tag{7.51}$$
$$U = \hat{p}_1 - \hat{p}_2 + z(\alpha/2)s(\hat{p}_1 - \hat{p}_2). \tag{7.52}$$

Example 7.14 *Two machines produce light bulbs. Of 400 produced by the first machine, 16 were defective; of 600 produced by the second machine, 60 were defective. Construct a 95% confidence interval for the difference $p_1 - p_2$.*

Solution The sample proportions are $\hat{p}_1 = 0.04$ and $\hat{p}_2 = 0.1$; therefore,

$$\hat{p}_1 - \hat{p}_2 = -0.06 \text{ and}$$
$$s(\hat{p}_1 - \hat{p}_2) = \sqrt{\frac{0.04 \times 0.96}{400} + \frac{0.1 \times 0.9}{600}} = 0.0157;$$
$$L = -0.06 - 1.96 \times 0.0157 = -0.0908;$$
$$U = -0.06 + 1.96 \times 0.0157 = -0.0292.$$

7.6   Some Methods of Estimation*

This section introduces two methods for constructing estimates: (i) the method of moments and (ii) the maximum likelihood method.

7.6.1   Method of Moments*

The kth moment μ_k of a random variable X is defined by $\mu_k = E(X^k)$; in terms of the distribution function the corresponding formulas are:

$$\mu_k = \sum_{i=1}^{\infty} x_i^k f(x_i) \text{ (discrete case)}; \tag{7.53}$$

$$\mu_k = \int_{-\infty}^{\infty} x^k f(x)\, dx \text{ (continuous case)}. \tag{7.54}$$

It is clear from the definition that $\mu_1 = \mu$ and it follows from Equation 5.74 that $\sigma_X^2 = \mu_2 - \mu_1^2$. It is not too difficult to show that if $X_1, \ldots, X_n$ is a random sample from a distribution $f(x)$ then the kth sample moment $\hat{\mu}_k$ defined by

$$\hat{\mu}_k = \frac{1}{n} \sum_{1 \le i \le n} X_i^k \tag{7.55}$$

is an unbiased estimate for the kth population moment μ_k. The verification of this fact is left to you as an exercise (see Problem 7.59). Note also that $\hat{\mu}_1 = \overline{X}$. If the distribution function of the random variable depends on one or more parameters then so will the moments.

1. **Exponential distribution** Suppose X is exponentially distributed with parameter λ; then

$$E(X) = \frac{1}{\lambda}; \text{ consequently, } \lambda = \frac{1}{\mu_1}.$$

The method of moments estimator (MME) is obtained by substituting $\hat{\mu}_1$ for μ_1 in the preceding equation, which yields the following estimator for λ:

$$\hat{\lambda} = \hat{\mu}_1^{-1} = \overline{X}^{-1}.$$

2. **Normal distribution** Suppose $X \overset{\mathcal{D}}{=} N(\mu, \sigma^2)$; in this case

$$\mu_1 = \mu; \ \mu_2 = \sigma^2 + \mu^2.$$

Solving these simultaneous equations for μ and σ^2 in terms of μ_1 and μ_2 yields

$$\mu = \mu_1 \text{ and } \sigma^2 = \mu_2 - \mu_1^2.$$

The method of moments estimator is obtained by substituting $\hat{\mu}_i$ for μ_i in these equations and thereby obtaining the following estimators for μ and σ^2:

$$\hat{\mu} = \hat{\mu}_1 = \overline{X} \text{ and}$$

$$\hat{\sigma}^2 = \hat{\mu}_2 - \hat{\mu}_1^2.$$

After some algebraic manipulations that we leave to the reader, it can be shown that

$$\hat{\sigma}^2 = \frac{1}{n} \sum_{1 \le i \le n} (x_i - \overline{x})^2 = \frac{n-1}{n} s^2; \text{ consequently,}$$

$$E(\hat{\sigma}^2) = \frac{n-1}{n} \sigma^2 < \sigma^2; \text{ thus}$$

the method of moments estimator for σ^2 is biased.

3. **Gamma distribution** Suppose X has a gamma distribution with parameters α, β; in this case

$$\mu_1 = \alpha\beta; \ \mu_2 = (\alpha + \alpha^2)\beta^2. \tag{7.56}$$

One can solve these equations for α, β in terms of μ_1, μ_2 obtaining the formulas:

$$\alpha = \frac{\mu_1^2}{\mu_2 - \mu_1^2} \text{ and } \beta = \frac{\mu_2 - \mu_1^2}{\mu_1}.$$

The method of moments estimator is obtained by substituting $\hat{\mu}_i$ for μ_i in these equations and thereby obtaining the following estimators for α and β:

$$\hat{\alpha} = \frac{\hat{\mu}_1^2}{\hat{\mu}_2 - \hat{\mu}_1^2} \text{ and } \hat{\beta} = \frac{\hat{\mu}_2 - \hat{\mu}_1^2}{\hat{\mu}_1}. \tag{7.57}$$

Example 7.15 *Estimating German tank production during World War II*

During World War II every tank produced by the Germans was stamped with a serial number indicating the order in which it was produced. The problem was to estimate the number N of tanks produced, based on on the serial numbers $\{x_1, \ldots, x_n\}$ of destroyed or captured German tanks. For this purpose it was assumed that $\{x_1, \ldots, x_n\}$ is a random sample taken from the (finite) population $\{1, 2, \ldots, N\}$. Its population mean is given by

$$\mu = \frac{1 + 2 + \cdots + N}{N} = \frac{N+1}{2}.$$

Solving for N in terms of μ yields $N = 2\mu - 1$; this yields the method of moments estimator

$$\hat{N} = 2\overline{X} - 1.$$

To illustrate the method (and its limitations!) suppose 10 tanks were captured and their serial numbers were recorded (in increasing order) as follows:

$$125, 135, 161, 173, 201, 212, 220, 240, 250, 502.$$

The sample mean and corresponding estimate for N are given by

$$\overline{X} = 221.9 \text{ and } \hat{N} = 2\overline{X} - 1 = 442.8.$$

Now it is immediately obvious that there is something seriously wrong with this estimate since common sense tells you that the number of tanks produced must exceed the largest captured serial number, so $N \geq 502$. Indeed, the "common sense estimator"

$$\hat{N}_1 = max\{125, 135, 161, 173, 201, 212, 220, 240, 250, 502\} = 502$$

is clearly superior to the MME estimator $2\overline{X} - 1$. In the next section we introduce an alternative method of constructing estimators that is in general much better than those obtained via other methods such as the method of moments. It is called the *method of maximum likelihood* and the estimators it produces are called *maximum likelihood* (MLE) estimators. In particular the estimator $\hat{N}_1$ defined above is an MLE estimator (see Example 7.16).

7.6.2   Maximum Likelihood Estimators*

The *maximum likelihood method* is a precise formulation of the intuitively appealing criterion that a good estimate of an unknown parameter is also the most plausible. The degree of plausibility is measured by the *likelihood function*, a concept that we now proceed to explain.

Definition 7.6 *Let $X_1, \ldots, X_n$ be a random sample of size n taken from a distribution $f(x; \theta)$. The* likelihood function *is the joint distribution function of the random sample, which is given by*

$$L(x_1, \ldots, x_n; \theta) = f(x_1; \theta) \cdots f(x_n; \theta). \tag{7.58}$$

When the distribution function depends on two or more parameters then so does the likelihood function:

$$L(x_1, \ldots, x_n; \theta_1, \theta_2) = f(x_1; \theta_1, \theta_2) \cdots f(x_n; \theta_1, \theta_2).$$

To simplify the notation we write the random sample and the set of parameters using **boldface** vector notation:

$$\boldsymbol{x} = (x_1, \ldots, x_n), \text{ and } \boldsymbol{\theta} = (\theta_1, \ldots, \theta_k).$$

The likelihood function is then written in the following more compact form:

$$L(x_1, \ldots, x_n; \theta_1, \theta_2) = L(\boldsymbol{x}; \boldsymbol{\theta}).$$

Definition 7.7 *Let $X_1, \ldots, X_n$ be a random sample from the distribution $f(x; \theta)$; then the maximum likelihood estimator(MLE) for θ is that value $\hat{\theta}$ which maximizes the likelihood function $L(\boldsymbol{x}; \theta)$, i.e.,*

$$L(\boldsymbol{x}, \hat{\theta}) \geq L(\mathbf{x}, \theta), \theta \in \Theta.$$

Remark 7.1 *For the purposes of computing* $\max L(\boldsymbol{x}; \theta)$ *it is sometimes more convenient to maximize the logarithm of the likelihood function instead of maximizing the likelihood function itself. This is because we compute the maximum by differentiation and the function*

$$\ln(\boldsymbol{x}; \theta) = \sum_i \ln f(x_i; \theta)$$

is easier to differentiate than $L(\boldsymbol{x}; \theta)$. *We obtain the same maximum no matter which function we choose to maximize since it is easily verified that*

$$L(\hat{\theta}) \geq L(\boldsymbol{x}; \theta), \theta \in \Theta \text{ if and only if } \ln L(\hat{\theta}) \geq \ln L(\boldsymbol{x}; \theta), \theta \in \Theta.$$

This will be illustrated in several of the examples below as well in some of the problems.

Examples

1. **Sampling from a Bernoulli distribution** We recall that a Bernoulli distribution is given by the following simple formula:

$$f(x; p) = p^x (1 - p)^{1-x}, \, x = 0, 1.$$

The likelihood function and its logarithm are given by

$$\begin{aligned}
L(\boldsymbol{x}; p) &= p^{x_1}(1 - p)^{1-x_1} \cdots p^{x_n}(1 - p)^{1-x_n} \\
&= p^{\sum_{1 \leq i \leq n} x_i}(1 - p)^{n - \sum_{1 \leq i \leq n} x_i} \\
&= p^T (1 - p)^{n-T}; \text{ where } T = \sum_{1 \leq i \leq n} x_i, \text{ and} \qquad (7.59)
\end{aligned}$$

$$\ln L(\boldsymbol{x}; p) = T \ln(p) + (n - T) \ln(1 - p).$$

To compute the maximum we differentiate $\ln L(\boldsymbol{x}; p)$ with respect to p, set it equal to zero, and solve for p, i.e.,

$$\frac{d}{dp} \ln L(\boldsymbol{x}; p) = \frac{T}{p} - \frac{n - T}{1 - p} = 0.$$

If $T = 0$ or n, then a direct calculation shows that $\hat{p} = 0, 1$, respectively. On the other hand, for $0 < T < n$, we have

$$\hat{p} = \frac{T}{n} = \frac{1}{n} \sum_{1 \leq i \leq n} x_i.$$

Consequently, the MLE for the binomial parameter is

$$\hat{p} = \overline{X}.$$

2. **Sampling from an exponential distribution** $f(x; \mu) = \mu e^{-\mu x}, x > 0; f(x; \mu) = 0, otherwise.$ The likelihood function and its logarithm are given by

$$\begin{aligned}
L(\boldsymbol{x}; \mu) &= \mu e^{-\mu x_1} \cdots \mu e^{-\mu x_n} \qquad (7.60) \\
&= \mu^n e^{-\mu \sum_{1 \leq i \leq n} x_i}, \, min(x_1, \ldots, x_n) > 0 \\
&= 0, \text{ otherwise, and}
\end{aligned}$$

$$\ln L(\boldsymbol{x}; \mu) = n \ln(\mu) - \mu T.$$

To compute the maximum we differentiate $\ln L(\boldsymbol{x}; \mu)$ with respect to μ, set it equal to zero, and solve for μ, i.e.,

$$\frac{d}{d\mu} \ln L(\boldsymbol{x}; \mu) = \frac{n}{\mu} - T = 0.$$

Thus $\hat{\mu} = n/T = 1/\overline{X}$; since $E(1/\overline{X}) \neq 1/E(\overline{X})$, it follows that $\hat{\mu}$ is a biased estimator of μ.

3. **Sampling from a Poisson distribution** $f(x; \lambda) = e^{-\lambda} \frac{\lambda^x}{x!}; \{x = 0, 1, \ldots, \}$. The likelihood function and its logarithm are given by

$$L(\boldsymbol{x}; \lambda) = \frac{e^{-n\lambda} \lambda^{\sum_{1 \leq i \leq n} x_i}}{x_1! \cdots x_n!}; \ln L(\boldsymbol{x}; \lambda) = -n\lambda + T \ln(\lambda) - \sum_{1 \leq i \leq n} \ln(x_i!). \tag{7.61}$$

To compute the maximum we differentiate $\ln L(\boldsymbol{x}; \lambda)$ with respect to λ, set it equal to zero, and solve for λ, i.e.,

$$\frac{d}{d\lambda} \ln L(\boldsymbol{x}; \lambda) = 0 = -n + \frac{T}{\lambda}, \text{ i.e.,}$$

$$\hat{\lambda} = \frac{T}{n}.$$

4. **Sampling from a normal distribution** Let $f(x; \mu, \sigma^2)$ denote the normal density with mean μ and variance σ^2. A straightforward calculation using Definition 4.35 gives

$$L(\boldsymbol{x}; \mu, \sigma^2) = \frac{1}{\sigma^n (2\pi)^{n/2}} e^{-(1/2\sigma^2) \sum_{1 \leq i \leq n} (x_i - \mu)^2}. \tag{7.62}$$

Taking logarithms of both sides yields

$$\ln L(\boldsymbol{x}; \mu, \sigma^2) = \ln\left(\frac{1}{\sigma^n (2\pi)^{n/2}}\right) - \frac{1}{2\sigma^2} \sum_{1 \leq i \leq n} (x_i - \mu)^2$$

$$= -n \ln(\sigma) - \frac{1}{2\sigma^2} \sum_{1 \leq i \leq n} (x_i - \mu)^2 - \frac{n}{2} \ln(2\pi).$$

Since $\ln L(\boldsymbol{x}; \mu, \sigma^2)$ is a function of two variables its maximum is obtained by computing the partial derivatives with respect to μ, σ and setting them equal to zero like so:

$$\frac{\partial \ln L}{\partial \mu} = \frac{1}{\sigma^2} \sum_{1 \leq i \leq n} (x_i - \mu) = 0;$$

$$\frac{\partial \ln L}{\partial \sigma} = -\frac{n}{\sigma} + \frac{1}{\sigma^3} \sum_{1 \leq i \leq n} (x_i - \mu)^2 = 0.$$

Solving these equations for μ and σ^2, respectively, yields

$$\hat{\mu} = \overline{x}; \tag{7.63}$$

$$\hat{\sigma}^2 = \frac{1}{n} \sum_{1 \leq i \leq n} (x_i - \overline{x})^2. \tag{7.64}$$

Notice that $\hat{\sigma}^2 = (n-1)/n \times s^2$ and therefore the MLE estimate for σ^2 is biased since

$$E(\hat{\sigma}^2) = (n-1)/n \times \sigma^2 \neq \sigma^2.$$

Example 7.16 *This is a continuation of Example 7.15.*

In order to estimate the number N of German tanks produced we assumed that the observed serial numbers $\{x_1, \ldots, x_n\}$ of the captured tanks were in effect a random sample taken from the (finite) population $\{1, 2, \ldots, N\}$. Consequently, the joint density function is given by

$$P(X_1 = x_1, \ldots, X_n = x_n) = \frac{1}{P_{N,n}}, \text{ provided } N \geq max\{x_1, \ldots, x_n\}$$
$$= 0, \text{ otherwise.}$$

This implies that the likelihood function is given by

$$L(\boldsymbol{x}; N) = \frac{1}{P_{N,n}}, \text{ provided } N \geq max\{x_1, \ldots, x_n\} \text{ and } L(\boldsymbol{x}; N) = 0, \text{ otherwise.} \quad (7.65)$$

Thus

$$L(\boldsymbol{x}; N) = \frac{1}{N(N-1)\cdots(N-n+1)}, \text{ provided } N \geq max\{x_1, \ldots, x_n\}.$$

Clearly, $L(\boldsymbol{x}; N)$ is maximized by choosing N as small as possible consistent with the constraint $N \geq max\{x_1, \ldots, x_n\}$; and this means choosing $\hat{N} = max\{x_1, \ldots, x_n\}$.

It is frequently the case that one wants to estimate a function $g(\theta)$ of the parameter. For example, if the data come from an exponential distribution with parameter λ then estimating the reliability at time t_0 means that one must estimate $\exp(-\lambda t_0)$, i.e., $g(\lambda) = \exp(-\lambda t_0)$. One of the nice properties of a MLE is that $g(\hat{\theta})$ is the MLE for $g(\theta)$. It is useful to restate this result more formally as a proposition:

Proposition 7.2 *Let $\hat{\theta}$ be the MLE of θ and let $g(\theta)$ denote any function of the parameter θ. Then $g(\hat{\theta})$ is the MLE of $g(\theta)$.*

Example 7.17 *Refer to the air conditioner failure times for the B8045 plane in Problem 1.37.*

Assuming the data come from an exponential distribution with parameter λ compute the maximum likelihood estimates for $\hat{\lambda}$ and $\hat{R}(t)$.

Solution It was shown that

$$\text{MLE } \hat{\lambda} = 1/\overline{X} = \frac{1}{82} = 1.22 \times 10^{-2}.$$

Consequently, the MLE for the reliability at time t is given by:

$$\hat{R}(t) = \exp(-\hat{\lambda}t) = \exp(-1.22 \times 10^{-2} \times t).$$

Setting $t = 50$ hours yields the estimate $\hat{R}(50) = 0.54$. This may be interpreted as follows: the probability that the air conditioning unit is still working after 50 hours is 0.54.

Under suitable hypotheses on $f(x; \theta)$, which are too technical to be stated here, MLEs enjoy the following properties:

1. An MLE is consistent, i.e.,

$$\lim_{n \to \infty} P(|\hat{\theta} - \theta| < d) = 1, \text{ for every } d > 0.$$

2. An MLE is *asymptotically unbiased*, i.e.,

$$\lim_{n \to \infty} E(\hat{\theta}) = \theta.$$

7.7   Chapter Summary

In this chapter we discussed several methods for *estimating* population parameters such as the mean μ and variance σ^2 of a distribution. The performance of an estimator depends on several factors including its bias, its standard error (or estimated standard error), and whether or not it is consistent. We then introduced the important concept of a confidence interval, which is an interval that contains the unknown parameter with high probability. The particular cases of the sample mean and sample proportion were then studied in some detail. We then discussed two basic methods for constructing estimators: the method of moments and the maximum likelihood method.

7.8   Problems

Section 7.2

Problem 7.1 *A random sample of size* $n = 20$ *from a normal distribution* $N(\mu, 16)$ *has mean* $\bar{x} = 75$. *Find:*
(a) a 90% confidence interval for μ.
(b) a lower 90% confidence interval for μ. *(c) a 95% confidence interval for* μ.
(d) a lower 95% confidence interval for μ.

Problem 7.2 *Refer to the data set of Problem 1.18. The sample mean and sample standard deviation of the* $n = 33$ *measurements of the lead level in the blood of the exposed group of children are* $\bar{x} = 31.85$, $s = 14.41$. *Assuming the data come from a normal distribution* $N(\mu, \sigma^2)$ *compute:*
(a) a 90% confidence interval for μ.
(b) a 99% confidence interval for μ.

Problem 7.3 *Refer to the data set of Problem 1.18. The sample mean and sample standard deviation of the* $n = 33$ *measurements of the lead level in the blood of the control group of children are* $\bar{x} = 15.88$, $s = 4.54$. *Assuming the data come from a normal distribution* $N(\mu, \sigma^2)$ *compute:*
(a) a 90% confidence interval for μ.
(b) a 99% confidence interval for μ.

Problem 7.4 *The price* $S(t)$ *(t measured in years) of a stock is governed by a geometric Brownian motion with parameters* $\mu = 0.12$, *volatility* $\sigma = 0.20$, *and initial price* $S_0 = \$75$. *Compute:*
(a) a 95% confidence interval for the stock price after 3 months.
(b) a 95% confidence interval for the stock price after 6 months.
(c) a 95% confidence interval for the stock price after 1 year.

Problem 7.5 *(a) Let* $X_1, \ldots, X_n$ *be a random sample from the Poisson distribution*

$$f(x; \lambda) = e^{-\lambda} \frac{\lambda^x}{x!}; \; x = 0, 1, \ldots.$$

Show that $\overline{X}$ is an unbiased estimator of λ.
(b) Compute the standard error $\sigma(\overline{X})$ and its estimated standard error $s(\overline{X})$.
(c) Refer to the data of Problem 1.72. Assuming the data come from a Poisson distribution with parameter λ, use the sample mean to estimate λ and compute the estimated standard error of your estimate.

Problem 7.6 *Let $X_1, \ldots, X_n$ be a random sample from a normal distribution $N(\mu, 4)$, with μ unknown. Is*

$$\hat{\theta}(X_1, \ldots, X_n) = \sum_{1 \leq i \leq n} (X_i - \mu)^2$$

a statistic? Justify your answer.

Problem 7.7 *Eight measurements were made on the viscosity of a motor oil blend yielding the values $\overline{x} = 10.23$, $s^2 = 0.64$. Assuming the data come from a normal distribution $N(\mu, \sigma^2)$ construct a 90% confidence interval for μ.*

Problem 7.8 *Suppose that $X_1, \ldots, X_{25}$ is a random sample from a normal population $N(\mu, \sigma^2)$ and that $\overline{x} = 148.30$.*
(a) Assume $\sigma = 4$; find a 95% confidence interval for μ.
(b) Assuming $s = 4$, i.e., σ^2 is unknown, obtain a 95% confidence interval for μ.

Problem 7.9 *From a sample of size 30 taken from a normal distribution we obtain*

$$\sum_{1 \leq i \leq 30} x_i = 675.2, \quad \sum_{1 \leq i \leq 30} x_i^2 = 17422.5.$$

(a) Find a 95% confidence interval for μ.
(b) Find a 95% confidence interval for σ^2.

Problem 7.10 *A sample of size $n = 6$ from a normal distribution with unknown variance produced $s^2 = 51.2$ Construct a 90% confidence interval for σ^2.*

Problem 7.11 *For a sample of 15 brand X automobiles the mean miles per gallon (mpg) was 29.3 with a sample standard deviation of 3.2; the gas tank capacity is 12 gallons.*
(a) Using these data determine a 95% one sided lower confidence interval for the mpg. Assume the data come from a normal distribution.
(b) The cruising range is defined to be the distance the automobile can travel on a full tank of gas. Find a 95% one sided lower confidence interval for the cruising range.

Problem 7.12 *The lifetimes, measured in miles, of 100 tires are recorded and the following results are obtained:*
$$\overline{x} = 38000 \ miles \ and \ s = 3200 \ miles.$$

(a) If the data cannot be assumed to come from a normal distribution what can you say about the distribution of $\overline{X}$?
(b) Construct a lower one sided 99% confidence interval for the mean lifetime.

Problem 7.13 *Refer to the reverse-bias collector current of Problem 1.28. Assuming the data come from a normal distribution construct a 95% confidence interval for μ.*

Problem 7.14 *An electronic parts factory produces resistors; statistical analysis of the output suggests that the resistances can be modeled by a normal distribution $N(\mu; \sigma^2)$. The following data gives the results of the resistances, measured in ohms, for 10 resistors.*
Resistances: {0.150, 0.154, 0.145, 0.148, 0.158, 0.152, 0.153, 0.157, 0.148, 0.156}
Using these data construct:
(a) a 95% confidence interval for the mean resistance μ.
(b) a 99% confidence interval for the mean resistance μ.

Problem 7.15 *(a) Compute a 95% confidence interval for the capacitances data in Problem 1.20.*
(b) Compute a 99% confidence interval.

Problem 7.16 *(a) Compute a 99% confidence interval for the mean level of cadmium dust and oxide fume for the data set of Problem 1.19. Assume the data come from a normal distribution.*
(b) Compute a 95% upper confidence interval.

Problem 7.17 *Suppose the viscosity of a motor oil blend is normally distributed with $\sigma = 0.2$.*
(a) A researcher wants to estimate the mean viscosity with a margin of error not to exceed 0.1 and a 95% level of confidence; how large must n be?
(b) Part (a) continued: Suppose he wants to estimate the mean viscosity with the same margin of error 0.1 but with a 99% level of confidence; how large must n be?
(c) Part (a) continued: Suppose he wants to estimate the mean viscosity with a margin of error not to exceed 0.05 and a 95% level of confidence; how large must n be?

Problem 7.18 *Suppose the atomic weight of a molecule is measured with two different instruments, each of which is subject to random error. More precisely, the recorded weight $W_i = \mu + e_i$, where μ is the true atomic weight of the molecule and e_i is the random variable that represents the error produced by the ith instrument. The accuracy of the instrument is determined by the variance $V(e_i) = \sigma_i^2$. Assume that $E(e_i) = 0, i = 1, 2$.*
(a) Show that $E(W_i) = \mu$, $i = 1, 2$.
(b) Show that $W_t = tW_1 + (1-t)W_2$ is an unbiased estimator for μ for every real number t.
(c) Assuming that the error terms e_1, e_2 are mutually independent derive a formula for $V(W_t)$ as a function of t.
(d) Find the value t^* for which $V(W_t)$ is a minimum and show that

$$V(W_{t^*}) = \frac{\sigma_1^2 \sigma_2^2}{\sigma_1^2 + \sigma_2^2} \leq \min(\sigma_1^2, \sigma_2^2).$$

This shows that we can reduce the variance of our estimate for μ by taking a suitable weighted average of the two measurements.

Problem 7.19 *Refer to the data set of Problem 1.20. Assuming the capacitance is normally distributed, compute a 95% confidence interval for the variance of the capacitance.*

Problem 7.20 *Refer to the data set of Problem 1.18. Assuming each variable is normally distributed:*
(a) Compute a 95% confidence interval for the variance of levels of lead in the blood of the exposed group.
(b) Compute a 95% confidence interval for the variance of levels of lead in the blood of the control group.

Problem 7.21 *Refer to the data set of Problem 1.19. Assuming the cadmium level is normally distributed, compute a 95% confidence interval for the variance of the cadmium levels.*

Problem 7.22 *Refer to the data on the reverse-bias current in Problem 1.28. Assuming that the data come from a normal distribution, compute a 95% confidence interval for the variance.*

Problem 7.23 *Refer to Example 7.8. The weekly mean return and volatility of GE for the year 1999 are 0.0091 and 0.0399, respectively (see Table 1.13). The sample size for these data is n = 51 (see Table 1.19). From these data estimate the mean return and volatility of GE's stock for:*
(a) one year. (b) one month. (c) one quarter (13 weeks).

Problem 7.24 *The weekly mean return and volatility of the S&P500 index for the year 1999 are 0.0031 and 0.0255, respectively (see Table 1.13). The sample size for these data is n = 51 (see Table 1.19). From these data estimate the mean return and volatility of the S&P500 index for:*
(a) one year. (b) six months.
(c) Find a 95% confidence interval for the annual volatility. Hint: Refer to Example 7.8.

Problem 7.25 *The weekly mean return and volatility of the bond mutual fund PTTRX for the year 1999 are −0.0011 and 0.0058, respectively (see Table 1.13). The sample size for these data is n = 51 (see Table 1.19). From these data estimate the mean return and volatility of the PTTRX mutual fund for:*
(a) one year. (b) one quarter (13 weeks).
(c) Find a 95% confidence interval for the annual volatility.

Problem 7.26 *The weekly mean return and volatility of the mutual fund BEARX for the year 1999 are −0.0046 and 0.0290, respectively (see Table 1.13). The sample size for these data is n = 51 (see Table 1.19). From these data find the mean return and volatility of the BEARX mutual fund for:*
(a) one year. (b) one quarter (13 weeks).
(c) Find a 95% confidence interval for the annual volatility.

Problem 7.27 *A portfolio has a weekly return of $\mu = 0.003$ and weekly volatility of $\sigma = 0.06$. The current value of the portfolio is $100,000.*
(a) Find the 99% VaR (rounded to the nearest dollar) for a one week period.
(b) Find the 99% VaR (rounded to the nearest dollar) for a two week period.

Problem 7.28 *An investor has $100,000 invested in the S&P500. Assuming its weekly return and volatility are $\mu = 0.0031$ and $\sigma = 0.0255$, respectively (see Table 1.13), find the 99% VaR (rounded to the nearest dollar) for:*
(a) a one week period. (b) a two week period.

Problem 7.29 *An investor has $100,000 invested in the PTTRX mutual fund. Assuming its weekly return and volatility are $\mu = -0.0011$ and $\sigma = 0.0058$, respectively (see Table 1.13), find the 99% VaR (rounded to the nearest dollar) for:*
(a) a one week period. (b) a two week period.

Problem 7.30 *An investor has $100,000 invested in the BEARX mutual fund. Assuming its weekly return and volatility are $\mu = -0.0046$ and $\sigma = 0.0290$, respectively (see Table 1.13), find the 99% VaR (rounded to the nearest dollar) for:*
(a) a one week period.   (b) a two week period.

Problem 7.31 *Derive Equation 7.33 for the $100(1 - \alpha)\%$ confidence value at risk.*

Section 7.4

Problem 7.32 *A random sample of size $n_1 = 16$ taken from a normal distribution $N(\mu_1, 36)$ has a mean $\overline{X}_1 = 30$; another random sample of size $n_2 = 25$ taken from a different normal distribution $N(\mu_2, 9)$ has a mean $\overline{X}_2 = 25$. Find:*
(a) a 99% confidence interval for $\mu_1 - \mu_2$.
(b) a 95% confidence interval for $\mu_1 - \mu_2$.
(c) a 90% confidence interval for $\mu_1 - \mu_2$.

Problem 7.33 *Two random samples of sizes $n_1 = 11$ and $n_2 = 13$, respectively, were taken from two normal distributions $N(\mu_1, \sigma^2)$ and $N(\mu_2, \sigma^2)$, and the following data were obtained: $\overline{X}_1 = 1.5$, $s_1 = 0.5$, $\overline{X}_2 = 2.1$, $s_2 = 0.6$. Find the values of:*
(a) s_p^2.   (b) $s(\overline{X}_1 - \overline{X}_2)$.   (c) a two sided 95% confidence interval for $\mu_1 - \mu_2$.

Problem 7.34 *The compression strengths, measured in kgms, of cylinders produced by two different manufacturers were tested and the following data were recorded: $\overline{X}_1 = 240$, $s_1 = 10$, $n_1 = 12$ and $\overline{X}_2 = 210$, $s_2 = 7$, $n_2 = 12$. Construct a 95% confidence interval for the difference in mean compression strengths. Assume that both samples come from normal populations with the same variance.*

Problem 7.35 *Two different weight loss programs are being compared to determine their effectiveness. Ten men were assigned to each program, so $n_1 = n_2 = 10$, and their weight losses are recorded below. Construct a 95% confidence interval for the difference in weight loss between the two diets. Do the data imply that one diet is more effective in reducing weight? Comment.*

| Diet 1 | 3.4 | 10.9 | 2.8 | 7.8 | 0.9 | 5.2 | 2.5 | 10.5 | 7.1 | 7.5 |
| Diet 2 | 11.9 | 13.1 | 11.6 | 6.8 | 6.8 | 8.8 | 12.5 | 8.6 | 17.5 | 10.3 |

Problem 7.36 *An experiment was conducted to compare the electrical resistances (measured in ohms) of resistors supplied by two manufacturers, denoted A and B, respectively. Six resistors from each manufacturer were randomly selected and their resistances measured; the following data were obtained:*

| Brand A | 0.140 | 0.138 | 0.143 | 0.142 | 0.144 | 0.137 |
| Brand B | 0.135 | 0.140 | 0.142 | 0.136 | 0.138 | 0.140 |

Assume both data sets come from normal populations. Construct a 95% confidence interval for the difference between the mean resistances.

Problem 7.37 *The skein strengths, measured in pounds, of two types of yarn are tested and the following results are reported.*

| Yarn A | 99 | 93 | 99 | 97 | 90 | 96 | 93 | 88 | 89 |
| Yarn B | 93 | 94 | 75 | 84 | 91 | | | | |

Construct a 95% confidence interval for the difference in means. Assume that both data sets come from a normal population.

Problem 7.38 *To determine the difference, if any, between two brands of radial tires, 12 tires of each brand are tested and the following mean lifetimes and sample standard deviations for each of the two brands were obtained:*

$$\bar{x}_1 = 38500 \text{ and } s_1 = 3100$$
$$\bar{x}_2 = 41000 \text{ and } s_2 = 3500$$

Assume the tire lifetimes are normally distributed. Find a 95% confidence interval for the difference in the mean lifetimes.

Problem 7.39 *With reference to the data set on lead levels in children's blood (Problem 1.18) find a $100(1 - \alpha)\%$ confidence interval for the differences in lead level between the exposed and control group of children when: (a) $\alpha = 0.10$. (b) $\alpha = 0.05$. (c) $\alpha = 0.01$.*

Problem 7.40 *Refer to the concentration of thiol in lysate data in Problem 1.62. (a) Derive a 95% confidence interval for the difference in the mean concentrations of thiol between the two groups. (b) Derive a 99% confidence interval for the difference in the mean concentrations of thiol between the two groups.*

Problem 7.41 *The following data are from Charles Darwin's study of cross- and self-fertilization. Pairs of seedlings of the same age, one produced by cross-fertilization and the other by self-fertilization, were grown together so that members of each pair were grown under nearly identical conditions. The data represent the final heights of the plants after a fixed period of time.*
(a) Compute a 95% confidence interval for the difference in the mean heights. State your model assumptions.
(b) Compute a 99% confidence interval for the difference in the mean heights.

Pair	Cross-fertilized	Self-fertilized
1	23.5	17.4
2	12.0	20.4
3	21.0	20.0
4	22.0	20.0
5	19.1	18.4
6	21.5	18.6
7	22.1	18.6
8	20.4	15.3
9	18.3	16.5
10	21.6	18.0
11	23.3	16.3
12	21.0	18.0
13	22.1	12.8
14	23.0	15.5
15	12.0	18.0

(Source: D.J.Hand et al. (1994), Small Data Sets, *Chapman & Hall, London.)*

Problem 7.42 *Refer to the dopamine b-hydroxylase (DBH) activity data set of Problem 1.40.*
(a) Construct a 95% confidence interval for the difference in the DBH activity between the two groups of patients.
(b) Construct a 95% confidence interval for the difference in the DBH activity between the two groups of patients.

Section 7.5

Problem 7.43 *Let $X_1, \ldots, X_n$ be a random sample from a Bernoulli distribution. Show that*

$$E(\hat{p}(1 - \hat{p})) = \frac{n - 1}{n} \times p(1 - p) < p(1 - p).$$

Deduce that $\hat{p}(1 - \hat{p})$ is a biased estimator of $p(1 - p)$.

Problem 7.44 *From a random sample of 800 TV monitors that were tested, 10 were found to be defective; construct a 99% one sided upper confidence interval for the proportion p of defective monitors.*

Problem 7.45 *A random sample of 1300 voters from a large population produced the following data: 675 in favor of proposition A and 625 opposed. Find a 95% confidence interval for the proportion p of voters:*
(a) in favor of proposition A.   (b) opposed to proposition A.

Problem 7.46 *From a sample of 400 items 5% were found to be defective. Find a*
(a) 95% confidence interval for the proportion p of defective items.
(b) 95% one sided upper confidence interval for p.

Problem 7.47 *(a) To determine the level of popular support for a political candidate a newspaper regularly surveys 600 voters. The results of the poll are reported as a 95% confidence interval. Determine the margin of error.*
(b) Calculate the margin of error in part (a) when the sample size is doubled to 1200.

Problem 7.48 *How large a sample size is needed if we want to be 95% confident that the sample proportion will be within 0.02 of the true proportion of defective items and:*
(a) the true proportion p is unknown;
(b) it is known that the true proportion $p \leq 0.1$.

Problem 7.49 *How large a sample size is needed if we want to be 90% confident that the sample proportion will be within 0.02 of the true proportion of defective items and:*
(a) the true proportion p is unknown;
(b) it is known that the true proportion $p \leq 0.1$.

Problem 7.50 *Two machines produce computer memory chips. Of 500 chips produced by machine 1, 20 were defective; of 600 chips produced by machine 2, 40 were defective. Construct 90% one sided upper confidence intervals for:*
(a) p_1, the proportion of defective chips produced by machine 1.
(b) p_2, the proportion of defective chips produced by machine 2.
(c) Construct a 95% two sided confidence interval for the difference $p_1 - p_2$. What conclusions can you draw if the confidence interval includes (or does not include) 0?

Problem 7.51 *To compare the effectiveness of two different methods of teaching arithmetic* 60 *students were divided into two groups of* 30 *students each. One group (the control group) was taught by traditional methods and the other group (the treatment group) by a new method. At the end of the term* 20 *students of the control group and* 25 *students of the treatment group passed the basic skills test.*

(a) Find an approximate 95% *confidence interval for* $p_1 - p_2$, *where* p_1 *and* p_2 *are the true proportions of students who pass the basic skills test for the control and treatment groups, respectively. What conclusions can you draw if the confidence interval includes (or does not include)* 0?

(b) Find an approximate 90% *confidence interval for* $p_1 - p_2$. *What conclusions can you draw if the confidence interval includes (or does not include)* 0?

Problem 7.52 *Samples of size* $n_1 = n_2 = 150$ *were taken from the weekly output of two factories. The number of defective items produced were* $x_1 = 10$ *and* $x_2 = 15$, *respectively. Construct an approximate* 95% *confidence interval for* $p_1 - p_2$, *the difference in the proportion of defective items produced by the two factories. What conclusions can you draw if the confidence interval includes (or does not include)* 0?

Section 7.6

Problem 7.53 *(a) Let* $X_1, \ldots X_n$ *be a random sample taken from an exponential distribution with parameter* λ. *Show that* $\overline{X}$ *is an unbiased estimator of* $1/\lambda$.
(b) Let $U = \min(X_1, \ldots, X_n)$. *Show that* U *is exponentially distributed with parameter* $n\lambda$. *Hint: Show that* $P(U > x) = \exp(-n\lambda x)$.
(c) Deduce from (a) that nU *is an unbiased estimator of* $1/\lambda$.
(d) Compute $V(nU)$ *and* $V(\overline{X})$. *Which estimator is better? Justify your answer by stating an appropriate criterion.*

Problem 7.54 *Ten measurements of the concentration (in* %) *of an impurity in an ore were taken and the following data were obtained.*
    3.8 3.5 3.4 3.9 3.7 3.7 3.6 3.7 4.0 3.9
(a) Assuming these data come from a normal distribution $N(\mu, \sigma^2)$ *compute the ML estimators for* μ, σ^2, σ.
(b) Which of the three estimators in part (a) are unbiased? Discuss.

Problem 7.55 *The breaking strengths of* 36 *steel wires were measured and the following results were obtained:* $\bar{x} = 9830$ *psi and* $s = 400$. *Assume that the measurements come from a normal distribution. The safe strength is defined to be the* 10*th percentile of the distribution. Compute the ML estimate of the safe strength.*

Problem 7.56 *The distribution of the lifetime of a component is assumed to have a gamma distribution with parameters* α, β. *The first two sample moments are* $\hat{\mu}_1 = 96$ *and* $\hat{\mu}_2 = 10368$. *Use the method of moments to estimate* α, β.

Problem 7.57 *Let*

$$f(x|\theta) = \theta x^{\theta-1}, \, 0 < x < 1; 0 < \theta$$
$$= 0, \, elsewhere.$$

(a) Show, via a suitable integration, that

$$\mu_1 = \frac{\theta}{\theta + 1}.$$

(b) Use the result from part (a) to derive the method of moments estimator for θ.

Problem 7.58 *In a political campaign $X = 185$ voters out of $n = 351$ voters polled are found to favor candidate A. Use these data to compute the ML estimate of p and $p(1 - p)$. Which of these estimators is unbiased? Discuss.*

Problem 7.59 *Verify that $\hat{\mu}_k$, as defined in Equation 7.55 is an unbiased estimate for μ_k.*

Problem 7.60 *It is sometimes necessary, e.g., in likelihood ratio tests, to compute the maximum value of the likelihood function $L(\boldsymbol{x}|\hat{\theta})$. In particular, with reference to the likelihood function displayed in Equation 7.62 show that*

$$L(\boldsymbol{x}|\hat{\mu}, \hat{\sigma}^2) = (\hat{\sigma}\sqrt{2\pi})^{-n} e^{-n/2}. \tag{7.66}$$

7.9 To Probe Further

We have not discussed *Bayesian estimation*. A good introduction to this important topic is given in

1. M. H. DeGroot (1986), *Probability and Statistics,* Addison–Wesley Publishing Co., Reading, MA.

   The reader will find a more advanced treatment of the topics treated here in

2. E.L. Lehman (1999), *Elements of Large Sample Theory,* Springer-Verlag.

3. P.J. Bickel and K.A. Doksum (1977), *Mathematical Statistics: Basic Ideas and Selected Topics,* Holden-Day. The proofs of the asymptotic normality and consistency of the sample quantiles appear on p. 400.

4. C. Howson and P. Urbach (1989), *Scientific Reasoning: The Bayesian Approach,* Open Court Publishing, La Salle, Illinois, is a penetrating comparison of the Bayesian approach to confidence intervals with the more traditional interpretation presented here.

Chapter 8

Hypothesis Testing

Doubt everything or believe everything; these are two equally convenient strategies. With either we dispense with the need for reflection.

Henri Poincaré (1854-1912), French mathematician

8.1 Orientation

A frequently occurring problem of no small importance is to determine whether or not a product meets a standard. The quality of the product is usually measured by a quantitative variable X defined on a population. Since the quality of the product is always subject to some kind of random variation, the standard is usually expressed as an assertion about the distribution of the variable X. For example, suppose X denotes the fuel efficiency of a car [measured in miles per gallon (mpg)], and the fuel efficiency standard is that the mean mpg μ be greater than μ_0. The assertion that $\mu > \mu_0$ is an example of a *statistical hypothesis*; it is a claim about the distribution of the variable X. For another example, consider the widely used Black-Scholes options pricing formula, which assumes that returns on a stock are normally distributed (see Section 5.9.1). Is this hypothesis consistent with the kinds of actual data listed in Tables 1.19 and 1.20? Problems of this sort are discussed in Section 8.4. In this chapter we learn how to use statistics such as the sample mean and sample variance to test the truth or falsity of a statistical hypothesis. We then extend these methods to testing hypotheses about two distributions where the goal is to study the difference between their means.

Organization of Chapter

1. Section 8.2: Tests of Statistical Hypotheses: Basic Concepts and Examples

2. Section 8.2.1: Significance Testing

3. Section 8.2.2: Power Function and Sample Size

4. Section 8.3: Comparing Two Populations

5. Section 8.3.1: The Wilcoxon Rank Sum Test for Two Independent Samples

6. Section 8.3.2: A Test of the Equality of Two Variances

7. Section 8.4: Normal Probability Plots

8. Section 8.5: Tests Concerning the Parameter p of a Binomial Distribution

9. Section 8.6: Chapter Summary

8.2   Tests of Statistical Hypotheses: Basic Concepts and Examples

The concepts of hypothesis testing are best introduced in the context of an example.

Example 8.1 *Testing a hypothesis about a population mean*

From previous experimental data it is known that the fuel efficiency of a taxi, measured in miles per gallon (mpg), is normally distributed with mean μ and standard deviation σ. In order to reduce fuel costs a taxi company plans to order a fleet of taxis, provided the fuel efficiency of the new fleet exceeds 27 mpg. To determine whether or not the new taxis meet the standard, nine of them are randomly selected and the average mpg was recorded as $\bar{x} = 28.5$. On the basis of these data, is it plausible to conclude that the taxis meet the fuel efficiency standard of 27 mpg? Clearly, there are only two possible conclusions: Either the taxis are not fuel efficient, so $\mu \leq 27$, or they are, so $\mu > 27$. To simplify the calculations we assume the standard deviation is known, with value $\sigma = 2$. This assumption is, of course, unrealistic, and we will study the situation when σ is unknown in Section 8.2.4. The assertion that $\mu \leq 27$ is an example of a statistical hypothesis called the *null hypothesis* or *no change hypothesis*. In plain English, the null hypothesis is that the taxis do not meet the new fuel efficiency standard. The particular value $\mu_0 = 27$ is called the *null value* and plays a special role in the performance analysis of a hypothesis test. The null hypothesis, in this case, is denoted $H_0 : \mu \leq \mu_0$, and the *alternative hypothesis* is denoted $H_1 : \mu > \mu_0$. An alternative hypothesis of the form $H_1 : \mu > \mu_0$ is said to be *one sided* since H_1 consists of those values of $\mu > \mu_0$, i.e., that deviate from μ_0 in one direction only. The distribution of the statistic $\overline{X}$ when $\mu = \mu_0$ is called the *null distribution*. To test this hypothesis it is natural use the sample mean $\overline{X}$ for our test statistic because it is intuitively clear that we would find $H_1 : \mu > \mu_0$ more likely (and H_0 less likely) to be true when the sample mean $\overline{X}$ is very much greater than μ_0. That is, we choose a number $c > \mu_0$, called the *cutoff value*, such that an observed value of $\overline{X} > c$ is highly unlikely if, in fact, H_0 be true. To determine which of the two hypotheses is more plausible statisticians use the following decision procedure:

$$\text{accept } H_0 \text{ if } \overline{X} \leq c; \text{ reject } H_0 \text{ if } \overline{X} > c.$$

A procedure for deciding whether to accept or reject the null hypothesis is called a *test of a statistical hypothesis*. It is a procedure based on the value of a statistic, usually the sample mean $\overline{X}$, for accepting or rejecting H_0. It is also a method for measuring the strength of the evidence against the null hypothesis contained in the experimental data. The set of values $\mathcal{C} = \{\bar{x} : \bar{x} > c\}$ is called the *rejection region* of the test, since H_0 is rejected when $\overline{X}$ falls in $\mathcal{C}$. A *type I error* occurs when one rejects H_0 when H_0 is true, and a *type II error* when one accepts H_0 when H_0 is false. Table 8.1 displays the alternatives, decision rule, and the type I and type II errors in the context of the taxis' fuel efficiency (with $\mu_0 = 27$).

Table 8.1 *Null and alternative hypothesis, decision rules, type I and type II errors*

	H_0 is true taxis are not fuel efficient $(\mu \leq \mu_0)$	H_0 is false taxis are fuel efficient $(\mu > \mu_0)$
Test Statistic $\overline{X}$		
$\overline{X} \leq c$	Correct decision	type II error
$\overline{X} > c$	type I error	Correct decision

The performance of the test is measured by how well it controls the type I and type II error probabilities. A good test will reject H_0 with small probability when it is true and will reject H_0 with probability close to one when it is false. Using the fact that $\overline{X}$ is normally distributed with mean μ and standard deviation $\sigma/\sqrt{n}$ we see that the type I and type II error probabilities are given by

$$P(\overline{X} > c|\mu_0) = P\left(\frac{\sqrt{n}(\overline{X} - \mu_0)}{\sigma} > \frac{\sqrt{n}(c - \mu_0)}{\sigma}\right), \ (H_0 : \mu \leq \mu_0)$$

$$= 1 - \Phi\left(\frac{\sqrt{n}(c - \mu_0)}{\sigma}\right) = \alpha \text{ (probability of type I error);} \qquad (8.1)$$

$$P(\overline{X} \leq c|\mu) = P\left(\frac{\sqrt{n}(\overline{X} - \mu)}{\sigma} \leq \frac{\sqrt{n}(c - \mu)}{\sigma}\right), \ (H_1 : \mu > \mu_0)$$

$$= \Phi\left(\frac{\sqrt{n}(c - \mu)}{\sigma}\right) = \beta(\mu) \text{ (probability of type II error).} \qquad (8.2)$$

Terminology: The *significance level of the test* is the probability α of a type I error; it is also called the *size of the rejection region*. The type I error probability for a test with rejection region $\mathcal{C} = \{\overline{x} : \overline{x} > c\}$ is given by

$$\alpha = P(\overline{X} > c|H_0 \text{ is true}) = P(\overline{X} > c|\mu = \mu_0).$$

We choose the value c such that the probability of a type I error is less than α, where α is typically chosen to be smaller than 0.05.

Constructing a level α test The cutoff value c corresponding to a level α test of $H_0 : \mu \leq \mu_0$ against $H_1 : \mu > \mu_0$ is given by

$$c = \mu_0 + z(\alpha)\frac{\sigma}{\sqrt{n}}. \qquad (8.3)$$

Equation 8.3 for the cutoff value c is a consequence of the relationship between α and c derived in Equation 8.1; in detail,

$$\alpha = P(\overline{X} > c|\mu = \mu_0) = 1 - \Phi\left(\frac{\sqrt{n}(c - \mu_0)}{\sigma}\right); \text{ therefore, } \frac{\sqrt{n}(c - \mu_0)}{\sigma} = z(\alpha).$$

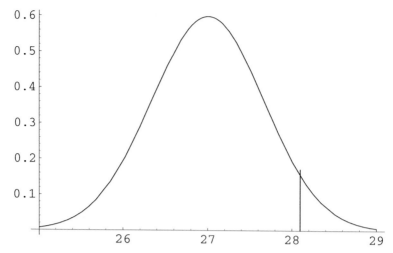

FIGURE 8.1: Rejection region based on sample mean $\overline{X} > 28.0967$

Example 8.2 *Continuation of Example 8.1. Constructing a level $\alpha = 0.05$ test.*

To test $H_0 : \mu \le 27$ against the alternative $H_1 : \mu > 27$ at the level $\alpha = 0.05$ we set $\mu_0 = 27$, $n = 9$ in Equation 8.1; consequently, $3(c - 27)/2 = 1.645$, and therefore $c = 27 + 1.645 \times (2/3) = 28.0967$. Figure 8.1 displays the graph of the null distribution $N(27, 4)$ and the rejection region $\{\overline{x} : \overline{x} > 28.0967\}$.

Computing the probability of a type II error at an alternative μ: The probability of a type II error at the alternative $\mu > \mu_0$, denoted by $\beta(\mu)$, is given by

$$\beta(\mu) = P(\overline{X} \le c | \mu > \mu_0) = \Phi\left(\frac{\sqrt{n}(c - \mu)}{\sigma}\right). \tag{8.4}$$

Example 8.3 *Continuation of Example 8.2. Computing $\beta(\mu)$.*

Suppose the alternative is $\mu = 28.5$ and $\overline{X} \le 28.0967$. In this case $c = 28.0967$, $n = 9$, $\sigma = 2$; inserting these values into Equation 8.4 yields $\beta(28.5) = \Phi(-0.605) = 0.2726$.

Power of a test

Since $\beta(\mu)$ is the probability that we accept H_0 when H_1 is true, $1 - \beta(\mu)$ is the probability that we reject H_0 when H_1 is true; which is why it is called the *power of the test* at the alternative μ. It is a measure of the test's ability to recognize that H_0 is false and decide, correctly, that H_1 is true.

Review: Constructing a level α test of $H_0 : \mu \le \mu_0$ against $H_1 : \mu > \mu_0$

1. Choose the rejection region $\mathcal{C} = \{\overline{x} : \overline{x} > c\}$ where

$$c = \mu_0 + z(\alpha)\frac{\sigma}{\sqrt{n}}.$$

With this choice of c, the probability of a type I error is α; more precisely, we have

$$P(\overline{X} > \mu_0 + z(\alpha)\frac{\sigma}{\sqrt{n}} | \mu_0) = \alpha.$$

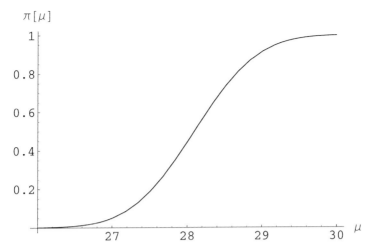

FIGURE 8.2: Power function of Example 8.2: $\mu_0 = 27$, $\alpha = 0.05$, $c = 28.0967$

2. Compute the sample mean $\overline{X}$ and

$$\text{accept } H_0 \text{ if } \overline{X} \le \mu_0 + z(\alpha)\frac{\sigma}{\sqrt{n}}; \text{ reject } H_0 \text{ if } \overline{X} > \mu_0 + z(\alpha)\frac{\sigma}{\sqrt{n}}. \tag{8.5}$$

A test of the form given in Equation 8.5 is called an *upper tailed test*. A *lower tailed test* is defined similarly (see Example 8.4). A test of either type is called a *one sided test*.

We summarize the performance of a test by computing its *power function* $\pi(\mu)$ defined as $\pi(\mu) = P(\text{reject } H_0|\mu)$. An explicit formula for the power function of a test of $H_0 : \mu \le \mu_0$ against $H_1 : \mu > \mu_0$ with rejection region $\mathcal{C} = \{\overline{x} : \overline{x} > c\}$ is given in Equation 8.6:

$$\pi(\mu) = P(\overline{X} > c|\mu) = 1 - \Phi\left(\frac{\sqrt{n}(c - \mu)}{\sigma}\right). \tag{8.6}$$

Note that the power function defined in Equation 8.6 is an increasing function of μ, that is, $\mu_1 < \mu_2$ implies $\pi(\mu_1) < \pi(\mu_2)$; in particular, $\pi(\mu_0) = \alpha$ and $\mu < \mu_0$ implies $\pi(\mu) < \alpha$. Consequently, for all $\mu \le \mu_0$ the type I error probability of this test never exceeds α. Problem 8.15(a) asks you to show that Equation 8.6 defines an increasing function of μ.

An Equivalent Test Based on a Standardized Test Statistic

The one sided test given in Equation 8.1 can be rewritten in an equivalent form using the standardized test statistic $Z_0 = \sqrt{n}(\overline{X} - \mu_0)/\sigma$. We note that the null distribution of the test statistic Z_0 is that of a standard normal random variable; that is,

$$Z_0 = \frac{\sqrt{n}(\overline{X} - \mu_0)}{\sigma} \stackrel{\mathcal{D}}{=} N(0; 1), \text{ when } \mu = \mu_0. \tag{8.7}$$

The rejection region in terms of the standardized test statistic Z_0 is the set $\{Z_0 > z(\alpha)\}$. We obtain this result by rewriting the original rejection region—refer back to Equation 8.5—in terms of Z_0; that is

$$\left(\overline{X} > \mu_0 + z(\alpha)\frac{\sigma}{\sqrt{n}}\right) = \left(\frac{\sqrt{n}(\overline{X} - \mu_0)}{\sigma} > z(\alpha)\right) = (Z_0 > z(\alpha)).$$

This leads to the following equivalent test of the hypothesis:

$$H_0 : \mu \le \mu_0 \text{ against the one sided (upper) alternative } H_1 : \mu > \mu_0 :$$

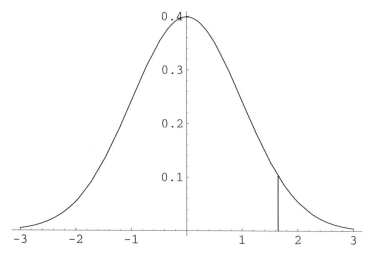

FIGURE 8.3:   Rejection region for Example 8.2 based on $Z_0 = (\overline{X} - 27)/(2/3) > 1.645$

$$\text{accept } H_0 \text{ if } \frac{\sqrt{n}(\overline{X} - \mu_0)}{\sigma} \leq z(\alpha); \text{ reject } H_0 \text{ if } \frac{\sqrt{n}(\overline{X} - \mu_0)}{\sigma} > z(\alpha). \qquad (8.8)$$

Figure 8.3 displays the graph of the distribution of the standardized test statistic $Z_0 = \sqrt{n}(\overline{X} - \mu_0)/\sigma$ and the rejection region $\{z : z > 1.645\}$ for Example 8.2.

The probabilities of making the four possible decisions (correct and incorrect) are listed in Table 8.2. Note that every entry in Table 8.2 can be expressed in terms of the power function.

Table 8.2 *Decision probabilities*

Test Statistic $\overline{X}$	H_0 *is true* $(\mu \leq \mu_0)$ *Decision probability*	H_1 *is true* $(\mu > \mu_0)$ *Decision probability*
$\overline{X} \leq c$	$\Phi(\sqrt{n}(c - \mu_0)/\sigma)$	$\beta = \Phi(\sqrt{n}(c - \mu)/\sigma)$
$\overline{X} > c$	$\alpha = 1 - \Phi(\sqrt{n}(c - \mu_0)/\sigma)$	$1 - \Phi(\sqrt{n}(c - \mu)/\sigma)$

Before proceeding further, we note that type I and type II errors are not of equal importance. A taxi company is much more concerned with the possibility of mistakenly ordering fuel inefficient taxis than with mistakenly rejecting fuel efficient ones. For this reason we focus our attention on controlling the type I error.

Confidence interval approach: We now introduce an alternative method, based on confidence intervals, for testing hypotheses about the mean. It is increasingly popular among engineers for reasons that will soon become apparent. In this example, we are concerned that the fuel efficiency of the new taxis is too low. For this purpose we observed that a one sided $100(1 - \alpha)\%$ lower confidence interval of the form

$$[\overline{X} - z(\alpha)\frac{\sigma}{\sqrt{n}}, \infty)$$

is appropriate (Equation 7.21). With reference to the data, we have $\bar{x} = 28.5, \sigma/\sqrt{n} = 2/3$, so the 95% lower confidence interval is given by

$$[\bar{x} - z(0.05)\frac{\sigma}{\sqrt{n}}, \infty) = 28.5 - 1.645 \times (2/3) = [27.4033 \, \infty).$$

Conclusion: These data suggest that the true mpg could be as low as 27.4033. In particular, we note that $27.4033 > 27 = \mu_0$ implies that the null value is not in the confidence interval. We interpret this to mean that the null hypothesis is not plausible. This leads to the following criterion for accepting H_0. *The null hypothesis $H_0 : \mu \leq \mu_0$ is accepted whenever μ_0 lies in the confidence interval.* Parameter values that lie outside the confidence interval are not plausible. Therefore, *we reject the null hypothesis $H_0 : \mu = \mu_0$ whenever μ_0 lies outside the confidence interval;* that is, we reject the null hypothesis when

$$\mu_0 < \overline{X} - z(\alpha)\frac{\sigma}{\sqrt{n}}.$$

Solving this inequality for $\overline{X}$ yields the rejection region for the level α test obtained earlier. That is, *the null hypothesis $H_0 : \mu = \mu_0$ is rejected whenever*

$$\overline{X} > \mu_0 + z(\alpha)\frac{\sigma}{\sqrt{n}}.$$

It is worth pointing out that when this test rejects $H_0 : \mu = \mu_0$ it will also reject $H_0 : \mu < \mu_0$. This is a consequence of the fact that whenever μ_0 lies outside the confidence interval, that is whenever μ_0 satisfies the inequality

$$\mu_0 < \overline{X} - z(\alpha)\frac{\sigma}{\sqrt{n}},$$

then all parameter values $\mu < \mu_0$ also satisfy the same inequality:

$$\mu < \mu_0 < \overline{X} - z(\alpha)\frac{\sigma}{\sqrt{n}}.$$

A Lower Tailed Test of a Hypothesis: We construct a lower tailed test of the null hypothesis

$$H_0 : \mu \geq \mu_0 \text{ against } H_1 : \mu < \mu_0$$

by modifying the reasoning previously used for an upper tailed test (see Example 8.1), so we content ourselves with a sketch. It is clear that we are very unlikely to observe a value of $\overline{X}$ that is very much smaller than the null value when $H_0 : \mu = \mu_0$ is true. The appropriate rejection region in this case for a level α lower tailed test is the set $\{\bar{x} : \bar{x} < c\}$, where

$$c = \mu_0 - z(\alpha)\frac{\sigma}{\sqrt{n}}. \tag{8.9}$$

The corresponding decision rule is

$$\text{accept } H_0 \text{ if } \overline{X} \geq \mu_0 - z(\alpha)\frac{\sigma}{\sqrt{n}};$$

$$\text{reject } H_0 \text{ if } \overline{X} < \mu_0 - z(\alpha)\frac{\sigma}{\sqrt{n}}.$$

The probability $\beta(\mu)$ of a type II error at the alternative $\mu < \mu_0$ is given by

$$\beta(\mu) = P(\overline{X} \geq c | \mu < \mu_0) = 1 - \Phi\left(\frac{\sqrt{n}(c - \mu)}{\sigma}\right). \tag{8.10}$$

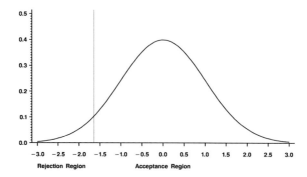

Rejection Region ($x < -1.645$) Based on Standardized Sample Mean
Lower Tailed Test, alpha = 0.05

FIGURE 8.4:   Rejection region based on Z_0

Using similar reasoning, we see that the power function is given by

$$\pi(\mu) = \Phi\left(\frac{\sqrt{n}(c-\mu)}{\sigma}\right). \tag{8.11}$$

The equivalent test based on the standardized statistic $Z_0 = \sqrt{n}(\overline{X} - \mu_0)/(\sigma)$ is given by:

$$\text{accept } H_0 \text{ if } \frac{\sqrt{n}(\overline{X} - \mu_0)}{\sigma} \geq -z(\alpha); \text{ reject } H_0 \text{ if } \frac{\sqrt{n}(\overline{X} - \mu_0)}{\sigma} < -z(\alpha). \tag{8.12}$$

The rejection region corresponding to the standardized statistic Z_0 is displayed in Figure 8.4.

Example 8.4 *Testing $H_0 : \mu \geq \mu_0$ against $H_1 : \mu < \mu_0$*

The problem of determining the concentration of arsenic in copper, measured as a percentage, is of considerable importance since the presence of even a small amount can significantly increase the copper's electrical resistance. A laboratory makes $n = 6$ measurements of the concentration of arsenic in copper, with the result $\overline{x} = 0.17\%$. We assume that the data come from a normal distribution with known standard deviation $\sigma = 0.04$. The copper cannot be used if the concentration of arsenic in copper exceeds 0.2%. (1) Construct a 5% level test of the null hypothesis $H_0 : \mu \geq 0.2$ against $H_1 : \mu < 0.2$ and draw the graph of the power function for this test. (2) Compute the type II error probability at the alternative $\mu = 0.17$ for the lower tailed test in part (1). (3) Test the null hypothesis at the level 5% level by constructing a 95% upper confidence interval.

Solution

1. Because this is a lower tailed test we substitute $n = 6$, $\mu_0 = 0.2$, $\sigma = 0.04$, $z(0.05) = 1.645$ into Equation 8.9 and obtain the cutoff value

$$c = 0.2 - 1.645 \times \frac{0.04}{\sqrt{6}} = 0.1731.$$

The rejection region is displayed in Figure 8.5.

Since $\overline{x} = 0.17 < 0.1731$, we reject H_0; that is, we conclude that the concentration of arsenic in the copper ore is below 0.2%.

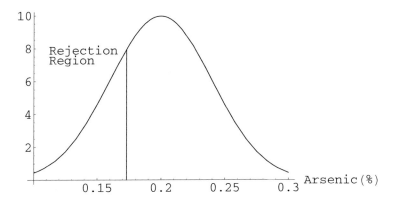

FIGURE 8.5: Rejection region $\mathcal{C} = \{x : x < 0.1731\}$ for Example 8.4

The same result could of course have been obtained using the standardized test statistic Z_0. To see this substitute $\overline{x} = 0.17$, $\sigma = 0.04$, $n = 6$, $-z(0.05) = -1.645$ into Equation 8.12. This yields the result $z = -1.83 < -1.645$; the null hypothesis is again rejected.

We obtain the power function corresponding to the rejection region $\mathcal{C} = \{\overline{x} : \overline{x} < 0.1731\}$ by setting $n = 6, \sigma = 0.04$, $c = 0.1731$ in Equation 8.11; the explicit formula follows:

$$\pi(\mu) = \Phi\left(\frac{\sqrt{6}(0.1731 - \mu)}{0.04}\right).$$

Its graph is displayed in Figure 8.6. Notice that the power function is decreasing; Problem 8.15(b) asks you to provide a detailed proof.

2. We obtain the type II error probability at the alternative $\mu = 0.17 < 0.20$ from the equation $\beta(\mu) = 1 - \pi(\mu)$.

Substituting $c = 0.1731$, $n = 6$, $\sigma = 0.04$, $\mu = 0.17$ in Equation 8.11 we obtain $\beta(0.17) = 1 - \pi(0.17) = 1 - 0.5753 = 0.4247$.

A one sided upper $100(1 - \alpha)\%$ confidence interval is given by

$$\left(-\infty, \overline{X} + z(\alpha)\frac{\sigma}{\sqrt{n}}\right]. \tag{8.13}$$

Substituting

$$\overline{x} = 0.17, \quad \frac{\sigma}{\sqrt{n}} = \frac{0.04}{\sqrt{6}} = 0.0163, \quad z(0.05) = 1.645$$

into Equation 8.13 yields the confidence interval $(-\infty, 0.1969]$. Since $0.1969 < 0.20$, we reject $H_0 : \mu = 0.20$ and conclude that the concentration of arsenic in the ore is less than 0.20%. Note that the confidence interval suggests that the concentration of copper could be as large as 0.1969%.

Two Sided Tests of a Hypothesis We construct a two sided test of the null hypothesis $H_0 : \mu = \mu_0$ against the alternative $H_1 : \mu \neq \mu_0$, by using the rejection region $\{\overline{x} : |\overline{x} - \mu_0| > c\}$. This is a reasonable choice since large values of $|\overline{X} - \mu_0|$ cast doubt on

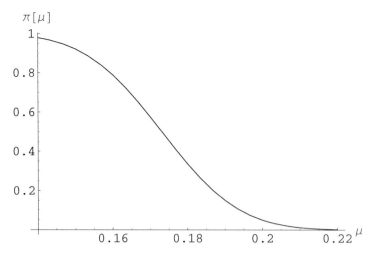

FIGURE 8.6:   Power function for Example 8.4: $\mu_0 = 0.20$; $\mathcal{C} = \{x : x > 0.1731\}$

the plausibility of H_0. The type I error probability α is given by

$$P(|\overline{X} - \mu_0| > c|H_0 \text{ is true}) = P(|\overline{X} - \mu_0| > c \text{ when } \mu = \mu_0)$$

$$= P\left(\frac{\sqrt{n}(|\overline{X} - \mu_0|)}{\sigma} > \frac{\sqrt{n}c}{\sigma}\right) = \alpha.$$

This leads to the following level α test:

$$\text{accept } H_0 \text{ if } |\overline{X} - \mu_0| \le z(\alpha/2)\sigma/\sqrt{n};$$
$$\text{reject } H_0 \text{ if } |\overline{X} - \mu_0| > z(\alpha/2)\sigma/\sqrt{n}. \tag{8.14}$$

An equivalent test based on the standardized statistic Z is given by:

$$\text{accept } H_0 \text{ if } \left|\frac{\sqrt{n}(\overline{X} - \mu_0)}{\sigma}\right| \le z(\alpha/2);$$

$$\text{reject } H_0 \text{ if } \left|\frac{\sqrt{n}(\overline{X} - \mu_0)}{\sigma}\right| > z(\alpha/2). \tag{8.15}$$

The power function is given by

$$\pi(\mu) = 1 - \Phi\left(\frac{\sqrt{n}(\mu_0 - \mu + c)}{\sigma}\right) + \Phi\left(\frac{\sqrt{n}(\mu_0 - \mu - c)}{\sigma}\right) \tag{8.16}$$

Example 8.5 *Testing $H_0 : \mu = \mu_0$ against $H_1 : \mu \ne \mu_0$*

The resistors produced by a manufacturer are required to have a resistance of $\mu_0 = 0.150$ ohms. Statistical analysis of the output suggests that the resistances can be approximated by a normal distribution $N(\mu; \sigma^2)$ with known standard deviation $\sigma = 0.005$. A random sample of $n = 10$ resistors is drawn and the sample mean is found to be $\overline{x} = 0.152$. (1) Test, at the 5% level, the hypothesis that the resistors are conforming. Graph the null distribution of $\overline{X}$ and identify the rejection region of the test. (2) Compute and graph the power function of this test. (3) Test this hypothesis at the 5% level by constructing a 95% two sided confidence interval.
    Solution

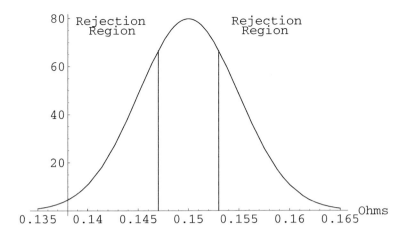

FIGURE 8.7: Rejection region for Example 8.5: $\mu_0 = 0.15$; $\mathcal{C} = \{x : |x - 0.15| > 0.003\}$

1. The null hypothesis in this case is that there is no difference between the specifications and the factory output, so the null hypothesis and its alternative are given by:

$$H_0 : \mu = 0.150 \text{ against } H_1 : \mu \neq 0.150.$$

We construct a test that is significant at the 5% level by substituting the values

$$n = 10, |\overline{x} - \mu_0| = |0.152 - 0.150| = 0.002, \sigma = 0.005, z(0.025) = 1.96$$

into Equation 8.14. This yields

$$z(\alpha/2)\sigma/\sqrt{n} = 1.96 \times \frac{0.005}{\sqrt{10}} = 0.003.$$

Since $|\overline{x} - 0.150| = 0.002 < 0.003$ we do not reject $H_0 : \mu = 0.150$. The null distribution of $\overline{X}$ and the rejection region for the two sided test are displayed in Figure 8.7.

2. The formula for the power function is obtained directly from Equation 8.16 and is given by

$$\pi(\mu) = 1 - \Phi\left(\frac{\sqrt{10}(0.15 - \mu + 0.003)}{0.005}\right) + \Phi\left(\frac{\sqrt{10}(0.15 - \mu - 0.003)}{0.005}\right).$$

Figure 8.8 displays the graph.

3. The 95% confidence interval is $0.152 \pm 0.0031 = [0.1489, 0.1551]$. The confidence interval contains the null value 0.150, so we accept the null hypothesis.

Table 8.3 summarizes the various hypotheses we have studied and the corresponding rejection regions based on the standardized test statistic $\sqrt{n}(\overline{X} - \mu_0)/\sigma$.

Table 8.3 *Rejection regions based on $\sqrt{n}(\overline{X} - \mu_0)/\sigma$*

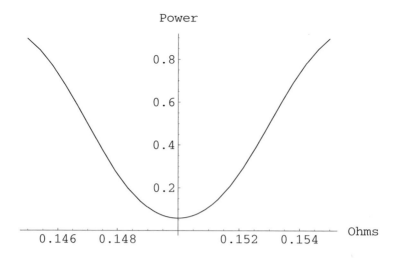

FIGURE 8.8: Power function for Example 8.5: $\mu_0 = 0.15$; $\mathcal{C} = \{x : |x - 0.15| > 0.003\}$

H_0	H_1	Rejection Region
$\mu \geq \mu_0$	$\mu < \mu_0$	$\frac{\sqrt{n}(\overline{X} - \mu_0)}{\sigma} < -z(\alpha)$
$\mu \leq \mu_0$	$\mu > \mu_0$	$\frac{\sqrt{n}(\overline{X} - \mu_0)}{\sigma} > z(\alpha)$
$\mu = \mu_0$	$\mu \neq \mu_0$	$\left\| \frac{\sqrt{n}(\overline{X} - \mu_0)}{\sigma} \right\| > z(\alpha/2)$

8.2.1   Significance Testing

In science we have to distinguish between mathematical propositions derived from hypotheses by logical deduction and *causal inference* based on empirical data. We can never prove the Pythagorean theorem by measuring the sides of sufficiently many right triangles. Mathematical truths cannot be refuted by experiments; and experiments cannot lead to mathematical certainty. Unfortunately, the terms reject and accept used in hypothesis testing imply a certitude that cannot be justified, since no scientific hypothesis is ever finally accepted or rejected in practice. We, therefore, interpret reject as meaning that the data provide strong evidence against the null hypothesis and we interpret accept to mean that the data do not. For this reason statisticians now recommend that one report the *P-value* of the test, which is a measure of how much agreement there is between the data and H_0.

Example 8.6 *The P-value of a test*

Return to Example 8.4. The null hypothesis, we recall, is $H_0 : \mu = 0.2$. We assume the data come form a normal distribution with $\sigma = 0.04$. The experimental data, based on a sample size $n = 6$, yielded a sample mean $\bar{x} = 0.17$. We are interested in the following question: If H_0 is true, so $\mu = 0.2\%$, what is the probability that the sample mean $\overline{X}$ would be as small as the observed value, 0.17%?

Solution   Using the given information ($\sigma = 0.04$ and sample size $n = 6$) we see that

$$P(\overline{X} < 0.17 | H_0 \text{ is true }) = P(\overline{X} < 0.17 | \mu = 0.2).$$

It follows that

$$P(\overline{X} < 0.17 | \mu = 0.2) = P\left(\frac{\sqrt{n}(\overline{X} - \mu_0)}{\sigma} < \frac{\sqrt{6}(0.17 - 0.20)}{0.04}\right)$$
$$= P(Z < -1.84) = \Phi(-1.84) = 0.033.$$

That is, the probability, assuming H_0 is true, of observing a sample mean $\overline{X}$ smaller than the observed value $\overline{x} = 0.17$ is 0.033; this probability, denoted by P, is called the P-value.

Interpreting the P-value

A small P-value indicates that H_0 is not plausible. Specifically, the test is said to be *statistically significant* at the level α if the P-value is smaller than α. Thus, instead of merely stating that H_0 was accepted (or rejected) at the significance level α one reports the P-value of the test and the significance of the result is left to the subjective judgement of the researcher. In practice, then, one proceeds as follows: Specify the significance level of the test, collect the data, and then compute the P-value P. A small value of P provides evidence against the null hypothesis; more precisely, the following criteria are suggested for weighing the evidence for and against the null hypothesis:

Strong evidence against H_0: $P < 0.01$;

Moderately strong evidence against H_0: $0.01 < P < 0.05$;

Relatively weak evidence against H_0: $P > 0.10$.

In words: The P-value (for a lower tailed test) is defined by

$$P\text{-}value = P\left(\begin{array}{c}\text{sample mean would be as small}\\\text{as the observed value}\\\text{if } H_0 \text{ is true.}\end{array}\right).$$

The significance testing approach just outlined is one that students must become familiar with since it is the one used in the current statistical software packages. Specifically, for hypothesis testing using Minitab or SAS, the output simply lists the observed P-value and leaves it to the user to judge the significance of the result.

Some suggestions for computing the P-value

Suppose that the data $\{x_1, \ldots, x_n\}$ come from a normal distribution $N(\mu, \sigma^2)$, where σ^2 is known and suppose the test is based on the value of the statistic

$$Z_0 = \frac{\sqrt{n}(\overline{X} - \mu_0)}{\sigma}.$$

When the observed value of the statistic Z_0 is z, then the P-value is

$$P\text{-}value = \left\{\begin{array}{l}\Phi(z) \text{ (lower tailed test)}\\1 - \Phi(z) \text{ (upper tailed test)}\\2(1 - \Phi(|z|)) \text{ (two tailed test)}\end{array}\right\}.$$

Example 8.7

Compute the P-value of the test of Example 8.5.

Solution In this case

$$\overline{x} = 0.152, \, n = 10, \, \sigma = 0.005.$$

Therefore,

$$z = \frac{\sqrt{n}(\overline{x} - \mu_0)}{\sigma} = \frac{\sqrt{10}(0.152 - 0.150)}{0.005} = 1.26.$$

This is a two sided test, so the P-value equals $2[1 - \Phi(1.26)] = 0.21$. This is relatively weak evidence against the the null hypothesis, so we do not reject it.

8.2.2 Power Function and Sample Size

When the sample size and significance level are fixed in advance by the statistician the type II error probability is determined. Consider, for instance, Example 8.4 where we showed that the power at the alternative $\mu = 0.17 < 0.20$ equals 0.5753. The probability of a type II error, then, equals $1 - 0.5753 = 0.4247$. Consequently, there is a considerable risk here of incurring unnecessary costs for reprocessing the ore. This raises the following question: Is there any way of reducing the type II error probability β? When the sample size n and significance level α are fixed in advance then the answer is no, since in this case the cutoff value c must satisfy the equation: $c = \mu_0 - z(\alpha)\frac{\sigma}{\sqrt{n}}$ (Equation 8.9). The only way, then, of increasing the power of the test at an alternative $\mu < \mu_0$ (without simultaneously increasing the type I error probability α) is to increase the sample size n. More precisely, the test will have significance level α and power $1 - \beta$ at the alternative μ if

$$\pi(\mu_0) = \alpha; \tag{8.17}$$
$$\pi(\mu) = 1 - \beta. \tag{8.18}$$

The sample size n that is required for the test to have power $1 - \beta$ (or type II error β) at the alternative μ is given by

$$n = \left(\frac{\sigma}{\mu_0 - \mu} \times (z(\alpha) + z(\beta)) \right)^2. \tag{8.19}$$

Derivation of Equation 8.19:
In this case the power function is

$$\pi(\mu) = \Phi\left(\frac{c - \mu}{\sigma/\sqrt{n}} \right), \quad -\infty < \mu < \infty.$$

Inserting this expression for $\pi(\mu)$ into Equations 8.17 and 8.18 yields two equations for the two unknowns c and n:

$$\Phi\left(\frac{c - \mu_0}{\sigma/\sqrt{n}} \right) = \alpha; \tag{8.20}$$
$$1 - \Phi\left(\frac{c - \mu}{\sigma/\sqrt{n}} \right) = \beta. \tag{8.21}$$

Solving these equations for $\sqrt{n}$ (Problem 8.21 asks you to supply the details) we obtain

$$\sqrt{n} = \frac{\sigma}{\mu_0 - \mu} \times (z(\alpha) + z(\beta)).$$

Squaring both sides of this expression yields Equation 8.19 for the sample size.
Equation 8.19 for the sample size is also valid for a upper tailed test.

Example 8.8

Suppose we wish to test the hypothesis $H_0 : \mu \geq 0.2$ against $H_1 : \mu < 0.2$ for the concentration of arsenic in copper at significance level $\alpha = 0.05$ and known $\sigma = 0.04$. Determine the minimum required sample size if the power of the test is to be 0.90 at the alternative $\mu = 0.17$.

Solution We compute the minimal sample size using Equation 8.19. That is, we set

$$\mu_0 = 0.2,\ \mu = 0.17,\ \sigma = 0.04,\ \alpha = 0.05,\ \text{and}\ 1 - \beta = 0.9.$$

This yields the result that the sample size equals $n = 3.9^2 = 15.21$, which we round up to $n = 16$.

Power and Sample Size: Two Sided Tests

When the alternative hypothesis is two sided the sample size n that is required to produce a test with significance level α and power $1 - \beta$ at the alternative μ is given by

$$n = \left(\frac{\sigma(z(\alpha/2) + z(\beta))}{\mu_0 - \mu} \right)^2. \tag{8.22}$$

The derivation is similar to that used to derive Equation 8.19, save for the fact that an explicit solution for n is not possible. However, one can show that Equation 8.22 is a useful approximation.

Example 8.9

Refer to Example 8.5. In this example the resistors are required to have a resistance of $\mu_0 = 0.150$ ohms. How large must n be if you want $\alpha = 0.05$ and the type II error probability at the alternative $\mu = 0.148$ is to be 0.2?

Solution Setting $z(\alpha/2) = z(0.05) = 1.96$, $z(\beta) = z(0.2) = 0.84$, and $\mu_0 - \mu = 0.002$ in Equation 8.21 yields the result that

$$\sqrt{n} = \left(\frac{0.005(1.96 + 0.84)}{0.002} \right) = 7; \text{ so } n = 49.$$

8.2.3 Large Sample Tests Concerning the Mean of an Arbitrary Distribution

When the distribution from which the sample is drawn is not known to be normal it is still possible to make inferences concerning the mean μ of the distribution, provided the sample size is sufficiently large so that the central limit theorem can be applied. More precisely, we drop the assumption that the random sample $X_1, \ldots, X_n$ is taken from a normal distribution; instead we assume that it comes from an arbitrary distribution with unknown mean μ and unknown variance σ^2. We make the additional assumption that the sample size n is large enough, ($n \geq 30$, say) so that the central limit theorem can be brought to bear; that is, assume that the distribution of standardized variable $Z = \sqrt{n}(\overline{X} - \mu)/\sigma$ has an approximate standard normal distribution $N(0, 1)$. This is still not useful since σ is unknown; we circumvent this difficulty by replacing σ with its estimate s, the sample standard deviation, since s will be close to σ with high probability for large values of n.

Proceeding by methods similar to those used earlier (Table 8.3) one can derive the following approximate level α one sided and two sided tests based on the standardized test statistic $\sqrt{n}(\overline{X} - \mu_0)/s$. They are displayed in Table 8.4.

Table 8.4 *Rejection regions (large samples)*

Alternative Hypothesis	Rejection Region
$H_1 : \mu < \mu_0$	$\dfrac{\sqrt{n}(\overline{X} - \mu_0)}{s} < -z(\alpha);$
$H_1 : \mu > \mu_0$	$\dfrac{\sqrt{n}(\overline{X} - \mu_0)}{s} > z(\alpha);$
$H_1 : \mu \neq \mu_0$	$\left| \dfrac{\sqrt{n}(\overline{X} - \mu_0)}{s} \right| > z(\alpha/2).$

Example 8.10

A laboratory makes $n = 30$ determinations of the concentration μ of an impurity in an ore and obtains the result that $\bar{x} = 0.09\%$ and $s = 0.03\%$. If $\mu \geq 0.10\%$ the ore cannot be used without additional processing. Consequently, we want to test

$$H_0 : \mu \geq 0.1 \text{ against } H_1 : \mu < 0.1.$$

How plausible is the null hypothesis given that $\bar{x} = 0.09$?

Solution   We shall solve this problem by first computing the observed value z of the standardized statistic Z and then computing its *approximate* P-value. We say approximate P-value because the distribution of the standardized statistic is only approximately normal. Now the observed value of Z equals -1.83, since

$$z = \frac{\sqrt{n}(\bar{x} - 0.1)}{s} = \frac{\sqrt{30} \times (-0.01)}{0.03} = -1.83.$$

Thus, the approximate P-value of the test is

$$\text{P-value} = P(Z \leq z) = P(Z \leq -1.83) = 0.03.$$

Since the P-value $= 0.03 < 0.05$ the observed value is significant at the 5% level. In other words, we reject the null hypothesis $H_0 : \mu \geq 0.1\%$.

8.2.4   Tests Concerning the Mean of a Distribution with Unknown Variance

When the variance σ^2 is unknown and the sample size n is small ($n < 30$, say) so the applicability of the central limit theorem is questionable, one can still test hypotheses about the mean μ, provided the distribution from which the random sample is drawn is approximately normal. We begin with an extension of the hypothesis testing methods of the previous sections to the important special case when the random sample $X_1, \ldots, X_n$ comes from a normal distribution $N(\mu, \sigma^2)$ with unknown variance σ^2. The fact that $\overline{X}$ is again normally distributed with mean μ and variance σ^2/n is not very useful since σ^2 is unknown. However, we do know (Theorem 6.5) that

$$T = \frac{\sqrt{n}(\overline{X} - \mu)}{s} \text{ is } t_{n-1} \text{ distributed ; consequently,}$$

$$P(\overline{X} \leq c|\mu) = P\left(\frac{\sqrt{n}(\overline{X} - \mu)}{s} \leq \frac{\sqrt{n}(c - \mu)}{s}\right)$$

$$= P\left(t_{n-1} \leq \frac{\sqrt{n}(c - \mu)}{s}\right).$$

It is this result that allows us to control the type I error probability using the t_{n-1} distribution instead of the normal distribution.

One sided test: Unknown variance

We now show how to construct a level α test

$$H_0 : \mu = \mu_0 \text{ against } H_1 : \mu < \mu_0$$

using the standardized test statistic

$$T = \frac{\sqrt{n}(\overline{X} - \mu_0)}{s}.$$

The method is similar to the one used to derive the lower tailed test given in Equation 8.12.

1. Choose a rejection region of the form

$$C = \left\{ \bar{x} : \frac{\sqrt{n}(\bar{x} - \mu_0)}{s} < -t_{n-1}(\alpha) \right\}, \tag{8.23}$$

so whenever the null hypothesis is true the probability of a type 1 error is α; that is,

$$\text{accept } H_0 \text{ if } \frac{\sqrt{n}(\overline{X} - \mu_0)}{s} \geq -t_{n-1}(\alpha);$$

$$\text{reject } H_0 \text{ if } \frac{\sqrt{n}(\overline{X} - \mu_0)}{s} < -t_{n-1}(\alpha).$$

2. The power function, which is not easy to compute, is given by

$$\pi(\mu) = P\left(\frac{\sqrt{n}(\overline{X} - \mu_0)}{s} < t \,\middle|\, \mu \right) = P(T_{n-1,\delta} < t), \tag{8.24}$$

where $T_{n-1,\delta}$ has a *non-central t distribution* with non-centrality parameter $\delta = \sqrt{n}(\mu - \mu_0)/\sigma$. To determine the minimal sample size required for a test of power $1 - \beta$ at the alternative μ statisticians use graphs (not given here) of the power function for selected values of n and α.

Example 8.11

This is a continuation of Example 8.4 except that we now assume that the variance is unknown. The reader will recall that the researcher was interested in determining if the concentration of arsenic in copper was greater than or equal to 0.2%. We change the problem slightly by assuming that the researcher made $n = 6$ measurements with the sample mean $\bar{x} = 0.17$ and sample standard deviation $s = 0.04$. Assuming that the data come from a normal distribution test (at the 5% level) the null hypothesis

$$H_0 : \mu = 0.20 \text{ against } H_1 : \mu < 0.20.$$

Solution We construct the rejection region using Equation 8.23 ; that is, we set

$$\mu_0 = 0.20, \, n = 6, \, t_5(0.05) = 2.015, \text{ and } s = 0.04$$

and obtain the rejection region

$$C = \left\{ \bar{x} : \frac{\sqrt{6}(\bar{x} - 0.20)}{0.04} < -2.015 \right\}.$$

When $\bar{x} = 0.17$ the value of the standardized test statistic T is given by

$$T = \frac{\sqrt{6}(0.17 - 0.20)}{0.04} = -1.837 > -2.015.$$

Consequently, we do not reject the null hypothesis and conclude that the concentration of arsenic in the copper ore is not less than 0.2%. Notice that our conclusion in this case differs from the conclusion obtained in Example 8.4. That is, a value of the sample mean $\overline{X} = 0.17$ is significant at the 5% level when the variance is known, and is not significant at the 5% level when the variance is unknown. Thus, knowing the value of the variance yields a test that is more sensitive to departures from the null hypothesis.

Two sided tests via confidence intervals: Variance unknown

A $100(1 - \alpha)\%$ confidence interval for the mean of a normal distribution with unknown variance has endpoints $[L, U]$ where

$$L = \overline{X} - t_{n-1}(\alpha/2)s/\sqrt{n}, \text{ and}$$
$$U = \overline{X} + t_{n-1}(\alpha/2)s/\sqrt{n}.$$

Keeping in mind the interpretation of the confidence interval as consisting of those parameter values μ that are consistent with the null hypothesis we arrive at the following two sided test:

Accept $H_0 : \mu = \mu_0$ when

$$\overline{X} - t_{n-1}(\alpha/2)\frac{s}{\sqrt{n}} \leq \mu_0 \leq \overline{X} + t_{n-1}(\alpha/2)\frac{s}{\sqrt{n}}; \tag{8.25}$$

and reject H_0 otherwise.

Example 8.12 *This is a continuation of Example 8.5 except that we now assume that the variance is unknown and that $s = 0.005$.*

Test at the 5% level the null hypothesis

$$H_0 : \mu = 0.15 \text{ against } H_1 : \mu \neq 0.15.$$

Solution We construct the 95% confidence interval by setting $n = 10$, $\mu_0 = 0.15$ and $\alpha = 0.05$, $t_9(0.025) = 2.262$ in Equation 8.25. This produces the confidence interval $[0.1489, 0.1551]$, which contains 0.150, so we do not reject the null hypothesis. It is once again worth noting that the confidence interval approach tells us that μ could be as small as 0.1489 or as large as 0.1551; which is clearly more informative than merely stating the null hypothesis is not rejected.

Table 8.5 summarizes the various hypotheses we have studied and the corresponding rejection regions based on the standardized test statistic $t_{n-1} = \sqrt{n}(\overline{X} - \mu_0)/s$.

Table 8.5 *Rejection regions based on the t distribution*

Alternative Hypothesis	*Rejection Region*
$(H_1 : \mu < \mu_0)$	$\dfrac{\sqrt{n}(\overline{X} - \mu_0)}{s} < -t_{n-1}(\alpha);$
$(H_1 : \mu > \mu_0)$	$\dfrac{\sqrt{n}(\overline{X} - \mu_0)}{s} > t_{n-1}(\alpha);$
$(H_1 : \mu \neq \mu_0)$	$\left\| \dfrac{\sqrt{n}(\overline{X} - \mu_0)}{s} \right\| > t_{n-1}(\alpha/2).$

Figure 8.9 displays the rejection regions for a two sided test based on the standardized t distribution with $n = 16$, so $t_{n-1} = t_{15}$, and $\alpha = 0.10, 0.05$. We use the fact that $t_{15}(0.05) = 1.753$ and $t_{15}(0.025) = 2.131$.

Remark The one sided and two sided level α tests listed in Table 8.5 are still approximately correct, provided one can safely assume that the distribution from which the random sample is drawn is at least approximately normal.

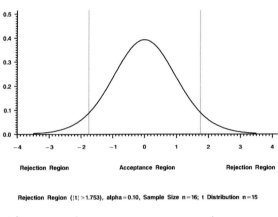

Rejection Region ($|t| > 1.753$), alpha = 0.10, Sample Size n = 16; t Distribution n = 15

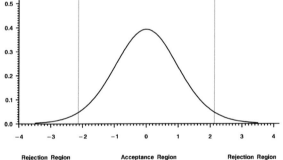

Rejection Region ($|t| > 2.131$), alpha = 0.05, Sample Size n = 16; t Distribution n = 15

FIGURE 8.9:   Rejection regions based on the t distribution, sample size $n = 16$

8.3   Comparing Two Populations

In many situations of practical interest we would like to compare two populations with respect to some numerical characteristic. For example, we might be interested in determining which of two steel cables has has the greater tensile strength, or which of two gasoline blends yields more miles per gallon (mpg). If we denote, say, the tensile strengths of the two types of cable by X_1 and X_2, respectively, then the quantity of interest will be their difference $X_1 - X_2$. The assumptions on X_1 and X_2 are the same as in Section 7.4, where we assumed that the distributions of the variables X_1 and X_2 are both normally distributed with (possibly) different means denoted by μ_1, μ_2, but with the same variances, i.e., $\sigma_1^2 = \sigma_2^2 = \sigma^2$. It is worth noting that the assumption of equal variances is itself a hypothesis that can be tested (see Section 8.3.2). When the assumption of equal variances is found to be questionable then the problem of testing hypotheses about the difference between the two means $\mu_1 - \mu_2$ is considerably more complicated; we refer the interested reader to G.W. Snedecor and W.G. Cochran (1980) *Statistical Methods,* 7th ed., Iowa State University Press, Ames, Iowa, for a discussion (with examples) of various remedies that have been proposed in the statistics literature.

We distinguish between *independent samples*, where it is assumed that the random samples selected from each of the two populations are mutually independent, and *paired samples* as discussed in Section 7.4.1.

Independent Samples

Consider the problem of comparing two normal distributions, with means μ_1, μ_2 and the same variance σ^2. In this context the null hypothesis of no difference between the means of the two populations is expressed as

$$H_0 : \Delta = \mu_1 - \mu_2 = 0.$$

There are various possibilities for the alternative hypothesis, such as:

$$H_1 : \Delta \neq 0, \text{ or}$$
$$H_1 : \Delta > 0, \text{ i.e., } \mu_1 > \mu_2.$$

Consider for example a comparison of the effectiveness of two weight loss programs, called diet 1 and diet 2, respectively. The null hypothesis states that there is no difference between the two diets. The alternative $H_1 : \Delta \neq 0$ asserts that there is a difference, whereas the alternative $H_1 : \Delta > 0$ asserts that diet 1 is more effective than diet 2. To construct a level α test of $H_0 : \Delta = 0$ against the two sided alternative $\Delta \neq 0$ we proceed as follows:

1. Take random samples of sizes n_1 and n_2 from the two populations, and denote their corresponding sample means and sample variances by $\overline{X}_i$ and s_i^2, $i = 1, 2$. Since the random samples selected from the two populations are assumed to be independent we have

$$V(\overline{X}_1 - \overline{X}_2) = \sigma^2 \left(\frac{1}{n_1} + \frac{1}{n_2} \right).$$

The estimated standard error is

$$s(\overline{X}_1 - \overline{X}_2) = s_p \sqrt{\left(\frac{1}{n_1} + \frac{1}{n_2} \right)},$$

where s_p^2 is the pooled estimate of σ^2 (Equation 7.34).

2. For our test statistic we use the fact (Theorem 7.1) that when H_0 is true

$$\frac{(\overline{X}_1 - \overline{X}_2) - (\mu_1 - \mu_2)}{s_p\sqrt{\frac{1}{n_1} + \frac{1}{n_2}}} \sim t_{n_1+n_2-2} \qquad (8.26)$$

has a student t distribution with n_1+n_2-2 degrees of freedom. Under the null hypothesis $\mu_1 = \mu_2$ it follows that

$$\frac{(\overline{X}_1 - \overline{X}_2)}{s_p\sqrt{\frac{1}{n_1} + \frac{1}{n_2}}} \sim t_{n_1+n_2-2}.$$

3. Choose the rejection region to be the set $\{t : |t| > t_{n_1+n_2-2}(\alpha/2)\}$; with this choice of the cutoff value the probability of a type I error is α. In other words you

$$\text{accept } H_0 \text{ if } \left| \frac{(\overline{X}_1 - \overline{X}_2)}{s_p\sqrt{\frac{1}{n_1} + \frac{1}{n_2}}} \right| \leq t_{n_1+n_2-2}(\alpha/2);$$

$$\text{reject } H_0 \text{ if } \left| \frac{(\overline{X}_1 - \overline{X}_2)}{s_p\sqrt{\frac{1}{n_1} + \frac{1}{n_2}}} \right| > t_{n_1+n_2-2}(\alpha/2).$$

Alternatively, one can proceed by constructing a two sided $100(1-\alpha)\%$ confidence interval of the form $[L, U]$ (using Equations 7.37 and 7.38), where

$$L = \overline{X}_1 - \overline{X}_2 - t_{n_1+n_2-2}(\alpha/2)s_p\sqrt{\frac{1}{n_1} + \frac{1}{n_2}};$$

$$U = \overline{X}_1 - \overline{X}_2 + t_{n_1+n_2-2}(\alpha/2)s_p\sqrt{\frac{1}{n_1} + \frac{1}{n_2}}.$$

We then accept H_0 if $0 \in [L, U]$ and reject H_0 otherwise, i.e.,

$$\text{accept } H_0 \text{ if } L \leq 0 \leq U; \qquad (8.27)$$
$$\text{reject } H_0 \text{ if } U < 0, \text{ or } L > 0. \qquad (8.28)$$

The rejection regions for lower and upper tailed tests are derived in a similar manner; the results are summarized in Table 8.6.

Table 8.6

Alternative Hypothesis	*Rejection Region*
$H_1 : \mu_1 - \mu_2 < 0,$	$\left(\dfrac{(\overline{X}_1 - \overline{X}_2)}{s_p\sqrt{\frac{1}{n_1} + \frac{1}{n_2}}} < -t_{n_1+n_2-2}(\alpha) \right);$
$H_1 : \mu_1 - \mu_2 > 0,$	$\left(\dfrac{(\overline{X}_1 - \overline{X}_2)}{s_p\sqrt{\frac{1}{n_1} + \frac{1}{n_2}}} > t_{n_1+n_2-2}(\alpha) \right);$
$H_1 : \mu_1 - \mu_2 \neq 0,$	$\left(\left\| \dfrac{(\overline{X}_1 - \overline{X}_2)}{s_p\sqrt{\frac{1}{n_1} + \frac{1}{n_2}}} \right\| > t_{n_1+n_2-2}(\alpha/2) \right).$

A test based on the statistic defined by Equation 8.26 is called a *two sample t test*.

Example 8.13

Refer to the weight loss data for the two diets in Problem 7.35. Is there a significant difference at the 5% level between the two diets?

Solution We use the confidence interval approach. We obtain a 95% confidence interval by computing the following quantities:

$$\bar{x}_1 = 5.86, \ \bar{x}_2 = 10.79, \ \bar{x}_1 - \bar{x}_2 = -4.93;$$
$$s_1^2 = 11.85, \ s_2^2 = 10.67;$$
$$s_p^2 = 11.26, \ s_p = 3.36;$$
$$t_{18}(0.025)s_p\sqrt{\frac{1}{10} + \frac{1}{10}} = 3.15;$$
$$L = -4.93 - 3.15 = -8.08;$$
$$U = -4.93 + 3.15 = -1.78.$$

Since the confidence interval $[-8.08, -1.78]$ does not contain 0 we do not accept the null hypothesis.

Paired sample t test

Before proceeding, it might be helpful to review the mathematical model of a paired sample as described in Definition 7.5. The key assumption made there is that the random variables $D_i = X_i - Y_i$, $(i = 1, \ldots, n)$ are mutually independent, normally distributed random variables with a common mean and variance denoted by $E(D_i) = \Delta$, $V(D_i) = \sigma_D^2$. As in the case of independent samples, a variety of combinations for the null and alternative hypotheses can be studied. Here we consider the two sided alternatives case only, that is $H_1 : \Delta \neq 0$. The necessary modifications for dealing with one sided alternatives are omitted. Our problem then is this: Use the confidence interval approach to construct a level α test of

$$H_0 : \Delta = 0 \text{ against } H_1 : \Delta \neq 0.$$

The following test is called a *paired sample t test*.

1. Collect the paired data (x_i, y_i), $i = 1, \ldots, n$ and compute $\overline{D}$ and s_D^2.

2. For our test statistic we use the fact that

$$\frac{(\overline{D} - \Delta)}{s_D/\sqrt{n}} \sim t_{n-1}. \tag{8.29}$$

Consequently, under the null hypothesis that $\Delta = 0$ we have

$$\frac{\overline{D}}{s_D/\sqrt{n}} \sim t_{n-1}.$$

3. Construct a two sided $100(1-\alpha)\%$ confidence interval of the form $[L, U]$ (Equations 7.18 and 7.19), where

$$L = \overline{D} - t_{n-1}(\alpha/2) \times s_D/\sqrt{n};$$
$$U = \overline{D} + t_{n-1}(\alpha/2) \times s_D/\sqrt{n}.$$

We then accept H_0 if $0 \in [L, U]$ and reject H_0 otherwise, that is

accept H_0 if $L \leq 0 \leq U$; reject H_0 if $U < 0$, or $L > 0$.

Example 8.14 *Comparing execution times of two processors*

Refer to Table 7.2 that reports the results of a study comparing the execution times of six different workloads on two different configurations of a processor. The parameter of interest is Δ, the difference between the mean execution times of the two processors. Is there a significant difference at the 5% level between the two processor configurations?

Solution We test the null hypothesis $(\Delta = 0)$ at the 5% level by constructing a 95% confidence interval for Δ, which is given by

$$L = -25.67 - t_5(0.025) \times \frac{10.41}{\sqrt{6}} = -36.60;$$

$$U = -25.67 + t_5(0.025) \times \frac{10.41}{\sqrt{6}} = -14.74.$$

The confidence interval $[-36.60, -14.74]$ does not contain 0, so the null hypothesis is not accepted. Of course, the data indicate much more, and that is that a one cache memory is significantly faster than a no cache memory.

8.3.1 The Wilcoxon Rank Sum Test for Two Independent Samples

The P-values and power of the t test change when the distribution of the data is non-normal, highly skewed, for instance. The *Wilcoxon rank sum test,* an alternative to the t test, is a test whose P-value does not depend explicitly on the distribution of the data and whose power remains high against the alternatives even when the data are non-normal.

1. **Population assumptions** The random samples $X_1, \ldots, X_{n_1}$ and $Y_1, \ldots, Y_{n_2}$ are independent, with continuous distribution functions F and G, respectively. When F and G are both normal with the same variance σ^2 but (possibly) different means μ_1 and μ_2, then their distribution functions are given by (see the discussion prior to Equation 4.42)

$$F(x) = \Phi\left(\frac{x - \mu_1}{\sigma}\right) \text{ and } G(x) = \Phi\left(\frac{x - \mu_2}{\sigma}\right).$$

It follows that

$$F(x) = \Phi\left(\frac{x - \mu_1}{\sigma}\right) = \Phi\left(\frac{[x - (\mu_1 - \mu_2)] - \mu_2}{\sigma}\right)$$
$$= G(x - (\mu_1 - \mu_2)) = G(x - \theta), \theta = \mu_1 - \mu_2.$$

In other words, when $\mu_1 < \mu_2$, $(\theta < 0)$ the probability density function of the $Y's$ is shifted to the right, and when $\mu_1 > \mu_2$, $(\theta > 0)$ the probability density function of the $Y's$ is shifted to the left. For a graphical display of this shift see Figure 8.10.

2. **The null and alternative hypotheses** The null hypothesis, as usual, is that both samples come from the same distribution, that is

$$H_0 : F(x) = G(x).$$

We shall consider three types of alternative hypotheses: $(\mu_1 < \mu_2)$, $(\mu_1 > \mu_2)$, and $(\mu_1 \neq \mu_2)$. Restated in terms of the parameter θ the three alternatives are:

$$H_1 : F(x) = G(x - \theta), \theta < 0 \, (\mu_1 < \mu_2);$$
$$H_1 : F(x) = G(x - \theta), \theta > 0 \, (\mu_1 > \mu_2);$$
$$H_1 : F(x) = G(x - \theta), \theta \neq 0 \, (\mu_1 \neq \mu_2).$$

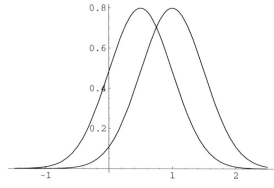

FIGURE 8.10: $f(x) = g(x - \theta)$

We interpret acceptance of the alternative $H_1 : \theta < 0$ or $H_1 : \theta > 0$ as implying that one population has a higher mean than the other; and acceptance of $H_1 : \theta \neq 0$ as implying that the two means differ.

3. **Convention**   Population 1 always denotes the smaller sample, population 2 the larger (so $n_1 \leq n_2$).

We will now illustrate the Wilcoxon rank sum test procedure, and the underlying theory, in the context of a specific example.

Example 8.15 *Comparing fuel efficiencies of two gasolines*

The data in Table 8.7 come from an experiment comparing the fuel efficiencies of two gasolines, a standard brand and a premium one. Six cars were driven with the premium gas and seven cars with the standard gas, so $n_1 = 6$, $n_2 = 7$. Test, at the 5% level, the null hypothesis that the premium blend is no more fuel efficient than the standard one against the one sided alternative that the premium blend is more fuel efficient.

Table 8.7

Premium (mpg)	Standard (mpg)
X	Y
26.6	22.7
26.0	28.1
28.4	24.0
27.5	26.8
30.9	25.5
29.0	24.7
	30.2

Solution   The null hypothesis is that the fuel efficiencies of the premium and standard gasolines are the same and the alternative hypothesis is that the premium brand is more fuel efficient, that is, we are testing

$$H_0 : F(x) = G(x) \text{ against } H_1 : F(x) = G(x - \theta), \theta = \mu_1 - \mu_2 > 0.$$

The Wilcoxon rank sum test begins by combining the observations from both populations into a single sample of size $6 + 7 = 13$, and arranging the data in increasing order (by computing the order statistics of the combined sample), preserving the population identity as shown in Table 8.8. The first column denotes the population from which the observation is drawn (premium $= 1$, standard $= 2$), the second column displays the order statistics of the combined sample, and the last column lists the rank of each observation in the combined sample.

In this example all the observations in the combined sample are distinct, so the ranks are uniquely defined. It may happen that two or more observations are *tied*. Suppose, for example that the recorded values for the mileages ranked 6 (26.6) and 7 (26.8) were tied with common value 26.7. Then the rank of each of these observations is the mean of the ranks that would have been assigned had they been different. In this case we would have assigned the rank 6.5 to each of them.

Table 8.8

Population	MPG	Rank
2	22.7	1
2	24.0	2
2	24.7	3
2	25.5	4
1	26.0	5
1	26.6	6
2	26.8	7
1	27.5	8
2	28.1	9
1	28.4	10
1	29.0	11
2	30.2	12
1	30.9	13

When H_0 is true, then the null distribution of the ranks assigned to the sample $X_1, \ldots, X_6$ is the same as if we had randomly selected 6 balls from $6+7 = 13$ balls numbered $1, 2, \ldots, 13$. Therefore, when the alternative $H_1 : \theta > 0$ is true, then the sum of the ranks assigned to the sample $X_1, \ldots, X_6$ will be larger than it would if H_0 were true. Looking at Table 8.8, we see that the four lowest ranking observations come from population 2 (the standard brand), suggesting that the lower ranks are associated with population 2 and the higher ranks are associated with population 1. A reasonable choice for our test statistic then, is W_1, the sum of the ranks associated with the X sample. For instance, with reference to Table 8.8 we see that the sum of the ranks associated with population 1 is

$$W_1 = 5 + 6 + 8 + 10 + 11 + 13 = 53.0.$$

We then choose a number c, so that an observed value of $W_1 \geq c$ implies that it is highly unlikely to have been observed if, in fact, H_0 is true.

It can be shown (we omit the proof) that under H_0 the mean and standard deviation of the rank sum W_1 (with n_1 X's and n_2 Y's) are

$$E(W_1) = \frac{n_1(n_1 + n_2 + 1)}{2}; \ \sigma(W_1) = \sqrt{\frac{n_1 n_2(n_1 + n_2 + 1)}{12}}. \tag{8.30}$$

With reference to the data set in Table 8.7, the expected value and standard deviation of the rank sum are given by:

$$E(W_1) = \frac{6 \times (6 + 7 + 1)}{2} = 42 \text{ and } \sigma(W_1) = \sqrt{\frac{6 \times 7(6 + 7 + 1)}{12}} = 7.0.$$

The rejection region and P-value are computed by using the result (we omit the proof) that the null distribution of W_1 is known to be approximately normal with mean $E(W_1)$ and standard deviation $\sigma(W_1)$ as given in Equation 8.30; that is,

$$\frac{W_1 - E(W_1)}{\sigma(W_1)} \approx Z,$$

where Z denotes a standard normal random variable.

The rejection region for an upper tailed Wilcoxon rank sum test (using the continuity correction to improve the accuracy of the normal approximation) is given by

$$\text{Rejection region: } \frac{W_1 - E(W_1) - 0.5}{\sigma(W_1)} > z(\alpha). \tag{8.31}$$

To illustrate the method we return to the fuel efficiency data (Table 8.8). We substitute

$$w_1 = 53, \; E(W_1) = 42, \; \sigma(W_1) = 7$$

into Equation 8.31. It follows that

$$\frac{W_1 - E(W_1) - 0.5}{\sigma(W_1)} = \frac{53 - 42 - 0.5}{7} = 1.5 < 1.645 = z(0.05).$$

The (approximate) P-value is given by

$$\text{P-value} = P(Z > 1.5) = 0.0668.$$

So the results are not significant at the 5% significance level. We do not reject H_0.

Table 8.9 lists the rejection regions for each of the three alternative hypotheses based on the normal approximation (with continuity correction) to the Wilcoxon rank sum statistic. Critical values for this statistic based on the exact distribution of W_1 can be found in Dixon and Massey (1969), *Introduction to Statistical Analysis,* 3rd ed., McGraw-Hill.

Table 8.9 *Rejection regions based on the Wilcoxon statistic*

Alternative hypothesis   Rejection region

$$H_1 : \mu_1 - \mu_2 < 0 \quad \frac{W_1 - E(W_1) + 0.5}{\sigma(W_1)} < -z(\alpha)$$

$$H_1 : \mu_1 - \mu_2 > 0 \quad \frac{W_1 - E(W_1) - 0.5}{\sigma(W_1)} > z(\alpha)$$

$$H_1 : \mu_1 - \mu_2 \neq 0 \quad \frac{|W_1 - E(W_1)| - 0.5}{\sigma(W_1)} > z(\alpha/2);$$

$$E(W_1) = \frac{n_1(n_1 + n_2 + 1)}{2}; \; \sigma(W_1) = \sqrt{\frac{n_1 n_2(n_1 + n_2 + 1)}{12}}.$$

Note: W_1 is the sum of the ranks associated with population 1 $(n_1 \leq n_2)$.

8.3.2   A Test of the Equality of Two Variances

Tests of hypotheses about the variance σ^2 are based on the statistic $(n-1)s^2/\sigma^2$, which is known to have a χ^2_{n-1} distribution (Theorem 6.4). Because this test is infrequently used, and this is an introductory text, we shall not pursue this matter further here. Of more importance is a test for the equality of two variances since, in our derivation of the two sample t test, we assumed that the variances of the two normal distributions being compared were equal. In this section we present a test of this hypothesis; readers interested in how to modify the two sample t test when $\sigma_1^2 \neq \sigma_2^2$ are advised to consult Snedecor and Cochran (1980).

1. Let s_1^2 and s_2^2 denote the sample variances corresponding to random samples of sizes n_1 and n_2 taken from two independent normal distributions with the same variance σ^2; then it follows from Theorems 6.4 and 6.7 that the ratio

$$F_0 = \frac{s_1^2}{s_2^2} \sim F_{n_1-1,n_2-1} \tag{8.32}$$

   has an F distribution with parameters $\nu_1 = n_1 - 1$, $\nu_2 = n_2 - 1$.

2. To test the null hypothesis that

$$H_0 : \sigma_1^2 = \sigma_2^2 \text{ against } H_1 : \sigma_1^2 \neq \sigma_2^2$$

   at the significance level α choose the rejection region

$$\{F : F < F_{n_1-1,n_2-1}(1-\alpha/2) \text{ or } F > F_{n_1-1,n_2-1}(\alpha/2)\}.$$

3. The critical values $F_{\nu_1,\nu_2}(1-(\alpha/2))$ are obtained from Table A.6 in the appendix by using the formula:

$$F_{\nu_1,\nu_2}(1-(\alpha/2)) = \frac{1}{F_{\nu_2,\nu_1}(\alpha/2)}. \tag{8.33}$$

   Equation 8.33 is a consequence of the relation

$$F_{\nu_1,\nu_2} \sim \frac{1}{F_{\nu_2,\nu_1}}.$$

To put it another way, we accept the null hypothesis when

$$\frac{1}{F_{n_2-1,n_1-1}(\alpha/2)} < F < F_{n_1-1,n_2-1}(\alpha/2). \tag{8.34}$$

Example 8.16

Refer to the weight loss data for the two diets in Problem 7.35. Is it reasonable to assume that both the diet 1 and diet 2 data come from normal distributions with the same variance?
   Solution In this case, as previously noted, we have $n_1 = n_2 = 10$ and $s_1^2 = 11.85$, $s_2^2 = 10.67$; so the F ratio equals $11.85/10.67 = 1.11$. Since the upper tail probabilities are given only for $\alpha = 0.05$ and $\alpha = 0.01$, we can only construct two sided tests with significance level $\alpha = 0.10$ or $\alpha = 0.02$. Thus, for instance, $F_{9,9}(0.05) = 3.18$, so our acceptance region is the set $\{F : 1/3.18 \leq F \leq 3.18\}$. Since $F = 1.11$ we do not reject $H_0 : \sigma_1^2 = \sigma_2^2$.

8.4 Normal Probability Plots

Many of the techniques used in statistical inference assume that the distribution $F(x)$ of the random sample $X_1, \ldots, X_n$ comes from a distribution of a particular type denoted by

$F_0(x)$. In hypothesis testing, for example, we have assumed throughout that the data are normally distributed, i.e., that

$$F_0(x) = \Phi\left(\frac{x - \mu}{\sigma}\right).$$

When this assumption is violated some of these tests perform poorly, and so it is important to determine whether it is reasonable to assume that the data come from a given distribution $F_0(x)$.

As another example, consider the data in Problem 1.8, which lists 24 measurements of the operating pressure (psi) of a rocket motor. The statistical methods used by these authors to study the reliability of the rocket motor assumed that the data came from a normal distribution. This is a special case of determining whether the data come from a given theoretical distribution $F_0(x)$ against the alternative that it does not; that is, we are interested in testing

$$H_0 : F(x) = F_0(x) \text{ against } H_1 : F(x) \neq F_0(x). \tag{8.35}$$

In this section we present two methods for studying the normality assumption:

1. Normal probability plots are also called Q–Q plots, which is an abbreviation for *quantile–quantile plot*. This is a two-dimensional graph with the property that if the random sample comes from a normal distribution, then the plotted points appear to lie in a straight line. Both SAS and MINITAB have built-in procedures for producing these plots. Our purpose here is to discuss the underlying theory, which is of independent interest.

2. The Shapiro–Wilk test is based on the fact that the correlation coefficient of the normal probability plot is a measure of the strength of the linear relationship between the empirical and normal quantiles. (A detailed study of the sample correlation coefficient is given in Chapter 10.) The hypothesis of normality is rejected if the correlation falls below a critical value that depends on the sample size. In the computer printout, the test statistic, denoted by W, and its P-value are printed side by side as in Equation 8.43.

A discussion of the theory underlying normal probability plots

The basic idea behind a Q–Q plot is that when the null hypothesis $F(x) = F_0(x)$ is true the empirical distribution function $\hat{F}_n(x)$ (see Chapter 1, Definition 1.5) of a random sample of size n drawn from the parent distribution $F_0(x)$ should approximate the null distribution $F_0(x)$. A precise statement of this result, which is known as the *fundamental theorem of mathematical statistics*, is beyond the scope of this text, so, we content ourselves with the following version:

Theorem 8.1 *Let $\hat{F}_n$ denote the empirical distribution function of a random sample taken from the continuous parent distribution function F_0; that is, assume the null hypothesis $H_0 : F(x) = F_0(x)$ is true. Then*

$$\lim_{n \to \infty} \hat{F}_n(x) = F_0(x). \tag{8.36}$$

Furthermore, the p quantiles of the empirical distribution function converge to the p quantiles of the parent distribution. That is,

$$\lim_{n \to \infty} Q_n(p) = Q(p), \tag{8.37}$$

where $Q_n(p)$ and $Q(p)$ denote the pth quantiles of $\hat{F}_n(x)$ and $F_0(x)$, respectively.

Proof We shall not give a detailed proof here except to remark that it is a consequence of the law of large numbers applied to the empirical distribution function (see Equation 5.42, which is a weaker version of Equation 8.36). Intuitively, Equation 8.36 states that if the random sample really comes from the distribution $F_0(x)$, then, for large values of n, the proportion of observations in the sample whose values are less than or equal to x is approximately equal to $F_0(x)$, which is a very reasonable result.

Applying this result to the special case where $F_0(x) = \Phi((x - \mu)/\sigma)$, we have

$$\lim_{n \to \infty} \hat{F}_n(x) = \Phi\left(\frac{x - \mu}{\sigma}\right). \tag{8.38}$$

Equation 8.37 implies that the p quantiles $Q_n(p)$ of the empirical distribution function should be approximately equal to the p quantiles $Q(p)$ of the normal distribution. This suggests that a plot of the quantiles of the empirical distribution function against the quantiles of the normal distribution should produce a set of points that appear to lie on a straight line. Intuitively, the jth order statistic is approximately the $100(j/n)$th percentile of the data. More precisely, we now claim that the jth order statistic is the $100(j - 0.5)/n$ percentile of the empirical distribution function; that is,

$$Q_n\left(\frac{j - 0.5}{n}\right) = x_{(j)}. \tag{8.39}$$

Proof To derive Equation 8.39 we use Equation 1.16 to verify that when

$$p = (j - 0.5)/n, \text{ then } j - 1 < np = j - 0.5 < j; \text{ consequently, } Q_n(p) = x_{(j)}.$$

Similarly, the $100(j - 0.5)/n$ percentiles of the normal distribution $N(\mu, \sigma^2)$ are given by

$$Q\left(\frac{j - 0.5}{n}\right) = \mu + \sigma z_j, \text{ where} \tag{8.40}$$

$$\Phi(z_j) = \frac{j - 0.5}{n}. \tag{8.41}$$

The n quantities $z_1, \ldots, z_n$ defined by Equation 8.41 are called the *standardized normal scores;* they are the $(j - 0.5)/n$ quantiles of the standard normal distribution. Since $P(X_i \leq \mu + \sigma z_j) = \Phi(z_j) = (j - 0.5)/n$, it follows at once that $\mu + \sigma z_j$ is the $((j - 0.5)/n)$ quantile of the normal distribution $N(\mu, \sigma^2)$, as stated in Equation 8.40.

Putting these results together yields

$$x_{(j)} \approx \mu + \sigma \times z_j; \tag{8.42}$$

since, as noted earlier, the two quantiles should be nearly equal.

The content of Equation 8.42 is this: If the data come from a normal distribution, then the points $(z_j, x_{(j)}), j = 1, \ldots, n$ will appear to lie on a straight line; this graph, as noted earlier, is called a *normal probability plot* or a Q–Q plot.

Definition 8.1 *To construct a Q–Q plot of the data set* $\{x_1, \ldots, x_n\}$ *compute:*
(1) the normal scores $z_j = \Phi^{-1}((j - 0.5)/n))$
(2) the order statistics $x_{(j)}$ *and then plot*
(3) $(z_j, x_{(j)}), j = 1, \ldots, n.$

Constructing a Q–Q plot and looking at the graph to see if the points lie on a straight line is one of the simplest methods for testing normality. Because computing the normal scores z_j can be quite tedious it is advisable to carry out the computations via a statistical software

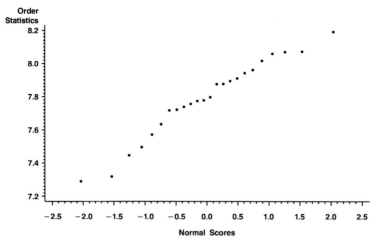

Q–Q Plot of Rocket Motor Data

FIGURE 8.11:   Q–Q plot of rocket motor data

package. To illustrate the method, we construct the Q–Q plot for the rocket motor data (Problem 1.8). Since the data points of the Q–Q plot appear to lie on a straight line we do not reject the hypothesis that the original data come from a normal distribution.

   If you do not have access to a computer then the computations can be carried out using a scientific calculator as illustrated in Table 8.10. The column (P) displays the order statistics of the rocket motor operating pressures data set of Problem 1.8 and column (Z) displays the normal scores z_j defined by $\Phi(z_j) = (j - 0.5)/24$, $j = 1, 2, \ldots, 24$.

Table 8.10 *Order statistics, normal scores for Problem 1.8 data*

P	Z	P	Z
7.2904	-2.03683	7.7981	0.05225
7.3196	-1.53412	7.8764	0.15731
7.4472	-1.25816	7.87785	0.26415
7.4965	-1.05447	7.89525	0.3741
7.5719	-0.88715	7.9112	0.48878
7.6357	-0.74159	7.9431	0.61029
7.71835	-0.61029	7.96195	0.74159
7.7227	-0.48878	8.01705	0.88715
7.7401	-0.3741	8.06055	1.05447
7.7575	-0.26415	8.0707	1.25816
7.7749	-0.15731	8.0736	1.53412
7.77925	-0.05225	8.1925	2.03683

   In addition to plotting and displaying the normal probability plot, a statistical software package will print out the value and P-value of the Shapiro–Wilk statistic, as in the following excerpt from a SAS printout.

$$\text{W:Normal} \qquad 0.961116 \qquad \text{Pr<W} \qquad 0.4674. \qquad (8.43)$$

The value of the Shapiro–Wilk statistic and its P-value are $W = 0.961116$ and $P = 0.4674$, respectively; so the hypothesis of normality is not rejected.

8.5   Tests Concerning the Parameter p of a Binomial Distribution

The effectiveness of a medical procedure is usually determined by the number of patients whose symptoms are relieved by the treatment. Because treatments using powerful drugs may have severe, if not fatal, side effects, the null hypothesis is that the treatment is ineffective; that is, it is up to the drug manufacturer to demonstrate the effectiveness of its drug to the Food and Drug Administration (FDA). It is known, for example, that approximately 25% of babies born to HIV infected mothers are themselves infected with the same virus. There was, therefore, justifiable excitement when the National Institute of Allergy and Infectious Diseases announced on Feb. 22, 1994 that giving the drug AZT to infected mothers reduced the infection rate to 8%—a three-fold reduction in the rate of infection! To put it another way, approximately 75% of babies born to HIV infected mothers do not have the HIV virus, but for those mothers treated with AZT the percentage increased to 92%. The results were regarded as so significant that clinics participating in the trials were advised to offer AZT to the women and babies in the control group.

We now place these results in the context of hypothesis testing. Suppose the natural recovery rate from the disease is p_0, and let p denote the recovery rate due to the treatment. Then the null hypothesis and its alternative will be of the form:

$$H_0 : p \leq p_0 \text{ against } H_1 : p > p_0.$$

We test this hypothesis by using the sample total X, where X is the number of patients out of n whose condition is improved as a result of the treatment. Under reasonable assumptions the null distribution of X is binomial with parameters n, p_0, so that evaluating the results of a clinical trial is equivalent to making inferences about the parameter p of a binomial distribution. In the clinical trials of AZT, for instance, we have $\hat{p} = 0.92 > 0.75 = p_0$, so the evidence against the null hypothesis appears to be quite strong.

Suppose we wish to make inferences concerning the (unknown) proportion of defective items produced by some manufacturing process or the proportion of patients who respond favorably to a new medical procedure. If the ith item is defective, we set the variable $X_i = 1$ and set $X_i = 0$ otherwise. Thus, the sample total $X = \sum_{1 \leq i \leq n} X_i$ counts the total number of defective items. The sample proportion $\hat{p} = X/n$ is an estimate of the proportion of defective items. If one can assume that conditions on the production line do not depend on the time at which the ith item is produced, then we can regard this as an example of a Bernoulli trial process (Definition 5.4); that is, it is an example of n independent repetitions of an experiment with only two outcomes, denoted by 0 and 1, respectively. Consequently, the sample total has a binomial distribution with parameters n, $p = P(X_i = 1)$. It is for this reason that a Bernoulli trials process is also called a *binomial experiment*. More formally, we can also regard $X_1, \ldots, X_n$ as a random sample taken from the Bernoulli distribution:

$$f(x; p) = p^x (1 - p)^{1-x}, \ x = 0, \ 1 \text{ with parameter space } \Theta = \{p : 0 \leq p \leq 1\}.$$

One Sided Upper Tailed Test of a Population Proportion

We suppose that our data come from a random sample $X_1, \ldots, X_n$ of size n taken from a Bernoulli distribution $f(x; p)$. We are interested in testing

$$H_0 : p \leq p_0 \text{ against } H_1 : p > p_0. \tag{8.44}$$

Intuitively, a large value of X provides evidence against the null hypothesis that the drug is ineffective. This suggests that we choose our rejection region to be the set $\mathcal{C} = \{x : x > c\}$. We have thus arrived at the following upper tailed test of hypothesis 8.44:

$$\text{accept } H_0 \text{ if } X \leq c; \tag{8.45}$$
$$\text{reject } H_0 \text{ if } X > c. \tag{8.46}$$

The power function of this test is given by

$$\pi(p) = P(X > c|p) = \sum_{c+1 \leq x \leq n} b(x; n, p) = 1 - B(c; n, p). \tag{8.47}$$

To compute the type I error probabilities, we use the fact that the power function $\pi(p)$ defined by Equation 8.47 is an *increasing* function of p (Proposition 4.6). Theorem 8.2 is a useful application of this fact to the context of hypothesis testing.

Theorem 8.2 *Suppose our test statistic X is binomial with parameters n, p and the power function $\pi(p)$ is given by*

$$\pi(p) = P(X > c|p) = \sum_{c+1 \leq x \leq n} b(x; n, p) = 1 - B(c; n, p).$$

Then $\pi(p)$ is an increasing function of p, so $p \leq p_0$ implies $\pi(p) \leq \pi(p_0)$. The significance level of the test is given by

$$\alpha = \pi(p_0) = P(X > c|p_0) = 1 - B(c; n, p_0). \tag{8.48}$$

Thus, to compute the significance level of the test we use Equation 8.48 and the binomial tables for $B(x; n, p_0)$.

We now put these formulas to good use, but first we make a few preliminary remarks. Testing a hypothesis about a population proportion is complicated by the fact that the computation of the power function, significance levels, and type I and type II errors depends on the sample size n; in particular, when $n \leq 25$ one can use the table of binomial probabilities; on the other hand, when $n > 25$ one must use the normal approximation, which yields only approximate values for the type I and type II errors. Another complication arises from the fact that X has a discrete distribution, so it will not always be possible to construct a test with a prescribed significance level α.

Upper Tailed Test of a Population Proportion (Small Sample)

Example 8.17

Assume that the natural recovery rate for a certain disease is 40%, so $p_0 = 0.4$. A new drug is developed and to test the manufacturer's claim that it boosts the recovery rate, a random sample of $n = 10$ patients, called the treatment group, is given the drug and the total number of recoveries X is recorded. Our problem is to find the cutoff value c that produces a test with significance level $\alpha \leq 0.05$.

Solution According to Equation 8.48 the test with rejection region $\mathcal{C} = \{x : x > c\}$ has significance level 0.05 when c satisfies the equation:

$$0.05 = \pi(0.40) = 1 - B(c; 10, 0.40).$$

From the binomial tables we see that

$$B(6; 10, 0.40) = 0.945 < 0.95 < 0.988 = B(7; 10, 0.40).$$

The choice of $c = 7$ yields $\alpha = 1 - 0.988 = 0.012$, and the choice of $c = 6$ yields $\alpha = 1 - 0.945 = 0.055$. If we insist on a significance level $\alpha \leq 0.05$, then we choose $c = 7$. Clearly, an exact 5% level test does not exist.

The type II error probability at the alternative $p = 0.7$ equals

$$\beta(0.7) = P(X \leq 7 | p = 0.7) = 1 - B(2; 10, 0.3)$$
$$= 1 - 0.383 = 0.617.$$

This is quite large because the sample size is small.

Upper Tailed Test of a Population Proportion (Large Sample)

We turn now to the problem of testing $H_0 : p = p_0$ against $H_1 : p > p_0$ when the sample size n is large enough so that the normal approximation is valid. The method is very similar to that just used except that it is more convenient to use the sample proportion $\hat{p} = X/n$ as our test statistic. It is intuitively clear that $\hat{p} > p_0$ provides evidence in favor of H_1 and this suggests, at least when the sample size is large enough, that we use a rejection region of the form

$$\mathcal{C} = \left\{ \hat{p} : \hat{p} > p_0 + z(\alpha) \sqrt{\frac{p_0(1 - p_0)}{n}} \right\} \tag{8.49}$$

to obtain an approximate level α test.

We can, of course, always use the standardized statistic

$$Z_0 = \frac{\hat{p} - p_0}{\sqrt{\frac{p_0(1-p_0)}{n}}}. \tag{8.50}$$

This yields the following large sample, approximate level α test of H_0 :

$$\text{accept } H_0 \text{ if } \frac{\hat{p} - p_0}{\sqrt{\frac{p_0(1-p_0)}{n}}} \leq z(\alpha); \tag{8.51}$$

$$\text{reject } H_0 \text{ if } \frac{\hat{p} - p_0}{\sqrt{\frac{p_0(1-p_0)}{n}}} > z(\alpha). \tag{8.52}$$

Example 8.18

Consider the following modification of the clinical trials experiment described in Example 8.17: The natural recovery rate from the discase is 40% and the treatment group now consists of $n = 200$ patients, 90 of whom recovered, so $\hat{p} = 0.45$. The question is this: Can we conclude, at the 5% significance level, that the drug is effective?

Solution Substituting $p_0 = 0.40$, $\hat{p} = 0.45$, $n = 200$ into the left hand side of Equation 8.52 yields the value

$$\frac{\hat{p} - p_0}{\sqrt{\frac{p_0(1-p_0)}{n}}} = \frac{0.05}{\sqrt{0.4 \times 0.6/200}} = 1.44 < 1.645 = z(0.05).$$

Since $Z_0 = 1.44 < 1.645$ the null hypothesis is not rejected. We arrive at the same conclusion using Equation 8.49 to construct the rejection region

$$\{\hat{p} : \hat{p} > c\} = \{\hat{p} : \hat{p} > 0.457\}.$$

Since $\hat{p} = 0.45 < 0.457$ we do not reject H_0.

The normal approximation to the binomial may also be used to approximate the power function and the type II error probability at the alternative p. The (approximate) power function and type II error probability for a one sided test with rejection region $\{\hat{p} : \hat{p} > c\}$ are given by

$$\pi(p) \approx 1 - \Phi\left(\frac{c-p}{\sqrt{p(1-p)/n}}\right); \tag{8.53}$$

$$\beta(p) \approx \Phi\left(\frac{c-p}{\sqrt{p(1-p)/n}}\right). \tag{8.54}$$

Derivation of 8.53:

$$\pi(p) = P(\hat{p} > c|p)$$

$$= P\left(\frac{\hat{p}-p}{\sqrt{p(1-p)/n}} > \frac{c-p}{\sqrt{p(1-p)/n}}\right)$$

$$\approx 1 - \Phi\left(\frac{c-p}{\sqrt{p(1-p)/n}}\right).$$

Example 8.19

Compute (1) the type II error at the alternative $p = 0.5$, and (2) the P-value of the observed test statistic $\hat{p} = 0.45$ of Example 8.18.

   Solution

1. The type II error probability at the alternative $p = 0.5$ is obtained by substituting $p = 0.5$, $n = 200$, $c = 0.457$ into Equation 8.54; thus,

$$\beta(0.5) = \Phi\left(\frac{0.457 - 0.5}{\sqrt{0.5 \times 0.5/200}}\right) = \Phi(-1.22) = 0.1112.$$

2. We recall that the P-value is the probability that the sample proportion $\hat{p}$ would be as large as the observed value $\hat{p} = 0.45$ if H_0 is true. We obtain the the P-value by setting $c = \hat{p} = 0.45$, $p = 0.4$, and $n = 200$ in Equation 8.53; thus,

$$P(\hat{p} > 0.45|p_0 = 0.4) = \pi(0.45) = 1 - \Phi(1.44) = 0.0749.$$

   Consequently, the P-value $= 0.0749$.

   Table 8.11 gives the level α tests, and the corresponding rejection regions for the one sided and two sided tests based on the standardized test statistic $Z_0 = (\hat{p}-p_0)/(\sqrt{p_0(1-p_0)/n})$.

Table 8.11 *Rejection regions based on $Z_0 = (\hat{p}-p_0)/(\sqrt{p_0(1-p_0)/n})$*

Alternative hypothesis	Rejection region
$H_1 : p < p_0;$	$\dfrac{\hat{p}-p_0}{\sqrt{\frac{p_0(1-p_0)}{n}}} < -z(\alpha);$
$H_1 : p > p_0;$	$\dfrac{\hat{p}-p_0}{\sqrt{\frac{p_0(1-p_0)}{n}}} > z(\alpha);$
$H_1 : p \neq p_0;$	$\left\|\dfrac{\hat{p}-p_0}{\sqrt{\frac{p_0(1-p_0)}{n}}}\right\| > z(\alpha/2).$

Controlling the type II error probabilities In Section 8.2.2 we showed how to choose the sample size n so that the corresponding level α test has type II error probability β at a specified alternative (Equations 8.19 and 8.21). Using similar methods here one can derive the following equations for the sample size n that has type II error probability β at the alternative p:

$$n = \left(\frac{z(\alpha)\sqrt{p_0(1-p_0)} + z(\beta)\sqrt{p(1-p)}}{p - p_0} \right)^2 \quad \text{(one sided test);} \qquad (8.55)$$

$$n = \left(\frac{z(\alpha/2)\sqrt{p_0(1-p_0)} + z(\beta)\sqrt{p(1-p)}}{p - p_0} \right)^2 \quad \text{(two sided test).} \qquad (8.56)$$

Example 8.20 *Determining minimal sample size*

Suppose we have to test $H_0 : p \leq 0.1$ against $H_1 : p > 0.1$ at the level $\alpha = 0.01$. What is the minimum sample size needed so that the type II error probability at the alternative $p = 0.25$ is $\beta = 0.1$?

Solution Since this is a one sided test we use Equation 8.55 and the given data, $z(0.01) = 2.33$, $p_0 = 0.1$, $p = 0.25$, $z(0.1) = 1.28$. The result is $n = 8.355^2$, which rounded up to the nearest integer, yields $n = 70$.

8.5.1 Tests of Hypotheses Concerning Two Binomial Distributions: Large Sample Size

Suppose we are interested in testing the null hypothesis that the proportion of defective items produced by two different manufacturing processes is the same. Denoting the corresponding proportions by p_1 and p_2, we are interested in testing

$$H_0 : p_1 = p_2 \text{ against } H_1 : p_1 \neq p_2.$$

When the samples are taken from two independent populations it follows that the sample proportions $\hat{p}_1$ and $\hat{p}_2$ are independent and approximately normal; consequently $\hat{p}_1 - \hat{p}_2$ is also approximately normal, with variance $V(\hat{p}_1 - \hat{p}_2)$, standard error $\sigma(\hat{p}_1 - \hat{p}_2)$, and estimated standard error $s(\hat{p}_1 - \hat{p}_2)$ given by

$$V(\hat{p}_1 - \hat{p}_2) = \frac{p_1(1-p_1)}{n_1} + \frac{p_2(1-p_2)}{n_2};$$

$$\sigma(\hat{p}_1 - \hat{p}_2) = \sqrt{\frac{p_1(1-p_1)}{n_1} + \frac{p_2(1-p_2)}{n_2}};$$

$$s(\hat{p}_1 - \hat{p}_2) = \sqrt{\frac{\hat{p}_1(1-\hat{p}_1)}{n_1} + \frac{\hat{p}_2(1-\hat{p}_2)}{n_2}}. \qquad (8.57)$$

To test this hypothesis we use the confidence interval approach; that is, we construct the $100(1-\alpha)\%$ two sided confidence interval $[L, U]$ for $p_1 - p_2$, where L and U are given by (see Section 7.5.1)

$$L = \hat{p}_1 - \hat{p}_2 - z(\alpha/2)s(\hat{p}_1 - \hat{p}_2); \quad U = \hat{p}_1 - \hat{p}_2 + z(\alpha/2)s(\hat{p}_1 - \hat{p}_2).$$

We reject the null hypothesis if 0 is not in $[L, U]$ and we accept it otherwise.

Example 8.21 *Continuation of Example 7.14*

In this example the following data were collected: Of 400 items produced by machine 1, 16 were defective; of 600 produced by machine 2, 60 were defective. Test the null hypothesis that there is no difference in the proportion of defectives produced by the two machines; use $\alpha = 0.05$.

Solution   From these data we compute

$$\hat{p}_1 - \hat{p}_2 = -0.06;$$
$$s(\hat{p}_1 - \hat{p}_2) = \sqrt{\frac{0.04 \times 0.96}{400} + \frac{0.1 \times 0.9}{600}} = 0.0157;$$
$$L = -0.06 - 1.96 \times 0.0157 = -0.0908;$$
$$U = -0.06 + 1.96 \times 0.0157 = -0.0292.$$

Since 0 is not in the confidence interval $[-0.0908, -0.0292]$, we reject the null hypothesis.

The two sided and one sided tests can also be expressed in terms of the standardized test statistic $Z_1 = (\hat{p}_1 - \hat{p}_2)/s(\hat{p}_1 - \hat{p}_2)$. The results are given in Table 8.12.

Table 8.12 *Rejection regions based on* $Z_1 = (\hat{p}_1 - \hat{p}_2)/s(\hat{p}_1 - \hat{p}_2)$

Alternative hypothesis	Rejection region
$H_1 : p_1 < p_2$;	$\dfrac{\hat{p}_1 - \hat{p}_2}{s(\hat{p}_1 - \hat{p}_2)} < -z(\alpha)$
$H_1 : p_1 > p_2$;	$\dfrac{\hat{p}_1 - \hat{p}_2}{s(\hat{p}_1 - \hat{p}_2)} > z(\alpha)$
$H_1 : p_1 \neq p_2$;	$\left\| \dfrac{\hat{p}_1 - \hat{p}_2}{s(\hat{p}_1 - \hat{p}_2)} \right\| > z(\alpha/2)$

8.6   Chapter Summary

Hypothesis testing (or significance testing) is a statistical method, based on the careful collection of experimental data, for evaluating the plausibility of a scientific model or for comparing two conflicting models. A statistical hypothesis is a claim about the distribution of a variable X defined on a population; it is called (somewhat arbitrarily) the null hypothesis. The denial of the claim is also a statistical hypothesis; it is called the alternative hypothesis. When the distribution of X is normal, the null hypothesis is usually stated in the form of an inequality, or equality, between the mean μ and a particular value μ_0 called the null value. In this chapter we studied primarily the following three types of null and alternative hypotheses:

$$H_0 : \mu \leq \mu_0 \text{ against } H_1 : \mu > \mu_0 \text{ (Example 8.1)};$$
$$H_0 : \mu \geq \mu_0 \text{ against } H_1 : \mu < \mu_0 \text{ (Example 8.4)};$$
$$H_0 : \mu = \mu_0 \text{ against } H_1 : \mu \neq \mu_0 \text{ (Example 8.5)}.$$

The alternative hypothesis $H_1 : \mu > \mu_0$ (or $H_1 : \mu < \mu_0$) is said to be one sided since it consists of those values of μ that deviate from μ_0 in one direction only. The alternative hypothesis $H_1 : \mu \neq \mu_0$ is said to be two sided since it consists of those values of μ that deviate from μ_0 in either of two directions, that is, either $\mu > \mu_0$ or $\mu < \mu_0$.

A test of a hypothesis is a procedure based on the value of a statistic, usually the sample mean $\overline{X}$, for accepting or rejecting H_0. A type I error occurs when one rejects H_0 when

H_0 is true. A type II error occurs when one accepts H_0 when H_0 is false. The performance of the test is measured by how well it controls the type I and type II error probabilities. A good test will reject H_0 with small probability when it is true and will reject H_0 with high probability when it is false. An important tool for analyzing the performance of a test is the power function, which is the probability of rejecting H_0 as a function of the parameter μ.

8.7 Problems

Section 8.2

Problem 8.1 *The Environmental Protection Agency (EPA) standard for the maximum contaminant level (MCL) of alpha emitters in drinking water is 15 pCi/L (Picocuries per liter). Picocuries is a meaure of radioactivity.*
(a) Anecdotal evidence suggests that a town's water supply fails to meet the EPA standard for radioactivity. To test this claim, the water supply will be regularly sampled and the radioactivity levels recorded. Identify the null hypothesis H_0, the null value μ_0, and the alternative hypothesis H_1 in terms of the population mean μ.
(b) If the null hypothesis is rejected, describe the situation for which this is the correct decision.
(c) If the null hypothesis is erroneously rejected, describe, in plain English, the public health consequences.

Problem 8.2 *Refer to the context of Problem 8.1. Suppose the water is regularly sampled for two weeks (so $n = 14$) and the observed radioactivity levels come from a normal distribution with unknown mean μ and known standard deviation $\sigma = 0.4$. The water is declared safe if $\overline{x} < 14.8$; that is, we use the rejection region $\mathcal{C} = \{\overline{x} : \overline{x} < 14.8\}$ for testing the hypothesis $H_0 : \mu \geq 15$ against $H_1 : \mu < 15$.*
(a) What is the significance level of this test?
(b) Compute the probability of a type I error if, in fact, $\mu = 15.2$.
(c) Compute the probability of a type II error if, in fact, $\mu = 14.9$.

Problem 8.3 *The EPA standard for the MCL of arsenic in drinking water is 50 parts per billion (ppb).*
(a) Suppose one tests H_0: $\mu \leq 50$ against $H_1 : \mu > 50$. Describe in this context the consequences of a type I and type II error. Which one has the more serious consequences?
(b) Suppose one tests H_0: $\mu \geq 50$ against $H_1 : \mu < 50$. Describe in this context the consequences of a type I and type II error. Which one has the more serious consequences?
(c) Given that drinking water with arsenic levels exceeding 50 ppb is a public health hazard, which of the hypotheses tests described in parts (a) and (b) would you recommend using? You can assume that both tests have significance levels α.

Problem 8.4 *The federal standard for cadmium dust in the workplace is $200\mu g/m^3$. To monitor the levels of cadmium dust an environmental engineer samples the air at 10 minute intervals, so each hourly average is the sample mean of 6 measurements. Assume that the level of cadmium dust is normally distributed with unknown mean μ and standard deviation $\sigma = 15\mu g/m^3$.*

(a) Suppose management considers the air quality to be unacceptable when the hourly average (denoted by $\bar{x}$) is greater than $200\mu g/m^3$; that is, it suspends production whenever $\bar{x} > 200$. The environmental engineer, therefore, uses the rejection region $C = \{\bar{x} : \bar{x} > 200\}$ for testing the hypothesis

$$H_0: \mu \le 200 \text{ against } H_1 : \mu > 200.$$

What is the consequence of a type I error on production? What is the effect of a type II error on a worker's health?

(b) (Part (a) continued) What is the significance level of this test? Compute the probability of a type I error if, in fact, $\mu = 190$. Hint: Use Equation 8.1. Compute the probability of a type II error at the alternative $\mu = 205$.

(c) Suppose we change the rejection region to the set $C = \{\bar{x} : \bar{x} > 210\}$. Compute the significance level of this test and the probability of a type II error at the alternative $\mu = 205$. Compare the type I and type II errors of this test with those obtained in part (b). Interpret the results.

Problem 8.5 *The Environmental Protection Agency (EPA) requires that when the level of lead (measured in micro-grams per liter ($\mu g/l$)) reaches $15\,\mu g/l$ in at least ten percent of the samples taken, then steps must be taken to reduce the levels of lead. Let $Q(p)$ denote the pth quantile of the distribution (assumed to be continuous) of the amount of lead X in a municipal water supply.*

Explain why the water supply meets the EPA standard when $Q(0.90) \le 15$ and does not meet the standard when $Q(0.90) > 15$.

Hint: Denote the amount of lead in a water sample by X. Show that $Q(0.90) \le 15$ implies that $P(X \le 15) \ge P(X \le Q(0.90)) = 0.90$.

Problem 8.6 *Suppose we use a sample of size $n = 9$ taken from a normal distribution with $\sigma = 4$ and the rejection region $C = \{\bar{x} : \bar{x} < 28\}$ to test the hypothesis*

$$H_0: \mu \ge 30 \text{ against } H_1 : \mu < 30.$$

(a) What is the significance level of this test? Compute the probability of a type II error at the alternative: $\mu = 25$.

(b) Suppose the cutoff value is changed to $c = 27.5$, so the rejection region is the set $C = \{\bar{x} : \bar{x} < 27.5\}$. Compute the significance level of this test; next, compute the probability of a type II error at the alternative $\mu = 25$. Compare the type I and type II errors of this test with the test of part (a). Interpret your results.

Problem 8.7 *The thickness of metal wires used in the manufacture of silicon wafers is assumed to be normally distributed with mean μ (microns) and standard deviation $\sigma = 1.0$. To monitor the production process the mean thickness of 4 wires taken from each wafer is computed. The output is considered unacceptable if the sample mean differs from the target value $\mu_0 = 10$ by more than 1.0 microns. We use the rejection region $C = \{\bar{x} : |\bar{x}-10| > 1.0\}$ to test $H_0: \mu = 10$ against $H_1 : \mu \ne 10$.*

(a) What is the consequence of a type I error on production? What is the effect of a type II error on the quality of the manufactured product?

(b) Compute the significance level α of this test.

(c) Compute the cutoff value c so that the significance level $\alpha = 0.01$.

Problem 8.8 *Suppose we use a sample of size $n = 25$ taken from a normal distribution with $\sigma = 10$ and the rejection region $C = \{\bar{x} : \bar{x} < c\}$ to test the hypothesis $H_0: \mu \ge$*

15 *against* $H_1 : \mu < 15$.

(a) Find the cutoff value c so that the test has significance level $\alpha = 0.1$. *Compute the probability of a type II error at the alternative* $\mu = 12.5$. *Hint: Use Equation 8.9.*

(b) Find the cutoff value c so that the test has significance level $\alpha = 0.05$. *Compute the probability of a type II error at the alternative* $\mu = 12.5$. *Compare the type I and type II errors of this test with the test of part (a). Discuss your conclusions.*

Problem 8.9 *The breaking strengths of steel wires used in elevator cables are assumed to come from a normal distribution with known* $\sigma = 400$. *Before accepting a large shipment of steel wires, an engineer wants to be confident that* μ *is greater than 10,000 pounds.*

(a) Identify the appropriate null and alternative hypotheses. Describe in this context, and in plain English, the consequences of a type I error and type II error.

(b) The breaking strengths of 16 steel wires were measured and the following result was obtained: $\bar{x} = 10,100$ *pounds. Using this data, test your null hypothesis at the 5% significance level and state your conclusions.*

(c) Use the confidence interval approach described to test your null hypothesis. What additional information concerning the value of the mean breaking strength μ *does the confidence interval approach provide the engineer? Discuss.*

Problem 8.10 *The concentration of an impurity in an ore, measured as a percentage, is assumed to be* $N(\mu, 0.05^2)$. *The ore cannot be used without additional processing if* $\mu \geq 0.1\%$. *To determine whether or not the ore is usable, a random sample of size* $n = 9$ *is drawn and the ore is declared unusable if* $\bar{x} \geq 0.08\%$.

(a) What are the null and alternative hypotheses in this case? What is the rejection region?

(b) Compute the significance level of this test.

(c) Compute the type II error if $\mu = 0.09\%$.

(d) Compute the P-value of the test if the observed value of the sample mean $\bar{x} = 0.09$.

Problem 8.11 *Suppose we use a sample of size* $n = 25$ *taken from a normal distribution with* $\sigma = 10$ *to test the hypothesis* $H_0: \mu = 50$ *against* $H_1 : \mu \neq 50$.

(a) Construct a test of H_0 *with significance level* $\alpha = 0.05$ *by constructing a 95% confidence interval for* μ.

(b) Describe the rejection region of the test in part (a).

(c) What is the P-value corresponding to $\bar{x} = 53$? $\bar{x} = 46$?

Problem 8.12 *Same problem as the previous one but this time assume* σ *is unknown and* $s = 10$.

Problem 8.13 *Suppose* $\pi(\mu) = 1 - \Phi(2(10.98 - \mu))$ *is the power function of a test of* $H_0 : \mu \leq 10$ *against* $H_1 : \mu > 10$. *Using this information find:*

(a) The significance level of the test.

(b) The probability of a type II error at the alternative $\mu = 11.5$.

Problem 8.14 *Suppose* $\pi(\mu) = \Phi((26.71 - \mu)/2)$ *is the power function of a test of* $H_0 : \mu \geq 30$ *against* $H_1 : \mu < 30$. *Using this information find:*

(a) The significance level of the test.

(b) The probability of a type II error at the alternative $\mu = 24.5$.

Problem 8.15 *(a) The graph of the power function in Figure 8.2 is an increasing function of* μ. *Show that this is a consequence of the fact that the power function defined by Equation*

8.6

$$\pi(\mu) = 1 - \Phi\left(\frac{\sqrt{n}(c-\mu)}{\sigma}\right)$$

is an increasing function of μ. Hint: Use the fact that $\Phi(x)$ is an increasing function of x.
(b) The graph of the power function in Figure 8.6 is a decreasing function of μ. Show that this is a consequence of the fact that the power function defined by Equation 8.11

$$\pi(\mu) = \Phi\left(\frac{\sqrt{n}(c-\mu)}{\sigma}\right)$$

is a decreasing function of μ.

Problem 8.16 *Twelve measurements of the ionization potential, measured in electron volts, of toluene were made and the following data were recorded:*
10.6, 9.51, 9.83, 10.11, 10.05, 10.31, 9.37, 10.44, 10.09, 10.55, 9.19, 10.09.
The published value of the ionization potential is 9.5. Assume that the data come from a normal distribution. Test the hypothesis that the new experimental data are consistent with the published value.

Problem 8.17 *Seven measurements of arsenic in copper yielded a sample mean of $\bar{x} = 0.17\%$ and sample standard deviation $s = 0.03\%$. Assuming the data come from a normal distribution $N(\mu, \sigma^2)$:*
(a) Describe the test statistic and rejection region that is appropriate for testing (at the level α) $H_0 : \mu \geq 0.2\%$ against $H_1 : \mu < 0.2\%$.
(b) Use the results of part (a) to test H_0 at the 5% significance level.
(c) Express the P-value in terms of the t distribution.
(d) Test the null hypothesis at the 5% level by constructing a suitable one sided confidence interval. What additional information concerning the concentration of arsenic in copper does the confidence interval provide?

Problem 8.18 *In a study of the effectiveness of a weight loss program the net weight loss, in lbs, of ten male subjects were recorded:*
8.4, 15.9, 7.8, 12.8, 5.9, 10.2, 7.5, 15.5, 12.1, 12.5.
The weight loss program is judged ineffective if the mean weight loss is less than or equal to 10 pounds. Test the null hypothesis that the weight loss program is ineffective at the 5% significance level.

Problem 8.19 *Suppose the lifetime of a tire is advertised to be 40000 miles. The lifetimes of 100 tires are recorded and the following results are obtained:*

$$\bar{x} = 39360 \ miles \ and \ s = 3200 \ miles.$$

(a) If we do not assume that the data come from a normal distribution what can you say about the distribution of $\overline{X}$?
(b) Describe the statistic you would choose to test $H_0 : \mu \geq 40000$ against $H_1 : \mu < 40000$.
(c) Compute the P-value of your test and determine whether or not it is significant at the 5% level.

Problem 8.20 *Statistical analysis of experimental data suggest that the compressive strengths of bricks can be approximated by a normal distribution $N(\mu, \sigma^2)$. A new (and cheaper) process for manufacturing these bricks is developed whose output is also assumed to be normally*

distributed. The manufacturer claims that the new process produces bricks whose mean compressive strengths exceed the standard $\mu_0 = 2500$ psi.
(a) Before purchasing these bricks, the customer needs to be convinced that the bricks' compressive strength exceeds the standard; that is, the customer wants to test the hypothesis $H_0 : \mu \leq 2500$ against $H_1 : \mu > 2500$. Describe in this context the consequences of type I and type II errors.
(b) To test the manufacturer's claim that the new process exceeds the standard, the compressive strengths of $n = 9$ bricks, selected at random, are measured with the following results reported: $\bar{x} = 2600$ and sample standard deviation $s = 75$. Describe the rejection region for testing, at the 5% significance level, your null hypothesis of part (a). State your conclusions.
(c) Test this hypothesis using the confidence interval approach.
(d) Compute the P-value of your test; state your conclusions.

Problem 8.21 *Derive the Equation 8.19 for the sample size n by solving the Equations 8.20 and 8.21.*

Problem 8.22 *Give the details of the derivation of Equation 8.16.*

Problem 8.23 *The graph of the power function in Figure 8.8 is symmetric about the null value μ_0, that is,*

$$\pi(\mu_0 + d) = \pi(\mu_0 - d).$$

Show that the power function defined by Equation 8.16 is symmetric about the null value. Hint: Use the fact that $\Phi(-x) = 1 - \Phi(x)$.

Problem 8.24 *In order to test $H_0 : \mu \leq 35$ against $H_1 : \mu > 35$ a researcher records the observed value $\bar{x}$ of the sample mean of a random sample of size $n = 9$ taken from a normal distribution $N(\mu, 49)$. Compute the P-value of the test if the observed value $\bar{x}$ is:*
(a) $\bar{x} = 36$. (b) $\bar{x} = 37$. (c) $\bar{x} = 39$.
(d) For which of the computed P-values in parts (a), (b), and (c), would the null hypothesis be rejected at the 5% level of significance?

Problem 8.25 *In order to test $H_0 : \mu \geq 10$ against $H_1 : \mu < 10$ a researcher computes the observed values $\bar{x}$ and s of the sample mean and sample standard deviation of random sample of size $n = 12$ taken from a normal distribution $N(\mu, \sigma^2)$. Compute the P-value of the test if the observed values $\bar{x}$ and s are:*
(a) $\bar{x} = 9$, $s = 3.0$.
(b) $\bar{x} = 8.5$, $s = 3.0$.
(c) $\bar{x} = 8.5$, $s = 2.5$.
(d) For which of the computed P-values in parts (a), (b), and (c) would the null hypothesis be rejected at the 5% level of significance?

Problem 8.26 *Compute and sketch the graph of the power function of the test with rejection region $\mathcal{C} = \{\bar{x} : \bar{x} > 210\}$ in part (c) of Problem 8.4.*

Problem 8.27 *Compute and sketch the graph of the power function of the test with rejection region $\mathcal{C} = \{\bar{x} : \bar{x} < 28\}$ in part (a) of Problem 8.6.*

Problem 8.28 *Compute and sketch the graph of the power function of the test with rejection region $\mathcal{C} = \{\bar{x} : |\bar{x} - 10| > 1\}$ of Problem 8.7.*

Problem 8.29 *Suppose you want to test* H_0: $\mu \le 50$ *against* $H_1 : \mu > 50$ *using a random sample of size n taken from a normal distribution $N(\mu, 81)$. Find the minimum sample size n and the cutoff value c if the probability of a type I error is to be $\alpha = 0.05$ and the probability of a type II error at the alternative $\mu = 55$ is to be $\beta(55) = 0.20$.*

Problem 8.30 *Suppose you want to test*

$$H_0 \colon \mu \ge 30 \ against \ H_1 : \mu < 30$$

using a random sample of size n taken from a normal distribution $N(\mu, 16)$. Find the minimum sample size n and the cutoff value c if the probability of a type I error is to be $\alpha = 0.05$ and the probability of a type II error at the alternative $\mu = 28$ is to be $\beta(28) = 0.20$.

Section 8.3

Problem 8.31 *The data set of Problem 7.36 comes from an experiment comparing the electrical resistances (measured in ohms) of two types of wire.*
(a) Test the null hypothesis that $\mu_1 = \mu_2$ against the alternative that $\mu_1 \ne \mu_2$ by constructing an appropriate two sided 95% confidence interval.
(b) What is the P-value of your test?

Problem 8.32 *Refer to the data set of Problem 7.37.*
(a) By constructing a suitable two sided 95% confidence interval, test the hypothesis that there is no difference between the mean skein strengths of the two types of yarn.
(b) What is the P-value of your test?
(c) Use the Wilcoxon rank sum statistic to test $H_0 : \mu_1 = \mu_2$ against $H_1 : \mu_1 \ne \mu_2$. Use $\alpha = 0.05$. What is the P-value of the test?

Problem 8.33 *Refer to Example 1.8 and Table 1.7. Denote the mean N-1-THB-ADE levels of the control and exposed group of workers by μ_1, μ_2, respectively. Use the Wilcoxon rank sum statistic to test $H_0 : \mu_1 \ge \mu_2$ against $H_1 : \mu_1 < \mu_2$. Use $\alpha = 0.05$. What is the P-value of the test?*

Problem 8.34 *To determine the difference, if any, between two brands of radial tires, 12 tires of each brand were tested and the following mean lifetimes and sample standard deviations for the two brands were obtained:*

$$\bar{x}_1 = 38500 \ and \ s_1 = 3100;$$
$$\bar{x}_2 = 41000 \ and \ s_2 = 3500.$$

Assuming that the tire lifetimes are normally distributed with identical variances, test the hypothesis that there is no difference between the two lifetimes. Compute the P-value of your test.

Problem 8.35 *In a code size comparison study, five determinations of the number of bytes required to code five workloads on two different processors were recorded; the data are displayed in the following table.*

System 1	System 2
101	130
144	180
211	141
288	374
72	302

(*Source: R. Jain (1991),* The Art of Computer Systems Performance Analysis, *John Wiley & Sons. Used with permission.*)

Do the data indicate a significant difference between the two systems? Use $\alpha = 0.05$. *(Hint: First decide whether the data come from two independent samples or from paired samples.)*

Problem 8.36 *Refer to the data in Problem 1.62.*
(a) Do the data indicate a significant difference in the mean concentration of thiol between the normal group and the rheumatoid arthritis group? Use the t test with significance level $\alpha = 0.05$. *Summarize your analysis by giving the P-value.*
(b) Repeat part (a) but use the Wilcoxon rank sum test.

Problem 8.37 *In a comparison study of a standard versus a premium blend of gasoline, the miles per gallon (mpg) for each blend was recorded for five compact automobiles.*

Car	Standard (mpg)	Premium (mpg)
1	26.9	27.7
2	22.5	22.2
3	24.5	25.3
4	26.0	26.8
5	28.5	29.7

Test, at the 5% significance level, the claim that the premium blend delivers more miles per gallon than the standard blend. Your null hypothesis is that "the premium blend is no more fuel efficient than the standard blend."

Problem 8.38 *Refer to Charles Darwin's data in Problem 7.41. Do the data point to a significant difference between the heights of the cross-fertilized and self-fertilized seedlings? Summarize your results by giving the P-value.*

Problem 8.39 *In a weight loss experiment* 10 *men followed diet A and* 7 *men followed diet B. The weight losses after one month are displayed in the following table.*

Diet A	Diet B	Diet A	Diet B
15.3	3.4	10	5.2
2.1	10.9	8.3	2.5
8.8	2.8	9.4	
5.1	7.8	12.5	
8.3	0.9	11.1	

(a) Use the two sample t test to determine if there is any significant difference between the two diets. Assume the data are normally distributed with a common variance; use $\alpha = 0.05$. *In detail: Describe the null and alternative hypotheses, and determine which diet is more effective in reducing weight.*
(b) Use the Wilcoxon rank sum test to test $H_0 : \mu_1 = \mu_2$ *against* $H_1 : \mu_1 > \mu_2$. *Use* $\alpha = 0.05$. *What is the P-value of the test? Remember that diet B is now population 1 because its sample size is the smaller one.*

Problem 8.40 *Refer to the data in Problem 1.40.*
(a) Do the data indicate a significant difference in the DBH activity between the non-psychotic and psychotic groups? Use the two sample t test and summarize your results

by stating the P-value.
(b) Repeat part (a) but use the Wilcoxon rank sum test.

Problem 8.41 *Refer to the data in Problem 1.64. Use the two sample t test to determine the effect, if any, of the word processor program on the length of time it takes to type a text file. Assume the data are normally distributed with a common variance; use $\alpha = 0.05$. In detail:*
(a) Compare the differences between word processing programs A and B. Describe the null and alternative hypotheses, and state your conclusions.
(b) Repeat part (a) but use the Wilcoxon rank sum test.

Problem 8.42 *In a study comparing the breaking strengths (measured in psi) of two types of fiber, random samples of sizes $n_1 = n_2 = 9$ of both types of fiber were tested, and the following summary statistics were obtained: $\bar{x}_1 = 40.1$, $s_1^2 = 2.5$; $\bar{x}_2 = 41.5$, $s_2^2 = 2.9$. Assume both samples come from normal distributions. Although fiber 2 is cheaper to manufacture it will not be used unless its breaking strength is found that to be greater than that of fiber 1. Thus, we want to test*

$$H_0 : \Delta = \mu_1 - \mu_2 = 0 \text{ against } H_1 : \Delta = \mu_1 - \mu_2 < 0.$$

Observe that the alternative hypothesis as formulated here asserts that the breaking strength of fiber 2 is greater than fiber 1. Controlling the type I error here protects you from using an inferior, although cheaper, product.
(a) Use the given data to test the null hypothesis at the levels: $\alpha = 0.05$; $\alpha = 0.01$.
(b) Test the null hypothesis by constructing an appropriate one sided $100(1-\alpha)\%$ confidence interval, where $\alpha = 0.01$.

Problem 8.43 *Refer to Lord Rayleigh's data that led to his discovery of argon (cf. Table 1.4 and Example 7.10). Are the data consistent with with Lord Rayleigh's claim that nitrogen gas obtained from air is heavier than nitrogen gas obtained from chemical decomposition? Use the two sample t-test.*

Problem 8.44 *Refer to the data in Problem 7.41. Let σ_1^2 and σ_2^2 denote the variances of the heights of the cross-fertilized and self-fertilized seedlings, respectively. Test for the equality of the variances at the 1% level; that is, test the following hypothesis:*

$$H_0 : \frac{\sigma_1^2}{\sigma_2^2} = 1 \text{ against } \frac{\sigma_1^2}{\sigma_2^2} \neq 1.$$

Problem 8.45 *In a comparison study of two independent normal populations the sample variances were $s_1^2 = 1.1$ and $s_2^2 = 3.4$, and the corresponding sample sizes were $n_1 = 6$, $n_2 = 10$. Is one justified in using the two sample t test? Justify your conclusions.*

Section 8.4: Problems (8.46-8.55) illustrate the importance of assessing the plausibility of the hypothesis of normality. You will need to use MINITAB, JMP, SAS, or the equivalent to apply the Shapiro–Wilk test for normality.

Problem 8.46 *(a) Refer to the data set of Problem 1.7. Construct a Q–Q plot to assess, informally, the plausibility of the normality hypothesis. Comment on whether or not the points $(z_j, x_{(j)})$ appear to lie along a straight line.*
(b) Test the data for normality by applying the Shapiro–Wilk test. State the alternatives, decision rule, and conclusion.

Problem 8.47 *Refer to the lead absorption data set of Problem 1.18. Construct a Q–Q plot to assess, informally, the plausibility of the normality hypothesis for the data in the first column (the exposed group). Comment on whether or not the points $(z_j, x_{(j)})$ appear to lie along a straight line.*
(b) Test the data for normality by applying the Shapiro–Wilk test. State the alternatives, decision rule,and conclusion.

Problem 8.48 *Refer to the lead absorption data set of Problem 1.18. Construct a Q–Q plot to assess, informally, the plausibility of the normality hypothesis for the data in the second column (control). Comment on whether or not the points $(z_j, x_{(j)})$ appear to lie along a line.*
(b) Test the data for normality by applying the Shapiro–Wilk test. State the alternatives, decision rule, and conclusion.

Problem 8.49 *Refer to the lead absorption data set of Problem 1.18. The paired sample t-test assumes that the data in the difference column are normally distributed. Test the third column (difference) for normality by applying the Shapiro–Wilk test. State the alternatives, decision rule, and conclusion.*

Problem 8.50 *(a) Refer to the radiation data of Table 1.3. The histogram (Figure 1.10) suggests that these data are not normally distributed. Construct a Q–Q plot of the data and use it as an heuristic tool to assess the plausibility of the hypothesis of normality.*
(b) Test the data for normality by applying the Shapiro–Wilk test. State the alternatives, decision rule, and conclusion. What is the P-value of the test?

Problem 8.51 *Refer to the cadmium levels data of Problem 1.19. Test the data for normality by (i) constructing a Q–Q plot of the data and (ii) applying the Shapiro–Wilk test. What is the P-value of the test?*

Problem 8.52 *Refer to the capacitor data set of Problem 1.20. Test the data for normality by (i) constructing a Q–Q plot of the data and (ii) applying the Shapiro–Wilk test. What is the P-value of the test?*

Problem 8.53 *The efficient market hypothesis, discussed in Section 5.5 implies that stock market returns are normally distributed. Refer to Example 1.17 and Figure 1.11 (histogram of the S&P returns (1999)), which suggest that the returns are normally distributed. Test the data for normality by applying the Shapiro–Wilk test. State the alternatives, decision rule, and conclusion.*

Problem 8.54 *Refer to the weekly 2002 return data for GE listed in Table 1.20. Use a statistical software package to:*
(a) Draw the graph of the Q–Q plot (normal probability plot) of the returns.
(b) Test for normality by computing the Shapiro–Wilk test statistic. State the alternatives, decision rule, and conclusion.

Problem 8.55 *Refer to the weekly 2002 return data for BEARX listed in Table 1.20. Use a statistical software package to:*
(a) Draw the graph of the Q–Q plot (normal probability plot) of the returns.
(b) Test for normality by computing the Shapiro–Wilk test statistic. State the alternatives, decision rule, and conclusion.

Problem 8.56 *Refer to the weekly 2002 return data for PTTRX listed in Table 1.20. Use a statistical software package to:*
(a) Draw the graph of the Q–Q plot (normal probability plot) of the returns.
(b) Test for normality by computing the Shapiro–Wilk test statistic. State the alternatives, decision rule, and conclusion.

Section 8.5

Problem 8.57 *Let X be the sample total from a random sample of size $n = 10$ from a Bernoulli distribution with parameter p. To test $H_0 : p \leq 0.25$ against $H_1 : p > 0.25$, we choose the rejection region $\mathcal{C} = \{x : x > c\}$.*
(a) What is the significance level of this test for $c = 4$? Determine the type II error if $p = 0.30$, $p = 0.50$.
(b) What is the significance level of this test for $c = 5$? Determine the type II error if $p = 0.30$, $p = 0.50$.

Problem 8.58 *Let X be the sample total from a random sample of size $n = 20$ from a Bernoulli distribution with parameter p. To test $H_0 : p \leq 0.40$ against $H_1 : p > 0.40$, we choose the rejection region $\mathcal{C} = \{x : x > c\}$.*
(a) Find the value of c such that the type I error probability $\alpha \leq 0.05$. What is the type II error if, in fact, $p = 0.7$?
(b) Find the value of c such that the type I error probability $\alpha \leq 0.1$. What is the type II error if, in fact, $p = 0.7$?

Problem 8.59 *Let X be the sample total from a random sample of size $n = 15$ taken from a Bernoulli distribution with parameter p. To test $H_0 : p = 0.90$ against $H_1 : p < 0.90$, we choose the rejection region $\mathcal{C} = \{x : x < c\}$, where the value of c is to be determined. How should we choose c so that the type I error probability $\alpha \leq 0.05$? Determine the type II error if, in fact, $p = 0.60$. Hint: To compute $B(x; n, p)$, $0.5 < p < 1$ we use Equation 3.41:*

$$B(x; n, p) = 1 - B(n - 1 - x; n, 1 - p).$$

Problem 8.60 *An urn contains two red balls, two white balls, and a fifth ball that is either red or white. Let p denote the probability of drawing a red ball. The null hypothesis is that the fifth ball is red.*
(a) Describe the parameter space Θ.
(b) To test this hypothesis we draw 10 balls at random, one at a time (with replacement), and let X equal the number of red balls drawn. Using a rejection region of the form $\mathcal{C} = \{X < c\}$, compute the power function $\pi(p)$.
(c) Use the result of part (b) to compute the type I and type II errors when $c = 3$.

Problem 8.61 *Suppose the natural recovery rate from a disease is $p_0 = 0.4$. A new drug is developed that, it is claimed, significantly improves the recovery rate. Consequently, the null hypothesis and its alternative are given by*

$$H_0 : p = 0.4 \ against \ H_1 : p > 0.4.$$

Consider the following test: A random sample of 15 patients are given the drug, and the number of recoveries is denoted by X. The null hypothesis will be rejected if $X > 8$, i.e.,

$$accept \ H_0 \ if \ X \leq 8 \ reject \ H_0 \ if \ X > 8.$$

(a) Derive the formula for the power function $\pi(p)$ of this test.
(b) Compute the significance level of the test.
(c) Compute the type II error β if, in fact, the drug boosts the recovery rate to 0.6.

Problem 8.62 *A garden supply company claims that its flower bulbs have an 80% germination rate. To challenge this claim, a consumer protection agency decides to test the following hypothesis:*

$$H_0 : p = 0.80 \ against \ H_1 : p < 0.80.$$

The following data were collected: $n = 100$ flower bulbs were planted, and 77 of them germinated. Compute the P-value of your test statistic and state whether or not it is significant at the 5% level.

Problem 8.63 *Before proposition A was put to the voters for approval, a random sample of 1300 voters produced the following data: 625 in favor of proposition A and 675 opposed. Test the null hypothesis that a majority of the voters will approve proposition A. Use $\alpha = 0.05$. What is the alternative hypothesis? What is the P-value of your test statistic?*

Problem 8.64 *Suppose that we have to test $H_0 : p = 0.2$ against $H_1 : p > 0.2$ at the level $\alpha = 0.05$. What is the minimum sample size needed so that the type II error at the alternative $p = 0.3$ is 0.1?*

Problem 8.65 *The natural recovery rate from a disease is 75%. A new drug is developed that, according to its manufacturer, increases the recovery rate to 90%. Suppose we test this hypothesis at the level $\alpha = 0.01$. What is the minimal sample size needed to detect this improvement with probability of at least 0.90?*

Problem 8.66 *In a comparison study of two machines for manufacturing memory chips the following data were collected: Of 500 chips produced by machine 1, 20 were defective; of 600 chips produced by machine 2, 40 were defective. Can you conclude from these results that there is no significant difference (at the level $\alpha = 0.05$) between the proportions of defective items produced by the respective machines? What is the P-value of your test statistic?*

Problem 8.67 *A random sample of 250 items from lot A contains 10 defective items and a random sample of 300 items from lot B is found to contain 18 defective items. What conclusions can you draw concerning the quality of the two lots?*

Problem 8.68 *The number of defective items out of a total production of 1125 is 28. The manufacturing process is then modified and the number of defective items out of a total production of 1250 is 22. Can you reasonably conclude that the modifications reduced the proportion of defective items?*

Problem 8.69 *Patients suffering cardiac arrest were divided into two groups: Group 1 consisted of n_1 patients who were given cardiac pulmonary resuscitation (CPR) by trained civilians until the arrival of an ambulance crew. Group 2 patients consisted of n_2 patients whose CPR was delayed until the ambulance crew arrived. Let X_i, $i = 1, 2$ denote the number of patients in group i that survived, and let $p_i, i = 1, 2$ denote the corresponding probabilities of survival.*
(a) Describe the distributions of X_1 and X_2; name them and identify the parameters.
(b) The null hypothesis is that CPR given to group 1 patients by trained civilians is not

effective; the alternative hypothesis is that it is. State the null and the alternative hypotheses in terms of the relevant population parameters.

(c) In a comparison study of the survival rate among patients suffering cardiac arrest the following data were collected. Group 1 patients consisted of 75 patients of whom 27 survived. Group 2 patients consisted of 556 patients of whom 43 survived. Using this data test the null hypothesis of part (b) by computing a 95% lower confidence interval for $p_1 - p_2$, the difference in survival rates between the group 1 and group 2 patients.

Problem 8.70 *Shingles is a virus infection of the nerves that causes a painful, blistering rash. It is the same virus that causes chickenpox. A new vaccine for shingles, Zostavax, was approved in May 2006 by the Food and Drug Administration (FDA). The approval was based on a large scale clinical trial with 38,500 people 60 or older, half of whom (group 1) received the vaccine and the other half (group 2) received a placebo. The vaccinated group had 315 cases of shingles, the placebo group had 642 cases. Let X_1, X_2 denote the number of people in the vaccinated group and unvaccinated group, respectively, that contracted shingles, and denote the corresponding shingles prevalence rate for the two groups by p_1, p_2.*

(a) Describe the distributions of X_1 and X_2; name them and identify the parameters.

(b) The null hypothesis is that the vaccine given to group 1 is not effective; the alternative hypothesis is that it is. State the null and the alternative hypotheses in terms of the relevant population parameters.

(c) Using the data from the clinical trial, test the null hypothesis of part (b)—that the vaccine given to group 1 is not effective—by computing a 95% upper confidence interval for $p_1 - p_2$, the difference in prevalence rates rates between the group 1 and group 2 patients.

8.8   To Probe Further

There is no "one size fits all" approach to statistical inference. The confidence interval approach to hypothesis testing is favored by a growing number of practitioners of statistics including G.E.P. Box, W.G. Hunter and J.S. Hunter (1978), *Statistics for Experimenters*, John Wiley & Sons, who write, "Significance testing in general has been a greatly overworked procedure, and in many cases where significance statements have been made it would have been better to provide an interval within which the value of the parameter would be expected to lie." The Wilcoxon rank sum test is an example of a *distribution free* or *non-parametric* statistical method. A useful introduction to and survey of these techniques is given in Randles and Wolfe (1979), *Introduction to The Theory of Non Parametric Statistics*, John Wiley & Sons. We omitted a formula for the (approximate) power function for the tests listed in Table 8.12, since it requires more technical details than are suitable for a first course in statistics.

Chapter 9

Statistical Analysis of Categorical Data

In God we trust; all others bring data.

Statistician's motto

9.1 Orientation

One of the most famous categorical data sets in biology is the plant breeding experiment of the Austrian monk and botanist G. Mendel (1822-1884) who classified peas according to their shape *(round (r) or wrinkled (w))* and color *(yellow (y) or green (g))*. Each seed was classified into one of four categories: $(ry) = (round, yellow)$, $(rg) = (round, green)$, $(wy) = (wrinkled, yellow)$, and $(wg) = (wrinkled, green)$. According to Mendel's theory the frequency counts of seeds of each type produced from this experiment occur in the following ratios: $9 : 3 : 3 : 1$. Thus, the ratio of the number of (ry) peas to (rg) peas should be equal to $9 : 3$, and similarly for the other ratios. Mendel's experiment raises the following fundamental scientific question: Are Mendel's data, displayed in Table 9.1, consistent with the predictions of his scientific model? To answer these and more complex questions arising in the study of categorical data, statisticians use the chi-square test.

Organization of Chapter

1. Section 9.2: Chi-Square Tests

2. Section 9.3: Contingency Tables

3. Section 9.4: Chapter Summary

9.2 Chi-Square Tests

The chi-square test is used to test hypotheses about the parameters of one or more multinomial distributions. Recall that a multinomial experiment is the mathematical model of n repetitions of an experiment with k mutually exclusive and exhaustive outcomes, denoted $C_1, \ldots, C_k$, with probabilities $p_i = P(C_i)$, where $\sum_i p_i = 1$. (The multinomial distribution is discussed in Section 5.3.) The outcomes C_i are called *categories* or *cells;* their probabilities $p_i = P(C_i)$ are called *cell probabilities.* In the simplest case the cell probabilities are completely specified, so the null hypothesis is given by:

$$H_0 : P(C_i) = p_i, \ i = 1, \ldots, k. \tag{9.1}$$

Example 9.1

One of the most famous multinomial experiments in the history of science was performed by G. Mendel, who classified $n = 556$ peas according to two traits: shape (round or wrinkled) and color (yellow or green). Each seed was put into one of $k = 4$ categories: $C_1 = ry$, $C_2 = rg$, $C_3 = wy$, $C_4 = wg$, where r, w, g, and y denote round, wrinkled, green, and yellow, respectively. The results of Mendel's breeding experiment are displayed in Table 9.1.

Table 9.1 *Mendel's data*

Seeds	Observed frequency
ry	315
wy	101
rg	108
wg	32
	$n = 556$ (total)

Mendel's theory predicted that the frequency counts of the seed types produced from his experiment should occur in the ratios: 9 : 3 : 3 : 1. Thus, the ratio of the number of ry peas to rg peas should be 9 : 3, and similarly for the other ratios. We can rewrite these proportions in terms of probabilities as follows:

$$p_1 = P(ry) = \frac{9}{16}, \ p_2 = P(rg) = \frac{3}{16},$$
$$p_3 = P(wy) = \frac{3}{16}, \ p_4 = P(wg) = \frac{1}{16}.$$

The null hypothesis corresponding to Mendel's theory is therefore

$$H_0 : p_1 = \frac{9}{16}, p_2 = \frac{3}{16}, p_3 = \frac{3}{16}, p_4 = \frac{1}{16}.$$

Do Mendel's theoretical predictions provide a good fit to the data?

We test the null hypothesis (9.1) by carrying out n independent repetitions of the experiment and then recording the observed frequencies $Y_1, \ldots, Y_k$ of each category. In Section 5.3 we showed that the random variables $Y_1, \ldots, Y_k$ have a *multinomial distribution*. Consequently, the cell frequency Y_i (also called the observed frequency and denoted O_i) has a binomial distribution with parameters n, p_i; therefore, the *expected frequency*, denoted E_i, of the ith category is given by $E(Y_i) = np_i$.

To test the null hypothesis that the observed data come from a multinomial distribution with given cell probabilities $P(C_i) = p_i$, $i = 1, \ldots, k$ we use the following important theorem (we omit the proof) due to the statistician K. Pearson (1857-1936).

Theorem 9.1 *Let $Y_1, \ldots, Y_k$ have a multinomial distribution with parameters $n, p_1, \ldots, p_k$. Then the distribution of the random variable χ^2, defined by*

$$\chi^2 = \sum_i \frac{(Y_i - np_i)^2}{np_i} = \sum_i \frac{(O_i - E_i)^2}{E_i} \tag{9.2}$$

has an approximate chi-square distribution with $k - 1$ degrees of freedom. The accuracy of the approximation improves as n increases.

Looking at Equation 9.2 we see that a small value of χ^2 indicates that the differences between the observed and expected frequencies are small, indicating a close agreement of the data with the model, while a large value of χ^2 indicates a poor fit. This leads directly to the χ^2 *goodness of fit* test.

Pearson's χ^2 Goodness of Fit Test (Large Sample)

1. Null hypothesis: $H_0 : P(C_i) = p_i$, $i = 1, \ldots, k$. Sample size: n.

2. Record the observed frequencies Y_i and compute the expected cell frequencies np_i. Then compute the χ^2 test statistic:

$$\chi^2 = \sum_{1 \leq i \leq k} \frac{(Y_i - np_i)^2}{np_i}.$$

3. Reject H_0 if $\chi^2 > \chi^2_{k-1}(\alpha)$, where k is the number of cells and α is the significance level of the test.

Solution

We test Mendel's predictions at the level $\alpha = 0.01$ by using the χ^2 test. Looking at Table 9.1, we see that the observed frequencies are $Y_1 = 315$, $Y_2 = 101$, $Y_3 = 108$, $Y_4 = 32$, and the sample size equals the sum of the cell frequencies: $n = 315 + 101 + 108 + 32 = 556$. Table 9.2 displays all the data needed to compute the χ^2 test statistic. The second, third, and fourth columns list the observed frequencies, cell probabilities, and expected cell frequencies, respectively.

Table 9.2 *Cell probabilities and expected cell frequencies for Mendel's data*

Seeds	Observed frequency	Cell probability	Expected frequency
ry	315	9/16	$556 \times 9/16 = 312.75$
wy	101	3/16	$556 \times 3/16 = 104.25$
rg	108	3/16	$556 \times 3/16 = 104.25$
wg	32	1/16	$556 \times 1/16 = 34.75$

Computation of the χ^2 statistic:

$$\chi^2 = \frac{(315 - 312.75)^2}{312.75} + \frac{(101 - 104.25)^2}{104.25} + \frac{(108 - 104.25)^2}{104.25} + \frac{(32 - 34.75)^2}{34.75}$$
$$= 0.47$$

Since $k - 1 = 3$ it follows that $\chi^2 \approx \chi^2_3$. From the chi-square tables (Table A.5) we see that $\chi^2 = 0.47 < \chi^2_3(0.05) = 7.815$; consequently, we do not reject the null hypothesis.

Some Remarks on the Chi-Square Approximation

The accuracy of the chi-square approximation depends on the values of the cell probabilities; the following conservative criteria are recommended:

1. To use this approximation, it is recommended that the expected cell frequency $np_i \geq 5$ for each i. This condition holds for Mendel's genetics data in Table 9.2.

2. Cells with expected frequencies less than 5 should be combined to meet this rule.

3. k is the number of cells *after* cells with small expected frequencies have been combined.

9.2.1   Chi-Square Tests When the Cell Probabilities Are Not Completely Specified

In the previous examples the cell probabilities were completely specified. We now consider the case when the cell probabilities depend on one or more parameters that are estimated from the data.

Example 9.2

A computer engineer conjectures that the number of packets X per unit of time arriving at a node in a computer network has a Poisson distribution with parameter λ, so $P(X = i) = e^{-\lambda}\lambda^i/i!$. To validate this model the number of arriving packets for each of 150 time intervals is recorded; the results are displayed in columns one and two of Table 9.3. Are these data consistent with the model assumption that X has a Poisson distribution?

Solution   Let C_i denote the event that i packets arrived in a time interval, and let f_i denote the number of times the event C_i occurred. Looking at Table 9.3, we see that C_0 occurred twice, so $f_0 = 2$; C_1 occurred 10 times, so $f_1 = 10$; and so forth. The total number of packet arrivals equals $600 = \sum_i i f_i$. The maximum likelihood estimator of the arrival rate λ equals the total number of arrivals divided by the total number of intervals, so $\hat{\lambda} = 600/150 = 4$. The estimated cell probabilities are given by

$$\hat{P}(X = i) = e^{-\hat{\lambda}}\hat{\lambda}^i/i! = e^{-4}4^i/i!, 0 \le i \le 9; \ \hat{P}(X \ge 10) = 0.0081,$$

and are displayed in the rightmost column of Table 9.3. Observe that the last entry in this column is $\hat{P}(X \ge 10) = 0.0081$ and not $\hat{P}(X = 10) = 0.00529$; this must be so, for otherwise the categories would not be exhaustive. The expected number of arrivals for each cell is computed by multiplying the cell probability by the sample size $n = 150$; these results are displayed in column three.

Table 9.3   *The number of packet arrivals to a node in a computer network*

No. of arrivals	Observed no. of time intervals with i arrivals	Expected no. of intervals with i arrivals	$i \times f_i$	Cell probabilities
i	f_i	np_i	$i \times f_i$	p_i
0	2	2.74735	0	0.01832
1	10	10.9894	10	0.0733
2	25	21.9788	50	0.1465
3	25	29.3050	75	0.1954
4	39	29.3050	156	0.1954
5	15	23.4440	75	0.1563
6	14	15.6293	84	0.1042
7	13	8.9311	91	0.0595
8	5	4.4655	40	0.02977
9	1	1.9847	9	0.01323
≥ 10	1	1.2150	10	0.0081
Totals	150		600	

Observe that the expected cell frequencies $np_0 = 2.75$, $np_8 = 4.47$, $np_9 = 1.98$, $np_{10} = 1.215$ are each less than 5; we therefore combine the first two cells and then the last three cells so that each expected cell frequency satisfies the condition $np_i \ge 5$.

The expected frequencies and cell probabilities corresponding to the grouped data are displayed in Table 9.4.

Table 9.4 *Table 9.3 after grouping of the data*

No. of arrivals	Observed no. of time intervals with i arrivals	Expected no. of intervals with i arrivals	Cell probabilities
≤ 1	12	13.7367	0.0916
2	25	21.9788	0.1465
3	25	29.305	0.1954
4	39	29.305	0.1954
5	15	23.444	0.1563
6	14	15.6293	0.1042
7	13	8.93105	0.0595
≥ 8	7	7.665	0.0511

Computing the χ^2 statistic and the degrees of freedom (DF)

In this example the cell probabilities and the expected cell frequencies are functions of the unknown parameter λ. Nevertheless, the χ^2 test is still applicable, *provided the unknown parameter is replaced by its maximum likelihood estimate.* More precisely, we have the following theorem:

Theorem 9.2 *When the cell probabilities depend on one or more unknown parameters, the χ^2 test is still applicable provided the unknown parameters are replaced by their maximum likelihood estimates and provided the degrees of freedom are reduced by the number of parameters estimated, i.e.,*

$$DF \text{ of } \chi^2 = k - 1 - (\text{no. of parameters estimated}). \tag{9.3}$$

We compute the chi-square statistic exactly as before except that we use the grouped data in Table 9.4. This yields the result $\chi^2 = 9.597$. The degrees of freedom, however, must be reduced by the number of parameters estimated from the data (Theorem 9.2). In the present instance, DF of $\chi^2 = k - 1 - (\text{no. of parameters estimated}) = 8 - 1 - 1 = 6$, so, $\chi^2 \approx \chi_6^2$. Because $\chi^2 = 9.597 < 12.592 = \chi_6^2(0.05)$ we conclude that the Poisson distribution provides a reasonably good fit to the data.

9.3 Contingency Tables

The concept of a *contingency table* (also called a *two way table*) is most easily understood in the context of a specific example.

Example 9.3 *A 2×3 contingency table*

The frequency data in Table 9.5 were obtained in a study of the causes of failure of vacuum tubes (VT) used in artillery shells during World War II. Each failure was classified according to two criteria or *attributes*: (1) *position* of the VT in the shell ($A_1 = $ top and $A_2 = $ bottom) and (2) *type of failure* (B_1, B_2, B_3). The position of a VT in the shell is an example of a *categorical variable,* with *two categories,* that is, the outcome of the experiment is a non-numerical value, *top* or *bottom,* as opposed to the numerical variables we have been studying heretofore. The type of failure is also a categorical variable with three categories corresponding to the three different failure types. The two classification criteria when applied simultaneously produce a partition of the sample space into the $2 \times 3 = 6$ cells denoted by $A_i \cap B_j (i = 1, 2; j = 1, 2, 3)$. For the 115 shells that had a type B_1 failure 75 VTs were in the top position and 40 were in the bottom position, so the cell frequency of $A_1 \cap B_1$ is 75; similarly, the cell frequency of $A_2 \cap B_1$ is 40 and so on. Summing the cell frequencies across the rows gives the *marginal row frequencies* and summing the cell frequencies down the columns gives the *marginal column frequencies*. For instance, the number of VTs that failed in the top position (category A_1) is 100, and there were 115 VTs classified as type B_1 failures. Table 9.5 is an example of a 2×3 *contingency table*.

Table 9.5 *Frequency data on the causes of failure of $n = 180$ vacuum tubes*

Position in Shell	Type of Failure B_1	B_2	B_3	Total
A_1 = Top	75	10	15	100
A_2 = Bottom	40	30	10	80
Total	115	40	25	180

(Source: M. G. Natrella (1963), *Experimental Statistics*, National Bureau of Standards Handbook 91.)

The General $r \times c$ Contingency Table

Table 9.5 consists of 2 rows and 3 columns because the categorical variables A (= position) and B (= failure type) have 2 and 3 categories, respectively. An $r \times c$ contingency table is a straightforward generalization; it consists of r rows and c columns that correspond to the r categories of the variable A and the c categories of the variable B. We denote the r categories of A by $A_1, \ldots, A_r$ and the c categories of B by $B_1, \ldots, B_c$. We let y_{ij} denote the observed frequency of the observations falling in the cell $A_i \cap B_j$. The resulting frequency counts are displayed in the $r \times c$ *contingency table* shown in Table 9.6. The marginal row and column frequencies are denoted by *dot subscript notation*, where a dot in place of the subscript i, say, indicates summation with respect to the i.

$$\text{Dot subscript notation: } y_{i.} = \sum_{1 \le j \le c} y_{ij}; y_{.j} = \sum_{1 \le i \le r} y_{ij}.$$

Table 9.6 *An $r \times c$ contingency table*

	B_1	$\ldots$	B_c	Total
A_1	y_{11}	y_{1j}	y_{1c}	$y_{1.}$
$\vdots$				$y_{i.}$
A_r	y_{r1}	$\ldots$	y_{rc}	$y_{r.}$
Total	$y_{.1}$	$y_{.j}$	$y_{.c}$	n

χ^2 test of independence

We use the data in a contingency table to test the hypothesis that the categorical variables A and B are independent. Looking at the data in Table 9.5, for example, we can ask whether the type of failure depends on the position of the vacuum tube in the shell. The null hypothesis is that the two variables are independent, that is,

$$H_0 : P(A_i \cap B_j) = P(A_i)P(B_j) \text{ for } i = 1, \ldots, r; j = 1, \ldots, c. \tag{9.4}$$

Estimating the parameters $p_{i.}$ and $p_{.j}$

The marginal probabilities $P(A_i)$ and $P(B_j)$ are obtained from the joint probabilities $p_{ij} = P(A_i \cap B_j)$ by computing the row and column sums in the usual way. Thus,

$$P(A_i) = p_{i.} = \sum_{1 \le j \le c} p_{ij}; \tag{9.5}$$

$$P(B_j) = p_{.j} = \sum_{1 \le i \le r} p_{ij}. \tag{9.6}$$

Using this notation, the null hypothesis of independence can be written as

$$H_0 : p_{ij} = p_{i.} p_{.j} \text{ for } i = 1, \ldots, r; \ j = 1, \ldots, c. \tag{9.7}$$

We estimate the probabilities $p_{i.} = P(A_i)$ and $p_{.j} = P(B_j)$ by computing their observed relative frequencies. These estimates are given by

$$\hat{p}_{i.} = \frac{y_{i.}}{n} \text{ and } \hat{p}_{.j} = \frac{y_{.j}}{n}. \tag{9.8}$$

Consequently, the estimated expected frequency count $\hat{e}_{ij}$ for the (i, j)th cell, when H_0 is true, is given by

$$\hat{e}_{ij} = n \hat{p}_{i.} \hat{p}_{.j} = \frac{y_{i.} y_{.j}}{n}.$$

We note that the marginal probabilities satisfy the equations

$$\sum_{1 \le i \le r} p_{i.} = 1 \text{ and } \sum_{1 \le j \le c} p_{.j} = 1. \tag{9.9}$$

This means that we have to estimate $r - 1 + c - 1 = r + c - 2$ parameters from the data. The number of degrees of freedom of χ^2 is given by (see Theorem 9.2 and Equation 9.3)

$$\text{DF of } \chi^2 = rc - 1 - (r + c - 2) = (r - 1)(c - 1). \tag{9.10}$$

Consequently, χ^2 has an approximate chi-square distribution with $(r - 1)(c - 1)$ degrees of freedom.

χ^2 Test of Independence

1. Null hypothesis: $H_0 : p_{ij} = p_{i.} p_{.j}$, $(i = 1, \ldots, r; \ j = 1, \ldots, c)$. First, compute the ith row sum $y_{i.}$ and the jth column sum $y_{.j}$. Then compute the estimated expected frequency of the i, jth cell using the formula $\hat{e}_{ij} = y_{i.} y_{.j} / n$.

2. Compute the test statistic χ^2

$$\chi^2 = \sum_{1 \le i \le r} \sum_{1 \le j \le c} \frac{(y_{ij} - \hat{e}_{ij})^2}{\hat{e}_{ij}}.$$

3. Reject H_0 if $\chi^2 > \chi^2_{(r-1)(c-1)}(\alpha)$, where α is the significance level of the test.

Example 9.4 *Continuation of Example 9.3*

Consider the following statistical problem arising from the data in Table 9.5: Is the type of VT failure independent of the position of the tube in the shell?

Solution The null hypothesis is that the VT position and its failure type are mutually independent, so we use the χ^2 test of independence. The observed and estimated expected frequencies derived from Table 9.5 are displayed in Table 9.7. The expected frequencies (or expected counts) are printed directly below the observed frequencies.

Table 9.7 *Observed and expected frequencies for the 2×3 contingency Table 9.5 (MINITAB output)*

```
Expected counts are printed below observed counts

             B1         B2         B3      Total
    1        75         10         15        100
            63.89      22.22      13.89

    2        40         30         10         80
            51.11      17.78      11.11

Total       115        40         25        180

ChiSq =  1.932 +  6.722 +  0.089 +
         2.415 +  8.403 +  0.111 = 19.673
df = 2
```

We illustrate the method for computing the expected frequencies by showing that $\hat{e}_{11} = 63.89$. From Table 9.5 we have

$$y_{1.} = 100 \text{ and } y_{.1} = 115, \text{ so} \frac{y_{1.}y_{.1}}{180} = 63.89, \text{ etc.}$$

The MINITAB output tells us that $\chi^2 = 19.673$ and that the number of degrees of freedom (df) is two. This follows from Equation 9.10, which states that $(r-1)(c-1) = 1 \times 2 = 2$. Thus $\chi_2^2 = 19.673 > 9.21 = \chi_2^2(0.01)$; in other words, the type of failure and position in the tube are not independent.

χ^2 **Test of Homogeneity**

In the previous example we considered a single population of elements that were classified according to two different criteria or attributes; we now consider the situation where the rows of the contingency table consist of independent samples taken from r different populations, with each population partitioned into c categories. We are, in effect, comparing two or more multinomial distributions since each row of the contingency table corresponds to a random sample taken from a multinomial distribution.

Example 9.5 *Is there a relationship between smoking and emphysema?*

The data in Table 9.8 were obtained in a study of the relationship between smoking and emphysema. The patients were classified according to two criteria: the number of packs of cigarettes smoked per day (A) and the severity of the patient's emphysema (B). The categorical variable A has three levels: A_1, A_2, A_3, where A_1 denotes those patients who smoked less than one pack a day, A_2 denotes those patients who smoked between 1 and 2 packs a day, and A_3 denotes those patients who smoked more than 2 packs a day. In this experiment 100 patients were randomly selected from category A_1, 200 patients were randomly selected from category A_2, and 100 patients were randomly selected from category A_3. The categorical variable B has 4 levels denoted by B_1, B_2, B_3, and B_4, where B_1 denotes the mildest form and B_4 denotes the most severe form of emphysema. The statistical problem is to determine if there is a link between the number of packs of cigarettes smoked per day and the severity of the emphysema. More precisely, are the proportions of patients in the four emphysema categories the same (homogeneous) for the three classes of smokers?

Table 9.8 *Cigarette smoking and emphysema data*

Packs of Cigarettes Smoked	Severity of Emphysema				
	B_1	B_2	B_3	B_4	Total
A_1	41	28	25	6	$n_1 = 100$
A_2	24	116	46	14	$n_2 = 200$
A_3	4	50	34	12	$n_3 = 100$
Total	69	194	105	132	$n = 400$

(Source: P. G. Hoel (1984), *Introduction to Mathematical Statistics*, 5th ed., John Wiley & Sons, New York. Used with permission.)

Solution   In this example, the sample size for each level of the categorical variable A is predetermined by the researcher, which is why this is called a contingency table with one margin fixed. We denote the sample size for the ith population by n_i; thus $n_1 = 100$, $n_2 = 200$, $n_3 = 100$, and the total sample size $n = n_1 + n_2 + n_3 = 400$. The four column totals, 69, 194, 105, and 132 are the cell frequencies of the four categories $B_1, \ldots, B_4$.

To proceed we first formulate the null hypothesis. Denote by p_{ij} the probability that an observation from the ith population falls into category B_j. The null hypothesis of homogeneity implies that p_{ij} depends not on the population (equivalently, it does not depend on i), but only on the category B_j. Our problem, then, is to test the hypothesis

$$H_0 : p_{1j} = p_{2j} = p_{3j} = p_j \text{ for } j = 1, 2, 3, 4.$$

Here p_j denotes the probability that a smoker with emphysema falls into severity category B_j. The null hypothesis is that this probability does not depend on the number of packs of cigarettes smoked per day. A hypothesis testing problem of this type is solved by the χ^2 *test of homogeneity*, which will now be explained.

1. First estimate the parameters p_j by computing the cell frequency of each category B_j and then dividing by the sample size:

$$\hat{p}_1 = \frac{69}{400}, \hat{p}_2 = \frac{194}{400}, \hat{p}_3 = \frac{105}{400}, \hat{p}_4 = \frac{132}{400}.$$

2. Next compute the expected cell frequencies via the formula

$$\hat{e}_{ij} = n_i \hat{p}_j = \frac{n_i y_{\cdot j}}{n} = \frac{i \text{ th row sum} \times j \text{ th column sum}}{n}.$$

The observed and expected cell frequencies are displayed in Table 9.9.

3. Compute the test statistic χ^2:

$$\chi^2 = \sum_{1 \leq i \leq r} \sum_{1 \leq j \leq c} \frac{(y_{ij} - \hat{e}_{ij})^2}{\hat{e}_{ij}}.$$

With reference to the data in Table 9.8 we obtain

$$\chi^2 = \frac{(41 - 17.25)^2}{17.25} + \ldots + \frac{(12 - 8)^2}{8} = 64.408.$$

4. Compute the degrees of freedom via Equation 9.10 as before: $DF = (r - 1)(c - 1) = 2 \times 3 = 6$. Since $\chi^2 = 64.408 > 12.59 = \chi_6^2(0.05)$, the null hypothesis is rejected.

   (Note: The assertion that $DF = (r - 1)(c - 1)$ will be justified next.)

Table 9.9 *MINITAB output: Expected frequency counts for the contingency Table 9.8*

```
Expected counts are printed below observed counts

             B1        B2        B3        B4      Total
    1         41        28        25         6        100
           17.25     48.50     26.25      8.00

    2         24       116        46        14        200
           34.50     97.00     52.50     16.00

    3          4        50        34        12        100
           17.25     48.50     26.25      8.00

Total         69       194       105        32        400

ChiSq = 32.699 +   8.665 +   0.060 +   0.500 +
         3.196 +   3.722 +   0.805 +   0.250 +
        10.178 +   0.046 +   2.288 +   2.000 = 64.408
df = 6
```

The $r \times c$ Contingency Table for the χ^2 Test of Homogeneity: The General Case

Before taking up the general case we make the following summary remarks. Table 9.8 is an example of an 3×4 contingency table with one margin fixed because, as noted earlier, specifying the sample size for each population determines the marginal row frequencies *prior* to the experiment. In particular, each row of Table 9.8 represents the results of a random sample, with predetermined sample size, taken from a population with each observation classified into one of four categories. We note that Table 9.8 is just a special case of the $r \times c$ contingency Table 9.10.

Table 9.10

	B_1	...	B_c	Total
A_1	y_{11}	y_{1j}	y_{1c}	n_1
$\vdots$				n_i
A_r	y_{r1}	...	y_{rc}	n_r
Total	$y_{.1}$	$y_{.j}$	$y_{.c}$	$n = \sum_i n_i$

The entry y_{ij} in the ith row represents the number of observations from the ith population that fall in category j, where $\sum_j y_{ij} = n_i$ is fixed. Denoting by p_{ij} the probability that an observation from the ith population falls into category j the null hypothesis of homogeneity takes the following form:

$$H_0 : p_{ij} = p_j (i = 1, \ldots, r; \ j = 1, \ldots, c).$$

Consequently, the cell frequencies $Y_{i1}, \ldots, Y_{ic}$ have a multinomial distribution with parameters $p_1, \ldots, p_c; n_i$, and therefore

$$\sum_{1 \leq j \leq c} \frac{(Y_{ij} - n_i p_j)^2}{n_i p_j} \approx \chi^2_{c-1}.$$

Thus, each row contributes $(c-1)$ degrees of freedom, and summing over the r rows contributes $r(c-1)$ degrees of freedom. This last assertion may be justified by noting that the row sums are mutually independent and approximately χ^2_{c-1} distributed; consequently, their sum has an approximate χ^2 distribution with degrees of freedom equal to their sum, which equals $(c-1)+\ldots+(c-1) = r(c-1)$ (Theorem 6.3). Finally, note that we estimate $(c-1)$ independent parameters since

$$\sum_{1 \leq j \leq c} \hat{p}_j = 1.$$

Thus $DF = r(c-1) - (c-1) = (r-1)(c-1)$, as claimed.

9.4 Chapter Summary

In this chapter we focused on the problem of analyzing and testing hypotheses about categorical data using the χ^2 test.

9.5 Problems

Problem 9.1 *A biology student repeating Mendel's experiment described in Example 9.1 obtained the following results:*

Seed type	ry	wy	rg	wg
Frequency	273	94	88	25

Are these data consistent with Mendel's theory that the frequency counts of seeds of each type produced from this experiment occur in the ratios: ry:rg:wy:wg=9:3:3:1?

Problem 9.2 *We define a fair die to mean that each face is equally likely; thus, there are six categories, and the null hypothesis we wish to test is*

$$H_0 : p_i = 1/6, \ i = 1,\ldots,6.$$

To determine if a die is fair, as opposed to the alternative that it is not, a die was thrown $n = 60$ times, with the results displayed in the following table. Complete columns 3 and 4, compute the χ^2 statistic, and test the null hypothesis at the level $\alpha = 0.05$

Face	Observed frequency y_i	Expected frequency np_{i0}	$y_i - np_{i0}$
1	10		
2	9		
3	11		
4	11		
5	8		
6	11		

Problem 9.3 *Let X denote the number of heads that appear when 5 coins are tossed. It is clear that X has a binomial distribution with $P(X = i) = b(i; 5, p), (i = 0, \ldots, 5)$. The following data record the frequency distribution of the number of heads in $n = 50$ tosses of the 5 coins.*

Number of heads	Observed frequency y_i	Expected frequency np_{i0}	$y_i - np_{i0}$
0	1		
1	17		
2	15		
3	10		
4	6		
5	1		

The null hypothesis is $H_0 : p = 0.5$.
(a) Compute the expected cell frequencies, and complete columns 3 and 4.
(b) After combining cells with small expected frequencies, compute the χ^2 statistic and compute the number of degrees of freedom.
(c) Use the value of χ^2 obtained in part (b) to test H_0 at the 5% level of significance.

Problem 9.4 *The number X of chocolate chips per cookie produced by a large commercial bakery is assumed to have a Poisson distribution with $\lambda = 5$. To test this hypothesis a random sample of 100 cookies is chosen and the frequency distribution of the number of chocolate chips per cookie is recorded below.*

No. of chocolate chips	Observed frequency O_i	Expected frequency E_i
0	0	
1	2	
2	6	
3	12	
4	22	
5	14	
6	17	
7	10	
8	8	
9	9	

The categorical variable C_i, , $i = 0, 1, \ldots$, denotes the event that the cookie had i chocolate chips, and the null hypothesis is that $P(C_i) = P(X = i), i = 0, 1, \ldots, 8$ and $P(C_9) = P(X \geq 9)$, where the distribution of X is Poisson with parameter $\lambda = 5$.
(a) Compute the expected cell frequencies and enter them in column 3.
(b) After combining cells with small expected frequencies, compute the χ^2 statistic and compute the number of degrees of freedom.
(c) Use the value of χ^2 obtained in part (b) to test H_0 at the 5% level of significance.

Problem 9.5 *The following table gives the results of a high-energy physics experiment that counts the number of electron-positron pairs observed in a hydrogen bubble chamber exposed to a beam of photons. The number of electron-positron pairs are recorded on a bubble chamber photograph. Theoretical physics predicts that the frequency distribution of bubble chamber photos with $0, 1, \ldots i$, electron-positron pairs is Poisson. Are these data consistent with this prediction? Use the appropriate χ^2 test at the level $\alpha = 0.01$.*

Number of electron-positron pairs on photograph i	Number of photos with k electron-positron pairs f_i
0	47
1	69
2	84
3	76
4	49
5	16
6	11
7	3

Problem 9.6 *Monte Carlo simulations of complex random processes are widely used in science, engineering, and finance. These simulations rely on a uniform random number generator (also called pseudorandom number generators), which is a computer program that produces, or is supposed to produce, an iid sequence $X_1, X_2, \ldots, X_n$ with common uniform distribution on $[0, 1]$. One way of testing whether or not the sequence is indeed uniform is to divide the unit interval into 10 intervals, say, of equal length, denoted by $c_i = [(i - 1)/10, i/10)$, $i = 1, \ldots, 10$, and count the number Y_i of observations that fall in the ith interval c_i. If the output is uniformly distributed as claimed then $P(X_j \in c_i) = 0.10$.*

(a) Assuming the output is an iid sequence with common uniform distribution, describe the joint distribution function of the random variables $Y_1, \ldots, Y_{10}$. Name it and identify the parameters.

(b) The data in the table below records the output of 100 observations produced by one such generator. Are these data consistent with the claim that they are uniformly distributed?

Interval	c1	c2	c3	c4	c5	c6	c7	c8	c9	c10
Frequency	11	9	4	7	12	6	13	11	9	18

Problem 9.7 *Show that the multinomial distribution reduces to the binomial distribution when $k = 2$. Here Y_1 and Y_2 denote the number of successes and failures, respectively; use the notation $p_1 = p$, $p_2 = 1 - p$.*

Problem 9.8 *The following 2×2 contingency table gives the pass/fail results for the boys and girls who took the advanced level test in pure mathematics with statistics in Great Britain for the 1974-75 year.*

Sex	Pass	Fail	Total
Boy	891	569	1460
Girl	666	290	956
Totals	1557	859	2416

(Source: J.R. Green and D. Margerison (1978), Statistical Treatment of Experimental Data, Elsevier Scientific Publishing Co. Used with permission.)

Use the data and the appropriate chi-square test to test the hypothesis that the passing rate does not depend on the sex of the candidate; use the level $\alpha = 0.01$.

Problem 9.9 *A political consultant surveyed 275 urban, 250 suburban, and 100 rural voters and obtained the following data concerning the public policy issue that is the most important to them.*

Voter	Health Care	Crime	Economy	Taxes	Totals
Urban	113	62	75	25	275
Suburban	88	37	75	50	250
Rural	60	10	20	10	100
Totals	261	109	170	85	625

Do these data indicate any differences in the responses of the three classes of voters with respect to the importance of these four public policy issues?

Problem 9.10 *The results of a newspaper poll of 1,000 voters to the question, "The Congress has passed and the President has signed a new tax cut. Overall, do you think this is a good thing or a bad thing?" are listed below. Do these data indicate any differences in the responses of the three classes of adults with respect to the tax cut?*

Voter	Good thing	Bad thing	No opinion
Democrat	152	204	44
Independent	85	70	25
Republican	294	67	59

Problem 9.11 *To determine the possible effects of the length of a heating cycle on the brittleness of nylon bars, 400 bars were subjected to a 30 second heat treatment and 400 were subjected to a 90 second heat treatment. The nylon bars were then classified as brittle or non-brittle.*

Length of Heating Cycle	Brittle	Non-Brittle	Total
30(sec)	77	323	400
90(sec)	177	223	400
Total	254	546	800

(Source: A. J. Duncan (1974), Quality Control and Industrial Statistics, 4th ed., Homewood, IL, R. D. Irwin.)

Test the hypothesis that the brittleness of the nylon bars is independent of the heat treatment.

Problem 9.12 *For each of 75 rockets fired the lateral deflection and range in yards were measured with the results displayed in the following 3×3 contingency table.*

	Lateral Deflection (Yards)			
Range	-250 to -51	-50 to 49	50 to 199	Total
0 − 1199	5	9	7	21
1200 − 1799	7	3	9	19
1800 − 2699	8	21	6	35
Total	20	33	22	75

(Source: NAVORD Report 3369 (1955); quoted in I. Guttman, S.S. Wilks, and J.S. Hunter (1982), Introductory Engineering Statistics, 3rd ed., John Wiley & Sons, Inc., New York.)

Test at the 5% level of significance that lateral deflection and range are independent.

Problem 9.13 *Each of 6805 pieces of moulded vulcanite made from a resinous powder was classified according two criteria: porosity and dimension. The results are displayed in the following* 2×2 *contingency table.*

	Porous	Nonporous	Total
With defective dimensions	142	331	473
Without defective dimensions	1233	5099	6332
Total	1375	5430	6805

(Source: A. Hald (1952), Statistical Theory with Engineering Applications, *John Wiley & Sons, Inc., New York.)*

 Test the hypothesis that the two classification criteria are independent.

Problem 9.14 *A survey was conducted to evaluate the effectiveness of a new flu vaccine. The vaccine was provided in a two-shot sequence over a period of two weeks. Some people received the two-shot sequence, some appeared for only the first shot, and others received neither. A survey of 1000 residents the following spring provided the information shown in the table below. The researchers were interested in testing the hypothesis that the vaccine was successful in reducing the number of flu cases.*

	No Vaccine	One Shot	Two Shot	Total
Flu	24	9	13	46
No Flu	289	100	565	954
Total	313	109	578	1000

(Source: Mendenhall, Beaver, and Beaver, 10th ed., Example 14.5, p. 623.)
(a) Compute the incidence of flu for each of the three groups.
(b) Describe in plain English the null and alternative hypotheses, and which χ^2 *test is appropriate here: the* χ^2 *test of independence or the* χ^2 *test of homogeneity?*
(c) Use the χ^2 *test described in part (b) to test the null hypothesis, at the level* $\alpha = 0.05$. *Describe the rejection region, decision rule, and conclusion.*

Problem 9.15 *The following data were collected in a study of the relationship between the blood pressures of children and their fathers. Blood pressures of the fathers were clasified as* A_1 *(below average),* A_2 *(average), and* A_3 *(above average). The childrens' blood pressures were classified similarly as* B_1, B_2, B_3. *The entries in the contingency table are the frequency counts for each category* $A_i \cap B_j$, $(i = 1, 2, 3;, j = 1, 2, 3)$.
(a) The researchers were interested in studying the association, if any, between the blood pressures of the fathers and their children. Give a careful statement of the null hypothesis and the appropriate χ^2 *test you will use to test it.*
(b) Use the χ^2 *test described in part (a) to test the null hypothesis, at the level* $\alpha = 0.05$. *Describe the rejection region, decision rule, and conclusion.*

	Child's blood pressure		
Father's blood pressure	B_1	B_2	B_3
A_1	14	11	8
A_2	11	11	9
A_3	6	10	12

Problem 9.16 *In a comparison study of two processes for manufacturing steel plates each plate is classified into one of three categories: no defects, minor defects, major defects. The results of the study are displayed in the following* 2×3 *contingency table.*

Process	No defects	Minor defects	Major defects
A	40	10	5
B	35	12	8

Test the hypothesis that there is no difference between the two processes; use $\alpha = 0.01$.

Problem 9.17 *To test the effectiveness of drug treatment for the common cold, 164 patients with colds were selected, half were given the drug treatment (the treatment group), and the other half (the control group) were given a sugar pill. The patient's reactions to the treatment and sugar pill are recorded in the following table.*

	Helped	Harmed	No effect	Total
Treatment	50	10	22	82
Control	42	12	28	82
Total	92	22	50	164

(Source: P. G. Hoel (1984), Introduction to Mathematical Statistics, *5th ed., John Wiley Press. Used with permission.)*

Test the hypothesis that there is no difference between the treatment and control groups; use $\alpha = 0.10$.

Problem 9.18 *In a comparative study of the effectiveness of two brands of motion sickness pills, brand A pills were given to 45 persons randomly selected from a group of 90 air travelers while the other 45 persons were given brand B pills. The responses were classified as: None (no motion sickness), Slight (slight motion sickness), etc.*

	None	Slight	Moderate	Severe	Total
Brand A	18	17	6	4	45
Brand B	11	14	14	6	45
Total	29	31	20	10	

(a) Give a careful description of the statistical test you would use to determine whether or not these two brands of pills differ significantly in effectiveness from one another. Identify the null and alternative hypotheses, the test statistic, its null distribution, and define the rejection region.
(b) Test your null hypothesis, described in part (a) above, at the level $\alpha = 0.05$.

9.6   To Probe Further

An important topic that we have not discussed here is Fisher's exact test, which concerns the critical region when one or both of the sample sizes are small (so that the central limit theorem cannot be applied). For further information on these topics we advise the reader to consult one of the standard references, such as A. Hald (1952), *Statistical Theory with Engineering Applications*, John Wiley & Sons, Inc., New York. See Section 21.13.

Chapter 10

Linear Regression and Correlation

> Most real-life statistical problems have one or more nonstandard features. There are
> no routine statistical questions; only questionable statistical routines.
> D. Cox, British statistician

10.1 Orientation

Scientists are frequently interested in studying the functional relationship between two
variables y and x. For example, in an agricultural experiment, the crop yield y may depend
on the amount of fertilizer x; in a metallurgical experiment one is interested in the shear
strength y of a spot weld as a function of the weld diameter x. The variable x is variously
called the *explanatory variable*, the *regressor variable,* or the *predictor variable*; y is called
the *dependent variable* and also the *response variable*.

We assume that the relationship between the explanatory and response variables is given
by a function $y = f(x)$, where f is unknown. In most cases the function is specified by
a theoretical model. For example, Hooke's law asserts that the force required to stretch a
spring by a distance x from its relaxed length is given by $y = \beta x$, where β is a constant
that is characteristic of the spring. In this example the relationship between the variables
is derived from the basic principles of mechanics. In most cases, however, the scientist is
uncertain as to the exact form of the functional relationship $y = f(x)$ and he proceeds to
study it by performing a series of experiments and drawing the scatter plot of the observed
values (x_i, y_i), $i = 1, \ldots n$, where y_i is the value of the response corresponding to the chosen
value x_i. The shape of the scatter plot will then provide some insight into the functional
relationship between y and x. Regression analysis is a method for determining the function
that best fits the observed data.

Organization of Chapter

1. Section 10.2: Method of Least Squares

2. Section 10.2.1: Fitting a Straight Line via Ordinary Least Squares

3. Section 10.3: The Simple Linear Regression Model

4. Section 10.3.1: The Sampling Distribution of $\hat{\beta}_1$, $\hat{\beta}_0$, SSE, and SSR

5. Section 10.3.2: Tests of Hypotheses Concerning the Regression Parameters

6. Section 10.3.3: Confidence Intervals and Prediction Intervals

7. Section 10.3.4: Displaying the Output of a Regression Analysis in an ANOVA Table

8. Section 10.3.5: Curvilinear Regression

9. Section 10.4: Model Checking

10. Section 10.5: Correlation Analysis

11. Section 10.6: Mathematical Details and Derivations

12. Section 10.7: Chapter Summary

10.2   Method of Least Squares

The first step in fitting a straight line to a set of bivariate data is to construct the *scatter plot*; that is, plot the points (x_i, y_i), $i = 1, \ldots n$ to see if the points appear to approximate a straight line. If they do not, the scientist must consider an alternative model for the data.

Example 10.1

Consider the data in Table 10.1, which records the shear strengths (y) of spot welds of two gauges (thicknesses) of steel, $0.040''$ and $0.064''$ thousandths of an inch, corresponding to the weld diameter (x) measured in units of thousandths of an inch. Notice that the response variable y is in the first column and the regressor variable is in the second column. The scatter plot for the $0.040''$ gauge steel is shown in Figure 10.1.

Table 10.1
Shear strength and weld diameter for two gauges of steel

0.040″ gauge		0.064″ gauge	
y	*x*	*y*	*x*
350	140	680	190
380	155	800	200
385	160	780	209
450	165	885	215
465	175	975	215
485	165	1025	215
535	195	1100	230
555	185	1030	250
590	195	1175	265
605	210	1300	250

(Source: A.J. Duncan (1985), Quality Control and Industrial Statistics, *5th ed., R. D. Irwin, Inc. Used with permission.)*

Notation The response variable Y and the explanatory variable X of the *regression model* are defined using the *model statement* format: Model $(Y = \text{shear strength} \mid X = \text{weld diameter})$.

From Figure 10.1 it appears that the shear strength y varies almost linearly with respect to the weld diameter x. We say almost because it is clear that it is impossible to construct a line of the form $y = b_0 + b_1 x$ which passes through all the observations (x_i, y_i), $i = 1, \ldots n$; which leads us to the following problem: How do we choose b_0 and b_1 so that the line $y = b_0 + b_1 x$ "best fits" the observations (x_i, y_i)? In Section 10.2.1 we shall solve this problem via the method of *least squares*.

Example 10.2 *Fuel efficiency vs. automobile weight*

Model(Y = shear strength ¦ X = weld diameter)

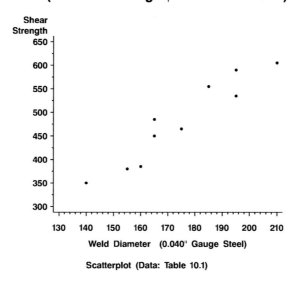

Scatterplot (Data: Table 10.1)

FIGURE 10.1:   Scatter plot of the data in Table 10.1

After the price of petroleum tripled in the early part of the 1970s, the fuel efficiencies of automobiles became an important concern for consumers, manufacturers, and governments. Table 10.2 records the weight (in units of 1000 lbs), fuel consumption in miles per gallon (MPG), engine displacement (DISP), and fuel consumption in gallons per 100 miles (GPM) for twelve 1992 model year automobiles. (Displacement is a measure of the size of the automobile engine.)

There are two different ways to measure fuel efficiency: (1) miles per gallon (MPG), which is the method used in the United States, and (2) gallons per hundred miles (GPM), as is used in some European countries. Another important goal of regression analysis is to determine the explanatory variable that best explains the observed response. Specifically, which is the better choice for the explanatory variable: weight or engine displacement? To study this question we first draw the scatter plots for the two models: Model $(Y = mpg|X = weight)$ (Figure 10.2) and Model $(Y = gpm|X = weight)$ (Figure 10.3).

Table 10.2
Fuel efficiency data for twelve 1992 model year automobiles

Automobile	WEIGHT	MPG	DISP	GPM
Saturn	2.495	32	1.9	3.12
Escort	2.530	30	1.8	3.34
Elantra	2.620	29	1.6	3.44
Camry V6	3.395	25	3.0	4.00
Camry4	3.030	27	2.2	3.70
Taurus	3.345	28	3.8	3.58
Accord	3.040	29	2.2	3.44
LeBaron	3.085	27	3.0	3.70
Pontiac	3.495	28	3.8	3.58
Ford	3.950	25	4.6	4.00
Olds88	3.470	28	3.8	3.58
Buick	4.105	25	5.7	4.00

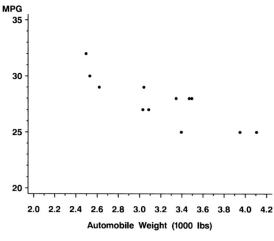

Model(Y = mpg | X = weight)

Scatter plot (Data: Table 10.2)

FIGURE 10.2:   Scatter plot of mpg vs. automobile weight (Table 10.2)

The choice of gpm as a measure fuel efficiency is justified by the following reasoning. The basic principles of mechanics suggest that the amount of gasoline G consumed is proportional to the work W expended in moving the vehicle (so $G = c_1 \times W$). Since $Work(W) = Force(F) \times Distance(D)$, it follows that $G = c_1 \times F \times D$; thus $G/D = c_1 \times F$. In addition, the force F is proportional to the *weight* (WGT), so

$$GPM = 100 \times \frac{G}{D} = 100 \times c_1 \times WGT = c_2 \times WGT.$$

These theoretical considerations imply that the functional relation between GPM and WGT is more likely to be a linear one. We will continue our study of this data set with the goal of exploring the following question: Which model, Model $(Y = mpg|X = weight)$ or Model $(Y = gpm|X = weight)$, best fits the data? We will develop some criteria that will help us to answer this question later in this chapter.

10.2.1   Fitting a Straight Line via Ordinary Least Squares

We now consider the problem of constructing a straight line with equation $y = b_0 + b_1 x$ that "best fits" the data $(x_1, y_1), \ldots, (x_n, y_n)$ in a sense that will be made precise.

The ith *fitted value*, (also called the ith *predicted value*) is denoted $\hat{y}_i$ and is defined as follows:

$$\hat{y}_i = b_0 + b_1 x_i, \ i = 1, \ldots n.$$

It is the y coordinate of the point on the line $y = b_0 + b_1 x$ with x coordinate x_i.

The ith *residual* is the difference between the ith observed value and the ith predicted value (see Figure 10.4). It is denoted e_i and is given by

$$i\text{th residual: } e_i = y_i - (b_0 + b_1 x_i).$$

Points of the scatter plot that lie below the line have negative residuals and points above the line have positive residuals. The sum of the squares of the residuals, denoted SSE (*error sum of squares*) is defined as follows:

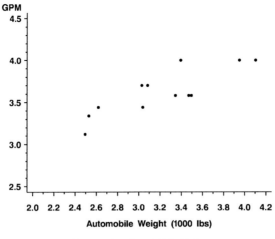

Model(Y = gpm ¦ X = weight)

Scatter plot (Data: Table 10.2)

FIGURE 10.3:   Scatter plot of gpm vs. automobile weight (Table 10.2)

$$SSE = \sum_{1 \leq i \leq n} e_i^2 = \sum_{1 \leq i \leq n} [y_i - (b_0 + b_1 x_i)]^2. \tag{10.1}$$

The error sum of squares is a measure of how well the line fits the data. For example, when all points of the scatter plot lie on the line $y = b_0 + b_1 x$ then each residual equals 0 and therefore SSE also equals 0.

The *method of least squares* is to find those values b_0, b_1 that minimize the sums of the squares of the residuals e_i. The error sum of squares is a function $Q(b_0, b_1)$ of the parameters b_0, b_1 and is given by the following formula:

$$SSE = \sum_{1 \leq i \leq n} [y_i - (b_0 + b_1 x_i)]^2 = Q(b_0, b_1). \tag{10.2}$$

Notational Convention: From now on we adopt the convention that a sum of the form $\sum x_i$, where the limits on the indices are not specified, is a convenient shorthand for $\sum_{1 \leq i \leq n} x_i$.

We minimize the sum of squares $Q(b_0, b_1)$ using an algebraic method that avoids the use of calculus and offers some insight into the least squares method itself. The basic idea is to partition Q into a sum of three squares (so that each summand is non-negative), from which the values that minimize it can be read off. The formula for the sum of squares uses the following notation, which is also useful for computing the minimizing values.

$$S_{xx} = \sum (x_i - \overline{x})^2 = (n-1)s_x^2 = \sum x_i^2 - n\overline{x}^2; \tag{10.3}$$

$$S_{yy} = \sum (y_i - \overline{y})^2 = (n-1)s_y^2 = \sum y_i^2 - n\overline{y}^2; \tag{10.4}$$

$$S_{xy} = \sum (x_i - \overline{x})(y_i - \overline{y}) = (n-1)s_{xy} = \sum x_i y_i - n\overline{x} \times \overline{y}; \tag{10.5}$$

$$\text{sample correlation coefficient: } r = \frac{S_{xy}}{\sqrt{S_{xx}}\sqrt{S_{yy}}}; \tag{10.6}$$

$$\text{sample covariance } s_{xy} = \frac{\sum (x_i - \overline{x})(y_i - \overline{y})}{n-1}. \tag{10.7}$$

$$\tag{10.8}$$

It can be shown (see the discussion following Equation 10.12) that $0 \le r^2 \le 1$ and therefore $-1 \le r \le 1$.

The partition of Q into a sum of squares: The sum of squares of Equation 10.2 can be partitioned into a sum of three squares given by

$$Q(b_0, b_1) = S_{yy}(1 - r^2) + (b_1\sqrt{S_{xx}} - r\sqrt{S_{yy}})^2 + n(\overline{y} - b_0 - b_1\overline{x})^2. \tag{10.9}$$

Derivation: The derivation is given in Section 10.6 at the end of this chapter. Looking at Equation 10.9 we see that only the second and third summands depend on the unknown parameters b_0 and b_1. We minimize $Q(b_0, b_1)$ by choosing these parameters so that these terms equal zero. Setting the second and third terms equal to zero yields the following two equations:

$$b_1\sqrt{S_{xx}} - r\sqrt{S_{yy}} = 0;$$
$$\overline{y} - b_0 - b_1\overline{x} = 0.$$

The equations for b_0 and b_1 can always be solved provided $S_{xx} \ne 0$. But this will always be true except for the uninteresting case when $x_1 = x_2 = \ldots = x_n$. Denoting the solutions by $\hat{\beta}_0$ and $\hat{\beta}_1$ we obtain the slope–intercept formula for the line that best fits the data in the sense of the least squares criterion:

$$\text{Least Squares Line: } y = \hat{\beta}_0 + \hat{\beta}_1 x, \text{ where} \tag{10.10}$$

$$\hat{\beta}_1 = \frac{r\sqrt{S_{yy}}}{\sqrt{S_{xx}}} = \frac{S_{xy}}{S_{xx}} \text{ and} \tag{10.11}$$

$$\hat{\beta}_0 = \overline{y} - \hat{\beta}_1\overline{x}. \tag{10.12}$$

The quantities $\hat{\beta}_0$ (intercept) $\hat{\beta}_1$ (slope) are called the *regression coefficients* or the *regression parameters*. From Equations 10.2 and 10.9 it follows that

$$SSE = S_{yy}(1 - r^2). \tag{10.13}$$

Equation 10.13 follows from the fact that the second and third summands on the right hand side of the partition of the sum of squares given in Equation 10.9 equal zero. This formula for SSE also implies that $0 \le r^2 \le 1$. This is because SSE and S_{yy} are always non-negative, since they are sums of squares. Therefore, $0 \le 1 - r^2$, so $0 \le r^2 \le 1$ as asserted earlier.

The fitted values and residuals associated with the least squares line, also denoted by $\hat{y}_i$, and e_i, are defined as before:

$$\hat{y}_i = \hat{\beta}_0 + \hat{\beta}_1 x_i, \ i = 1, \ldots n; \tag{10.14}$$

$$e_i = y_i - \hat{y}_i = y_i - \hat{\beta}_0 - \hat{\beta}_1 x_i. \tag{10.15}$$

The least squares Equation 10.10 is also called the *estimated regression function* of y on x. The use of the adjective estimated may appear strange since the calculations do not require any statistical assumptions. We will resolve this mystery in the next section when we introduce a statistical model for the bivariate data set (x_i, y_i), $i = 1, \ldots, n$. The regression parameters are then interpreted as estimators of the true regression line.

Some suggestions for computing the regression coefficients The regression coefficients are most conveniently computed by using a statistical software package such as MINITAB or SAS. If you do not have access to a statistical software package, the regression coefficients can be computed using a calculator as shown in the next example.

Example 10.3 *Continuation of Example 10.1 (0.040″ gauge steel)*

Compute the estimated regression parameters and the estimated regression function for the Model (Y = shear strength $|X$ = weld diameter) problem of Example 10.1 (0.040″ gauge steel).

Solution We first compute $\bar{x}$, $\bar{y}$, S_{xx}, S_{xy}, S_{yy}. A routine computation using Equations 10.3-10.5 yields

$$\bar{x} = 174.5, \ \bar{y} = 480, \ S_{xx} = 4172.5, \ S_{xy} = 16650, \ S_{yy} = 73450.$$

It follows from Equations 10.11 and 10.12 that the slope $\hat{\beta}_1$ and intercept $\hat{\beta}_0$ of the regression line are given by

$$\hat{\beta}_1 = \frac{S_{xy}}{S_{xx}} = 3.990413; \ \hat{\beta}_0 = \bar{y} - \hat{\beta}_1 \bar{x} = -216.3271.$$

Note: The final results were taken from the computer printout and may differ slightly from those you obtained due to round off error.

The next step in the regression analysis is to plot the estimated regression line and the scatter plot on the same set of axes, as shown in Figure 10.4. Table 10.3 displays a SAS printout of the observations $y_1, \ldots y_n$ (in the column headed *Dep Var*), the fitted values $\hat{y}_1, \ldots, \hat{y}_n$ (in the column headed *Predict*), and the *residuals* for the data displayed in Table 10.1 for (0.040″ gauge steel). We obtain the first residual by computing $y_1 - \hat{y}_1 = 350 - (-216.3271 + 3.990413 \times 140) = 7.6692$. Again, because of round off error, the final answer obtained here differs slightly from the value in the computer printout.

Interpreting the regression coefficients The estimated slope $\hat{\beta}_1 = 3.990413$ implies that each 0.001″ increase in the weld diameter produces an increase of 3.990413 in the shear strength. The estimate $\hat{\beta}_0 = -216.3271$ states that the estimated regression line intercepts the y axis at -216.3271. This clearly is without any physical significance since the shear strength can never be negative. This should serve as a warning against using the estimated regression line to predict the response y when x lies far outside the range of the initial data, which in this case is the interval [140, 210]. We can, however, use the estimated regression line to predict the value of the shear strength corresponding to a weld diameter of 180; in this case the predicted value is $-216.3271 + 3.990413 \times 180 = 501.9472$.

An interpretation of the sign of r When the sample correlation is positive, then so is the estimated slope of the estimated regression line; therefore, an increase in x tends to increase the value of the response variable y. A negative value for the sample correlation, on the other hand, implies that the estimated regression line has a negative slope; in other words, an increase in x tends to decrease the value of the response variable y.

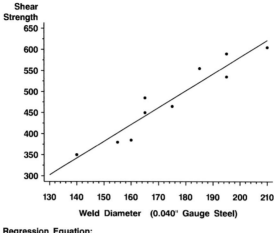

Model(Y = shear strength ¦ X = weld diameter)

Regression Equation:
Y = −216.3271 + 3.990413*X
Scatterplot and Regression Line (Data: Table 10.1)

FIGURE 10.4:   Scatter plot and regression line for shear strength vs. weld diameter (Table 10.1)

Table 10.3 *Shear strength, predicted shear strength, and residual*

Obs.	Shear	Predict value	Residual
1	350.0	342.3	7.6693
2	380.0	402.2	−22.1869
3	385.0	422.1	−37.1390
4	450.0	442.1	7.9089
5	465.0	482.0	−16.9952
6	485.0	442.1	42.9089
7	535.0	561.8	−26.8035
8	555.0	521.9	33.1007
9	590.0	561.8	28.1965
10	605.0	621.7	−16.6597

The Coefficient of Determination

The square of the sample correlation coefficient is called the *coefficient of determination* and is denoted by R^2. It plays an important role in *model checking*. This refers to a variety of analytical and graphical methods for evaluating how well the estimated regression line fits the data; it is discussed in Section 10.4.

We now derive an alternative formula for the coefficient of determination that gives us a deeper understanding of its role in model checking. It is based on the sum of squares partition given in Equation 10.16 below, which is absolutely fundamental for understanding the results of a regression analysis:

$$\sum (y_i - \bar{y})^2 = \sum (\hat{y}_i - \bar{y})^2 + \sum (y_i - \hat{y}_i)^2. \tag{10.16}$$

The sum of squares partition (Equation 10.16) is most easily derived and understood via a matrix approach to regression given in Chapter 11. The decomposition 10.16 is also written in the form:

$$\text{SST} = \text{SSR} + \text{SSE}, \tag{10.17}$$

where,

$$\text{SST} = \sum (y_i - \bar{y})^2 = S_{yy}, \text{ (by Equation 10.4)}, \quad \text{SSR} = \sum (\hat{y}_i - \bar{y})^2$$
$$\text{SSE} = \sum (y_i - \hat{y}_i)^2 = \text{SST}(1 - R^2) \text{ (by Equation 10.13)}.$$

The quantity SST is called the *total sum of squares;* it is a measure of the total variability of the original observations. The quantity SSR is called the *regression sum of squares;* it is a measure of the total variability of the fitted values. It is also a measure of the total variability explained by the regression model. The quantity SSE is the error sum of squares, defined earlier; it is a measure of the unexplained variability.

The partition (Equation 10.17) of the total sum of squares leads to an alternative formula for the coefficient of determination that gives us a deeper understanding of its role in model checking. Dividing both sides of Equation 10.17 by SST, we see that

$$1 = \frac{\text{SSR}}{\text{SST}} + \frac{\text{SSE}}{\text{SST}}.$$

From Equation 10.17 it follows that

$$SSR = SST - SSE = SST - SST(1 - R^2) = R^2 SST.$$

Dividing both sides by SST yields the following formula for R^2:

$$R^2 = \frac{SSR}{SST}. \tag{10.18}$$

Thus, the quantity $R^2 = SSR/SST$ represents the *proportion of the total variability that is explained by the model,* and SSE/SST represents the proportion of the unexplained variability. A value of R^2 close to one implies that most of the variability is explained by the regression model; a value of R^2 close to zero indicates that the regression model is not appropriate.

Example 10.4 *Continuation of Example 10.2*

(i) Compute the regression parameters, (ii) the coefficient of determination, and (iii) graph the estimated regression line for the regression Model $(Y = mpg|X = weight)$ of Example 10.2.

Solution (i) These quantities were computed using a computer software package; the following results (rounded to two decimal places) were obtained:

$$\bar{x} = 3.21, \bar{y} = 27.75, S_{xx} = 2.95, S_{xy} = -10.13, S_{yy} = 50.25; \text{ consequently,}$$
$$\hat{\beta}_1 = \frac{S_{xy}}{S_{xx}} = \frac{-10.13}{2.95} = -3.43, \hat{\beta}_0 = \bar{y} - \hat{\beta}_1\bar{x} = 27.75 + 3.43 \times 3.21 = 38.77.$$

The slope is negative since an increase in weight leads to a decrease in mpg.
(ii) From the definitions of SST, SSR, and SSE just given we obtain the following values:

$$SST = 50.25, SSR = 34.77, SSE = 15.48.$$

Consequently,

$$R^2 = \frac{34.77}{50.25} = 0.69, r = -0.83.$$

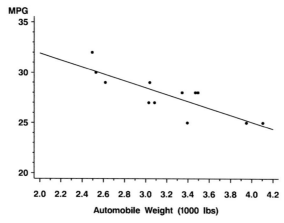

Regression Equation:
MPG = 38.78474 − 3.434047*WEIGHT
Scatter Plot and Estimated Regression Line   (Data: Table 10.2)

FIGURE 10.5:   Scatter plot and estimated regression line for mpg vs. automobile weight (Table 10.2)

(iii) The scatter plot and the estimated regression line are shown in Figure 10.5.

Interpreting the regression coefficients The estimated slope $\hat{\beta}_1 = -3.43$ implies that each 1000 lb increase in the automobile's weight produces a *decrease* of 3.43 in the fuel efficiency as measured in miles per gallon. The estimate $\hat{\beta}_0 = 38.78$ states that the estimated regression line intercepts the y axis at 38.78. Note that the predicted fuel efficiency for an automobile weighing 12,000 lbs is $38.78 - 3.43 \times 12 = -2.38$ mpg, which is absurd since mpg is always non-negative. Once again, this should serve as a warning against using the estimated regression line to "predict" the response y when x lies far outside the range of the initial data. On the other hand, the fitted mpg value for an automobile weighing 2495 lbs is given by $38.78 - 3.43 \times 2.495 = 30.22$ mpg. The residual equals $32 - 30.22 = 1.78$.

Some properties of the residuals It can be shown that the residuals and the fitted values satisfy the following conditions:

$$\sum e_i = \sum (y_i - \hat{y}_i) = 0; \tag{10.19}$$

$$\sum e_i x_i = \sum (y_i - \hat{y}_i)x_i = 0; \tag{10.20}$$

$$\frac{1}{n} \sum \hat{y}_i = \bar{y}. \tag{10.21}$$

The derivations of these equations are left as Problem 10.14.

10.3   The Simple Linear Regression Model

The *simple linear regression model* assumes that the functional relationship between the response variable and the explanatory variable is a linear function plus an error term denoted

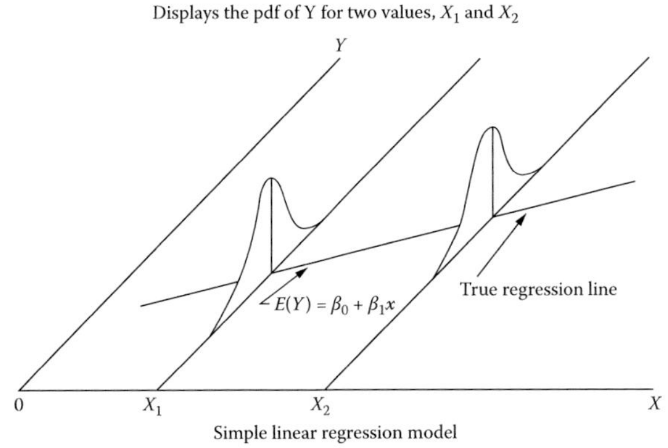

Displays the pdf of Y for two values, X_1 and X_2

FIGURE 10.6: Simple linear regression model

by ϵ; that is,

$$Y = \beta_0 + \beta_1 x + \epsilon. \tag{10.22}$$

The variable ϵ includes all the factors other than the explanatory variable that can also produce changes in Y.

From these considerations we are naturally led to a statistical model for the bivariate data sets of the sort we have analyzed previously and that consists of the following elements:

1. Y_i denotes the value of the response corresponding to the value x_i of the explanatory variable, where $x_i, i = 1, \ldots n$, are the scientist's particular choices of the explanatory variable x.

2. We assume that

$$Y_i = \beta_0 + \beta_1 x_i + \epsilon_i. \tag{10.23}$$

Our statistical model assumes that the random variables $\epsilon_1, \ldots, \epsilon_n$ are mutually independent and normally distributed with zero means and variances σ^2, so

$$E(\epsilon_i) = 0 \text{ and } Var(\epsilon_i) = \sigma^2. \tag{10.24}$$

The parameters β_0 and β_1, called the *intercept* and *slope* respectively, define the *true regression function* $y = \beta_0 + \beta_1 x$.

We shall refer to conditions 1 and 2 as the *standard assumptions*.

3. The standard assumptions imply that the random variables $Y_1, \ldots, Y_n$ are mutually independent and normally distributed with

$$E(Y_i) = \beta_0 + \beta_1 x_i \text{ and } Var(Y_i) = \sigma^2.$$

Notation To emphasize the dependence of the expected value of Y on x we use the notation: $E(Y|x) = \beta_0 + \beta_1 x$. Some authors use the notation: $\mu_{Y|x} = \beta_0 + \beta_1 x$. Figure 10.6 displays the pdf of Y for two values, x_1 and x_2.

10.3.1 The Sampling Distribution of $\hat{\beta}_1$, $\hat{\beta}_0$, SSE, and SSR

The simple linear regression model has *three* unknown parameters: β_0, β_1, and σ^2. We now estimate these parameters via the least squares method described in the previous section. Thus, we obtain the *least squares estimators* for β_0 and β_1 by substituting Y for y in Equations 10.11 and 10.12. Consequently, the least squares estimators of the regression parameters are given by

$$\hat{\beta}_1 = \frac{S_{xY}}{S_{xx}} = \frac{\sum (x_i - \overline{x})(Y_i - \overline{Y})}{S_{xx}}; \tag{10.25}$$

$$\hat{\beta}_0 = \overline{Y} - \hat{\beta}_1 \overline{x}. \tag{10.26}$$

An estimator for σ^2 will be given shortly. In the meantime, it is important to note that in the context of the simple linear regression model the least squares estimators $\hat{\beta}_0$ and $\hat{\beta}_1$ are now random variables, and that in order to make statistical inferences concerning the unknown parameters we must first derive their sampling distributions. Theorem 10.1 summarizes the essential facts; its proof, which follows directly from Equations 10.25 and 10.26, is sketched in Section 10.6 at the end of this chapter.

Theorem 10.1 *1. The distribution of the least squares estimator for the slope β_1 is normal with mean, variance, and second moment given by*

$$E(\hat{\beta}_1) = \beta_1; \tag{10.27}$$

$$V(\hat{\beta}_1) = \frac{\sigma^2}{S_{xx}}; \tag{10.28}$$

$$E(\hat{\beta}_1^2) = \frac{\sigma^2}{S_{xx}} + \beta_1^2. \tag{10.29}$$

2. The distribution of the least squares estimator for the intercept β_0 is normal with mean and variance given by

$$E(\hat{\beta}_0) = \beta_0; \tag{10.30}$$

$$V(\hat{\beta}_0) = \sigma^2 \left[\frac{1}{n} + \frac{\overline{x}^2}{S_{xx}} \right]. \tag{10.31}$$

It follows from Equation 10.27 that $\hat{\beta}_1$ is an unbiased estimator of the slope. Similarly, Equation 10.30 implies that $\hat{\beta}_0$ is an unbiased estimator of the intercept.

The parameter σ^2 is an important one because it is a measure of the variability of the response Y; a small value of σ^2 indicates that the observations (x_i, y_i) lie close to the true regression line and a large value of σ^2 indicates that they are more widely dispersed about the true line. To estimate σ^2 we use the statistic

$$s^2 = \frac{\sum (Y_i - \hat{Y}_i)^2}{n - 2} = \frac{SSE}{n - 2}. \tag{10.32}$$

The next theorem completes our description of the sampling distributions of the regression parameters; its proof is somewhat technical and lies outside the scope of this text.

Theorem 10.2 *The statistic s^2 is an unbiased estimator of σ^2; that is*

$$E(s^2) = E\left(\frac{SSE}{n - 2} \right) = \sigma^2. \tag{10.33}$$

The random variable $(n - 2)s^2/\sigma^2$ has a chi-square distribution with $n - 2$ degrees of freedom; that is,

$$\frac{(n - 2)s^2}{\sigma^2} \stackrel{D}{=} \chi^2_{n-2}.$$

In addition, s^2 is independent of both $\hat{\beta}_0$ and $\hat{\beta}_1$.

A direct proof of 10.33 is givn in Section 10.6.

We will also need the following result concerning the expected value of SSR:

$$E(SSR) = \sigma^2 + \beta_1^2 S_{xx}. \tag{10.34}$$

We give the proof in Section 10.6.

Degrees of freedom associated with the sum of squares $\sum (Y_i - \hat{Y}_i)^2$ The divisor $n - 2$ is the number of *degrees of freedom* associated with the sum of squares $\sum (Y_i - \hat{Y}_i)^2$. Recall (see Equations 10.19 and 10.20) that the residuals satisfy the following two equations:

$$\sum e_i = \sum (Y_i - \hat{Y}_i) = 0;$$

$$\sum e_i x_i = \sum (Y_i - \hat{Y}_i) x_i = 0.$$

Each of these equations allows us to eliminate one summand from the sum of squares $\sum (Y_i - \hat{Y}_i)^2$; consequently, there are only $n - 2$ independent quantities $Y_i - \hat{Y}_i$.

Confidence Intervals for the Regression Coefficients

We now derive confidence intervals for the parameters β_0, β_1, and σ^2 from the sampling distributions of $\hat{\beta}_0$, $\hat{\beta}_1$, and s^2. We begin with the sampling distribution of $\hat{\beta}_1$ since this is the most important parameter in the regression model. Using Theorems 10.1 and 10.2 and then standardizing the random variable $\hat{\beta}_1$ in the usual way, we obtain

$$\frac{\hat{\beta}_1 - \beta_1}{\sigma/\sqrt{S_{xx}}} = Z \overset{\mathcal{D}}{=} N(0,1); \quad \frac{(n-2)s^2}{\sigma^2} = V \overset{\mathcal{D}}{=} \chi_{n-2}^2; \text{ consequently,}$$

$$\frac{\hat{\beta}_1 - \beta_1}{s/\sqrt{S_{xx}}} = \frac{Z}{\sqrt{V/(n-2)}} \overset{\mathcal{D}}{=} t_{n-2}. \tag{10.35}$$

The quantity in the denominator of Equation 10.35 is called the *estimated standard error* of the estimate;

$$\text{Estimated standard error: } s(\hat{\beta}_1) = \frac{s}{\sqrt{S_{xx}}}. \tag{10.36}$$

The standard error for the estimate of β_0 is obtained by noting that

$$V(\hat{\beta}_0) = \sigma^2 \left[\frac{1}{n} + \frac{\bar{x}^2}{S_{xx}} \right];$$

Replacing σ^2 with its estimate s^2 yields the following formula for the estimated standard error for β_0:

$$s(\hat{\beta}_0) = s \sqrt{\left[\frac{1}{n} + \frac{\bar{x}^2}{S_{xx}} \right]}. \tag{10.37}$$

The sampling distribution of $\hat{\beta}_0$ is derived in a similar way, leading to the result

$$\frac{\hat{\beta}_0 - \beta_0}{s(\hat{\beta}_0)} \overset{\mathcal{D}}{=} t_{n-2}. \tag{10.38}$$

We can now construct confidence intervals for β_1 and β_0 using methods similar to those used to derive confidence intervals in Chapter 7. Thus, a $100(1 - \alpha)\%$ confidence interval for β_1 is given by

$$\hat{\beta}_1 \pm t_{n-2}(\alpha/2) s(\hat{\beta}_1). \tag{10.39}$$

Similarly, a $100(1 - \alpha)\%$ confidence interval for β_0 is given by

$$\hat{\beta}_0 \pm t_{n-2}(\alpha/2) s(\hat{\beta}_0). \tag{10.40}$$

Example 10.5 *Confidence intervals for the slope and intercept*

Construct 95% confidence intervals for the slope and intercept parameters of the regression Model (Y = shear strength$|X$ = weld diameter) (Example 10.1).

Solution   We computed the quantities $\bar{x}$, S_{xx}, SSE, $\hat{\beta}_0$, and $\hat{\beta}_1$ earlier (Example 10.3) and we restate them here for your convenience:

$$n = 10; \ \bar{x} = 174.5; \ S_{xx} = 4172.5;$$
$$SSE = 7009.6165; \ \hat{\beta}_0 = -216.3271; \ \hat{\beta}_1 = 3.990413.$$

We estimate σ^2 using Equation 10.32; thus,

$$s^2 = \frac{SSE}{8} = \frac{7009.6165}{8} = 876.2021.$$

The estimated standard error for $\hat{\beta}_1$ is therefore

$$s(\hat{\beta}_1) = \frac{s}{\sqrt{S_{xx}}} = \sqrt{\frac{876.2021}{4172.5}} = 0.4583.$$

Similarly,

$$s(\hat{\beta}_0) = \sqrt{876.2021}\sqrt{\frac{1}{10} + \frac{174.5^2}{4172.5}} = 80.5109.$$

To construct a 95% confidence interval for β_1 we choose $t_8(0.025) = 2.306$ and compute $3.9904 \pm 2.306 \times 0.4583 = 3.990413 \pm 1.0568$. The confidence interval is $[2.9336, 5.0473]$.

The 95% confidence interval for β_0 is given by $-216.3271 \pm 2.306 \times 80.5109 = -216.3271 \pm 185.6581$, with confidence interval is $[-401.9852, -30.6690]$.

Testing Hypotheses About the Regression Parameters Using Confidence Intervals

We also use confidence intervals for the regression parameters to test hypotheses about them. Thus, we accept the null hypothesis $H_0 : \beta_1 = 0$ against $H_1 : \beta_1 \neq 0$ if the confidence interval for β_1 contains 0; otherwise we reject it. For example, the confidence interval for the slope in Example 10.5 is $[2.9336, 5.0472]$, which does not contain 0, thus we reject $H_0 : \beta_1 = 0$. We present another approach in the next section.

10.3.2   Tests of Hypotheses Concerning the Regression Parameters

The concepts and methods of hypothesis testing developed in Chapter 8 are easily adapted to make inferences about the slope β_1 (and also the intercept β_0). This is because under the standard assumptions we have shown (see Equation 10.35) that

$$\frac{\hat{\beta}_1 - \beta_1}{s/\sqrt{S_{xx}}} \overset{\mathcal{D}}{=} t_{n-2}.$$

Suppose, for example, we wish to test the hypothesis $H_0 : \beta_1 = 0$ against $H_1 : \beta_1 \neq 0$. As noted earlier, when $\hat{\beta}_1 = 0$, then $SSR = 0$, so the regression model explains none of the variation. Thus $\beta_1 = 0$ implies that varying x has no effect on the response variable y. It follows from Equation 10.35 that the null distribution of $\hat{\beta}_1/(s/\sqrt{S_{xx}})$ is a t distribution with $n - 2$ degrees of freedom. Consequently, a (two sided) level α test of H_0 is

$$\text{reject } H_0 \text{ if } \frac{|\hat{\beta}_1|}{s/\sqrt{S_{xx}}} \geq t_{n-2}(\alpha/2). \tag{10.41}$$

Example 10.6 *Continuation of Example 10.5*

Refer back to Example 10.5. Test the null hypothesis $H_0 : \beta_1 = 0$ against $H_1 : \beta_1 \neq 0$ for the regression Model (Y = shear strength $|X$ = weld diameter) at the 5% significance level.

Solution The regression parameters, S_{xx}, s^2, and $\bar{x}$ were computed earlier, and the following results were obtained:

$$\hat{\beta}_0 = -216.3271; \ \hat{\beta}_1 = 3.990413; \ s^2 = 876.202, \ S_{xx} = 4172.5; \bar{x} = 174.5, \ n = 10.$$

The value of the test statistic is given by

$$\frac{|\hat{\beta}_1|}{s/\sqrt{S_{xx}}} = \frac{3.990413}{\sqrt{876.202/4172.5}} = 8.708 \geq t_8(0.025) = 2.306.$$

This is significant at the 5% level, so we reject the null hypothesis.

Hypothesis testing via the computer The result we just obtained is contained in the edited SAS computer printout of the regression analysis displayed in Table 10.4. The column labeled Parameter Estimate displays the estimated regression coefficients $\hat{\beta}_0$ (INTERCEP), $\hat{\beta}_1$ (SLOPE). The column Standard Error displays the estimated standard errors $s(\hat{\beta}_0)$, $s(\hat{\beta}_1)$. The next to last column titled T for HO is an abbreviation for "Test of the null hypothesis $H_0 : \beta_i = 0$, $(i = 1, 2)$ using the t test." It contains the value of the t statistic used to test H_0. The P-value for each test is printed in the last column, Prob > |T|.

Table 10.4

| | Parameter Estimates | | | |
| | Parameter | Standard | T for HO: | |
Variable	Estimate	Error	Parameter=0	Prob > |T|
INTERCEP	-216.327142	80.51090203	-2.687	0.0276
SLOPE	3.990413	0.45825157	8.708	0.0001

10.3.3 Confidence Intervals and Prediction Intervals

In this section we consider two estimation problems that are easily confused because they are so closely related.

1. The first problem is to derive point and interval estimates for the mean response $E(Y|x)$, which is a numerical parameter.

2. The second problem, which appears to be similar to the first, is to *predict* the response Y corresponding to a new observation taken at the value x. Note that, in this case, Y is *not* a parameter but a random variable.

Confidence Intervals for the Mean Response $E(Y|x)$

In order to construct a confidence interval for $E(Y|x)$ we need some facts concerning the sampling distribution of the estimator $\hat{\beta}_0 + \hat{\beta}_1 x$, as summarized in the following theorem.

Theorem 10.3 *The point estimator $\hat{\beta}_0 + \hat{\beta}_1 x$ of the mean response $\beta_0 + \beta_1 x$ is unbiased and normally distributed with mean, variance, and estimated standard error given by*

$$E(\hat{\beta}_0 + \hat{\beta}_1 x) = \beta_0 + \beta_1 x;$$

$$V(\hat{\beta}_0 + \hat{\beta}_1 x) = \sigma^2 \left[\frac{1}{n} + \frac{(x - \bar{x})^2}{S_{xx}} \right];$$

$$s(\hat{\beta}_0 + \hat{\beta}_1 x) = s \sqrt{\left[\frac{1}{n} + \frac{(x - \bar{x})^2}{S_{xx}} \right]}.$$

Remark 10.1 *Note that $s(\hat{\beta}_0 + \hat{\beta}_1 x)$ depends on x; in particular, the further x deviates from $\bar{x}$ the greater the estimated standard error. This underscores the warning about extrapolating the model beyond the range of the original data.*

Invoking Theorems 10.1 and 10.2 again, we can show that

$$\frac{\hat{\beta}_0 + \hat{\beta}_1 x - (\beta_0 + \beta_1 x)}{s(\hat{\beta}_0 + \hat{\beta}_1 x)} \overset{\mathcal{D}}{=} t_{n-2}. \tag{10.42}$$

Proceeding in a now familiar way, we see that a $100(1 - \alpha)\%$ confidence interval for the mean response $E(Y|x)$ is given by

$$\hat{\beta}_0 + \hat{\beta}_1 x \pm t_{n-2}(\alpha/2)s(\hat{\beta}_0 + \hat{\beta}_1 x). \tag{10.43}$$

Example 10.7 *Continuation of Example 10.4*

Refer to the fuel efficiency data of Table 10.2. Compute a 95% confidence interval for the expected mpg for a car weighing 3495 lbs.

Solution We first convert the car's weight to 3.495, corresponding to the units 3495 used in the data set. All the quantities required for the following computations can be found in in Example 10.4. The point estimate (final results rounded to two decimal places) is

$$\hat{\beta}_0 + \hat{\beta}_1 x = 38.78 - 3.43 \times 3.495 = 26.78 \text{ mpg}.$$

We obtain the confidence interval by substituting the values

$$s = 1.2442, \, n = 12, \, x = 3.495, \, \bar{x} = 3.21$$

into the definition of $s(\hat{\beta}_0 + \hat{\beta}_1 x)$ given in the last line of Theorem 10.3. Next substitute the values

$$t_{10}(0.025) = 2.228 \text{ and } s(\hat{\beta}_0 + \hat{\beta}_1 x) = 0.413$$

into Equation 10.43. We obtain the confidence interval $[25.86, 27.70]$.

The scatter plot, estimated regression line, and 95% confidence limits for $E(Y|X = x)$ for the fuel efficiency data (Table 10.2) are displayed in Figure 10.7.

Prediction Interval

According to the hypotheses of the simple linear regression model the future response is given by $Y = \beta_0 + \beta_1 x + \epsilon$ and the predicted future response is given by $\hat{Y} = \hat{\beta}_0 + \hat{\beta}_1 x$, where ϵ is an $N(0, \sigma^2)$ random variable that is independent of $\hat{Y}$. Consequently,

$$V(Y - \hat{Y}) = V(Y) + V(\hat{Y}) = \sigma^2 + \sigma^2 \left[\frac{1}{n} + \frac{(x - \bar{x})^2}{S_{xx}} \right] = \sigma^2 \left[1 + \frac{1}{n} + \frac{(x - \bar{x})^2}{S_{xx}} \right].$$

Thus, the estimated standard error $s(Y - \hat{Y})$ is given by

$$s(Y - \hat{Y}) = s\sqrt{\left[1 + \frac{1}{n} + \frac{(x - \bar{x})^2}{S_{xx}} \right]} \tag{10.44}$$

Model(Y = mpg|X = weight)

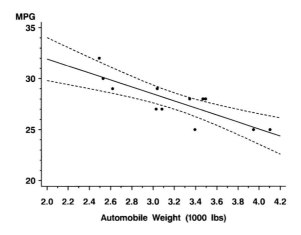

MPG

Automobile Weight (1000 lbs)

Estimated Regression Line and
95% Confidence Limits for Mean Response E(Y|X=x) (Data: Table 10.2)

FIGURE 10.7: Scatter plot, estimated regression line, and 95% confidence limits for the mean response $E(Y|X = x)$ (Table 10.2)

and the distribution of the random variable $(Y - \hat{Y})/s(Y - \hat{Y})$ is t_{n-2}, that is,

$$\frac{Y - (\hat{\beta}_0 + \hat{\beta}_1 x)}{s\sqrt{\left[1 + \frac{1}{n} + \frac{(x-\bar{x})^2}{S_{xx}}\right]}} \overset{\mathcal{D}}{=} t_{n-2}.$$

It follows in the usual way that the corresponding $100(1-\alpha)\%$ *prediction interval* is given by

$$\hat{\beta}_0 + \hat{\beta}_1 x \pm t_{n-2}(\alpha/2)s\sqrt{\left[1 + \frac{1}{n} + \frac{(x-\bar{x})^2}{S_{xx}}\right]}. \tag{10.45}$$

Example 10.8 *Computing a prediction interval*

Compute a 95% prediction interval for the mpg for a car weighing 3495 lbs. (Refer to Table 10.2 for the fuel efficiency data.)

Solution We compute a 95% prediction interval for the fuel efficiency (measured in mpg) in the same way that we computed the confidence interval except that we now use Equation 10.45 in place of 10.43. We compute $s(Y - \hat{Y})$ by making the following substitutions into Equation 10.44:

$$s = 1.2442, \, n = 12, \, x = 3.495, \, \bar{x} = 3.21, \, S_{xx} = 2.95.$$

Therefore,

$$1.2442\sqrt{\left[1 + \frac{1}{12} + \frac{(3.495 - 3.21)^2}{2.95}\right]} = 1.3111.$$

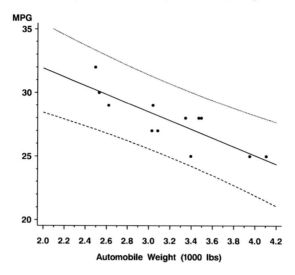

Estimated Regression Line and
95% Prediction Limits for Future Response Y (Data: Table 10.2)

FIGURE 10.8:   Scatter plot, estimated regression line, and 95% prediction intervals for the fuel efficiency data of Table 10.2

Consequently, the prediction interval is given by

$$26.78 \pm 2.228 \times 1.311 = [23.86, 29.70].$$

The scatter plot, estimated regression line, and 95% prediction intervals for the fuel efficiency data of Table 10.2 are displayed in Figure 10.8.

10.3.4   Displaying the Output of a Regression Analysis in an ANOVA Table

Regression analysis is a computationally intensive process that is best carried out via a statistical software package. The computer output, however, is displayed in the form of an *analysis of variance (ANOVA)* table, a concept involving some new ideas of independent interest.   The ANOVA approach to regression analysis is based on the sum of squares decomposition $SST = SSR + SSE$ (Equation 10.17) and a partition of the *degrees of freedom* (DF) associated with the sum of squares SST. The following elementary example illustrates the basic idea: Suppose $X_1, \ldots, X_n$ are iid $N(0,1)$ random variables. Then it is well known (see Theorems 6.2 and 6.4) that

1. $\sum X_i^2$ has a chi-square distribution with n degrees of freedom;

2. $\sum (X_i - \overline{X})^2$ has a chi-square distribution with $n-1$ degrees of freedom;

3. $n\overline{X}^2$ has a chi-square distribution with 1 degree of freedom.

The partition of the sum of squares $\sum X_i^2 = n\overline{X}^2 + \sum (X_i - \overline{X})^2$ yields the following partition of the degrees of freedom: $n = 1 + (n-1)$. Similarly, the sum of squares partition

$$SST = \sum (y_i - \overline{y})^2 = \sum (\hat{y}_i - \overline{y})^2 + \sum (y_i - \hat{y}_i)^2 = SSR + SSE$$

yields the following partition of the degrees of freedom: $n - 1 = 1 + (n - 2)$. In detail: The SST term on the left hand side has $n - 1$ degrees of freedom, and the SSE term on the right has $n - 2$ degrees of freedom. This is a consequence of the fact that the random variables SSR and SSE are mutually independent. (We omit the technical details.)

A *mean square* (MS) is a sum of squares divided by its degrees of freedom. Thus, the *mean square due to regression* (MSR) and *mean square error* (MSE) are calculated by dividing SSR and SSE by their respective degrees of freedom. Their ratio, MSR/MSE, is a random variable denoted F because it has an F distribution, as will be explained shortly. These quantities play an important role in statistical inferences about the regression coefficients.

The sum of squares partition and the partition of the degrees of freedom are conveniently summarized in an ANOVA table as shown in Table 10.5.

Table 10.5 *ANOVA table for the simple linear regression model*

Source	DF	Sum of Squares	Mean Square	F value	*Prob > F*
Model	1	SSR	MSR=SSR/1	MSR/MSE	
Error	$n - 2$	SSE	MSE=SSE/$(n - 2)$		
Total	$n - 1$	SST			

We use the F ratio MSR/MSE to test the null hypothesis $H_0 : \beta_1 = 0$ against $H_1 : \beta_1 \neq 0$. This follows from the fact that $MSR = SSR$ in the simple linear regression model and Equation 10.34, which states that

$$E(MSR) = \sigma^2 + \beta_1^2 S_{xx}.$$

Consequently, when $H_1 : \beta_1 \neq 0$ is true, we have

$$E(MSR) = \sigma^2 + \beta_1^2 S_{xx} > \sigma^2 = E(MSE).$$

We therefore reject H_0 at the level α when $F > F_{1,n-2}(\alpha)$. This test is in fact equivalent to the one based on the t distribution discussed previously (Equation 10.41). This is a consequence of the result that $t_\nu^2 = F_{1,\nu}$ (Corollary 6.2).

Example 10.9

Display the results of the regression analysis of Example 10.1 in the format of an ANOVA table.

Solution The ANOVA table of the edited SAS printout is shown in Table 10.6.

Table 10.6 *Anova table for Example 10.1*

```
                   Analysis of Variance
                     Sum of         Mean
     Source     DF    Squares       Square      F Value     Prob>F
     Model       1  66440.38346  66440.38346     75.828     0.0001
     Error       8   7009.61654    876.20207
     C Total     9  73450.00000
          Root MSE   29.60071     R-square      0.9046
```

Interpreting the output In Table 10.6 the names for SSR, SSE, and SST are listed in the first column, titled *Source*, which stands for the source of the variation. SSR is called *Model*, SSE is called *Error*, and SST is called *C Total*, which is short for *corrected total*[1]. The second column, titled DF, indicates the *degrees of freedom*. The values of SSR, SSE, and SST appear in column three, *Sum of Squares*. The value of the F ratio appears in the column titled F value, and the last column (Prob > F) contains the P value, which in this case equals $P(F_{1,8} > 75.828) < 0.001$. Refer back to Table 10.4, where the value of the t statistic for testing $H_0 : \beta_1 = 0$ is calculated as 8.708, and note that $75.828 = (8.708)^2$, a consequence of the fact that $F_{1,\nu} = t_\nu^2$. In other words, the F test in the ANOVA table and the t test for $\beta_1 = 0$ are equivalent.

The term Root MSE in the table is the square root of the mean square error; that is, $\sqrt{MSE} = \sqrt{876.20207} = 29.60071$. The coefficient of determination is printed as R–square in the table; in this case $R^2 = 0.9046$.

Example 10.10 *Continuation of Example 10.2*

Display the results of the regression analysis of Model $(Y = mpg|X = weight)$ of Example 10.2 in the format of an ANOVA table.

Solution The SAS printout of the ANOVA table is displayed in Table 10.7.

Table 10.7 *Anova table for Example 10.2*

Analysis of Variance

Source	DF	Sum of Squares	Mean Square	F Value	Prob>F
Model	1	34.76972	34.76972	22.461	0.0008
Error	10	15.48028	1.54803		
C Total	11	50.25000			

	Root MSE	1.24420	R-square	0.6919

10.3.5 Curvilinear Regression

It sometimes happens that the points of the scatter plot appear to lie on a curve, ruling out the simple linear regression model. Nevertheless, it may still be possible to fit a curve to these data by transforming them so that the least squares method can be brought to bear on the transformed data. The following example illustrates the technique.

Example 10.11

The data in Table 10.8 come from a series of experiments to determine the relationship between the viscosity v (measurement units omitted) of the compound heptadecane as a function of the temperature t (measured in Kelvins). The scatter plot of (t_i, v_i), shown in Figure 10.9, indicates clearly that the data points lie on a curve, so the simple linear regression model is not appropriate.

---

[1]In SAS terminology the sum $\sum (y_i - \bar{y})^2$ is called C Total to distinguish it from the sum $\sum y_i^2$, the uncorrected sum of squares.

Table 10.8

t	v	$y = \ln v$	$1/t$	$\ln t$
303.15	*3.291*	*1.19119*	*0.0032987*	*5.71423*
313.50	*2.652*	*0.97531*	*0.0031898*	*5.74780*
323.15	*2.169*	*0.77427*	*0.0030945*	*5.77812*
333.15	*1.829*	*0.60377*	*0.0030017*	*5.80859*
343.15	*1.557*	*0.44276*	*0.0029142*	*5.83817*
353.15	*1.340*	*0.29267*	*0.0028317*	*5.86689*
363.15	*1.161*	*0.14928*	*0.0027537*	*5.89482*
373.15	*1.014*	*0.01390*	*0.0026799*	*5.92198*
393.15	*0.794*	*-0.23067*	*0.0025436*	*5.97419*
413.15	*0.655*	*-0.42312*	*0.0024204*	*6.02381*
433.15	*0.546*	*-0.60514*	*0.0023087*	*6.07108*
453.15	*0.460*	*-0.77653*	*0.0022068*	*6.11622*
473.15	*0.392*	*-0.93649*	*0.0021135*	*6.15941*
493.15	*0.339*	*-1.08176*	*0.0020278*	*6.20081*
513.15	*0.296*	*-1.21740*	*0.0019487*	*6.24057*
533.15	*0.260*	*-1.34707*	*0.0018756*	*6.27880*
553.15	*0.229*	*-1.47403*	*0.0018078*	*6.31563*
573.15	*0.203*	*-1.59455*	*0.0017447*	*6.35115*

(Source: D. S. Viswanath and G. Natarjan (1989), *Data Book on the Viscosity of Liquids*, p. 363, Hemisphere Publishing Corporation, New York.)

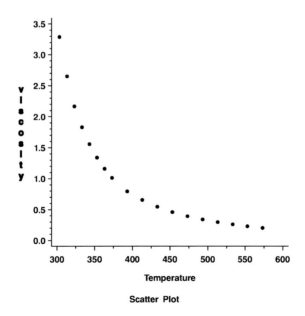

FIGURE 10.9:   Scatter plot of viscosity data (Table 10.8)

Chemists believe that a curve defined by either $v = a \exp(b/t)$, or $v = at^b$ (where a and b are to be determined), provides a good fit to the observed data. We convert this curve fitting problem to a linear regression problem by taking logarithms:

$$v = a \exp(b/t), \text{ so } \ln v = \ln a + b\frac{1}{t}; \tag{10.46}$$

$$v = at^b, \text{ so } \ln v = \ln a + b \ln t. \tag{10.47}$$

Thus, to fit the curve defined by Equation 10.46 we note that the transformed variables $y = \ln v$ and $x = 1/t$ satisfy the linear equation

$$y = \beta_0 + \beta_1 x, \ (\beta_0 = \ln a, \ \beta_1 = b).$$

(Models of the type $v = a \exp bt$, or $v = at^b$, are called *intrinsically linear* because they can be transformed into linear models by means of a suitable change of variables.) Therefore, we use the least squares method to fit a straight line to the transformed data, shown in columns 3 and 4 of Table 10.8. That is, we solve the curvilinear regression problem by first fitting a straight line to the transformed data $y_i = \ln v_i$, $x_i = 1/t_i$, as in Section 10.2.1, and then estimate the parameters of the curve $v = a \exp(b/t)$ using the inverse transformation $a = \exp \beta_0$ and $b = \beta_1$; thus, $\hat{a} = \exp(\hat{\beta}_0)$ and $\hat{b} = \hat{\beta}_1$.

Solution The regression model statement for the transformed variables is Model (Y= $\ln v | x = 1/t$). That is, we assume

$$Y_i = \beta_0 + \beta_1 x_i + \epsilon_i, \text{ where, } Y_i = \ln v_i \text{ and } x_i = \frac{1}{t_i}.$$

In other words, we assume the simple linear regression model of Equation 10.22 is valid for the transformed data. The scatter plot and estimated regression line for the transformed data, shown in Figure 10.3.5, suggest a strong linear relation between the transformed variables. This is confirmed by the formal results of the regression analysis as summarized in Table 10.9, which shows the ANOVA table followed by the estimates for the intercept and slope:

$$\hat{\beta}_0 = -4.654467 \text{ and } \hat{\beta}_1 = 1754.551654;$$
$$\hat{a} = \exp(-4.654467) = 0.00961273 \text{ and } \hat{b} = 1754.551654.$$

The coefficient of determination equals 0.9993.

Table 10.9 *Anova table and parameter estimates for Model* $(Y = ln(v)|x = 1/t)$

Analysis of Variance

Source	DF	Sum of Squares	Mean Square	F Value	Prob>F
Model	1	13.36161	13.36161	21593.985	0.0001
Error	16	0.00990	0.00062		
C Total	17	13.37151			

Root MSE	0.02487	R-square	0.9993

Parameter Estimates

Variable	Parameter Estimate	Standard Error	T for H0: Parameter=0	Prob > |T|
INTERCEP	-4.654467	0.03026500	-153.790	0.0001
SLOPE	1754.551654	11.93987456	146.949	0.0001

We now mention several problems arising in curvilinear regression that will not be dealt with here, but are discussed in R.H. Myers (1990), *Classical and Modern Regression with Applications,* 2nd ed., PWS-Kent, Boston.

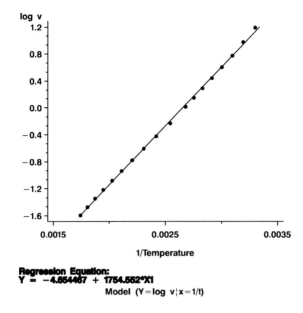

FIGURE 10.10: Scatter plot and estimated regression line for the transformed data of Example 10.11

1. There is in general no "best" solution to fitting a curve to a set of empirical data. For example, one can sometimes find a polynomial of sufficiently high degree that passes through all the points of the scatter plot.

2. It is important to note that fitting a straight line to the transformed data is not equivalent to applying the least squares method to the original data. For instance, minimizing the sum of squares

$$\sum_i [v_i - a \exp(bu_i)]^2$$

produces estimates for a and b that are not equal to the transforms of the least squares estimates. Thus, $\exp(\hat{\beta}_0) \neq \hat{a}$ and $\hat{\beta}_1 \neq \hat{b}$.

3. The form of the error term for an intrinsically linear model is also transformed. Suppose, for example, the transformed model

$$\ln y = \beta_0 + \beta_1 x + \epsilon$$

satisfies the standard conditions of the simple linear regression model. Taking exponentials of both sides yields the following model for the original data: $y = a \exp(bx) \exp(\epsilon)$, $(a = \exp(\beta_0), b = \beta_1)$. This is not the same as the model $y = a \exp(bx) + \epsilon$.

10.4 Model Checking

An important part of any regression analysis is determining whether or not the standard assumptions of the simple linear regression model are satisfied. We recall that the relationship between the response and explanatory variables is assumed to be of the form

$Y = f(x) + \epsilon$, where $f(x) = \beta_0 + \beta_1 x$ is a (non-random) function of x, and ϵ represents the experimental error. It is instructive to think of $f(x)$ as a *signal,* ϵ as the *noise,* and Y as the observed, or received, signal. *The goal of regression analysis is to extract the signal $f(x)$ from the noise.* Model checking refers to a collection of graphical displays, called *residual plots,* that help the statistician evaluate the validity of the linear regression model.

Residual Plots It can now be shown that the residuals are normally distributed with mean, variance, and estimated standard error given by:

$$E(e_i) = 0; \ V(e_i) = \sigma^2 \left(1 - \left[\frac{1}{n} + \frac{(x_i - \bar{x})^2}{S_{xx}} \right] \right);$$

$$s(e_i) = s \sqrt{\left(1 - \left[\frac{1}{n} + \frac{(x_i - \bar{x})^2}{S_{xx}} \right] \right)}.$$

The residuals are not, however, mutually independent since $\sum_i e_i = 0$. It can be shown, however, that the vector of residuals $(e_1, \ldots, e_n)$ and the vector of fitted values $(\hat{y}_1, \ldots, \hat{y}_n)$ are mutually independent. These results imply that when the regression model is valid, the residual plot $(\hat{y}_i, e_i)$ should appear to be a set of points, randomly scattered about the line $y = 0$ and such that approximately 95% of them lie within $\pm 2s$ of zero. Another plot, closely related to the one just described, is the *studentized residual plot* consisting of $(\hat{y}_i, e_i^*)$ where

$$\text{Studentized residual: } e_i^* = \frac{e_i}{s(e_i)}.$$

The studentized residuals should appear to be randomly scattered about the line $y = 0$ and approximately 95% of them should lie within ± 2 of zero. Because of the tedious nature of the computations, you are advised to graph these plots using a computer.

Example 10.12 *Continuation of Example 10.1*

Draw the plots of (1) the residuals and (2) the studentized residuals for the regression Model $(Y = $ shear strength $| X = $ weld diameter$)$ of Example 10.1. Are the residual plots consistent with the hypotheses of the simple linear regression model?

   Solution The residual and studentized residual plots are shown in Figures 10.11 and 10.12. Looking at the plot of the studentized residuals (Figure 10.12) we note that the points appear to be randomly scattered within a horizontal band of width 2 centered at about zero; the same is true for the residuals plot, Figure 10.11, since $\pm 2s = \pm 2 \times 29.6 = \pm 59.2$. This is consistent with the scatter plot and estimated regression line shown in Figure 10.4. We conclude that the data do not conflict with the hypotheses of the simple linear regression model.

Example 10.13 *Continuation of Example 10.2*

Draw the plot of the studentized residuals for the regression Model$(Y = $ MPG $| X = $ weight$)$ of Example 10.2. Do the assumptions of the simple linear regression model appear to be valid?

   Solution The plot of the studentized residuals for Model$(Y = $ MPG $| X = $ weight$)$ is shown in Figure 10.13. All studentized residuals appear to be randomly scattered within a horizontal band of width 2 centered about 0, so the model appears to be consistent with the hypotheses of the simple linear regression model.

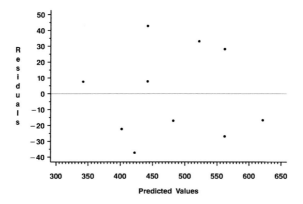

FIGURE 10.11: Plot of residuals against predicted values (data from Table 10.1)

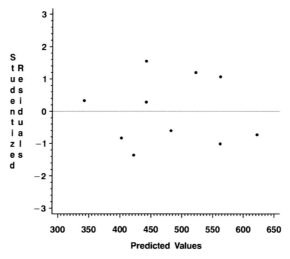

FIGURE 10.12: Plot of studentized residuals against predicted values (data from Table 10.1)

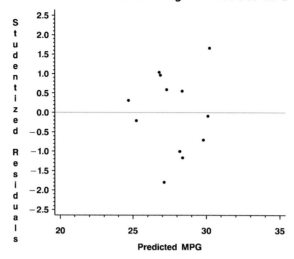

FIGURE 10.13:   Plot of studentized residuals against predicted mpg (fuel efficiency data in Table 10.2)

Example 10.14 *A residual plot reveals non-constant variances*

Refer to the data set of Problem 10.21 and Problem 10.22. Is the simple linear regression model appropiate in this case? Looking at Figure 10.14, which is a plot of the residuals for Model (Y = time $|X$ = data bytes) (ARGUS DATA) of Problem 10.21, suggests that the the linear regression model is not appropriate. Notice that the variances of the residuals seem to increase with $\hat{y}_i$; the condition that the variances be constant appears to be violated in this case. This suggests that a linear regression model is not appropriate here. Additional evidence for this conclusion comes from a normal probability plot of the residuals, as will now be explained.

Normal Probability Plots of the Residuals We use normal probability plots (defined in Section 8.4) of the residuals to check the assumption that the error terms $\epsilon_1, \ldots, \epsilon_n$ are iid normal random variables.

Example 10.15 *Continuation of Example 10.14*

Draw the normal probability plot of the residuals for the Model (Y = time $|X$ = data bytes) (ARGUS DATA) of Problem 10.21. What departures from the simple linear regression model are suggested by the normal probability plot?

Solution The normal probability plot is shown in Figure 10.15. The points do not appear to lie on a straight line, which strengthens the conclusion obtained earlier that the simple linear regression model is inappropriate.

We conclude with a brief summary of the criteria that statisticians recommend in order to assess the validity of the simple linear regression model.

The simple linear regression model is inappropriate if

1. the scatter plot does not exhibit, at least approximately, a linear relationship.

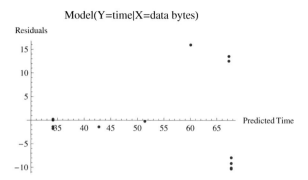

FIGURE 10.14: Plot of residuals against predicted time: Example 10.14

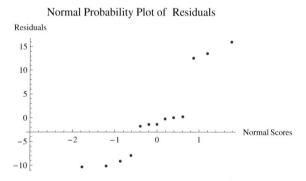

FIGURE 10.15: Normal probability plot of residuals: Example 10.15

2. the residuals exhibit a pattern, that is, if the residuals do not appear to be randomly scattered about the line $y = 0$ or approximately 95% of them do not lie within $\pm 2s$ of zero, then the linear regression model might not be appropriate.

3. the error variances $V(\epsilon_1), \ldots, V(\epsilon_n)$ are not, at least approximately, equal. For an example of where this might not be the case, see Figure 10.14.

4. the error terms $\epsilon_1, \ldots, \epsilon_n$ are not normally distributed; use normal probability plots.

5. the error terms are not independent. This may occur if the measurements are taken in a time sequence during which the experimental apparatus is affected by the physical conditions under which previous measurements were taken. To detect this statisticians recommend a plot of the residuals versus the *time order* in which the observations are taken. If the independence assumption is valid, there should be no pattern, e.g., quadratic or linear, in the plot.

10.5   Correlation Analysis

Correlation analysis studies the *linear* relationship between two variables (X, Y) defined on the *same* population using methods similar to those used in simple linear regression (SLR); the underlying data structures, however, are quite different, as illustrated in Example 10.16. The main goal of SLR is estimating the parameters β_0, β_1 of the linear regression function $\beta_0 + \beta_1 x$, whereas the primary focus of correlation analysis is estimating the correlation coefficient ρ (Equation 10.6). The two problems are, of course, linked, since the coefficient of determination R^2 equals r^2, the square of the sample correlation; the difference between them is that R^2 is a measure of how well the model fits the data, while r is an estimate of the correlation coefficient ρ, a population parameter.

With these preliminaries out of the way, some examples of correlation analysis, which occur frequently in modern portfolio theory, will now be given.

Example 10.16 *Does the size of a mutual fund influence its rate of return?*

The bivariate data set in Table 10.10 (Source: *NY Times,* Feb. 3, 1991) records the size, in billions of dollars, and the five year rate of return of 25 General Equity (stock) mutual funds. The population consists of all such stock funds; the variables are size (X, billions of dollars) and return (Y, in percent).

Table 10.10 *Size and five year rate of return for 25 mutual funds*

Size	Return	Size	Return	Size	Return	Size	Return
12.325	104	2.328	79	3.191	86	3.426	77
6.534	62	2.198	92	2.786	78	3.21	74
5.922	97	2.171	62	2.525	68	1.897	84
5.606	88	2.154	60	2.514	73	1.729	120
4.356	61	2.1172	93	1.69	98		
3.925	48	2.06	99	1.578	97		
3.759	52	1.908	111	1.575	65		

Does the size of a mutual fund influence its rate of return? In particular, do larger funds have higher rates of return? To answer this question we first draw the scatter plot $(x_i, y_i), i = 1, 2, \ldots, 25$, where (x_i, y_i) denote the size and return of the ith mutual fund (Figure 10.16). The scatter plot does not indicate a positive, or a negative, association between the size of a mutual fund and its rate of return. In this case $r = 0.029$, which means that the size of the fund should not be an important factor for an investor. Before continuing with our analysis note the difference between these bivariate data and those in Table 10.1, where X, the weld diameter, is specified by the engineer and Y, the shear strength, is the response.

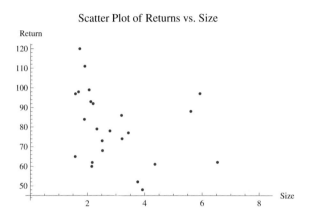

FIGURE 10.16: Scatter plot of mutual fund's rate of return vs. size (billions of dollars); data from Table 10.10

10.5.1 Computing the Market Risk of a Stock

Comparing the risks and returns of alternative investment strategies is an important task for investors. An important parameter for evaluating the riskiness of a stock—and whether or not a stock should be added to a portfolio—is the stock's β, which we discussed earlier in Example 1.30. We continue that discussion here, but from the more advanced perspective of the simple linear regression model. In particular, we note that β is the slope of the regression line for the Model (Y = stock return $|X$ = market return) (this is equivalent to Definition 1.10). More precisely, we consider the *market model*

$$Y = \beta_0 + \beta X + \epsilon, \text{ where,} \tag{10.48}$$

$$Y = \text{ return on asset for some fixed period;}$$

$$X = \text{ return on market portfolio for the same fixed period.}$$

Intuitively, β measures the sensitivity of a stock's future returns to the returns of the market portfolio (defined here as the S&P500). It plays an important role in the capital asset pricing model (CAPM), which describes the relationship between a stock's beta and its expected return; in other words, it helps us to decide whether or not a stock's price is justified in view of the risks incurred investing in it. The market model of Equation 10.48 and the role of β in CAPM and are discussed in G.J. Alexander, W.F. Sharpe, J.V. Bailey (2001), *Fundamentals of Investments*, 3rd ed., Prentice-Hall, p. 157 and p. 190.

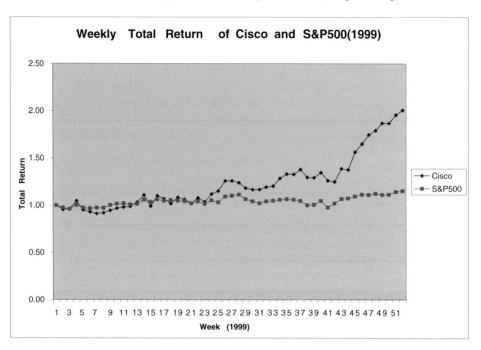

FIGURE 10.17:   Weekly total returns of Cisco and S&P500 (1999)

Example 10.17 *Measuring the beta (β) of a stock*

The concepts of a stock's β is best understood in the context of an example based on real stock market data. Consider, then, Figure 10.17, which displays the time series of the weekly total returns for Cisco Systems and the S&P500 for 1999 (the last year of the 1990s bull market). These data are listed in Table 10.11. **Note:** Cisco Prices are adjusted for stock splits, etc. The sequence of weekly total returns is the time series $S_i/S_0\ i = 1, 2\ldots$, where S_0 is the initial price and S_i is the ith weekly closing price. For example, the initial and final price of Cisco Systems on Jan. 4, 1999, and Dec. 27, 1999, were $S_0 = \$26.67$ and $S_{52} = \$53.56$, respectively; yielding a remarkable 201% total return—a confirmation of the "irrational exuberance" description of the 1990s bull market. By way of comparison, the total return for the same period of the S&P500 was 115.2%, and its annual return (excluding dividends and commissions) a more modest 15.2%. Figure 10.18, by way of contrast, shows the relative performance of Cisco versus the S&P500 for the year 2002. These data are listed in Table 10.12. During this year, a classic example of a bear market, the S&P500 lost 25%, but the losses for Cisco's investors were far more severe; they lost 37%. This example is a reminder of the fact, which too many investors sometimes forget, that the law of gravity remains valid, even during bull markets.

Looking at Figures 10.17 and 10.18 we see that in a bull market Cisco's gains, are significantly greater than the S&P500 index, but, in a bear market, its losses are much greater. In other words, it's a much riskier investment. Of course, a portion of Cisco's return must be attributed to the performance of the S&P500, and the rest must be attributed to the unique features of Cisco itself. To study the relation between the returns of Cisco Systems (Y) and the S&P500 (X) we proceed as in Example 1.30; that is, we draw the scatter plot (x_i, y_i), where x_i and y_i are the ith weekly returns of the S&P500 and Cisco Systems, respectively, and the least squares fitted line (Figure 10.19). In this example, the estimates for the intercept and slope are $\beta_0 = 0.0106$, $\beta = 1.445$, with sample correlation $r = 0.717$.

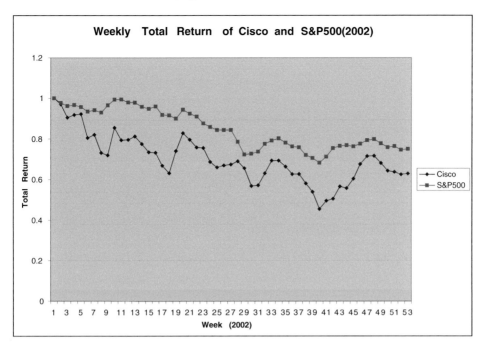

FIGURE 10.18: Weekly total returns of Cisco and S&P500 (2002)

It is to be observed, however, that in this example both X and Y are random variables, consequently, X is not a predictor variable, nor is Y a response variable, in the sense of the simple linear regression model discussed in Section 10.3.

A Partition of Total Risk

Equation 10.48 yields a partition of the *total risk* of a stock into two terms, both of great importance to investors. First, some notation: Denote the market risk by $\sigma_M^2 = Var(X)$ and the variance of the error term by σ_ϵ^2. Assuming the error term ϵ is independent of the market return X, Equation 10.48 implies that

$$\sigma_Y^2 = \beta^2 \sigma_M^2 + \sigma_\epsilon^2 \tag{10.49}$$

where,

1. $\beta^2 \sigma_M^2$ is the market (or systematic) risk of the stock, and

2. σ_ϵ^2 is the *unique risk* of the stock.

Interpretation of β We begin with the observation that the market portfolio's $\beta = 1$, since it is perfectly correlated with itself, that is, the points of the scatter plot (x_i, x_i), $i = 1, \ldots, n$ lie on the straight line with slope 1. The market model for Cisco—based on the stock market data—is:

$$\text{Cisco return} = 0.0106 + 1.44 \times \text{S\&P500 return}.$$

Consequently, if the S&P index has an increase of 10% then the expected increase of Cisco is 14.4%; if the S&P500 index suffers a 10% decrease then the expected decrease of Cisco is -14.4%. Consider now the bond fund PTTRX whose returns are listed in Table 1.19. This is an example of a portfolio with $\beta = 0.102 < 1$ (Problem 10.44 asks you to verify this.) In this case, when the market declines 10% of PTTRX's expected return will be

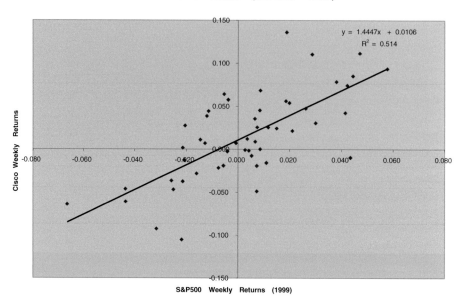

Scatter Plot of Returns (S&P500, Cisco)

FIGURE 10.19:   Scatter plot, least squares fitted line of Cisco's returns vs. S&P returns (1999); $\beta = 1.44$

down by about 1%. Stocks with $\beta < 1$ are called *defensive stocks* and stocks with $\beta > 1$ are called *aggressive stocks*. In other words, stocks with $\beta < 1$ have less risk and stocks with $\beta > 1$ have more risk than the market portfolio. In the current finance theory literature β is expressed in the following equivalent form:

$$\beta = \frac{Cov(X, Y)}{\sigma_X^2}. \tag{10.50}$$

This is easily derived from Equation 10.11; we leave the details to the reader.

A Statistical Model for Correlation Analysis The statistical analysis of a bivariate data set, such as the weekly returns of the S&P500 and Cisco Systems, assumes that the data come from a random sample drawn from a bivariate normal population. In particular, we assume that the observations $(X_1, Y_1), \ldots, (X_n, Y_n)$ come from a bivariate normal distribution with parameters μ_X, μ_Y, σ_X, σ_Y, ρ as defined by Equation 5.143. The assumption that (X, Y) has a bivariate normal distribution can be justified on mathematical grounds—e.g., the multidimensional central limit theorem—and by empirical data since the frequency histograms of returns often follow a bell shaped curve. Recall that, in this case, the regression of Y on X (see Equation 5.148) is given by

$$E(Y|X = x) = \mu_{Y|x} = \mu_Y + \rho \frac{\sigma_Y}{\sigma_X}(x - \mu_X). \tag{10.51}$$

For example, the point estimate for the correlation coefficient between the S&P500 and Cisco returns scores is $r = 0.717$ (Example 10.17) and the correlation coefficient between the size and return of a mutual fund is $r = 0.029$ (Example 10.16). Obtaining a confidence interval for ρ is complicated since the exact distribution of r is known only in the case when $\rho = 0$. For this reason, we shall content ourselves with discussing how to test the null hypothesis $H_0 : \rho = 0$.

The Sampling Distribution of r The following theorem gives the distribution of the sample correlation coefficient under the null hypothesis.

Theorem 10.4 *When $H_0 : \rho = 0$ is true, then*

$$T = \frac{r\sqrt{n-2}}{\sqrt{1-r^2}} \tag{10.52}$$

has a t distribution with $n-2$ degrees of freedom.

Example 10.18 *Continuation of Example 10.16*

Based on the data of Example 10.16 and the T statistic of Equation 10.52, does the size of a mutual fund determine its rate of return?

Solution In this case there are several choices for the alternative hypothesis; for example, we might conjecture that a larger fund has a tendency to perform better, which suggests that we choose the alternative hypothesis to be $H_1 : \rho > 0$; that is, we will reject the null hypothesis at the level α if $T > t_{n-2}(\alpha)$. Choosing $\alpha = 0.05$ and $n = 25$ produces the critical region $\{T > t_{23}(0.05) = 1.714\}$. The value of the T statistic in this case is given by

$$T = \frac{0.029\sqrt{23}}{\sqrt{1-0.029^2}} = 0.1391 < 1.714.$$

Consequently, the null hypothesis cannot be rejected in this case. In other words, a larger fund size does not necessarily produce a larger rate of return.

10.5.2 The Shapiro–Wilk Test for Normality

In Section 8.4 we introduced the Q–Q plot as a useful graphical tool for checking the normality of a given data set. Recall that the idea behind the Q–Q plot is this: If the data $\{x_1, \ldots, x_n\}$ come from a normal distribution, then the points of the Q–Q plot $(z_j, x_{(j)}), j = 1, \ldots, n$, will appear to lie on a straight line; here $x_{(1)}, \ldots, x_{(n)}$ denote the order statistics and z_j denotes the normal scores, i.e., $\Phi(z_j) = (j - 0.5)/n$). We remark that the mean and variance of the order statistics are the same as the mean and variance of the original data set, since the order statistics are just a permutation of the original data set. Now, one way of measuring the straightness of the Q–Q plot is to compute its correlation coefficient, denoted in the SAS computer package by W, i.e., compute

$$W = \frac{\sum (x_{(j)} - \bar{x})(z_j - \bar{z})}{\sqrt{S_{xx}}\sqrt{S_{zz}}}.$$

We reject the hypothesis of normality if $W < w$. The idea of basing a test of normality on this statistic is due to Shapiro and Wilk, and this test is available as an option in most computer software packages (e.g., PROC UNIVARIATE of SAS).

Example 10.19

Draw the normal probability plots and compute the Shapiro–Wilk statistic for the studentized residuals of (1) Model ($Y = $ mpg $|X =$ weight) and (2) Model ($Y = $ gpm $|X =$ weight).
    Solution

1. The normal probability plot is shown in Figure 10.20. Using SAS (PROC UNIVARIATE), we obtained the value $W = 0.9793$ with a P-value $P(W < 0.9793) = 0.9519$. The hypothesis of normality is not rejected.

Normal Probability Plot of Studentized Residuals

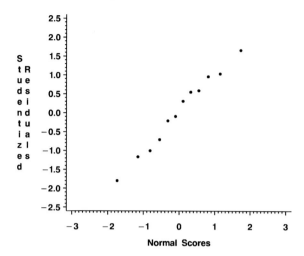

Model (Y=mpg¦X=weight) (Data: Table 10.2)

FIGURE 10.20:   Normal probability plot of studentized residuals (fuel efficiency data, Model $(Y = mpg|X = weight)$, Table 10.2)

2. The normal probability plot for the regression Model $(Y = \text{gpm} \,|X =\text{weight})$ is shown in Figure 10.21. Using SAS (PROC UNIVARIATE) we obtain the value $W = 0.9549$ for the Shapiro-Wilk statistic with a P-value $P(W < 0.9549) = 0.6579$. This is not strong evidence against the hypothesis of normality.

We conclude that both models appear to satisfy the conditions of the simple linear regression model. So which one is better? Comparing the coefficients of determination of the two models gives a slight edge to Model $(Y = \text{gpm} \,|X = \text{weight})$, since its coefficient of determination equals 0.7052, while the coefficient of determination for the Model $(Y = \text{mpg} \,|X = \text{weight})$ equals 0.6919.

10.6   Mathematical Details and Derivations

The details of the derivations of Equation 10.9 and the sampling distributions for $\hat{\beta}_0$, $\hat{\beta}_1$, and related random variables will now be given.

Derivation of the Sum of Squares Partition (10.9)

We begin with the following algebraic identity:

$$y_i - b_0 - b_1 x_i = (y_i - \overline{y}) - b_1(x_i - \overline{x}) + (\overline{y} - b_0 - b_1\overline{x}).$$

The corresponding sum of squares $Q(b_0, b_1)$ takes the form

$$Q(b_0, b_1) = \sum [(y_i - \overline{y}) - b_1(x_i - \overline{x}) + (\overline{y} - b_0 - b_1\overline{x})]^2.$$

This can be simplified by noting that

$$\sum (x_i - \overline{x}) = \sum (y_i - \overline{y}) = 0$$

Normal Probability Plot of Studentized Residuals

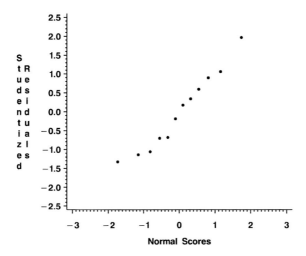

Model (Y=gpm¦X=weight) (Data: Table 10.2)

FIGURE 10.21: Normal probability plot of studentized residuals (fuel efficiency data, Model $(Y = gpm|X = weight)$, Table 10.2)

and therefore

$$\sum (y_i - \overline{y})\overline{y} = 0;$$

$$\sum (y_i - \overline{y})(\overline{y} - b_0 - b_1\overline{x}) = 0;$$

$$\sum b_1(x_i - \overline{x})(\overline{y} - b_0 - b_1\overline{x}) = 0.$$

Consequently,

$$Q(b_0, b_1) = \sum (y_i - \overline{y})^2 - 2 \sum b_1(x_i - \overline{x})(y_i - \overline{y}) + b_1^2 \sum (x_i - \overline{x})^2$$
$$+ n(\overline{y} - b_0 - b_1\overline{x})^2$$
$$= S_{yy} - 2b_1 S_{xy} + b_1^2 S_{xx} + n(\overline{y} - b_0 - b_1\overline{x})^2$$
$$= S_{yy}(1 - r^2) + (b_1\sqrt{S_{xx}} - r\sqrt{S_{yy}})^2 + n(\overline{y} - b_0 - b_1\overline{x})^2,$$

where the second summand in the last equation comes from "completing the square" of the term $-2b_1 S_{xy} + b_1^2 S_{xx}$.

Proof of Theorem 10.1

The most efficient way to derive the sampling distribution for $\hat{\beta}_1$ is to rewrite it as a linear combination of mutually independent, normally distributed random variables Y_i of the sort discussed in conjunction with Theorem 6.1. The key idea is the representation

$$\hat{\beta}_1 = \sum d_i Y_i, \text{ where} \tag{10.53}$$

$$d_i = \frac{x_i - \overline{x}}{S_{xx}}. \tag{10.54}$$

The representation 10.53 is a consequence of the following alternative expression for S_{xy}:

$$S_{xy} = \sum (x_i - \overline{x})y_i. \tag{10.55}$$

Equation 10.55 is a consequence of the following sequence of algebraic manipulations:

$$S_{xy} = \sum (x_i - \overline{x})(y_i - \overline{y}) = \sum (x_i - \overline{x})y_i - \sum (x_i - \overline{x})\overline{y} = \sum (x_i - \overline{x})y_i,$$

where we used the fact that $\sum (x_i - \overline{x})\overline{y} = \overline{y} \sum (x_i - \overline{x}) = 0$.

Incidentally, the same argument shows that $S_{xx} = \sum x_i(x_i - \overline{x})$. Thus, $S_{xY} = \sum (x_i - \overline{x})Y_i$ and Equation 10.25 together imply that

$$\hat{\beta}_1 = \frac{S_{xY}}{S_{xx}}$$

$$= \sum \frac{(x_i - \overline{x})}{S_{xx}} Y_i.$$

This completes the derivation of Equation 10.53.

Applying Theorem 6.1 and using the facts that $E(Y_i) = \beta_0 + \beta_1 x_i$ and $V(Y_i) = \sigma^2$ yield the result that $\hat{\beta}_1$ is normally distributed with mean

$$E(\beta_1) = \sum \left(\frac{x_i - \overline{x}}{S_{xx}} \right) (\beta_0 + \beta_1 x_i)$$

$$= \frac{\beta_0}{S_{xx}} \sum (x_i - \overline{x}) + \frac{\beta_1}{S_{xx}} \sum x_i(x_i - \overline{x})$$

$$= \frac{\beta_0}{S_{xx}} \times 0 + \frac{\beta_1}{S_{xx}} \times S_{xx} = \beta_1,$$

and variance

$$V(\hat{\beta}_1) = \sum \left(\frac{x_i - \overline{x}}{S_{xx}} \right)^2 \times \sigma^2$$

$$= \frac{S_{xx}}{S_{xx}^2} \times \sigma^2 = \frac{\sigma^2}{S_{xx}}.$$

The sampling distribution of $\hat{\beta}_0$ is derived in a similar way; we only sketch the outlines, leaving the details as an exercise.

A routine calculation shows that

$$E(\overline{Y}) = \beta_0 + \beta_1 \overline{x} \text{ and therefore}$$
$$E(\hat{\beta}_0) = E(\overline{Y} - \hat{\beta}_1 \overline{x})$$
$$= E(\overline{Y}) - E(\hat{\beta}_1)\overline{x} = \beta_0 + \beta_1 \overline{x} - \beta_1 \overline{x} = \beta_1.$$

This shows that $\hat{\beta}_0$ is an unbiased estimator of β_0, as claimed. The next step is to represent $\hat{\beta}_0$ as a linear combination of $Y_1, \dots, Y_n$:

$$\hat{\beta}_0 = \overline{Y} - \hat{\beta}_1 \overline{x} = \sum c_i Y_i \qquad (10.56)$$

where

$$c_i = \frac{1}{n} - \overline{x} \left(\frac{x_i - \overline{x}}{S_{xx}} \right). \qquad (10.57)$$

Since the random variables $Y_1, \dots, Y_n$, are independent, it follows that

$$V\left(\sum c_i Y_i \right) = \sum c_i^2 \sigma^2 = \sigma^2 \sum c_i^2.$$

We leave it to you as a problem to show that

$$\sum c_i^2 = \sum \left[\frac{1}{n} - \bar{x} \left(\frac{x_i - \bar{x}}{S_{xx}} \right) \right]^2 = \frac{1}{n} + \frac{\bar{x}^2}{S_{xx}}; \text{ therefore} \tag{10.58}$$

$$V(\hat{\beta}_0) = \sigma^2 \left[\frac{1}{n} + \frac{\bar{x}^2}{S_{xx}} \right]. \tag{10.59}$$

This completes the proof of Theorem 10.1.

Derivation of Equation 10.34

Using the fact that the fitted values and $\bar{y}$ are given by $\hat{y}_i = \hat{\beta}_0 + \hat{\beta}_1 x_i$, and $\bar{y} = \hat{\beta}_0 + \hat{\beta}_1 \bar{x}$, we see that

$$E(SSR) = E\left(\sum (\hat{y}_i - \bar{y})^2 \right) = E\left(\sum \hat{\beta}_1^2 (x_i - \bar{x})^2 \right)$$

$$= E(\hat{\beta}_1^2) S_{xx} = \left(\frac{\sigma^2}{S_{xx}} + \beta_1^2 \right) S_{xx} = \sigma^2 + \beta_1^2 S_{xx}.$$

The Sampling Distribution of $\hat{\beta}_0 + \hat{\beta}_1 x^*$, Where x^* is a Specified Value of the Explanatory Variable

It follows from Equations 10.56 and 10.53 that

$$\hat{\beta}_0 + \hat{\beta}_1 x^* = \sum (c_i + d_i x^*) Y_i$$

$$= \sum \left(\frac{1}{n} + \frac{(x^* - \bar{x})(x_i - \bar{x})}{S_{xx}} \right) Y_i. \tag{10.60}$$

Thus, we are able to represent the random variable $\hat{\beta}_0 + \hat{\beta}_1 x^*$ as a linear combination of mutually independent, normally distributed random variables Y_i of the sort discussed in the context of Theorem 6.1.

Applying Theorem 6.1 and using the facts that $E(Y_i) = \beta_0 + \beta_1 x_i$ and $V(Y_i) = \sigma^2$ yield the result that $\hat{\beta}_0 + \hat{\beta}_1 x^*$ is normally distributed with mean, variance, and estimated standard error given by

$$E(\hat{\beta}_0 + \hat{\beta}_1 x^*) = \beta_0 + \beta_1 x^*, \text{ and variance} \tag{10.61}$$

$$V(\hat{\beta}_0 + \hat{\beta}_1 x^*) = \sigma^2 \left[\frac{1}{n} + \frac{(x^* - \bar{x})^2}{S_{xx}} \right]; \tag{10.62}$$

$$s(\hat{\beta}_0 + \hat{\beta}_1 x^*) = s\sqrt{\left[\frac{1}{n} + \frac{(x^* - \bar{x})^2}{S_{xx}} \right]}. \tag{10.63}$$

Derivation of Equation 10.33

Equation 10.33 asserts that $s^2 = SSE/(n-2)$ is an unbiased estimator of σ^2. It is equivalent to

$$E(SSE) = E\left(\sum (Y_i - \hat{Y}_i)^2 \right) = (n-2)\sigma^2. \tag{10.64}$$

Derivation of Equation 10.64

We begin with the representation (the proof is omitted)

$$SSE = S_{YY} - \hat{\beta}_1^2 S_{xx}.$$

Taking expectations of both sides yields

$$E(SSE) = E(S_{YY}) - E(\hat{\beta}_1^2) S_{xx}, \text{ where}$$

$$E(\hat{\beta}_1^2) = V(\hat{\beta}_1) + E(\hat{\beta}_1)^2 = \frac{\sigma^2}{S_{xx}} + \beta_1^2.$$

Similarly,

$$E(S_{YY}) = E(\sum (Y_i - \overline{Y})^2)$$
$$= E(\sum Y_i^2) - nE(\overline{Y})^2, \text{ where}$$
$$E(Y_i^2) = \sigma^2 + (\beta_0 + \beta_1 x_i)^2 \text{ and}$$
$$E(\overline{Y}^2) = \frac{\sigma^2}{n} + (\beta_0 + \beta_1 \overline{x})^2.$$

Therefore,

$$E(S_{YY}) = n\sigma^2 + \sum (\beta_0 + \beta_1 x_i)^2 - n \left(\frac{\sigma^2}{n} + (\beta_0 + \beta_1 \overline{x})^2 \right)$$
$$= (n-1)\sigma^2 + \sum (\beta_0 + \beta_1 x_i)^2 - n(\beta_0 + \beta_1 \overline{x})^2$$
$$= (n-1)\sigma^2 + \sum [(\beta_0 + \beta_1 x_i) - (\beta_0 + \beta_1 \overline{x})]^2$$
$$= (n-1)\sigma^2 + \beta_1^2 \sum (x_i - \overline{x})^2$$
$$= (n-1)\sigma^2 + \beta_1^2 S_{xx}. \text{ Consequently,}$$
$$E(\text{SSE}) = E(S_{YY}) - E(\hat{\beta}_1^2)S_{xx}$$
$$= (n-1)\sigma^2 + \beta_1^2 S_{xx} - \left(\frac{\sigma^2}{S_{xx}} + \beta_1^2 \right) S_{xx} = (n-2)\sigma^2.$$

10.7   Chapter Summary

The regression model is used to study the relationship between a response variable y and an explanatory variable x. We assume that the relationship between the explanatory and response variables is given by a function $y = f(x)$, where f is unknown. A linear regression model assumes that $f(x) = \beta_0 + \beta_1 x$, and a curvilinear regression model assumes that the function $f(x)$ is non-linear. When the function is intrinsically linear then we can reduce a curvilinear regression model to a linear one by making a suitable change of variables. The validity of the model hypotheses is examined by graphing scatter plots and residual plots. Correlation analysis studies the linear relationship between two random variables X and Y. The Shapiro–Wilk test for normality is a correlation test for assessing the linearity of the points of a Q–Q plot.

10.8   Problems

Section 10.2

Problem 10.1
(a) Draw the scatter plot for the following artificial data.

x:	0	0	1	1	2	2
y:	0	1	1	3	2	3

(b) Sketch a straight line that seems to satisfactorily fit these points. Denoting the equation of your line by $y = b_0 + b_1 x$ find the values of b_0, b_1.
(c) Calculate the least squares line $y = \hat{\beta}_0 + \hat{\beta}_1 x$.
(d) Is the line you obtained in part (b) close to the least squares line obtained in part (c)?

Problem 10.2
(a) Draw the scatter plot corresponding to the shear strength data for the 0.064" gauge steel given in Table 10.1.
(b) Calculate the least squares estimates of $\hat{\beta}_0$ and $\hat{\beta}_1$ and draw the graph of the estimated regression line $y = \hat{\beta}_0 + \hat{\beta}_1 x$. Use the same axes on which you graphed the scatter plot.
(c) Compute the sample correlation, coefficient of determination, and error sum of squares.
(d) Compute the predicted values and the residuals, and display the results in a format similar to that of Table 10.3.
Save your computations.

Problem 10.3 *The following data record the amount of water (x), in centimeters, and the yield of hay (y), in metric tons per hectare, on an experimental farm.*

water (x)	30	45	60	75	90	105	120
yield (y)	2.11	2.27	2.5	2.88	3.21	3.48	3.37

(a) Draw the scatter plot (x_i, y_i).
(b) Calculate the least squares estimates $\hat{\beta}_0$ and $\hat{\beta}_1$ and draw the graph of the estimated regression line $y = \hat{\beta}_0 + \hat{\beta}_1 x$. Use the same axes on which you graphed the scatter plot.
(c) Compute the sample correlation, coefficient of determination, and error sum of squares.
(d) Compute the predicted value and the residual for $x = 30$, $x = 75$.

Problem 10.4 *The following data represent the number of disk I/Os (input-output) and CPU times for seven programs.*
Disk I/Os: 14, 16, 27, 42, 39, 50, 83
CPU times: 2, 5, 7, 9, 10, 13, 20
(Source: R. Jain (1991), The Art of Computer Systems Performance Analysis, *John Wiley & Sons, New York. Used with permission.)*
(a) Draw the scatter plot for the regression Model $(Y = CPU\ time | X = Disk\ I/Os)$.
(b) Calculate the least squares estimates $\hat{\beta}_0$ and $\hat{\beta}_1$ and draw the graph of the estimated regression line $y = \hat{\beta}_0 + \hat{\beta}_1 x$. Use the same axes on which you graphed the scatter plot.
(c) Compute the sample correlation, coefficient of determination, and error sum of squares.
(d) Compute the predicted values, the residuals, and display the results in a format similar to that of Table 10.3.
Save your computations.

Problem 10.5 *Refer to the fuel efficiency data (Table 10.2).*
(a) Draw the scatter plot for the regression Model$(Y = mpg | X = disp)$.
(b) Calculate the least squares estimates of $\hat{\beta}_0$ and $\hat{\beta}_1$ and draw the graph of the estimated regression line $y = \hat{\beta}_0 + \hat{\beta}_1 x$. Use the same axes on which you graphed the scatter plot.
(c) Compute the sample correlation, coefficient of determination, and the error sum of squares.
(d) Compute the predicted value and residual for $x = 1.6$, $x = 4.6$.
(e) Which of the two variables, weight or engine displacement, is a better predictor of fuel efficiency (mpg)? (Hint: Compare the coefficients of determination.)

Problem 10.6 *Refer to the fuel efficiency data (Table 10.2).*
(a) Calculate the least squares estimates $\hat{\beta}_0$ and $\hat{\beta}_1$ for the regression Model($Y = gpm$ $|X = weight$). Copy the scatter plot from Figure 10.3 and draw the graph of the estimated regression line $y = \hat{\beta}_0 + \hat{\beta}_1 x$. Use the same axes on which you graphed the scatter plot.
(b) Compute the sample correlation, coefficient of determination, and error sum of squares.
(c) Compute the residuals and display the results in a format similar to that of Table 10.3.

Problem 10.7 *Refer to the fuel efficiency data (Table 10.2).*
(a) Draw the scatter plot for the regression Model($Y = gpm$ $|X = disp$).
(b) Calculate the least squares estimates $\hat{\beta}_0$ and $\hat{\beta}_1$ and draw the graph of the estimated regression line $y = \hat{\beta}_0 + \hat{\beta}_1 x$. Use the same axes on which you graphed the scatter plot.
(c) Compute the sample correlation, coefficient of determination, and error sum of squares.
(d) Compute the predicted value and residual for $x = 1.6$, $x = 4.6$.
(e) Which of the two variables, weight or engine displacement, is a better predictor of fuel efficiency (gpm)? (Hint: Compare the coefficients of determination.)

Problem 10.8 *Refer to the fuel efficiency data for 18 2004-05 model year automobiles of Problem 1.76.*
(a) Draw the scatter plot for the regression Model($Y = mpg$ $|X = weight$).
(b) Calculate the least squares estimates of $\hat{\beta}_0$ and $\hat{\beta}_1$ and draw the graph of the estimated regression line $y = \hat{\beta}_0 + \hat{\beta}_1 x$. Use the same axes on which you graphed the scatter plot.
(c) Compute the sample correlation, coefficient of determination, and the error sum of squares.
(d) Compute the predicted values and residuals for $x = 2.78$, $x = 3.64$, $x = 5.59$.

Problem 10.9 *(Continuation of Problem 10.8) Refer to the automobile fuel efficiency data of Problem 1.76.*
(a) Draw the scatter plot for the regression Model($Y = gpm$ $|X = weight$).
(b) Calculate the least squares estimates $\hat{\beta}_0$ and $\hat{\beta}_1$ for the regression Model($Y = gpm$ $|X = weight$).
(c) Compute the sample correlation, coefficient of determination, and error sum of squares.
(d) Compute the predicted values and residuals for $x = 2.78$, $x = 3.64$, $x = 5.59$.
(e) Which of the two models, Model ($Y = mpg$ $|X = weight$) or Model ($Y = gpm$ $|X = weight$), best fits the data? (Hint: Compare their coefficients of determination.)

Problem 10.10 *(Continuation of Problem 10.8) Refer to the automobile fuel efficiency data of Problem 1.76.*
(a) Draw the scatter plot for the regression Model($Y = mpg$ $|X = disp$).
(b) Calculate the least squares estimates of $\hat{\beta}_0$ and $\hat{\beta}_1$ and draw the graph of the estimated regression line $y = \hat{\beta}_0 + \hat{\beta}_1 x$. Use the same axes on which you graphed the scatter plot.
(c) Compute the sample correlation, coefficient of determination, and the error sum of squares.
(d) Compute the predicted values and residuals for $x = 2.0$, $x = 3.2$, $x = 5.3$.
(e) Which of the two variables, weight or engine displacement, is a better predictor of fuel efficiency (mpg)? (Hint: Compare the coefficients of determination.)

Problem 10.11 *(Continuation of Problem 10.9) Refer to the automobile fuel efficiency data of Problem 1.76.*
(a) Draw the scatter plot for the regression Model($Y = gpm$ $|X = disp$).

(b) *Calculate the least squares estimates $\hat{\beta}_0$ and $\hat{\beta}_1$ and draw the graph of the estimated regression line $y = \hat{\beta}_0 + \hat{\beta}_1 x$. Use the same axes on which you graphed the scatter plot.*
(c) *Compute the sample correlation, coefficient of determination, and error sum of squares.*
(d) *Compute the predicted value and residual for $x = 2.0$, $x = 3.2$, $x = 5.3$.*
(e) *Which of the two variables, weight or engine displacement, is a better predictor of fuel efficiency (gpm)? (Hint: Compare the coefficients of determination.)*

Problem 10.12 *The following table lists 10 measurements of three ground water quality parameters from city wells in Fresno, California: specific electrical conductivity (SEC), total dissolved solids (TDS), and silica (SiO_2).*

SEC (ms/cm)	SiO_2 (mg/l)	TDS (mg/l)
422	56	275
470	48	305
841	48	535
714	55	456
749	59	484
450	61	293
214	63	173
213	65	175
229	36	147
370	54	243

(Source: B. Day and H. Nightingale (1984), Ground Water, vol. 22, no. 1, pp. 80-85.)

These variables are routinely measured in a laboratory. The SEC measurement procedure is simple, rapid, and precise. Measuring TDS is a time consuming procedure that requires several days for evaporation and drying of a known volume of filtered water under constant and standard laboratory conditions. One study used the regression Model $(Y = TDS \,|X = SEC)$ to estimate TDS from SEC measurements.
(a) *Draw the scatter plot for the regression Model $(Y = TDS \,|X = SEC)$.*
(b) *Calculate the least squares estimates of $\hat{\beta}_0$ and $\hat{\beta}_1$ and draw the graph of the estimated regression line $y = \hat{\beta}_0 + \hat{\beta}_1 x$. Use the same axes on which you graphed the scatter plot.*
(c) *Compute the sample correlation, coefficient of determination, and error sum of squares.*

Problem 10.13 *Suppose $y_i = a + bx_i$, $i = 1, \ldots n$. Show that $r = +1$ when $b > 0$ and $r = -1$ when $b < 0$.*

Problem 10.14
(a) *Verify Equation 10.19, which asserts that $\sum (y_i - \hat{y}_i) = 0$.*
(b) *Verify Equation 10.20, which asserts that $\sum (y_i - \hat{y}_i)x_i = 0$.*
(c) *Verify Equation 10.21, which asserts that the arithmetic mean of the observed values equals the arithmetic mean of the fitted values; that is,*

$$\frac{1}{n} \sum y_i = \frac{1}{n} \sum \hat{y}_i.$$

Section 10.3

Problem 10.15 *In Table 10.1 the weld diameter is the same for observations 4 and 6 ($x = 165$) but the shear strengths Y are different. How does the regression model (10.22) account for this?*

Problem 10.16 *A student gave the following expression for the fitted regression line of Figure 10.4:* $Y = -216.3271 + 3.990413 \times x + \epsilon$. *Explain why this expression is incorrect.*

Problem 10.17
(a) Display the results of the regression analysis of Problem 10.2 in the format of an ANOVA table (Table 10.5).
(b) What proportion of the variability is explained by the model?
(c) Compute a 95% confidence interval for the regression coefficients.
(d) Compute a 95% confidence interval for the mean shear strength when $x = 220$.

Problem 10.18
(a) Display the results of the regression analysis of Problem 10.4 in the format of an ANOVA table (Table 10.5).
(b) What proportion of the variability is explained by the model?
(c) Compute a 95% confidence interval for the regression coefficients.
(d) Compute a 95% confidence interval for the mean CPU time when $x = 30$.
(e) Compute a 95% prediction interval for the CPU time when $x = 30$.

Problem 10.19 *(This is a continuation of the fuel efficiency Example 10.2.)*
(a) Using the computer output displayed in the Table 10.7 (and the result that $\bar{x} = 3.21$, $S_{xx} = 2.95$), *construct 95% confidence intervals for* β_0 *and* β_1.
(b) Compute a 95% confidence interval for the mpg when the automobile weighs 2530 lbs.
(c) Compute a 95% prediction interval for the mpg Y *of a 1992 Escort that also weighs 2530 lbs. Compare the predicted value to the actual value, which is 30 mpg.*

Problem 10.20 *(This is a continuation of the fuel efficiency Example 10.2.) The ANOVA table for the Model* $(Y = gpm|X = weight)$ *follows.*

Analysis of Variance

Source	DF	Sum of Squares	Mean Square	F Value	Prob>F
Model	1	0.59506	0.59506	23.916	0.0006
Error	10	0.24881	0.02488		
C Total	11	0.84387			

Root MSE   0.15774   R-square   0.7052

Parameter Estimates

Variable	Parameter Estimate	Standard Error	T for H0: Parameter=0	Prob > |T|
INTERCEP	2.179754	0.29867670	7.298	0.0001
SLOPE	0.449247	0.09186264	4.890	0.0006

(a) Using the computer output displayed in the ANOVA table (and the result that $\bar{x} = 3.21$, $S_{xx} = 2.95$), *construct 95% confidence intervals for* β_0 *and* β_1.
(b) Compute a 95% confidence interval for the mean gpm of an automobile that weighs 2530 lbs.
(c) Compute a 95% prediction interval for the gpm of a 1992 Escort that also weighs 2530 lbs. Compare the predicted value to the actual value, which is 3.34 gpm.

Problem 10.21 *The following data were obtained in a comparative study of a remote procedure call (RPC) on two computer operating systems, UNIX and ARGUS.*

UNIX DATA BYTES	TIME	ARGUS DATA BYTES	TIME
64	26.4	92	32.8
64	26.4	92	34.2
64	26.4	92	32.4
64	26.2	92	34.4
234	33.8	348	41.4
590	41.6	604	51.2
846	50.0	860	76.0
1060	48.4	1074	80.8
1082	49.0	1074	79.8
1088	42.0	1088	58.6
1088	41.8	1088	57.6
1088	41.8	1088	59.8
1088	42.0	1088	57.4

(Source: R. Jain (1991), The Art of Computer Systems Performance Analysis, *John Wiley & Sons. Used with permission.)*
(a) Draw the scatter plot for Model $(Y = time \mid X = data\ bytes)$ (UNIX DATA).
(b) Display the results of the regression analysis in the format of Table 10.5.
(c) What proportion of the variability is explained by the model?
(d) Compute 95% confidence intervals for the regression coefficients.
Save your computations.

Problem 10.22 *(Continuation of Problem 10.21)*
(a) Draw the scatter plot for Model $(Y = time \mid X = data\ bytes)$ (ARGUS DATA).
(b) Display the results of the regression analysis in the format of Table 10.5.
(c) What proportion of the variability is explained by the model?
(d) Compute 95% confidence intervals for the regression coefficients.
Save your computations.

Problem 10.23 *Twelve batches of of plastic are made, and from each batch one test item was molded and its brinell hardness y was measured at time x. The results are recorded in the following table.*

Time(x)	Hardness(y)	Time(x)	Hardness(y)
32	230	40	248
48	262	48	279
72	323	48	267
64	298	24	214
48	255	80	359
16	199	56	305

(Source: J. Neter, W. Wasserman, and M.H. Kutner (1985), Applied Linear Statistical Models, *Richard D. Irwin, Inc., 2nd ed., Homewood, IL. Used with permission.)*
(a) Draw the scatter plot for Model $(Y = hardness \mid X = time)$.
(b) Display the results of your regression analysis in the format of Table 10.5.
(c) Plot the estimated regression line and comment on how well the estimated regression line fits the data.
(d) What proportion of the variability is explained by the model?
(e) Compute 95% confidence intervals for the regression coefficients.

(f) Give a point estimate of the mean hardness when $x = 48$.
Save your computations.

Problem 10.24 *(This is a continuation of Problem 10.8.)*
(a) Display the results of your regression analysis in the format of Table 10.5.
(b) Compute 95% confidence intervals for the regression coefficients.
(c) Compute a 95% confidence interval for the mean mpg of an automobile that weighs 3655 lbs.
(d) Compute a 95% prediction interval for the mpg of a 2004-05 Chevy Impala that also weighs 3655 lbs. Compare the predicted value to the actual value, which is 24 mpg.

Problem 10.25 *(This is a continuation of Problem 10.9.)*
(a) Display the results of your regression analysis in the format of Table 10.5.
(b) Compute 95% confidence intervals for the regression coefficients.
(c) Compute a 95% confidence interval for the mean gpm (gallons per hundred miles) of an automobile that weighs 3655 lbs.
(d) Compute a 95% prediction interval for the gpm of a 2004-05 Chevy Impala with engine displacement 3.8. Compare the predicted value to the actual value, which is 4.167 gpm.

Problem 10.26 *(This is a continuation of Problem 10.10.)*
(a) Display the results of your regression analysis in the format of Table 10.5.
(b) Compute 95% confidence intervals for the regression coefficients.
(c) Compute a 95% confidence interval for the mean mpg of an automobile that weighs 3655 lbs.
(d) Compute a 95% prediction interval for the mpg of a 2004-05 Chevy Impala with engine displacement 3.8. Compare the predicted value to the actual value, which is 24 mpg.

Problem 10.27 *(This is a continuation of Problem 10.11.)*
(a) Display the results of your regression analysis in the format of Table 10.5.
(b) Compute 95% confidence intervals for the regression coefficients.
(c) Compute a 95% confidence interval for the mean gpm (gallons per hundred miles) of an automobile with engine displacement 3.8.
(d) Compute a 95% prediction interval for the gpm of a 2004-05 Chevy Impala with engine displacement 3.8. Compare the predicted value to the actual value, which is 4.167 gpm.

Section 10.3.5

Problem 10.28 *Refer to Example 10.11 and Table 10.8. Fit the curve of Equation 10.47 by carrying out a linear regression analysis on the transformed data $y = \ln v$, $x = \ln t$. Graph the scatter plot and estimated regression line for the transformed data.*

Problem 10.29 *Kepler's third law of planetary motion states that the square of the period (P) of revolution of a planet is proportional to the cube of its mean distance (D) from the sun; that is, $P^2 = kD^3$. Taking the square root of both sides yields the formula $P = cD^{3/2}$; $c = \frac{2\pi}{\sqrt{\gamma M}}$, where M is the mass of the sun and γ is the gravitational constant. We simplify Kepler's third law by taking the unit period as one earth year and the unit of distance to be the earth's mean distance to the sun (these are also called astronomical units). Thus, $P = R^{3/2}$. The periods of revolution and the distances for all 9 planets are shown in the following table. Use the transformation $y = \ln P$, $x = \ln D$ to find the curve $P = aD^b$ that best fits the data. Are your results consistent with Kepler's predicted values: $a = 1$, $b = 1.5$?*

Planet	Distance	Period
Mercury	0.387	0.241
Venus	0.723	0.615
Earth	1.0	1.0
Mars	1.524	1.881
Jupiter	5.202	11.862
Saturn	9.555	29.458
Uranus	19.218	84.01
Neptune	30.109	164.79
Pluto	39.44	248.5

(Source: H. Skala, The College Mathematics Journal, vol. 27, no. 3, May 1996, pp. 220-223. Note: There is a typo in Professor Skala's table; the headings for the period and distances appear to have been interchanged.)

Problem 10.30 *The data in the following table come from measuring the average speeds (s) (in meters/minute) for racing shells seating different numbers of rowers (n). From calculations based on fluid mechanics, it is known that the speed is proportional to the 1/9th power of the number of rowers; that is, $s = cn^{1/9}$. Use the transformation $y = \ln s$, $x = \ln n$ to find the curve $s = an^b$ that best fits the data. Compare your results with the theoretical prediction $b = 1/9$.*

n	s	n	s
1	279	4	316
1	276	4	312
1	275	4	309
1	279	4	326
2	291	8	341
2	289	8	338
2	287	8	343
2	295	8	349

(Source: H. Skala, The College Mathematics Journal, vol. 27, no. 3, May 1996, pp. 220-223.)

Problem 10.31
(a) The data in the following table are pressures, in atmospheres, of oxygen gas kept at 25 degrees C and made to occupy various volumes, measured in liters. Draw the scatter plot of (v_i, p_i). Is the simple linear regression model appropriate?
(b) Chemists believe that the relation between the pressure and volume of a gas at constant temperature is given by $p = av^b$. Draw the scatter plot for the transformed data (x_i, y_i), where $y = \ln p$ and $x = \ln v$. Is the linear regression model appropriate for the transformed data?
(c) Fit a straight line to the transformed data by the method of least squares and determine the point estimates for the parameters a and b.
(d) Are these results consistent with Boyle's law for an ideal gas, which states that $pv = k$? That is, the product of pressure by volume is constant at constant temperature.

Volume (v)	Pressure (P)
3.25	7.34
5.00	4.77
5.71	4.18
8.27	2.88
11.50	2.07
14.95	1.59
17.49	1.36
20.35	1.17
22.40	1.06

(Source: R.G.D. Steel and J.H. Torrie (1980), Principles and Procedures of Statistics, *2nd ed., p. 458, McGraw-Hill. Used with permission.)*

Problem 10.32

(a) A biologist believes that the curve defined by the function $v = \frac{u}{a+bu}$ provides a good fit to a data set. Show that the transformed variables $x = 1/u$, $y = 1/v$ lie on a straight line $y = \beta_0 + \beta_1 x$ and express β_0 and β_1 in terms of a and b.
(b) Sketch the curve $v = u/(1+u)$ for $0 \leq u \leq 2$.

Section 10.4

Problem 10.33 *Refer back to Problem 10.23.*
(a) Plot the studentized residuals against predicted hardness for the regression Model(Y = hardness |X = time).
(b) Draw the normal probability plots of the studentized residuals.
(c) Are these plots consistent with the hypotheses of the simple linear regression model? Justify your conclusions.

Problem 10.34 *Consider the following two bivariate data sets each consisting of eleven (x, y) pairs. The two data sets have the same x values but different y values.*

	Dataset I	Dataset II
x	y	y
4	4.26	3.10
5	5.68	4.74
6	7.24	6.13
7	4.82	7.26
8	6.95	8.14
9	8.81	8.77
10	8.04	9.14
11	8.33	9.26
12	10.84	9.13
13	7.58	8.74
14	9.96	8.10

(Source: F. J. Anscombe (1973), Graphs in Statistical Analysis, The American Statistician, *vol. 27, pp. 17–21. Used with permission.)*
(a) Draw the scatter plots of data sets I and II. Using only the information contained in these scatter plots comment on the suitability of the simple linear regression model for each

of the data sets.
(b) Compute the estimated regression line, coefficient of determination, and sample correlation coefficient for each of the data sets.
(c) Draw the residual plots for each of the data sets.
(d) Explain why only one of the linear regression models is appropriate even though the regression analysis for data sets I and II lead to the same parameter estimates, the same coefficient of determination, and the same standard errors.

Problem 10.35 *Refer back to Problem 10.12. Do you agree with the authors' conclusion that TDS can be estimated from the regression models with acceptable error? In particular, are the data consistent with the hypotheses of the simple linear regression model? Justify your conclusions by model checking, that is, write a brief report based upon an analysis of the scatter, residual, and normal probability plots for the regression Model(Y = TDS |X = SEC).*

Problem 10.36 *Refer back to the UNIX data of Problem 10.21.*
(a) Draw the studentized residual and normal probability plots for Model (Y = time |X = data bytes) (UNIX DATA).
(b) Do these plots and the scatter plot indicate any departures from the simple linear regression model? Summarize your conclusions in a brief paragraph or two.

Section 10.5

Problem 10.37 *Calculate the sample correlation coefficient r for the following artificial data:*

x   1   2   3   4   5   6
y   2   2   3   5   4   6

Problem 10.38 *Refer to the StatLab data set in Table 1.16. Psychologists are interested in exploring the relationship between the girls' scores on the Raven test (Y) and the Peabody test (X). The Peabody test measures verbal facility; the Raven test measures a different sort of mental ability, related to geometrical intuition and spatial organization. The Raven test uses no words, only geometrical figures. It is an interesting problem to determine whether there is any relationship between the verbal facility measured by the Peabody test and the geometrical–spatial abilities measured by the Raven test.*
(a) Draw the scatter plot of (x, y).
(b) Compute the correlation coefficient r.

Problem 10.39 *Refer to the girls' heights X and weights Y shown in the StatLab data set (Table 1.16).*
(a) Draw the scatter plot and guess the value of r.
(b) Calculate r.

Problem 10.40 *Refer to the fathers' heights X and weights Y shown in the StatLab data set (Table 1.16).*
(a) Draw the scatter plot and guess the value of r.
(b) Calculate r.

Problem 10.41 *Refer back to of the CO_2 emissions data in Problem 1.17. In this problem we explore the relationships between some of the variables $X_1, \ldots, X_4$.*
(a) Draw the scatter plot of (X_3, X_2). In this context would you expect a positive, zero, or

negative correlation? Justify your answer.
(b) Compute the sample correlation coefficient r between X_2 and X_3, and discuss its usefulness in explaining the relationship between X_2 and X_3.
(c) Draw the scatter plot of (X_3, X_4). In this context, would you expect a positive, zero, or negative correlation? Justify your answer.
(d) Compute the sample correlation coefficient r between X_3 and X_4, and discuss its usefulness in explaining the relationship between X_3 and X_4.

Problem 10.42 *Table 10.13 in Section 10.9 lists the number of vehicles per 10,000 population (VPP), the number of deaths per 10,000 vehicles (DPV), and the number of deaths per 10,000 population for twelve countries for the year 1987. In this problem we apply the techniques of correlation analysis to explore the relationships between these variables.*
(a) Draw the scatter plot of (X_1, X_2). In this context, would you expect a positive, zero, or negative correlation? Justify your answer.
(b) Compute the sample correlation coefficient r between X_1 and X_2, and discuss its usefulness in explaining the relationship between X_1 and X_2.
(c) Draw the scatter plot of (X_1, X_3). In this context would you expect a positive, zero, or negative correlation? Justify your answer.
(d) Compute the correlation coefficient r between X_1 and X_3, and discuss its usefulness in explaining the relationship between X_1 and X_3.

Problem 10.43 *Cirrhosis, a liver disease, is a side effect of excessive alcohol consumption. Table 10.14 in Section 10.9, based on statistics compiled by the Commodity Board of the Spirits Industry, records the rate of consumption of pure alcohol per capita for various countries and the cirrhosis death rates per 100,000 men in 1992, or the most recent year of data. Draw the scatter plot of deaths Y against alcohol consumption X. Do the data support the existence of a positive association between the two variables? Are there any data points that depart from this pattern? Discuss.*

Problem 10.44 *Refer to the 1999 weekly returns for the bond fund PTTRX listed in Table 1.19. Compute PTTRX's β for the year 1999.*

Problem 10.45 *Show that $E(\overline{Y}) = \beta_0 + \beta_1 \overline{x}$.*

Problem 10.46 *Verify the representation for $\hat{\beta}_0$ displayed in Equation 10.56.*

Problem 10.47 *Perform the necessary algebraic manipulations in order to derive Equation 10.58.*

10.9 Large Data Sets

Table 10.11 *Prices and total returns for Cisco Systems and the S&P500, 1999*

Cisco Price	Cisco Total Return	S&P Price	S&P Total Return	Cisco Price	Cisco Total Return	S&P Price	S&P Total Return
26.67	1.00	1275.09	1.00	33.53	1.26	1403.28	1.10
25.42	0.95	1243.26	0.98	33	1.24	1418.78	1.11
25.7	0.96	1225.19	0.96	31.47	1.18	1356.94	1.06
27.89	1.05	1279.64	1.00	31.07	1.16	1328.72	1.04
25.31	0.95	1239.4	0.97	31.12	1.17	1300.29	1.02
24.76	0.93	1230.13	0.96	31.78	1.19	1327.68	1.04
24.28	0.91	1239.22	0.97	32.06	1.20	1336.61	1.05
24.45	0.92	1238.33	0.97	34.25	1.28	1348.27	1.06
25.2	0.94	1275.47	1.00	35.47	1.33	1357.24	1.06
25.81	0.97	1294.59	1.02	35.38	1.33	1351.66	1.06
26.12	0.98	1299.29	1.02	36.75	1.38	1335.42	1.05
26.3	0.99	1282.8	1.01	34.5	1.29	1277.36	1.00
27.49	1.03	1293.72	1.01	34.44	1.29	1282.81	1.01
29.53	1.11	1348.35	1.06	35.9	1.35	1336.02	1.05
26.42	0.99	1319	1.03	33.6	1.26	1247.41	0.98
29.34	1.10	1356.85	1.06	33.28	1.25	1301.65	1.02
28.51	1.07	1335.18	1.05	37	1.39	1362.93	1.07
27.12	1.02	1345	1.05	36.72	1.38	1370.23	1.07
28.86	1.08	1337.8	1.05	41.72	1.56	1396.06	1.09
28.31	1.06	1330.29	1.04	44.06	1.65	1422	1.12
27.25	1.02	1301.84	1.02	46.6	1.75	1416.62	1.11
28.72	1.08	1327.75	1.04	47.78	1.79	1433.3	1.12
27.67	1.04	1293.64	1.01	49.9	1.87	1417.04	1.11
29.84	1.12	1342.84	1.05	49.85	1.87	1421.03	1.11
30.66	1.15	1315.31	1.03	52.22	1.96	1458.34	1.14
33.53	1.26	1391.22	1.09	53.56	2.01	1469.25	1.15

Table 10.12 *Prices and total returns for Cisco Systems and the S&P500, 2002*

Cisco Price	Cisco Total Return	S&P Price	S&P Total Return	Cisco Price	Cisco Total Return	S&P Price	S&P Total Return
20.83	1	1172.51	1	14.05	0.67	989.03	0.84
20.21	0.97	1145.6	0.98	14.38	0.69	921.39	0.79
18.85	0.9	1127.58	0.96	13.65	0.66	847.75	0.72
19.13	0.92	1133.28	0.97	11.82	0.57	852.84	0.73
19.21	0.92	1122.2	0.96	11.89	0.57	864.24	0.74
16.76	0.8	1096.22	0.93	13.12	0.63	908.64	0.77
17.09	0.82	1104.18	0.94	14.45	0.69	928.77	0.79
15.24	0.73	1089.84	0.93	14.45	0.69	940.86	0.8
15	0.72	1131.78	0.97	13.82	0.66	916.07	0.78
17.8	0.85	1164.31	0.99	13.03	0.63	893.92	0.76
16.54	0.79	1166.16	0.99	13.05	0.63	889.81	0.76
16.57	0.8	1148.7	0.98	12.08	0.58	845.39	0.72
16.93	0.81	1147.39	0.98	11.23	0.54	827.37	0.71
16.15	0.78	1122.73	0.96	9.46	0.45	800.58	0.68
15.3	0.73	1111.01	0.95	10.32	0.5	835.32	0.71
15.26	0.73	1125.17	0.96	10.51	0.5	884.39	0.75
13.91	0.67	1076.32	0.92	11.78	0.57	897.65	0.77
13.14	0.63	1073.43	0.92	11.61	0.56	900.96	0.77
15.42	0.74	1054.99	0.9	12.56	0.6	894.74	0.76
17.25	0.83	1106.59	0.94	14.08	0.68	909.83	0.78
16.57	0.8	1083.82	0.92	14.89	0.71	930.55	0.79
15.78	0.76	1067.14	0.91	14.92	0.72	936.31	0.8
15.73	0.76	1027.53	0.88	14.18	0.68	912.23	0.78
14.3	0.69	1007.27	0.86	13.4	0.64	889.48	0.76
13.74	0.66	989.14	0.84	13.27	0.64	895.76	0.76
13.95	0.67	989.82	0.84	13.01	0.62	875.4	0.75
14.05	0.67	989.03	0.84	13.1	0.63	879.82	0.75

Table 10.13 *International comparisons (1987) of vehicle ownership and death rates*

COUNTRY	VPP X1	DPV X2	DPP X3
USA	7322	2.6	1.9
NZ	5950	4.0	2.4
UK	3701	2.5	0.9
FRG	4855	2.7	1.3
Japan	4078	1.9	0.8
Finland	2219	3.0	1.2
Denmark	3367	3.7	1.4
Kuwait	3014	4.8	1.5
Jordan	771	17.7	1.4
Thailand	50	46.7	0.2
Niger	278	13.0	0.4
Pakistan	40	121.4	0.5

(Source: K. S. Jadaan, *An Overview of Road Safety in New Zealand,* ITE Journal, April, 1993.)

Table 10.14 *Country, alcohol consumption, cirrhosis deaths per 100,000*

Country	Alcohol Consumption	Cirrhosis Deaths	Country	Alcohol Consumption	Cirrhosis Deaths
Luxembourg	13.3	27.0	Ireland	8.8	3.2
France	12.2	23.9	Rumania	8.5	50.1
Austria	11.1	41.6	Netherlands	8.4	6.2
Germany	11.0	33.1	Australia	7.9	8.8
Portugal	11.0	40.6	Argentina	7.8	15.0
Hungary	10.8	104.5	Britain	7.7	7.2
Denmark	10.6	18.8	New Zealand	7.7	5.1
Switzerland	10.6	28.9	Finland	7.2	13.7
Spain	10.6	28.9	USA	7.2	13.7
Greece	9.7	13.2	Japan	7.0	19.1
Belgium	9.6	14.5	Canada	6.9	11.1
Czech	9.4	14.5	Poland	6.7	15.6
Italy	9.1	34.8	Sweden	5.6	10.0
Bulgaria	8.8	28.9			

10.10 To Probe Further

The following well-written texts contain a wealth of additional information on regression analysis, including a more careful discussion of model checking and non-linear regression. The Alexander et al. text is a useful source of information concerning modern portfolio theory.

1. R.H. Myers (1990), *Classical and Modern Regression with Applications,* 2nd ed., PWS-Kent, Boston.

2. J. Neter, M.H. Kutner, C.J. Nachtseim, and W. Wasserman (1996), *Applied Linear Statistical Models,* 4th ed., Irwin, Chicago.

3. G.J. Alexander, W.F. Sharpe, and J.V. Bailey (2001), *Fundamentals of Investments,* 3rd ed., Prentice Hall.

Chapter 11

*Multiple Linear Regression**

Data analysis without guidance is foolish, but uncritical belief in a specific model is dangerously foolish.

J.W. Tukey and M.B. Wilk, American statisticians

11.1 Orientation

Sometimes the fit of a simple linear regression model can be improved by adding one or more explanatory variables. For instance, the crop yield may depend not only on the amount of fertilizer but also on the amount of water. In this instance we have one response variable, the yield Y, and two explanatory variables, the amount of fertilizer x_1 and the amount of water x_2. Denote the relation between the expected yield and x_1, x_2 by $f(x_1, x_2)$. The object of regression analysis is to describe the unknown response function $f(x_1, x_2)$. In many cases, the response function can be approximated by a function that is linear in the variables; that is, we assume that the functional relation between the expected crop yield and the amounts of fertilizer and water is

$$E(Y | \text{fertilizer } = x_1; \text{ water } = x_2) = \beta_0 + \beta_1 x_1 + \beta_2 x_2 = f(x_1, x_2).$$

Here, the values of the variables x_1, x_2 are known to the researcher while the *regression parameters* β_0, β_1, β_2 are not. *Multiple linear regression* is a statistical procedure for estimating and making inferences about the regression parameters. Since this procedure leads to tedious and complex calculations when done by hand we assume the student has access to a computer equipped with a standard statistical software package. Our primary goal in this chapter is to present those basic concepts of multiple linear regression that are essential for understanding the computer printouts. The most convenient and efficient way to reach this goal is to use vectors and matrices to represent the response variable, the explanatory variables, and the relations among them. For this reason we assume the student is familiar with the following basic concepts from linear algebra: vector space, subspace, vectors, matrices, matrix multiplication, and some additional concepts that will be introduced as needed. It is instructive to begin with the matrix formulation of the simple linear regression model studied in Chapter 10. This will not only ease the transition to the more complicated multiple linear regression models discussed later, but will also give us some additional insights into the simple linear regression model itself.

Organization of Chapter

1. Section 11.2: The Matrix Approach to Simple Linear Regression

2. Section 11.2.1: Sampling Distribution of the Least Squares Estimators

3. Section 11.2.2: Geometric Interpretation of the Least Squares Solution

4. Section 11.3: The Matrix Approach to Multiple Linear Regression

5. Section 11.3.2: Testing Hypotheses About the Regression Model

6. Section 11.3.3: Model Checking

7. Section 11.3.4: Confidence Intervals and Prediction Intervals

8. Section 11.4: Mathematical Details and Derivations

9. Section 11.5: Chapter Summary

11.2   The Matrix Approach to Simple Linear Regression

In the simple linear regression model, discussed in Chapter 10, we assume that the relation between the value of the response variable Y_i and the value x_i of the explanatory variable is given by

$$Y_i = \beta_0 + \beta_1 x_i + \epsilon_i, \ (i = 1, \ldots n), \ \text{where } \epsilon_i \overset{\mathcal{D}}{=} N(0, \sigma^2),$$

and the random variables $\epsilon_1, \ldots, \epsilon_n$ are mutually independent.

In the matrix approach to the simple linear regression model we represent the n responses $Y_1, \ldots, Y_n$, the two regression parameters β_0 and β_1, and the n random variables $\epsilon_1, \ldots, \epsilon_n$ as the column vectors[1]

$$\boldsymbol{Y} = \begin{pmatrix} Y_1 \\ \vdots \\ Y_n \end{pmatrix}; \ \boldsymbol{\beta} = \begin{pmatrix} \beta_0 \\ \beta_1 \end{pmatrix}; \ \boldsymbol{\epsilon} = \begin{pmatrix} \epsilon_1 \\ \vdots \\ \epsilon_n \end{pmatrix}.$$

The matrix formulation of the simple linear regression model is

$$\boldsymbol{Y} = \boldsymbol{X}\boldsymbol{\beta} + \boldsymbol{\epsilon} \text{ where } \boldsymbol{X} = \begin{pmatrix} 1 & x_1 \\ \vdots & \vdots \\ 1 & x_n \end{pmatrix}. \tag{11.1}$$

The $(n \times 2)$ matrix $\boldsymbol{X}$ is called the *design matrix*. Observe that the number of columns in the design matrix equals the number of regression parameters and the number of rows equals the number of observations. The n fitted values $\hat{Y}_i = \hat{\beta}_0 + \hat{\beta}_1 x_i, \ (i = 1, \ldots, n)$ and residuals $e_i = Y_i - \hat{Y}_i = Y_i - (\hat{\beta}_0 + \hat{\beta}_1 x_i), \ (i = 1, \ldots, n)$ can also be represented using matrix notation as follows:

$$\hat{\boldsymbol{Y}} = \begin{pmatrix} \hat{Y}_1 \\ \vdots \\ \hat{Y}_n \end{pmatrix} = \begin{pmatrix} \hat{\beta}_0 + \hat{\beta}_1 x_1 \\ \vdots \\ \hat{\beta}_0 + \hat{\beta}_1 x_n \end{pmatrix} = \boldsymbol{X}\hat{\boldsymbol{\beta}};$$

$$\boldsymbol{e} = \begin{pmatrix} Y_1 - \hat{Y}_1 \\ \vdots \\ Y_n - \hat{Y}_n \end{pmatrix} = \boldsymbol{Y} - \boldsymbol{X}\hat{\boldsymbol{\beta}}.$$

---

[1]Vectors and matrices are denoted in boldface type.

We observed in Chapter 10 (Equations 10.19 and 10.20) that the residuals satisfy the following two linear equations:

$$\sum_i e_i = 0; \sum_i x_i e_i = 0. \tag{11.2}$$

These two equations can be written as the single matrix equation

$$\begin{pmatrix} 1 & \cdots & 1 \\ x_1 & \cdots & x_n \end{pmatrix} \begin{pmatrix} e_1 \\ \vdots \\ e_n \end{pmatrix} = \begin{pmatrix} 0 \\ 0 \end{pmatrix}.$$

The matrix $\boldsymbol{X}'$ defined by

$$\boldsymbol{X}' = \begin{pmatrix} 1 & \cdots & 1 \\ x_1 & \cdots & x_n \end{pmatrix}$$

is obtained from the design matrix by interchanging its rows and columns; it is called the *transpose* of $\boldsymbol{X}$. Similarly, the transpose of a column vector $\boldsymbol{x}$, a $(n \times 1)$ matrix, is the row vector $\boldsymbol{x}' = (x_1, \ldots, x_n)$, a $(1 \times n)$ matrix. In general the transpose of a matrix with n rows and p columns is a matrix with p rows and n columns. Using this notation we write the Equation 11.2 in the following more useful form:

$$\boldsymbol{X}'(\boldsymbol{Y} - \boldsymbol{X}\hat{\boldsymbol{\beta}}) = 0. \tag{11.3}$$

Applying the distributive law for matrix multiplication to the left hand side of Equation 11.3 we see that $\hat{\boldsymbol{\beta}}$ satisfies the *normal equations:*

$$\boldsymbol{X}'\boldsymbol{X}\hat{\boldsymbol{\beta}} = \boldsymbol{X}'\boldsymbol{Y}. \tag{11.4}$$

Some Additional Properties of the Transpose and Inverse of a Matrix

We list for future reference some useful facts concerning the transpose and inverse of a matrix. Proofs are omitted.

$$(\boldsymbol{A}')' = \boldsymbol{A} \tag{11.5}$$
$$(\boldsymbol{A}\boldsymbol{B})' = \boldsymbol{B}'\boldsymbol{A}' \tag{11.6}$$
$$(\boldsymbol{A}^{-1})^{-1} = \boldsymbol{A} \tag{11.7}$$
$$(\boldsymbol{A}\boldsymbol{B})^{-1} = \boldsymbol{B}^{-1}\boldsymbol{A}^{-1} \tag{11.8}$$
$$(\boldsymbol{A}')^{-1} = (\boldsymbol{A}^{-1})' \tag{11.9}$$

Matrix Solution to the Normal Equations

In the simple linear regression case the determinant of $\boldsymbol{X}'\boldsymbol{X}$ is nS_{xx} (the proof is left to you as a problem). Because $S_{xx} \neq 0$, except for the uninteresting case when $x_1 = \ldots = x_n$, we can always assume that $det(\boldsymbol{X}'\boldsymbol{X}) \neq 0$. Consequently, the inverse matrix $(\boldsymbol{X}'\boldsymbol{X})^{-1}$ exists and the solution to the normal equation in matrix form is given by

$$\hat{\boldsymbol{\beta}} = (\boldsymbol{X}'\boldsymbol{X})^{-1}\boldsymbol{X}'\boldsymbol{Y}. \tag{11.10}$$

The matrix representation of the vector of fitted values $\boldsymbol{Y}$ is given by

$$\hat{\boldsymbol{Y}} = \boldsymbol{X}\hat{\boldsymbol{\beta}} = \boldsymbol{X}(\boldsymbol{X}'\boldsymbol{X})^{-1}\boldsymbol{X}'\boldsymbol{Y} = \boldsymbol{H}\boldsymbol{Y} \tag{11.11}$$

where

$$\boldsymbol{H} = \boldsymbol{X}(\boldsymbol{X}'\boldsymbol{X})^{-1}\boldsymbol{X}'. \tag{11.12}$$

The matrix $\boldsymbol{H}$ is called the *hat matrix;* it plays an important role in describing the sampling distribution of the least squares estimators and in model checking. The residual vector $\boldsymbol{e}$ can also be written in terms of the hat matrix as follows:

$$\boldsymbol{e} = \boldsymbol{Y} - \hat{\boldsymbol{Y}} = \boldsymbol{Y} - \boldsymbol{HY} = (\boldsymbol{I} - \boldsymbol{H})\boldsymbol{Y}, \tag{11.13}$$

where $\boldsymbol{I}$ denotes the identity matrix. The hat matrix plays an important role in detecting violations of the regression model assumptions. For future reference we list some useful properties of the hat matrix. We leave the derivations to the reader as an exercise (Problem 11.7).

$$\boldsymbol{H}' = \boldsymbol{H} \text{ (Symmetric matrix)} \tag{11.14}$$
$$\boldsymbol{H}^2 = \boldsymbol{H} \text{ (Idempotent matrix)} \tag{11.15}$$
$$\boldsymbol{H}(\boldsymbol{I} - \boldsymbol{H}) = \boldsymbol{0} \tag{11.16}$$
$$(\boldsymbol{I} - \boldsymbol{H})^2 = (\boldsymbol{I} - \boldsymbol{H}) \tag{11.17}$$

Example 11.1

Re-derive Equations 10.11 and 10.12 for the regression coefficients obtained in Chapter 10 by solving the normal Equation 11.4.

Solution It suffices to show that

$$\hat{\beta}_0 + \hat{\beta}_0 \overline{x} = \overline{Y} \text{ and } \hat{\beta}_1 = \frac{S_{xY}}{S_{xx}}.$$

In this case the normal equations are two linear equations in two unknowns. To see this, compute the matrix products $\boldsymbol{X}'\boldsymbol{X}$ and $\boldsymbol{X}'\boldsymbol{Y}$:

$$\boldsymbol{X}'\boldsymbol{X} = \begin{pmatrix} n & \sum x_i \\ \sum x_i & \sum x_i^2 \end{pmatrix}; \ \boldsymbol{X}'\boldsymbol{Y} = \begin{pmatrix} \sum_i Y_i \\ \sum_i x_i Y_i \end{pmatrix}. \tag{11.18}$$

We therefore obtain two equations for the two unknowns $\hat{\beta}_0$, $\hat{\beta}_1$:

$$n\hat{\beta}_0 + \left(\sum_i x_i\right) \hat{\beta}_1 = \sum_i Y_i;$$

$$\left(\sum_i x_i\right) \hat{\beta}_0 + \left(\sum_i x_i^2\right) \hat{\beta}_1 = \sum_i x_i Y_i.$$

Using the relations that $\sum_i x_i = n\overline{x}$ and $\sum_i Y_i = n\overline{Y}$ we transform these equations into the equivalent form:

$$n\hat{\beta}_0 + n\overline{x}\hat{\beta}_1 = n\overline{Y};$$

$$n\overline{x}\hat{\beta}_0 + \left(\sum_i x_i^2\right) \hat{\beta}_1 = \sum_i x_i Y_i.$$

Dividing both sides of the first equation by n yields the familiar relation

$$\hat{\beta}_0 + \hat{\beta}_1 \overline{x} = \overline{Y}.$$

Solving for $\hat{\beta}_0 = \overline{Y} - \hat{\beta}_1\overline{x}$ and substituting into the second equation yields the previously derived equation for $\hat{\beta}_1$:

$$n\overline{x}(\overline{Y} - \hat{\beta}_1\overline{x}) + \left(\sum_i x_i^2\right)\hat{\beta}_1 = \sum_i x_i Y_i. \text{ Rearranging terms yields}$$

$$\hat{\beta}_1\left(\sum_i x_i^2 - n\overline{x}^2\right) = \sum_i x_i Y_i - n\overline{x}\overline{Y}; \text{ equivalently,}$$

$$\hat{\beta}_1 S_{xx} = S_{xY}, \text{ and therefore } \hat{\beta}_1 = \frac{S_{xY}}{S_{xx}}.$$

Example 11.2

Refer to Table 10.2 of Example 10.2 (fuel efficiency data of 12 automobiles). Write the response variable as a column vector and compute the design matrix for the Model($Y = mpg|X = weight$). Write down, but do not solve, the matrix form of the normal equations.

Solution There are $n = 12$ observations, so the response vector has 12 rows and the design matrix has 12 rows and 2 columns given by

$$Y = \begin{pmatrix} 32 \\ 30 \\ 29 \\ 25 \\ 27 \\ 28 \\ 29 \\ 27 \\ 28 \\ 25 \\ 28 \\ 25 \end{pmatrix}; X = \begin{pmatrix} 1 & 2.495 \\ 1 & 2.530 \\ 1 & 2.620 \\ 1 & 3.395 \\ 1 & 3.030 \\ 1 & 3.345 \\ 1 & 3.040 \\ 1 & 3.085 \\ 1 & 3.495 \\ 1 & 3.950 \\ 1 & 3.470 \\ 1 & 4.105 \end{pmatrix}.$$

The matrix product $X'X$ and the vector $X'Y'$ are given by

$$X'X = \begin{pmatrix} 12 & 38.56 \\ 38.56 & 126.8546 \end{pmatrix} \text{ and } X'Y = \begin{pmatrix} 333 \\ 1059.915 \end{pmatrix}.$$

Computational details To compute $X'X$ we use Equation 11.18 and the fact that $\sum_i x_i = 38.56$ and $\sum_i x_i^2 = 126.8546$. We compute the vector $X'Y$ in the same way, using Equation 11.18 and the fact that $\sum_i Y_i = 333$ and $\sum_i x_i Y_i = 1059.915$.

Therefore, the matrix form of the normal equations is given by

$$\begin{pmatrix} 12 & 38.56 \\ 38.56 & 126.8546 \end{pmatrix} \times \begin{pmatrix} \hat{\beta}_0 \\ \hat{\beta}_1 \end{pmatrix} = \begin{pmatrix} 333 \\ 1059.915 \end{pmatrix}. \tag{11.19}$$

In practice we never solve the normal equations by inverting the matrix $X'X$; we use, instead, one of the standard statistical software packages.

Random Vectors and Random Matrices

446   *Introduction to Probability and Statistics for Science, Engineering, and Finance*

A vector with components that are random variables is called a *random vector*. The response vector $\boldsymbol{Y} = \boldsymbol{X\beta} + \boldsymbol{\epsilon}$ in the regression model is an example of random vector. The *mean vector* $E(\boldsymbol{Y})$ is the vector of means, defined by

$$E(\boldsymbol{Y}) = \begin{pmatrix} E(Y_1) \\ \vdots \\ E(Y_n) \end{pmatrix}. \tag{11.20}$$

The mean vector for the response variable is

$$E(\boldsymbol{Y}) = \boldsymbol{X\beta}. \tag{11.21}$$

The Variance–Covariance Matrix of a Random Vector

The *variance–covariance matrix* of the random vector $\boldsymbol{Y}' = (Y_1, \ldots, Y_n)$ is the $n \times n$ matrix defined as follows:

$$Cov(\boldsymbol{Y}) = \begin{pmatrix} V(Y_1) & \ldots & Cov(Y_1, Y_n) \\ Cov(Y_2, Y_1) & \ldots & Cov(Y_2, Y_n) \\ \vdots & \vdots & \vdots \\ Cov(Y_n, Y_1) & \ldots & V(Y_n) \end{pmatrix}. \tag{11.22}$$

The diagonal elements of the matrix are the variances and the off diagonal elements are the covariances. Recall that $V(Y_i) = Cov(Y_i, Y_i)$, consequently the (i, j) entry is $Cov(Y_i, Y_j)$. The matrix $Cov(\boldsymbol{Y})$ is symmetric because $Cov(Y_i, Y_j) = Cov(Y_j, Y_i)$.

Example 11.3 *Compute the variance–covariance matrix of the response vector in the simple linear regression model.*

Solution Because Y_i and Y_j are independent random variables when $i \neq j$, it follows that $Cov(Y_i, Y_j) = 0$, $(i \neq j)$; that is, the off diagonal elements are 0. Moreover, $V(Y_i) = \sigma^2$, so the diagonal elements are σ^2. Therefore

$$Cov(\boldsymbol{Y}) = \begin{pmatrix} \sigma^2 & 0 & \ldots & 0 \\ 0 & \sigma^2 & \ldots & 0 \\ 0 & \vdots & \vdots & 0 \\ 0 & 0 & \ldots & \sigma^2 \end{pmatrix} = \sigma^2 I. \tag{11.23}$$

A matrix whose entries are random variables is called a *random matrix*. The product of the column vector $\boldsymbol{Y} - E(\boldsymbol{Y})$ and the row vector $(\boldsymbol{Y} - E(\boldsymbol{Y}))'$ is the $n \times n$ symmetric random matrix

$$(\boldsymbol{Y} - E(\boldsymbol{Y}))(\boldsymbol{Y} - E(\boldsymbol{Y}))' = \begin{pmatrix} (Y_1 - E(Y_1))(Y_1 - E(Y_1)) & \ldots & (Y_1 - E(Y_1))(Y_n - E(Y_n)) \\ (Y_2 - E(Y_2))(Y_1 - E(Y_1)) & \ldots & (Y_2 - E(Y_2))(Y_n - E(Y_n)) \\ \vdots & \vdots & \vdots \\ (Y_n - E(Y_n))(Y_1 - E(Y_1)) & \ldots & (Y_n - E(Y_n))(Y_n - E(Y_n)) \end{pmatrix}.$$

Observe that the expected value of the random variable in the ith row and jth column is the covariance of the random variables Y_i and Y_j; that is, $Cov(Y_i, Y_j) = E[(Y_i - E(Y_i))(Y_j - E(Y_j))]$. It follows that the variance–covariance matrix $Cov(\boldsymbol{Y})$ can be expressed as the expected value of the random matrix $(\boldsymbol{Y} - E(\boldsymbol{Y}))(\boldsymbol{Y} - E(\boldsymbol{Y}))'$. That is,

$$Cov(\boldsymbol{Y}) = E[(\boldsymbol{Y} - E(\boldsymbol{Y}))(\boldsymbol{Y} - E(\boldsymbol{Y}))']. \tag{11.24}$$

11.2.1 Sampling Distribution of the Least Squares Estimators

The sampling distributions of the least squares estimators can also be derived via matrix methods. The next theorem (we omit the proof) is useful for computing mean vectors and variance covariance matrices for the vector of fitted values, the regression parameters, and the residual vector.

Theorem 11.1 *Suppose the random vector $\boldsymbol{W}$ is obtained by multiplying the random vector $\boldsymbol{Y}$ by the matrix $\boldsymbol{A}$, that is $\boldsymbol{W} = \boldsymbol{A}\boldsymbol{Y}$. Then*

$$E(\boldsymbol{W}) = \boldsymbol{A}E(\boldsymbol{Y}); \tag{11.25}$$
$$Cov(\boldsymbol{W}) = \boldsymbol{A}Cov(\boldsymbol{Y})A'. \tag{11.26}$$

Our first application of this theorem is to compute the mean vector and variance–covariance matrix of $\hat{\boldsymbol{\beta}}$.

Theorem 11.2 *Under the assumptions of the linear regression model*

$$E(\hat{\boldsymbol{\beta}}) = \boldsymbol{\beta}; \tag{11.27}$$
$$Cov(\hat{\boldsymbol{\beta}}) = \sigma^2(\boldsymbol{X}'\boldsymbol{X})^{-1}. \tag{11.28}$$

Proof It follows from Equation 11.25 (with $\boldsymbol{A} = (\boldsymbol{X}'\boldsymbol{X})^{-1}\boldsymbol{X}'$ and $E(\boldsymbol{Y}) = \boldsymbol{X}\boldsymbol{\beta}$) that

$$E(\hat{\boldsymbol{\beta}}) = (\boldsymbol{X}'\boldsymbol{X})^{-1}\boldsymbol{X}'E(\boldsymbol{Y}) = (\boldsymbol{X}'\boldsymbol{X})^{-1}(\boldsymbol{X}'\boldsymbol{X})\boldsymbol{\beta} = \boldsymbol{I}\boldsymbol{\beta} = \boldsymbol{\beta}.$$

In other words, the least squares estimator $\hat{\boldsymbol{\beta}} = (\boldsymbol{X}'\boldsymbol{X})^{-1}\boldsymbol{X}'\boldsymbol{Y}$ is an unbiased estimator of the vector of regression parameters $\boldsymbol{\beta}$.

It follows from Equation 11.26 (with $\boldsymbol{A} = (\boldsymbol{X}'\boldsymbol{X})^{-1}\boldsymbol{X}'$, $\boldsymbol{A}' = \boldsymbol{X}(\boldsymbol{X}'\boldsymbol{X})^{-1}$, and $Cov(\boldsymbol{Y}) = \sigma^2\boldsymbol{I}$) that

$$Cov(\hat{\boldsymbol{\beta}}) = \boldsymbol{A}\sigma^2\boldsymbol{I}\boldsymbol{A}' = \sigma^2(\boldsymbol{X}'\boldsymbol{X})^{-1}\boldsymbol{X}'\boldsymbol{X}(\boldsymbol{X}'\boldsymbol{X})^{-1} = \sigma^2(\boldsymbol{X}'\boldsymbol{X})^{-1}.$$

When we substitute the unbiased estimator $s^2 = MSE = SSE/(n-2)$ for σ^2 in Equation 11.28 we obtain the *estimated variance–covariance matrix*:

$$s^2(\hat{\boldsymbol{\beta}}) = s^2(\boldsymbol{X}'\boldsymbol{X})^{-1}. \tag{11.29}$$

Example 11.4 *Continuation of Example 11.2*

Compute the estimated standard errors $s(\hat{\beta}_0)$ and $s(\hat{\beta}_1)$ from the diagonal entries of the matrix $s^2(\boldsymbol{X}'\boldsymbol{X})^{-1}$.

Solution

The error sum of squares was computed in Example 10.4. Using the definition $s^2 = SSE/(n-2)$ it follows that $s^2 = 1.54803$. (Note: The following calculations were performed on a computer and the final values were rounded to 4 places. Because of round off errors, these answers will differ slightly from those obtained using a scientific calculator.) The inverse matrix

$$(\boldsymbol{X}'\boldsymbol{X})^{-1} = \begin{pmatrix} 3.5854 & -1.0899 \\ -1.0899 & 0.3392 \end{pmatrix}.$$

Consequently,

$$s^2(\boldsymbol{X}'\boldsymbol{X})^{-1} = \begin{pmatrix} 5.5503 & -1.6871 \\ -1.6871 & 0.5250 \end{pmatrix}.$$

Thus, $s(\hat{\beta}_0) = \sqrt{5.5503} = 2.3559$ and $s(\hat{\beta}_1) = \sqrt{0.5250} = 0.7246$. (Problem 11.6 asks you to compute the estimated standard errors directly using the methods of Chapter 10.)

Matrix Expressions for Confidence Intervals and Prediction Intervals

Point estimates and confidence intervals for the mean response and prediction intervals for a future response can also be expressed using matrix notation. The mean response for a specified value x_0 of the explanatory variable is $E(Y|x_0) = \beta_0 + \beta_1 x_0$. The estimated mean response, denoted $\hat{Y}(x_0)$, can be written as the matrix product

$$\hat{Y}(x_0) = \hat{\beta}_0 + \hat{\beta}_1 x_0 = (1, x_0)\begin{pmatrix}\hat{\beta}_0\\\hat{\beta}_1\end{pmatrix} = \boldsymbol{x}_0'\hat{\boldsymbol{\beta}}$$

where $\boldsymbol{x}_0' = (1, x_0)$. It follows from Equation 11.10 that $\boldsymbol{x}_0'\hat{\boldsymbol{\beta}} = \boldsymbol{AY}$ where

$$\boldsymbol{A} = \boldsymbol{x}_0'(\boldsymbol{X'X})^{-1}\boldsymbol{X'} \text{ and } \boldsymbol{A'} = \boldsymbol{X}(\boldsymbol{X'X})^{-1}\boldsymbol{x}_0' \qquad (11.30)$$

and from Equation 11.26 (with $\boldsymbol{A}$ defined as in Equation 11.30 and $Cov(\boldsymbol{Y}) = \sigma^2\boldsymbol{I}$) that

$$V(\hat{Y}(x_0)) = \sigma^2\boldsymbol{x}_0'(\boldsymbol{X'X})^{-1}\boldsymbol{x}_0. \qquad (11.31)$$

We obtain the estimated *standard error of prediction*, denoted $s(\hat{Y}(x_0))$, by replacing σ^2 with its estimate $s^2 = MSE = SSE/(n-2)$; thus,

$$s(\hat{Y}(x_0)) = s\sqrt{\boldsymbol{x}_0'(\boldsymbol{X'X})^{-1}\boldsymbol{x}_0}. \qquad (11.32)$$

A $100(1-\alpha)\%$ confidence interval for the mean response at the value x_0 is given by

$$\boldsymbol{x}_0'\hat{\boldsymbol{\beta}} \pm t_{n-2}(\alpha/2)s\sqrt{\boldsymbol{x}_0'(\boldsymbol{X'X})^{-1}\boldsymbol{x}_0}. \qquad (11.33)$$

Similarly, the matrix expression for the estimated standard error of the difference between a future response $Y(x_0) = \beta_0 + \beta_1 x_0 + \epsilon$ and the predicted future response $\hat{Y}(x_0) = \hat{\beta}_0 + \hat{\beta}_1 x_0$ at the value x_0 of the explanatory variable, denoted $s(Y(x_0) - \hat{Y}(x_0))$, is given by

$$s(Y(x_0) - \hat{Y}(x_0)) = s\sqrt{1 + \boldsymbol{x}_0'(\boldsymbol{X'X})^{-1}\boldsymbol{x}_0}. \qquad (11.34)$$

The corresponding prediction interval for a future response at the value x_0 is

$$\boldsymbol{x}_0'\hat{\boldsymbol{\beta}} \pm t_{n-2}(\alpha/2)s\sqrt{1 + \boldsymbol{x}_0'(\boldsymbol{X'X})^{-1}\boldsymbol{x}_0}. \qquad (11.35)$$

Matrix Representation of the Sums of Squares SST, SSR, and SSE

Recall that the computer output of a regression analysis is displayed in an ANOVA table that is based on the sum of squares decomposition

$$SST = SSR + SSE. \qquad (11.36)$$

We now give a derivation of Equation 11.36 using a matrix method that extends without change to the multiple linear regression context.

We begin by deriving matrix expressions for the sums of squares SST, SSR, and SSE.

$$SST = \sum_i (Y_i - \overline{Y})^2 = \sum_i Y_i^2 - n\overline{Y}^2 = \boldsymbol{Y'Y} - n\overline{Y}^2;$$

$$SSR = \sum_i \hat{Y}_i^2 - n\overline{Y}^2 = \hat{\boldsymbol{Y}}'\hat{\boldsymbol{Y}} - n\overline{Y}^2;$$

$$SSE = \sum_i e_i^2 = \boldsymbol{e'e}.$$

Recall that the residual vector is orthogonal to the column space of the design matrix (Equation 11.3). It follows that

$$\hat{\boldsymbol{Y}}'\boldsymbol{e} = (\boldsymbol{X}\hat{\boldsymbol{\beta}})'\boldsymbol{e} = \hat{\boldsymbol{\beta}}'\boldsymbol{X}'\boldsymbol{e} = \boldsymbol{0} \text{ and } \boldsymbol{e}'\hat{\boldsymbol{Y}} = (\hat{\boldsymbol{Y}}'\boldsymbol{e})' = \boldsymbol{0}' = \boldsymbol{0}.$$

Consequently,

$$\boldsymbol{Y}'\boldsymbol{Y} = (\hat{\boldsymbol{Y}} + \boldsymbol{e})'(\hat{\boldsymbol{Y}} + \boldsymbol{e}) = \hat{\boldsymbol{Y}}'\hat{\boldsymbol{Y}} + \boldsymbol{e}'\hat{\boldsymbol{Y}} + \hat{\boldsymbol{Y}}'\boldsymbol{e} + \boldsymbol{e}'\boldsymbol{e} = \hat{\boldsymbol{Y}}'\hat{\boldsymbol{Y}} + \boldsymbol{e}'\boldsymbol{e}.$$

Therefore, $\boldsymbol{Y}'\boldsymbol{Y} - n\overline{Y}^2 = (\hat{\boldsymbol{Y}}'\hat{\boldsymbol{Y}} - n\overline{Y}^2) + \boldsymbol{e}'\boldsymbol{e}$, which demonstrates that $SST = SSR + SSE$.

11.2.2 Geometric Interpretation of the Least Squares Solution

The essence of the least squares method depends on the concept of the distance between two points in n dimensional space. We begin with the concept of the *length* of a vector, denoted by $|\boldsymbol{x}|$, and defined as follows:

$$|\boldsymbol{x}| = \sqrt{\sum_i x_i^2}.$$

The *distance* between the vectors $\boldsymbol{x}$ and $\boldsymbol{y}$ is the length of their difference $\boldsymbol{x} - \boldsymbol{y} = (x_1 - y_1, \ldots, x_n - y_n)$, which equals

$$|\boldsymbol{x} - \boldsymbol{y}| = \sqrt{\sum_i (x_i - y_i)^2}.$$

In order to free ourselves from the tyranny of subscripts it is useful to know that the square of the length of a vector can also be expressed in terms of matrix multiplications. For instance, the matrix product $\boldsymbol{x}'\boldsymbol{x}$ equals the square of the length of $\boldsymbol{x}$, since

$$\boldsymbol{x}'\boldsymbol{x} = \sum_i x_i^2 = |\boldsymbol{x}|^2.$$

More generally, the matrix product of a row vector and column vector, called the *scalar product* and denoted $\boldsymbol{x}'\boldsymbol{y}$, is a real number. Recall that in 2- and 3-dimensional space the condition $\boldsymbol{x}'\boldsymbol{y} = 0$ implies that the angle between the two vectors is a right angle. In the more general context of linear algebra we say that two vectors are *orthogonal* when $\boldsymbol{x}'\boldsymbol{y} = 0$. The matrix form of the normal equations (Equation 11.3) implies that the residual vector $\boldsymbol{Y} - \boldsymbol{X}\hat{\boldsymbol{\beta}}$ is orthogonal to each column of the design matrix. This leads to a geometric characterization of the least squares solution that provides additional insight and understanding, as will now be explained.

We recall that the least squares solution for the best fitting line was obtained by determining the values of the parameters $(b_0, b_1) = \boldsymbol{b}'$ that minimize the following sum of squares:

$$Q(b_0, b_1) = \sum_{1 \le i \le n} (Y_i - (b_0 + b_1 x_i))^2. \tag{11.37}$$

To proceed further we use matrix notation to represent the sum of squares $Q(b_0, b_1)$ as the square of the distance between the vectors $\boldsymbol{Y}$ and $\boldsymbol{X}\boldsymbol{b}$. Thus,

$$\boldsymbol{Xb} = \begin{pmatrix} b_0 + b_1 x_1 \\ \vdots \\ b_0 + b_1 x_n \end{pmatrix}, \text{ so } \boldsymbol{Y} - \boldsymbol{Xb} = \begin{pmatrix} Y_1 - (b_0 + b_1 x_1) \\ \vdots \\ Y_n - (b_0 + b_1 x_n) \end{pmatrix}.$$

Therefore,

$$|\boldsymbol{Y} - \boldsymbol{Xb}|^2 = \sum_{1 \le i \le n} (Y_i - (b_0 + b_1 x_i))^2 = Q(b_0, b_1). \tag{11.38}$$

The set of vectors $y = \boldsymbol{Xb}$ is called the *column space* of the design matrix. It is the *subspace* consisting of all linear combination of the columns of the design matrix, that is

$$y = \boldsymbol{Xb} = \begin{pmatrix} b_0 + b_1 x_1 \\ \vdots \\ b_0 + b_1 x_n \end{pmatrix} = b_0 \begin{pmatrix} 1 \\ \vdots \\ 1 \end{pmatrix} + b_1 \begin{pmatrix} x_1 \\ \vdots \\ x_n \end{pmatrix}.$$

The *least squares problem* is to find the vector $\boldsymbol{Xb}$ in the column space of the design matrix that minimizes the distance to $\boldsymbol{Y}$. The least squares solution now has a simple and elegant geometric interpretation as the minimum distance from the response vector to the column space of the design matrix. The reader will recall from Euclidean geometry that the minimum distance from a point to a given line is obtained by measuring the length of the perpendicular line segment from the point to the line. This perpendicular line segment is the geometric representation of the residual vector. In the regression context we interpret this to mean that $\hat{\boldsymbol{\beta}}$ is the least squares solution provided the residual vector $\boldsymbol{e} = \boldsymbol{Y} - \boldsymbol{X}\hat{\boldsymbol{\beta}}$ is orthogonal to the column space of the design matrix. Expressed in matrix form we obtain the condition that $\boldsymbol{X}'(\boldsymbol{Y} - \boldsymbol{X}\hat{\boldsymbol{\beta}}) = 0$, which the reader will recognize as Equation 11.3.

To repeat: The normal equations are a consequence of the fact that the vector of residuals must be orthogonal to the columns of the design matrix. No calculus is required to obtain this fundamental result. In Section 11.4 we show that any solution of the normal equations minimizes the sum of squares.

11.3   The Matrix Approach to Multiple Linear Regression

The *multiple linear regression model* with two explanatory variables assumes that the statistical relationship between the response variable and the explanatory variables is of the form

$$Y_i = \beta_0 + \beta_1 x_{i1} + \beta_2 x_{i2} + \epsilon_i, \tag{11.39}$$
$$\epsilon_i \sim N(0, \sigma^2)$$

where ϵ_i denotes an iid sequence of independent, normal random variables with zero means and common variance σ^2. The matrix representation for this regression model is exactly the same as in the simple linear regression case except that one more column representing the new variable is adjoined to the design matrix; that is,

$$\boldsymbol{Y} = \boldsymbol{X}\boldsymbol{\beta} + \boldsymbol{\epsilon} \tag{11.40}$$

where

$$\boldsymbol{Y} = \begin{pmatrix} Y_1 \\ \vdots \\ Y_n \end{pmatrix} ; \boldsymbol{X} = \begin{pmatrix} 1 & x_{11} & x_{12} \\ \vdots & \vdots & \vdots \\ 1 & x_{n1} & x_{n2} \end{pmatrix}$$

$$\boldsymbol{\beta} = \begin{pmatrix} \beta_0 \\ \beta_1 \\ \beta_2 \end{pmatrix} ; \boldsymbol{\epsilon} = \begin{pmatrix} \epsilon_1 \\ \vdots \\ \epsilon_n \end{pmatrix}.$$

Notation   We use the *model statement* format
Model $(Y = y | X_1 = x_1, X_2 = x_2)$ to specify the response variable Y and the explanatory variables X_1, X_2.

Example 11.5 *Continuation of Example 10.4*

Refer to Table 10.2 of Example 10.2 (fuel efficiency data of 12 automobiles). It was shown (see Example 10.4) that for the Model$(Y = mpg | X_1 = weight)$ the coefficient of determination $R^2 = 0.69$. In an effort to fit a better model to the data the statistician decided to fit the multiple linear regression Model $(Y = mpg | X_1 = weight, X_2 = displacement)$. Write the response variable as a column vector and compute the design matrix for this model.

   Solution

$$\boldsymbol{Y} = \begin{pmatrix} 32 \\ 30 \\ 29 \\ 25 \\ 27 \\ 28 \\ 29 \\ 27 \\ 28 \\ 25 \\ 28 \\ 25 \end{pmatrix} ; \boldsymbol{X} = \begin{pmatrix} 1 & 2.495 & 1.9 \\ 1 & 2.530 & 1.8 \\ 1 & 2.620 & 1.6 \\ 1 & 3.395 & 3.0 \\ 1 & 3.030 & 2.2 \\ 1 & 3.345 & 3.8 \\ 1 & 3.040 & 2.2 \\ 1 & 3.085 & 3.0 \\ 1 & 3.495 & 3.8 \\ 1 & 3.950 & 4.6 \\ 1 & 3.470 & 3.8 \\ 1 & 4.105 & 5.7 \end{pmatrix}.$$

Polynomial Regression Models
   Polynomial regression deals with the problem of fitting a polynomial to a data set. Suppose the statistical relation between the response Y and the explanatory variable x is of the form

$$Y = \beta_0 + \beta_1 x + \beta_2 x^2 + \epsilon \text{ where } \epsilon \overset{\mathcal{D}}{=} N(0, \sigma^2). \tag{11.41}$$

   It follows that the mean response $E(Y|x) = \beta_0 + \beta_1 x + \beta_2 x^2$ is a quadratic polynomial of the explanatory variable x. Although the mean response is a non-linear function we can reduce it to a multiple linear regression problem by defining the explanatory variables $x_1 = x$ and $x_2 = x^2$. The polynomial regression model Equation 11.41 is transformed into the multiple linear regression model

$$Y = \beta_0 + \beta_1 x + \beta_2 x_2 + \epsilon \text{ where } \epsilon \overset{\mathcal{D}}{=} N(0, \sigma^2). \tag{11.42}$$

Example 11.6 *Determining the fuel efficiency of a light truck*

The data listed in Table 11.1 came from an experiment to determine the fuel efficiency of a light truck equipped with an experimental overdrive gear. The response variable Y is miles per gallon (mpg) and the explanatory variable x is the constant speed in miles per hour (mph) on the test track. Graph the scatter plot of mpg (Y) against speed (x) to see if the simple linear regression model is appropriate. Calculate the design matrix for the quadratic polynomial Model ($Y = mpg|x_1 = speed, x_2 = (speed)^2$).

Table 11.1 *Fuel efficiency data of Example 11.6*

MPG (Y)	Speed (x)
22	35
20	35
28	40
31	40
37	45
38	45
41	50
39	50
34	55
37	55
27	60
30	60

(Source: J. Neter, M.H. Kutner, C.J. Nachtseim, and W. Wasserman (1996), *Applied Linear Statistical Models*, 4th ed., R. D. Irwin, Chicago. Used with permission.)

Solution The scatter plot is shown in Figure 11.3. Clearly, the simple linear regression model is inappropriate.

The design matrix is

$$X = \begin{pmatrix} 1 & 35 & 1225 \\ 1 & 35 & 1225 \\ 1 & 40 & 1600 \\ 1 & 40 & 1600 \\ 1 & 45 & 2025 \\ 1 & 45 & 2025 \\ 1 & 50 & 2500 \\ 1 & 50 & 2500 \\ 1 & 55 & 3025 \\ 1 & 55 & 3025 \\ 1 & 60 & 3600 \\ 1 & 60 & 3600 \end{pmatrix}.$$

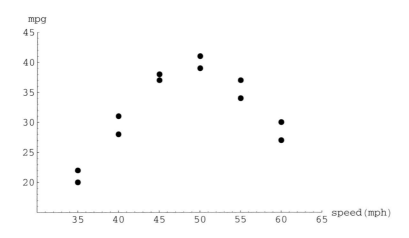

FIGURE 11.1: Scatter plot for the data in Table 11.1

11.3.1   Normal Equations, Fitted Values, and ANOVA Table for the Multiple Linear Regression Model

Consider the general multiple linear regression model

$$Y = X\beta + \epsilon$$

where

$$Y = \begin{pmatrix} Y_1 \\ \vdots \\ Y_n \end{pmatrix}; \quad X = \begin{pmatrix} 1 & x_{11} & \dots & x_{1k} \\ \vdots & \vdots & \vdots & \vdots \\ 1 & x_{n1} & \dots & x_{nk} \end{pmatrix}$$

$$\beta = \begin{pmatrix} \beta_0 \\ \beta_1 \\ \vdots \\ \beta_k \end{pmatrix}; \quad \epsilon = \begin{pmatrix} \epsilon_1 \\ \vdots \\ \epsilon_n \end{pmatrix}$$

and ϵ_i denotes, as usual, an iid sequence of independent, normal random variables with zero means and common variance σ^2. Thus the ith response Y_i is

$$Y_i = \beta_0 + \beta_1 x_{i1} + \dots + \beta_k x_{ik} + \epsilon_i. \tag{11.43}$$

It follows that the mean response as a function of the explanatory variables is

$$E(Y|x_1, \dots, x_k) = \beta_0 + \beta_1 x_1 + \dots + \beta_k x_k. \tag{11.44}$$

Interpretation of the Regression Coefficients

The regression coefficient β_i measures the change in the expected response of Y when the explanatory variable x_i is increased by one unit and all the other explanatory variables are held constant.

Notation We use the model statement format Model $(Y = y|X_1 = x_1, \dots, X_k = x_k)$ to specify the response variable Y and the k explanatory variables $X_1, \dots, X_k$. The multiple linear regression model has $k+1$ regression parameters $\beta_0, \beta_1, \dots, \beta_k$ and the design matrix has n rows and $p = k + 1$ columns, corresponding to the number of regression parameters.

The Least Squares Solution

The least squares solution and its geometric interpretation are the same as in the simple linear regression case; that is, the residual vector $e = Y - X\hat{\beta}$ is orthogonal to the columns of the design matrix. Consequently, Equation 11.3

$$X'(Y - X\hat{\beta}) = 0$$

remains valid, as well as the matrix form of the normal equations derived earlier (Equation 11.4)

$$X'X\hat{\beta} = X'Y.$$

When the inverse matrix $(X'X)^{-1}$ exists, the solutions to the normal equations, the vector of fitted values and the residual vector in matrix form are given by

$$\hat{\beta} = (X'X)^{-1}X'Y;$$
$$\hat{Y} = X\hat{\beta} = X(X'X)^{-1}X'Y = HY;$$
$$e = Y - \hat{Y} = Y - HY = (I - H)Y;$$

where

$$H = X(X'X)^{-1}X'.$$

So much for the theory. In practice, the sums of squares SST, SSR, SSE, the estimated regression parameters $\hat{\beta}_i$, and their estimated standard errors $s(\hat{\beta}_i)$ are listed in the computer printouts as shown in Table 11.3. For this reason we omit the tedious computational details needed for calculating these terms by hand.

ANOVA Table for the Multiple Linear Regression Model

The computer printout of a regression analysis is displayed in the format of the ANOVA Table 11.2. The entries in the third column are computed from the sum of squares decomposition $SST = SSR + SSE$ derived earlier (see the discussion following Equation 11.36) and the entries in the other columns are computed using theorems concerning the distributions of SSR and SSE (we omit the derivations).

Table 11.2 *ANOVA table for the multiple linear regression model*

Source	DF	Sum of Squares	Mean Square	F value	Prob>F
Model	$p-1$	SSR	MSR=SSR/$(p-1)$	MSR/MSE	
Error	$n-p$	SSE	MSE=SSE/$(n-p)$		
Total	$n-1$	SST			

Degrees of Freedom Associated with SSE

It can be shown that the number of degrees of freedom associated with the error sum of squares is $(n-p)$, and the *mean square error* $MSE = SSE/(n-p)$ (the error sum of squares divided by its degrees of freedom) is an unbiased estimator of σ^2; that is,

$$E(MSE) = E\left(\frac{SSE}{n-p}\right) = \sigma^2.$$

The terms $n-p$, SSE, and MSE are listed in the second, third, and fourth columns of Table 11.2.

Estimated Standard Errors of the Estimated Regression Parameters

Equations 11.28 and 11.29 for the variance–covariance and estimated variance–covariance matrices remain valid for the multiple linear regression model; that is

$$Cov(\hat{\boldsymbol{\beta}}) = \sigma^2(\boldsymbol{X'X})^{-1};$$
$$s^2(\hat{\boldsymbol{\beta}}) = s^2(\boldsymbol{X'X})^{-1}, \ (s^2 = MSE).$$

We use the ANOVA table to test the null hypothesis

$$H_0 : \beta_1 = \ldots \beta_k = 0 \text{ against } H_1 : \text{ not all } \beta_i = 0, \ (i = 1, \ldots, k).$$

When H_0 is true there is no regression relation between the response variable Y and the explanatory variables $X_1, \ldots, X_k$; consequently, the regression model is not a useful one. In addition, assuming H_0 is true, it can be shown that the distribution of SSR/σ^2 is chi-square with $(p-1)$ degrees of freedom, so the number of degrees of freedom associated with SSR is $(p-1)$. This number is listed in the second column (DF) of Table 11.2. The *mean square due to regression*, denoted MSR, is $SSR/(p-1)$. The terms SSR and MSR are listed in the third and fourth columns respectively. The SST term, as usual, has $n-1$ degrees of freedom. The partition of the number of degrees of freedom

$$n - 1 = (p - 1) + (n - p) \tag{11.45}$$

(displayed in the second column of Table 11.2) corresponds to the decomposition $SST = SSR + SSE$.

Suppose we wish to test the null hypothesis

$$H_0 : \beta_0 = \beta_1 = \ldots \beta_k = 0.$$

It can be shown, when H_0 is true, that the ratio MSR/MSE has an F distribution with $(p-1)$ degrees of freedom in the numerator and $(n-p)$ degrees in the denominator; that is

$$F = \frac{MSR}{MSE} \sim F_{p-1,n-p}.$$

The value of the F ratio is listed in the fifth column. The P-value of the F test is

$$\text{P-value} = P(F_{p-1,n-p} > F)$$

and it is listed in the last column of Table 11.2.

Coefficient of Multiple Determination

The *coefficient of multiple determination*, denoted R^2, is defined as follows:

$$R^2 = \frac{SSR}{SST}. \tag{11.46}$$

We obtain an interpretation of R^2 by noting that $1 = \frac{SSR}{SST} + \frac{SSE}{SST}$. (This equation is obtained by dividing both sides of Equation 11.36 by SST). It follows that the term $R^2 = SSR/SST$ represents (as in the simple linear regression model) the *proportion of the total variability that is explained by the model,* and SSE/SST represents the proportion of the unexplained variability. A value of R^2 close to one implies that most of the variability is explained by the regression model; a value of R^2 close to zero indicates that the regression model is not appropriate. It can be shown (the details are left as a problem) that

$$F = \frac{MSR}{MSE} = \left(\frac{n-p}{p-1}\right)\frac{R^2}{1-R^2}. \tag{11.47}$$

Consequently, the F ratio is large when R^2 is close to one.

Example 11.7 *Continuation of Example 11.5*

Refer to Model($Y = mpg | X_1 = weight, X_2 = displacement$) of Example 11.5. Fit a multiple linear regression model to the data using a computer software package. State the estimated regression function. Compute the coefficient of multiple determination and test the null hypothesis

$$H_0 : \beta_1 = \beta_2 = 0 \text{ against } H_1 : \text{ not all } \beta_i = 0, (i = 1, 2).$$

Solution Table 11.3 is the SAS computer printout. The estimated regression parameters are referred to by name; thus, $\hat{\beta}_0$ is called INTERCEP (which is the word intercept reduced to 8 characters), $\hat{\beta}_1$ is WGT (weight), and $\hat{\beta}_2$ is DISP (displacement). The estimated regression parameters, estimated regression function, denoted $\hat{E}(Y|x_1, x_2)$, and coefficient of multiple of determination (R-square in the computer printout) are

$$\hat{\beta}_0 = 46.351255, \hat{\beta}_1 = -7.477042, \hat{\beta}_2 = 1.740633;$$
$$\hat{E}(Y|x_1, x_2) = 46.351255 - 7.477042 \times x_1 + 1.740633 \times x_2;$$
$$R^2 = 0.7878.$$

The estimated standard errors for the estimated regression parameters $\hat{\beta}_i$, $(i = 0, 1, 2)$ appear in the third column. The entries in the last two columns are the results of testing hypotheses about the individual regression parameters and are discussed in the next section.

There are $n = 12$ observations and $p = 3$ regression parameters. Consequently, the F ratio has $p - 1 = 2$ degrees of freedom in the numerator and $n - p = 9$ degrees of freedom in the denominator. The F ratio is $16.706 > F_{2,9}(0.01) = 8.02$. The P-value is 0.0009, so we reject the null hypothesis.

Table 11.3 *Anova table for Example 11.7*

```
                    Analysis of Variance
                    Sum of          Mean
    Source    DF    Squares        Square     F Value      Prob>F
    Model     2     39.58691       19.79346    16.706      0.0009
    Error     9     10.66309        1.18479
    C Total   11    50.25000
         Root MSE   1.08848     R-square     0.7878
                    Parameter Estimates
                    Parameter      Standard   T for H0:
    Variable  DF    Estimate       Error      Parameter=0  Prob > |T|

    INTERCEP  1     46.351255      4.28124784   10.827      0.0001
    WGT       1     -7.477042      2.10287560   -3.556      0.0062
    DISP      1      1.740633      0.86323760    2.016      0.0746
```

11.3.2 Testing Hypotheses about the Regression Model

When the null hypothesis of no regression relation is rejected we would like to determine which regression parameters differ from zero. Suppose, for example, we want to test the null hypothesis

$$H_0 : \beta_i = 0 \text{ against } H_1 : \beta_i \neq 0.$$

It can be shown that

$$\frac{\hat{\beta}_i - \beta_i}{s(\hat{\beta}_i)} \sim t_{n-p}. \tag{11.48}$$

Confidence intervals for the regression parameters can now be derived in the usual way. For example, a $100(1 - \alpha)\%$ confidence interval for $\hat{\beta}_i$ is

$$\hat{\beta}_i \pm s(\hat{\beta}_i)t_{n-p}(\alpha/2).$$

When the null hypothesis $H_0 : \beta_i = 0$ is true, it follows from Equation 11.48 that $\hat{\beta}_i/s(\hat{\beta}_i)$ has a t distribution with $(n - p)$ degrees of freedom. We therefore reject H_0, at the level α if

$$\left| \frac{\hat{\beta}_i}{s(\hat{\beta}_i)} \right| > t_{n-p}(\alpha/2).$$

Example 11.8 *Continuation of Example 11.7*

Compute 95% confidence intervals for the regression parameters β_1, β_2. Test at the 5% level the null hypotheses $H_0 : \beta_i = 0$, $(i = 1, 2)$.

Solution From Table 11.3 we see that $s(\hat{\beta}_1) = 2.10287560$ and $s(\hat{\beta}_2) = 0.86323760$. Moreover, $n - p = 12 - 3 = 9$, so $t_{n-p}(0.025) = t_9(0.025) = 2.262$. It follows that the 95% confidence interval for β_1 (answers are rounded to 4 decimal places) is

$$-7.477042 \pm 2.262 \times 2.10287560 = [-12.2337, -2.7203].$$

Similarly, the 95% confidence interval for β_2 is

$$1.740633 \pm 2.262 \times 0.86323760 = [-0.2120, 3.6933].$$

Notice that the confidence interval for β_2 contains 0; this suggests that we cannot reject (at the 5% level) the null hypothesis that $\beta_2 = 0$. This is confirmed by computing the ratio

$$\left| \frac{\hat{\beta}_2}{s(\hat{\beta}_2)} \right| = \frac{1.740633}{0.86323760} = 2.0164 < 2.262 = t_9(0.025).$$

The ratio $\hat{\beta}_2/s(\hat{\beta}_2) = 2.016$ is listed in column 5 (T for HO:) of Table 11.3 and the P-value $P(|t_9| > 2.016) = 0.0746$ is listed in column 6 ($Prob > |T|$).

Building the Regression Model

Adding a variable to a regression model can lead to results that appear strange and that cast doubts on the validity of the model itself. To illustrate, let us pursue further the multiple linear regression Model($Y = mpg|X_1 = weight$, $X_2 = displacement$) of Example 11.7. The estimated regression coefficient $\hat{\beta}_1 = -7.477042$ implies that an increase of 1000 lb in the car's weight (holding the other variable constant) results in a mean decrease of 7.477042 miles per gallon. Similarly, a one liter increase in the displacement of the engine produces a mean increase of 1.74063 miles per gallon. The latter result is unusual because a larger engine should decrease fuel efficiency. The problem is that the car's weight and engine size are highly correlated (it can be shown that the sample correlation $\rho(X_1, X_2) = 0.95348$); in particular, heavier cars tend to have larger engines, so it is not really possible to increase one variable and hold the other constant. This is an example of *multicollinearity,* which occurs when an explanatory variable is highly correlated with one or more other variables already in the model or when the estimated regression coefficient, as in Example 11.7, has the wrong sign. Detecting and grappling with multicollinearity is a serious problem in multiple regression; it is treated at great length in the references listed at the end of this chapter.

The *partial F test* is a general statistical procedure for evaluating the contribution of additional explanatory variables to the regression model. To fix our ideas, let us continue our study of some problems that arise when we add the explanatory variable X_2 (displacement) to the simple linear regression Model($Y = mpg|X_1 = weight$); we call this the *reduced model*. We call Model($Y = mpg|X_1 = weight$, $X_2 = displacement$) the *full model*. The sums of squares associated with the full model are denoted SSR(f) and SSE(f); similarly, SSR(r) and SSE(r) denote the sums of squares associated with the reduced model. The total sum of squares is the same for all models, consequently,

$$SST = SSR(f) + SSE(f) = SSR(r) + SSE(r)$$
therefore
$$SSR(f) - SSR(r) = SSE(r) - SSE(f).$$

The difference SSR(f)−SSR(r) is called the *extra sum of squares*. It measures the increase in the regression sum of squares when the explanatory variable X_2 is added to the model. This difference is always non-negative, since adding more variables to the model always results in an increase in the total variability explained by the model. It follows that adding an explanatory variable reduces the unexplained variability. Thus the extra sum of squares also measures the reduction in the error sum of squares that results when one adds an explanatory variable to the model.

Example 11.9 *Continuation of Example 11.2*

Compute the extra sum of squares when the variable X_2 (displacement) is added to the simple linear regression $\text{Model}(Y = mpg|X = weight)$ of Example 11.2.

Solution The analysis of variance for the reduced and full models are shown in Tables 10.7 and 11.3, respectively. It follows that

$$SSR(f) - SSR(r) = 39.58691 - 34.76792 = 4.81719.$$

Observe that

$$SSE(r) - SSE(f) = 15.48028 - 10.66309 = 4.81719$$

as predicted by the theory.

Consider the problem of determining when the extra sum of squares obtained by adding an explanatory variable to the model is large enough to justify including it in the model; that is, we want to test the null hypothesis

$$H_0 : \beta_2 = 0 \text{ against } H_1 : \beta_2 \neq 0.$$

When H_0 is accepted this means that we have not found it useful to add X_2 to the model; that is, the full model performs no better (and may actually be worse) than the reduced model. Denote the number of degrees of freedom of SSE(f) and SSE(r) by $DF(f)$ and $DF(r)$, respectively. Under the null hypothesis it can be shown that the ratio

$$F = \frac{[SSE(r) - SSE(f)]/[DF(r) - DF(f)]}{SSE(f)/DF(f)} \tag{11.49}$$

has an F distribution with $DF(r) - DF(f)$ degrees of freedom in the numerator and $DF(f)$ in the denominator. Consequently, the decision rule is

$$\text{accept } H_0 : \quad \text{if } F \leq F_{DF(r)-DF(f),DF(f)}(\alpha); \tag{11.50}$$

$$\text{reject } H_O : \quad \text{if } F > F_{DF(r)-DF(f),DF(f)}(\alpha). \tag{11.51}$$

Example 11.10 *Continuation of Example 11.9*

Use the partial F test to determine whether or not the variable $X_2 = $ displacement should be dropped from the model.

Solution In the solution to Example 11.9 we showed that the extra sum of squares SSE(r)−SSE(f) is 4.81719. It follows that

$$SSE(r) - SSE(f) = 4.81719 \text{ and } DF(r) - DF(f) = 10 - 9 = 1.$$

$$\text{Thus, } F = \frac{[SSE(r) - SSE(f)]/[DF(r) - DF(f)]}{SSE(f)/DF(f)} = \frac{4.81719/1}{10.66309/9} = 4.065867.$$

Because $F = 4.065867 < 5.12 = F_{1,9}(0.05)$ we do not reject $H_0 : \beta_2 = 0$. It follows from Corollary 6.2 (which states that $t_n^2 = F_{1,n}$) that in this case the partial F test is equivalent to the t test. Indeed, $t_9^2 = (2.016)^2 = 4.0643$, which equals the F ratio (except for round off error).

Adjusted R^2 Criterion

Although adding a variable to the model always reduces the unexplained variability it does not necessarily reduce the mean square error because the number of degrees of freedom of the error sum of squares for the full model is less than the corresponding number of degrees of freedom for the reduced model. More precisely,

$$DF(r) - DF(f) = \text{number of additional variables in the full model.}$$

Consequently $MSE(f) = SSE(f)/DF(f)$ could be larger than $MSE(r) = SSE(r)/DF(r)$. This leads to the *adjusted R-square*, denoted R_a^2 and defined as follows:

$$R_a^2 = 1 - \left(\frac{SSE/(n-p)}{SST/(n-1)} \right) = 1 - \frac{MSE}{SST/(n-1)}. \qquad (11.52)$$

Choosing a model with the largest adjusted R-square is equivalent to choosing the model with the smallest mean square error, since $SST/(n-1)$ is the same for all models. If the reduced model has a larger adjusted R-square then it will have a smaller estimated standard error of prediction. Consequently, the reduced model has shorter confidence and prediction intervals. The reader should consult the references at the end of the chapter for a detailed discussion of various methods for selecting the best model.

11.3.3   Model Checking

The assumptions of the multiple linear regression model (refer back to Equation 11.43) should always be checked by analyzing the residuals. When the model is valid the statistical properties of the residuals should resemble those of the error terms ϵ_i, $(i = 1, \ldots n)$, where $E(\epsilon_i) = 0$ and $V(\epsilon_i) = \sigma^2$. Two particularly useful methods for detecting violations of model assumptions are the residual plots and normal probability plots described previously in Chapter 10.

Residual Plots

It can be shown that the mean and variance of the ith residual are given by

$$E(e_i) = 0 \text{ and } V(e_i) = \sigma^2(1 - h_{ii}), \qquad (11.53)$$

where h_{ii} is the ith diagonal element of the hat matrix (Equation 11.12). We obtain the *estimated standard error for the ith residual*, denoted $s(e_i)$, replacing σ^2 with its estimate s^2. Thus

$$s(e_i) = s\sqrt{1 - h_{ii}}. \qquad (11.54)$$

It follows that the residuals do not have constant variance. For this reason it is more useful to plot the *studentized residuals*, denoted e_i^*, defined as follows:

$$e_i^* = \frac{e_i}{s(e_i)}. \qquad (11.55)$$

When the assumptions of the multiple linear regression model hold, the joint distribution of the studentized residuals is approximately equal to a random sample taken from a normal population with zero mean and variance one. Moreover, it can be shown that the fitted values are independent of the studentized residuals. Therefore the points $(\hat{Y}_i, e_i^*)$, $(i = 1, \ldots, n)$ of the residual plot should appear to be randomly scattered about the line $y = 0$, with most of the points within ± 2 of the line $y = 0$. The plot of the studentized residuals shown in Figure 11.2 appear to be randomly scattered within a horizontal band of width 2 centered about the line $y = 0$.

Normal Probability Plots

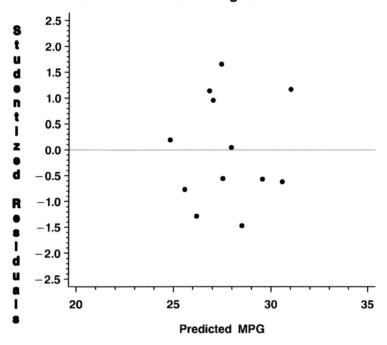

Plot of Studentized Residuals Against Predicted MPG

Model(Y = mpg ¦ X1 = weight, X2 = Disp) (Data: Table 10.2)

FIGURE 11.2:   Plot of studentized residuals against predicted mpg (data in Table 10.2)

Normal Probability Plot of Studentized Residuals

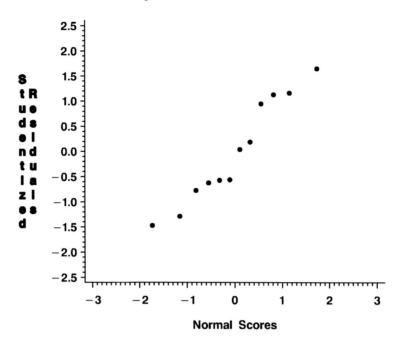

Model (Y = gpm ¦ X1 = weight, X2 = disp) (Data: Table 10.2)

FIGURE 11.3:   Normal probability plot of studentized residual (data in Table 10.2)

Normal probability plots of the studentized residuals are useful for checking the assumption that the error terms ϵ_i, $(i = 1, \ldots n)$ are iid normal random variables with zero means. To construct a *normal probability plot of the studentized residuals* we follow the procedure outlined in Definition 8.1. That is, compute
(1) the normal scores $z_i = \Phi^{-1}((i - 0.5)/n))$, $(i = 1, \ldots n)$;
(2) the order statistics $e^*_{(i)}$, $(i = 1, \ldots n)$;
(3) plot the points $(z_i, e^*_{(i)})$, $(i = 1, \ldots, n)$.

The model assumption that ϵ_i, $(i = 1, \ldots n)$ are iid normal random variables with zero means is doubtful when the points of the normal probability plot do not appear to lie nearly on a straight line.

The normal probability plot of the studentized residuals shown in Figure 11.3 reveals no serious violation of the normality assumption.

It would take too much time and space to fully exploit all the information we can obtain from a careful analysis of the residuals. The reader is advised to consult the references at the end of this chapter for a more comprehensive treatment of this subject.

11.3.4   Confidence Intervals and Prediction Intervals in Multiple Linear Regression

Matrix expressions for confidence intervals on the mean response and prediction intervals on a future response for the simple linear regression model were derived earlier in this chapter (refer to Equations 11.33 and 11.35 in Section 11.2.1). The formulas for the confidence intervals and prediction intervals in a multiple linear regression (assuming the error terms ϵ_i are iid, normal random variables with zero means) are similar except that the number of degrees of freedom for the t distribution is $(n - p)$. We omit the derivations.

The vector $\boldsymbol{x}_0' = (1, x_{01}, \ldots, x_{0k})$ denotes the values of the explanatory variables; that is $X_1 = x_{01}, \ldots, X_k = x_{0k}$. A $100(1 - \alpha)\%$ confidence interval for the mean response is given by

$$\boldsymbol{x}_0'\hat{\boldsymbol{\beta}} \pm t_{n-p}(\alpha/2)s\sqrt{\boldsymbol{x}_0'(\boldsymbol{X}'\boldsymbol{X})^{-1}\boldsymbol{x}_0}. \tag{11.56}$$

The term

$$s\sqrt{\boldsymbol{x}_0'(\boldsymbol{X}'\boldsymbol{X})^{-1}\boldsymbol{x}_0}$$

is called the *standard error of prediction* and is usually listed in the computer printouts as shown in Table 11.4.

Similarly, a $100(1 - \alpha)\%$ prediction interval for a future response

$$Y(\boldsymbol{x}_0) = \beta_0 + \beta_1 x_{01} + \ldots + \beta_k x_{0k} + \epsilon_0$$

is given by

$$\boldsymbol{x}_0'\hat{\boldsymbol{\beta}} \pm t_{n-p}(\alpha/2)s\sqrt{1 + \boldsymbol{x}_0'(\boldsymbol{X}'\boldsymbol{X})^{-1}\boldsymbol{x}_0} \, (p = k + 1). \tag{11.57}$$

Example 11.11 *Continuation of Example 11.7*

Compute 95% confidence intervals and prediction intervals for each observation using a computer software package.

Solution These values are shown in Table 11.4 and were obtained using SAS. The second column lists the observed value Y_i, the third column (Predict/Value) lists the fitted value

$$\hat{Y}_i = \boldsymbol{x}'\hat{\boldsymbol{\beta}} = \hat{\beta}_0 + \hat{\beta}_1 x_1 + \ldots + \hat{\beta}_k x_k$$

the fourth column lists the standard error of prediction (Std Err/Predict), columns 5 and 6 list the lower and upper limits for the 95% confidence intervals, and columns 7 and 8 list the lower and upper limits for the prediction intervals. Consider observation 1 (refer to the design matrix shown in the solution to Example 11.5). The values of the explanatory variables are $X_1 = 2.495$ and $X_2 = 1.9$, so

$$\boldsymbol{x}_0' = (1, \, 2.495, \, 1.9).$$

The observed value Y_1 is 32.00 and the fitted value $\hat{Y}_1$ is 31.0032. The standard error of prediction is 0.677. The 95% confidence interval is [29.4720, 32.5345] and the 95% prediction interval is [28.1036, 33.9029].

Table 11.4 *Confidence and prediction intervals for Example 11.7*

Obs	Dep Var MPG	Predict Value	Std Err Predict	Lower95% Mean	Upper95% Mean	Lower95% Predict	Upper95% Predict
1	32.0000	31.0032	0.677	29.4720	32.5345	28.1036	33.9029
2	30.0000	30.5675	0.584	29.2467	31.8883	27.7733	33.3617
3	29.0000	29.5464	0.504	28.4052	30.6876	26.8325	32.2603
4	25.0000	26.1886	0.573	24.8927	27.4845	23.4061	28.9711
5	27.0000	27.5252	0.540	26.3033	28.7471	24.7764	30.2740
6	28.0000	27.9550	0.460	26.9136	28.9963	25.2815	30.6284

7	29.0000	27.4504	0.555	26.1956	28.7053	24.6868	30.2141
8	27.0000	28.5065	0.360	27.6912	29.3217	25.9127	31.1002
9	28.0000	26.8334	0.362	26.0139	27.6529	24.2383	29.4285
10	25.0000	24.8239	0.596	23.4752	26.1725	22.0164	27.6313
11	28.0000	27.0203	0.362	26.2020	27.8387	24.4256	29.6151
12	25.0000	25.5796	0.783	23.8074	27.3518	22.5458	28.6134

11.4   Mathematical Details and Derivations

We will now show that any solution of the normal equations minimizes the sum of squares. More precisely, we will show that for any $\boldsymbol{b}$ the following inequality holds:

$$|\boldsymbol{Y} - \boldsymbol{X}\boldsymbol{b}|^2 \geq |\boldsymbol{Y} - \boldsymbol{X}\hat{\boldsymbol{\beta}}|^2. \tag{11.58}$$

Proof Let $\boldsymbol{b} = \hat{\boldsymbol{\beta}} + \boldsymbol{d}$, $\boldsymbol{d} \neq \boldsymbol{0}$. The inequality 11.58 is a consequence of the following equation:

$$|\boldsymbol{Y} - \boldsymbol{X}\boldsymbol{b}|^2 = |\boldsymbol{Y} - \boldsymbol{X}\hat{\boldsymbol{\beta}}|^2 + |\boldsymbol{X}\boldsymbol{d}|^2. \tag{11.59}$$

We obtain Equation 11.59 by substituting $\hat{\boldsymbol{\beta}} + \boldsymbol{d} = \boldsymbol{b}$ into the left hand side of Equation 11.59, expanding the matrix products and noting that the cross product terms vanish; that is,

$$(\boldsymbol{X}\boldsymbol{d})'(\boldsymbol{Y} - \boldsymbol{X}\hat{\boldsymbol{\beta}}) = \boldsymbol{0};\ (\boldsymbol{Y} - \boldsymbol{X}\hat{\boldsymbol{\beta}})'\boldsymbol{X}\boldsymbol{d} = \boldsymbol{0}.$$

In detail,

$$\begin{aligned}
\boldsymbol{Y} - \boldsymbol{X}\boldsymbol{b}|^2 &= \boldsymbol{Y} - \boldsymbol{X}(\hat{\boldsymbol{\beta}} + \boldsymbol{d})|^2 = (\boldsymbol{Y} - \boldsymbol{X}(\hat{\boldsymbol{\beta}} + \boldsymbol{d}))'(\boldsymbol{Y} - \boldsymbol{X}(\hat{\boldsymbol{\beta}} + \boldsymbol{d})) \\
&= (\boldsymbol{Y} - \boldsymbol{X}\hat{\boldsymbol{\beta}})'(\boldsymbol{Y} - \boldsymbol{X}\hat{\boldsymbol{\beta}}) - (\boldsymbol{X}\boldsymbol{d})'(\boldsymbol{Y} - \boldsymbol{X}\hat{\boldsymbol{\beta}}) \\
&\quad -(\boldsymbol{Y} - \boldsymbol{X}\hat{\boldsymbol{\beta}})'\boldsymbol{X}\boldsymbol{d} + (\boldsymbol{X}\boldsymbol{d})'(\boldsymbol{X}\boldsymbol{d}) \\
&= (\boldsymbol{Y} - \boldsymbol{X}\hat{\boldsymbol{\beta}})'(\boldsymbol{Y} - \boldsymbol{X}\hat{\boldsymbol{\beta}}) + (\boldsymbol{X}\boldsymbol{d})'(\boldsymbol{X}\boldsymbol{d}) \\
&= \boldsymbol{Y} - \boldsymbol{X}\hat{\boldsymbol{\beta}}|^2 + |\boldsymbol{X}\boldsymbol{d}|^2.
\end{aligned}$$

We complete the proof by showing that the cross product terms vanish. This follows from Equation 11.3 for the residual vector, which implies that

$$(\boldsymbol{X}\boldsymbol{d})'(\boldsymbol{Y} - \boldsymbol{X}\hat{\boldsymbol{\beta}}) = \boldsymbol{d}'\boldsymbol{X}'(\boldsymbol{Y} - \boldsymbol{X}\hat{\boldsymbol{\beta}}) = \boldsymbol{0};$$
$$(\boldsymbol{Y} - \boldsymbol{X}\hat{\boldsymbol{\beta}})'\boldsymbol{X}\boldsymbol{d} = ((\boldsymbol{X}\boldsymbol{d})'(\boldsymbol{Y} - \boldsymbol{X}\hat{\boldsymbol{\beta}}))' = \boldsymbol{0}.$$

11.5   Chapter Summary

A multiple linear regression model is used to study the relationship between a response variable y and two or more explanatory variables, $x_1, \ldots, x_k$. We assume that the relationship between the explanatory and response variables is given by a function $y = f(x_1, \ldots, x_k) + \epsilon$, where f is unknown. A multiple linear regression model assumes that $f(x_1, \ldots, x_k) = \beta_0 + \beta_1 x_+ \ldots + \beta_k x_k$. It is difficult, although not impossible, to explain the concepts and methodology of multiple linear regression without using matrix algebra. In addition to freeing ourselves from the tyranny of subscripts, the geometric interpretation of the response variable and the design matrix (as a vector and subspace of a vector space) gives us a deeper understanding of the least squares method itself, even in the case of simple linear regression.

11.6 Problems

Section 11.2

Problem 11.1 *Refer to Example 11.2. Show by direct substitution that $\hat{\beta}_0 = 38.78474$, $\hat{\beta}_1 = -3.434047$ is a solution to the normal Equation 11.19.*

Problem 11.2 *Refer to Example 10.3 of Chapter 10.*
(a) State the matrix formulation of the the Model ($Y =$ shear strength $|X =$ weld diameter).
(b) Write down the normal equations for this model in the matrix format 11.4.
(c) Verify that $\hat{\beta}_0 = -216.325$, $\hat{\beta}_1 = 3.9904$ satisfy the normal equations.
(d) Compute the estimated variance–covariance matrix $s^2(\boldsymbol{X}'\boldsymbol{X})^{-1}$.

Problem 11.3 *Refer to Problem 10.4 of Chapter 10.*
(a) State the matrix formulation of the the Model ($Y =$ CPU time $|X =$ Disk I/Os).
(b) State the normal equations for this model in the matrix format 11.4.
(c) Verify that $\hat{\beta}_0 = -0.008282$ and $\hat{\beta}_1 = 0.243756$ satisfy the normal equations given in part (b).
(d) Compute the estimated variance–covariance matrix $s^2(\boldsymbol{X}'\boldsymbol{X})^{-1}$.

Problem 11.4 *Refer to Problem 10.12 of Chapter 10.*
(a) State the matrix formulation of the the Model ($Y =$ TDS $|X =$ SEC).
(b) State the normal equations for this model in the matrix format 11.4.
(c) Verify that $\hat{\beta}_0 = 29.268569$ and $\hat{\beta}_1 = 0.597884$ satisfy the normal equations given in part (b).
(d) Compute the estimated variance–covariance matrix $s^2(\boldsymbol{X}'\boldsymbol{X})^{-1}$.

Problem 11.5 *Show that in the case of simple linear regression*

$$det(\boldsymbol{X}'\boldsymbol{X}) = nS_{xx}.$$

Problem 11.6 *Refer to Example 11.4. Compute the estimated standard errors using the equations (derived in Chapter 10)*

$$s(\hat{\beta}_0) = s\sqrt{\left[\frac{1}{n} + \frac{\bar{x}^2}{S_{xx}}\right]};$$

$$s(\hat{\beta}_1) = \frac{s}{\sqrt{S_{xx}}}.$$

Your results should agree with those obtained in the solution to Example 11.4.

Problem 11.7 *Derive the properties of the hat matrix $\boldsymbol{H}$ listed in Equations 11.14-11.17.*

Problem 11.8 *Refer to Problem 10.23 of Chapter 10.*
(a) State the matrix formulation of the the Model ($Y =$ hardness $|X =$ time).
(b) State the normal equations for this model in the matrix format 11.4.
(c) Verify that $\hat{\beta}_0 = 153.916667$ and $\hat{\beta}_1 = 2.416667$ satisfy the normal equations given in part (b).
(d) Compute the estimated variance-covariance matrix $s^2(X'X)^{-1}$.

Section 11.3

Problem 11.9 *Refer to the ground water quality data set of Problem 10.12.*
(a) Fit the multiple linear regression Model ($Y = TDS \,|\, X_1 = SEC$, $X_2 = SiO_2$) and state the estimated regression function.
(b) Estimate the change in the total dissolved solids (TDS) when the specific electrical conductivity (SEC) is increased by 100 ms/cm and the amount of silica (SiO_2) is increased by 5.
(c) What is the fitted value and residual corresponding to $X_1 = 841$ and $X_2 = 48$?
(d) Compute the ANOVA table, R^2, and the adjusted R-square.
(e) Use the partial F test to decide if X_2 should be dropped from the regression model.

Problem 11.10 *Refer to Example 11.6 and the associated data set Table 11.1.*
(a) Fit the quadratic polynomial Model ($Y = MPG \,|\, X_1 = speed$, $X_2 = speed^2$) and state the estimated regression function.
(b) Draw the scatter plot and the graph of the fitted polynomial on the same coordinate axes.
(c) Compute the fitted value, 95% confidence interval and prediction interval for $x'_0 = (1, 35, 1225)$.

Problem 11.11 *The following data come from an experiment to determine the relation between Y, the shear strength (psi) of a rubber compound, and X, the cure temperature ($^\circ F$).*
(a) Draw the scatter plot to see if the simple linear regression model is appropriate.
(b) Fit the quadratic polynomial Model ($Y =$ shear strength $|X_1 = x$, $X_2 = x^2$) to these data.
(c) Draw the graph of the fitted polynomial on the same set of axes you used for the scatter plot.
(d) Test the hypothesis that $\hat{\beta}_2 = 0$.
(e) Check the assumptions of the model by (1) graphing the studentized residual plot and (2) graphing the normal probability plot of the studentized residuals.

y	x
770	280
800	284
840	292
810	295
735	298
640	305
590	308
560	315

(Source: A.F. Dutka and F.J. Ewens (1971), A Method for Improving the Accuracy of Polynomial Regression Analysis, Journal of Quality Technology, pp. 149-155. Used with permission.)

Problem 11.12 *The data in the next table were obtained in a forestry study of certain characteristics of a stand of pine trees including A, age of a particular pine stand; HD, the average height of dominant trees in feet; N, the number of pine trees per acre at age A; and MDBH, the average diameter measured at breast height (4.5 feet above ground). A theoretical analysis suggested that the following model is appropriate:*

Model $(Y = MDBH | X_1 = HD, X_2 = A \times N, X_3 = HD/N)$.

(a) Fit this model to the data. Compute R^2 and R_a^2 (the adjusted R-square–see Equation 11.52).

(b) Test the hypothesis $H_0 : \beta_3 = 0$ against $H_1 : \beta_3 \neq 0$. State your conclusions.

(c) Fit the reduced Model $(Y = MDBH | X_1 = HD, X_2 = A \times N)$. Compute R^2 and R_a^2 for the reduced model.

(d) For each model prepare a table of fitted values, standard error of prediction, etc., similar in format to that of Table 11.4. Compare the full and reduced models by comparing the standard errors of prediction. Which model performs better?

A	HD	N	MDBH
19	51.5	500	7.0
14	41.3	900	5.0
11	36.7	650	6.2
13	32.2	480	5.2
13	39.0	520	6.2
12	29.8	610	5.2
18	51.2	700	6.2
14	46.8	760	6.4
20	61.8	930	6.4
17	55.8	690	6.4
13	37.3	800	5.4
21	54.2	650	6.4
11	32.5	530	5.4
19	56.3	680	6.7
17	52.8	620	6.7
15	47.0	900	5.9
16	53.0	620	6.9
16	50.3	730	6.9
14	50.5	680	6.9
22	57.7	480	7.9

(Source: R.H. Myers (1990), Classical and Modern Regression with Applications, *2nd ed., PWS-Kent, Boston. Used with permission.)*

Problem 11.13 *The data listed in the next table come from an experiment to study the effects of three environmental variables on exhaust emissions of light-duty diesel trucks. Fit the Model $(Y = NO_x | X_1 = $ Humidity, $X_2 = $ Temperature, $X_3 = $ Barometric pressure). Test the null hypothesis*

$$H_0 : \beta_i = 0, \ (i = 1, 2, 3) \text{ against } H_1 : \text{ at least one } \beta_i \neq 0.$$

NO_x (ppm)	Humidity (%)	Temperature ($^\circ F$)	Barometric Pressure (in. Hg)
0.70	96.5	78.1	29.08
0.79	108.72	87.93	29.98
0.95	61.37	68.27	29.34
0.85	91.26	70.63	29.03
0.79	96.83	71.02	29.05
0.77	95.94	76.11	29.04
0.76	83.61	78.29	28.87
0.79	75.97	69.35	29.07
0.77	108.66	75.44	29.00
0.82	78.59	85.67	29.02
1.01	33.85	77.28	29.43
0.94	49.20	77.33	29.43
0.86	75.75	86.39	29.06
0.79	128.81	86.83	28.96
0.81	82.36	87.12	29.12
0.87	122.60	86.20	29.15
0.86	124.69	87.17	29.09
0.82	120.04	87.54	29.09
0.91	139.47	87.67	28.99
0.89	105.44	86.12	29.21

(Source: R.L. Mason, R.F. Gunst, and J.L. Hess, (1989), Statistical Design and Analysis of Experiments, *John Wiley and Sons, New York. Used with permission.)*

Problem 11.14 *Show that the variance–covariance matrix of the residual vector is*

$$Cov(e) = \sigma^2(I - H)$$

where H is the hat matrix. Hint: Use the representation $e = (I - H)Y$, Equation 11.26, and Equation 11.16.

11.7   To Probe Further

The following texts contain a wealth of additional information on multiple linear regression including multicollinearity, model selection, model checking, and regression diagnostics.

1. R.H. Myers (1990), *Classical and Modern Regression with Applications,* 2nd ed., PWS-Kent, Boston.

2. J. Neter, M.H. Kutner, C.J. Nachtseim, and W. Wasserman (1996), *Applied Linear Statistical Models,* 4th ed., R. D. Irwin, Chicago.

Chapter 12

Single Factor Experiments: Analysis of Variance

To consult a statistician after an experiment is finished is often merely to ask him to conduct a post mortem examination. He can perhaps say what the experiment died of.

Sir Ronald A. Fisher (1890-1962), British statistician

12.1 Orientation

In this chapter we introduce the *analysis of variance* (ANOVA) method, a powerful technique of great generality for estimating and comparing the means of two or more populations, and in particular for testing the null hypothesis that the means are the same. This is an extension of the problem of testing hypotheses about the means of two populations discussed in Section 8.3. When the null hypothesis of no difference among the means is rejected, the next step in the analysis is to determine which means differ significantly from one another. This naturally leads us to study the problem of multiple comparisons of the population means.

Organization of Chapter

1. Section 12.2: The Single Factor ANOVA Model

2. Section 12.2.1: Estimating the ANOVA Model Parameters

3. Section 12.2.2: Testing Hypotheses about the Model Parameters

4. Section 12.2.3: Model Checking via Residual Plots

5. Section 12.3: Confidence Intervals for the Treatment Means; Contrasts

6. Section 12.3.1: Multiple Comparisons of Treatment Means

7. Section 12.4: Random Effects Model

8. Section 12.6: Chapter Summary

12.2 The Single Factor ANOVA Model

We develop the basic concepts and methods of the single factor ANOVA model in the context of an experiment to compare different methods of teaching arithmetic.

Example 12.1

Table 12.1 lists the numerical grades on a standard arithmetic test given to 45 students divided randomly into five equal sized groups. Groups 1 and 2 were taught by the current method; groups 3, 4, and 5 were taught together for a number of days. On each day group 3 students were praised publicly for their previous work, group 4 students were criticized publicly, and group 5 students, while hearing the praise and criticism of groups 3 and 4, were ignored.

Table 12.1 *Arithmetic grades for five groups of children*

Method	Arithmetic test grades									Means
1 (Control)	17	14	24	20	24	23	16	15	24	19.67
2 (Control)	21	23	13	19	13	19	20	21	16	18.33
3 (Praised)	28	30	29	24	27	30	28	28	23	27.44
4 (Criticized)	19	28	26	26	19	24	24	23	22	23.44
5 (Ignored)	21	14	13	19	15	15	10	18	20	16.11

(Source: G.B. Wetherill (1967), *Elementary Statistical Methods*, Methuen & Co. Ltd, p. 263.)

The response variable in this experiment is the student's grade, and the explanatory variable is the teaching group, which is an example of a non-numerical variable called a *factor*. A particular value of the factor is called a *factor level*; it is also called a *treatment*. The different groups correspond to five factor levels. For each treatment there are 9 students; we say that there are 9 *replicates* for each treatment. Since the treatments are distinguished from one another by the different levels of a single factor, the experiment is called a *single-factor experiment;* the subsequent analysis is called *one-way ANOVA*. We define the model using the model statement format: Model ($Y =$ arithmetic grade $|X = group$).

An experiment is called a *completely randomized design* when, as in Example 12.1, the different treatments are randomly assigned to the experimental units. The purpose of *randomization* is to guard against bias. For instance, if one put the best students into the third group, then their better than average performance could not be attributed primarily to the teaching method. This is an example of an *experimental design*.

The goal of ANOVA is to study the mean of the response variable as a function of the factor levels. In this sense ANOVA is similar to regression analysis, except that the explanatory variable may be non-numeric. In the context of Example 12.1, there are two questions of primary interest: (1) What effect does the teaching method have on the students' performance? (This is an estimation problem.) (2) Are there significant differences among the teaching methods? (This is, of course, a hypothesis testing problem in which we are interested in comparing the means of five populations based upon independent samples drawn from of each of them.)

The null hypothesis here is that there is no difference in the mean grades produced by these teaching methods, thus the null and alternative hypotheses are given by

$$H_0 : \mu_1 = \mu_2 = \mu_3 = \mu_4 = \mu_5 \text{ against } H_1 : \text{ at least two of the means are different.}$$

A Visual Comparison of the Five Teaching Methods: Side-by-Side Box Plots

Before turning to more advanced methods of analysis, we first look at the data using some of the exploratory data analysis techniques discussed in Chapter 1. The side-by-side box plot shown in Figure 12.1 is a very informative graphic display that reveals several important features of the data:

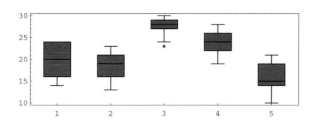

FIGURE 12.1:   Side-by-side box plots of teaching methods data (Table 12.1)

1. The width of each box gives a rough indication of the extent to which the variability of the grades depends on the teaching method.

2. The horizontal lines in each box (which represent the sample medians) indicate the extent to which the row medians depend on the teaching method. The box plots indicate some variability in the medians, particularly for method 3, which appears to be significantly better than the other four. (The * in the box plot for method 3 indicates an outlier, corresponding to observation 23.)

3. The box plots suggest that it is very unlikely that these samples were drawn from the same population; in other words, the null hypothesis that these teaching methods are equally effective appears to be highly unlikely. We return to this point shortly.

Notation and Model Assumptions Before proceeding further we introduce some notation and assumptions.

1. **Treatments and replications** It is assumed that there are I treatments and that the number of replications, denoted J, is the same for each treatment; so the total number of observations is $n = I \times J$. The jth observation at the ith factor level is denoted by y_{ij}. The data are displayed in the format of Table 12.2.

Table 12.2 *Data format for a one–way ANOVA*

Treatment	Observation			
1	y_{11}	y_{12}	$\cdots$	y_{1J}
2	y_{21}	y_{22}	$\cdots$	y_{2J}
$\vdots$	$\vdots$	$\vdots$	$\vdots$	$\vdots$
I	y_{I1}	y_{I2}	$\cdots$	y_{IJ}

For an example refer back to Table 12.1. The five teaching methods correspond to factor levels 1, 2, 3, 4, and 5, respectively. In particular, $y_{1,1} = 17$, $y_{1,2} = 14$, $y_{5,1} = 21$, etc.

Since the sample sizes corresponding to each treatment are equal, the experimental design is said to be *balanced.* The case of unequal sample sizes will be dealt with later.

2. **Random allocations** We assume that the experimental units have been randomly allocated to the treatments.

3. **The statistical model for the response variable** We assume that the response of the jth experimental unit to the ith factor level is a random variable Y_{ij} defined by

$$Y_{ij} = \mu_i + \epsilon_{ij} \tag{12.1}$$

where the random variables ϵ_{ij} are iid $N(0, \sigma^2)$, so $E(Y_{ij}) = \mu_i$ and $V(Y_{ij}) = \sigma^2$. In particular, we assume that the variance of the response variable is the same for all factor levels. The parameter μ_i is called the *ith treatment mean.*

4. **The overall mean and treatment effects** It is convenient to specify the statistical model in terms of new parameters μ, called the *overall mean*, and $\alpha_i = \mu_i - \mu$, which is the difference between the ith treatment mean and the overall mean; it is called the *ith treatment effect.* In detail:

$$\text{Overall mean } \mu = \frac{1}{I} \sum_{i=1}^{I} \mu_i; \; i\text{th treatment effect: } \alpha_i = \mu_i - \mu. \text{ Consequently,} \sum_{i=1}^{I} \alpha_i = 0.$$

In terms of the new parameters, the response variables are:

$$Y_{ij} = \mu + \alpha_i + \epsilon_{ij}; \text{ thus, } E(Y_{ij}) = \mu_i = \mu + \alpha_i.$$

5. **"Dot" subscript notation** is useful for expressing summation with respect to one index with the other fixed. Replacing the subscript j with a dot means summation with respect to the subscript j. Replacing each subscript i and j with a dot means summation with respect to both subscripts i and j. Thus,

$$y_{i.} = \sum_{1 \le j \le J} y_{ij}; \quad \overline{y}_{i.} = \frac{y_{i.}}{J} \; (i\text{th row mean});$$

$$y_{..} = \sum_{i} \sum_{j} y_{ij}; \quad \overline{y}_{..} = \frac{y_{..}}{n} \; (\text{grand mean}).$$

The quantity $\overline{y}_{..}$ is called the *grand mean* and the quantity $\overline{y}_{i.}$ is called the *ith row mean.*

Example 12.2 *Continuation of Example 12.1*

Refer back to Example 12.1 (comparing the five methods for teaching arithmetic). Compute the grand mean and the five treatment means. Express the results using the dot subscript notation.

Solution With reference to Table 12.1, it can be shown that the grand mean $\overline{y}_{..} = 21$, which means that the average grade of all students tested was 21. Similarly, it can be shown that the average grades for teaching methods 1 through 5 are:

$$\overline{y}_{1.} = 19.67, \; \overline{y}_{2.} = 18.33, \; \overline{y}_{3.} = 27.44, \; \overline{y}_{4.} = 23.44, \; \overline{y}_{5.} = 16.11.$$

The results indicate that the average grade for method 3 is greater than the average grades for the other four methods. Notice, however, that $y_{1,9} = 24 > 23 = y_{3,9}$; not every student in group 3 graded higher than all students in group 1. The fact that the box plots of groups 3 and 5 do not overlap indicates that there are major differences between these two teaching methods. On the other hand, there is considerable overlap between the box plots of groups 1 and 2, indicating that the differences between their means is probably not statistically significant. We will discuss these preliminary inferences shortly from a more sophisticated perspective.

12.2.1 Estimating the ANOVA Model Parameters

In this section we use the method of least squares to compute the values of the parameters $\mu, \alpha_1, \ldots, \alpha_I$ of the ANOVA model that best fits the data; the method is similar to the one used in regression analysis. We then develop a criterion for measuring how well the model fits the data; it is similar to the coefficient of determination used in regression analysis, and is in fact obtained from a similar partition of the total variability into two parts: the variability explained by the model and the unexplained variability.

It can be shown, see Section 12.5 for the details, that the least squares estimates for the model parameters μ, α_i, μ_i, denoted by $\hat{\mu}$, $\hat{\alpha}_i$, $\hat{\mu}_i$, are given by

$$\hat{\mu} = \overline{Y}_{..}, \ \hat{\alpha}_i = \overline{Y}_{i.} - \overline{Y}_{...} \ \text{Consequently,} \ \hat{\mu}_i = \hat{\mu} + \hat{\alpha}_i = \overline{Y}_{i.}. \tag{12.2}$$

These results are intuitively reasonable. The overall mean is estimated by the grand mean, the ith treatment effect is estimated by the difference between the ith row mean and the grand mean, and the ith treatment mean is estimated by the ith row mean.

Notice that the assumption of normality was not used to derive the least squares estimates; the normality assumption, however, will be used to derive confidence intervals for the model parameters.

Example 12.3 *Continuation of Example 12.1*

Refer back to the teaching methods data (Table 12.1) of Example 12.1. Compute the least squares estimators for the grand mean, treatment means, and treatment effects.

Solution Using Equation 12.2 we obtain the following results:

$$\text{Grand mean } \hat{\mu} = \overline{y}_{..} = 21.$$

Treatment means	Least squares estimate
$\hat{\mu}_1 = \overline{y}_{1.} = 19.67$	
$\hat{\mu}_2 = \overline{y}_{2.} = 18.33$	
$\hat{\mu}_3 = \overline{y}_{3.} = 27.44$	
$\hat{\mu}_4 = \overline{y}_{4.} = 23.44$	
$\hat{\mu}_5 = \overline{y}_{5.} = 16.11$	

The corresponding least squares estimates for the treatment effects are listed in the next table.

$$
\begin{array}{cc}
\text{Treatment} & \text{Least squares} \\
\text{effects} & \text{estimate}
\end{array}
$$

$$\hat{\alpha}_1 = \overline{y}_{1.} - \overline{y}_{..} = -1.33$$
$$\hat{\alpha}_2 = \overline{y}_{2.} - \overline{y}_{..} = -2.67$$
$$\hat{\alpha}_3 = \overline{y}_{3.} - \overline{y}_{..} = 6.44$$
$$\hat{\alpha}_4 = \overline{y}_{4.} - \overline{y}_{..} = 2.44$$
$$\hat{\alpha}_5 = \overline{y}_{5.} - \overline{y}_{..} = -4.89$$

These results indicate that the average score of the students taught using method 3 (praised) was 6.44 points above the grand mean (21), while the average score of the students taught using method 5 (ignored) was 4.89 points below.

Fitted Values and Residuals As in regression analysis, we define the fitted (predicted) values and the residuals as follows:

$$\textbf{Fitted value: } \hat{y}_{ij} = \overline{y}_{i.}; \textbf{ Residual: } e_{ij} = y_{ij} - \hat{y}_{ij} = y_{ij} - \overline{y}_{i.}.$$

Partitioning the Total Variability of the Data To study how well the model produced by the least squares estimates fits the data, we use a partition of the total sum of squares $SST = \sum \sum (y_{ij} - \overline{y}_{..})^2$ analogous to that used in regression analysis. Adding and subtracting $\overline{y}_{i.}$ we obtain the algebraic identity

$$y_{ij} - \overline{y}_{..} = (\overline{y}_{i.} - \overline{y}_{..}) + (y_{ij} - \overline{y}_{i.}); \tag{12.3}$$

that is,

$$\text{Observation} - \left(\begin{array}{c}\text{grand} \\ \text{mean}\end{array}\right) = \left(\begin{array}{c}\text{difference due to} \\ \text{treatment}\end{array}\right) + (\text{residual}).$$

This corresponds to the model $Y_{ij} - \mu = \alpha_i + \epsilon_{ij}$. Proceeding in a manner similar to that used in regression analysis (refer to Equation 10.17) we partition the total sum of squares SST into the sum

$$SST = SSTr + SSE, \tag{12.4}$$

where,

$$SST = \sum_{i=1}^{I}\sum_{j=1}^{J}(y_{ij} - \overline{y}_{..})^2 = \sum_{i=1}^{I}\sum_{j=1}^{J} y_{ij}^2 - \frac{y_{..}^2}{n};$$

$$SSTr = \sum_{i=1}^{I}\sum_{j=1}^{J}(\overline{y}_{i.} - \overline{y}_{..})^2 = J\sum(\overline{y}_{i.} - \overline{y}_{..})^2 = \frac{1}{J}\sum_{i=1}^{I} y_{i.}^2 - \frac{y_{..}^2}{n};$$

$$SSE = \sum_i\sum_j(y_{ij} - \overline{y}_{i.})^2 = SST - SSTr = \sum_i\sum_j e_{ij}^2.$$

$SSTr$ is called the *treatment sum of squares*. It represents the variability explained by the ANOVA model. SSE, the *error sum of squares*, represents the unexplained variability. The *proportion of the total variability explained by the model* is denoted R^2, where

$$R^2 = \frac{SSTr}{SST}. \tag{12.5}$$

Derivation of Equation 12.4 We obtain Equation 12.4 by squaring both sides of Equation 12.3, summing over the indices i, j and using the fact that the cross product term vanishes, that is,

$$\sum_i \sum_j (\overline{y}_{i.} - \overline{y}_{..})(y_{ij} - \overline{y}_{i.}) = 0.$$

Example 12.4 *Continuation of Example 12.1*

Refer back to the teaching methods data of Example 12.1. Compute the sums of squares SST, SSTr, SSE, and R^2.

Solution Based on the side-by-side box plots (Figure 12.1), it is clear that there are two sources of variability: the variability due to the different teaching methods and the variability arising from the differences between the individual students. The sums of squares and the proportion of variability explained by the model are given by

$$SST = 1196.00, \ SSTr = 722.67, \ SSE = 473.33, \ R^2 = \frac{722.67}{1196} = 0.60.$$

The model explains 60% of the total variability of the data, although you must keep in mind that the results of a statistical analysis of this type of experiment cannot be reduced to a single number.

Computing the Degrees of Freedom

Associated with each sum of squares appearing in Equation 12.4 is its number of *degrees of freedom (DF)*, a count of the independent summands appearing in its definition. As a general rule,

$$\begin{pmatrix} \text{DF of a} \\ \text{SS} \end{pmatrix} = \begin{pmatrix} \text{Number of summands} \\ \text{in SS} \end{pmatrix} - \begin{pmatrix} \text{Number of linear equations} \\ \text{satisfied by the summands} \end{pmatrix}.$$

In particular,

1. the number of degrees of freedom associated with SST is $n - 1$ since $SST = \sum \sum (Y_{ij} - \overline{Y}_{..})^2$ contains $n = IJ$ summands and the single constraint $\sum \sum (Y_{ij} - \overline{Y}_{..}) = 0$.

2. the number of degrees of freedom associated with SSTr is $I - 1$ since $SSTr = J \sum (\overline{Y}_{i.} - \overline{Y}_{..})^2$ contains I summands satisfying the one constraint $\sum (\overline{Y}_{i.} - \overline{Y}_{..}) = 0$.

3. the number of degrees of freedom associated with SSE is $n - I$ since $SSE = \sum \sum (Y_{ij} - \overline{Y}_{i.})^2$ contains $n = IJ$ summands satisfying the I constraints $\sum (Y_{ij} - \overline{Y}_{i.}) = 0$.

4. **Partitioning the degrees of freedom:** Notice that the partition (12.4) of the sum of squares SST yields a corresponding partition of the degrees of freedom: $n - 1 = I - 1 + (n - I)$. We will use this result shortly.

12.2.2 Testing Hypotheses about the Parameters

In addition to estimating the treatment means the researcher will also be interested in determining the differences among them. For instance, do the data on the five methods for teaching arithmetic indicate significant differences among them? Let us reformulate this as a hypothesis testing problem. Recall that the ANOVA model assumes that the response

Y_{ij} is of the form $Y_{ij} = \mu + \alpha_i + \epsilon_{ij}$, where the random variables ϵ_{ij} are iid $N(0, \sigma^2)$; thus $E(Y_{ij}) = \mu + \alpha_i$ and $V(Y_{ij}) = \sigma^2$. This suggests that we formulate the null hypothesis as follows:

$$H_0 : \alpha_1 = \ldots = \alpha_5 = 0 \text{ against } H_1 : \alpha_i \neq 0 \text{ for at least one } i = 1, \ldots, 5.$$

The single factor ANOVA test of H_0 is based on the partition of the sum of squares SST given in Equation 12.4 and some results concerning the sampling distribution of SSE, SSTr, and SST, as will now be explained.

Sampling Distribution of SSE, SSTr, SST

The statistic on which we base our test of H_0 is defined by the F *ratio:*

$$F = \frac{SSTr/(I-1)}{SSE/(n-I)} = \frac{MSTr}{MSE},$$

where

1. MSE is the *mean square error* computed as the error sum of squares divided by its degrees of freedom:
$$MSE = \frac{SSE}{n-I}.$$

It can be shown that MSE is always an unbiased estimator of σ^2 whether H_0 is true or not. Thus

$$E(MSE) = E\left(\frac{SSE}{n-I}\right) = \sigma^2. \tag{12.6}$$

2. MSTr is the *mean square treatment* computed as the treatment sum of squares divided by its degrees of freedom:

$$MSTr = \frac{SSTr}{I-1}.$$

It can also be shown that
$$E(MSTr) = \sigma^2 + \frac{J\sum \alpha_i^2}{I-1}. \tag{12.7}$$

Consequently, when H_0 is true, so that $\alpha_1 = \ldots = \alpha_I = 0$, then $E(MSTr) = \sigma^2 = E(MSE)$; however, when H_0 is false, which means that some of the α_i's are different from zero, then Equation 12.7 implies that $E(MSTr) > E(MSE)$. This suggests that we use the ratio $F = MSTr/MSE$ to test H_0; in particular, large values of F make H_0 less plausible. Theorem 12.1 is a more precise version of this statement.

3. Advanced statistical theory tells us that, when H_0 is true, the treatment sum of squares and error sum of squares are mutually independent χ^2 distributed random variables with $I-1$ and $n-I$ degrees of freedom, respectively; that is,

$$\frac{SSTr}{\sigma^2} \overset{\mathcal{D}}{=} \chi_{I-1}^2 \text{ and } \frac{SSE}{\sigma^2} \overset{\mathcal{D}}{=} \chi_{n-I}^2.$$

Thus, the distribution of the F ratio $MSTr/MSE$ is $F_{I-1, n-I}$.

Our test of the equality of the treatment means is based on the following theorem.

Theorem 12.1 *The null distribution of the ratio MSTr/MSE has an F distribution with $I-1$ degrees of freedom in the numerator and $n-I$ degrees of freedom in the denominator; that is,*

$$\text{F ratio: } F = \frac{MSTr}{MSE} \overset{D}{=} F_{I-1,n-I}$$

where F_{ν_1,ν_2} denotes the F distribution with ν_1, ν_2 degrees of freedom, respectively. The critical points of the F distribution are given in Table A.6 of the appendix.

The F test for the equality of treatment means We now show how to use the F ratio to test

$$H_0: \mu_1 = \mu_2 = \ldots = \mu_I \text{ against } H_1: \text{ at least two of the means are different.}$$

The Decision Rule When H_0 is true $F = MSTr/MSE$ is distributed as $F_{I-1,n-I}$, so large values of F support the alternative H_1. Consequently, our decision rule is

accept H_0 if $F \leq F_{I-1,n-I}(\alpha)$, and

reject H_0 if $F > F_{I-1,n-I}(\alpha)$.

The P-value of the test is the probability $P(F_{I-1,n-I} > F)$ and usually appears in the computer printout of the results.

Example 12.5 *Continuation of Example 12.1*

Use the F ratio to test the null hypothesis that there is no difference among the five teaching methods in Example 12.1.

Solution We previously computed $SSTr = 722.67$ and $SSE = 473.33$. We compute the F ratio by noting that $I = 5$, $J = 9$, so the degrees of freedom are given by $I - 1 = 4$ and $n - I = 45 - 5 = 40$. Therefore,

$$MSTr = \frac{SSTr}{I-1} = \frac{SSTr}{4} = 180.67;$$

$$MSE = \frac{SSE}{n-I} = \frac{SSE}{40} = 11.83;$$

$$F = \frac{MSTr}{MSE} = 15.27.$$

Since $15.27 > F_{4,40}(0.01) = 3.83$, we see that the P-value is less than 0.01; this is pretty strong evidence against the null hypothesis. These calculations are conveniently summarized in the single factor ANOVA table (also called the one-way ANOVA) shown in Table 12.3.

Table 12.3 *Single factor ANOVA table*

Source	DF	SS	MS	F
Model	$I-1$	$SSTr$	$MSTR = \frac{SSTr}{I-1}$	$\frac{MSTR}{MSE}$
Error	$n-I$	SSE	$MSE = \frac{SSE}{n-I}$	
Total	$n-1$	SST		

Example 12.6 *Continuation of Example 12.1*

Plot of Residuals Against Predicted Math Scores

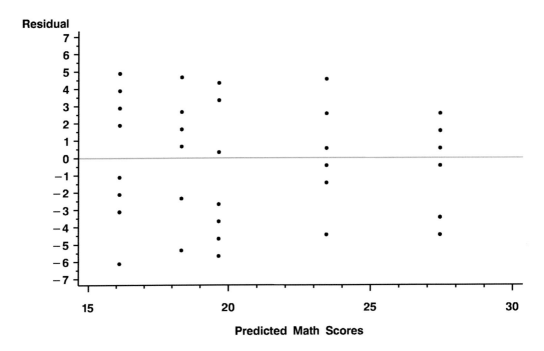

FIGURE 12.2:   Plot of residuals against predicted math scores (Example 12.1)

Construct the one-way ANOVA table for the teaching methods data (Table 12.1) of Example 12.1.

Solution The ANOVA table is shown in Table 12.4.

Table 12.4 *Single factor ANOVA table for Example 12.1*

Source	DF	SS	MS	F
Model	4	722.67	180.67	15.27
Error	40	473.33	11.83	
Total	44	1196.00		

12.2.3   Model Checking via Residual Plots

The assumptions of equal variances and normality should always be checked with various residual plots similar to those used in Chapter 10. Figure 12.2, which is the graph of $(\hat{y}_{ij}, e_{ij})$, is a plot of the residuals against the predicted values. Looking at this graph we do not see any evidence that the variances of the residuals depend on the predicted values.

The normal probability plot for the residuals is shown in Figure 12.3. It suggests that the residuals are not normally distributed. Moreover, the P-value of the Shapiro–Wilk statistic equals 0.0107, which provides additional evidence against the normality hypothesis. Fortunately, many of the statistical tests for comparing means, such as the t test and ANOVA tests, are relatively insensitive to moderate violations of the normality assumption (see G.E.P. Box, W.G. Hunter, and J.S. Hunter (1978), *Statistics for Experimenters*, John Wiley & Sons, New York, p. 91).

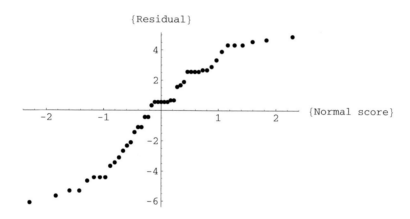

FIGURE 12.3: Normal probability plot of residuals for Example 12.1

12.2.4   Unequal Sample Sizes

It is not always possible to arrange an experiment so that there is the same number of observations for each treatment.

Example 12.7 *Comparing the silver content of Byzantine coins*

Consider the data in Table 12.5, which records the silver content (% Ag) of 27 coins discovered in Cyprus. The coins were grouped into four classes, corresponding to four different coinages during the reign of King Manuel I, Comnenus (1143-1180).

Table 12.5

Coinages			
1	2	3	4
5.9	6.9	4.9	5.3
6.8	9.0	5.5	5.6
6.4	6.6	4.6	5.5
7.0	8.1	4.5	5.1
6.6	9.3		6.2
7.7	9.2		5.8
7.2	8.0		5.8
6.9			
6.2			

(Source: M.F. Hendy and J.A. Charles (1970), The Production Techniques, Silver Content and Circulation History of the Twelfth–Century Byzantine Trachy, Archaeometry, vol. 12, pp. 13-21. Quoted in D. J. Hand et al. (1994), Small Data Sets, Chapman & Hall, London.)

Archaeologists were interested in whether there were significant differences in the silver content of coins minted early and late in King Manuel's reign.

In the preceding example there are $I = 4$ treatments, corresponding to the four different coinages. Note that the sample sizes are *not* equal; this is an example of an *unbalanced design*. The ANOVA procedure in this case is a straightforward modification of the equal sample sizes case.

The F Test for the Equality of Treatment Means (Unbalanced Design)

Let $J_i (i = 1, \ldots, I)$ denote the sample size corresponding to the ith treatment; the total number of observations is then $n = J_1 + \cdots + J_I$. Looking at Table 12.5 we see that $J_1 = 9$, $J_2 = 7$, $J_3 = 4$, $J_4 = 7$ and that the total number of observations $n = 27$. The sum of squares partition

$$SST = SSTr + SSE$$

still holds, where

$$SST = \sum_{i=1}^{I} \sum_{j=1}^{J_i} (y_{ij} - \overline{y}_{..})^2;$$

$$SSTr = \sum_{i=1}^{I} \sum_{j=1}^{J_i} (\overline{y}_{i.} - \overline{y}_{..})^2 = \sum_{i=1}^{I} J_i (\overline{y}_{i.} - \overline{y}_{..})^2;$$

$$SSE = \sum_{i=1}^{I} \sum_{j=1}^{J_i} (y_{ij} - \overline{y}_{i.})^2.$$

The mean square treatment, mean square error, and F ratio are defined by

$$MSTR = \frac{SSTr}{I-1}; \ MSE = \frac{SSE}{n-I}; \ F = \frac{MSTR}{MSE}.$$

The quantities $I-1$ and $n-I$ are the number of degrees of freedom associated with the sum of squares SSTr and SSE, respectively. (A more detailed explanation of this point will be given shortly.) There is also an analogue of Theorem 12.1:

Theorem 12.2 *The null distribution of the F ratio (unbalanced design) is $F_{I-1,n-I}$; that is,*

$$F = \frac{MSTr}{MSE} \overset{\mathcal{D}}{=} F_{I-1,n-I}$$

where $n = \sum_i J_i$.

When H_0 is true $F = MSTr/MSE$ is distributed as $F_{I-1,n-I}$. A large F value provides evidence against the null hypothesis; consequently our decision rule is

accept H_0 if $F \le F_{I-1,n-I}(\alpha)$, and reject H_0 if $F > F_{I-1,n-I}(\alpha)$.

The P-value of the test is the probability $P(F_{I-1,n-I} > F)$.

Example 12.8 *Continuation of Example 12.7*

Refer back to the coinage data in Table 12.5. (1) Compute the grand mean, the treatment means, and the components SSTr, and SSE of the sum of squares partition of SST. (2) Test the null hypothesis (no difference in the percentage of silver in the coins during the reign of King Manuel I) by computing the F ratio.
   Solution

1. The quantities $\overline{y}_{..}$, $\overline{y}_{i.}$, SSTr, SSE, and SST are

$$\overline{y}_{..} = 6.54; \overline{y}_{1.} = 6.74; \overline{y}_{2.} = 8.16; \overline{y}_{3.} = 4.88; \overline{y}_{4.} = 5.61;$$
$$SSTr = 35.77;$$
$$SSE = 10.90;$$
$$SST = 46.67.$$

2. We compute the F ratio after computing the mean square treatment and the mean square error:

$$MSTr = \frac{SSTr}{I-1} = 11.923, \ MSE = \frac{SSE}{n-I} = 0.474, \ F = \frac{MSTr}{MSE} = 25.170.$$

In this case $F \stackrel{\mathcal{D}}{=} F_{3,23}$; since $F = 25.170 > F_{3,23}(0.01) = 4.76$, the P-value of the test is < 0.01; the null hypothesis is not accepted.

The Single Factor ANOVA Table (Unbalanced Design)

To construct the ANOVA table it is necessary to calculate the degrees of freedom associated with the sums of squares SSE and SSTr.

1. The number of degrees of freedom associated with SSTr is $I - 1$ since $SSTr = \sum J_i(\overline{Y}_{i.} - \overline{Y}_{..})^2$ contains I summands satisfying the one constraint $\sum J_i(\overline{Y}_{i.} - \overline{Y}_{..}) = 0$;

2. The number of degrees of freedom associated with SSE is $n - I$ since $SSE = \sum\sum_{j=1}^{J_i}(Y_{ij} - \overline{Y}_{i.})^2$ contains n summands satisfying the I constraints $\sum_{j=1}^{J_i}(Y_{ij} - \overline{Y}_{i.}) = 0$.

3. The number ofdegrees of freedom associated with SST is $n - 1$ since $SST = \sum\sum(Y_{ij} - \overline{Y}_{..})^2$ contains n summands and the single constraint $\sum_{i=1}^{I}\sum_{j=1}^{J_i}(Y_{ij} - \overline{Y}_{..}) = 0$.

4. Notice that once again that the partition $SST = SSTr + SSE$ yields a corresponding partition of the degrees of freedom:

$$n - 1 = (I - 1) + (n - I).$$

The single factor ANOVA table for unequal sample sizes is displayed in Table 12.6.

Table 12.6   *Single factor ANOVA table (unequal sample sizes)*

Source	DF	SS	MS	F
Model	$I - 1$	SSTr	$MSTr = \frac{SSTr}{I-1}$	$\frac{MSTr}{MSE}$
Error	$n - I$	SSE	$MSE = \frac{SSE}{n-I}$	
Total	$n - 1$	SST		

The single factor ANOVA table for the data set in Table 12.5 is displayed in Table 12.7.

Table 12.7   *Single factor ANOVA table for the data in Table 12.5*

Source	DF	SS	MS	F
Model	3	37.75	12.58	26.27
Error	23	11.02	0.48	
Total	26	48.76		

12.3   Confidence Intervals for the Treatment Means; Contrasts

In Section 12.2.1 we derived point estimates for the treatment means $\mu, \mu_1, \ldots, \mu_I$. We now turn our attention to deriving confidence intervals for these parameters and their differences $\mu_i - \mu_k$. For instance, when the null hypothesis of no difference among the treatment means is rejected, the experimenter would still like to know which means, or linear combinations of means, are significantly different from one another. In particular, the experimenter might be interested in constructing a confidence interval for the parameter $\theta = \mu_i - \mu_k$ with the goal of testing the hypothesis

$$H_0 : \mu_i - \mu_k = 0 \text{ against } H_1 : \mu_i - \mu_k \neq 0.$$

Of course, this is equivalent to testing

$$H_0 : \mu_i = \mu_k \text{ against } H_1 : \mu_i \neq \mu_k.$$

The experimenter will conclude that the treatment means are significantly different if and only if the confidence interval does not contain 0. Before proceeding further, however, we assume that there is the same number of replicates, denoted J, for each treatment; this will simplify the calculations. The modifications to be made when the design is unbalanced will be mentioned briefly at the end of this section. The pairwise comparison $\mu_i - \mu_k$ of two treatment means is an example of a *contrast,* which refers to any linear combination of two or more treatment means of the form

$$\theta = \sum_i c_i \mu_i, \text{ where } \sum_i c_i = 0 \text{ (balanced design)}. \tag{12.8}$$

Example 12.9 *Continuation of Example 12.1*

Refer again to the teaching methods study (Example 12.1). In this experiment groups 1 and 2 (the control groups) were taught using the current method; the other three groups were taught with three different methods. This suggests that we study the differences between the test scores for the current teaching method and those for the other three. Express this comparison as a contrast.

Solution The null hypothesis in this case is that there is no difference between the current method of teaching arithmetic and the alternative methods. Consequently, we test the null hypothesis

$$H_0 : \frac{\mu_1 + \mu_2}{2} = \frac{\mu_3 + \mu_4 + \mu_5}{3}.$$

The contrast in this case is:

$$\frac{1}{2}\mu_1 + \frac{1}{2}\mu_2 - \frac{1}{3}\mu_3 - \frac{1}{3}\mu_4 - \frac{1}{3}\mu_5 = 0;$$

$$\text{Thus } c_1 = c_2 = \frac{1}{2} \text{ and } c_3 = c_4 = c_5 = -\frac{1}{3}.$$

Confidence Intervals for Treatment Means and Their Differences

To test a null hypothesis of the form

$$H_0 : \sum_i c_i \mu_i = 0, \text{ where } \sum_i c_i = 0, \tag{12.9}$$

we construct a confidence interval for the parameter $\theta = \sum_i c_i \mu_i$ and reject H_0 if the confidence interval does not contain 0.

To illustrate the method we consider the contrast $\theta = \mu_i - \mu_k$. We use the statistics $\overline{Y}_{i.}$ and $\overline{Y}_{i.} - \overline{Y}_{k.}$ to construct $100(1 - \alpha)\%$ confidence intervals for the parameters μ_i and their differences $\mu_i - \mu_k$. Using Equation 12.6 we obtain the following estimate for σ:

$$\hat{\sigma} = \sqrt{MSE} = \sqrt{\frac{SSE}{n - I}}. \tag{12.10}$$

We note that $V(\overline{Y}_{i.}) = \sigma^2/J$ and $V(\overline{Y}_{i.} - \overline{Y}_{k.}) = 2\sigma^2/J$; therefore, the corresponding estimated standard errors are given by

$$s(\overline{Y}_{i.}) = \sqrt{\frac{MSE}{J}}, \; s(\overline{Y}_{i.} - \overline{Y}_{k.}) = \sqrt{\frac{2MSE}{J}}.$$

Expressions for the $100(1-\alpha)\%$ confidence intervals for μ_i and $\mu_i - \mu_k$ are derived in the usual way and are given in Equations 12.11 and 12.12.

$$(100(1-\alpha)\% \text{ confidence limits for } \mu_i\text{:}) \quad \overline{Y}_{i.} \pm t_{n-I}(\alpha/2)\sqrt{\frac{MSE}{J}}; \qquad (12.11)$$

$$(100(1-\alpha)\% \text{ confidence limits for } \mu_i - \mu_k\text{:}) \quad \overline{Y}_{i.} - \overline{Y}_{k.} \pm t_{n-I}(\alpha/2)\sqrt{\frac{2MSE}{J}}. \quad (12.12)$$

If the design is unbalanced, the corresponding formulas are given by

$$\overline{Y}_{i.} \pm t_{n-I}(\alpha/2)\sqrt{\frac{MSE}{J_i}};$$

$$\overline{Y}_{i.} - \overline{Y}_{k.} \pm t_{n-I}(\alpha/2)\sqrt{MSE\left(\frac{1}{J_i} + \frac{1}{J_k}\right)}.$$

The *least significant difference*, abbreviated *LSD*, is defined by the following equation:

$$LSD = t_{n-I}(\alpha/2)\sqrt{2MSE/J}. \qquad (12.13)$$

Example 12.10 *Continuation of Example 12.1*

Refer back to the study of the five methods of teaching arithmetic (Example 12.1). Using the F test, we concluded that these teaching methods are not equally effective. More precisely, we did not accept, at the 5% level, the null hypothesis that the mean score on the arithmetic test is the same for each of the five teaching methods. This leads to the next step in the analysis: Determine which of these teaching methods differ significantly from one another and by how much. In particular, determine a 95% confidence interval for the difference between teaching methods 3 and 4.

Solution   We compute the confidence interval using Equation 12.12. Refer to Table 12.4 for the arithmetic grades for groups of children taught by the five methods. In this case $n - I = 40$, $J = 9$, and $MSE = 11.83$; consequently, choosing $\alpha = 0.05$ yields the values

$$t_{n-I}(\alpha/2)\sqrt{2MSE/J} = 2.021 \times 1.62 = 3.28.$$

Using this criterion we see that the difference between the mean responses for methods 3 and 4 is significant at the 5% level because $\overline{y}_3 - \overline{y}_4 = 27.44 - 23.44 = 4 > 3.28$.

Testing the Significance of a General Contrast   In the preceding discussion we considered contrasts of the form $\theta = \mu_i - \mu_k$. We now consider testing the significance of the contrast $\theta = \sum_i c_i\mu_i$, where $\sum_i c_i = 0$. To construct a confidence interval for the contrast we use the statistic

$$\hat{\theta} = \sum_i c_i\overline{Y}_{i.}, \qquad (12.14)$$

which is an unbiased estimate of θ. Any linear combination of the sample treatment means of the form (12.14) is also called a *contrast*. The null and alternative hypotheses corresponding to a contrast are

$$H_0 : \sum_i c_i\mu_i = 0 \text{ against } H_1 : \sum_i c_i\mu_i \neq 0, \text{ where } \sum_i c_i = 0.$$

To test this hypothesis we use the result that the random variable $\hat{\theta}$ defined by $\hat{\theta} = \sum_i c_i \overline{Y}_i$ is normally distributed with mean and variance given by:

$$E(\hat{\theta}) = \theta; \tag{12.15}$$

$$\sigma^2(\hat{\theta}) = \sigma^2 \sum_i \frac{c_i^2}{J}. \tag{12.16}$$

An unbiased estimator of σ^2 is MSE; consequently

$$s^2(\hat{\theta}) = MSE \sum_i \frac{c_i^2}{J}$$

is an unbiased estimator of $\sigma^2(\hat{\theta})$. It follows that $\frac{\hat{\theta}-\theta}{s(\hat{\theta})}$ is t_{n-I} distributed and the $100(1-\alpha)\%$ confidence limits for θ are

$$\hat{\theta} \pm t_{n-I}(\alpha/2)s(\hat{\theta}). \tag{12.17}$$

Example 12.11 *Continuation of Example 12.1*

Refer back to the teaching methods study (Example 12.1). Construct a 95% confidence interval for the contrast

$$\theta = \frac{1}{2}\mu_1 + \frac{1}{2}\mu_2 - \frac{1}{3}\mu_3 - \frac{1}{3}\mu_4 - \frac{1}{3}\mu_5.$$

This corresponds to comparing the averages of groups 1 and 2, who were taught by the current method, to the averages of groups $3, 4$, and 5, who were taught by different methods.

   Solution A straightforward calculation using Equation 12.17 yields $\hat{\theta} = -3.33$, $s(\hat{\theta}) = 1.0467$, $t_{40}(0.025) = 2.021$. The 95% confidence interval is therefore $-3.33 \pm 2.021 \times 1.0467 = -3.33 \pm 2.12$. Since the interval does not include 0, we conclude that there is a significant difference between teaching methods 1 and 2 and teaching methods $3, 4$, and 5.

12.3.1 Multiple Comparisons of Treatment Means

   In the previous section we constructed $100(1 - \alpha)\%$ confidence intervals of the form

$$\overline{Y}_{i.} - \overline{Y}_{k.} - t_{n-I}(\alpha/2)\sqrt{\frac{2MSE}{J}} \le \mu_i - \mu_k \le \overline{Y}_{i.} - \overline{Y}_{k.} + t_{n-I}(\alpha/2)\sqrt{\frac{2MSE}{J}}$$

for each of the $C_{I,2}$ paired differences $\mu_i - \mu_k$ that can be formed from the I treatment means. We then used these confidence intervals to test whether the differences between ith and kth treatment means are statistically significant. For example, the difference between ith and kth treatment means are considered significant at level α if

$$|\overline{y}_{i.} - \overline{y}_{k.}| \ge t_{n-I}(\alpha/2)\sqrt{2MSE/J} = LSD. \tag{12.18}$$

   There are, however, serious objections to using the least significant difference as a criterion for testing whether or not the difference between two treatment means is statistically significant. To see why, consider the case of 5 treatments with $C_{5,2} = 10$ paired comparisons. Suppose, in fact, that the null hypothesis is true, so that all treatment means are equal. Then, as Snedecor and Cochran point out (see reference at the end of this chapter), the probability of at least one of the paired differences exceeding the LSD bound 12.18 (with $\alpha = 0.05$) is approximately 0.30. Thus, when this experiment is repeated many times the largest observed paired difference will exceed the LSD bound nearly 30% of the time! Consequently, the significance level is really 0.30 and not, as the researcher might think, 0.05. In general, the probability under the null hypothesis of finding a pair of treatment means whose difference is significant increases with I; in other words, using this method to detect differences in the means increases the type I error probabilities.

Tukey's Simultaneous $100(1-\alpha)\%$ Confidence Intervals for the Differences between Two Means (Balanced Design)

We now present a method for comparing the treatment means that avoids the aforementioned difficulties; it is called *Tukey's multiple comparison procedure*. We give the formula for the $100(1-\alpha)\%$ confidence interval first, followed by the level α test for the difference between two means:

$$\overline{Y}_{i.} - \overline{Y}_{k.} - q(\alpha; I, n-I)\sqrt{\frac{MSE}{J}} \le \mu_i - \mu_k \tag{12.19}$$

$$\le \overline{Y}_{i.} - \overline{Y}_{k.} + q(\alpha; I, n-I)\sqrt{\frac{MSE}{J}}.$$

Here $q(\alpha; I, \nu)$ denotes the critical value of the *studentized range* distribution. Table A.7 in the appendix lists some of these critical values for $\alpha = 0.05$ and selected values of I and ν.

The corresponding test for rejecting the null hypothesis that $\mu_i = \mu_k$, $i \ne k$ is given by

$$|\overline{y}_{i.} - \overline{y}_{k.}| \ge q(\alpha; I, n-I)\sqrt{MSE/J} = HSD. \tag{12.20}$$

The quantity HSD appearing on the right hand side of Equation 12.20 is called the *honest significant difference*.

Example 12.12 *Continuation of Example 12.1*

Refer back to the teaching methods study (Example 12.1). Construct a 95% confidence interval for the difference between teaching methods 3 and 4; use the HSD criterion.

Solution Refer again to Table 12.1 on the arithmetic grades for groups of children taught by the five methods. In this case $n - I = 40$, $J = 9$, $MSE = 11.83$; consequently, choosing $\alpha = 0.05$ yields the values

$$HSD = q(0.05; 5, 40)\sqrt{MSE/J} = 4.04 \times \sqrt{1.3148} = 4.63.$$

Notice that $HSD = 4.63 > 3.23 = LSD$; in other words, the width of the simultaneous confidence interval is larger than the one obtained via the t-test. Using the HSD criterion we see that the difference in mean response between methods 3 and 4 is not significant at the 5% level because

$$\overline{y}_3 - \overline{y}_4 = 27.44 - 23.44 = 4 < 4.63.$$

Observe that the difference between two treatment means is significant only if the corresponding confidence interval does not contain zero. A useful way to visualize the relationships between the various means is to arrange the means in increasing order and identify those means that differ by less than HSD; the following table illustrates how this is done for the comparison of the teaching methods.

$\overline{y}_{5.}$	$\overline{y}_{2.}$	$\overline{y}_{1.}$	$\overline{y}_{4.}$	$\overline{y}_{3.}$
16.11	18.33	19.67	23.44	27.44

Notice that, according to Tukey's HSD criterion, there are fewer significant differences. For instance, the difference $\mu_3 - \mu_4$ is not significant according to Tukey's criterion, but is significant based on the LSD criterion.

Interpretation of Tukey's Multiple Comparisons

1. The confidence intervals produced by Tukey's method (Equation 12.19) include *all* differences $\mu_i - \mu_k$ with probability at least $1 - \alpha$.

2. Similarly, if we test for differences between some or all pairs of treatment means using the criterion in Equation 12.20, the probability of a type I error is less than α. Since the error rate α refers to all pairs of treatment means, it is also called the *experiment wise error rate*.

Tukey's Method for Unbalanced Designs

When the sample sizes for the treatments are not equal, but are not too much different from one another, then Tukey's method, with the following modifications, can still be used. The $100(1 - \alpha)\%$ confidence intervals produced by this method, however, are only approximately valid.

$$\overline{Y}_{i.} - \overline{Y}_{k.} \pm q(\alpha; I, n - I)\sqrt{\frac{MSE}{2}\left(\frac{1}{J_i} + \frac{1}{J_k}\right)} \qquad (12.21)$$

12.4   Random Effects Model

In the fixed effects model the factor levels the researcher chooses in order to compare the means of the response variable as a function of these specific factor levels. In many situations, however, the factor levels are not chosen in this way, but are instead randomly selected from a much larger population. The next example illustrates how such problems arise.

Example 12.13 *Does the analysis of ground water quality depend on the laboratory?*

Total dissolved solids (TDS), measured in milligrams per liter (mg/L), is a ground water quality parameter that is routinely measured in the laboratory. There is a justifiable concern that the measured TDS value depends on the laboratory performing the test. Excessive variability in the results reported by different laboratories would support the need for regulating the conditions under which these tests are performed. In addition to determining whether or not the labs differ in their measurements of TDS levels, researchers are interested in studying how much of the variability in these measurements is due to the variability among the labs.

To study the variability in the measurements, the following experiment is performed: $I = 4$ laboratories are selected at random from the large population of federal, state, and commercial water-quality laboratories. One sample of well water is divided into four parts— one for each lab—and each lab makes $J = 5$ measurements of the TDS level in the sample. The data, which are fictitious, are displayed in Table 12.8. Does the measured value of total dissolved solids in the water depend on the laboratory performing the test?

Table 12.8 *Measured values (mg/L) of total dissolved solids (TDS) recorded by four laboratories*

Laboratory	TDS levels (mg/L)				
1	112	190	145	65	21
2	246	134	121	136	79
3	131	122	146	193	129
4	137	231	277	211	157

Solution In the random effects model we assume that the jth measurement of the TDS level made by the ith laboratory is a random variable Y_{ij}, written as

$$Y_{ij} = \mu + A_i + \epsilon_{ij}$$

where the random variables A_i $(i = 1, \ldots, I)$ and ϵ_{ij} are mutually independent, normally distributed, random variables with zero means and variances σ_A^2, σ^2, respectively; in symbols we write

$$A_i \overset{\mathcal{D}}{=} N(0, \sigma_A^2) \text{ and } \epsilon_{ij} \overset{\mathcal{D}}{=} N(0, \sigma^2); \text{ thus } V(Y_{ij}) = \sigma_A^2 + \sigma^2.$$

The quantities σ_A^2 and σ^2 are called the *components of variance*.

Note that $\sigma_A^2 = 0$ implies that $A_i = 0\,(i = 1, \ldots, I)$; thus, the null hypothesis (of no treatment effects due to the different factor levels) and the alternative are written as

$$H_0 : \sigma_A^2 = 0 \text{ against } H_1 : \sigma_A^2 > 0.$$

Although this model is different from the fixed effects model previously discussed, the construction of the ANOVA table is carried out in exactly the same way, so we content ourselves with a sketch. The quantities SST, SSTr, MSTr, SSE, MSE, and F=MSTr/MSE are defined exactly as before and the sum of squares decomposition

$$SST = SSTr + SSE$$

remains valid. It can be shown (we omit the tedious details) that

$$E(MSE) = E\left(\frac{SSE}{IJ - I}\right) = \sigma^2; \tag{12.22}$$

$$E(MSTr) = \sigma^2 + J\sigma_A^2; \tag{12.23}$$

$$F = \frac{MSTr}{MSE} \overset{\mathcal{D}}{=} F_{I-1, IJ-I}. \tag{12.24}$$

From Equations 12.22 and 12.23 we see that if H_0 is true then $E(MSTr) = E(MSE)$, otherwise, if H_0 is false then $E(MSTr) > E(MSE)$. Thus, we do not accept H_0 at the level α if $F > F_{I-1, IJ-I}(\alpha)$.

An unbiased estimate of σ_A^2 is given by

$$\hat{\sigma}_A^2 = \frac{MSTr - MSE}{J}. \tag{12.25}$$

It is not too difficult to show (the details are left as a problem) that the variance and estimated variance of the grand mean are given by:

$$V(\overline{Y}_{..}) = \frac{J\sigma_A^2 + \sigma^2}{n}; \tag{12.26}$$

$$s^2(\overline{Y}_{..}) = \frac{MSTr}{n}. \tag{12.27}$$

Equation 12.27 asserts that the estimated variance of the grand mean $\overline{Y}_{..}$ is the *mean square treatment divided by the total number of observations.*

The ANOVA table derived from the TDS data in Table 12.8 is displayed in Table 12.9, from which the following conclusions can be drawn:

1. $F = 2.58 < 3.24 = F_{3,16}(0.05)$; consequently, the null hypothesis is not rejected.

2. The estimates of the variance components are

$$s^2(\overline{Y}_{..}) = \frac{MSTr}{n} = \frac{7878.85}{20} = 393.94;$$

$$\hat{\sigma}_A^2 = \frac{MSTr - MSE}{J} = \frac{7878.85 - 3057.125}{5} = 964.35;$$

$$\hat{\sigma}^2 = 3057.125.$$

Table 12.9 *Anova table for TDS data (Table 12.8)*

Source	DF	Sum of Squares	Mean Square	F Value	Pr > F
Model	3	23636.550	7878.850	2.58	0.0899
Error	16	48914.000	3057.125		
Corrected Total	19	72550.550			

Remark It sometimes happen that the ANOVA estimate $\hat{\sigma}_A^2$ (Equation 12.25) is negative! When this occurs there is no hard and fast rule about what to do next. One possibility, but not the only one, is to conclude that the true value of σ_A^2 is 0.

12.5 Mathematical Derivations and Details

We recall that the least squares method is to find those values $\mu, \alpha_1, \ldots, \alpha_I$ that minimize the error sum of squares

$$Q(\mu, \alpha_1, \ldots, \alpha_I) = \sum\sum e_{ij}^2 = \sum\sum (y_{ij} - \mu - \alpha_i)^2.$$

We minimize the function Q using calculus; that is, we first compute the partial derivatives of Q with respect to each of the parameters $\mu, \alpha_1, \ldots, \alpha_I$ and set them equal to zero. This yields the following $I + 1$ linear equations for the parameters:

$$\frac{\partial Q}{\partial \mu} = -2\sum\sum (y_{ij} - \mu - \alpha_i) = 0,$$

$$\frac{\partial Q}{\partial \alpha_i} = -2\sum_j (y_{ij} - \mu - \alpha_i) = 0, \; i = 1, \ldots, I,$$

subject to the constraint

$$\sum_i \alpha_i = 0.$$

Solving these linear equations yields the least squares estimates for the model parameters given in Equation 12.2.

12.6   Chapter Summary

A single factor experiment analyzes the Model (Y = response variable $|X$ = factor level), where X denotes a (non-numeric) variable called a factor. The different values of the factor are called factor levels or treatments. Some examples of factor levels are teaching methods, types of fabric, and diet methods. The goal of ANOVA is to study the mean of the response variable as a function of the factor levels. Side-by-side box plots help us understand how the mean of the response variable depends on the factor levels. ANOVA is a statistical method for estimating and comparing the means of two or more populations, and, in particular, for testing the null hypothesis that the means are the same. We test the model assumptions by graphing residual plots. When the null hypothesis of no difference among the means is rejected, the next step in the analysis is to determine which means differ significantly from one another. One method to accomplish this is Tukey's method of multiple comparisons.

12.7   Problems

Section 12.2

Problem 12.1 *The side-by-side box plot displayed in Figure 12.1 reveals an outlier in the group of students taught by method 3 (praised). (Outliers are discussed in Section 1.5.) Verify that 23 is the outlier by showing that it lies outside the interval*

$$(Q_1 - 1.5 \times IQR, \; Q_3 + 1.5 \times IQR),$$

where IQR is the interquartile range and Q_1, Q_3 are the lower and upper quartiles of the arithmetic scores of the students taught by method 3.

Problem 12.2 *The following table lists the tensile strengths of a rubber compound produced by three different methods, denoted A, B, and C.*

Treatment	Tensile strength			
A	4038	4364	4410	4242
B	4347	3859	3432	3720
C	3652	3829	4071	4232

(a) Compute the three treatment means and the grand mean $\overline{y}_{1.}, \overline{y}_{2.}, \overline{y}_{3.}, \overline{y}_{..}$.
(b) Compute SSTr, SSE, and SST, and verify that $SST = SSTr + SSE$.
(c) Construct the ANOVA table, and test the null hypothesis, at the 5% level, that there are no significant differences among the three methods of producing the rubber compound.

Problem 12.3
(a) Table 12.1 indicates that the first two students taught by method 3 obtained different scores (28 and 30) on the arithmetic test. What feature of the statistical model (Equation 12.1) accounts for this?
(b) As shown in Table 12.1, three students taught by method 1 scored a 24 on the arithmetic test and one student in group 3 scored a 23, even though the group 3 mean (27.44) is larger than the group 1 mean (19.67). What feature of the statistical model (12.1) accounts for this?

Problem 12.4 *The effect of four different mixing techniques on the compressive strengths of concrete is investigated and the following data are collected.*

Mixing technique	Compressive strength			
1	2868	2913	2844	2887
2	3250	3262	3110	2996
3	2838	3049	3030	2832
4	2847	2721	2530	2598

Does the mixing technique affect the compressive strength? State the null hypothesis and the alternative hypothesis. Construct the ANOVA table. Test the null hypothesis at the level $\alpha = 0.05$. Summarize your conclusions. Save your computations.

Problem 12.5 *The data in the following table are the results of a test in which the burst strengths of cardboard boxes made with four different types of wood pulp were studied; three measurements were taken for each wood pulp mixture.*

Wood pulp	Burst strength (psi)		
1	237	219	209
2	205	202	201
3	144	220	167
4	256	246	245

(a) Compute the four treatment means and the grand mean $(\bar{y}_{1.}, \ldots \bar{y}_{4.}, \bar{y}_{..})$.
(b) Compute SSTr, SSE, and SST.
(c) Construct the ANOVA table and test at the level $\alpha = 0.05$ whether the four types of boxes have the same mean burst strength.

Problem 12.6 *The following data give the burn times, in seconds, of four fabrics tested for flammability.*

Fabric	Burn times (sec)					
1	18	17	18	17	14	18
2	12	11	11	11	12	11
3	15	9	13	7	12	9
4	14	12	8	13	15	9

Construct the ANOVA table and test at the level $\alpha = 0.05$, the null hypothesis that there is no difference in the flammability of the four fabrics. What conclusions can you draw from this experiment?
Save your computations.

Problem 12.7 *The following data give four measurements of the ionization potential, measured in electron volts (eV), of a gaseous ion for three different methods: photoionization (PI), extrapolated voltage difference (EVD), and semi log plot (SL).*

Method	Ionization potential (eV)			
PI	10.68	9.22	8.47	8.22
EVD	10.53	9.92	11.10	10.22
SL	15.03	14.72	15.39	15.04

*Test the null hypothesis that there is no significant difference among the three measure-
ment methods by constructing the ANOVA table and performing the F test. Use $\alpha = 0.05$.*

Problem 12.8 *The following data give the results of an experiment to study the effects of
three different word processing programs on the length of time (measured in minutes) to type
a text file. A group of 24 secretaries were randomly divided into 3 groups of 8 secretaries
each, and each group was randomly assigned to one of the three word processing programs.*

Program	Typing time (min)							
1	15	13	11	19	14	19	16	19
2	13	12	9	18	13	17	15	20
3	18	17	13	18	15	22	16	22

*Test the hypothesis that the word processing program has no effect on the length of time it
takes to type a text file; use $\alpha = 0.05$. Summarize your results in an ANOVA table.*

Problem 12.9 *In a weight loss experiment men were given one of three different diets: 10
men followed diet A, 7 men followed diet B, and 8 men followed diet C. The weight losses
after one month are displayed in the following table. Notice that the observations at each
factor level now appear as columns instead of rows. The question to be decided is whether
or not the mean weight loss is the same for each of these diets.*

	Weight loss (lbs)	
Diet A	Diet B	Diet C
15.3	3.4	11.9
2.1	10.9	13.1
8.8	2.8	11.6
5.1	7.8	6.8
8.3	0.9	6.8
10	5.2	8.8
8.3	2.5	12.5
9.4		17.5
12.5		
11.1		

(a) Identify the factor, the number of levels, and the response variable.
(b) State the null hypothesis.
*(c) Construct the ANOVA table and test at the level $\alpha = 0.05$ whether the three diets produce
the same weight loss.*
Save your computations.

Problem 12.10 *The analytical methods committee of the Royal Society of Chemistry reported the following results on the levels of tin recovered after samples (taken from the same
product) were boiled with hydrochloric acid for 30, 45, 60, and 75 minutes, respectively.*

Refluxing time (min)	Tin found (mg/kg)					
30	55	57	59	56	56	59
45	57	59	58	57	57	56
60	58	57	56	57	58	59
75	57	55	58	59	59	59

(*Source:* The Determination of Tin in Organic Matter by Atomic Absorption Spectrometry, *The Analyst* (1983), vol. 108, pp. 109-115, Table VIII. Used with permission.)
Test the hypothesis that the amount of tin recovered is the same for all four boiling times. Use $\alpha = 0.05$. Summarize your results in an ANOVA table.

Problem 12.11
(a) Show that $E(\overline{y}_{i.}) = \mu + \alpha_i$.
(b) Show that $E(\overline{y}_{i.} - \overline{y}_{..}) = \alpha_i$, i.e., $\overline{y}_{i.} - \overline{y}_{..}$ is an unbiased estimator of the ith treatment mean.

Section 12.3

Problem 12.12 *The analytical methods committee of the Royal Society of Chemistry reported the following results on the levels of tin recovered from samples of blended fish–meat product after boiling with hydrochloric acid for 30, 45, 60, and 70 minutes, respectively.*

Refluxing time (min)	Tin found (mg/kg)
30	57 57 55 56 56 55 56 55
45	59 56 55 55 57 51 50 50
60	58 56 52 56 57 56 54 50
70	51 60 48 32 46 58 56 51

(*Source:* The Determination of Tin in Organic Matter by Atomic Absorption Spectrometry, *The Analyst* (1983), vol. 108, pp. 109-115, Table VII. Used with permission.)

During the experiment the authors noted "that maintaining boiling for 70 minutes without sputtering and enforced cooling was difficult. The effect is shown in a lower mean recovery and much greater variability of results." Write a brief report justifying these conclusions.

Problem 12.13 *Continuation of Problem 12.4.*
(a) Construct 95% confidence intervals for the treatment means.
(b) Use an appropriate t statistic to construct 95% confidence intervals for all differences $\mu_i - \mu_k$.
(c) Use Tukey's method to identify those differences in the treatment means that are significant. Use $\alpha = 0.05$.

Problem 12.14 *Continuation of Problem 12.6.*
(a) Construct 95% confidence intervals for each of the treatment means.
(b) Use Tukey's method to identify those differences in the treatment means that are significant. Use $\alpha = 0.05$.

Problem 12.15 *Continuation of Problem 12.7.*
(a) Construct 95% confidence intervals for each of the measurement methods.
(b) Use Tukey's method to identify those differences in the measurement methods that are significant. Use $\alpha = 0.05$.

Problem 12.16 *Continuation of Problem 12.8.*
(a) Construct 95% confidence intervals for the treatment means.
(b) Use Tukey's method to determine those differences in the treatment means that are significant. Use $\alpha = 0.05$.

Problem 12.17 *Continuation of Problem 12.9.*
(a) Construct 95% confidence intervals for each of the treatment means.
(b) Use Tukey's method to determine those differences in the treatment means that are significant. Use $\alpha = 0.05$.

Problem 12.18 *A high voltage electric cable consists of 12 wires. The following data are the tensile strengths of the 12 wires in each of 9 cables.*
(a) Use a computer to draw side-by-side box plots for the nine cables. Do the box plots indicate that the strength of the wires varies among the cables?
(b) Cables 1 to 4 were made from one lot of raw material and cables 5 to 9 came from another lot. Do the box plots indicate that the tensile strength depends on the lot?
(c) Perform a one–way ANOVA to test the null hypothesis that the tensile strength does not vary among the cables.
(d) Use Tukey's method to determine those differences in the treatment means that are significant. Are these results consistent with the information that cables 1 to 4 came from one lot of raw material and cables 5 to 9 came from another lot?
(e) Express the comparison between the two lots as a contrast, and test whether the tensile strength depends on the lot.

Cable number

1	2	3	4	5	6	7	8	9
345	329	340	328	347	341	339	339	342
327	327	330	344	341	340	340	340	346
335	332	325	342	345	335	342	347	347
338	348	328	350	340	336	341	345	348
330	337	338	335	350	339	336	350	355
334	328	332	332	346	340	342	348	351
335	328	335	328	345	342	347	341	333
340	330	340	340	342	345	345	342	347
337	345	336	335	340	341	341	337	350
342	334	339	337	339	338	340	346	347
333	328	335	337	330	346	336	340	348
335	330	329	340	338	347	342	345	341

(Source: A. Hald (1952), Statistical Theory with Engineering Applications, *John Wiley & Sons, New York. Used with permission.)*

Problem 12.19 *Verify that, under the assumptions of the single factor ANOVA model, the random variable*

$$\hat{\theta} = \sum_i c_i \overline{Y}_i$$

is normally distributed with mean and variance given by Equations 12.15 and 12.16.

Section 12.4

Problem 12.20 *The following data give the tensile strengths (psi) of cotton yarn produced by four different looms selected at random from a large group of such looms at a textile mill.*

Loom	Tensile strength (psi)					
1	18	18	20	16	14	17
2	16	14	15	20	16	14

| 3 | | 11 | 8 | 8 | 5 | 13 | 8 |
| 4 | | 24 | 23 | 20 | 19 | 16 | 19 |

(a) Is this a fixed effects model or a random effects model? Justify your answer.
(b) Construct the ANOVA table to test at the level $\alpha = 0.05$ whether the tensile strength is the same for each loom.
(c) Estimate the components of the variance.

Problem 12.21 *The following data give the polysilicon resistivities measured at three different positions for each of six randomly selected silicon wafers.*

	Wafer 1	Wafer 2	Wafer 3	Wafer 4	Wafer 5	Wafer 6
Resistivity	47.337	56.053	44.594	44.430	47.093	46.825
Resistivity	46.815	55.640	42.759	47.790	47.529	47.239
Resistivity	46.221	56.126	43.670	46.313	48.056	47.830

(a) Construct the ANOVA table and test the hypothesis that there is no significant difference in the resistivities among the silicon wafers; use $\alpha = 0.05$.
(b) Estimate the components of variance.

Problem 12.22 *The following data give the amount, in parts per million (ppm), of a toxic substance as measured by four randomly selected laboratories. Each laboratory made six measurements of the toxic substance. The question to be studied was how much of the variability is due to the different laboratory techniques and how much is due to random error.*

Laboratory	Toxic level (ppm)
1	53.2 54.5 52.8 49.3 50.4 53.8
2	51 40.5 50.8 51.5 52.4 49.9
3	47.4 46.2 46 45.3 48.2 47.1
4	51 51.5 48.8 49.2 48.3 49.8

(Source: J.S. Milton and J.C. Arnold, Introduction to Probability and Statistics, 2nd ed., McGraw–Hill.)
(a) Explain why the random effects model is appropriate.
(b) Construct the appropriate ANOVA table, and test at the level $\alpha = 0.05$ the null hypothesis that there is no significant differences in the toxic levels as measured by the different laboratories.
(c) Estimate the variance components.

Problem 12.23 *A manufacturing engineer wants to study the amount of time required to assemble a piece of equipment; he is particularly interested in how the assembly time varies among workers. To study this question he selects four workers at random and records the assembly times for each of the workers.*

	Assembly Time (min)		
Worker 1	Worker 2	Worker 3	Worker 4
20	16	18	21
22	23	20	17
21	20	23	18

22	*18*	*21*	*17*
22	*19*	*24*	*18*
24	*19*	*19*	*23*

(a) Construct the ANOVA table and test the hypothesis that there is no significant variability in assembly time between workers; use $\alpha = 0.05$.
(b) Estimate the components of variance.

Problem 12.24 *Derive Equation 12.26.*

Problem 12.25
(a) Show that $E(\overline{Y}_{i.}) = \mu + \alpha_i$.
(b) Show that $E(\overline{Y}_{i.} - \overline{Y}_{..}) = \alpha_i$, i.e., $\overline{Y}_{i.} - \overline{Y}_{..}$ is an unbiased estimator of the ith treatment mean.

12.8   To Probe Further

The following references contain a wealth of useful information on the practical applications of many statistical techniques, including ANOVA.

1. G. W. Snedecor and W. G. Cochran (1980), *Statistical Methods*, 7th ed., Iowa State University Press, Ames, Iowa.

2. G. E. P. Box, W. G. Hunter, and J. S. Hunter (1978), *Statistics for Experimenters*, John Wiley & Sons, New York.

Chapter 13

Design and Analysis of Multi-Factor Experiments

Probably the most important aim in experimental design is to obtain a given degree of accuracy at the least cost.

W. G. Cochran, English statistician

13.1   Orientation

We now turn our attention to the study of *two-way factorial* experiments; that is, experiments where the mean response is a function of two factors: factor A with a levels denoted by $i = 1, \ldots, a$, and factor B with b levels denoted by $j = 1, \ldots, b$. We define the (i, j)th *cell* to be the set of observations taken at the level i of factor A and the level j of factor B. We assume that n experiments (called replicates) have been performed at each possible combination of the factor-levels. For instance, a computer engineer might be interested in the execution times (the response variable) of five different workloads (factor A, $a = 5$) on three different configurations of a processor (factor B, $b = 3$). The goal here is to study how the processor performance is affected by the different processor configurations.

In a two-way factorial experiment, also called two-way ANOVA, the factor A and factor B effects are of equal interest to the experimenter. There is, however, an important subclass of experiments in which it is the factor A that is of primary interest and the factor B represents an extraneous variable whose effects are of little or no interest to the experimenter. This is the so-called *randomized complete block design* (RCBD). This is one of the most frequently used experimental designs and is also one of the simplest to analyze. We then consider (in increasing level of complexity) the two factor experiment with $n > 1$ observations at each combination of factor-levels. The case where the number of observations in each cell are not the same is more complicated and will not be dealt with here. Finally, we consider 2^k factorial experiments. These are experiments with k factors each of which has two levels, denoted high and low. For instance, in an agricultural experiment the researcher might be interested in how the the crop yield is affected by the amounts of herbicide, pesticide, and fertilizer applied to a field. Since there are many possible factor combinations the researcher might begin by studying the crop yields with each factor at high and low levels. If the crop yields do not change much when the factor levels are varied from low to high then it is unlikely that these factors are important and can be dropped from the research program.

Organization of Chapter

1. Section 13.2: Randomized Complete Block Designs

2. Section 13.2.1: Confidence Intervals and Multiple Comparison Procedures

3. Section 13.2.2: Model Checking via Residual Plots

4. Section 13.3: Two Factor Experiments with $n > 1$ Observations per Cell

5. Section 13.3.1: Confidence Intervals and Multiple Comparison Procedures

6. Section 13.4: 2^k Factorial Designs

13.2   Randomized Complete Block Designs

The precision of an experiment to compare two or more population means can be adversely affected by an extraneous variable in which the scientist has no interest. As an example consider the results of a study of lead absorption in children of employees who worked in a factory where lead is used to make batteries described in Problem 1.18 of Chapter 1. In this study the authors matched 33 such children (the exposed group) from different families to 33 children (the control group) of the same age and neighborhood whose parents were employed in other industries not using lead. The advantages of pairing the children in this way are clear; it eliminates the variability in the data due to the influence of the extraneous variables of age and neighborhood; therefore, the differences in the lead concentration between the two groups will not be masked by these variables. For instance, a child living close to a major highway may have a high lead concentration in his blood even though his father does not work in a battery factory. The partition of the experimental units into homogeneous groups so that the experimental units within the groups have certain common characteristics such as age, sex, weight, etc., is an example of a method of *experimental design* called *blocking*; the groups themselves are called *blocks* and the criteria used to define the blocks are called *blocking variables*. The purpose of the blocking, then, is to reduce the experimental error by eliminating the variability due to an extraneous variable.

Example 13.1 *Testing the effectiveness of an insecticide*

To illustrate the advantages of blocking consider the data set in Table 13.1, which records the mortality rates (measured in percent) for four different genetic strains (B) of fruit flies subjected to three different dosages of an insecticide (A). The experimental unit is the group of flies to which the insecticide is applied; the response variable is the mortality rate. The blocking variable is the genetic strain and there are three factor levels corresponding to the three dosage levels. Observe that while the experimenter randomly assigns the fly to one of three dosages of the insecticide, she cannot similarly assign the genetic strain. Are the mortality rates for the fruit flies influenced by the dosage levels of the insecticide?

Table 13.1 *Mortality rates (%) of four genetic strains of fruit flies from three dosage levels of insecticide*

	Blocks (B)				
Treatment (A)	1	2	3	4	Treatment means
1	66	55	43	32	49.00
2	71	57	44	37	52.25
3	79	63	51	44	59.25
Block means	72.00	58.33	46.00	37.67	

(Source: J. L. Gill (1978), Design and Analysis of Experiments in the Animal and Medical Sciences, vol. 2, p. 25, Iowa State University Press, Ames, IA. Used with permission.)

Solution We can model this experiment in two different ways. For instance, one can ignore the genetic strain and consider the dosage level as the explanatory variable and the mortality rate as the response variable. The model format statement in this case is Model (Y = mortality rate $|A$ = dosage level). This is, of course, a one-way ANOVA. Problem 13.1 asks you to analyze this model. There is also the possibility that the mortality rate is affected by the genetic strain. The model statement in this case is Model (Y = mortality rate $|A$ = dosage level, B = genetic strain). It is this model, called randomized complete block design (RCBD), to which we shall now turn our attention.

1. **A statistical model for the randomized complete block design** We assume that there are a levels of factor A and b blocks. Within each block we randomly assign the experimental units to each treatment; so there are a treatments within each block. This experimental design is called a *randomized complete block design (RCBD)* because each treatment is represented in each block. The data in Table 13.1 is an example of a randomized complete block design consisting of three factor-levels and four blocks, so $a = 3$, $b = 4$. We assume that the response of the jth experimental unit in the jth block at the ith factor-level is a random variable Y_{ij} such that

$$Y_{ij} = \mu_{ij} + \epsilon_{ij}, (i = 1, \ldots, a; \ j = 1, \ldots, b), \text{ where}$$

the random variables ϵ_{ij} are iid $N(0, \sigma^2)$. Consequently,

$$E(Y_{ij}) = \mu_{ij} \text{ and } V(Y_{ij}) = \sigma^2, \ (i = 1, \ldots, a; \ j = 1, \ldots, b).$$

The random variables ϵ_{ij} represent all factors other than A or B that can also produce changes in the response Y_{ij}.

2. **Data format** In the general case of a RCBD experiment with a levels for factor A and b levels for factor B we display the data in the format of an $a \times b$ matrix as in Table 13.2.

Table 13.2 *Data format for a randomized complete block design*

	Blocks (B)			
Treatment (A)	1	2	...	b
1	y_{11}	y_{12}	$\cdots$	y_{1b}
2	y_{21}	y_{22}	$\cdots$	y_{2b}
$\vdots$	$\vdots$	$\vdots$	$\vdots$	$\vdots$
a	y_{a1}	y_{a2}	$\vdots$	y_{ab}

3. **The treatment effects and the block effects** We use the symbols μ, $\overline{\mu}_{i.}$, and $\overline{\mu}_{.j}$ to denote the overall mean, the mean response when the treatment is at level i, and the mean response when the experimental unit is in block j, respectively. In particular,

$$\text{Overall mean: } \mu = \frac{1}{ab} \sum_{i=1}^{a} \sum_{j=1}^{b} \mu_{ij};$$

$$i\text{th treatment mean } (A): \overline{\mu}_{i.} = \frac{1}{b} \sum_{j=1}^{b} \mu_{ij} \ (i = 1, \ldots, a);$$

$$j\text{th block mean } (B): \overline{\mu}_{.j} = \frac{1}{a} \sum_{i=1}^{a} \mu_{ij} \ (j = 1, \ldots, b).$$

Similarly, we denote the treatment effects and block effects by the symbols α_i, β_j, where

$$\alpha_i = \overline{\mu}_{i.} - \mu \text{ and } \beta_j = \overline{\mu}_{.j} - \mu.$$

More precisely, α_i is the effect of the ith treatment and β_j is the effect of the jth block. It is easy to verify that these effects satisfy the following conditions:

$$\sum_{i=1}^{a} \alpha_i = 0; \; \sum_{j=1}^{b} \beta_j = 0.$$

4. **A parameterization of the randomized complete block design** For the RCBD we assume that the treatment effects and the block effects are *additive*. This means that that the mean response at the (i, j)th factor combination level is additive with respect to the two factors, that is,

$$\mu_{ij} = \mu + \alpha_i + \beta_j, \text{ where} \tag{13.1}$$

$$\sum_{i=1}^{a} \alpha_i = 0 \text{ and } \sum_{j=1}^{b} \beta_j = 0. \tag{13.2}$$

Note: The plausibility of the additive effects model 13.1 can be checked by plotting a suitable graph. We will return to this point shortly (see Figures 13.1 and 13.2).

5. **Homogeneity of the variances** We assume that the variance of the response variable is the same for all factor-levels. Serious violations of normality and homogeneity of variances can be checked by plotting the residuals. We shall return to this point later (see Section 13.2.2).

6. **Least squares estimates of the model parameters** Using the least squares method as in Section 12.2.1 (we omit the details) we obtain the following least squares estimate for the overall mean, treatment effects, and block effects:

$$\hat{\mu} = \overline{Y}_{..} \text{ (grand mean)}; \tag{13.3}$$

$$\hat{\alpha}_i = \overline{Y}_{i.} - \overline{Y}_{..}; \hat{\beta}_j = \overline{Y}_{.j} - \overline{Y}_{..}. \tag{13.4}$$

Thus, the overall mean is estimated by the grand mean, the factor A effect at the level i is estimated by the ith row mean minus the grand mean, and the effect of the jth block is estimated by the jth column mean minus the grand mean.

The fitted values and the residuals are:

Fitted value: $\hat{y}_{ij} = \hat{\mu} + \hat{\alpha}_i + \hat{\beta}_j$;

Residual: $e_{ij} = y_{ij} - \hat{y}_{ij} = y_{ij} - \hat{\mu} - (\overline{y}_{i.} - \overline{y}_{..}) - (\overline{y}_{.j} - \overline{y}_{..})$

$\qquad\qquad = y_{ij} - \overline{y}_{i.} - \overline{y}_{.j} + \overline{y}_{..}.$

It can be shown (the details are left to you as a problem) that

$$\sum_{i=1}^{a} e_{ij} = 0, \; \sum_{j=1}^{b} e_{ij} = 0. \tag{13.5}$$

The corresponding estimates for the ith treatment mean and jth block mean are given by:

$$i\text{th treatment mean} = \overline{Y}_{i.}; \; j\text{th block mean} = \overline{Y}_{.j}.$$

Example 13.2 *Continuation of Example 13.1*

Refer to the fruit fly mortality rate data in Table 13.1. Compute the least squares estimates of the overall mean, treatment effects, and block effects.

Solution We calculate the least squares estimates using formula 13.4 and the treatment means and block means that are displayed in the last column and bottom row of Table 13.1.

$$
\begin{aligned}
&\text{Parameter} \qquad \text{Least Squares Estimate} \\
&\hat{\mu} = \overline{y}_{..} = 53.50 \\
&\hat{\alpha}_1 = \overline{y}_{1.} - \overline{y}_{..} = 49 - 53.50 = -4.50 \\
&\hat{\alpha}_2 = \overline{y}_{2.} - \overline{y}_{..} = 52.25 - 53.50 = -1.25 \\
&\hat{\alpha}_3 = \overline{y}_{3.} - \overline{y}_{..} = 59.25 - 53.50 = 5.75 \\
&\hat{\beta}_1 = \overline{y}_{.1} - \overline{y}_{..} = 72 - 53.50 = 18.50 \\
&\hat{\beta}_2 = \overline{y}_{.2} - \overline{y}_{..} = 58.33 - 53.50 = 4.83 \\
&\hat{\beta}_3 = \overline{y}_{.3} - \overline{y}_{..} = 46 - 53.50 = -7.50 \\
&\hat{\beta}_4 = \overline{y}_{.4} - \overline{y}_{..} = 37.67 - 53.50 = -15.83
\end{aligned}
$$

7. **How well does the model fit the data?** We judge the performance of the model by first partitioning the total variability SST into the sum of SSA (the variability explained by the treatments), SSB (the variability explained by the blocks), and SSE (the unexplained variability). The coefficient of determination R^2 is the proportion of the total variability explained by the model. Thus

$$SST = SSA + SSB + SSE, \ DF = ab - 1, \tag{13.6}$$

$$R^2 = \frac{SSA + SSB}{SST}, \tag{13.7}$$

where the sums of squares and their degrees of freedom are defined by

$$SSA = \sum_{i=1}^{a}\sum_{j=1}^{b}(\overline{y}_{i.} - \overline{y}_{..})^2, \ DF = a - 1$$

$$= b\sum_{i=1}^{a}(\overline{y}_{i.} - \overline{y}_{..})^2;$$

$$SSB = \sum_{i=1}^{a}\sum_{j=1}^{b}(\overline{y}_{.j} - \overline{y}_{..})^2, \ DF = b - 1$$

$$= a\sum_{j=1}^{b}(\overline{y}_{.j} - \overline{y}_{..})^2;$$

$$SSE = \sum_{i}\sum_{j}(y_{ij} - \hat{y}_{ij})^2, \ DF = (a - 1)(b - 1).$$

Equation 13.6 for SST is a consequence of the algebraic identity

$$y_{ij} - \overline{y}_{..} = (\overline{y}_{i.} - \overline{y}_{..}) + (\overline{y}_{.j} - \overline{y}_{..}) + (y_{ij} - \overline{y}_{i.} - \overline{y}_{.j} + \overline{y}_{..}), \tag{13.8}$$

which corresponds to the model assumption $Y_{ij} - \mu = \alpha_i + \beta_j + \epsilon_{ij}$. We obtain the partition 13.6 by squaring both sides of the algebraic identity 13.8, adding all the terms, and noting that the cross product terms equal zero; in detail,

$$\sum_i \sum_j (\overline{y}_{i.} - \overline{y}_{..})(\overline{y}_{.j} - \overline{y}_{..}) = 0;$$

$$\sum_i \sum_j (\overline{y}_{i.} - \overline{y}_{..})(y_{ij} - \overline{y}_{i.} - \overline{y}_{.j} + \overline{y}_{..}) = 0;$$

$$\sum_i \sum_j (\overline{y}_{.j} - \overline{y}_{..})(y_{ij} - \overline{y}_{i.} - \overline{y}_{.j} + \overline{y}_{..}) = 0.$$

We omit the algebraic details.

Example 13.3 *Continuation of Example 13.1*

Refer to the fruit fly mortality rate data in Table 13.1. Compute the sum of squares and the proportion of the total variability explained by the RCBD model.

Solution

$$SSA = 219.5; \; SSB = 2017.67; \; SSE = 11.83; \; SST = 2249.00.$$

The proportion of the total variability explained by the RCBD model is given by:

$$R^2 = \frac{SSA + SSB}{SST} = \frac{2237.17}{2249.00} = 0.99.$$

Thus 99% of the total variability is accounted for by the model.

8. **Computing the degrees of freedom** The next step in the analysis is to compute the degrees of freedom (DF) associated with the sums of squares SSA, SSB, SSE, and SST.

   (a) There are $a-1$ degrees of freedom associated with SSA since $SSA = b \sum_i (\overline{Y}_{i.} - \overline{Y}_{..})^2$ contains a summands satisfying the one constraint $\sum_i (\overline{Y}_{i.} - \overline{Y}_{..}) = 0$;

   (b) There are $b - 1$ degrees of freedom associated with SSB since $SSB = a \sum_j (\overline{Y}_{.j} - \overline{Y}_{..})^2$ contains b summands satisfying the one constraint $\sum_j (\overline{Y}_{.j} - \overline{Y}_{..}) = 0$;

   (c) There are $(a - 1)(b - 1)$ degrees of freedom associated with SSE. To see this, write the error sum of squares in terms of the residuals as follows:

$$SSE = \sum_i \sum_j e_{ij}^2.$$

Therefore, we need to specify, say, only the first $(b - 1)$ entries for each row: $e_{i1}, \ldots, e_{i(b-1)}$, and the first $(a - 1)$ entries for each column: $e_{1j}, \ldots, e_{(a-1)j}$. Then use the fact that the residuals satisfy the conditions 13.5. Therefore, we need to specify, say, only the first $(b - 1)$ entries for each row:

$$e_{i1}, \ldots, e_{i(b-1)};$$

and the first $(a - 1)$ entries for each column:

$$e_{1j}, \ldots, e_{(a-1)j}.$$

Consequently, knowing these $(a - 1)(b - 1)$ residuals suffices to determine all the others.

Mean Response For Each Block: Additive Model

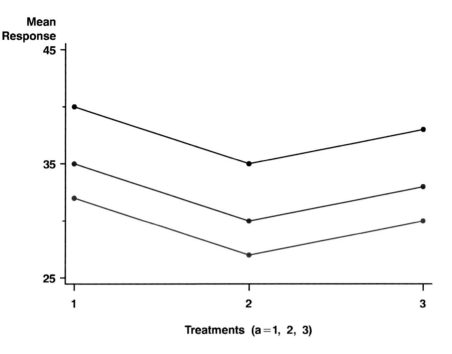

FIGURE 13.1: Mean response for each block vs. treatment $I = 1, 2, 3$; no interactions between treatments and blocks

(d) the degrees of freedom associated with SST is $ab-1$ since $SST = \sum_i \sum_j (Y_{ij} - \overline{Y}_{..})^2$ contains ab summands and the single constraint $\sum_i \sum_j (Y_{ij} - \overline{Y}_{..}) = 0$.

(e) Notice that the partition 13.6 of the sum of squares SST yields a corresponding partition of the degrees of freedom:

$$ab - 1 = (a - 1) + (b - 1) + (a - 1)(b - 1).$$

9. **Consequences of the additive effects hypothesis** The additive effects hypothesis $(\mu_{ij} = \mu + \alpha_i + \beta_j)$ implies

$$\mu_{i'j} - \mu_{ij} = \alpha_{i'} - \alpha_i,$$

which is independent of the jth block. This means that the change in the mean response due to a change in the level of factor A is independent of the block. When this condition holds, we say that there is no *interaction* between the treatments and the blocks. The additivity of the treatment and block effects implies that the plots obtained by connecting the points $(i, \mu_{ij})(i = 1, \ldots, a)$ by straight line segments are parallel to one another as in Figure 13.1.

Because the values μ_{ij} are unknown, statisticians recommend plotting the points (i, y_{ij}), $(i = 1, \ldots, a)$ and connecting them by straight line segments as in Figure 13.2, which uses the data on fruit fly mortality rates (Table 13.1). For instance, we see that the differences in mortality rates between the second and first treatments for blocks 1, 2, 3 and 4 are given by

$$y_{21} - y_{11} = 5, \; y_{22} - y_{12} = 2, \; y_{23} - y_{13} = 1, \; y_{24} - y_{14} = 5.$$

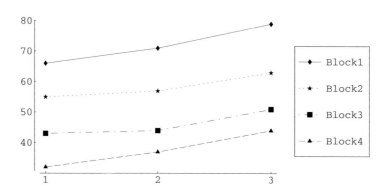

FIGURE 13.2:   Mortality rate for each block vs. treatment $I = 1, 2, 3$ (Example 13.1)

The fact that these graphs are not exactly parallel is due to the fact that the response is a random variable and not necessarily because the hypothesis of additivity is false. In any event, Figure 13.2 does not indicate a gross violation of the hypothesis that the treatment and block effects are additive.

Testing Hypotheses about the Model Let us return to the problem of testing the null hypothesis that there are no differences in the fruit fly mortality rates due to the different dosages of the insecticide (Table 13.1). That is, our main interest is in testing the equality of the treatment means; consequently, the null hypothesis and its alternative have the following form:

$$H_0 : \alpha_1 = \ldots = \alpha_a = 0 \text{ against } H_1 : \alpha_i \neq 0 \text{ for at least one } i = 1, \ldots, a.$$

The basic idea is to compare the variability explained by factor A to the unexplained variability; that is, we compute the F ratio:

$$F = \frac{SSA/(a-1)}{SSE/(a-1)(b-1)}. \tag{13.9}$$

A large value of the F ratio suggests that differences among the treatment effects are not due to chance (thus providing strong evidence against H_0); a small F value provides weak evidence against H_0. The quantities $(a-1)$ and $(a-1)(b-1)$ are the degrees of freedom associated with the sums of squares SSA and SSE. Theorem 13.1 (to follow) gives a formula for the distribution of F; the F test for the equality of the treatment means appears immediately afterward.

Sampling Distribution of SST, SSA, SSB, and SSE Assuming that the error terms in the model are iid $N(0, \sigma^2)$ it is possible to derive the following results on the sampling distribution of SST, SSA, F, SSB, and SSE. The statistic on which we base our test of H_0 is the F ratio, defined in Equation 13.9. To see why this is the "right" statistic, we list without proof some facts concerning the distributions of SSE, SSA, and SST.

1. The **mean square error** (MSE) is the error sum of squares divided by its number of degrees of freedom, which is $(a-1)(b-1)$. Consequently,

$$MSE = \frac{SSE}{(a-1)(b-1)}.$$

It can also be shown that MSE is an unbiased estimator of σ^2; thus,

$$E(MSE) = E\left(\frac{SSE}{(a-1)(b-1)}\right) = \sigma^2. \tag{13.10}$$

2. The quantity MSA denotes the sum of squares SSA divided by its number of degrees of freedom, which is $(a-1)$; thus,

$$MSA = \frac{SSA}{a-1}.$$

It can also be shown that

$$E(MSA) = \sigma^2 + b\frac{\sum_i \alpha_i^2}{a-1}. \tag{13.11}$$

Similarly, we define $MSB = SSB/(b-1)$ and note that

$$E(MSB) = \sigma^2 + a\frac{\sum_j \beta_j^2}{b-1}. \tag{13.12}$$

3. We next observe that when H_0 is true, that is, $\alpha_1 = \ldots = \alpha_a = 0$, then $E(MSA) = \sigma^2 = E(MSE)$. When H_0 is not true, which means that some of the α_i's are different from zero, then Equation 12.25 implies that $E(MSA) > E(MSE)$. This suggests that the ratio $F = MSA/MSE$ be used to test H_0; in particular, large values of F would make H_0 less plausible. Theorem 13.1 is a more precise version of this statement.

Theorem 13.1 *The null distribution of the F ratio MSA/MSE has an F distribution with $(a-1)$ degrees of freedom in the numerator and $(a-1)(b-1)$ degrees of freedom in the denominator; that is,*

$$F = \frac{MSA}{MSE} \sim F_{a-1,(a-1)(b-1)}.$$

The F test for the equality of treatment means
We now show how to use the F ratio to test:

$$H_0 : \mu_1 = \mu_2 = \ldots = \mu_a \text{ against } H_1 : \text{ at least two of the means are different.}$$

The decision rule When H_0 is true, Theorem 13.1 tells us that $F = MSA/MSE$ is distributed as $F_{a-1,(a-1)(b-1)}$ and that large values of F support the alternative H_1. Consequently, our decision rule is to:

accept H_0 if $F \le F_{a-1,(a-1)(b-1)}(\alpha)$, and reject H_0 if $F > F_{a-1,(a-1)(b-1)}(\alpha)$.

The P-value of the test is the probability P-value $= P(F_{a-1,(a-1)(b-1)} > F)$. Your final results should be summarized in the format of the ANOVA Table 13.3.

Table 13.3 *ANOVA table for a randomized complete block design*

Source	DF	SS	MS	F
A	$a-1$	SSA	$MSA = \frac{SSA}{a-1}$	$\frac{MSA}{MSE}$
B	$b-1$	SSB	$MSB = \frac{SSB}{b-1}$	$\frac{MSB}{MSE}$
Error	$(a-1)(b-1)$	SSE	$MSE = \frac{SSE}{(a-1)(b-1)}$	
Total	$ab-1$	SST		

Example 13.4 *Continuation of Example 13.3*

Refer to the the fruit fly mortality rate data in Table 13.1. Compute the ANOVA table and use the F ratio to test whether the mean mortality rate depends on the dosage level of the insecticide. State the null hypothesis, the alternatives, decision rule, and conclusion.

Solution The sums of squares SSA, SSB, SSE, and SST were computed in Example 13.3. The corresponding mean squares are given by

$$MSA = \frac{SSA}{a-1} = \frac{219.5}{2} = 109.75;$$

$$MSB = \frac{SSB}{b-1} = \frac{2017.67}{3} = 672.56;$$

$$MSE = \frac{SSE}{(a-1)(b-1)} = \frac{11.83}{6} = 1.97;$$

$$F = \frac{MSA}{MSE} = \frac{126.33}{2.67} = 55.65.$$

The ANOVA table for the fruit fly mortality data (Table 13.1) is displayed in Table 13.4.

Table 13.4 *ANOVA table for the fruit fly mortality rate data (Table 13.1) analyzed as a randomized complete block design*

Source	DF	Sum of Squares	Mean Square	F Value	Pr > F
A	2	219.500000	109.750000	55.65	0.0001
Block (B)	3	2017.666667	672.555556	341.01	0.0001
Error	6	11.833333	1.972222		
Corrected Total	11	2249.000000			

Conclusion The null hypothesis is that the mean mortality rates do not depend on the dosage levels of the insecticide. There are three dosage levels so

$$H_0 : \mu_1 = \mu_2 = \mu_3 \text{ against}$$

$$H_1 : \mu_i \ne \mu_j \text{ for at least one pair } i,j.$$

Our decision rule is based on the P-value of the F ratio. Since $F = 55.65 > F_{2,6}(0.01) = 10.92$, we reject the null hypothesis at the 1% level. In fact, the computer printout shows that the P-value is less than 0.0001. Consequently, the three dosage levels differ significantly in their effects.

13.2.1 Confidence Intervals and Multiple Comparison Procedures

When one rejects the null hypothesis of no differences among the treatment means the next step in the analysis of the experiment is to obtain confidence intervals for the individual treatment means and for their differences. We first derive confidence intervals for the treatment means.

The standard error of a treatment mean for a randomized complete block design is

$$s(\overline{Y}_{i.}) = \sqrt{\frac{MSE}{b}}.$$

Therefore, a $100(1-\alpha)\%$ confidence interval for the ith treatment mean is given by:

$$\overline{Y}_{i.} \pm t_\nu(\alpha/2)\sqrt{\frac{MSE}{b}}; \nu = (a-1)(b-1). \tag{13.13}$$

Tukey's simultaneous $100(1-\alpha)\%$ confidence intervals for the differences of two means (randomized complete block design): We now turn our attention to determining which pairs of treatment means are significantly different. We do this in the usual way by noting that a $100(1-\alpha)\%$ confidence interval for the difference between the ith and kth treatment means is given by:

$$\overline{Y}_{i.} - \overline{Y}_{k.} \pm q(\alpha; a, \nu)\sqrt{\frac{MSE}{b}}; \nu = (a-1)(b-1). \tag{13.14}$$

$$\tag{13.15}$$

We therefore reject the null hypothesis that $\mu_i = \mu_k$, $i \neq k$ if:

$$|\overline{y}_{i.} - \overline{y}_{k.}| > q(\alpha; a, \nu)\sqrt{2MSE/b}. \tag{13.16}$$

Example 13.5 *Continuation of Example 13.3*

Refer to the fruit fly mortality rate data in Table 13.1 and the ANOVA Table 13.4. Use Tukey's method to find the significant differences between the treatment means. Use $\alpha = 0.05$.

Solution

$$\nu = 6, a = 3, b = 4, MSE = 1.972, \text{ so}$$
$$q(0.05; a, \nu)\sqrt{2MSE/b} = 4.31.$$

The treatment means in increasing order are $49 < 52.25 < 59.25$. The difference between the first and second treatments is

$$|\overline{y}_{1.} - \overline{y}_{2.}| = 3.25 < 4.31,$$

which is not significant. On the other hand, difference between the third and second treatments equals

$$|\overline{y}_{3.} - \overline{y}_{2.}| = 7.00 > 4.31,$$

which is significant.

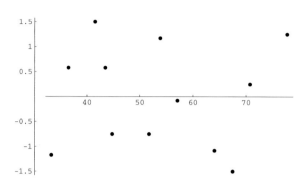

FIGURE 13.3:   Plot of residuals versus predicted values (Example 13.1)

13.2.2   Model Checking via Residual Plots

Residual and normal probability plots are highly useful for checking model assumptions. For instance, a plot of the residuals against the fitted values is very useful for detecting a gross violation of the hypothesis of equal variances. Looking at Figure 13.3 we see that the assumption of equal variances for Example 13.1 does not appear to be violated. The normal probability plot for the studentized residuals is displayed in Figure 13.4. The plot does not reveal any gross violation of normality, which is confirmed by the fact that the P-value of the Shapiro–Wilk test for normality is $p = 0.41$.

13.3   Two Factor Experiments with $n > 1$ Observations per Cell

We now consider experiments where the response variable depends on two factors with both factors of equal interest. The statistical analysis is necessarily more complex because there are several competing models that the scientist must consider before deciding which one best fits the data. We shall consider in detail the *main effects* model (no interactions), and the model with interactions. A thorough analysis requires a combination of analytical and graphical methods that are best understood in the context of a concrete example.

Example 13.6 *Analysis of fungicide and insecticides effects*

The data in Table 13.5 record the results of a study made to determine the effects of a fungicide and two insecticides on the egg production of pheasants over a five week period.

Table 13.5 *The effects of a fungicide and two insecticides on the egg production of pheasants over a five week period*

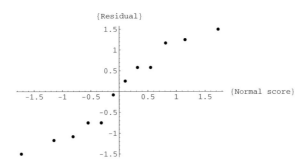

FIGURE 13.4: Normal probability plot of residuals (Example 13.1)

Fungicide (A)	Pesticide (B)					
	None		Dieldrin		Diazinon	
None	15	29	22	32	30	25
	18	30	24	7	28	21
	25	21	2	15	17	19
	26	24	9	18	16	30
Captan	37	12	1	13	9	7
	38	5	1	10	0	1
	24	21	6	18	0	4
	10	6	20	25	4	9

(Source: J. L. Gill (1978), *Design and Analysis of Experiments in the Animal and Medical Sciences*, vol. 1, Iowa State University Press, Ames, IA. Used with permission.)

This is a two factor experiment with two levels of factor A (no fungicide, Captan) and three levels of factor B (no pesticide, Dieldrin, and Diazinon). There are also 8 observations at each combination of the factor-levels; consequently in this experiment we have: $a = 2$, $b = 3$, $n = 8$. This is an example of a *complete $a \times b$ factorial experiment,* which refers to the fact that the experiment was performed at all possible combinations of the factor-levels.

Statistical analysis of the two factor experiment

1. **The statistical model for the response variable** We assume that we have performed $n > 1$ experiments at each combination of the factor-levels. We denote the value of the kth response at the (i, j)th combination of the factor-levels by Y_{ijk}, where

$$Y_{ijk} = \mu_{ij} + \epsilon_{ijk}, (i = 1, \ldots, a; j = 1, \ldots, b; k = 1, \ldots, n).$$

We assume the random variables ϵ_{ijk} are iid $N(0, \sigma^2)$; thus, their means and variances are given by

$$E(Y_{ijk}) = \mu_{ij} \text{ and } V(Y_{ijk}) = \sigma^2.$$

The ab parameters μ_{ij} are called the *cell means*. We use the symbols μ, $\overline{\mu}_{i.}$, and $\overline{\mu}_{.j}$ to denote the overall mean, the mean response when factor A is at level i (row mean), and the mean response when factor B is at level j (column mean), respectively. In particular,

$$\text{Overall mean response: } \mu = \frac{1}{ab}\sum_{i=1}^{a}\sum_{j=1}^{b}\mu_{ij};$$

$$\text{mean response at } i\text{th level of A: } \overline{\mu}_{i.} = \frac{1}{b}\sum_{j=1}^{b}\mu_{ij}\ (i=1,\ldots,a);$$

$$\text{mean response at } j\text{th level of B: } \overline{\mu}_{.j} = \frac{1}{a}\sum_{i=1}^{a}\mu_{ij}\ (j=1,\ldots,b).$$

Similarly, the quantities α_i and β_j, defined by the equations

$$\alpha_i = \overline{\mu}_{i.} - \mu \text{ and } \beta_j = \overline{\mu}_{.j} - \mu,$$

are called treatment effects. In particular, α_i is the effect of the ith level of factor A and β_j is the effect of the jth level of factor B. More generally, an *effect* is a change in the mean response caused by a change in the factor-level combination. It is easy to verify that the treatment effects satisfy the following conditions:

$$\sum_{i=1}^{a}\alpha_i = 0;\ \sum_{j=1}^{b}\beta_j = 0. \tag{13.17}$$

2. **Data format** We denote the kth observation in the (i,j)th cell by

$$y_{ijk},\ (i=1,\ldots,a;\ j=1,\ldots,b;\ k=1,\ldots,n).$$

We group these observations together by cell and arrange the ab cells in an $a \times b$ matrix consisting of a rows and b columns as shown in Table 13.6.

Table 13.6 *Data format for a two-factor design*

Factor A	*Factor B* 1	...	b
1	$y_{111},\ldots,y_{11n}$	...	$y_{1b1},\ldots,y_{1bn}$
⋮	⋮	⋮	⋮
a	$y_{a11},\ldots,y_{a1n}$	...	$y_{ab1},\ldots,y_{abn}$

3. **The dot subscript notation** We denote the sample cell means, factor A (row) means,

factor B (column) means, and the grand mean by the dot subscript notation; thus,

$$\bar{y}_{ij.} = \frac{1}{n} \sum_{k=1}^{n} y_{ijk};$$

$$\bar{y}_{i..} = \frac{1}{bn} \sum_{j=1}^{b} \sum_{k=1}^{n} y_{ijk};$$

$$\bar{y}_{.j.} = \frac{1}{an} \sum_{i=1}^{a} \sum_{k=1}^{n} y_{ijk};$$

$$\bar{y}_{...} = \frac{1}{abn} \sum_{i=1}^{a} \sum_{j=1}^{b} \sum_{k=1}^{n} y_{ijk} \text{ (grand mean)}.$$

Example 13.7 *Continuation of Example 13.6*

Refer to the data set 13.5. Compute the cell means, grand mean, and the factor A and factor B means, and use the dot subscript notation to express your results.

Solution

The cell means, grand mean, factor A means, and factor B means are given by

$$\bar{y}_{11.} = 23.5, \bar{y}_{12.} = 16.125, \bar{y}_{13.} = 23.25;$$
$$\bar{y}_{21.} = 19.125, \bar{y}_{22.} = 11.75, \bar{y}_{23.} = 4.25;$$
$$\bar{y}_{...} = 16.33, \bar{y}_{1..} = 20.96, \bar{y}_{2..} = 11.71;$$
$$\bar{y}_{.1.} = 21.32, \bar{y}_{.2.} = 13.94, \bar{y}_{.3.} = 13.75.$$

4. **The main effects and the interaction effects** There are now two models to consider depending on whether or not there are *interactions* between the two factors.

$$\text{No interactions: } \mu_{ij} = \mu + \alpha_i + \beta_j (i = 1, \ldots, a; j = 1, \ldots, b);$$
$$\text{Interactions : } \mu_{ij} \neq \mu + \alpha_i + \beta_j \text{ for at least one pair } (i, j).$$

The main effects model (also called the no interaction model) assumes that

$$\mu_{ij} = \mu + \alpha_i + \beta_j (i = 1, \ldots, a; j = 1, \ldots, b).$$

The model statement format for the two factor experiment with no interactions is Model $(Y = y | A = i, B = j)$.

As noted earlier, the main effects model implies that

$$\mu_{i'j} - \mu_{ij} = \alpha_{i'} - \alpha_i.$$

That is, the effect of a change in a level of factor A is the same for all levels of factor B. In particular, if a change (from i to i', say) in the temperature of a reaction increases the yield at one level of the pressure, then it will produce the same increase in the yield at all pressures. Of course, if there is an interaction effect this will not be the case. This consequence of the main effects model can be visualized by drawing the following graphs:

$$(i, \mu_{ij}), (i = 1, \ldots, a),$$

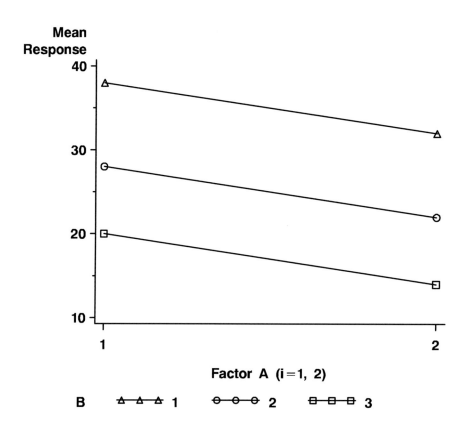

Mean response for each factor B level vs factor A level (i=1, 2)
No interaction between factors A and B

FIGURE 13.5:   Mean response for each factor B level vs. factor A level ($i = 1, 2$); no interaction between factors A and B

where there are b graphs, one for each level of factor B. Figure 13.5 displays these graphs for the case $a = 2$ and $b = 3$. Looking at this figure you see that the three graphs are obtained from one another by a vertical translation; in particular the three line segments are parallel to one another.

Of course, the cell means are not known and so we have to content ourselves with replacing them with their estimates $\overline{y}_{ij.}$; that is we graph the points

$$(i, \overline{y}_{ij.}), \ (i = 1, \ldots, a).$$

Figure 13.6 displays the graph obtained by plotting the cell means of the data in Table 13.5. Note that the magnitude of the decrease in egg production as we change the level of factor A from no fungicide ($i = 1$) to Captan ($i = 2$) is much greater when the pesticide is Diazinon (B at level $j = 3$) than it is for the case when no pesticide is used (B at level $j = 1$). An *interaction* effect occurs when the A effect is not the same for each level of B. Similarly, we say that an interaction effect exists when there is a change in the magnitude and/or direction (up or down) of a response to a change in a level of the B factor that is not the same for each level of the A factor. The graphs strongly suggest that there is an interaction effect, which will be confirmed when we compute the two-way ANOVA table.

5. The (i, j)th *interaction effect* is denoted by

$$(\alpha\beta)_{ij} = \mu_{ij} - (\mu + \alpha_i + \beta_j).$$

This quantity is a measure of the discrepancy between the no interaction model and the one with interactions. It is easy to verify that

$$\sum_{i=1}^{a} (\alpha\beta)_{ij} = 0; \ \sum_{j=1}^{b} (\alpha\beta)_{ij} = 0. \tag{13.18}$$

6. **The two factor model with interactions** To allow for the possibility of interactions we assume the mean response at the (i, j) factor-level combination is given by

$$\mu_{ij} = \mu + \alpha_i + \beta_j + (\alpha\beta)_{ij}.$$

The statistical model of the kth response at the (i, j)th factor level is given by:

$$Y_{ijk} = \mu + \alpha_i + \beta_j + (\alpha\beta)_{ij} + \epsilon_{ijk},$$

where the random variables ϵ_{ijk} are iid $N(0, \sigma^2)$. The corresponding model statement format for the two factor experiment with interactions is Model ($Y = y | A = i, B = j, A * B = (i, j)$).

7. **Least squares estimates of the model parameters for the two factor model with interactions** It can be shown that the least squares estimates of the overall mean, the factor A and B, and interaction effects are:

$$\hat{\mu} = \overline{Y}_{...}; \tag{13.19}$$

$$\hat{\alpha}_i = \overline{Y}_{i..} - \overline{Y}_{...}; \tag{13.20}$$

$$\hat{\beta}_j = \overline{Y}_{.j.} - \overline{Y}_{...}; \tag{13.21}$$

$$\hat{(\alpha\beta)}_{ij} = \overline{Y}_{ij.} - \overline{Y}_{i..} - \overline{Y}_{.j.} + \overline{Y}_{....}. \tag{13.22}$$

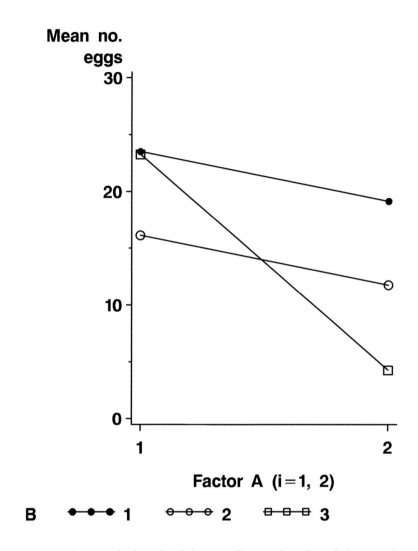

Mean response for each level of factor B   vs levels of factor A
Example 13.6; Table 13.5

FIGURE 13.6: Mean response for each factor B level vs. factor A level ($i = 1, 2$) of Example 13.6

In other words, the overall mean is estimated by the grand mean, the factor A effect at the level i is estimated by the ith row mean minus the grand mean, the factor B effect at the level j is estimated by the jth column mean minus the grand mean, and the (i, j)th interaction is estimated by the (i, j)th cell mean minus the (i, j)th additive effect. Similarly, the fitted values and the residuals are:

$$\textbf{Fitted value: } \hat{y}_{ijk} = \hat{\mu} + \hat{\alpha}_i + \hat{\beta}_j + (\hat{\alpha\beta})_{ij} = \overline{Y}_{ij\cdot};$$
$$\textbf{Residual: } e_{ijk} = Y_{ijk} - \overline{Y}_{ij\cdot}.$$

Thus, the fitted value is the (i, j)th estimated cell mean, and the (i, j, k)th residual is the (i, j, k)th observation minus the (i, j)th estimated cell mean.

It can be shown (the details are left to you as a problem) that

$$\sum_{k=1}^{n} e_{ijk} = 0. \tag{13.23}$$

8. **How well does the model fit the data?** We analyze the model goodness of fit by first partitioning the total variability SST into the sum of SSA (the variability explained by factor A), SSB (the variability explained by factor B), SSAB (the variability explained by the interaction of factors A and B), and SSE (the unexplained variability). The coefficient of determination R^2 is the proportion of the total variability explained by the model; it is given by

$$R^2 = \frac{SSA + SSB + SSAB}{SST}.$$

The definitions of these sums of squares and their corresponding degrees of freedom will now be given. We then arrange these quantities in the format of a two-way ANOVA table. This will allow us to test hypotheses about the model parameters.

Partitioning the Total Variability of the Data

We begin with the algebraic identity

$$y_{ijk} - \overline{y}_{\cdots} = (\overline{y}_{i\cdot\cdot} - \overline{y}_{\cdots}) + (\overline{y}_{\cdot j\cdot} - \overline{y}_{\cdots}) + (\overline{y}_{ij\cdot} - \overline{y}_{i\cdot\cdot} - \overline{y}_{\cdot j\cdot} + \overline{y}_{\cdots}) + (y_{ijk} - \overline{y}_{ij\cdot})$$
$$= \hat{\alpha}_i + \hat{\beta}_j + (\hat{\alpha\beta})_{ij} + e_{ijk},$$

which corresponds to the model $Y_{ijk} - \mu = \alpha_i + \beta_j + (\alpha\beta)_{ij} + \epsilon_{ijk}$.

Squaring both sides and summing over the indices i, j, k yields the following partition of the total sum of squares (SST) and their associated degrees of freedom (DF):

$$SST = SSA + SSB + SSAB + SSE, \tag{13.24}$$

where

$$SSA = \sum_{i=1}^{a}\sum_{j=1}^{b}\sum_{k=1}^{n} (\overline{y}_{i..} - \overline{y}_{...})^2; \ DF = a - 1$$

$$= nb\sum_{i=1}^{a} (\overline{y}_{i..} - \overline{y}_{...})^2;$$

$$SSB = \sum_{j=1}^{b}\sum_{i=1}^{a}\sum_{k=1}^{n} (\overline{y}_{.j.} - \overline{y}_{...})^2; \ DF = b - 1$$

$$= na\sum_{j=1}^{b} (\overline{y}_{.j.} - \overline{y}_{...})^2;$$

$$SSAB = \sum_{i=1}^{a}\sum_{j=1}^{b}\sum_{k=1}^{n} (\overline{y}_{ij.} - \overline{y}_{i..} - \overline{y}_{.j.} + \overline{y}_{...})^2; \ DF = (a-1)(b-1)$$

$$= n\sum_{i=1}^{a}\sum_{j=1}^{b} (\overline{y}_{ij.} - \overline{y}_{i..} - \overline{y}_{.j.} + \overline{y}_{...})^2$$

$$SSE = \sum_{i=1}^{a}\sum_{j=1}^{b}\sum_{k=1}^{n} (y_{ijk} - \overline{y}_{ij.})^2; \ DF = ab(n-1)$$

$$= SST - SSA - SSB - SSAB;$$

$$SST = \sum_{i=1}^{a}\sum_{j=1}^{b}\sum_{k=1}^{n} (y_{ijk} - \overline{y}_{...})^2; \ DF = nab - 1.$$

It can be shown by means of a tedious algebraic calculation that the sums of the cross product terms equal zero; we omit the details. We do not give computational formulas for the sums of squares because we assume the reader has access to one of the standard statistical software packages.

Sampling Distribution of SST, SSA, SSB, SSAB, SSE

(a) The **mean square error** is the random variable MSE defined by the equation:

$$MSE = \frac{SSE}{ab(n-1)}.$$

Note: When there is only one observation per cell (so $n = 1$) we cannot estimate the experimental error, since the denominator is 0 and so the quotient is undefined. This follows from the fact that the ab parameters of the model with interactions equals the number of observations.

The random variable MSE is an unbiased estimator of σ^2. Thus,

$$E(MSE) = E\left(\frac{SSE}{ab(n-1)}\right) = \sigma^2. \tag{13.25}$$

(b) **The mean squares** MSA, MSB, and MSAB are the sums of squares divided by their associated degrees of freedom. Their expected values are given by:

$$MSA = \frac{SSA}{a-1}; \ E(MSA) = \sigma^2 + \frac{nb\sum_i \alpha_i^2}{a-1};$$

$$MSB = \frac{SSB}{b-1}; \ E(MSB) = \sigma^2 + \frac{na\sum_j \beta_j^2}{b-1};$$

$$MSAB = \frac{SSAB}{(a-1)(b-1)}; \ E(MSAB) = \sigma^2 + \frac{n\sum_i\sum_j (\alpha\beta)_{ij}^2}{(a-1)(b-1)}.$$

(c) **Partitioning the degrees of freedom** The total sum of squares SST has $nab-1$ degrees of freedom. The degrees of freedom associated with the sums of squares SSA, SSB, $SSAB$, and SSE are

$$(a-1),\ (b-1),\ (a-1)(b-1) \text{ and } ab(n-1), \text{ respectively.}$$

We therefore have the following partition of the degrees of freedom:

$$nab - 1 = (a-1) + (b-1) + (a-1)(b-1) + ab(n-1).$$

These calculations are conveniently summarized in the two-way ANOVA Table 13.7.

Table 13.7 *Two-way ANOVA table*

Source	DF	SS	MS	F
A	$a-1$	SSA	$MSA = \frac{SSA}{a-1}$	$\frac{MSA}{MSE}$
B	$b-1$	SSB	$MSB = \frac{SSB}{b-1}$	$\frac{MSB}{MSE}$
AB	$(a-1)(b-1)$	$SSAB$	$MSAB = \frac{SSAB}{(a-1)(b-1)}$	$\frac{MSAB}{MSE}$
Error	$ab(n-1)$	SSE	$MSE = \frac{SSE}{ab(n-1)}$	
Total	$abn-1$	SST		

9. **Testing hypotheses about the parameters** In a multi-factor experiment statisticians recommend that one should first check for the presence of an interaction effect, otherwise one might erroneously conclude that there are no main effects (see Example 13.9). Our main interest therefore is in testing the following hypotheses:

(a) There are no interactions between the A and B factors:

$$H_{(\alpha\beta)} : (\alpha\beta)_{ij} = 0, \ (i = 1,\ldots a; \ j = 1,\ldots b).$$

(b) There are no factor A effects:

$$H_\alpha : \alpha_1 = \ldots = \alpha_a = 0.$$

(c) There are no factor B effects:

$$H_\beta : \beta_1 = \ldots = \beta_b = 0.$$

Testing the null hypotheses $H_{\alpha\beta}$, H_α, H_β

To test for the existence of interactions and the main effects we proceed in the usual way by computing the F ratios defined by

$$\frac{MSAB}{MSE}, \ \frac{MSA}{MSE}, \ \frac{MSB}{MSE}.$$

In detail:

(a) **A test for the existence of an interaction** If there is no interaction effect then the null hypothesis

$$H_{(\alpha\beta)} : (\alpha\beta)_{ij} = 0, \ (i = 1, \ldots a; \ j = 1, \ldots b)$$

is true and therefore $E(MSAB) = \sigma^2 = E(MSE)$. On the other hand if it is false then $E(MSAB) > \sigma^2 = E(MSE)$. It can be shown (under the null hypothesis) that the ratio $MSAB/MSE$ has the F distribution with numerator and denominator degrees of freedom given by $\nu_1 = (a-1)(b-1)$, $\nu_2 = ab(n-1)$. Thus,

$$F = \frac{MSAB}{MSE} \sim F_{(a-1)(b-1), ab(n-1)}.$$

Therefore our decision rule is to:

accept H_0 if $F \le F_{(a-1)(b-1), ab(n-1)}(\alpha)$, and reject H_0 if $F > F_{(a-1)(b-1), ab(n-1)}(\alpha)$.

(b) **Tests for the existence of main effects** To detect the existence of a factor A effect we consider the F ratio MSA/MSE, which is distributed as

$$F = \frac{MSA}{MSE} \sim F_{(a-1), ab(n-1)}.$$

The corresponding decision rule is to:

accept H_0 if $F \le F_{(a-1), ab(n-1)}(\alpha)$, and reject H_0 if $F > F_{(a-1)(b-1), ab(n-1)}(\alpha)$.

To detect the existence of a factor B effect we proceed just as we did in the previous case, only we now use the F ratio MSB/MSE, which is distributed as

$$F = \frac{MSB}{MSE} \sim F_{(b-1), ab(n-1)}.$$

The corresponding decision rule is to:

accept H_0 if $F \le F_{(b-1), ab(n-1)}(\alpha)$, and reject H_0 if $F > F_{(a-1)(b-1), ab(n-1)}(\alpha)$.

Example 13.8 *Continuation of Example 13.6*

Refer to the pheasant egg production data in Table 13.5. Compute the two-way ANOVA table and the proportion of the total variability explained by the model with interaction. Use the F ratio to test whether there is an interaction between the fungicide (A) and pesticide (B). State the null hypothesis, the alternatives, decision rule, and conclusion. What further conclusions can you draw from the ANOVA table?

Solution The two-way ANOVA table for the data set in Table 13.5 is displayed in Table 13.8.

Table 13.8 *Two-way ANOVA table for the data set in Table 13.5*

Source	DF	SS	MS	F Value	Pr > F
FUNGICIDE (A)	1	1026.7500000	1026.7500000	14.53	0.0004
PESTICIDE (B)	2	595.2916667	297.6458333	4.21	0.0215
AB	2	570.3750000	285.1875000	4.04	0.0249
Error	42	2968.2500000	70.6726190		
Corrected Total	47	5160.6666667			

Conclusion The proportion of the total variability explained by the model is given by the coefficient of determination

$$R^2 = \frac{SSA + SSB + SSAB}{SST} = \frac{2192.42}{5160.67} = 0.42.$$

Consequently 42% of the total variability is explained by the model. The null hypothesis is that there is no interaction effect between the levels of the fungicide and pesticide. The alternative is that there is an interaction effect. Our decision rule is to reject the no interaction hypothesis when the P-value of the F ratio $MSAB/MSE$ is less than 0.05.

Looking at Table 13.8 we see that the F ratio $MSAB/MSE$ has an $F_{2,42}$ distribution, which is not in the Table A.6 in the appendix. From Table A.6 in the appendix it follows that $3.23 > F_{2,42}(0.05) > 3.15$ because

$$F_{2,40}(0.05) = 3.23 > F_{2,42}(0.05) > F_{2,60}(0.05) = 3.15.$$

Since $F = 4.04 > 3.23$ we conclude that there is an interaction effect. The computer printout tells us that the F ratio has a P-value equal to 0.0249; that is $P(F_{2,42} > 4.04) = 0.0249$. The P-values for the tests of fungicide effects, pesticide effects, and interaction effects are listed in the rightmost column. The conclusion is that each of these effects is significant.

10. **Equal error variances** In our model we make the assumption that the variances of the error terms are equal: $V(\epsilon_{ijk}) = \sigma^2$, $(i = 1, \ldots, a; j = 1, \ldots, b; k = 1, \ldots, n)$.

Serious violations of this assumption are frequently detected by plotting the residuals against the fitted values. Figure 13.7 displays the graph of the residual plot for the data in Table 13.5. You will recall that the fitted values are the estimated cell means $\bar{y}_{ij.}$, which

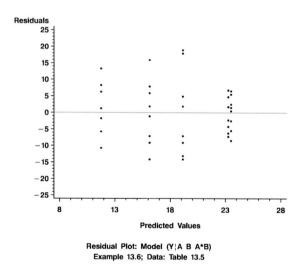

Residual Plot: Model (Y¦A B A*B)
Example 13.6; Data: Table 13.5

FIGURE 13.7: Residual plot: Model($Y|A$, B, $A * B$); Example 13.6

we computed in Example 13.7. When listed in increasing order of magnitude the estimated cell means are: 4.25, 11.75, 16.125, 19.125, 23.25, 23.50. The *funnel shape* of the residual plot suggests that the variances are not constant. At this point the reader might well ask how are our conclusions affected by the fact that the variances appear to be unequal. When this appears to be the case, statisticians recommend a transformation of the data of the form $u = f(y)$ to stabilize the variance (see Snedecor and Cochran, op.cit., for a discussion and specific suggestions). When, as in this case, the observed values in each cell comes from counting, then a square root transformation is often appropriate. In detail, it is recommended that you replace each observation y_{ijk} by $u_{ijk} = \sqrt{1 + y_{ijk}}$, and then calculate the two-way ANOVA table using the transformed data (see Problem 13.9).

Example 13.9 *A factor A effect masked by an interaction effect*

The data set displayed in Table 13.9 (modified from a real data set for the purposes of simplifying the calculations) indicates how an interaction effect can mask the effect of a factor. The entries in each cell are the estimated cell means of a 3×3 complete factorial experiment. The data have been modified so that $\overline{Y}_{i..} = \overline{Y}_{...} = 55$, which implies that $SSA = 0$. The naive analyst might conclude that A has no effect on the response. That A does have an effect is clear once we look at the data more closely. For instance, you will note that as level A varies from 1 to 3 in column 2 the response variable is increasing. We say that the factor A has a positive effect on the response at the second level of factor B. On the other hand, as level A varies from 1 to 3 in column 3 the response variable is decreasing. Consequently, the factor A has a negative effect on the response at the third level of factor B. The existence of an interaction is made clear by looking at Figure 13.8, which is a plot of the estimated cell means $\overline{y}_{ij.}$ against the three levels of factor A.

Table 13.9 *Hypothetical data set in which the existence of a factor A effect is masked by the interaction effect*

| | Estimated Cell Means | | | |
| | Factor (B) | | | |
Factor (A)	1	2	3	Row means
1	60	40	65	55
2	50	59	56	55
3	60	63	42	55
Column means	56.67	54.0	54.33	$\overline{y}_{...} = 55$

13.3.1   Confidence Intervals and Multiple Comparisons

When the null hypothesis of no differences among the factor A means (or among the factor B means) is rejected the researcher will be interested in confidence intervals for each of the factor means as well as determining which pairs of means are significantly different. Unbiased estimators of the factor A means $\overline{\mu}_{i.}$ ($i = 1, \ldots, a$) and the factor B means $\overline{\mu}_{.j}$ ($j = 1, \ldots, b$) are given by $\overline{Y}_{i..}$ and $\overline{Y}_{.j.}$. The $\overline{Y}_{i..}$ term is the sample mean of bn independent summands with the same variance σ^2 so $Var(\overline{Y}_{i..}) = \sigma^2/bn$. Similarly, $Var(\overline{Y}_{.j.}) = \sigma^2/an$. We obtain unbiased estimators for these variances by using MSE to estimate σ^2. The estimated standard error of the estimators are

$$s(\overline{Y}_{i..}) = \sqrt{\frac{MSE}{bn}}; \; s(\overline{Y}_{.j.}) = \sqrt{\frac{MSE}{an}}.$$

The confidence limits for the factor A and factor B means are given by

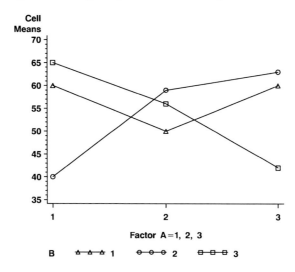

Estimated B Level Cell Means vs. Factor A Levels
Example 13.9; Data: Table 13.9

FIGURE 13.8: Estimated factor B level cell means vs. factor A level cell means; Table 13.9

$$\overline{Y}_{i..} \pm t_{ab(n-1)}(\alpha/2)\sqrt{\frac{MSE}{bn}}, \tag{13.26}$$

$$\overline{Y}_{.j.} \pm t_{ab(n-1)}(\alpha/2)\sqrt{\frac{MSE}{an}}, \tag{13.27}$$

where the degrees of freedom in the t distribution comes from the $ab(n-1)$ degrees of freedom associated with the MSE term. **Tukey's simultaneous multiple comparisons** The simultaneous confidence intervals for the differences among all pairs of factor A and factor B means are given by:

$$(\overline{y}_{i..} - \overline{y}_{i'..}) \pm q(\alpha; a, ab(n-1))\sqrt{\frac{MSE}{bn}}; \ (\overline{y}_{.j.} - \overline{y}_{.j'.}) \pm q(\alpha; b, ab(n-1))\sqrt{\frac{MSE}{an}}.$$

The quantities

$$HSD_A = q(\alpha; a, ab(n-1))\sqrt{\frac{MSE}{bn}}, \ HSD_B = q(\alpha; b, ab(n-1))\sqrt{\frac{MSE}{an}}$$

are called Tukey's *honest significant difference*. As noted earlier in the previous chapter, Tukey's HSD provides simultaneous $100(1-\alpha)\%$ confidence intervals for all pairs of differences of treatment means.

Example 13.10 *Continuation of Example 13.6*

Compute Tukey's HSD for the two factor experiment on the egg production of pheasants (Table 13.5).

Solution   In this case, $a = 2, b = 3, n = 8, ab(n - 1) = 42$, and $q(0.05; 2, 42) = 2.854$ and $q(0.05; 3, 42) = 3.436$. Consequently,

$$HSD_A = q(0.05; 2, 42)\sqrt{\frac{70.673}{24}} = 4.897;$$

$$HSD_B = q(0.05; 3, 42)\sqrt{\frac{70.673}{16}} = 7.221.$$

Since $|\overline{y}_{.2.} - \overline{y}_{.3.}| = 5.811 < 7.221$, we conclude that the difference between the Dieldrin and Diazinon treatments is not significant.

13.4 2^k Factorial Designs

A 2^k factorial experiment is an experimental design used to study the effects of k factors with two levels for each factor. We call the two levels of the factor "high" and "low." Such designs are frequently used when the number of factors and levels of an experiment are so large that running a complete factorial design would be prohibitively expensive in time, labor, or material. A major goal of this experiment would be to determine those factors that have a significant effect on the response variable and screen out those that do not. The total number of factor-level combinations equals 2^k and with n replicates at each factor-level combination there are $2^k n$ experimental runs. However, it is not only the number of experimental runs that increases rapidly with k but the complexity of the analysis as well. Theoretically, there are $2^k - 1$ interactions that are of potential importance, although in practice the higher order interactions are usually not significant. For this reason we will primarily focus our attention on the simplest cases: $k = 2, 3$.

The 2^2 experiment   We denote the two factors by A and B. Statisticians represent the high levels of factors A and B by the symbols a and b and the low levels by 1. More precisely, we represent the factor-level combinations by the formal product

$$a^{l_1} b^{l_2}, \; l_i = 0, 1,$$

where $l_i = 0$ denotes the low level of the factor and $l_i = 1$ denotes the high level. By convention we set $a^0 = b^0 = 1$. For instance, the symbol $ab = a^1 b^1$ represents the factor-level combination with both factors at the high level ($l_1 = l_2 = 1$); the symbol $a = a^1 b^0$ represents the factor-level combination with factor A at the high level and factor B at the low level ($l_1 = 1, l_2 = 0$). The symbol $1 = a^0 b^0$ denotes the factor-level combination with all factors at the low level. Summing up, the four factor-level combinations are denoted by: $1, a, b, ab$.

Notation and Model Assumptions

1. **Data format**   We represent the kth replicate at the i, j factor-level combination by y_{ijk}, where

$$y_{ijk}, (i = 1, 2; \; j = 1, 2; \; k = 1, \ldots, n).$$

With this notation the factor-level combination with both factors at the low level corresponds to ($i = 1, j = 1$). Similarly, ($i = 2, j = 1$) corresponds to the factor-level combination with A at the high level and B at the low level.

In the context of 2^2 factorial experiments statisticians arrange the data in columns by first writing the factor-level combinations in the *standard order*: $1, a, b, ab$. The n replicates at each factor-level combination are then listed in the next column. A third column, containing the sample totals for each factor-level combination, is sometimes added to the display as in Table 13.10. We use the *special notation* (1), (a), (b), (ab) to denote the sample totals for each factor-level combination. In terms of the dot subscript notation the sample totals and sample means for factor-level combination are given by

$$y_{11.} = (1), \ \overline{y}_{11.} = \frac{(1)}{n};$$

$$y_{21.} = (a), \ \overline{y}_{21.} = \frac{(a)}{n};$$

$$y_{12.} = (b), \ \overline{y}_{12.} = \frac{(b)}{n};$$

$$y_{22.} = (ab), \ \overline{y}_{22.} = \frac{(ab)}{n}.$$

Example 13.11 *Performance of a computer workstation*

Table 13.10 records the result of an experiment to measure the performance of a computer workstation, measured in million instructions per second (MIPS), as a function of memory size (factor A) and cache size (factor B). Memory size has two levels: 4 M-bytes (low) and 16 M-bytes (high); and cache size has two levels: 1 K-bytes (low) and 2 K-bytes (high). This is a 2^2 experiment with three replicates ($n = 3$) at each factor-level combination, resulting in a total sample size of $2^2 \times 3 = 12$. We shall analyze this data in more detail shortly (see Examples 13.13 and 13.14).

Table 13.10 *Performance of a computer workstation with two memory sizes and two cache sizes*

Factor-level Combination	Replicates	Sample Total
1	15, 18, 12	$(1) = 45$
a	25, 28, 19	$(a) = 72$
b	45, 48, 51	$(b) = 144$
ab	75, 75, 81	$(ab) = 231$

(Source: R. Jain (1991), *The Art of Computer Systems Performance Analysis*, p. 294, John Wiley & Sons, New York. Used with permission.)

2. **The statistical model for the response variable** We assume that we have performed n experiments at each combination of the factor-levels. We represent the value of the kth response at the (i, j)th combination of the factor-levels by Y_{ijk}, where

$$Y_{ijk} = \mu_{ij} + \epsilon_{ijk} \ (i = 1, 2; \ j = 1, 2; \ k = 1, \ldots, n), \text{ where}$$
$$\mu_{ij} = \mu + \alpha_i + \beta_j + (\alpha\beta)_{ij}, \tag{13.28}$$

and the random variables ϵ_{ijk} are iid $N(0, \sigma^2)$.

3. **Main effects and interaction effects** The *effect of the factor A at the low level of B* is the difference in the mean response when the factor A is changed from the low to the high level with the factor B fixed at its low level. It is denoted by $\mu_{21} - \mu_{11}$. Similarly,

we call $\mu_{22} - \mu_{12}$ the *effect of A at the high level of B*. We define the *main effect* of A to be the average of the effects of A at the low and high levels of B; thus, the

$$\text{main effect of } A = \frac{1}{2}(\mu_{22} - \mu_{12} + \mu_{21} - \mu_{11}).$$

Interchanging the roles of 1 and 2 we obtain the analogous formulas for the factor B effects. Thus $\mu_{12} - \mu_{11}$ is the *effect of B at the low level of A* and $\mu_{22} - \mu_{21}$ is the *effect of B at the high level of A*. The *main effect* of B equals the average of the effects of B at the low and high levels of A; thus, the

$$\text{main effect of } B = \frac{1}{2}(\mu_{22} - \mu_{21} + \mu_{12} - \mu_{11}).$$

The reader will recall that no interaction effect exists when the effect of A is independent of the level of B; equivalently, there is no interaction when the effect of A at the high level of B equals the effect of A at the low level of B. Consequently, there is no AB interaction when $\mu_{21} - \mu_{11} = \mu_{22} - \mu_{12}$. Therefore, we define the AB interaction effect to be the average of the difference between the effect of A at the high and low levels of B; thus, the

$$\text{AB interaction effect } = \frac{1}{2}(\mu_{22} - \mu_{12} - \mu_{21} + \mu_{11}).$$

Notice that each main effect and interaction is defined by a linear combination of the cell means whose coefficients sum to zero; that is, the main effects and interactions are contrasts.

Estimates for the main effects and the interactions   As just noted, the main effects and interactions are linear combinations of the cell means. Consequently, their estimates will also be linear combinations of the estimates of the cell means. These estimators (using the special notation) are given by

$$\hat{\mu}_{11} = \frac{(1)}{n} \quad \hat{\mu}_{12} = \frac{(b)}{n};$$
$$\hat{\mu}_{21} = \frac{(a)}{n} \quad \hat{\mu}_{22} = \frac{(ab)}{n}.$$

This leads to the next three formulas for computing the estimated main effects for A and B, and their AB interaction effects:

$$\text{estimator of main effect of } A = \frac{1}{2n}[(ab) - (b) + (a) - (1)];$$
$$\text{estimator of main effect of } B = \frac{1}{2n}[(ab) + (b) - (a) - (1)];$$
$$\text{estimator of AB interaction } = \frac{1}{2n}[(ab) - (b) - (a) + (1)].$$

It is an interesting and useful fact that each quantity in square brackets of the preceding expressions for the main and interaction effects is a contrast in the sample totals. In particular, we represent the main effects of A and B and the AB interaction effect in terms of the contrasts L_A, L_B, and L_{AB} defined as follows:

$$L_A = (ab) - (b) + (a) - (1); \tag{13.29}$$
$$L_B = (ab) + (b) - (a) - (1); \tag{13.30}$$
$$L_{AB} = (ab) - (b) - (a) + (1). \tag{13.31}$$

4. **Computing the contrast coefficients** In Table 13.11 we display the sign patterns for each contrast as a column. An algorithm for computing the sign pattern for each contrast will now be given. Begin by first writing down the factor-level combinations in the *standard order:* (1), (a), (b), (ab). These are listed in the first column. We ignore for the moment the column I. In each of the A and B columns you enter -1 when the factor occurs at its low level and enter $+1$ when it occurs at its high level in the corresponding contrast. For instance, the pattern $-1, +1, -1, +1$ in column A corresponds to the *low, high, low, high* coefficients of the factor-level combinations in the contrast L_A. Each coefficient for the AB contrast (column AB) is the product of the corresponding coefficients in columns A and B. The column I, which consists only of $+1$'s, is not a contrast since its coefficients do not sum to zero. It plays the role of the identity with respect to the operation of column multiplication; thus $IA = A$, $IAB = AB$, etc. Note also that the product of each column with itself equals the identity I. Thus, $A^2 = B^2 = (AB)^2 = I$.

Table 13.11 *Table of contrast coefficients for a 2^2 factorial experiment*

Factor-level Combination	I	A	B	AB
1	$+1$	-1	-1	$+1$
a	$+1$	$+1$	-1	-1
b	$+1$	-1	$+1$	-1
ab	$+1$	$+1$	$+1$	$+1$
Contrasts	$\bar{y}$	L_A	L_B	L_{AB}

Example 13.12 *Estimating main effects and interactions*

Refer to Table 13.10 (computer performance data). Compute the estimates of the main effects and interactions.

Solution We first compute the contrasts, using Equations 13.29-13.31, and the sample totals listed in Table 13.10. Then we compute the effects.

Computing the contrasts

$$L_A = 231 - 144 + 72 - 45 = 114$$
$$L_B = 231 + 144 - 72 - 45 = 258$$
$$L_{AB} = 231 - 144 - 72 + 45 = 60$$

Computing the effects

$$\text{Main effect of A} = \frac{L_A}{2n} = \frac{114}{6} = 19$$
$$\text{Main effect of B} = \frac{L_A}{2n} = \frac{258}{6} = 43$$
$$\text{Interaction effect AB} = \frac{L_{AB}}{2n} = \frac{60}{6} = 10$$

5. **Partitioning the total variability of the data** We analyze the model in the usual way by partitioning the total variability of the data into a portion explained by the model and the unexplained variability. These computations are much simpler than in the

general case because each sum of squares SSA, SSB, $SSAB$ equals the corresponding contrast sum of squares divided by $4n$; that is,

$$SST = SSA + SSB + SSAB + SSE, \text{ where} \tag{13.32}$$

$$SSA = \frac{L_A^2}{4n}; \tag{13.33}$$

$$SSB = \frac{L_B^2}{4n}; \tag{13.34}$$

$$SSAB = \frac{L_{AB}^2}{4n}. \tag{13.35}$$

Problem 13.28 asks the reader to derive Equations 13.33, 13.34, and 13.35.

Summing up: We compute SSE by first computing SST and then use the formulas:

$$SST = \sum_{i=1}^{2}\sum_{j=1}^{2}\sum_{k=1}^{n} y_{ijk}^2 - \frac{y_{...}^2}{4n};$$

$$SSE = SST - SSA - SSB - SSAB.$$

Example 13.13 *Continuation of Example 13.11*

Compute the partition of the total variability of the data in Table 13.10 (computer performance data).

Solution Using the values for the contrasts computed earlier,

$$L_A = 114, \ L_B = 258, \ L_{AB} = 60,$$

a straightforward application of Equations 13.33, 13.34, and 13.35 yields the following values for the sums of squares:

$$SSA = \frac{(114)^2}{12} = 1083, \ SSB = \frac{(258)^2}{12} = 5547, \ SSAB = \frac{(60)^2}{12} = 300,$$
$$SST = 7032, \ SSE = 7032 - 1083 - 5547 - 300 = 102.$$

6. **Partitioning the degrees of freedom** A 2^2 factorial experiment with n replicates for each factor-level combination is a special case of the two-way ANOVA with $a = 2$, $b = 2$, and $4n$ observations. Consequently, the sums of squares SSA, SSB, SSAB each have one degree of freedom, while SSE has $4(n-1)$ degrees of freedom.

   We therefore have the following partition of the degrees of freedom:

$$4n - 1 = 1 + 1 + 1 + 4(n-1).$$

We have thus reduced the task of computing the two-way ANOVA table for the 2^2 experiment (and this is also true for the 2^k experiment) to computing the contrasts in the sample totals for each factor-level combination.

Example 13.14 *Analysis of computer performance data*

Compute the ANOVA table for the computer performance data (Table 13.10) and test for the existence of main effects and interactions. Use $\alpha = 0.05$.

Solution The analysis of variance is displayed in Table 13.12. The entries in the sums of squares column (SS) were computed in Example 13.13.

Conclusions Since $F_{1,8}(0.01) = 11.26$ we conclude that the main effects and the AB interaction effects are all significant at the 0.01 level and *a fortiori* at the 5% level.

Table 13.12 *Two-way ANOVA table for 2^2 factorial experiment Example 13.11*

Source	DF	SS	MS	F
A	1	1083	1083	84.94
B	1	5547	5547	435.06
AB	1	300	300	23.53
Error	8	102	12.75	
Total	11	7032		

Measuring the Importance of an Effect

The preceding analysis is somewhat misleading because the effects are not of equal importance even though they are all judged to be significant. We measure the importance of an effect by computing the proportion of the total variability of the data that is explained by that effect. Thus:

$$\text{proportion of total variability explained by } A = \frac{SSA}{SST};$$
$$\text{proportion of total variability explained by } B = \frac{SSB}{SST};$$
$$\text{proportion of total variability explained by } AB = \frac{SSAB}{SST}.$$

Example 13.15 *Continuation of Example 13.11*

Refer to Table 13.10 (computer performance data). Compute the proportion of variability that is explained by each main and interaction effect.

    Solution

$$\text{proportion of total variability explained by } A = \frac{1083}{7032} = 0.15;$$
$$\text{proportion of total variability explained by } B = \frac{5547}{7032} = 0.79;$$
$$\text{proportion of total variability explained by } AB = \frac{300}{7032} = 0.04.$$

    Conclusions   Clearly, the most important factor affecting the computer performance is factor B (cache size), since it accounts for 79% of the total variability.

Example 13.16 *Analysis of a complex manufacturing process*

The manufacture of an integrated circuit is a complex process depending on many control parameters. The first step is to grow an epitaxial layer with a specified thickness on a silicon wafer. The data in Table 13.13 come from a carefully planned experimental design in which the researchers were interested in controlling the mean and reducing the variance of the thickness of the epitaxial layer. The epitaxial thickness is specified to be between 14 and 15 micrometers thick with a target value of 14.5 micrometers. Each experimental run produces 14 wafers. The epitaxial thickness is measured at 5 different positions on each wafer yielding a total of $70 = 14 \times 5$ measurements of the epitaxial thickness. The Y1 column records the mean epitaxial thickness and the Y2 column records $\log s^2$, the logarithm of the sample variance, a measure of the variation in the manufacturing process.[1] Reducing the variation Y2 while keeping Y1 on target would result in a smaller proportion of non-conforming output and thus produce a higher yield. The 8 control parameters labeled X1-X8 are 0 or 1 according as the factor is at the low or high level. The two response variables are Y1 and Y2. An exploratory analysis of the data indicated that the effects of the factors X1 (rotation method) and X8 (nozzle position) on Y2 was significant, but their effects on Y1 were not. Verify these conclusions by computing the two-way ANOVA table for each of the following two models: Model (Y2 | X1, X8, X1*X8) and Model (Y1 | X1, X8, X1*X8).

Key to the variables:
X1= Susceptor-rotation method
X2= Wafer code
X3= Deposition temperature
X4= Deposition time
X5= Arsenic gas flow rate
X6= Hydrochloric acid etch temperature
X7= Hydrochloric acid flow rate
X8= Nozzle position
Y1= Epitaxial thickness
Y2= $\log s^2$

Table 13.13 *Parameter settings for the control variables X1-X8 (0 = low, 1 = high) for each of 16 experimental runs*

EXPT	X1	X2	X3	X4	X5	X6	X7	X8	Y1	Y2
1	0	0	0	0	0	0	0	0	14.821	-0.4425
2	0	0	0	0	1	1	1	1	14.888	-1.1989
3	0	0	1	1	0	0	1	1	14.037	-1.4307
4	0	0	1	1	1	1	0	0	13.880	-0.6505
5	0	1	0	1	0	1	0	1	14.165	-1.4230
6	0	1	0	1	1	0	1	0	13.860	-0.4969
7	0	1	1	0	0	1	1	0	14.757	-0.3267
8	0	1	1	0	1	0	0	1	14.921	-0.6270
9	1	0	0	1	0	1	1	0	13.972	-0.3467
10	1	0	0	1	1	0	0	1	14.032	-0.8563
11	1	0	1	0	0	1	0	1	14.843	-0.4369
12	1	0	1	0	1	0	1	0	14.415	-0.3131
13	1	1	0	0	0	0	1	1	14.878	-0.6154
14	1	1	0	0	1	1	0	0	14.932	-0.2292
15	1	1	1	1	0	0	0	0	13.907	-0.1190
16	1	1	1	1	1	1	1	1	13.914	-0.8625

(Source: R.N. Kackar and A. Shoemaker (1986), *Robust Design: A Cost Effective Method For Improving Manufacturing Processes*, AT&T Technical Journal, March/April, vol. 65, issue 2, pp. 39-50. Used with permission.)

Solution

1. **Analysis of variance for the Model (Y2 | X1, X8, X1*X8)** The first step is to drop all the other variables from the model except X1, X8, Y1, and Y2 and sort the

---
[1]We use the symbols Y1, X1, etc., instead of Y_1, X_1 to make our notation consistent with the computer printouts.

data in the standard order with A=X1 and B=X8. This is done in Table 13.14. Table 13.15 is the data for the model rearranged in the standard order, remembering that X1 and X8 play the roles of the factors A and B, respectively. Table 13.16 is the analysis of variance table for this model. Since this table has a different format from a standard two-way ANOVA table (for example Table 13.12) a brief explanation is called for. The sums of squares SSX1, SSX8, and SSX1X8 appear in the lower left portion of the table. The variability explained by the model, denoted SSm, is defined to be

$$SSm = SSX1 + SSX8 + SSX1X8 = 0.49600 + 1.28035 + 0.06249 = 1.83884.$$

This sum appears in the "Sum of Squares Column." It has three degrees of freedom because it is a sum of three squares, each of which has one degree of freedom. The error sum of squares SSE represents the unexplained variability; it is given by $SSE = SST - SSm = 0.63983382$. The *overall F ratio* equals $F = (SSm/3)/(SSE/12) = 11.50$. The overall F ratio tests the null hypothesis of no X1 effects, no X8 effects, and no X1*X8 interactions. The P-value for the F ratio is 0.0008, so we reject the null hypothesis. In other words the method of susceptor rotation and the nozzle position have significant effects on the process variation.

Table 13.14 *The data for the Model (Y2 | X1, X8, X1*X8) sorted in the standard order: (X1=0, X8=0), (X1=1, X8=0), (X1=0, X8=1), (X1=1, X8=1)*

EXPT	X1	X8	Y1	Y2
1	0	0	14.821	-0.4425
4	0	0	13.880	-0.6505
6	0	0	13.860	-0.4969
7	0	0	14.757	-0.3267
9	1	0	13.972	-0.3467
12	1	0	14.415	-0.3131
14	1	0	14.932	-0.2292
15	1	0	13.907	-0.1190
2	0	1	14.888	-1.1989
3	0	1	14.037	-1.4307
5	0	1	14.165	-1.4230
8	0	1	14.921	-0.6270
10	1	1	14.032	-0.8563
11	1	1	14.843	-0.4369
13	1	1	14.878	-0.6154
16	1	1	13.914	-0.8625

Table 13.15 *The data for the Model (Y2 | X1, X8, X1*X8) rearranged in the standard order*

Treatment Combination	Replicates	Total
1	$-0.4425, -0.6505, -0.4969, -0.3267$	$(1) = -1.9166$
a	$-0.3467, -0.3131, -0.2292, -0.1190$	$(a) = -1.0080$
b	$-1.1989, -1.4307, -1.4230, -0.6270$	$(b) = -4.6796$
ab	$-0.8563, -0.4369, -0.6154, -0.8625$	$(ab) = -2.7710$

Table 13.16 *Two-way ANOVA table for the model (Y2 | X1, X8, X1*X8)*

Source	DF	Sum of Squares	Mean Square	F Value	Pr > F
Model	3	1.83883960	0.61294653	11.50	0.0008
Error	12	0.63983382	0.05331948		
Corrected Total	15	2.47867342			

	R-Square	C.V.	Root MSE		Y2 Mean
	0.741864	-35.60921	0.230910		-0.648456

Source	DF	Anova SS	Mean Square	F Value	Pr > F
X1	1	0.49600328	0.49600328	9.30	0.0101
X8	1	1.28034883	1.28034883	24.01	0.0004
X1*X8	1	0.06248750	0.06248750	1.17	0.3003

2. **Analysis of variance for the Model (Y1 | X1, X8, X1*X8)** We analyze this model in the same way by first rearranging the data in the standard order as in Table 13.17 and then performing the analysis of variance, which is displayed in Table 13.18.

Conclusions Looking at the ANOVA Table 13.16 we see that for the Model (Y2 | X1, X8, X1*X8) the main effects of X1 and X8 are significant, while the interaction effect is not. The proportion of variability accounted for by this model is $R^2 = 0.741864$, that is, this model accounts for 74% of the total variability. The results for the Model (Y1 | X1, X8, X1*X8) are quite different. Looking at the ANOVA Table 13.18 we see that none of the main effects are significant (the P-value of the overall F ratio is 0.9407). The proportion of the total variability explained by this model is $R^2 = 0.031372$, which is negligible. This means that setting the X1 and X8 control parameters to minimize the process variance will have a negligible effect on the epitaxial thickness.

Table 13.17 *The data for the Model (Y1 | X1, X8, X1*X8) rearranged in the standard order*

Treatment Combination	Replicates	Total
1	14.821, 13.880, 13.860, 14.757,	$(1) = 57.318$
a	13.972, 14.415, 14.932, 13.907	$(a) = 57.226$
b	14.888, 14.037, 14.165, 14.921	$(b) = 58.011$
ab	14.032, 14.843, 14.878, 13.914	$(ab) = 57.667$

Table 13.18 *Two-way ANOVA table for the model (Y1 | X1, X8, X1*X8)*

Source	DF	Sum of Squares	Mean Square	F Value	Pr > F
Model	3	0.09622225	0.03207408	0.13	0.9407
Error	12	2.97090150	0.24757512		
Corrected Total	15	3.06712375			

	R-Square	C.V.	Root MSE		Y1 Mean
	0.031372	3.458013	0.497569		14.38888

(continued on next page)

Source	DF	Anova SS	Mean Square	F Value	Pr > F
X1	1	0.01188100	0.01188100	0.05	0.8303
X8	1	0.08037225	0.08037225	0.32	0.5793
X1*X8	1	0.00396900	0.00396900	0.02	0.9013

The 2^3 factorial experiment

In a 2^3 experiment the researcher is interested in studying the response variable as a function of three factors with two levels for each factor.

1. **Notation** We denote the three factors by A, B, C and represent the 2^3 factor-level combinations by the formal product

$$a^{l_1} b^{l_2} c^{l_3},$$

where $l_i = 0$ denotes the low level of the factor and $l_i = 1$ denotes the high level. Thus $1 = a^0 b^0 c^0$ represents the treatment combination with all factors at their low levels; the symbol (1) represents the sample total at this factor-level combination. Similarly, $bc = a^0 b^1 c^1$ represents the factor-level combination *(low, high, high)*; the corresponding sample total is denoted (bc). With n replicates per factor-level combination there are a total of $2^3 n$ observations. There are 3 main effects (A, B, C), 3 *two* factor interactions (AB, AC, BC), and 1 three factor interaction ABC.

2. **Data format** We arrange the data from a 2^3 factorial experiment in columns by first writing the factor-level combinations in the *standard order*:

$$1, a, b, ab, c, ac, bc, abc.$$

The sample totals for each factor-level combination are denoted by

$$(1), (a), (b), (ab), (c), (ac), (bc), (abc).$$

The n replicates at each factor-level combination are then listed in the next column. Sometimes, a third column containing the sample totals for each factor-level combination, is added to the display. Table 13.19 is an example of how the data from such an experiment are displayed.

Table 13.19 *Data format for a 2^3 experiment with n replicates for each treatment combination*

Factor-level Combination	Replicates	Sample Total
1	$y_{1111}, \ldots, y_{111n}$	(1)
a	$y_{2111}, \ldots, y_{211n}$	(a)
b	$y_{1211}, \ldots, y_{121n}$	(b)
ab	$y_{2211}, \ldots, y_{221n}$	(ab)
c	$y_{1121}, \ldots, y_{112n}$	(c)
ac	$y_{2121}, \ldots, y_{212n}$	(ac)
bc	$y_{1221}, \ldots, y_{122n}$	(bc)
abc	$y_{2221}, \ldots, y_{222n}$	(abc)

3. **Main effects and interactions** We estimate the main effects and interactions by constructing the corresponding contrasts, following the same procedure used to construct Table 13.11. First write down sample totals for the factor-level combinations in the standard order: $\{(1), (a), (b), (ab), (c), (ac), (bc), (abc)\}$. These are listed in the first column of Table 13.20. Columns A, B, and C list the patterns of $\pm$ signs corresponding to the coefficients in the contrast used to estimate the main effects. We will give the method for deriving these patterns below. Each coefficient for the AB contrast (column AB) is the product of the corresponding coefficients in columns A and B; and each coefficient for the ABC contrast (column ABC) is the product of the corresponding coefficients in columns A, B, and C. The entries for the columns AC and BC are obtained by multiplying the corresponding columns.

Table 13.20 *Table of contrast coefficients for a 2^3 factorial experiment*

Treatment Totals	Main Effects and Interactions							
	I	A	B	C	AB	AC	BC	ABC
(1)	+1	−1	−1	−1	+1	+1	+1	−1
(a)	+1	+1	−1	−1	−1	−1	+1	+1
(b)	+1	−1	+1	−1	−1	+1	−1	+1
(ab)	+1	+1	+1	−1	+1	−1	−1	−1
(c)	+1	−1	−1	+1	+1	−1	−1	+1
(ac)	+1	+1	−1	+1	−1	+1	−1	−1
(bc)	+1	−1	+1	+1	−1	−1	+1	−1
(abc)	+1	+1	+1	+1	+1	+1	+1	+1
Contrasts	$\overline{y}$	L_A	L_B	L_C	L_{AB}	L_{AC}	L_{BC}	L_{ABC}

Derivation of the estimates for the main effects via contrasts

Looking at column A of Table 13.20 we see that the A contrast L_A equals:

$$L_A = -(1) + (a) - (b) + (ab) - (c) + (ac) - (bc) + (abc). \qquad (13.36)$$

The estimates for an effect and its corresponding sum of squares are given by:

$$\text{Estimate of an effect} = \frac{L_{effect}}{4n}; \qquad (13.37)$$

$$SS(effect) = \frac{L^2_{effect}}{8n}. \qquad (13.38)$$

Derivation of Equation 13.37: It suffices to consider the main effect A, since the derivations for the other effects are similar. In particular we want to show that the

$$\text{estimate of the main effect of A} = \frac{L_A}{4n}.$$

We first compute the effect of A at each of the four possible factor-level combinations of B and C. You will recall that this equals the change in the response when only the factor A is changed from its low to high level, with all the other factors held constant. These computations are listed in Table 13.21.

Table 13.21 *Computing the effects of A at each of the four factor-level combinations of B and C in a 2^3 factorial experiment*

B	C	Effect of A
low	low	$\frac{(a)-(1)}{n}$
high	low	$\frac{(ab)-(b)}{n}$
low	high	$\frac{(ac)-(c)}{n}$
high	high	$\frac{(abc)-(bc)}{n}$

The estimated effect of A is defined to be the average of these four effects, which you can easily verify equals $L_A/4n$. Using the same reasoning you can show that the estimated B effect equals $L_B/4n$, etc.

4. **Partitioning the total variability of the data** To each main effect and interaction there corresponds a sum of squares that measures the contribution of that effect to the total variability of the data. In a three-way ANOVA the partition of SST takes the following form:

$$SST = SSA + SSB + SSC + SSAB + SSAC + SSBC + SSABC + SSE. \quad (13.39)$$

We use Equation 13.38 to compute the sums of squares for the main effects and interactions. We compute the total sum of squares in the usual way

$$SST = \sum_i \sum_j \sum_k \sum_{1 \le l \le n} (y_{ijkl} - \bar{y}_{....})^2$$
$$= \sum_i \sum_j \sum_k \sum_{1 \le l \le n} y_{ijkl}^2 - N\bar{y}_{....}^2,$$

where $N = n2^k$ is the total number of observations. The formulas for computing the sums of squares for the general three-way ANOVA are not given because these calculations are best done via one of the standard statistical software packages such as MINITAB or SAS.

5. **Partitioning the degrees of freedom** Corresponding to the partition of the total variability given in Equation 13.39 we have the following partition of the degrees of freedom. The total sum of squares SST has $2^3 n - 1$ degrees of freedom and there are $2^3 - 1 = 7$ single degree of freedom contrasts; therefore,

$$2^3 n - 1 = (2^3 - 1) + 2^3(n - 1).$$

Consequently, the error sum of squares SSE has $2^3(n - 1)$ degrees of freedom. The preceding formula for the partition of the degrees of freedom is a special case of the following more general result that holds for a complete 2^k experiment with n replicates for each treatment combination. The total sum of squares SST has $2^k n - 1$ degrees of freedom and there are $2^k - 1$ single degree of freedom contrasts; therefore, SSE has $2^k(n - 1)$ degrees of freedom and the partition of the degrees of freedom is given by:

$$2^k n - 1 = (2^k - 1) + 2^k(n - 1).$$

The sum of squares for each effect $SS(effect) = L^2_{effect}/8n$ (refer back to Equation 13.38) has an F distribution with parameters $\nu_1 = 1$ and $\nu_2 = 2^k(n-1)$.

Example 13.17 *A 2^3 factorial experiment*

The following data come from a 2^3 factorial experiment with three replicates for each treatment combination. Estimate the main effects and all multi-factor interactions. Then analyze the data by constructing the ANOVA table.

Table 13.22 *Data from a complete 2^3 factorial experiment with three replicates for each treatment combination*

Treatment Combination	Replicates	Total
1	14, 24, 11	49
a	18, 26, 20	64
b	32, 20, 21	73
ab	30, 21, 36	87
c	21, 33, 27	81
ac	20, 24, 24	68
bc	32, 30, 39	101
abc	38, 33, 26	97

Solution We first compute the contrasts, the main effects and all multi-factor interactions using Equation 13.37. We compute the corresponding sums of squares using Equation 13.38. The results are summarized in Table 13.23.

Table 13.23 *Table of contrasts and estimates of the main effects and interactions for the data in Table 13.22*

Contrasts	Estimates of the effects and interactions
$L_A = 12,$	$\dfrac{L_A}{12} = 1.00$
$L_B = 96,$	$\dfrac{L_B}{12} = 8.00$
$L_{AB} = 8,$	$\dfrac{L_{AB}}{12} = 0.67$
$L_C = 74,$	$\dfrac{L_C}{12} = 6.17$
$L_{AC} = -46,$	$\dfrac{L_{AC}}{12} = -3.83$
$L_{BC} = 2,$	$\dfrac{L_{BC}}{12} = 0.17$
$L_{ABC} = 10,$	$\dfrac{L_{ABC}}{12} = 0.83$

Preliminary conclusions

Looking at the contrasts only it is easy to see that the most significant main effects are due to the factors B and C, with A being relatively insignificant. The higher order interactions, with the possible exception of the AC interaction, do not appear to be significant. These preliminary findings are confirmed by examining the ANOVA Table 13.24 below. Note that the entries in column three (SS) are computed via the formula $SS(effect) = L^2_{effect}/8n$. Thus $SSA = 12^2/24 = 6.00$.

Table 13.24 *ANOVA table for a 2^3 factorial experiment; data set: Table 13.22; edited version of SAS printout*

Source	DF	SS	Mean Square	F Value	Pr > F
A	1	6.0000000	6.0000000	0.18	0.6761
B	1	384.0000000	384.0000000	11.59	0.0036
A*B	1	2.6666667	2.6666667	0.08	0.7803
C	1	228.1666667	228.1666667	6.89	0.0184
A*C	1	88.1666667	88.1666667	2.66	0.1223
B*C	1	0.1666667	0.1666667	0.01	0.9443
A*B*C	1	4.1666667	4.1666667	0.13	0.7275
Error	16	530.0000000	33.1250000		
Total	23	1243.3333333			

Conclusions Looking at the last column of the ANOVA table we see that only the P-values for the B and C main effects are significant at the 5% level. Another way of evaluating the various effects and the interactions is to observe that the factor B accounts for $31\%[= (384/1243) \times 100\%]$ of the total variability, whereas the factor A accounts for only $0.48\%[= (6/1243) \times 100\%]$ of the total variability. This suggests that the experimenter should concentrate her future research on factors B and C because the other factors and their interactions appear to be negligible.

Fractional Replication of a 2^k Factorial Experiment

The results of the preceding analysis of the data in Table 13.22 imply that the factor A and the higher order interactions are negligible; only the factors B and C were judged to be significant. In other words, most of the observed variability is due to a small number of factors and/or low order interactions. This suggests that we performed more experiments than were necessary to reach these conclusions. Yet another consideration of no small importance is that a complete 2^k factorial experiment with n replicates for each treatment combination requires $2^k n$ experimental runs. This number is rapidly increasing even for small values of k and n. For instance, a complete factorial experiment with four factors and three replicates for each treatment combination requires $2^4 \times 3 = 48$ experimental runs; a complete factorial experiment with five factors and two replicates for each treatment combination requires $2^5 \times 2 = 64$ experimental runs. If the researcher believes (as in the previous example) that only the main effects are important, and that the higher order interactions are negligible, then useful information concerning these effects can still be obtained by performing only a fraction of the experiments needed for the complete factorial. This is the driving force behind the idea of a *fractional replication* of a 2^k factorial experiment. This topic, which is of considerable importance in modern manufacturing, is treated at great length in the references cited in Section 13.5.

We will not attempt a comprehensive introduction here since it is very easy to get lost in intricate technical details that obscure the basic ideas. For this reason we shall study only the simplest example, the so-called *half fraction of the 2^k factorial*. In this design the experimenter makes half the experimental runs of the complete factorial, and this is why we denote the half fraction design by the symbol 2^{k-1} because it equals one-half of 2^k. For instance, a half fraction of a complete 2^3 factorial experiment consists of $2^{3-1} = 4$ treatment combinations.

Although there are many possible factors that can affect the final results, most scientists have a pretty good idea (based on theory and experience) that some factors and higher order interactions are not as important as others. It is on this basis that the senior scientist and her staff determine which treatment combinations to include in the experimental runs.

Choosing the Treatment Combinations for a 2^{3-1} Design

A half fraction of a 2^3 factorial experiment is run with $2^{3-1} = 4$ treatment combinations. We will now present an algebraic method that statisticians have devised for selecting the treatment combinations. Pick a contrast, for example

$$L_{ABC} = -(1) + (a) + (b) - (ab) + (c) - (ac) - (bc) + (abc),$$

and choose the treatment combinations with a plus sign for the experimental runs. This means we perform the experiments on the four treatment combinations: (a), (b), (c), (abc).

Table 13.25 is obtained by rearranging the rows of Table 13.20 so that the treatment combinations appearing with a plus sign in the contrast L_{ABC} occupy the first four rows.

Table 13.25 *Table of contrast coefficients for a 2^3 factorial experiment*

Treatment	Main Effects and Interactions							
Totals	I	A	B	C	AB	AC	BC	ABC
(a)	+1	+1	−1	−1	−1	−1	+1	+1
(b)	+1	−1	+1	−1	−1	+1	−1	+1
(c)	+1	−1	−1	+1	+1	−1	−1	+1
(abc)	+1	+1	+1	+1	+1	+1	+1	+1
(ab)	+1	+1	+1	−1	+1	−1	−1	−1
(ac)	+1	+1	−1	+1	−1	+1	−1	−1
(bc)	+1	−1	+1	+1	−1	−1	+1	−1
(1)	+1	−1	−1	−1	+1	+1	+1	−1
Contrasts	$\overline{y}$	L_A	L_B	L_C	L_{AB}	L_{AC}	L_{BC}	L_{ABC}

Computing the Contrasts for a 2^{3-1} Design with Defining Relation $I = ABC$

Looking at the first four rows of Table 13.25 we see that the contrast coefficients for column $I =$ column ABC, so we write $I = ABC$ and call it the *defining relation* of our design. L_{ABC} is called the *defining contrast,* and ABC is called the *generator* of this design.

Let us agree to call a contrast coming from a fractional replicate a *fractional contrast* and denote it by the symbols $l_A, l_B, \ldots$. Looking at Table 13.25 we see that the fractional contrasts are given by:

$$l_A = (a) - (b) - (c) + (abc) = l_{BC},$$
$$l_B = -(a) + (b) - (c) + (abc) = l_{AC},$$
$$l_C = -(a) - (b) + (c) + (abc) = l_{BC}.$$

When two fractional contrasts are identical we say that they are *aliases* of one another. Thus A is aliased with BC since $l_A = l_{BC}$; we write $A \equiv BC$. Define the *sum of the contrasts* A and BC to be the sum of the corresponding contrast coefficients in columns A and BC. This leads to another interpretation of the phenomenon of aliasing. Looking at Table 13.25

we see that

$$l_A = \frac{1}{2}(L_A + L_{BC}),$$

$$l_B = \frac{1}{2}(L_B + L_{AC}),$$

$$l_C = \frac{1}{2}(L_C + L_{AB}).$$

Therefore the corresponding estimate $l_A/2n$ of the main effect of A is really an estimate of the main effect of A *plus* the main effect of BC.

Example 13.18 *Analysis of data using a 2^{3-1} design*

Refer to Table 13.22. Analyze these data using a 2^{3-1} design with defining relation $I = ABC$.

Solution The estimates for the main effects for the half fraction factorial and complete factorial are listed in Table 13.26.

Table 13.26 *Estimates of the main effects based on fractional contrasts and complete contrasts; data: Table 13.22*

Fractional factorial	Complete factorial
$\dfrac{l_A}{2n} = 1.17$	$\dfrac{L_A}{4n} = 1$
$\dfrac{l_B}{2n} = 4.17$	$\dfrac{L_B}{4n} = 8$
$\dfrac{l_C}{2n} = 6.83$	$\dfrac{L_C}{4n} = 6.17$

The results indicate clearly that factor A is much less important than the factors B and C. This is consistent with the results obtained from the complete factorial experiment obtained by running the complete factorial, but were obtained by running four experiments instead of eight. This illustrates the cost savings that are possible using a suitable fractional factorial. A formal analysis of variance (not given here) for the Model $(Y|A, B, C)$ uses the formulas $SSA = l_A^2/2^2 n$ and so on to compute the sums of squares for the main effects. Each of these has one degree of freedom while SST has $2^2 n - 1$ degrees of freedom and $SSE = SST - SSA - SSB - SSC$ has $2^2 n - 4$ degrees of freedom.

Confounding in a 2^k Factorial Experiment

Confounding occurs when it is impossible to run all 2^k factor-level combinations under the same experimental conditions. Consider a 2^2 agricultural experiment to study the effects of nitrogen (factor A) and phosphorus (factor B) on crop yield. Suppose we have two separate plots of land, each one of which is large enough to run only two factor-level combinations. Each plot of land is a *block* and the experimenter must decide which two treatments are assigned to each block. This is an example of a factorial experiment in *incomplete blocks*, since not all factor-level combinations can be run in each block. This complicates the analysis somewhat since the experimenter must now include an additional parameter in the model to account for a possible block effect. Table 13.27 shows one possible way of allocating the factor-level combinations to each of the two blocks.

Table 13.27 *Format for a 2^2 factorial experiment partitioned into two incomplete blocks*

Block 1	Block 2
1	a
ab	b

To estimate the block effect we compute the difference between the mean response of block 1 to that of block 2; that is we compute the following contrast:

$$\frac{(1) + (ab)}{2n} - \frac{(a) + (b)}{2n} =$$

$$= \frac{1}{2n}[(1) + (ab) - (a) - (b)]$$

$$= \frac{L_{AB}}{2n}.$$

You will recognize that the contrast used to estimate the block effect is the same as the contrast used to estimate the AB interaction! In other words, it is not possible to separate the AB interaction effect from the block effect. We say that the AB interaction is *completely confounded with blocks*. The quantities $(1) + (ab)$ and $(a) + (b)$ are called *block totals*.

Similarly, confounding occurs in a 2^3 factorial experiment when it is impossible to run the experiment for all eight factor-level combinations under the same experimental conditions. For instance, suppose each experiment requires two hours, which means your research staff can only run four experiments per shift (assuming your staff works an eight hour shift). This is yet another example of a factorial experiment in incomplete blocks. In this case each shift plays the role of a block.

The method for assigning treatments to two blocks is similar to that used for selecting the treatments in a half replicate 2^{k-1} design. In detail, choose a contrast and assign the treatment combinations with a plus sign to one block and assign the remaining treatments to the other block. Table 13.28 shows the block assignments corresponding to the contrast L_{ABC}.

Table 13.28 *Format for a 2^3 factorial experiment divided into two incomplete blocks*

Block 1	Block 2
1	a
ab	b
ac	c
bc	abc

The contrast for estimating the block effect is the difference between the block totals which equals

$$(a) + (b) + (c) + (abc) - [(1) + (ab) + (ac) + (bc)] = L_{ABC}.$$

So the block effects are confounded with the ABC interaction and it is therefore not possible to separate the interaction from the block effect. Call the block containing the treatment combination 1 the *principal block*. It is worth noting that the treatments in the other block can be obtained by choosing any treatment combination not in the principal block and multiplying it by the treatment combinations in the principal block. For instance the treatment combinations in block 2 of Table 13.28 are (a, b, c, abc); they can also be obtained by choosing a, which is not in the principal block, and multiplying it according the rule that $a^2 = b^2 = c^2 = 1$; thus,

$$a \times 1 \equiv a;$$
$$a \times ab \equiv a^2 b \equiv b;$$
$$a \times ac \equiv a^2 c \equiv c;$$
$$a \times bc \equiv abc.$$

Constructing the ANOVA Table for a 2^3 Factorial Experiment Divided into Two blocks

The formal analysis of variance is the same as in the non-confounded case. That is, you compute the sums of squares exactly as before and calculate the sum of squares due to the blocks (denoted SSBl), by adding the sum of squares of all effects confounded with the blocks, which in this case yields $SSBl = SSABC$.

Example 13.19

The data in Table 13.29 came from a 2^3 factorial experiment divided into two blocks. Compute the ANOVA table and determine which effects are significant at the 5% level. In addition compute the proportion of variability accounted for by each main effect.

Table 13.29 *Data from a 2^3 factorial experiment divided into two blocks with three replicates per treatment combination*

Block 1	Block 2
1: 7, 17, 8	a: 29, 20, 30
ab: 69, 66, 66	b: 43, 53, 51
ac: 28, 26, 32	c: 11, 8, 9
bc: 51, 54, 53	abc: 72, 77, 80

Solution We used SAS (PROC GLM) to compute the ANOVA Table 13.30 in the usual way and then edited the output to take into account the fact that the ABC interaction is confounded with the blocks.

Table 13.30 *ANOVA table for a 2^3 factorial experiment divided into two blocks; data set: Table 13.29*

Source	DF	Type I SS	Mean Square	F Value	Pr > F
A	1	2204.16667	2204.16667	144.54	0.0001
B	1	10837.50000	10837.50000	710.66	0.0001
A*B	1	16.66667	16.66667	1.09	0.3114
C	1	73.50000	73.50000	4.82	0.0432
A*C	1	32.66667	32.66667	2.14	0.1627
B*C	1	54.00000	54.00000	3.54	0.0782
Block	1	1.50000	1.50000	0.10	0.7579
Error	16	244.00000	15.25000		
Total	23	13464.00000			

540 Introduction to Probability and Statistics for Science, Engineering, and Finance

Conclusions Only the main effects are significant at the 5% level, and all two-factor interactions are not. The factor B is clearly the most important factor since it accounts for $80\% = (10837/13464) \times 100\%$ of the total variability. The factor C, even though it is judged statistically significant, accounts for only 0.55% of the total variability.

13.5 Chapter Summary

The validity of a scientific investigation depends in a critical way on carefully designed and analyzed experiments. This means identifying and reducing sources of extraneous variation so that observed differences will be mostly due to the factor-level combinations in which the researcher is interested. Excessive variation in the production process is a major cause of poor quality output. Identifying and eliminating the source of excessive variation is an essential first step toward reducing the costs of scrap work, repair, and warranties. In this chapter we have presented some of the basic statistical concepts underlying the design and analysis of experiments depending on two or more factors. The basic philosophy is to construct a statistical model for the data and to check its validity by performing an analysis of variance. A comprehensive treatment of this important topic is simply not possible within the confines of an introductory text so we have contented ourselves with the simplest examples: randomized complete block designs, two-way ANOVA, and 2^k experiments.

13.6 Problems

Section 13.2

Problem 13.1 *With reference to the fruit fly mortality data (Table 13.1) perform a one-way ANOVA for the Model (Y = mortality rate |A = dosage level). That is, drop the blocking variable B (genetic strain) from the model and analyze the data using a one-way ANOVA model to determine whether or not the mortality rates of the fruit flies are affected by the dosage levels of the insecticide. Display your results in the format of a single factor ANOVA table (Table 12.3). Compare your conclusions concerning the existence of a treatment effect with the results obtained in the text using the RCBD model. Discuss in particular the effects of blocking.*

Problem 13.2 *In the table below the counts of the numbers of insects (*Leptinotarsa dec-imlineata*) were recorded after four treatments were applied to six areas of a field.*

	Area					
Treatment	1	2	3	4	5	6
1	492	410	475	895	401	330
2	111	67	233	218	28	18
3	58	267	283	278	392	141
4	4	1	53	14	138	11

(Source: G. Beall (1942), Biometrika, vol. 32, pp. 243-262; quoted in Tukey (1977), Exploratory Data Analysis, Addison–Wesley, Reading, MA.)

(a) Estimate the overall mean, the treatment means, and the area means.
(b) Compare the four treatments and test for differences among them by computing the ANOVA table. Use $\alpha = 0.05$.
(c) Check the model assumptions by plotting the residuals.

Problem 13.3 *The table below lists the final weight (in grams) of 18 hamsters fed three different diets varying in fat content. The hamsters were blocked by weight in groups of three; that is, the hamsters in each block had the same weights at the start of the experiment.*
(a) Estimate the overall mean, the treatment means, and the block means.
(b) Compare the three diets and test for differences among them by computing the ANOVA table. Use $\alpha = 0.05$.
(c) Check the model assumptions by plotting the residuals.

			Block			
Treatment	1	2	3	4	5	6
1	96	96	94	99	99	102
2	103	101	103	105	101	107
3	103	104	106	108	109	110

(Source: J. L. Gill (1978), Design and Analysis of Experiments in the Animal and Medical Sciences, *vol. 2, Iowa State University Press, Ames, IA. Used with permission.)*

Problem 13.4 *Show that the residuals satisfy both conditions displayed in Equation 13.5.*

Problem 13.5 *The data below come from a comparison study of the execution times (measured in milliseconds) of five different work loads on three different configurations of a processor. The goal was to study the effect of processor configuration on computer performance.*
(a) Identify the experimental units, the response variable, and the two factors.
(b) Calculate the ANOVA table and test whether there are differences among the three configurations. Use $\alpha = 0.05$.
(c) Which treatment means are significantly different from one another? Use Tukey's multiple comparison method with $\alpha = 0.05$.
(d) Check the model assumptions by graphing suitable plots of the residuals.

	Configurations		
Workload	Two Cache	One Cache	No Cache
ASM	54.0	55.0	106.0
TECO	60.0	60.0	123.0
SIEVE	43.0	43.0	120.0
DRHYSTONE	49.0	52.0	111.0
SORT	49.0	50.0	108.0

(Source: R. Jain (1991), The Art of Computer Systems Performance Analysis, *John Wiley & Sons, New York. Used with permission.)*

Problem 13.6 *Two types of fertilizer are being compared against a control (no fertilizer). The response variable is the yield (tons) in sugar beets. The results are displayed in the next table.*

(a) Calculate the ANOVA table and test whether there are differences among the treatments. Use $\alpha = 0.05$.
(b) Check the model assumptions.

	Fertilizer Applied		
	None	PO_4	PO_4 NO_3
Area			
1	12.45	6.71	6.48
2	2.25	5.44	7.11
3	4.38	4.92	5.88
4	4.35	5.23	7.54
5	3.42	6.74	6.61
6	3.27	4.74	8.86

(Source: G. W. Snedecor (1946), Statistical Methods, *University of Iowa Press, Ames, IA. Used with permission. Quoted in Tukey (1977), p. 371.)*

Problem 13.7 *The data below come from a larger experiment to determine the effectiveness of blast furnace slags (A) as agricultural liming materials on three types of soil (B): sandy loam (1), sandy clay loam (2), and loamy sand (3). The treatments were all applied at 4000 lbs per acre and the response variable was the corn yield in bushels per acre.*
(a) Calculate the ANOVA table and use your results to test for differences among the treatment means. Use $\alpha = 0.05$.
(b) Plot the residuals against the fitted values. Do you see any non-random patterns?
(c) Calculate the normal probability plot of the residuals and test the residuals for normality using the Shapiro–Wilk test.
(d) Plot $(i, y_{ij})(i = 1, \ldots, 7)$ for each soil type; follow the pattern used to obtain Figure 13.2. Does the graph support or refute the assumption that there are no treatment–block interactions?

	Soil (B)		
Furnace slags (A)	1	2	3
None	11.1	32.6	63.3
Coarse slag	15.3	40.8	65.0
Medium slag	22.7	52.1	58.8
Agricultural slag	23.8	52.8	61.4
Agricultural limestone	25.6	63.1	41.1
Agricultural slag + minor elements	31.2	59.5	78.1
Agricultural limestone + minor elements	25.8	55.3	60.2

(Source: D.E. Johnson and F.A. Graybill (1972), Journal of the American Statistical Association, *vol. 67, pp. 862-868. Used with permission.)*

Problem 13.8 *The data below come from an experiment to compare four methods for manufacturing penicillin. Note that the blocks are displayed as rows and the treatments are displayed as columns. The raw material consisted of $b = 5$ blends consisting of enough material so that all $a = 4$ methods could be tested on each blend. It was known that the variability between blends was not negligible. The response variable is the yield (the units* *were not given).*

(a) Calculate the ANOVA table and determine whether there are differences among the four methods for manufacturing penicillin.

(b) Check the model assumptions by graphing suitable residual plots.

	Methods (A)			
Blends (B)	1	2	3	4
1	89	88	97	94
2	84	77	92	79
3	81	87	87	85
4	87	92	89	84
5	79	81	80	88

(Source: Box, Hunter, and Hunter(1978), Statistics for Experimenters, p. 209, John Wiley & Sons, New York. Used with permission.)

Section 13.3

Problem 13.9 *Refer to the data of Table 13.5. The residuals plot 13.7 indicates that the hypothesis of equal variances is invalid. To stabilize the variance we use the transformation*

$$u = \sqrt{1 + y}.$$

The transformed data set is given in the next table.

(a) Compute the two-way ANOVA table for the transformed data and test for the existence of interactions and main effects.

(b) Plot the residuals against the fitted values and compare this plot with the one obtained in the text using the original data set (Figure 13.7). Do you detect any improvement?

	$u = \sqrt{1+y}$ of Table 13.5					
	Pesticide (B)					
Fungicide (A)	None		Dieldrin		Diazinon	
None	4.00	5.48	4.80	5.74	5.57	5.10
	4.36	5.57	5.00	2.83	5.39	4.69
	5.10	4.69	1.73	4.00	4.24	4.47
	5.20	5.00	3.16	4.36	4.12	5.57
Captan	6.16	3.61	1.41	3.74	3.16	2.83
	6.24	2.45	1.41	3.32	1.00	1.41
	5.00	4.69	2.65	4.36	1.00	2.24
	3.32	2.65	4.58	5.10	2.24	3.16

Problem 13.10 *Refer to Problem 13.9.*

(a) Compute Tukey's HSD_A, HSD_B. Use $\alpha = 0.05$.

(b) Using Tukey's HSD determine all pairs of means that are significantly different.

Problem 13.11 *Show that $\overline{Y}_{i..}$ and $\overline{Y}_{.j.}$ are unbiased estimators of the factor A means $\overline{\mu}_{i.}$ $(i = 1, \ldots, a)$ and the factor B means $\overline{\mu}_{.j}$ $(j = 1, \ldots, b)$, respectively.*

Problem 13.12
(a) Show that $\overline{Y}_{ij\cdot}$ is an unbiased estimator of the (i,j)th cell mean μ_{ij} $(i = 1,\ldots,a)$; $(j = 1,\ldots,b)$.
(b) Compute $Var(\overline{Y}_{ij\cdot})$.

Problem 13.13 *Verify that interaction effects $(\alpha\beta)_{ij}$ satisfy Equation 13.18.*

Problem 13.14 *Verify Equation 13.23.*

Problem 13.15 *The data in the next table are the results of an experiment to investigate the sources of variability in testing the strength of Portland cement. A sample of cement was divided into small samples for testing. The cement was "gauged" (mixed with water) by three different men called gaugers, and then it was cast into cubes. Three men, called breakers, later tested the cubes for compressive strength. The measurements are in pounds per square inch. Each gauger gauged 12 cubes, which were then divided into three sets of four, and each breaker tested one set of four cubes from each gauger. All the tests were carried out on the same machine. The purpose of the experiment was to identify the source of and measure the variability among gaugers and breakers.*
(a) Compute the estimated cell means and display your results in the format of Table 13.9.
(b) Plot the estimated cell means of the breakers (factor B) against the the three levels of gaugers (factor A) (see Figure 13.6 for an example). Does the graph suggest the existence of an interaction?
(c) Test for the existence of main effects and interactions by constructing the ANOVA table.
(d) Check the model assumptions by graphing suitable residual plots.

	Breaker (B)					
Gauger (A)	Breaker 1		Breaker 2		Breaker 3	
Gauger 1	5280	5520	4340	4400	4160	5180
	4760	5800	5020	6200	5320	4600
Gauger 2	4420	5280	5340	4880	4180	4800
	5580	4900	4960	6200	4600	4480
Gauger 3	5360	6160	5720	4760	4460	4930
	5680	5500	5620	5560	4680	5600

(Source: O.L. Davies and P.L. Goldsmith (eds.) (1972), Statistical Methods in Research and Production, 4th ed., Edinburgh, Oliver and Boyd, p. 154; quoted in D.J. Hand (1994), p. 24.)

Problem 13.16 *The data in the table below record the service time (in minutes) required to service disk drives from three different manufacturers by three different technicians. Each technician was randomly assigned to repair five disk drives for each brand of disk drive.*
(a) Compute the estimated cell means and display your results in the format of Table 13.9.
(b) Plot the estimated cell means $(i, \overline{y}_{ij\cdot})(i = 1,\ldots,3)$ for each brand in the format of Figure 13.8. Does the graph support or refute the assumption that there are no interactions between the technicians and the brand of the disk drive?
(c) Test for the existence of interactions and main effects by computing the two-way ANOVA table.

(d) Compute a 95% confidence interval for the time to service each brand of drive.
(e) Compute a 95% confidence interval for the time for each technician to service a drive.

Technician (A)	Brand of Drive (B)		
	Brand 1	Brand 2	Brand 3
Technician 1	62	57	59
	48	45	53
	63	39	67
	57	54	66
	69	44	47
Technician 2	51	61	55
	57	58	58
	45	70	50
	50	66	69
	39	51	49
Technician 3	59	58	47
	65	63	56
	55	70	51
	52	53	44
	70	60	50

(*Source: Neter, Wasserman, and Kutner (1990),* Applied Linear Statistical Models, *p. 723, R. D. Irwin, Homewood, IL. Used with permission.*)

Problem 13.17 *The data in the table below come from an experiment to determine the amount of warping (the response variable) in a metal plate as a function of the following two factors: (A) temperature (measured in degrees Celsius), and (B) copper content (measured as a percentage).*
(a) *Compute the estimated cell means and display your results in the format of Table 13.9.*
(b) *Plot the estimated cell means* $(i, \bar{y}_{ij.})(i = 1, \ldots, 4)$ *for each concentration of copper in the format of Figure 13.8. Does the graph support or refute the assumption that there are no interactions between the temperature and copper content of the metal plate?*
(c) *Test for the existence of interactions and main effects by computing the two-way ANOVA table.*

Temperature (°C) (A)	Copper Content (%) (B)							
	40		60		80		100	
50	17	20	16	21	24	22	28	27
75	12	9	18	13	17	12	27	31
100	16	12	18	21	25	23	30	23
125	21	17	23	21	23	22	29	31

(*Source: N. L. Johnson and F. C. Leone (1964),* Statistics and Experimental Design, *vol. II, p. 98, John Wiley Press, Inc. Used with permission.*)

Problem 13.18 *Refer to Problem 13.17.*
(a) *Compute Tukey's* HSD_A, HSD_B. *Use* $\alpha = 0.05$.
(b) *Using Tukey's HSD determine all pairs of means that are significantly different.*

Section 13.4

Problem 13.19 *Refer to the Model (Y2 | X1, X8, X1*X8) of Example 13.16.*
(a) Estimate the main effects and interactions by first computing the contrasts L_{X1}, L_{X8},
L_{X1X8}.
(b) Compute SSX1, SSX8, and SSX1X8. Your answers should agree with those in Table
13.16. Hint: Use Equations 13.33, 13.34, and 13.35.
(c) How would you set the parameter value X8 (0 or 1) so as to minimize the process
variation? Justify your answer. Hint: Compute the average value of Y2 at the levels
X8 = 0 and X8 = 1.

Problem 13.20 *Refer to the Model (Y1 | X1, X8, X1*X8) of Example 13.16.*
(a) Estimate the main effects and interactions by first computing the contrasts L_{X1}, L_{X8},
L_{X1X8}.
(b) Compute SSX1, SSX8, and SSX1X8. Your answers should agree with those in Table
13.18. Hint: Use Equations 13.33, 13.34, and 13.35.
(c) Which variable has the largest effect on Y1 (the epitaxial thickness)?

Problem 13.21 *Refer to Example 13.16. An exploratory analysis of the data indicated*
that the effect of the factor X4 (deposition time) on Y1 was significant, but its effect on Y2
was not. This suggests that we study both Model (Y1 | X4, X8, X4*X8) and Model (Y2 |
X4, X8, X4*X8).
(a) Rearrange the data for Model (Y1 | X4, X8, X4*X8) in the standard order; use the for-
mat of Table 13.17.
(b) Perform the analysis of variance and test for the existence of main effects and interac-
tions. Use $\alpha = 0.05$.
(c) Consider the Model (Y2 | X4, X8, X4*X8). Rearrange the data for this model in the
standard order; use the format of Table 13.15.
(d) Perform the analysis of variance and test for the existence of main effects and interac-
tions. Use $\alpha = 0.05$.
(e) Combining the results from parts (b) and (d) is it reasonable to conclude that the effect
of deposition time on epitaxial thickness is significant and that its effect on Y2 (process
variance) is not? Justify your answer.

Problem 13.22 *The following data came from a complete 2^3 factorial experiment with*
three replicates for each treatment combination.

Treatment Combination	Replicates
1	48, 37, 58
a	94, 90, 107
b	94, 91, 93
ab	160, 156, 154
c	70, 75, 68
ac	117, 114, 110
bc	123, 105, 126
abc	187, 170, 182

(a) Prepare a table similar in format to Table 13.23 listing all the contrasts and estimates
of the main effects and interactions.

(b) Using the results obtained in part (a) compute the sums of squares for each main and interaction effect.

(c) Given the information that $SST = 39811.96$ and $SSE = 863.33$ compute the proportion of the total variability explained by each main effect and interaction. Rank the effects in decreasing order of importance as measured by $SS(effect)/SST$.

(d) Using the information obtained in parts (b) and (c) compute the ANOVA table in a format similar to Table 13.24. Are your final results consistent with the preliminary conclusions you obtained in part (c)? Comment.

Problem 13.23 *Analyze the data of the preceding problem as a half replicate with defining relation $I = ABC$. That is, compute the fractional replicates l_A, l_B, l_C and the corresponding sum of squares; then compute SST and SSE for this half replicate. Determine the proportion of variability that is explained by each main effect and comment on whether or not your results are consistent with those obtained from the complete factorial.*

Problem 13.24 *The data in the following table come from an experiment to study the effects of temperature (factor A), processing time (factor B), and the rate of temperature rise (factor C) on the amount dye (the response variable) left in the residue bath of a dyeing process. The experiment was run at two levels for each factor, with two replicates at each factor-level combination.*

Treatment Combination	Replicates
1	19.9, 18.6
a	25.0, 22.8
b	17.4, 16.8
ab	19.5, 18.3
c	14.5, 16.1
ac	27.7, 18.0
bc	16.3, 14.6
abc	28.3, 26.2

(Source: Milton and Arnold (1990), Introduction to Probability and Statistics, *2nd. ed., McGraw-Hill, New York. Used with permission.)*

(a) Prepare a table similar in format to Table 13.23 listing all the contrasts and estimates of the main effects and interactions.

(b) Using the results obtained in part (a) compute the sums of squares for each main and interaction effect.

(c) Given the information that $SST = 313.88$ and $SSE = 56.14$ compute the proportion of the total variability explained by each main effect and interaction. Rank the effects in decreasing order of importance as measured by $SS(effect)/SST$.

(d) Using the information obtained in parts (b) and (c) compute the ANOVA table in a format similar to Table 13.24. Are your final results consistent with the preliminary conclusions you obtained in part (c)? Comment.

Problem 13.25 *Analyze the data of the preceding problem as a half replicate with defining relation $I = ABC$. That is, compute the fractional replicates l_A, l_B, l_C and the corresponding sum of squares; then compute SST and SSE for this half replicate. Determine the proportion of variability that is explained by each main effect and comment on whether or not your results are consistent with those obtained from the complete factorial.*

Problem 13.26 *The data in the following table come from a complete 2^3 factorial experiment to study the effect of cutting speed (A), tool geometry (B), and cutting angle (C) on the life of a machine tool. Two levels of each factor were chosen and there were three replicates for each factor-level combination.*

Treatment Combination	Replicates
1	22, 31, 25
a	32, 43, 29
b	35, 34, 50
ab	55, 47, 46
c	44, 45, 38
ac	40, 37, 36
bc	60, 50, 54
abc	39, 41, 47

(Source: D. C. Montgomery (1984), Design and Analysis of Experiments, *2nd ed. p. 292, John Wiley Press, Inc. Used with permission.)*
(a) Given that $SSE = 482.67$, $SST = 2095.33$ analyze the data by computing the ANOVA table and determine which main effects and interactions are significant at the level $\alpha = 0.05$.
(b) Using the results in part (a) compute the proportion of the total variability explained by each main effect and interaction. Rank the effects in decreasing order of importance as measured by $SS(effect)/SST$.

Problem 13.27 *Consider a 2^4 agricultural experiment with four factors A, B, C, D. The experiment is to be carried out on two plots of land each of which is large enough to run only eight factor-level combinations. The scientists choose to confound the ABCD interaction with the block effect. Prepare a table, similar to Table 13.28, listing the factor-level combinations for each block. Hint: First prepare the table of contrast coefficients similar to Table 13.20.*

Problem 13.28 *In this problem we outline a derivation of the formula 13.33 for the sum of squares SSA. The proof is easily modified to derive the formulas 13.34 and 13.35 for SSB and SSAB.*
(a) Show that

$$\hat{\alpha}_1 + \hat{\alpha}_2 = (\overline{y}_{1..} - \overline{y}_{...}) + (\overline{y}_{2..} - \overline{y}_{...}) = 0.$$

(b) Use the result in part (a) to show that $SSA = 4n\hat{\alpha}_2^2$.
(c) Show that $\hat{\alpha}_2 = L_A/4n$ and therefore $SSA = L_A^2/4n$ as claimed. Hint: These formulas are applications of the equations for the sums of squares displayed in the text just after Equation 13.24. The sums simplify when one uses the fact that $1 \leq i \leq 2$; $1 \leq j \leq 2$.

13.7   To Probe Further

Students interested in further applications and generalizations should consult the following references:

1. G.E.P. Box, W.G. Hunter, and J.S. Hunter (1978), *Statistics for Experimenters*, John Wiley & Sons, Inc, New York.

2. R.L. Mason, R.F. Gunst, and J.L. Hess (1989), *Statistical Design and Analysis of Experiments*, John Wiley & Sons, Inc., New York.

3. D.C. Montgomery (1984), *Design and Analysis of Experiments*, 2nd ed., John Wiley & Sons, Inc., New York.

Chapter 14

Statistical Quality Control

The aim in production should be not just to get statistical control, but to shrink variation. Costs go down as variation is reduced. It is not enough to meet specifications.

W.E. Deming (1982), *Out of the Crisis*, MIT Press, p. 334

14.1   Orientation

The fluctuations in the quality of a manufactured product, such as integrated circuits, steel bars, and cotton fabric, are a result of the variations in the raw materials, machines, and workers used to manufacture them. We measure quality by recording the value of a numerical variable X defined on the population of manufactured objects. Examples of such variables include the thickness of an epitaxial layer deposited on a silicon wafer, the compressive strength of concrete, the tensile strength of a steel cable, etc. Sometimes the variable is categorical as when we classify each item inspected as conforming or non-conforming. In all cases the product is required to meet a standard usually stated in the form of lower and upper control limits on the variable X.

The primary goal of a quality control program is to save money by reducing or eliminating the production of defective products since, if uncorrected, this can only result in declining sales, loss of market share, layoffs, and bankruptcy. The following example is cited in Kazmierski (1995), (*Statistical Problem Solving in Quality Engineering*, McGraw-Hill, New York):

A small company had $10 million in sales over a given period. During that period, warranty costs were $1,550,000. Scrap and rework were $150,000. The inspection budget was $250,000, and the prevention budget was $50,000. The total cost of poor quality was $2 million (20% of sales). Reducing the cost of poor quality by 50% would increase profit by $ 1 million.

Among the most important techniques for detecting variation in the quality of a manufactured product is the *control chart*. We will study several types of control charts depending on whether the variable being measured is numerical or categorical. The $\bar{x}$ chart monitors the fluctuations about the target value μ and the R chart monitors the fluctuations in the process variation. Two other charts are the p chart, which monitors the proportion of inspected items rejected as non-conforming, and the c chart, which monitors the number defects in a non-conforming item.

Organization of Chapter

1. Section 14.2: $\bar{x}$ and R Control Charts

2. Section 14.2.1: Detecting a Shift in the Process Mean

3. Section 14.3: p Charts and c Charts

4. Section 14.4: Chapter Summary

14.2 $\bar{x}$ and R Control Charts

The concept of the control chart rests on the assumption that the manufacturing process is in *statistical control.* Informally, this means that the variability in product quality comes from a statistical model of random variation. The formal mathematical model of a process in statistical control follows.

Definition 14.1 *The manufacturing process is said to be in statistical control when the measurements $X_1, X_2, \ldots, X_n$ taken on a sequence of successively drawn objects behave as if they come from a random sample of size n taken from a common distribution F. We then say that the pattern of variation is* stable.

Quality control engineers emphasize the importance of first partitioning the variation in the manufacturing process into two parts as indicated in Equation 14.1:

$$\begin{pmatrix} \text{Total process} \\ \text{variation} \end{pmatrix} = \begin{pmatrix} \text{common cause} \\ \text{variation} \end{pmatrix} + \begin{pmatrix} \text{special cause} \\ \text{variation} \end{pmatrix}. \tag{14.1}$$

Deming's red bead experiment (described in Example 1.3) provides an entertaining example of a *common cause of variation,* which refers to the variations due to random fluctuations within the system itself, as opposed to a *special cause variation,* which refers to variations caused by external, possibly non-random events such as computer software bugs, defective raw materials, misaligned gauges, workers not familiar with the equipment, and so forth. From the engineering point of view the variation due to special causes can be reduced or eliminated, for example, by using only fully trained workers, properly maintained equipment and high quality raw materials. Historically, it was Shewhart, Deming, and their disciples who were among the first to point out that the type of action required to improve the quality of a manufactured product depends first on identifying the causes of the observed variation. "Until special causes have been identified and eliminated," wrote Deming, "one dare not predict what the process will produce the next hour." The common cause variation is what remains after eliminating the variation due to special causes.

$\bar{x}$ and R Control Charts for a Process in Statistical Control with Known Mean and Known Variance

Assume that the manufacturing process is in statistical control and that the k samples of size n come from a normal distribution with known mean μ and known variance σ^2. Consequently, the sample mean $\overline{X}$ has a normal distribution with mean μ and standard deviation $\sigma(\overline{X}) = \sigma/\sqrt{n}$. It follows that

$$P(\mu - 3\sigma/\sqrt{n} \le \overline{X} \le \mu + 3\sigma/\sqrt{n}) = 0.9974.$$

It is, therefore, very unlikely for a sample average to fall outside the *lower control limit* and *upper control limit*, defined by

$$LCL_{\overline{X}} = \mu - \frac{3\sigma}{\sqrt{n}} \quad \text{and} \quad UCL_{\overline{X}} = \mu + \frac{3\sigma}{\sqrt{n}}. \tag{14.2}$$

It is worth pointing out that the sample average will be approximately normal even if the parent distribution from which the sample is drawn is not normal.

1. **Three-sigma control chart for averages ($\bar{x}$ chart)** Take k successive samples of size n and compute their sample means $\bar{x}_1, \ldots, \bar{x}_k$. The $\bar{x}$ chart is obtained by first plotting the horizontal *center line (CL)* at the target value μ, and the horizontal lower and upper control lines at the values $LCL_{\overline{X}}$, $UCL_{\overline{X}}$ defined in Equation 14.2. We then plot the points $(i, \bar{x}_i)$, $(i = 1, \ldots, k)$. Points outside the control limits suggest that this is due to a *special cause variation* (malfunctioning machinery, a change in personnel) and not due to the system itself. We will illustrate the method in Example 14.1. The corresponding $\bar{x}$ chart is the top graph in Figure 14.1.

2. **Control chart for the sample ranges (R chart)** One of the main causes of poor quality is excessive variation in the manufacturing process. The R control chart is a graphical method for monitoring the process variation. We graph an R chart as follows: Take k successive samples of size n and for each sample compute the sample range defined by

$$R = \max(X_1, \ldots, X_n) - \min(X_1, \ldots, X_n).$$

The ith sample range is denoted by R_i. It can be shown (we omit the proof) that the mean and standard deviation of the sample range R of a sequence of n independent and identically distributed normal random variables is proportional to σ; thus,

$$E(R) = d_2(n)\sigma, \tag{14.3}$$
$$\sigma(R) = d_3(n)\sigma, \tag{14.4}$$

where the constants of proportionality depend only on the sample size n. These constants are listed in Table A.8. In the tables these constants are denoted d_2, d_3 and the explicit dependence of these constants on n is understood. Consequently, when σ is known, the center line (CL) of the R chart is defined to be the horizontal line at the value $\overline{R}$, defined by

$$\overline{R} = d_2\sigma, \ (\sigma \text{ known}). \tag{14.5}$$

The horizontal lower and upper control lines at the values LCL_R, UCL_R are defined by

$$LCL_R = D_1\sigma \text{ and } UCL_R = D_2\sigma, \ (\sigma \text{ known}), \tag{14.6}$$

where the values of D_1, D_2 are listed in Table A.10. Note that $D_1 = 0$ for $n \leq 6$. The final step is to plot the points (i, R_i), $(i = 1, \ldots, k)$. We illustrate the method in Example 14.1. The corresponding R chart is the bottom graph in Figure 14.1.

Example 14.1 *Computing lower and upper control limits for $\bar{x}$ and R charts*

At regular time intervals a sample of five parts is drawn from a manufacturing process and the width (measured in millimeters) of a critical gap in each part is recorded in Table 14.1. The ith row lists the sample number, the five measurements of the gap widths, the sample mean, and sample range. Here, $n = 5$ and $k = 21$. When the process is in control the distribution of the gap width is normal with mean $\mu = 200$ and $\sigma = 10$. Compute the lower and upper control limits for both the $\bar{x}$ and R charts, the sample means, and sample ranges. Then plot the three-sigma $\bar{x}$ chart and the R chart.

Solution We compute the lower and upper control limits by substituting $\mu = 200$, $\sigma = 10$, $n = 5$ into Equation 14.2; consequently,

$$LCL_{\overline{X}} = 200 - \frac{3 \times 10}{\sqrt{5}} = 186.6 \text{ and } UCL_{\overline{X}} = 200 + \frac{3 \times 10}{\sqrt{5}} = 213.4.$$

The last two columns of Table 14.1 record the ith sample mean $\bar{x}_i$ and the ith sample range R_i. Let us now compute the sample average and range for the first sample of five measurements (the final results are rounded to one decimal place). We then compute $\bar{R}$ and the lower and upper control limits for the R chart.

$$\bar{x}_1 = \frac{212 + 206 + 184 + 198 + 185}{5} = 197.0,$$
$$R_1 = 212 - 184 = 28,$$
$$\bar{R} = d_2\sigma = 2.326 \times 10 = 23.3,$$
$$D_1 = 0 \text{ and } D_2 = 4.92, \text{ so}$$
$$LCL_R = 0 \text{ and } UCL_R = D_2\sigma = 4.92 \times 10 = 49.2.$$

Figure 14.1 displays the $\bar{x}$ and R charts. The central line in an $\bar{x}$ chart is always denoted by $\bar{\bar{x}}$ even when μ is known. Looking at these charts we see that all the plotted points lie within the control limits, so we conclude that the process is in control. However, a process that is in statistical control is not necessarily satisfactory from the standpoint of quality control. Suppose, for example, that the product specifications are that the gap width be 200 ± 10. Call 190 and 210 *tolerance limits*. A total of 27 parts are outside the tolerance limits in Table 14.1. That is, 25.7% of the parts are non-conforming, even though none of the sample averages fell outside the control limits! In this case the control chart is telling management that the excessive variation is in the system and will not be improved by firing workers, the "get tough" policy.

Table 14.1 *Gap widths, sample means, and sample ranges for Example 14.1*

Sample	Gap width data					$\bar{x}$	R
1	212	206	184	198	185	197.0	28
2	196	195	199	209	207	201.2	14
3	190	203	201	186	210	198.0	24
4	204	194	208	203	207	203.2	14
5	217	196	206	204	207	206.0	21
6	194	211	219	199	204	205.4	25
7	197	200	192	200	197	197.2	8
8	206	208	184	188	195	196.2	24
9	200	196	209	207	195	201.4	14
10	207	207	202	209	194	203.8	15
11	209	181	194	206	204	198.8	28
12	200	200	192	199	192	196.6	8
13	174	192	212	211	184	194.6	38
14	180	190	194	203	216	196.6	36
15	201	212	184	196	210	200.6	28
16	212	177	195	218	203	201.0	41
17	202	196	200	201	203	200.4	7
18	219	218	206	202	204	209.8	17
19	202	208	196	200	208	202.8	12
20	207	200	175	186	196	192.8	32
21	211	207	205	214	210	209.4	9

We now show how to construct control charts for the process mean and process variation when the parameters μ and σ^2 are unknown. In this case we replace the previous control limits with their estimates determined from the recorded data. In detail we proceed in the

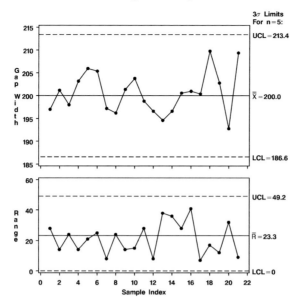

FIGURE 14.1: $\overline{x}$ and R control charts for a process in statistical control (Table 14.1)

following way: Assume that the manufacturing process is in statistical control and that the k samples of size n come from a normal distribution with unknown mean μ and unknown variance σ^2. We estimate μ by computing the grand mean defined by

$$\overline{\overline{x}} = \frac{\sum_{1 \leq j \leq k} \overline{x}_j}{k}. \tag{14.7}$$

Take k successive samples of size n and denote the ith sample range by R_i.

The estimated central line for the R chart, which is also denoted by $\overline{R}$, is the average of the k sample ranges and is given by

$$\overline{R} = \frac{\sum_{1 \leq i \leq k} R_i}{k}. \tag{14.8}$$

The average $\overline{R}$ of the sample ranges together with Equations 14.3 and 14.4 can be used to derive the following unbiased estimates of σ and the standard deviation of the sample range, denoted $\sigma(R)$:

$$\hat{\sigma} = \frac{\overline{R}}{d_2} \text{ and } \hat{\sigma}(R) = \frac{d_3 \overline{R}}{d_2}. \tag{14.9}$$

It follows that $\frac{3\overline{R}}{d_2 \sqrt{n}}$ is an unbiased estimator of $\frac{3\sigma}{\sqrt{n}}$. Consequently, the estimated lower and upper control limits are given by

$$LCL_{\overline{X}} = \overline{\overline{x}} - A_2 \overline{R} \text{ and } UCL_{\overline{X}} = \overline{\overline{x}} + A_2 \overline{R}, \left(A_2 = \frac{3}{d_2 \sqrt{n}} \right). \tag{14.10}$$

The values A_2 are given in Table A.9.

Similarly, the lower and upper control limits for the R chart are given by

$$LCL_R = \overline{R} - 3\hat{\sigma}(R) = \overline{R}\left(1 - \frac{3d_3}{d_2}\right) = D_3\overline{R}, \qquad (14.11)$$

$$UCL_R = \overline{R} + 3\hat{\sigma}(R) = \overline{R}\left(1 + \frac{3d_3}{d_2}\right) = D_4\overline{R}, \qquad (14.12)$$

where the constants D_3, D_4 are given in Table A.9 For $n \leq 6$ we set $D_3 = 0$ because in these cases $(1 - 3d_3/d_2) < 0$.

1. **Three-sigma control chart for averages ($\overline{x}$ chart)** Take k successive samples of size n and compute their sample means $\overline{x}_1, \ldots, \overline{x}_k$. The $\overline{x}$ chart is obtained by first plotting the horizontal *center line (CL)* at the value $\overline{\overline{x}}$, and the horizontal lower and upper control lines at the values $LCL_{\overline{X}}, UCL_{\overline{X}}$ defined in Equation 14.10. We then plot the points $(i, \overline{x}_i)$, $(i = 1, \ldots, k)$. We will illustrate the method in Example 14.2.

2. **Control chart for the ranges (R chart)** Take k successive samples of size n and for each sample compute the sample range. The R chart is obtained by first plotting the horizontal *center line (CL)* at the value $\overline{R}$, and the horizontal lower and upper control lines at the values LCL_R, UCL_R defined by Equations 14.11 and 14.12. We will illustrate the method in Example 14.2.

Example 14.2 *Control limits, three-sigma $\overline{x}$, and R charts*

Every hour a sample of five fittings to be used for an aircraft hydraulic system is drawn from from the production line and the pitch diameter of the threads was measured. The pitch diameter was specified as 0.4037 ± 0.0013. The data were coded by subtracting 0.4000 from each observation and then expressed in units of 0.0001 inch. Thus, the observation 0.4036 was recorded as 36. The measurements are displayed in Table 14.2. The ith row lists the sample number, the five measurements of the pitch diameter, the sample mean, and sample range. Here, $n = 5$ and $k = 20$. We now drop the assumption that the parameters μ and σ^2 are known. Compute the lower and upper control limits for both the $\overline{x}$ and R charts. Then plot the three-sigma $\overline{x}$ chart and the R chart.

   Solution We compute the lower and upper control limits for the $\overline{x}$ chart by substituting $\overline{\overline{x}} = 33.55, \overline{R} = 6.2, A_2 = 0.58$ into Equation 14.10. Consequently,

$$LCL_{\overline{X}} = 33.55 - 0.58 \times 6.2 = 29.95; \ UCL_{\overline{X}} = 33.55 + 0.58 \times 6.2 = 37.15.$$

These results differ from those obtained by the computer only in the second decimal place and are due to round off errors.

   The lower and upper control limits for the R chart are obtained by substituting $\overline{R} = 6.2, D_3 = 0, D_4 = 2.11$ into Equations 14.11 and 14.12. This yields the result (rounded to one decimal)

$$LCL_R = D_3\overline{R} = 0; \ UCL_R = D_4\overline{R} = 2.11 \times 6.2 = 13.1.$$

   Looking at the $\overline{x}$ chart in Figure 14.2 we see that three samples (10, 12, and 18) are outside the control limits. There are also two samples (9 and 13) outside the control limits of the R chart.

Table 14.2 *Sample number, pitch diameter of threads on aircraft fittings (5 per hour), sample mean, and sample range*

Sample	Pitch diameters					Average	Range
1	36	35	34	33	32	34.0	4
2	31	31	34	32	30	31.6	4
3	30	30	32	30	32	30.8	2
4	32	33	33	32	35	33.0	3
5	32	34	37	37	35	35.0	5
6	32	32	31	33	33	32.2	2
7	33	33	36	32	31	33.0	5
8	23	33	36	35	36	32.6	13
9	43	36	35	24	31	33.8	19
10	36	35	36	41	41	37.8	6
11	34	38	35	34	38	35.8	4
12	36	38	39	39	40	38.4	4
13	36	40	35	26	33	34.0	14
14	36	35	37	34	33	35.0	4
15	30	37	33	34	35	33.8	7
16	28	31	33	33	33	31.6	5
17	33	30	34	33	35	33.0	5
18	27	28	29	27	30	28.2	3
19	35	36	29	27	32	31.8	9
20	33	35	35	39	36	35.6	6

(Source: E.L. Grant and R.S. Leavenworth (1996), Statistical Quality Control, *7th ed., McGraw-Hill. Used with permission.)*

14.2.1 Detecting a Shift in the Process Mean

A control chart is not infallible. It can fail to detect quickly an out of control process, or it can send, incorrectly, an out of control signal. The similarity with hypothesis testing should be noted. The null hypothesis is that the process is in a state of statistical control. A failure to detect an out of control process is a type I error, and an incorrect out of control signal is a type II error.

We begin our analysis of the performance of the $\bar{x}$ chart by computing the expected number of samples until an incorrect out of control signal is indicated. Under the null hypothesis (that the process is in statistical control) the sample means $\overline{X}_1, \ldots, \overline{X}_k$ are a random sample taken from a normal distribution with mean μ_0 and variance $\sigma/\sqrt{n}$. Let T denote the first time a sample mean falls outside the control limits. This is another example of a waiting time problem of the sort discussed in Chapter 3. Consequently, the random variable T has the geometric distribution

$$P(T = i) = q^{i-1}p, (i = 1, 2, \ldots),$$

where p (the probability that a single sample falls outside the control limits) is given by

$$p = P\left(|\overline{X} - \mu_0| > \frac{3\sigma}{\sqrt{n}}\right) = 0.0026 \text{ and } q = 1 - p.$$

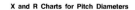

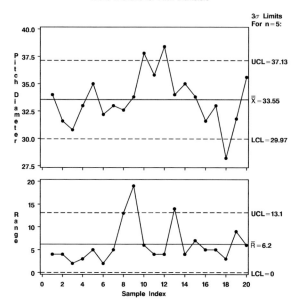

FIGURE 14.2: $\overline{x}$ and R charts for pitch diameters (Example 14.2)

Since T has a geometric distribution it follows that

$$P(T \le i) = 1 - (1-p)^i; \tag{14.13}$$

$$E(T) = \frac{1}{p}. \tag{14.14}$$

Equation 14.13 gives the probability that an incorrect out of control signal occurs at the ith sample or earlier; Equation 14.14 gives the expected number samples required. Since $p = 0.0026$ we see that $E(T) = 384.62 \approx 385$. In other words the average number of samples before an incorrect out of control signal occurs is approximately 385.

Example 14.3 *Probability of detecting an out of control process*

Suppose a process assumed to be in control with process mean μ_0 and variance σ is thrown out of control because the process mean has shifted to $\mu_0 + \delta\sigma$, where δ is a constant (positive or negative). What is the probability that the three-sigma control chart will detect this shift no later than the ith sample? What is the expected number of samples required in order to detect the shift?

Solution   Under the hypothesis that the process mean has shifted to $\mu_0 + \delta\sigma$ the sample means $\overline{X}_1, \dots, \overline{X}_k$ are a random sample taken from a normal distribution with mean $\mu_0 + \delta\sigma$ and variance $\sigma/\sqrt{n}$. Let T denote the first time a sample mean falls outside the control limits. Consequently, the distribution of T is again geometric where p (the probability that a single sample falls outside the control limits) is given by

$$p = P\left(|\overline{X} - \mu_0| > \frac{3\sigma}{\sqrt{n}} | \mu = \mu_0 + \delta\sigma\right)$$
$$= \Phi(-3 - \delta\sqrt{n}) + (1 - \Phi(3 - \delta\sqrt{n})). \tag{14.15}$$

The derivation of Equation 14.15 is left to the reader as Problem 14.3.

To illustrate the formula, suppose the process mean of Example 14.1 is shifted to 207; the process standard deviation, however, remains fixed at $\sigma = 10$. What is the probability the shift is detected no later than the third sample? What is the expected number of steps until the shift is detected? The shift will be detected when a sample mean lies outside the the control limits. Denoting the number of steps required to detect this shift by T we see that it too has a geometric distribution with probability p given by Equation 14.15. A shift in the process mean from 200 to 207 means that $200 + \delta\sigma = 207$. Consequently, $\delta = 0.7$, since $\sigma = 10$. Inserting the values $\delta = 0.7$, $n = 5$ into Equation 14.15 we obtain the value $p = 0.0764$. Therefore,

$$P(T \le 3) = 1 - (1 - 0.0764)^3 = 0.2121;$$
$$E(T) = \frac{1}{0.0764} = 13.09 \approx 13.$$

In other words, it will take 13 samples, on average, for the control chart to detect this shift in the process mean. Even after three samples have been taken the probability is still approximately 0.79 that the shift in the process mean remains undetected.

14.3 p Charts and c Charts

Another important measure of quality is the proportion of defective (non-conforming) items produced. At regular time intervals $i = 1, \ldots, k$ the quality control inspector takes a sample of size n and records the number of defective items for the ith sample, denoted X_i, and also the proportion of defective items $\hat{p}_i = X_i/n$. The sample proportion is the total number of defective items divided by the total number of items sampled; it is denoted $\bar{p}$, where

$$\bar{p} = \frac{\sum_{1 \le i \le k} X_i}{kn}.$$

When the process is in statistical control the distribution of X_i is binomial with parameters n, p, and, when n is large ($n > 30$ say), the ith sample proportion has an approximate normal distribution with mean p and variance $p(1 - p)/n$. Consequently,

$$P\left(p - 3\sqrt{\frac{p(1-p)}{n}} \le \hat{p}_i \le p + 3\sqrt{\frac{p(1-p)}{n}}\right) \approx 0.9974.$$

When the true proportion of defective items is unknown we replace it with its estimate $\bar{p}$; it follows that

$$P\left(\bar{p} - 3\sqrt{\frac{\bar{p}(1-\bar{p})}{n}} \le \hat{p}_i \le \bar{p} + 3\sqrt{\frac{\bar{p}(1-\bar{p})}{n}}\right) \approx 0.9974.$$

Reasoning as we did before in the case of the $\bar{x}$ chart we see that it is very unlikely for a sample proportion to fall outside the *lower control limit* and *upper control limit*, defined by

$$LCL_{\bar{p}} = \bar{p} - 3\sqrt{\frac{\bar{p}(1-\bar{p})}{n}} \text{ and } UCL_{\bar{p}} = \bar{p} + 3\sqrt{\frac{\bar{p}(1-\bar{p})}{n}}. \tag{14.16}$$

We call $LCL_{\bar{p}}$, $UCL_{\bar{p}}$ the three-sigma control limits. The p chart monitors the proportion of defective items by drawing the center line at $\bar{p}$ and the lower and upper control limits

$LCL_{\overline{p}}$, $UCL_{\overline{p}}$ defined by Equation 14.16. (Note: It sometimes happens that the lower control limit is negative; when this occurs we set it equal to zero.) We then plot the points $(i, \hat{p}_i)$, $i = 1, \ldots, k$. A lack of control is indicated when one or more sample proportions lie outside the control limits.

Example 14.4 *Deming's red bead experiment*

This example is a simulation of Deming's red bead experiment (Example 1.3) described in Chapter 1. We recall the details. Suppose 4000 beads are placed in a bowl. Of these, 3800 are white and 800 are red. The red beads represent the defective items. Each worker stirs the beads and then, while blindfolded, inserts a special tool into the mixture with which he draws out exactly 50 beads. The aim of this process is to produce white beads; red beads, as previously noted, are defective and will not be accepted by the customers. When he is done the sampled beads are replaced, the beads are thoroughly stirred for the next worker who repeats the process. The results for 6 workers over a 5 day period are shown in Table 14.3. Compute $\overline{p}$, $LCL_{\overline{p}}$, $UCL_{\overline{p}}$, graph the p chart and determine if the process is in control.

   Solution   The total number of items inspected equals $kn = 30 \times 50 = 1500$, and the total number of defective items is 307. So $\overline{p} = 307/1500 = 0.204$ (rounded to three decimal places). Therefore the lower and upper control limits are

$$LCL_{\overline{p}} = 0.204 - 0.171 = 0.033;$$
$$UCL_{\overline{p}} = 0.204 + 0.171 = 0.375.$$

The p chart is shown in Figure 14.3. No points lie outside the control limits so we conclude that the process is in control. Notice that worker 1 "produced" the fewest number of defectives on day 1, but produced the largest number of defectives on day 2. What the control chart tells us is that the excessive proportion of defectives is due to the system and not due to the workers.

Table 14.3

Day	Worker 1	Worker 2	Worker 3	Worker 4	Worker 5	Worker 6
1	3	9	10	11	6	10
2	15	9	11	9	10	10
3	15	13	6	10	17	10
4	6	12	9	15	8	9
5	6	14	12	9	12	11

The c chart for the number of defects per unit
   The c chart is used when it is the number of defects per item that is of interest rather than the proportion of defective items. Consider, for example, an aircraft fuselage assembled with hundreds of rivets. One defective rivet is highly unlikely to compromise aircraft safety, but 10 or more might. The variable of interest here is C the number of defective rivets per fuselage. A c chart is a control chart for the number of defects per item. The control limits are based on the assumption that the the number of defects per item is a Poisson random variable with parameter λ. This assumption is a reasonable one for an airplane fuselage consisting of hundreds of rivets each of which has a small probability of being defective.

   Control limits for the c control charts   Let C_i, $i = 1, \ldots, k$ denote the number of defects found on the ith item. We assume that C_i has a Poisson distribution with parameter

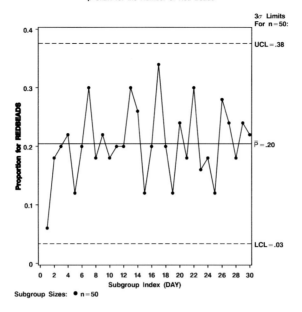

p Chart for the Number of Red Beads

3σ Limits For n=50:

UCL = .38

$\bar{P}$ = .20

LCL = .03

Proportion for REDBEADS

Subgroup Index (DAY)

Subgroup Sizes: ● n=50

FIGURE 14.3: *p* chart for the number of read beads (Example 14.4)

λ. Therefore, it has mean value λ and standard deviation $\sqrt{\lambda}$. The theoretical three-sigma control limits are

$$LCL = \lambda - 3\sqrt{\lambda} \text{ and } UCL = \lambda + 3\sqrt{\lambda}.$$

When λ is unknown we estimate it by dividing the total number of defects by the total number of items sampled; we denote this quantity by $\bar{c}$. Thus,

$$\bar{c} = \frac{\sum_{1 \leq i \leq k} C_i}{k}.$$

The estimated control limits are given by

$$LCL_{\bar{c}} = \bar{c} - 3\sqrt{\bar{c}} \text{ and } UCL_{\bar{c}} = \bar{c} + 3\sqrt{\bar{c}}. \tag{14.17}$$

Example 14.5 *Control limits for a c chart*

The following table is a record of the number of defective rivets (missing, misaligned, etc.) on 14 aircraft fuselages.

$$2, 9, 10, 11, 6, 10, 16, 7, 5, 5, 6, 8, 6, 5$$

Compute $\bar{c}$, the control limits, and plot the *c* chart.

Solution The total number of defects is 106, so $\bar{c} = 106/14 = 7.57$. Therefore, the lower and upper control limits are

$$LCL_{\bar{c}} = 7.57 - 8.25 = -0.68 \text{ and } UCL_{\bar{c}} = 7.57 + 8.25 = 15.82.$$

Since $\bar{c} - 3\sqrt{\bar{c}} < 0$ we set $LCL_{\bar{c}} = 0$. The *c* chart is displayed in Figure 14.4. There is one point (airplane 7 with $C_7 = 16$) outside the control limits, so the system is not in control.

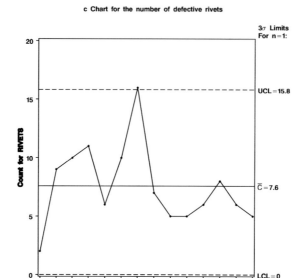

FIGURE 14.4:   *c* chart for number of defective rivets (Example 14.5)

14.4   Chapter Summary

In this chapter we have given a brief introduction to statistical quality control. We pointed out that no improvements in product quality are possible until the manufacturing process is in a state of statistical control and this means that one must first identify and eliminate variation due to special causes. Control charts are used to detect a lack of control. We noted a similarity between hypothesis testing and statistical control. The null hypothesis is that the process is in a state of statistical control. A failure to detect an out of control process is a type I error, and an incorrect out of control signal is a type II error.

14.5   Problems

Problem 14.1 *The thickness of an epitaxial layer on a silicon wafer is specified to be* 14.5 ± 0.5 *micrometers. Assume the manufacturing process is in statistical control where the distribution of the thickness is normal with mean* $\mu = 14.5$ *and standard deviation* $\sigma = 0.4$. *To monitor the production process the mean thickness of four wafers is determined at regularly spaced time intervals.*
(a) Compute the lower and upper control limits $LCL_{\overline{X}}$, $UCL_{\overline{X}}$ *for a three-sigma control chart for the process mean.*
(b) Compute the lower and upper control limits LCL_R, UCL_R *for a three-sigma control chart for the process variation.*
(c) Determine the probability that a sample mean (based on a sample of size $n = 4$*) falls*

outside the tolerance limits 14.5 ± 0.5.

Problem 14.2 *Suppose the process mean in Problem 14.1 shifts to* $\mu = 15.0$.
(a) What is the probability that this shift will be detected on the first sample after the shift has occurred?
(b) What is the probability that this shift will be detected by the third sample?
(c) What is the expected number of samples required to detect this shift?

Problem 14.3 *Derive the formula for p (the probability that a single sample falls outside the control limits) stated in Equation 14.15.*

Problem 14.4 *The diameters (measured in millimeters) of ball bearings are monitored by* $\bar{x}$ *and R charts. After 30 samples of size* $n = 6$ *were taken the following values were recorded:* $\sum_i \bar{x}_i = 150mm$ *and* $\sum_i \bar{R}_i = 12.0$.
(a) Assuming the process is in statistical control calculate the $\bar{x}$ *and R control limits.*
(b) Assuming that the ball bearing diameters are normally distributed estimate σ.

Problem 14.5 *A dimension of a rheostat knob was specified to be* 0.140 ± 0.003 *inches. Otherwise the knob would not fit into another part. Five knobs from each each hour's production were inspected and the dimension was recorded (in units of* 0.001 *inches) in the following table. The sample means and sample ranges are recorded in the last two columns.*
(a) Draw the histogram of the raw data using the class marks:
132, 134, 136, 138, 140, 142, 144, 146, 148. How many rheostats are non-conforming?
(b) Is the process in statistical control? Draw the $\bar{x}$ *and R charts to find out.*

Sample	Dimension of Rheostat Part (0.001 in.)					$\bar{x}$	R
1	140	143	137	134	135	137.8	9
2	138	143	143	145	146	143.0	8
3	139	133	147	148	139	141.2	15
4	143	141	137	138	140	139.8	6
5	142	142	145	135	136	140.0	10
6	136	144	143	136	137	139.2	8
7	142	147	137	142	138	141.2	10
8	143	137	145	137	138	140.0	8
9	141	142	147	140	140	142.0	7
10	142	137	145	140	132	139.2	13
11	137	147	142	137	135	139.6	12
12	137	146	142	142	140	141.4	9
13	142	142	139	141	142	141.2	3
14	137	145	144	137	140	140.6	8
15	144	142	143	135	144	141.6	9
16	140	132	144	145	141	140.4	13
17	137	137	142	143	141	140.0	6
18	137	142	142	145	143	141.8	8
19	142	142	143	140	135	140.4	8
20	136	142	140	139	137	138.8	6
21	142	144	140	138	143	141.4	6
22	139	146	143	140	139	141.4	7
23	140	145	142	139	137	140.6	8
24	134	147	143	141	142	141.4	13
25	138	145	141	137	141	140.4	8

| 26 | 140 | 145 | 143 | 144 | 138 | 142.0 | 7 |
| 27 | 145 | 145 | 137 | 138 | 140 | 141.0 | 8 |

(Source: E.L. Grant and R.S. Leavenworth (1996), Statistical Quality Control, *7th ed., McGraw-Hill. Used with permission.)*

Problem 14.6 *Items are shipped in cartons containing several hundreds of them. From each carton 100 are selected at random and inspected for defects. In a shipment of 12 cartons the number of defective items found in the sample for each carton were recorded as follows:*

$$0, 1, 1, 2, 0, 2, 4, 1, 2, 1, 7, 5.$$

Calculate the three-sigma control limits for the p chart. Is the process in statistical control? If not, list the values that are out of control.

Problem 14.7 *The proportion of defective items produced by a process in control is known to be $p = 0.02$. The process is monitored by taking samples of size 50 and determining the proportion of defective items in the sample.*
(a) Compute the lower and upper control limits for the proportion defective.
(b) Suppose the proportion of defective items produced increases to 0.04. Let T denote the first sample for which the proportion defective lies above the upper control limit obtained in part (a). Describe the distribution of T. Name it and identify its parameters. Compute $E(T)$.

Problem 14.8 *A c chart is used to monitor the number of typos per page in a printed document. Based on past experience the number of typos is assumed to have a Poisson distribution with $\bar{c} = 1.5$.*
(a) Assuming the process is in statistical control find the three-sigma control limits.
(b) Use Table A.2 (Poisson distribution) to determine the probability that a point on the control chart will fall outside the control limits.
(c) Assuming the process is in statistical control determine the expected number of pages one has to sample before an incorrect out of control signal is given by the c chart.

Problem 14.9 *The number of defects in a bolt of cloth is to be monitored by means of a c chart. Defects include flaws in the weave and irregularities in the dye. Before actually implementing the control chart a trial run using simulated data was carried out. The first 6 observations in the data set below come from a Poisson distribution with parameter $\lambda_1 = 6$, the next 8 observations come from a Poisson distribution with parameter $\lambda_2 = 9$, and the final 6 observations come from a Poisson distribution with parameter $\lambda_3 = 6.5$. Consequently, the simulated data do not come from a process in statistical control. Calculate the three-sigma control limits using Equation 14.2 and plot the c control chart. Does the control chart detect a lack of control?*

$$1, 5, 6, 7, 3, 6, 11, 9, 8, 10$$
$$12, 10, 9, 8, 4, 6, 5, 6, 4, 7$$

Problem 14.10 *An alternative method for constructing lower and upper control limits $LCL_{\bar{c}}, UCL_{\bar{c}}$ for a c chart is to choose them so that the inequalities*

$$P(C < LCL_{\bar{c}}) < \frac{\alpha}{2} \text{ and } P(C > UCL_{\bar{c}}) < \frac{\alpha}{2}$$

are satisfied. It is assumed that C has a Poisson distribution. When $LCL_{\bar{c}} = 0$ then an out of control signal occurs only when $C > UCL_{\bar{c}}$.
(a) Find the control limits when $\lambda = 4$ and $\alpha = 0.01$.
(b) Find the control limits when $\lambda = 6$ and $\alpha = 0.01$.
(c) Find the control limits when $\lambda = 6$ and $\alpha = 0.05$.

14.6 To Probe Further

The common practice in manufacturing, according to Grant and Leavenworth (see the reference at the end of this paragraph), is to set the control chart limits far enough apart so that type I errors are rare on the grounds that "it seldom pays to hunt for trouble without a strong basis for confidence that trouble is really there." On the other hand, the Shewhart control chart, because it only looks at one sample at a time, may not detect quickly enough a lack of control caused by a shift in the process mean. There are several methods for reducing the type II error probability (failure to detect an out of control process). The *cumulative sum control chart* is a useful method for coping with this type of problem. These and other refinements and extensions are discussed in E. L. Grant and R. S. Leavenworth (1996), *Statistical Quality Control*, 7th ed., McGraw–Hill, New York, which is the classic text on this subject. Part Four contains an extensive account of the economic aspects, historical origins, current trends, and future directions of the quality assurance movement in the United States and Japan.

Appendix A

Tables

- Table A.1 Cumulative Binomial Distribution

- Table A.2 Cumulative Poisson Distribution

- Table A.3 Standard Normal Probabilities

- Table A.4 Critical Values $t_\nu(\alpha)$ of the t Distribution

- Table A.5 Quantiles $Q_\nu(p) = \chi_\nu^2(1 - p)$ of the χ^2 Distribution

- Table A.6 Critical Values of the $F_{\nu_1,\nu_2}(\alpha)$ Distribution

- Table A.7 Critical Values of the Studentized Range $q(\alpha; n, \nu)$

- Table A.8 Factors for Estimating σ, $\overline{s}$, or $\overline{\sigma}_{RMS}$ and σ_R from $\overline{R}$

- Table A.9 Factors for Determining from $\overline{R}$ the Three-Sigma Control Limits for $\overline{X}$ and R Charts

- Table A.10 Factors for Determining from σ the Three-Sigma Control Limits for $\overline{X}$, R, and s or $\overline{\sigma}_{RMS}$ Charts

A.1   Cumulative Binomial Distribution

$$B(x; n, p) = \sum_{0 \leq y \leq x} b(y; n, p)$$

n	x	0.05	0.10	0.15	0.2	0.25	0.3	0.35	0.4	0.45	0.5
5	0	0.774	0.590	0.444	0.328	0.237	0.168	0.116	0.078	0.050	0.031
	1	0.977	0.919	0.835	0.737	0.633	0.528	0.428	0.337	0.256	0.187
	2	0.999	0.991	0.973	0.942	0.896	0.837	0.765	0.683	0.593	0.500
	3	1.000	1.000	0.998	0.993	0.984	0.969	0.946	0.913	0.869	0.812
	4	1.000	1.000	1.000	1.000	0.999	0.998	0.995	0.990	0.982	0.969
	5	1.000	1.000	1.000	1.000	1.000	1.000	1.000	1.000	1.000	1.000
10	0	0.599	0.349	0.197	0.107	0.056	0.028	0.013	0.006	0.003	0.001
	1	0.914	0.736	0.544	0.376	0.244	0.149	0.086	0.046	0.023	0.011
	2	0.988	0.930	0.820	0.678	0.526	0.383	0.262	0.167	0.100	0.055
	3	0.999	0.987	0.950	0.879	0.776	0.650	0.514	0.382	0.266	0.172
	4	1.000	0.998	0.99	0.967	0.922	0.850	0.751	0.633	0.504	0.377
	5	1.000	1.000	0.999	0.994	0.98	0.953	0.905	0.834	0.738	0.623
	6	1.000	1.000	1.000	0.999	0.996	0.989	0.974	0.945	0.898	0.828
	7	1.000	1.000	1.000	1.000	1.000	0.998	0.995	0.988	0.973	0.945
	8	1.000	1.000	1.000	1.000	1.000	1.000	0.999	0.998	0.995	0.989
	9	1.000	1.000	1.000	1.000	1.000	1.000	1.000	1.000	1.000	0.999
	10	1.000	1.000	1.000	1.000	1.000	1.000	1.000	1.000	1.000	1.000
15	0	0.463	0.206	0.087	0.035	0.013	0.005	0.002	0.000	0.000	0.000
	1	0.829	0.549	0.319	0.167	0.080	0.035	0.014	0.005	0.002	0.000
	2	0.964	0.816	0.604	0.398	0.236	0.127	0.062	0.027	0.011	0.004
	3	0.995	0.944	0.823	0.648	0.461	0.297	0.173	0.091	0.042	0.018
	4	0.999	0.987	0.938	0.836	0.686	0.515	0.352	0.217	0.120	0.059
	5	1.000	0.998	0.983	0.939	0.852	0.722	0.564	0.403	0.261	0.151
	6	1.000	1.000	0.996	0.982	0.943	0.869	0.755	0.610	0.452	0.304
	7	1.000	1.000	0.999	0.996	0.983	0.950	0.887	0.787	0.654	0.500
	8	1.000	1.000	1.000	0.999	0.996	0.985	0.958	0.905	0.818	0.696
	9	1.000	1.000	1.000	1.000	0.999	0.996	0.988	0.966	0.923	0.849
	10	1.000	1.000	1.000	1.000	1.000	0.999	0.997	0.991	0.975	0.941
	11	1.000	1.000	1.000	1.000	1.000	1.000	1.000	0.998	0.994	0.982
	12	1.000	1.000	1.000	1.000	1.000	1.000	1.000	1.000	0.999	0.996
	13	1.000	1.000	1.000	1.000	1.000	1.000	1.000	1.000	1.000	1.000
	14	1.000	1.000	1.000	1.000	1.000	1.000	1.000	1.000	1.000	1.000
	15	1.000	1.000	1.000	1.000	1.000	1.000	1.000	1.000	1.000	1.000

Table A.1 (continued)

n	x	0.05	0.10	0.15	0.2	0.25	0.3	0.35	0.4	0.45	0.5
20	0	0.358	0.122	0.039	0.012	0.003	0.001	0.000	0.000	0.000	0.000
	1	0.736	0.392	0.176	0.069	0.024	0.008	0.002	0.001	0.000	0.000
	2	0.925	0.677	0.405	0.206	0.091	0.035	0.012	0.004	0.001	0.000
	3	0.984	0.867	0.648	0.411	0.225	0.107	0.044	0.016	0.005	0.001
	4	0.997	0.957	0.830	0.630	0.415	0.238	0.118	0.051	0.019	0.006
	5	1.000	0.989	0.933	0.804	0.617	0.416	0.245	0.126	0.055	0.021
	6	1.000	0.998	0.978	0.913	0.786	0.608	0.417	0.250	0.130	0.058
	7	1.000	1.000	0.994	0.968	0.898	0.772	0.601	0.416	0.252	0.132
	8	1.000	1.000	0.999	0.990	0.959	0.887	0.762	0.596	0.414	0.252
	9	1.000	1.000	1.000	0.997	0.986	0.952	0.878	0.755	0.591	0.412
	10	1.000	1.000	1.000	0.999	0.996	0.983	0.947	0.872	0.751	0.588
	11	1.000	1.000	1.000	1.000	0.999	0.995	0.980	0.943	0.869	0.748
	12	1.000	1.000	1.000	1.000	1.000	0.999	0.994	0.979	0.942	0.868
	13	1.000	1.000	1.000	1.000	1.000	1.000	0.998	0.994	0.979	0.942
	14	1.000	1.000	1.000	1.000	1.000	1.000	1.000	0.998	0.994	0.979
	15	1.000	1.000	1.000	1.000	1.000	1.000	1.000	1.000	0.998	0.994
	16	1.000	1.000	1.000	1.000	1.000	1.000	1.000	1.000	1.000	0.999
	17	1.000	1.000	1.000	1.000	1.000	1.000	1.000	1.000	1.000	1.000
	18	1.000	1.000	1.000	1.000	1.000	1.000	1.000	1.000	1.000	1.000
	19	1.000	1.000	1.000	1.000	1.000	1.000	1.000	1.000	1.000	1.000
	20	1.000	1.000	1.000	1.000	1.000	1.000	1.000	1.000	1.000	1.000
25	0	0.277	0.072	0.017	0.004	0.001	0.000	0.000	0.000	0.000	0.000
	1	0.642	0.271	0.093	0.027	0.007	0.002	0.000	0.000	0.000	0.000
	2	0.873	0.537	0.254	0.098	0.032	0.009	0.002	0.000	0.000	0.000
	3	0.966	0.764	0.471	0.234	0.096	0.033	0.010	0.002	0.000	0.000
	4	0.993	0.902	0.682	0.421	0.214	0.090	0.032	0.009	0.002	0.000
	5	0.999	0.967	0.838	0.617	0.378	0.193	0.083	0.029	0.009	0.002
	6	1.000	0.991	0.930	0.780	0.561	0.341	0.173	0.074	0.026	0.007
	7	1.000	0.998	0.975	0.891	0.727	0.512	0.306	0.154	0.064	0.022
	8	1.000	1.000	0.992	0.953	0.851	0.677	0.467	0.274	0.134	0.054
	9	1.000	1.000	0.998	0.983	0.929	0.811	0.630	0.425	0.242	0.115
	10	1.000	1.000	1.000	0.994	0.970	0.902	0.771	0.586	0.384	0.212
	11	1.000	1.000	1.000	0.998	0.989	0.956	0.875	0.732	0.543	0.345
	12	1.000	1.000	1.000	1.000	0.997	0.983	0.940	0.846	0.694	0.500
	13	1.000	1.000	1.000	1.000	0.999	0.994	0.975	0.922	0.817	0.655
	14	1.000	1.000	1.000	1.000	1.000	0.998	0.991	0.966	0.904	0.788
	15	1.000	1.000	1.000	1.000	1.000	1.000	0.997	0.987	0.956	0.885
	16	1.000	1.000	1.000	1.000	1.000	1.000	0.999	0.996	0.983	0.946
	17	1.000	1.000	1.000	1.000	1.000	1.000	1.000	0.999	0.994	0.978
	18	1.000	1.000	1.000	1.000	1.000	1.000	1.000	1.000	0.998	0.993
	19	1.000	1.000	1.000	1.000	1.000	1.000	1.000	1.000	1.000	0.998
	20	1.000	1.000	1.000	1.000	1.000	1.000	1.000	1.000	1.000	1.000
	21	1.000	1.000	1.000	1.000	1.000	1.000	1.000	1.000	1.000	1.000
	22	1.000	1.000	1.000	1.000	1.000	1.000	1.000	1.000	1.000	1.000
	23	1.000	1.000	1.000	1.000	1.000	1.000	1.000	1.000	1.000	1.000
	24	1.000	1.000	1.000	1.000	1.000	1.000	1.000	1.000	1.000	1.000
	25	1.000	1.000	1.000	1.000	1.000	1.000	1.000	1.000	1.000	1.000

A.2   Cumulative Poisson Distribution

$$P(x; \lambda) = \sum_{0 \leq y \leq x} p(y; \lambda)$$

	λ								
x	0.1	0.2	0.3	0.4	0.5	0.6	0.7	0.8	0.9
0	0.905	0.819	0.741	0.67	0.607	0.549	0.497	0.449	0.407
1	0.995	0.982	0.963	0.938	0.910	0.878	0.844	0.809	0.772
2	1.000	0.999	0.996	0.992	0.986	0.977	0.966	0.953	0.937
3	1.000	1.000	1.000	0.999	0.998	0.997	0.994	0.991	0.987
4	1.000	1.000	1.000	1.000	1.000	1.000	0.999	0.999	0.998
5	1.000	1.000	1.000	1.000	1.000	1.000	1.000	1.000	1.000
6	1.000	1.000	1.000	1.000	1.000	1.000	1.000	1.000	1.000

	λ								
x	1	1.5	2	2.5	3	3.5	4	4.5	5
0	0.368	0.223	0.135	0.082	0.050	0.030	0.018	0.011	0.007
1	0.736	0.558	0.406	0.287	0.199	0.136	0.092	0.061	0.040
2	0.920	0.809	0.677	0.544	0.423	0.321	0.238	0.174	0.125
3	0.981	0.934	0.857	0.758	0.647	0.537	0.433	0.342	0.265
4	0.996	0.981	0.947	0.891	0.815	0.725	0.629	0.532	0.440
5	0.999	0.996	0.983	0.958	0.916	0.858	0.785	0.703	0.616
6	1.000	0.999	0.995	0.986	0.966	0.935	0.889	0.831	0.762
7	1.000	1.000	0.999	0.996	0.988	0.973	0.949	0.913	0.867
8	1.000	1.000	1.000	0.999	0.996	0.990	0.979	0.960	0.932
9	1.000	1.000	1.000	1.000	0.999	0.997	0.992	0.983	0.968
10	1.000	1.000	1.000	1.000	1.000	0.999	0.997	0.993	0.986
11	1.000	1.000	1.000	1.000	1.000	1.000	0.999	0.998	0.995
12	1.000	1.000	1.000	1.000	1.000	1.000	1.000	0.999	0.998
13	1.000	1.000	1.000	1.000	1.000	1.000	1.000	1.000	0.999
14	1.000	1.000	1.000	1.000	1.000	1.000	1.000	1.000	1.000

Table A.2 (continued)

x	\multicolumn{9}{c}{λ}								
	6	7	8	9	10	11	12	13	14
0	0.002	0.001	0.000	0.000	0.000	0.000	0.000	0.000	0.000
1	0.017	0.007	0.003	0.001	0.000	0.000	0.000	0.000	0.000
2	0.062	0.030	0.014	0.006	0.003	0.001	0.001	0.000	0.000
3	0.151	0.082	0.042	0.021	0.010	0.005	0.002	0.001	0.000
4	0.285	0.173	0.100	0.055	0.029	0.015	0.008	0.004	0.002
5	0.446	0.301	0.191	0.116	0.067	0.038	0.020	0.011	0.006
6	0.606	0.450	0.313	0.207	0.130	0.079	0.046	0.026	0.014
7	0.744	0.599	0.453	0.324	0.220	0.143	0.090	0.054	0.032
8	0.847	0.729	0.593	0.456	0.333	0.232	0.155	0.100	0.062
9	0.916	0.830	0.717	0.587	0.458	0.341	0.242	0.166	0.109
10	0.957	0.901	0.816	0.706	0.583	0.460	0.347	0.252	0.176
11	0.980	0.947	0.888	0.803	0.697	0.579	0.462	0.353	0.260
12	0.991	0.973	0.936	0.876	0.792	0.689	0.576	0.463	0.358
13	0.996	0.987	0.966	0.926	0.864	0.781	0.682	0.573	0.464
14	0.999	0.994	0.983	0.959	0.917	0.854	0.772	0.675	0.570
15	0.999	0.998	0.992	0.978	0.951	0.907	0.844	0.764	0.669
16	1.000	0.999	0.996	0.989	0.973	0.944	0.899	0.835	0.756
17	1.000	1.000	0.998	0.995	0.986	0.968	0.937	0.890	0.827
18	1.000	1.000	0.999	0.998	0.993	0.982	0.963	0.930	0.883
19	1.000	1.000	1.000	0.999	0.997	0.991	0.979	0.957	0.923
20	1.000	1.000	1.000	1.000	0.998	0.995	0.988	0.975	0.952
21	1.000	1.000	1.000	1.000	0.999	0.998	0.994	0.986	0.971
22	1.000	1.000	1.000	1.000	1.000	0.999	0.997	0.992	0.983
23	1.000	1.000	1.000	1.000	1.000	1.000	0.999	0.996	0.991
24	1.000	1.000	1.000	1.000	1.000	1.000	0.999	0.998	0.995
25	1.000	1.000	1.000	1.000	1.000	1.000	1.000	0.999	0.997
26	1.000	1.000	1.000	1.000	1.000	1.000	1.000	1.000	0.999
27	1.000	1.000	1.000	1.000	1.000	1.000	1.000	1.000	0.999
28	1.000	1.000	1.000	1.000	1.000	1.000	1.000	1.000	1.000

A.3   Standard Normal Probabilities

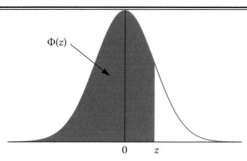

z	0	0.01	0.02	0.03	0.04	0.05	0.06	0.07	0.08	0.09
-3.4	0.0003	0.0003	0.0003	0.0003	0.0003	0.0003	0.0003	0.0003	0.0003	0.0002
-3.3	0.0005	0.0005	0.0005	0.0004	0.0004	0.0004	0.0004	0.0004	0.0004	0.0003
-3.2	0.0007	0.0007	0.0006	0.0006	0.0006	0.0006	0.0006	0.0005	0.0005	0.0005
-3.1	0.0010	0.0009	0.0009	0.0009	0.0008	0.0008	0.0008	0.0008	0.0007	0.0007
-3.0	0.0013	0.0013	0.0013	0.0012	0.0012	0.0011	0.0011	0.0011	0.0010	0.0010
-2.9	0.0019	0.0018	0.0018	0.0017	0.0016	0.0016	0.0015	0.0015	0.0014	0.0014
-2.8	0.0026	0.0025	0.0024	0.0023	0.0023	0.0022	0.0021	0.0021	0.0020	0.0019
-2.7	0.0035	0.0034	0.0033	0.0032	0.0031	0.0030	0.0029	0.0028	0.0027	0.0026
-2.6	0.0047	0.0045	0.0044	0.0043	0.0041	0.0040	0.0039	0.0038	0.0037	0.0036
-2.5	0.0062	0.0060	0.0059	0.0057	0.0055	0.0054	0.0052	0.0051	0.0049	0.0048
-2.4	0.0082	0.0080	0.0078	0.0075	0.0073	0.0071	0.0069	0.0068	0.0066	0.0064
-2.3	0.0107	0.0104	0.0102	0.0099	0.0096	0.0094	0.0091	0.0089	0.0087	0.0084
-2.2	0.0139	0.0136	0.0132	0.0129	0.0125	0.0122	0.0119	0.0116	0.0113	0.0110
-2.1	0.0179	0.0174	0.0170	0.0166	0.0162	0.0158	0.0154	0.0150	0.0146	0.0143
-2.0	0.0228	0.0222	0.0217	0.0212	0.0207	0.0202	0.0197	0.0192	0.0188	0.0183
-1.9	0.0287	0.0281	0.0274	0.0268	0.0262	0.0256	0.0250	0.0244	0.0239	0.0233
-1.8	0.0359	0.0351	0.0344	0.0336	0.0329	0.0322	0.0314	0.0307	0.0301	0.0294
-1.7	0.0446	0.0436	0.0427	0.0418	0.0409	0.0401	0.0392	0.0384	0.0375	0.0367
-1.6	0.0548	0.0537	0.0526	0.0516	0.0505	0.0495	0.0485	0.0475	0.0465	0.0455
-1.5	0.0668	0.0655	0.0643	0.0630	0.0618	0.0606	0.0594	0.0582	0.0571	0.0559
-1.4	0.0808	0.0793	0.0778	0.0764	0.0749	0.0735	0.0721	0.0708	0.0694	0.0681
-1.3	0.0968	0.0951	0.0934	0.0918	0.0901	0.0885	0.0869	0.0853	0.0838	0.0823
-1.2	0.1151	0.1131	0.1112	0.1093	0.1075	0.1056	0.1038	0.1020	0.1003	0.0985
-1.1	0.1357	0.1335	0.1314	0.1292	0.1271	0.1251	0.1230	0.1210	0.1190	0.1170
-1.0	0.1587	0.1562	0.1539	0.1515	0.1492	0.1469	0.1446	0.1423	0.1401	0.1379
-0.9	0.1841	0.1814	0.1788	0.1762	0.1736	0.1711	0.1685	0.1660	0.1635	0.1611
-0.8	0.2119	0.2090	0.2061	0.2033	0.2005	0.1977	0.1949	0.1922	0.1894	0.1867
-0.7	0.2420	0.2389	0.2358	0.2327	0.2296	0.2266	0.2236	0.2206	0.2177	0.2148
-0.6	0.2743	0.2709	0.2676	0.2643	0.2611	0.2578	0.2546	0.2514	0.2483	0.2451
-0.5	0.3085	0.3050	0.3015	0.2981	0.2946	0.2912	0.2877	0.2843	0.2810	0.2776
-0.4	0.3446	0.3409	0.3372	0.3336	0.3300	0.3264	0.3228	0.3192	0.3156	0.3121
-0.3	0.3821	0.3783	0.3745	0.3707	0.3669	0.3632	0.3594	0.3557	0.3520	0.3483
-0.2	0.4207	0.4168	0.4129	0.4090	0.4052	0.4013	0.3974	0.3936	0.3897	0.3859
-0.1	0.4602	0.4562	0.4522	0.4483	0.4443	0.4404	0.4364	0.4325	0.4286	0.4247
-0.0	0.5000	0.4960	0.4920	0.4880	0.4840	0.4801	0.4761	0.4721	0.4681	0.4641

Table A.3 (continued)

z	0.0	0.01	0.02	0.03	0.04	0.05	0.06	0.07	0.08	0.09
0.0	0.5000	0.5040	0.5080	0.5120	0.5160	0.5199	0.5239	0.5279	0.5319	0.5359
0.1	0.5398	0.5438	0.5478	0.5517	0.5557	0.5596	0.5636	0.5675	0.5714	0.5753
0.2	0.5793	0.5832	0.5871	0.5910	0.5948	0.5987	0.6026	0.6064	0.6103	0.6141
0.3	0.6179	0.6217	0.6255	0.6293	0.6331	0.6368	0.6406	0.6443	0.6480	0.6517
0.4	0.6554	0.6591	0.6628	0.6664	0.6700	0.6736	0.6772	0.6808	0.6844	0.6879
0.5	0.6915	0.6950	0.6985	0.7019	0.7054	0.7088	0.7123	0.7157	0.7190	0.7224
0.6	0.7257	0.7291	0.7324	0.7357	0.7389	0.7422	0.7454	0.7486	0.7517	0.7549
0.7	0.7580	0.7611	0.7642	0.7673	0.7704	0.7734	0.7764	0.7794	0.7823	0.7852
0.8	0.7881	0.7910	0.7939	0.7967	0.7995	0.8023	0.8051	0.8078	0.8106	0.8133
0.9	0.8159	0.8186	0.8212	0.8238	0.8264	0.8289	0.8315	0.8340	0.8365	0.8389
1.0	0.8413	0.8438	0.8461	0.8485	0.8508	0.8531	0.8554	0.8577	0.8599	0.8621
1.1	0.8643	0.8665	0.8686	0.8708	0.8729	0.8749	0.8770	0.8790	0.8810	0.8830
1.2	0.8849	0.8869	0.8888	0.8907	0.8925	0.8944	0.8962	0.8980	0.8997	0.9015
1.3	0.9032	0.9049	0.9066	0.9082	0.9099	0.9115	0.9131	0.9147	0.9162	0.9177
1.4	0.9192	0.9207	0.9222	0.9236	0.9251	0.9265	0.9279	0.9292	0.9306	0.9319
1.5	0.9332	0.9345	0.9357	0.9370	0.9382	0.9394	0.9406	0.9418	0.9429	0.9441
1.6	0.9452	0.9463	0.9474	0.9484	0.9495	0.9505	0.9515	0.9525	0.9535	0.9545
1.7	0.9554	0.9564	0.9573	0.9582	0.9591	0.9599	0.9608	0.9616	0.9625	0.9633
1.8	0.9641	0.9649	0.9656	0.9664	0.9671	0.9678	0.9686	0.9693	0.9699	0.9706
1.9	0.9713	0.9719	0.9726	0.9732	0.9738	0.9744	0.9750	0.9756	0.9761	0.9767
2.0	0.9772	0.9778	0.9783	0.9788	0.9793	0.9798	0.9803	0.9808	0.9812	0.9817
2.1	0.9821	0.9826	0.9830	0.9834	0.9838	0.9842	0.9846	0.9850	0.9854	0.9857
2.2	0.9861	0.9864	0.9868	0.9871	0.9875	0.9878	0.9881	0.9884	0.9887	0.9890
2.3	0.9893	0.9896	0.9898	0.9901	0.9904	0.9906	0.9909	0.9911	0.9913	0.9916
2.4	0.9918	0.9920	0.9922	0.9925	0.9927	0.9929	0.9931	0.9932	0.9934	0.9936
2.5	0.9938	0.9940	0.9941	0.9943	0.9945	0.9946	0.9948	0.9949	0.9951	0.9952
2.6	0.9953	0.9955	0.9956	0.9957	0.9959	0.9960	0.9961	0.9962	0.9963	0.9964
2.7	0.9965	0.9966	0.9967	0.9968	0.9969	0.9970	0.9971	0.9972	0.9973	0.9974
2.8	0.9974	0.9975	0.9976	0.9977	0.9977	0.9978	0.9979	0.9979	0.9980	0.9981
2.9	0.9981	0.9982	0.9982	0.9983	0.9984	0.9984	0.9985	0.9985	0.9986	0.9986
3.0	0.9987	0.9987	0.9987	0.9988	0.9988	0.9989	0.9989	0.9989	0.9990	0.9990
3.1	0.9990	0.9991	0.9991	0.9991	0.9992	0.9992	0.9992	0.9992	0.9993	0.9993
3.2	0.9993	0.9993	0.9994	0.9994	0.9994	0.9994	0.9994	0.9995	0.9995	0.9995
3.3	0.9995	0.9995	0.9995	0.9996	0.9996	0.9996	0.9996	0.9996	0.9996	0.9997
3.4	0.9997	0.9997	0.9997	0.9997	0.9997	0.9997	0.9997	0.9997	0.9997	0.9998

A.4   Critical Values $t_\nu(\alpha)$ of the t Distribution

$$P(t_\nu > t_\nu(\alpha)) = \alpha$$

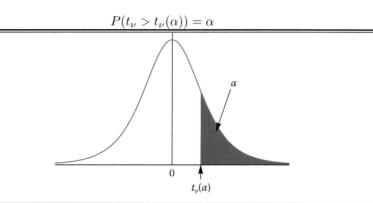

ν	α				
	0.100	0.050	0.025	0.010	0.005
1	3.078	6.314	12.706	31.821	63.657
2	1.886	2.920	4.303	6.965	9.925
3	1.638	2.353	3.182	4.541	5.841
4	1.533	2.132	2.776	3.747	4.604
5	1.476	2.015	2.571	3.365	4.032
6	1.440	1.943	2.447	3.143	3.707
7	1.415	1.895	2.365	2.998	3.499
8	1.397	1.860	2.306	2.896	3.355
9	1.383	1.833	2.262	2.821	3.250
10	1.372	1.812	2.228	2.764	3.169
11	1.363	1.796	2.201	2.718	3.106
12	1.356	1.782	2.179	2.681	3.055
13	1.350	1.771	2.160	2.650	3.012
14	1.345	1.761	2.145	2.624	2.977
15	1.341	1.753	2.131	2.602	2.947
16	1.337	1.746	2.120	2.583	2.921
17	1.333	1.740	2.110	2.567	2.898
18	1.330	1.734	2.101	2.552	2.878
19	1.328	1.729	2.093	2.539	2.861
20	1.325	1.725	2.086	2.528	2.845
21	1.323	1.721	2.080	2.518	2.831
22	1.321	1.717	2.074	2.508	2.819
23	1.319	1.714	2.069	2.500	2.807
24	1.318	1.711	2.064	2.492	2.797
25	1.316	1.708	2.060	2.485	2.787
26	1.315	1.706	2.056	2.479	2.779
27	1.314	1.703	2.052	2.473	2.771
28	1.313	1.701	2.048	2.467	2.763
29	1.311	1.699	2.045	2.462	2.756
30	1.310	1.697	2.042	2.457	2.750
40	1.303	1.684	2.021	2.423	2.704
60	1.296	1.671	2.000	2.390	2.660
∞	1.282	1.645	1.960	2.326	2.576

A.5 Quantiles $Q_\nu(p) = \chi_\nu^2(1-p)$ of the χ^2 Distribution

$$P\left(\chi_\nu^2 \le Q_\nu(p)\right) = p$$

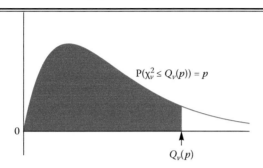

| | | | | | | p | | | | |
ν	0.005	0.010	0.025	0.050	0.100	0.900	0.950	0.975	0.990	0.995
1	0.000	0.000	0.001	0.004	0.016	2.706	3.841	5.024	6.635	7.879
2	0.010	0.020	0.051	0.103	0.211	4.605	5.991	7.378	9.210	10.597
3	0.072	0.115	0.216	0.352	0.584	6.251	7.815	9.348	11.345	12.838
4	0.207	0.297	0.484	0.711	1.064	7.779	9.488	11.143	13.277	14.860
5	0.412	0.554	0.831	1.145	1.610	9.236	11.070	12.833	15.086	16.750
6	0.676	0.872	1.237	1.635	2.204	10.645	12.592	14.449	16.812	18.548
7	0.989	1.239	1.690	2.167	2.833	12.017	14.067	16.013	18.475	20.278
8	1.344	1.646	2.180	2.733	3.490	13.362	15.507	17.535	20.090	21.955
9	1.735	2.088	2.700	3.325	4.168	14.684	16.919	19.023	21.666	23.589
10	2.156	2.558	3.247	3.940	4.865	15.987	18.307	20.483	23.209	25.188
11	2.603	3.053	3.816	4.575	5.578	17.275	19.675	21.920	24.725	26.757
12	3.074	3.571	4.404	5.226	6.304	18.549	21.026	23.337	26.217	28.300
13	3.565	4.107	5.009	5.892	7.042	19.812	22.362	24.736	27.688	29.819
14	4.075	4.660	5.629	6.571	7.790	21.064	23.685	26.119	29.141	31.319
15	4.601	5.229	6.262	7.261	8.547	22.307	24.996	27.488	30.578	32.801
16	5.142	5.812	6.908	7.962	9.312	23.542	26.296	28.845	32.000	34.267
17	5.697	6.408	7.564	8.672	10.085	24.769	27.587	30.191	33.409	35.718
18	6.265	7.015	8.231	9.390	10.865	25.989	28.869	31.526	34.805	37.156
19	6.844	7.633	8.907	10.117	11.651	27.204	30.144	32.852	36.191	38.582
20	7.434	8.260	9.591	10.851	12.443	28.412	31.410	34.170	37.566	39.997
21	8.034	8.897	10.283	11.591	13.240	29.615	32.671	35.479	38.932	41.401
22	8.643	9.542	10.982	12.338	14.041	30.813	33.924	36.781	40.289	42.796
23	9.260	10.196	11.689	13.091	14.848	32.007	35.172	38.076	41.638	44.181
24	9.886	10.856	12.401	13.848	15.659	33.196	36.415	39.364	42.980	45.559
25	10.520	11.524	13.120	14.611	16.473	34.382	37.652	40.646	44.314	46.928
26	11.160	12.198	13.844	15.379	17.292	35.563	38.885	41.923	45.642	48.290
27	11.808	12.879	14.573	16.151	18.114	36.741	40.113	43.195	46.963	49.645
28	12.461	13.565	15.308	16.928	18.939	37.916	41.337	44.461	48.278	50.993
29	13.121	14.256	16.047	17.708	19.768	39.087	42.557	45.722	49.588	52.336
30	13.787	14.953	16.791	18.493	20.599	40.256	43.773	46.979	50.892	53.672
40	20.707	22.164	24.433	26.509	29.051	51.805	55.758	59.342	63.691	66.766
50	27.991	29.707	32.357	34.764	37.689	63.167	67.505	71.420	76.154	79.490
60	35.534	37.485	40.482	43.188	46.459	74.397	79.082	83.298	88.379	91.952

A.6   Critical Values of the $F_{\nu_1,\nu_2}(\alpha)$ Distribution

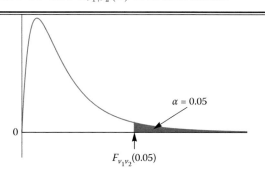

$$\alpha = 0.05$$

$$\nu_1$$

ν_2	1	2	3	4	5	6	7	8	9
1	161.45	199.50	215.71	224.58	230.16	233.99	236.77	238.88	240.54
2	18.51	19.00	19.16	19.25	19.30	19.33	19.35	19.37	19.38
3	10.13	9.55	9.28	9.12	9.01	8.94	8.89	8.85	8.81
4	7.71	6.94	6.59	6.39	6.26	6.16	6.09	6.04	6.00
5	6.61	5.79	5.41	5.19	5.05	4.95	4.88	4.82	4.77
6	5.99	5.14	4.76	4.53	4.39	4.28	4.21	4.15	4.10
7	5.59	4.74	4.35	4.12	3.97	3.87	3.79	3.73	3.68
8	5.32	4.46	4.07	3.84	3.69	3.58	3.50	3.44	3.39
9	5.12	4.26	3.86	3.63	3.48	3.37	3.29	3.23	3.18
10	4.96	4.10	3.71	3.48	3.33	3.22	3.14	3.07	3.02
11	4.84	3.98	3.59	3.36	3.20	3.09	3.01	2.95	2.90
12	4.75	3.89	3.49	3.26	3.11	3.00	2.91	2.85	2.80
13	4.67	3.81	3.41	3.18	3.03	2.92	2.83	2.77	2.71
14	4.60	3.74	3.34	3.11	2.96	2.85	2.76	2.70	2.65
15	4.54	3.68	3.29	3.06	2.90	2.79	2.71	2.64	2.59
16	4.49	3.63	3.24	3.01	2.85	2.74	2.66	2.59	2.54
17	4.45	3.59	3.20	2.96	2.81	2.70	2.61	2.55	2.49
18	4.41	3.55	3.16	2.93	2.77	2.66	2.58	2.51	2.46
19	4.38	3.52	3.13	2.90	2.74	2.63	2.54	2.48	2.42
20	4.35	3.49	3.10	2.87	2.71	2.60	2.51	2.45	2.39
21	4.32	3.47	3.07	2.84	2.68	2.57	2.49	2.42	2.37
22	4.30	3.44	3.05	2.82	2.66	2.55	2.46	2.40	2.34
23	4.28	3.42	3.03	2.80	2.64	2.53	2.44	2.37	2.32
24	4.26	3.40	3.01	2.78	2.62	2.51	2.42	2.36	2.30
25	4.24	3.39	2.99	2.76	2.60	2.49	2.40	2.34	2.28
26	4.23	3.37	2.98	2.74	2.59	2.47	2.39	2.32	2.27
27	4.21	3.35	2.96	2.73	2.57	2.46	2.37	2.31	2.25
28	4.20	3.34	2.95	2.71	2.56	2.45	2.36	2.29	2.24
29	4.18	3.33	2.93	2.70	2.55	2.43	2.35	2.28	2.22
30	4.17	3.32	2.92	2.69	2.53	2.42	2.33	2.27	2.21
40	4.08	3.23	2.84	2.61	2.45	2.34	2.25	2.18	2.12
60	4.00	3.15	2.76	2.53	2.37	2.25	2.17	2.10	2.04
120	3.92	3.07	2.68	2.45	2.29	2.18	2.09	2.02	1.96

Table A.6 (continued)

					$\alpha = 0.05$				
					ν_1				
ν_2	10	12	15	20	24	30	40	60	120
1	241.88	243.91	245.95	248.01	249.05	250.10	251.14	252.20	253.25
2	19.40	19.41	19.43	19.45	19.45	19.46	19.47	19.48	19.49
3	8.79	8.74	8.70	8.66	8.64	8.62	8.59	8.57	8.55
4	5.96	5.91	5.86	5.80	5.77	5.75	5.72	5.69	5.66
5	4.74	4.68	4.62	4.56	4.53	4.50	4.46	4.43	4.40
6	4.06	4.00	3.94	3.87	3.84	3.81	3.77	3.74	3.70
7	3.64	3.57	3.51	3.44	3.41	3.38	3.34	3.30	3.27
8	3.35	3.28	3.22	3.15	3.12	3.08	3.04	3.01	2.97
9	3.14	3.07	3.01	2.94	2.90	2.86	2.83	2.79	2.75
10	2.98	2.91	2.85	2.77	2.74	2.70	2.66	2.62	2.58
11	2.85	2.79	2.72	2.65	2.61	2.57	2.53	2.49	2.45
12	2.75	2.69	2.62	2.54	2.51	2.47	2.43	2.38	2.34
13	2.67	2.60	2.53	2.46	2.42	2.38	2.34	2.30	2.25
14	2.60	2.53	2.46	2.39	2.35	2.31	2.27	2.22	2.18
15	2.54	2.48	2.40	2.33	2.29	2.25	2.20	2.16	2.11
16	2.49	2.42	2.35	2.28	2.24	2.19	2.15	2.11	2.06
17	2.45	2.38	2.31	2.23	2.19	2.15	2.10	2.06	2.01
18	2.41	2.34	2.27	2.19	2.15	2.11	2.06	2.02	1.97
19	2.38	2.31	2.23	2.16	2.11	2.07	2.03	1.98	1.93
20	2.35	2.28	2.20	2.12	2.08	2.04	1.99	1.95	1.90
21	2.32	2.25	2.18	2.10	2.05	2.01	1.96	1.92	1.87
22	2.30	2.23	2.15	2.07	2.03	1.98	1.94	1.89	1.84
23	2.27	2.20	2.13	2.05	2.01	1.96	1.91	1.86	1.81
24	2.25	2.18	2.11	2.03	1.98	1.94	1.89	1.84	1.79
25	2.24	2.16	2.09	2.01	1.96	1.92	1.87	1.82	1.77
26	2.22	2.15	2.07	1.99	1.95	1.90	1.85	1.80	1.75
27	2.20	2.13	2.06	1.97	1.93	1.88	1.84	1.79	1.73
28	2.19	2.12	2.04	1.96	1.91	1.87	1.82	1.77	1.71
29	2.18	2.10	2.03	1.94	1.90	1.85	1.81	1.75	1.70
30	2.16	2.09	2.01	1.93	1.89	1.84	1.79	1.74	1.68
40	2.08	2.00	1.92	1.84	1.79	1.74	1.69	1.64	1.58
60	1.99	1.92	1.84	1.75	1.70	1.65	1.59	1.53	1.47
120	1.91	1.83	1.75	1.66	1.61	1.55	1.50	1.43	1.35

Table A.6 (continued)

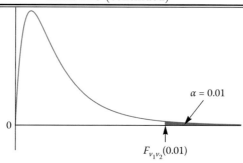

$F_{\nu_1 \nu_2}(0.01)$

$$\alpha = 0.01$$

ν_2	ν_1								
	1	2	3	4	5	6	7	8	9
1	4052.18	4999.50	5403.35	5624.58	5763.65	5858.99	5928.36	5981.07	6022.47
2	98.50	99.00	99.17	99.25	99.30	99.33	99.36	99.37	99.39
3	34.12	30.82	29.46	28.71	28.24	27.91	27.67	27.49	27.35
4	21.20	18.00	16.69	15.98	15.52	15.21	14.98	14.80	14.66
5	16.26	13.27	12.06	11.39	10.97	10.67	10.46	10.29	10.16
6	13.75	10.92	9.78	9.15	8.75	8.47	8.26	8.10	7.98
7	12.25	9.55	8.45	7.85	7.46	7.19	6.99	6.84	6.72
8	11.26	8.65	7.59	7.01	6.63	6.37	6.18	6.03	5.91
9	10.56	8.02	6.99	6.42	6.06	5.80	5.61	5.47	5.35
10	10.04	7.56	6.55	5.99	5.64	5.39	5.20	5.06	4.94
11	9.65	7.21	6.22	5.67	5.32	5.07	4.89	4.74	4.63
12	9.33	6.93	5.95	5.41	5.06	4.82	4.64	4.50	4.39
13	9.07	6.70	5.74	5.21	4.86	4.62	4.44	4.30	4.19
14	8.86	6.51	5.56	5.04	4.69	4.46	4.28	4.14	4.03
15	8.68	6.36	5.42	4.89	4.56	4.32	4.14	4.00	3.89
16	8.53	6.23	5.29	4.77	4.44	4.20	4.03	3.89	3.78
17	8.40	6.11	5.18	4.67	4.34	4.10	3.93	3.79	3.68
18	8.29	6.01	5.09	4.58	4.25	4.01	3.84	3.71	3.60
19	8.18	5.93	5.01	4.50	4.17	3.94	3.77	3.63	3.52
20	8.10	5.85	4.94	4.43	4.10	3.87	3.70	3.56	3.46
21	8.02	5.78	4.87	4.37	4.04	3.81	3.64	3.51	3.40
22	7.95	5.72	4.82	4.31	3.99	3.76	3.59	3.45	3.35
23	7.88	5.66	4.76	4.26	3.94	3.71	3.54	3.41	3.30
24	7.82	5.61	4.72	4.22	3.90	3.67	3.50	3.36	3.26
25	7.77	5.57	4.68	4.18	3.85	3.63	3.46	3.32	3.22
26	7.72	5.53	4.64	4.14	3.82	3.59	3.42	3.29	3.18
27	7.68	5.49	4.60	4.11	3.78	3.56	3.39	3.26	3.15
28	7.64	5.45	4.57	4.07	3.75	3.53	3.36	3.23	3.12
29	7.60	5.42	4.54	4.04	3.73	3.50	3.33	3.20	3.09
30	7.56	5.39	4.51	4.02	3.70	3.47	3.30	3.17	3.07
40	7.31	5.18	4.31	3.83	3.51	3.29	3.12	2.99	2.89
60	7.08	4.98	4.13	3.65	3.34	3.12	2.95	2.82	2.72
120	6.85	4.79	3.95	3.48	3.17	2.96	2.79	2.66	2.56

Table A.6 (continued)

	$\alpha = 0.01$								
	ν_1								
ν_2	10	12	15	20	24	30	40	60	120
1	6055.85	6106.32	6157.28	6208.73	6234.63	6260.65	6286.78	6313.03	6339.39
2	99.40	99.42	99.43	99.45	99.46	99.47	99.47	99.48	99.49
3	27.23	27.05	26.87	26.69	26.60	26.50	26.41	26.32	26.22
4	14.55	14.37	14.20	14.02	13.93	13.84	13.75	13.65	13.56
5	10.05	9.89	9.72	9.55	9.47	9.38	9.29	9.20	9.11
6	7.87	7.72	7.56	7.40	7.31	7.23	7.14	7.06	6.97
7	6.62	6.47	6.31	6.16	6.07	5.99	5.91	5.82	5.74
8	5.81	5.67	5.52	5.36	5.28	5.20	5.12	5.03	4.95
9	5.26	5.11	4.96	4.81	4.73	4.65	4.57	4.48	4.40
10	4.85	4.71	4.56	4.41	4.33	4.25	4.17	4.08	4.00
11	4.54	4.40	4.25	4.10	4.02	3.94	3.86	3.78	3.69
12	4.30	4.16	4.01	3.86	3.78	3.70	3.62	3.54	3.45
13	4.10	3.96	3.82	3.66	3.59	3.51	3.43	3.34	3.25
14	3.94	3.80	3.66	3.51	3.43	3.35	3.27	3.18	3.09
15	3.80	3.67	3.52	3.37	3.29	3.21	3.13	3.05	2.96
16	3.69	3.55	3.41	3.26	3.18	3.10	3.02	2.93	2.84
17	3.59	3.46	3.31	3.16	3.08	3.00	2.92	2.83	2.75
18	3.51	3.37	3.23	3.08	3.00	2.92	2.84	2.75	2.66
19	3.43	3.30	3.15	3.00	2.92	2.84	2.76	2.67	2.58
20	3.37	3.23	3.09	2.94	2.86	2.78	2.69	2.61	2.52
21	3.31	3.17	3.03	2.88	2.80	2.72	2.64	2.55	2.46
22	3.26	3.12	2.98	2.83	2.75	2.67	2.58	2.50	2.40
23	3.21	3.07	2.93	2.78	2.70	2.62	2.54	2.45	2.35
24	3.17	3.03	2.89	2.74	2.66	2.58	2.49	2.40	2.31
25	3.13	2.99	2.85	2.70	2.62	2.54	2.45	2.36	2.27
26	3.09	2.96	2.81	2.66	2.58	2.50	2.42	2.33	2.23
27	3.06	2.93	2.78	2.63	2.55	2.47	2.38	2.29	2.20
28	3.03	2.90	2.75	2.60	2.52	2.44	2.35	2.26	2.17
29	3.00	2.87	2.73	2.57	2.49	2.41	2.33	2.23	2.14
30	2.98	2.84	2.70	2.55	2.47	2.39	2.30	2.21	2.11
40	2.80	2.66	2.52	2.37	2.29	2.20	2.11	2.02	1.92
60	2.63	2.50	2.35	2.20	2.12	2.03	1.94	1.84	1.73
120	2.47	2.34	2.19	2.03	1.95	1.86	1.76	1.66	1.53

A.7   Critical Values of the Studentized Range $q(\alpha; n, \nu)$

	$\alpha = 0.05$								
	n								
ν	2	3	4	5	6	7	8	9	10
1	17.97	26.98	32.82	37.08	40.41	43.12	45.40	47.36	49.07
2	6.08	8.33	9.80	10.88	11.74	12.44	13.03	13.54	13.99
3	4.50	5.91	6.82	7.50	8.04	8.48	8.85	9.18	9.46
4	3.93	5.04	5.76	6.29	6.71	7.05	7.35	7.60	7.83
5	3.64	4.60	5.22	5.67	6.03	6.33	6.58	6.80	6.99
6	3.46	4.34	4.90	5.30	5.63	5.90	6.12	6.32	6.49
7	3.34	4.16	4.68	5.06	5.36	5.61	5.82	6.00	6.16
8	3.26	4.04	4.53	4.89	5.17	5.40	5.60	5.77	5.92
9	3.20	3.95	4.41	4.76	5.02	5.24	5.43	5.59	5.74
10	3.15	3.88	4.33	4.65	4.91	5.12	5.30	5.46	5.60
11	3.11	3.82	4.26	4.57	4.82	5.03	5.20	5.35	5.49
12	3.08	3.77	4.20	4.51	4.75	4.95	5.12	5.27	5.39
13	3.06	3.73	4.15	4.45	4.69	4.88	5.05	5.19	5.32
14	3.03	3.70	4.11	4.41	4.64	4.83	4.99	5.13	5.25
15	3.01	3.67	4.08	4.37	4.59	4.78	4.94	5.08	5.20
16	3.00	3.65	4.05	4.33	4.56	4.74	4.90	5.03	5.15
17	2.98	3.63	4.02	4.30	4.52	4.70	4.86	4.99	5.11
18	2.97	3.61	4.00	4.28	4.49	4.67	4.82	4.96	5.07
19	2.96	3.59	3.98	4.25	4.47	4.65	4.79	4.92	5.04
20	2.95	3.58	3.96	4.23	4.45	4.62	4.77	4.90	5.01
24	2.92	3.53	3.90	4.17	4.37	4.54	4.68	4.81	4.92
30	2.89	3.49	3.85	4.10	4.30	4.46	4.60	4.72	4.82
40	2.86	3.44	3.79	4.04	4.23	4.39	4.52	4.63	4.73
60	2.83	3.40	3.74	3.98	4.16	4.31	4.44	4.55	4.65
120	2.80	3.36	3.68	3.92	4.10	4.24	4.36	4.47	4.56
∞	2.77	3.31	3.63	3.86	4.03	4.17	4.29	4.39	4.47

n = number of treatments; ν = degrees of freedom.

Table A.7 (continued)

	$\alpha = 0.05$									
	n									
v	11	12	13	14	15	16	17	18	19	20
1	50.59	51.96	53.20	54.33	55.36	56.32	57.22	58.04	58.83	59.56
2	14.39	14.75	15.08	15.38	15.65	15.91	16.14	16.37	16.57	16.77
3	9.72	9.95	10.15	10.35	10.52	10.69	10.84	10.98	11.11	11.24
4	8.03	8.21	8.37	8.52	8.66	8.79	8.91	9.03	9.13	9.23
5	7.17	7.32	7.47	7.60	7.72	7.83	7.93	8.03	8.12	8.21
6	6.65	6.79	6.92	7.03	7.14	7.24	7.34	7.43	7.51	7.59
7	6.30	6.43	6.55	6.66	6.76	6.85	6.94	7.02	7.10	7.17
8	6.05	6.18	6.29	6.39	6.48	6.57	6.65	6.73	6.80	6.87
9	5.87	5.98	6.09	6.19	6.28	6.36	6.44	6.51	6.58	6.64
10	5.72	5.83	5.93	6.03	6.11	6.19	6.27	6.34	6.40	6.47
11	5.61	5.71	5.81	5.90	5.98	6.06	6.13	6.20	6.27	6.33
12	5.51	5.61	5.71	5.80	5.88	5.95	6.02	6.09	6.15	6.21
13	5.43	5.53	5.63	5.71	5.79	5.86	5.93	5.99	6.05	6.11
14	5.36	5.46	5.55	5.64	5.71	5.79	5.85	5.91	5.97	6.03
15	5.31	5.40	5.49	5.57	5.65	5.72	5.78	5.85	5.90	5.96
16	5.26	5.35	5.44	5.52	5.59	5.66	5.73	5.79	5.84	5.90
17	5.21	5.31	5.39	5.47	5.54	5.61	5.67	5.73	5.79	5.84
18	5.17	5.27	5.35	5.43	5.50	5.57	5.63	5.69	5.74	5.79
19	5.14	5.23	5.31	5.39	5.46	5.53	5.59	5.65	5.70	5.75
20	5.11	5.20	5.28	5.36	5.43	5.49	5.55	5.61	5.66	5.71
24	5.01	5.10	5.18	5.25	5.32	5.38	5.44	5.49	5.55	5.59
30	4.92	5.00	5.08	5.15	5.21	5.27	5.33	5.38	5.43	5.47
40	4.82	4.90	4.98	5.04	5.11	5.16	5.22	5.27	5.31	5.36
60	4.73	4.81	4.88	4.94	5.00	5.06	5.11	5.15	5.20	5.24
120	4.64	4.71	4.78	4.84	4.90	4.95	5.00	5.04	5.09	5.13
∞	4.55	4.62	4.68	4.74	4.80	4.85	4.89	4.93	4.97	5.01

Table A.7 (continued)

	$\alpha = 0.01$								
	n								
v	2	3	4	5	6	7	8	9	10
1	90.03	135.0	164.3	185.6	202.2	215.8	227.2	237.0	245.6
2	14.04	19.02	22.29	24.72	26.63	28.20	29.53	30.68	31.69
3	8.26	10.62	12.17	13.33	14.24	15.00	15.64	16.20	16.69
4	6.51	8.12	9.17	9.96	10.58	11.10	11.55	11.93	12.27
5	5.70	6.98	7.80	8.42	8.91	9.32	9.67	9.97	10.24
6	5.24	6.33	7.03	7.56	7.97	8.32	8.61	8.87	9.10
7	4.95	5.92	6.54	7.01	7.37	7.68	7.94	8.17	8.37
8	4.75	5.65	6.20	6.62	6.96	7.24	7.47	7.68	7.86
9	4.60	5.43	5.96	6.35	6.66	6.91	7.13	7.33	7.49
10	4.48	5.27	5.77	6.14	6.43	6.67	6.87	7.05	7.21
11	4.39	5.15	5.62	5.97	6.25	6.48	6.67	6.84	6.99
12	4.32	5.05	5.50	5.84	6.10	6.32	6.51	6.67	6.81
13	4.26	4.96	5.40	5.73	5.98	6.19	6.37	6.53	6.67
14	4.21	4.89	5.32	5.63	5.88	6.08	6.26	6.41	6.54
15	4.17	4.84	5.25	5.56	5.80	5.99	6.16	6.31	6.44
16	4.13	4.79	5.19	5.49	5.72	5.92	6.08	6.22	6.35
17	4.10	4.74	5.14	5.43	5.66	5.85	6.01	6.15	6.27
18	4.07	4.70	5.09	5.38	5.60	5.79	5.94	6.08	6.20
19	4.05	4.67	5.05	5.33	5.55	5.73	5.89	6.02	6.14
20	4.03	4.64	5.02	5.29	5.51	5.69	5.84	5.97	6.09
24	3.96	4.55	4.91	5.17	5.37	5.54	5.69	5.81	5.92
30	3.89	4.45	4.80	5.05	5.24	5.40	5.54	5.65	5.76
40	3.82	4.37	4.70	4.93	5.11	5.26	5.39	5.50	5.60
60	3.76	4.28	4.59	4.82	4.99	5.13	5.25	5.36	5.45
120	3.70	4.20	4.50	4.71	4.87	5.01	5.12	5.21	5.30
∞	3.64	4.12	4.40	4.60	4.76	4.88	4.99	5.08	5.16

Table A.7 (continued)

$$\alpha = 0.01$$

| v | \multicolumn{10}{c}{n} |
	11	12	13	14	15	16	17	18	19	20
1	253.2	260.0	266.2	271.8	277.0	281.8	286.3	290.4	294.3	298.0
2	32.59	33.40	34.13	34.81	35.43	36.00	36.53	37.03	37.50	37.95
3	17.13	17.53	17.89	18.22	18.52	18.81	19.07	19.32	19.55	19.77
4	12.57	12.84	13.09	13.32	13.53	13.73	13.91	14.08	14.24	14.40
5	10.48	10.70	10.89	11.08	11.24	11.40	11.55	11.68	11.81	11.93
6	9.30	9.48	9.65	9.81	9.95	10.08	10.21	10.32	10.43	10.54
7	8.55	8.71	8.86	9.00	9.12	9.24	9.35	9.46	9.55	9.65
8	8.03	8.18	8.31	8.44	8.55	8.66	8.76	8.85	8.94	9.03
9	7.65	7.78	7.91	8.03	8.13	8.23	8.33	8.41	8.49	8.57
10	7.36	7.49	7.60	7.71	7.81	7.91	7.99	8.08	8.15	8.23
11	7.13	7.25	7.36	7.46	7.56	7.65	7.73	7.81	7.88	7.95
12	6.94	7.06	7.17	7.26	7.36	7.44	7.52	7.59	7.66	7.73
13	6.79	6.90	7.01	7.10	7.19	7.27	7.35	7.42	7.48	7.55
14	6.66	6.77	6.87	6.96	7.05	7.13	7.20	7.27	7.33	7.39
15	6.55	6.66	6.76	6.84	6.93	7.00	7.07	7.14	7.20	7.26
16	6.46	6.56	6.66	6.74	6.82	6.90	6.97	7.03	7.09	7.15
17	6.38	6.48	6.57	6.66	6.73	6.81	6.87	6.94	7.00	7.05
18	6.31	6.41	6.50	6.58	6.65	6.73	6.79	6.85	6.91	6.97
19	6.25	6.34	6.43	6.51	6.58	6.65	6.72	6.78	6.84	6.89
20	6.19	6.28	6.37	6.45	6.52	6.59	6.65	6.71	6.77	6.82
24	6.02	6.11	6.19	6.26	6.33	6.39	6.45	6.51	6.56	6.61
30	5.85	5.93	6.01	6.08	6.14	6.20	6.26	6.31	6.36	6.41
40	5.69	5.76	5.83	5.90	5.96	6.02	6.07	6.12	6.16	6.21
60	5.53	5.60	5.67	5.73	5.78	5.84	5.89	5.93	5.97	6.01
120	5.37	5.44	5.50	5.56	5.61	5.66	5.71	5.75	5.79	5.83
∞	5.23	5.29	5.35	5.40	5.45	5.49	5.54	5.57	5.61	5.66

A.8   Factors for Estimating σ, $\bar{s}$, or $\bar{\sigma}_{RMS}$ and σ_R from $\bar{R}$

Number of observations in subgroup, n	Factor $d_2, d_2 = \frac{\bar{R}}{\sigma}$	Factor $d_3, d_3 = \frac{\sigma_R}{\sigma}$	Factor $c_2, c_2 = \frac{\bar{\sigma}_{RMS}}{\sigma}$	Factor $c_4, c_4 = \frac{\bar{S}}{\sigma}$
2	1.128	0.8525	0.5642	0.7979
3	1.693	0.8884	0.7236	0.8862
4	2.059	0.8798	0.7979	0.9213
5	2.326	0.8641	0.8407	0.9400
6	2.534	0.8480	0.8686	0.9515
7	2.704	0.8332	0.8882	0.9594
8	2.847	0.8198	0.9027	0.9650
9	2.970	0.8078	0.9139	0.9693
10	3.078	0.7971	0.9227	0.9727
11	3.173	0.7873	0.9300	0.9754
12	3.258	0.7785	0.9359	0.9776
13	3.336	0.7704	0.9410	0.9794
14	3.407	0.7630	0.9453	0.9810
15	3.472	0.7562	0.9490	0.9823
16	3.532	0.7499	0.9523	0.9835
17	3.588	0.7441	0.9551	0.9845
18	3.640	0.7386	0.9576	0.9854
19	3.689	0.7335	0.9599	0.9862
20	3.735	0.7287	0.9619	0.9869
21	3.778	0.7242	0.9638	0.9876
22	3.819	0.7199	0.9655	0.9882
23	3.858	0.7159	0.9670	0.9887
24	3.895	0.7121	0.9684	0.9892
25	3.931	0.7084	0.9696	0.9896
30	4.086	0.6926	0.9748	0.9914
35	4.213	0.6799	0.9784	0.9927
40	4.322	0.6692	0.9811	0.9936
45	4.415	0.6601	0.9832	0.9943
50	4.498	0.6521	0.9849	0.9949
55	4.572	0.6452	0.9863	0.9954
60	4.639	0.6389	0.9874	0.9958
65	4.699	0.6337	0.9884	0.9961
70	4.755	0.6283	0.9892	0.9964
75	4.806	0.6236	0.9900	0.9966
80	4.854	0.6194	0.9906	0.9968
85	4.898	0.6154	0.9912	0.9970
90	4.939	0.6118	0.9916	0.9972
95	4.978	0.6084	0.9921	0.9973
100	5.015	0.6052	0.9925	0.9975

Estimate of $\sigma = \bar{R}/d_2$ or $\bar{S}/c_4$ or $\bar{\sigma}_{RMS}/C_2$; $\sigma_R = \bar{R}/d_3$. These factors assume sampling from a normal distribution.

Source: Reproduced from E. L. Grant and R. S. Leavenworth (1996), *Statistical Quality Control*, 7th ed., New York, McGraw-Hill. Used with permission.

A.9 Factors for Determining from $\bar{R}$ the Three-Sigma Control Limits for $\bar{X}$ and R Charts

Number of observations in subgroup, n	Factor for $\bar{X}$ chart, A_2	Factors for R chart	
		Lower control limit, D_3	Upper control limit, D_4
2	1.88	0	3.27
3	1.02	0	2.57
4	0.73	0	2.28
5	0.58	0	2.11
6	0.48	0	2.00
7	0.42	0.08	1.92
8	0.37	0.14	1.86
9	0.34	0.18	1.82
10	0.31	0.22	1.78
11	0.29	0.26	1.74
12	0.27	0.28	1.72
13	0.25	0.31	1.69
14	0.24	0.33	1.67
15	0.22	0.35	1.65
16	0.21	0.36	1.64
17	0.20	0.38	1.62
18	0.19	0.39	1.61
19	0.19	0.40	1.60
20	0.18	0.41	1.59

Upper control limit for $\bar{X} = UCL_{\bar{X}} = \bar{\bar{X}} + A_2\bar{R}$

Lower control limit for $\bar{X} = LCL_{\bar{X}} = \bar{\bar{X}} - A_2\bar{R}$

(If aimed-at or standard value $\bar{X}_0$ is used rather than $\bar{\bar{X}}$ as the central line on the control chart, $\bar{X}_0$ should be substituted for $\bar{\bar{X}}$ in the preceding formulas.)

Upper control limit for $R = \text{UCL}_R = D_4\bar{R}$

Lower control limit for $R = \text{UCL}_R = D_3\bar{R}$

All factors in the table are based on the normal distribution.

There is no lower control limit for R chart where n is less than 7. *Source:* Reproduced from E. L. Grant and R. S. Leavenworth (1996), *Statistical Quality Control*, 7th ed., New York, McGraw-Hill. Used with permission.

A.10   Factors for Determining from σ the Three-Sigma Control Limits for $\overline{X}$, R, and s or $\overline{\sigma}_{RMS}$ Charts

Number of observations in subgroup, n	Factors for $\overline{X}$ chart, A	Factors for R chart		Factors for σ_{RMS} chart		Factors for s chart	
		Lower control limit, D_1	Upper control limit, D_2	Lower control limit, B_1	Upper control limit, B_2	Lower control limit, B_5	Upper control limit, B_6
2	2.12	0	3.69	0	1.84	0	2.61
3	1.73	0	4.36	0	1.86	0	2.28
4	1.50	0	4.70	0	1.81	0	2.09
5	1.34	0	4.92	0	1.76	0	1.96
6	1.22	0	5.08	0.03	1.71	0.03	1.87
7	1.13	0.20	5.20	0.10	1.67	0.11	1.81
8	1.06	0.39	5.31	0.17	1.64	0.18	1.75
9	1.00	0.55	5.39	0.22	1.61	0.23	1.71
10	0.95	0.69	5.47	0.26	1.58	0.28	1.67
11	0.90	0.81	5.53	0.30	1.56	0.31	1.64
12	0.87	0.92	5.59	0.33	1.54	0.35	1.61
13	0.83	1.03	5.65	0.36	1.52	0.37	1.59
14	0.80	1.12	5.69	0.38	1.51	0.40	1.56
15	0.77	1.21	5.74	0.41	1.49	0.42	1.54
16	0.75	1.28	5.78	0.43	1.48	0.44	1.53
17	0.73	1.36	5.82	0.44	1.47	0.46	1.51
18	0.71	1.43	5.85	0.46	1.45	0.48	1.50
19	0.69	1.49	5.89	0.48	1.44	0.49	1.48
20	0.67	1.55	5.92	0.49	1.43	0.50	1.47
21	0.65			0.50	1.42	0.52	1.46
22	0.64			0.52	1.41	0.53	1.45
23	0.63			0.53	1.41	0.54	1.44
24	0.61			0.54	1.40	0.55	1.43
25	0.60			0.55	1.39	0.56	1.42

Table A.10 (continued)

Number of observations in subgroup, n	Factors for $\overline{X}$ chart, A	Factors for R chart		Factors for σ_{RMS} chart		Factors for s chart	
		Lower control limit, D_1	Upper control limit, D_2	Lower control limit, B_1	Upper control limit, B_2	Lower control limit, B_5	Upper control limit, B_6
30	0.55			0.59	1.36	0.60	1.38
35	0.51			0.62	1.33	0.63	1.36
40	0.47			0.65	1.31	0.66	1.33
45	0.45			0.67	1.30	0.68	1.31
50	0.42			0.68	1.28	0.69	1.30
55	0.40			0.70	1.27	0.71	1.28
60	0.39			0.71	1.26	0.72	1.27
65	0.37			0.72	1.25	0.73	1.26
70	0.36			0.74	1.24	0.74	1.25
75	0.35			0.75	1.23	0.75	1.24
80	0.34			0.75	1.23	0.76	1.24
85	0.33			0.76	1.22	0.77	1.23
90	0.32			0.77	1.22	0.77	1.22
95	0.31			0.77	1.21	0.78	1.22
100	0.30			0.78	1.20	0.78	1.21

$$UCL_{\overline{X}} = \mu + A\sigma \qquad LCL_{\overline{X}} = \mu - A\sigma$$

(If actual average is to be used rather than standard or aimed-at average, $\overline{\overline{X}}$ should be substituted for μ in the preceding formulas.)

$$UCL_R = D_2\sigma \qquad UCL_s = B_6\sigma \qquad UCL_{\sigma_{RMS}} = B_2\sigma$$
Central line$_R = d_2\sigma$ Central line$_s = c_4\sigma$ Central line$_{RMS} = c_2\sigma$
$$LCL_R = D_1\sigma \qquad LCL_s = B_5\sigma \qquad LCL_{\sigma_{RMS}} = B_1\sigma$$

There is no lower control for dispersion chart when n is less than 6.

Source: Reproduced from E. L. Grant and R. S. Leavenworth (1996), *Statistical Quality Control*, 7th ed., New York, McGraw-Hill. Used with permission.

Answers to Selected
Odd-Numbered Problems

Chapter 1

Data Analysis

Problem 1.1 *(a)*

```
0  114555799999
1  00000000022222255
2  0058
3  000023
4  05
5
6  0
```

Problem 1.3 *The histogram is clearly more symmetric with a well defined center at the class mark -1.8.*

Problem 1.5 *(a)* $146 < 147 < 148 < 149 < 151 < 152 < 153 \leq 153 < 154 < 157$
(b) $\hat{F}_{10}(145) = 0$, $\hat{F}_{10}(150) = 0.4$, $\hat{F}_{10}(150.5) = 0.4$
$\hat{F}_{10}(152.9) = 0.6$, $\hat{F}_{10}(153.1) = 0.8$, $\hat{F}_{10}(158) = 1$

Problem 1.7

$$(c) \quad 15.30 < 16.00 \leq 16.00 < 16.05 < 16.10 \leq 16.10$$
$$< 16.20 < 16.30 < 16.40 < 16.50 < 16.60 < 16.75$$
$$\leq 16.75 < 17.10 \leq 17.10 < 17.50 \leq 17.50$$

(d) 35.29%; 76.47% *(e)* 52.49%; 52.94% *(f)* 76.47%; 35.29%

Problem 1.9 *(a)*

year	2001	2002	2003	2004	2005
value of portfolio	9,096	8,3459	10,526	12,946	14,436

(b) Ms. Allen's annualized rate of return for the five year period 2001-2005:
$r = \exp(\ln(1.43570)/5) - 1 = 0.0762 (= 7.62\%)$.
(c) Average rate of return $= 8.67\%$.

Problem 1.11 *(b) The GE data are approximately symmetric, but heavier tailed.*

Problem 1.13 *(b) Both histograms are symmetrically distributed around a well defined center.*

Problem 1.15 *(c) Both histograms are symmetrically distributed around a well defined center.*
(d) Clearly, PTTRX investors had many more reasons to rejoice in 2002 than they did in 1999.

Problem 1.17 *(a)*

COUNTRY	X1	X2	X3	X4
GDR	88.1	5.40	11300	4.9
USA	1328.3	5.37	18529	2.8
Canada	124.3	4.73	15160	2.9
Australi	70.3	4.22	11103	3.6
Czechosl	61.8	3.95	9280	4.5
USSR	1038.2	3.62	8375	4.3
Poland	120.3	3.15	1926	17.6
FRG	175.1	2.86	14399	2.1
UK	155.1	2.70	10419	2.6
Romania	57.9	2.50	6030	4.2
Japan	284.0	2.31	15764	1.3
SAfrica	76.0	2.20	1870	12.5
Italy	106.4	1.86	10355	1.7
France	97.5	1.74	12789	1.4
ROK	60.3	1.42	2689	4.4
Spain	55.5	1.42	5972	2.1
Mexico	87.3	1.01	1825	5.4
China	651.9	0.59	294	19.0
Brazil	56.5	0.38	2021	1.9
India	177.9	0.21	311	6.1

(b)

COUNTRY	X1	X2	X3	X4
China	651.9	0.59	294	19.0
Poland	120.3	3.15	1926	17.6
SAfrica	76.0	2.20	1870	12.5
India	177.9	0.21	311	6.1
Mexico	87.3	1.01	1825	5.4
GDR	88.1	5.40	11300	4.9
Czechosl	61.8	3.95	9280	4.5
ROK	60.3	1.42	2689	4.4
USSR	1038.2	3.62	8375	4.3
Romania	57.9	2.50	6030	4.2
Australi	70.3	4.22	11103	3.6
Canada	124.3	4.73	15160	2.9
USA	1328.3	5.37	18529	2.8
UK	155.1	2.70	10419	2.6
FRG	175.1	2.86	14399	2.1
Spain	55.5	1.42	5972	2.1
Brazil	56.5	0.38	2021	1.9
Italy	106.4	1.86	10355	1.7
France	97.5	1.74	12789	1.4
Japan	284.0	2.31	15764	1.3

(c) The USA produces more CO_2 per capita than China because it produces much more GNP per capita; however, it produces much less CO_2 per unit of GNP.

Problem 1.19 *(a)*

```
Stem Leaf                        #           Boxplot
 208 0                           1              0
 206 0                           1              |
 204 00                          2              |
 202 00                          2              |
 200 00000                       5              |
 198 00000000                    8           +-----+
 196 0000000                     7           *--+--*
 194 0000000                     7           |     |
 192 00000000                    8           +-----+
 190 000                         3              |
 188 000                         3              |
 186
 184
 182
 180 0                           1              0
      ----+----+----+----+
```

(c) The level of cadmium dust is greater or equal to the federal standard $22.92\% = (11/48) \times 100\%$ of the time. Clearly, this is not a safe workplace.

Problem 1.21 *(a)* $1.9 < 3.0 < 4.0 < 4.7 < 4.9 < 5.8 < 6.1 \leq 6.1 < 7.1 < 7.3$
(b) median $= (4.9 + 5.8)/2 = 5.35$
(c) $Q_1 = x_{(3)} = 4.0; Q_3 = x_{(8)} = 6.1; IQR = 6.1 - 4.0 = 2.1$

Problem 1.23 *(a) Sample median* $\tilde{x} = 50.0$
(b) $Q_1 = 40; Q_3 = 75; IQR = 35$ *(c) Outliers:* $\{220, 285\}$

Problem 1.25 *(a) Sample range* $= 1270 - 1017 = 253, \tilde{x} = 1078.5$
(b) $Q_1 = 1028, Q_3 = 1138, IQR = 110.$ *There are no outliers.*

Problem 1.27 *(a)* $Q(0.10) = 2962.5, Q(0.50) = 8803.5$
(b) Range $= 15411, Q_1 = 4379.5, Q_3 = 12226.5, IQR = 7847$
(c) No outliers.

Problem 1.29 *(b)* $Q_1 = 0.05, Q_3 = 0.18$

Problem 1.31 *(a) Sample range* $= 11, \tilde{x} = 151.5, Q_1 = 148, Q_3 = 153, IQR = 5$

Problem 1.33 *(a)* $\tilde{x} = 684.5; Q_1 = 598; Q_3 = 767; IQR = 169.$ *No outliers.*
(b) $\tilde{x} = 946; Q_1 = 903; Q_3 = 981; IQR = 78.$ *Outliers:* $\{745, 1202\}.$

Problem 1.35 *(a)* $\tilde{x} = 1.0, Q_1 = 0.3, Q_3 = 0.43, IQR = 4.0.$ *Outliers* $= \{12.5, 18, 25\}.$
(b) $\tilde{x} = 0.1, Q_1 = 0.1, Q_3 = 1.8, IQR = 1.7.$ *No outliers.*

Problem 1.37 *(a) Range* $= 260; median = 22; Q_1 = 12; Q_3 = 87; IQR = 75.$ *There are three outliers:* 261, 246, 225.
(b) Range $= 215; median = 63; Q_1 = 18; Q_3 = 111; IQR = 93.$ *There are no outliers.*
(c) Range $= 207; median = 41.5; Q_1 = 18.5; Q_3 = 92.5; IQR = 74.$ *There is one outlier:* 210.
(d) Range $= 216; median = 60; Q_1 = 33; Q_3 = 118; IQR = 85.$ *There are no outliers.*

Problem 1.39 *Median* $= 0.50055; Q_1 = 0.49515; Q_3 = 0.5069; IQR = 0.001175.$ *There are no outliers.*

Problem 1.41 *(a)* $F(1400) = 4;$ *that is, 4 light bulbs failed. So, the company loses \$4. The company earns \$6 on those light bulbs that last more than 1400 hours. Therefore, net revenue* $R = \$2.$
(b) $F(1500) = 6.$ *Reasoning as in part (a), the net revenue* $R = -\$2.$

Problem 1.43 *(a) From air:* $\overline{x} = 2.3101, s = 0.0001426.$
(b) By chemical decomposition: $\overline{x} = 2.2994725, s = 0.0013792.$

Problem 1.45 $\overline{x} = 76.07, s^2 = 6258.38, s = 79.11$

Problem 1.47 $\overline{x} = 1092.9, s^2 = 5872.1, s = 76.63$

Problem 1.49 *(a)*

$$\text{Method 1: } \overline{x} = 191.1667; \; s = 97.9752$$
$$\text{Method 2: } \overline{x} = 208.6667; \; s = 104.2892$$
$$\text{Difference: } \overline{x} = -17.5; \; s = 53.6563$$

(c) Because the box plot for the difference includes 0 within its interquartile range we conclude that there is no important difference between the two methods for measuring nitrates.

Problem 1.51 $\overline{x} = 70.19$, $s = 2.88$

Problem 1.53

$$\overline{x} = 8568, \; s^2 = 19,771,046, \; s = 4446.46$$

Problem 1.55 *(a)* $\overline{x} = 0.338$; $s^2 = 0.03$; $s = 0.17$
(b) $\overline{x} = 0.549$; $s^2 = 0.016$; $s = 0.13$

Problem 1.57 $\overline{x} = 151$, $s = 3.4641$

Problem 1.59 $\overline{x} = 0.50$, $s^2 = 0.000075$, $s = 0.0086$

Problem 1.61 *The sample mean, variance, and standard deviation of the difference variable $d = exposed - control$ are $\overline{d} = 15.97$, $s^2 = 251.66$, $s = 15.86$. The difference variable has a positive sample mean confirming once again that the lead levels of the exposed group are larger than the lead levels of the control group. In Chapter 8 we study whether this difference is too large to be due to chance. The data strongly suggest that there are real differences in the lead levels between the two groups.*

Problem 1.63 *(a)* $\overline{x} = 0.01643$, $s = 0.00470$ *(b)* $\overline{x} = 0.02426$, $s = 0.00514$

Problem 1.65 $\overline{x} = 196.13$, $s^2 = 25.69$, $s = 5.07$

Problem 1.67 *(a)* $\overline{x} = 0.001$ *(b)* $s = 0.006$

Problem 1.69 *(b) The inefficient portfolios are: A, D, G.*

Problem 1.71 *(a) The summary statistics for each diet are given by*

$$\text{diet } A: \; \overline{x} = 9.04, \; s = 3.67$$
$$\text{diet } B: \; \overline{x} = 5.86, \; s = 3.44$$
$$\text{diet } C: \; \overline{x} = 10.79, \; s = 3.27$$

There are two outliers in diet A: 2.1 and 15.3 pounds.
(c) Diet C is most effective, followed by diets A and B (least effective). However, diet C does not appear to be much better than diet A.

Problem 1.73 *(a) If $a > 0$ and $x_{(i)} \leq x_{(i+1)}$ then it follows that $ax_{(i)} + b \leq ax_{(i+1)} + b$. Consequently,*

$$ax_{(1)} + b \leq ax_{(2)} + b \leq \ldots \leq ax_{(n)} + b.$$

Therefore $y_{(i)} = ax_{(i)} + b$.

Problem 1.75 *(c)* $2.3378 = 1.5975 + 0.01645 \times 45$
(d) Yield is increased by 0.1645 tons per hectare.

Problem 1.77 *(c) Fitted values: 3.732, 4.444, 6.035*
(d) Fuel consumption would increase by 0.524 gallons per hundred miles.

Problem 1.79 *(c) Fitted values: 3.122, 4.370, 5.411*
(d) It consumes an additional 0.374 gallons per 100 miles.

Problem 1.81 *(a)* $r = 0.1852$ *(b)* $r = 0.4996$ *(c) The daughter's height is much more closely correlated with the midparent height than with the mother's height.*

Problem 1.83 *(a)* $r = -0.4720$ *(b)* $\beta = -0.096$

Chapter 2

Probability Theory

Problem 2.1
(a) $A_0 = TTTT$; $P(A_0) = 1/16$
(b) $A_1 = HTTT, THTT, TTHT, TTTH$; $P(A_1) = 4/16$
(c) $B_3 = HHHT, HHTH, HTHH, THHH, HHHH$; $P(B_3) = 5/16$
(d) $B_4 = HHHH$; $P(B_4) = 1/16$
(e) $A_4 = HHHH$; $P(A_4) = 1/16$

Problem 2.3
(a) B_2 (b) A_1
(c) A_4 (d) A_0

Problem 2.5

$$(a) \quad c = (1 + 1/2 + 1/3 + 1/4 + 1/5)^{-1} = 60/137 = 0.438$$
$$(b) \quad P(s_0) = 0.438 \text{ and } 1 - P(s_0) = 0.5620$$

Problem 2.9
(a) $A_3 \cup A_5 \cup A_7 \cup A_9 \cup A_{11}$
(b) $A_2 \cup A_3 \cup A_4 \cup A_5 \cup A_6$
(c) $A_7 \cup A_8 \cup A_9 \cup A_{10} \cup A_{11} \cup A_{12}$

Problem 2.11 *(a) 0.9, (b) 0.1, (c) 0.8, (d) 0.2, (e) 0.5, (f) 0.2, (g) 0.8*

Problem 2.13 *(a) T, (b) F, (c) F*

Problem 2.15 *Assume $A_1 \cap A_2 = \emptyset$, which implies that $(A_1 \cap B) \cap (A_2 \cap B) = \emptyset$. We have to show that*
$P'(A_1 \cup A_2) = P'(A_1) + P'(A_2)$. In detail:

$$
\begin{aligned}
P'(A_1 \bigcup A_2) &= P(A_1 \bigcup A_2 | B) \\
&= \frac{P((A_1 \bigcup A_2) \cap B)}{P(B)} \\
&= \frac{P((A_1 \cap B) \bigcup ((A_2 \cap B))}{P(B)} \\
&= \frac{P(A_1 \cap B)}{P(B)} + \frac{P(A_2 \cap B)}{P(B)} \\
&= P(A_1 | B) + P(A_2 | B) = P'(A_1) + P'(A_2).
\end{aligned}
$$

Problem 2.17 *(a) $10, 90, 720$*
(b) Using the recurrence formula $P_{n,k} = P_{n,k-1} \times (n - k + 1)$, we see that $P_{9,4} = 504 \times 6 = 3024$. Using similar reasoning we see that $P_{9,5} = 15, 120$, $P_{9,6} = 60, 480$.

Problem 2.19 *(a)* $S = \{(i,j) : i \neq j, i \in \{1,2,3,4,5\}, j \in \{1,2,3,4,5\}$. *There are* $P_{5,2} = 20$ *sample points in* S. *So the equally likely probability measure assigns to each sample point the probability* $1/20$.
(b) $2/20$ *(c)* $1-(2/20)=0.9$

.

Problem 2.21 *The first 24 numbers (in the ordered list) begin with the digit 1, the 25th through 48th numbers on the list begin with the digit 2, the 49th through 72nd numbers begin with the digit 3, so the 73rd number in the list is 41235.*

Problem 2.23

$$\binom{6}{k} = 6, 15, 20, 15, 6, 1 \text{ for } k = 1, \ldots, 6$$

Problem 2.25 *(a)* $126x^2$ *(b)* $84x^3$

Problem 2.27 *(a) Both numbers must be odd or both numbers must be even. There are* $C_{50,2} = 1,225$ *ways of choosing two odd numbers and, similarly, there are* $1,225$ *of choosing two even numbers; so the total number of ways of getting an even sum is* $2,450$.
(b) All three numbers must be even, and there are $C_{50,3} = 19,600$ *of them; or, two of them must be odd and the third must be even, and there are* $C_{50,2} \times 50 = 61,250$ *of them. Consequently, there are* $80,850$ *ways of getting an even sum.*

Problem 2.29 *(a)* 4 *(b)* 2

Problem 2.31 $10 \times 10 \times 10 \times 26 \times \times 26 \times 26 = 17,576,000$

Problem 2.33

$$\frac{1}{\binom{36}{6}} = 5.134 \times 10^{-7}$$

Problem 2.35 *(a)* $1/12$ *(b)* $1/4$ *(c)* $1/4$

Problem 2.37

$$\left(\frac{12!}{4!4!4!}\right)/3! = 34650/6 = 5775$$

Problem 2.39 *(a)* 0.0175 *(b)* 0.3193 *(c)* 0.4789 *(d)* 0.1842

Problem 2.41 *(a) There are* 12 *outcomes:* $(1,2),(1,3),(1,4),(2,1),(2,3),(2,4),(3,1),(3,2),$
$(3,4),(4,1),(4,2),(4,3)$
(b) 0.5 *(c)* $1/3$ *(d)* $1/3$

Problem 2.43 *(a)* $4/\binom{52}{5} = 1.5391 \times 10^{-6}$

$$(b) \ \frac{13 \times \binom{4}{3} \times 12 \times \binom{4}{2}}{\binom{52}{5}} = 0.0014$$

Problem 2.45 *(a)* $C_{20,12}/C_{30,12} = 0.0015$ *(b)* $C_{20,2}/C_{30,12} = 2.1967 \times 10^{-6}$

Problem 2.47
(a) $P(A \cap B) = 0.09$ *(b)* $P(A \cup B) = 0.61$ *(c)* $P(A \cap B') = 0.31$ *(d)* $P(B|A) = 0.225$

Problem 2.49 *(a)* 0.4 *(b)* 0.4

Problem 2.51 *Let M, W, E denote the event that the student selected is man, woman, engineering student, respectively. We solve the problem by computing the conditional probability $P(M|E)$. So we must compute $P(E)$ and $P(M \cap E)$.*

$$P(E) = 0.4 \times 0.7 + 0.6 \times 0.3 = 0.46; \ \ so$$
$$P(M|E) = \frac{0.4 \times 0.7}{0.46} = 0.61$$

Problem 2.53 *(a)* $P(A^+) = 0.109$ *(b)* $P(W|A^+) = 0.088$

Problem 2.55 *From the information given we see that: $P(S) = 0.001$, $P(T^+|S') = 1600/9990 = 0.16$, $P(T^+|S) = 0.8$. Therefore, (a) $P(T^+) = 0.1608$. (b) $P(S|T^+) = 0.001$*

Problem 2.57

$$P(D|T^-) = \frac{0.01 \times 0.0002}{0.98} = 2.04 \times 10^{-6}$$
$$P(D'|T^-) = \frac{0.98 \times 0.9998}{0.98} = 0.98$$

Problem 2.59

$$P(both \ are \ defective \mid at \ least \ one \ is \ defective) = \frac{3/45}{24/45} = \frac{1}{8}$$

Problem 2.61 *(a)* 0.33 *(b)* 0.5

Problem 2.63 *(a)* $0.98 \times 0.95 = 0.931$ *(b)* $1 - 0.931 = 0.069$

Problem 2.65 $P(W) = 0.891$

Problem 2.67 *(a) Since $\mathcal{S} \cap A = A$ it follows that $P(\mathcal{S} \cap A) = P(A)$. On the other hand $P(\mathcal{S}) \times P(A) = 1 \times P(A) = P(A)$.*
(b) $P(\emptyset \cap A) = P(\emptyset) = 0 = 0 \times P(A)$

Problem 2.69

$$P(A) = P(A \cap A) = P(A)P(A),$$
$$so \ P(A) = P(A)^2 \ and \ therefore$$
$$P(A)(1 - P(A)) = 0.$$

Problem 2.71 *We must show that $P(A \cap B') = P(A)P(B')$, which is an immediate consequence of the fact that $P(A \cap B') = P(A) - P(A \cap B) = P(A) - P(A)P(B) = P(A)(1 - P(B)) = P(A)P(B')$.*

Chapter 3

Discrete Random Variables and Their Distribution Functions

Problem 3.1

(a) $P(X \leq 0) = 10/15$	(b) $P(X < 0) = 5/15$	(c) $P(X \leq 1) = 14/15$
(d) $P(X \leq 1.5) = 14/15$	(e) $P(|X| \leq 1) = 13/15$	(f) $P(|X| < 1) = 5/15$

Problem 3.3 (a) $c = 1/(6 + (7 + 5) + (8 + 4)) = 1/30$
(b)

$$F(-2.5) = 0 \quad F(-0.5) = 15/30, \ F(0) = 21/30$$
$$F(1.5) = 26/30 \quad F(1.7) = 26/30, \ F(3) = 1$$

Problem 3.5

x	1	5	10	15
$f(x)$	0.4	0.3	0.2	0.1

Problem 3.7 *There are* $C_{5,2} = 10$ *sample points given by*
$(12), (13), (14), (15), (23), (24), (25), (34), (35), (45)$.

w	3	4	5	6	7	8	9
$f(w)$	0.1	0.1	0.2	0.2	0.2	0.1	0.1

Problem 3.9

$$f(x) = \frac{\binom{4}{x}\binom{46}{5-x}}{\binom{50}{5}}, \ x = 0, 1, 2, 3, 4$$

Problem 3.11 (a)

$$\sum_{1 \leq x \leq n} cx = c\frac{n(n+1)}{2} = 1; \ so$$

$$c = \frac{2}{n(n+1)}.$$

(b)

$$F(x) = \frac{2}{n(n+1)} \sum_{1 \leq i \leq x} i$$

$$= \frac{2}{n(n+1)} \frac{x(x+1)}{2} = \frac{x(x+1)}{n(n+1)}$$

Problem 3.13

$$(a)\ P(Y > a) = \sum_{a+1 \leq x < \infty} pq^{x-1} = pq^a(1 + q + q^2 + \ldots)$$

$$= pq^a \left(\frac{1}{1-q} \right) = q^a$$

$$(b)\ P(Y > a + b | Y > b) = \frac{P(Y > a + b)}{P(Y > b)} = \frac{q^{a+b}}{q^b} = q^a$$

(c) We use the fact that every odd integer $x = 2k - 1$, $k = 1, 2, 3, \ldots$. Therefore,

$$P(Y \text{ is odd }) = \sum_{1 \leq k < \infty} pq^{(2k-1)-1}$$

$$= p(\sum_{1 \leq k < \infty} q^{2k-2}) = p(1 + q^2 + q^4 + \ldots)$$

$$= p \left(\frac{1}{1-q^2} \right) = \frac{p}{(1-q)(1+q)} = \frac{1}{1+q}.$$

(d) Note that $(-1)^Y = +1$ when Y is even and $(-1)^Y = -1$ when Y is odd. Therefore, by part (c)

$$P((-1)^Y = -1) = \frac{1}{1+q} \text{ and } P((-1)^Y = 1) = \frac{q}{1+q}.$$

Problem 3.15 *(a) $c = 0.438$ (b) $E(X) = 1.1899$, $E(X^2) = 3.1898$, $V(X) = 1.7739$*

Problem 3.17   *(a) $E(X) = 1.5$ (b) $E(X^2) = 3$, so $V(X) = 3 - (1.5)^2 = 0.75$.*

Problem 3.19
   $E(X) = 1.25$; $V(X) = E(X^2) - 1.25^2 = 0.8633$

Problem 3.21
(a) $E(X) = 3$ (b) $E(X^2) = 11$ (c) $V(X) = 11 - 3^2 = 2$ (d) $E(2^X) = 12.4$
(e) $E(\sqrt{X}) = 1.68$ (f) $E(2X + 5) = 2E(X) + 5 = 2 \times 3 + 5 = 11$
(g) $V(2X + 5) = 2^2 V(X) = 4 \times 2 = 8$

Problem 3.23 *(a) 2 (b) 3 (c) 9*

Problem 3.25   *(a) $E(X) = 0$ (b) $E(X^2) = 16/15$*
(c) $V(X) = 16/15$ (d) $E(2^X) = 1.28$
(e) $E(2^{-X}) = 1.28$ (f) $E(\cos(\pi X)) = -1/15$

Problem 3.27   *(a) $E(X) = -0.33$ (b) $E(X^2) = 2$ (c) $V(X) = 1.89$*
(d) $E(2^X) = 1.32$ (e) $E(2^{-X}) = 1.85$ (f) $E(\cos(\pi X)) = 0.2$

Problem 3.29 $E(X) = \frac{2n+1}{3}$

Problem 3.31   *The graph of $p(1 - p)$ is a parabola, which assumes its maximum value at $p = 1/2$. Therefore, $p(1 - p) \leq 1/4$. The variance of a Bernoulli random variable is given by $V(X) = p(1 - p)$, so $V(X) \leq \frac{1}{4}$.*

Problem 3.33

(a) See the answers to part (b) below.

$$(b)\ h(0) = \frac{\binom{14}{4}}{\binom{20}{4}} = 0.2066;$$

$$R_h(x) = \frac{(4-x)(6-x)}{(x+1)(10+x+1)},\ (x = 0, 1, 2, 3),\ so$$

$$h(1) = R_h(0) \times h(0) = 0.4508$$
$$h(2) = R_h(1) \times h(1) = 0.2817$$
$$h(3) = R_h(2) \times h(2) = 0.0578$$
$$h(4) = R_h(3) \times h(3) = 0.0031.$$

Problem 3.35

$$(a)\ n = 5,\ N = 50,\ D = 1$$

$$P(X = 0) = \frac{\binom{49}{5}}{\binom{50}{5}} = 0.90$$

$$(b)\ n = 10,\ N = 100,\ D = 2$$

$$P(X = 0) = \frac{\binom{98}{10}}{\binom{100}{10}} = 0.8091$$

$$(c)\ n = 20,\ N = 200,\ D = 4$$

$$P(X = 0) = \frac{\binom{196}{20}}{\binom{200}{20}} = 0.6539$$

Problem 3.37

$$P(Y = 1) = \frac{2}{10},$$
$$P(Y = x) = \frac{P_{8,x-1}}{P_{10,x-1}} \times \frac{2}{10-(x-1)},\ (x = 2, \ldots, 9)$$

Problem 3.39 *Notation:* $(U = 2) = (WW)$ *denotes the event that the first two draws produced white balls,* $(U = 3) = (WRW) \cup (RWW)$ *is the event that both white balls were drawn on the third draw, etc. Consequently,*

$$P(U = 3) = P(RWW) + P(WRW)$$
$$= \frac{4 \times 2}{6 \times 5 \times 4} + \frac{2 \times 4}{6 \times 5 \times 4}$$
$$= \frac{2 \times 4 \times 2}{6 \times 5 \times 4} = \frac{2}{15}.$$

The remaining probabilities are computed in a similar fashion.

$$P(U = 2) = \frac{2 \times 1}{6 \times 5} = \frac{1}{15}$$

$$P(U = 3) = \frac{2 \times 4 \times 2}{6 \times 5 \times 4} = \frac{2}{15}$$

$$P(U = 4) = \frac{3 \times 4 \times 3 \times 2}{6 \times 5 \times 4 \times 3} = \frac{3}{15}$$

$$P(U = 5) = \frac{4 \times 4 \times 3 \times 2 \times 2}{6 \times 5 \times 4 \times 3 \times 2} = \frac{4}{15}$$

$$P(U = 6) = \frac{5 \times 4 \times 3 \times 2 \times 2}{6 \times 5 \times 4 \times 3 \times 2} = \frac{5}{15}$$

Problem 3.41

$$(a)\ B(4; 10; 0.1) = 0.998$$
$$B(4; 10; 0.4) = 0.633$$
$$B(4; 10; 0.6) = 1 - B(5; 10; 0.4) = 1 - 0.834 = 0.166$$
$$B(4; 10; 0.8) = 1 - B(5; 10; 0.2) = 1 - 0.994 = 0.006$$
$$(b)\ 1 - B(9; 20; 0.2) = 1 - 0.997 = 0.003$$
$$1 - B(9; 20; 0.5) = 1 - 0.412 = 0.588$$
$$1 - B(9; 20; 0.7) = B(10; 20; 0.3) = 0.983$$
$$1 - B(9; 20; 0.9) = B(10; 20; 0.1) = 1.00$$

Problem 3.43 *The probability of getting x heads in n tosses of a coin with probability p for getting a head equals the probability of getting $n - x$ tails (in n tosses) with probability $1 - p$ of getting a tail.*

Problem 3.45 *The probability that the student guesses correctly is $1/4 = 0.25$. Consequently, the number of correct guesses has a binomial distribution with parameters $n = 10$, $p = 0.25$.*

$$P(X \geq 6) = 1 - B(5; 10; 0.25) = 1 - 0.980 = 0.02$$

Problem 3.47
(a) *The net revenue, denoted R, for a four seat van is given by $R = 10N - 20$, where N is the number of passengers in the van. Consequently, $E(R) = 10E(N) - 20$. The pf of N is given by*

n	0	1	2	3	4
$f(n)$	0.006	0.040	0.121	0.215	0.618

  Note that $P(N = 4) = P(X \geq 4)$.
  Now $E(N) = 3.399$, so $E(R) = 10 \times 3.399 - 20 = \$13.99$.
(b) *The net revenue, denoted R, for a five seat van is given by $R = 10N - 25$, where N is the number of passengers in the van. Consequently, $E(R) = 10E(N) - 25$. The pf of N is given by*

n	0	1	2	3	4	5
$f(n)$	0.006	0.040	0.121	0.215	0.251	0.367

Note that $P(N = 5) = P(X \geq 5)$. Now $E(N) = 3.766$, so $E(R) = 10 \times 3.766 - 25 = \$12.66$. (c) A four seat van has greater expected net revenue, so it is more profitable to operate.

Problem 3.49 *We justify the binomial approximation to the hypergeometric by noting that $n/N = 20/1000 = 0.02 < 0.05$. The number of defective items in the sample has an approximate binomial distribution with parameters $n = 20$, $p = 50/1000 = 0.05$.*

$$(a) \ P(X \leq 2) = B(2; 20; 0.05) = 0.925$$
$$(b) \ P(X \leq 1) = B(1; 20; 0.05) = 0.736$$
$$(c) \ P(X = 0) = B(0; 20; 0.05) = 0.358$$

Problem 3.51 *In this case R is binomially distributed with parameters $n = 200$ and $p = 0.4$ (if the drug is worthless) and $p = 0.8$ if the recovery rate for patients given the drug is indeed 80%.*
(a) $E(R) = 200 \times 0.4 = 80$ and $V(R) = 200 \times 0.4 \times 0.6 = 48$
(b) $E(R) = 200 \times 0.8 = 160$ and $V(R) = 200 \times 0.8 \times 0.2 = 32$

Problem 3.53

$$(a) \ P(225 \leq X \leq 275) = P(|\hat{p} - 0.5| \leq 0.05)$$
$$\geq 1 - \frac{1}{4 \times 500 \times (0.05)^2} = 0.8$$
$$(b) \ P(80 \leq X \leq 120) = P(|\hat{p} - 0.5| \leq 0.1)$$
$$\geq 1 - \frac{1}{4 \times 200 \times (0.1)^2} = 0.875$$
$$(c) \ P(45 \leq X \leq 55) = P(|\hat{p} - 0.5| \leq 0.05)$$
$$\geq 1 - \frac{1}{4 \times 100 \times (0.05)^2} = 0$$

Problem 3.55 $P(42 \leq X \leq 58) = P(|X - 50| \leq 8) \geq 1 - (\sigma_X^2/d^2) = 1 - (16/64) = 0.75$

Problem 3.57 *Insert the values $n = 4$, $d = 0.99$, $u = 1.02$, $p = 0.52, 1 - p = 0.48$ into the formula $E(return) = (d(1 - p) + pu)^n - 1 = (0.99 \times 0.48 + 0.52 \times 1.02)^4 - 1 = 0.023$. Similarly,*

$$\sqrt{V(R_4)} = \sqrt{(d^2(1 - p) + pu^2)^n - (d(1 - p) + pu)^{2n}} = 0.0305.$$

Problem 3.59 *Insert the values $n = 4$, $d = 0.985$, $u = 1.015$, $p = 0.52, 1 - p = 0.48$ into the formula $E(return) = (d(1 - p) + pu)^n - 1 = (0.985 \times 0.48 + 0.52 \times 1.015)^4 - 1 = 0.0024$. Similarly,*

$$\sqrt{V(R_4)} = \sqrt{(d^2(1 - p) + pu^2)^n - (d(1 - p) + pu)^{2n}} = 0.0300.$$

Problem 3.61

Face	1	2	3	4	5	6
Frequency	3407	3631	3176	2916	3448	3422

(a) $p = 1/6 = 0.1667$, $\hat{p} = 3448/20000 = 0.1724$, *so* $|\hat{p} - p| = 0.0057$. *Therefore,*

$$P(|\hat{p} - p| \geq 0.0057) \leq \frac{1/6 \times 5/6}{20000 \times (0.0057)^2} = 0.21.$$

(b) $p = 1/6 = 0.1667$, $\hat{p} = 2916/20000 = 0.1458$, *so* $|\hat{p} - p| = 0.0209$. *Therefore,*

$$P(|\hat{p} - p| \geq 0.0209) \leq \frac{1/6 \times 5/6}{20000 \times (0.0209)^2} = 0.0159.$$

Problem 3.63 *The number X of disk drives that last 6000 or more hours is binomial with $n = 20000$ and $p = 0.80$. Consequently, $E(X) = 20000 \times 0.8 = 16000$.*

Problem 3.65

$$p(x + 1; \lambda) = \frac{\lambda^{x+1}}{(x + 1)!} \times \exp(-\lambda)$$

$$= \frac{\lambda}{x + 1} \frac{\lambda^x}{x!} \times \exp(-\lambda)$$

$$= \frac{\lambda}{x + 1} p(x; \lambda), \quad so$$

$$\frac{p(x + 1; \lambda)}{p(x; \lambda)} = \frac{\lambda}{x + 1}.$$

Problem 3.67

$$\exp(-1.5) = 0.2231$$
$$1.5 \times \exp(-1.5) = 0.3347$$
$$\frac{(1.5)^2}{2} \times \exp(-1.5) = 0.2510$$
$$\frac{(1.5)^3}{6} \times \exp(-1.5) = 0.1255$$

(b) $p(0; 1.5) = 0.223$
$p(1; 1.5) = P(1; 1.5) - P(0; 1.5) = 0.558 - 0.223 = 0.335$
$p(2; 1.5) = P(2; 1.5) - P(1; 1.5) = 0.809 - 0.558 = 0.251$
$p(3; 1.5) = P(3; 1.5) - P(2; 1.5) = 0.934 - 0.809 = 0.125$

Problem 3.69 *Let X denote the number of unvaccinated students. It has a binomial distribution with parameters $n = 50$, $p = 0.01$. We approximate it with the Poisson distribution with parameter $\lambda = 50 \times 0.01 = 0.5$.*

(a) $P(X = 0) = p(0; 0.5) = 0.6065$
(b) $P(X = 1) = p(1; 0.5) = 0.3033$
(c) $P(X \geq 2) = 1 - P(X \leq 1) = 0.0902$

Problem 3.71 *(a)* $P(X = 0; 4) = e^{-4} = 0.0183$ *(b)* $P(X \geq 7; 4) = 1 - P(X \leq 6; 4) = 0.111$.

Problem 3.73 *(a) The distribution of X is binomial with $n = 20,000$, $p = 0.0001$.*
(b) $E(X) = 20,000 \times 0.0001 = 2$
(c) We use the Poisson approximation with $\lambda = 2$. $P(X \geq 3; 2) = 1 - 0.667$.

Problem 3.75 *Let X denote the number of defective items in the sample. It has an approximate binomial distribution with parameters $n = 100$, $p = D/10000$. We approximate the binomial distribution by the Poisson distribution with parameter $\lambda = 100 \times D/10000 = D/100$. So, $P(X \leq x) \approx P(x; D/100)$.*
(a) Here $D = 50$, so $P(X \leq 2) \approx P(2; 0.5) = 0.986$.
(b) $D = 100$; $P(X \leq 2) \approx P(2; 1) = 0.92$
(c) $D = 200$; $P(X \leq 2) \approx P(2; 2) = 0.677$

Problem 3.77

$$(a)\ E(e^{sX}) = \sum_{1 \leq j < \infty} e^{sj} pq^{j-1} = \sum_{0 \leq j < \infty} e^{s(j+1)} pq^j$$

$$= pe^s \sum_{0 \leq j < \infty} (qe^s)^j = pe^s (1 - qe^s)^{-1}$$

$$(b)\ M_X'(s) = pe^s(1 - qe^s)^{-1} + pqe^{2s}(1 - qe^s)^{-2}$$

$$M_X''(s) = pe^s(1 - qe^s)^{-1} + 3pqe^{2s}(1 - qe^s)^{-2} + 2pq^2 e^{3s}(1 - qe^s)^{-3}$$

$$M_X'(0) = \frac{1}{p} = \mu_1$$

$$M_X''(0) = \frac{1+q}{p^2} = \mu_2;\ V(X) = \frac{1+q}{p^2} - \frac{1}{p^2} = \frac{q}{p^2}$$

Chapter 4

Continuous Random Variables and Their Distribution Functions

Problem 4.1
(a) $P(1 < Y < 2) = 0.2492$ (b) $P(Y > 3) = 0.1054$
(c) Solve $1 - \exp(-0.75q(0.95)) = 0.95$; the solution is $q(0.95) = 3.9943$.

Problem 4.3

$$(b) \quad P(0.2 < Y < 0.5) = \frac{(0.5)^3}{8} - \frac{(0.2)^3}{8} = 0.0146$$

$$(b) \quad P(Y > 0.6) = 1 - \frac{(0.6)^3}{8} = 0.9730$$

$$(c) \quad f(y) = F'(y) = \frac{3y^2}{8}, \, 0 < y < 2,$$

$$f(y) = 0, \; elsewhere$$

Problem 4.5 Solve $1 - \exp(-q(0.10)/3.5) = 0.10$; therefore, $q(0.10) = 0.3688$ hours, which equals 22.13 minutes.

Problem 4.7 (a) It is easy to see that $F(0) = 1/2$, so the median $= 0$.
(b) Solving $(1 + e^{-x})^{-1} = 0.1$ and $(1 + e^{-x})^{-1} = 0.9$, respectively, yields
$Q(0.10) = -\ln(9)$, $Q(0.90) = \ln(9)$.
(c) Solving $(1 + e^{-x})^{-1} = 0.05$ and $(1 + e^{-x})^{-1} = 0.95$, respectively, yields
$Q(0.05) = -\ln(19)$, $Q(0.90) = \ln(19)$.

Problem 4.9

$$(b) \quad F(x) = 0, \, x \le 0$$
$$F(x) = 3x^2 - 2x^3, \, 0 \le x \le 1$$
$$F(x) = 1, \, x > 1$$
$$(c) \quad 10cm = 0.1m, \; so \; probability \; of \; hitting \; bull's \; eye \; = F(0.1) = 0.028.$$

Problem 4.11

(b) $f(t) = 0, t \leq 1$

$$f(t) = F'(t) = \frac{2}{t^3}, t > 1$$

(c) $P(T > 100) = \dfrac{1}{100^2} = 0.0001$

(d) $P(T > 150 | T > 100) = \dfrac{P(T > 150)}{P(T > 100)} = 0.44$

Problem 4.13

(a) $P(T < 5000) = 1 - \exp(-5000/28700) = 0.1599$
(b) $P(T < 8000) = 1 - \exp(-8000/28700) = 0.2433$

Problem 4.15 $P(T > 1) = \exp(-1/43.3) = 0.9770$

Problem 4.17

(a) F *is a continuous df because*

$f(x) = F'(x) = 0, x \leq 0,$

$f(x) = F'(x) = \dfrac{1}{(1 + x)^2} > 0, x > 0,$ *so*

$f(x) \geq 0;$

$$\int_{-\infty}^{\infty} f(t)\,dt = \int_{0}^{\infty} \frac{1}{(1 + t)^2}\,dt$$

$$\lim_{x \to \infty} \int_{0}^{x} \frac{1}{(1 + t)^2}\,dt = \lim_{x \to \infty} \frac{x}{x + 1} = 1.$$

(b) F *is a continuous df because*

$f(x) = F'(x) = 0, x \leq 2,$

$f(x) = F'(x) = \dfrac{3}{x^2} > 0, x > 2,$ *so*

$f(x) \geq 0;$

$$\int_{-\infty}^{\infty} f(t)\,dt = \int_{2}^{\infty} \frac{3}{t^2}\,dt$$

$$= \lim_{x \to \infty} \int_{2}^{x} \frac{3}{t^2}\,dt = \lim_{x \to \infty} \left(1 - \frac{4}{x^2}\right) = 1.$$

(c) F *is not a continuous df because*

$F(x) = \sin x < 0, -\pi/2 < x < 0.$

Problem 4.19

(a) $E(X) = -\dfrac{1}{6};\ E(X^2) = \dfrac{1}{3};\ V(X) = \dfrac{11}{36}$

(b) $E(X) = \dfrac{\pi}{2};\ E(X^2) = \dfrac{\pi^2 - 4}{2};\ V(X) = \dfrac{\pi^2}{4} - 2$

(c) $E(X) = 0.25;\ E(X^2) = 0.1;\ V(X) = 0.0375$

(d) $E(X) = 1;\ E(X^2) = 1.167;\ V(X) = 0.167$

(e) $E(X) = 0;\ E(X^2) = 0.167,\ V(X) = 0.167$

(f) $E(X) = 0;\ E(X^2) = 2;\ V(X) = 2$

Problem 4.21

$$(a)\ E(X) = \int_a^b \frac{x}{b-a}\,dx = \frac{b^2 - a^2}{2(b-a)} = \frac{a+b}{2}$$

$$(b)\ E(X^2) = \int_a^b \frac{x^2}{b-a}\,dx = \frac{b^2 + ab + a^2}{3(b-a)};$$

therefore,

$$V(X) = E(X^2) - E(X)^2 = \frac{b^2 - 2ab + a^2}{12} = \frac{(b-a)^2}{12}.$$

Problem 4.23 *Since $0 \le U \le 1$ it follows that $a \le X \le a + (b-a) = b$. For $a \le t \le b$ we have*

$$P(X \le t) = P(a + (b-a)U \le t) = P\left(U \le \frac{t-a}{b-a}\right) = \frac{t-a}{b-a}.$$

Problem 4.25 $E(X) = 0$, $E(X^2) = 0.6$, *and* $V(X) = 0.6$.

Problem 4.27 *Integrating by parts we obtain the formula*

$$\int \alpha t(1+t)^{-(\alpha+1)}\,dt = -t(1+t)^{-\alpha} + \frac{(1+t)^{-\alpha+1}}{-\alpha+1}.$$

Consequently,

$$E(T) = \int_0^\infty \alpha t(1+t)^{-(\alpha+1)}\,dt = \lim_{a \to \infty} \int_0^\infty \alpha t(1+t)^{-(\alpha+1)}\,dt$$

$$= \lim_{a \to \infty} \left(-t(1+t)^{-\alpha} + \frac{(1+t)^{-\alpha+1}}{-\alpha+1}\right)\Big|_0^a$$

$$= \frac{1}{\alpha - 1},\ (1 < \alpha < 2).$$

To show that $E(T^2) = \infty$ one must show that the definite integral that defines it is a divergent improper integral; that is:

$$E(T^2) = \int_0^\infty \alpha t^2(1+t)^{-(\alpha+1)}\,dt = \infty.$$

For $(1 < \alpha < 2)$ it is easy to see that $t^2/(1+t)^{\alpha+1} > t^2/(1+t)^3 > C/t$ (for some constant $0 < C < 1$) as $t \to \infty$. As is well known $\int_1^\infty 1/t\,dt = \infty$; therefore $E(T^2) = \infty$.

Problem 4.29

(a) $P(Z < 1) = 0.8413$	*(b)* $P(Z < -1) = 0.1587$	*(c)* $P(|Z| < 1) = 0.6826$
(d) $P(Z < -1.64) = 0.0505$	*(e)* $P(Z > 1.64) = 0.0505$	*(f)* $P(|Z| < 1.64) = 0.8990$
(g) $P(Z > 2) = 0.0228$	*(h)* $P(|Z| > 2) = 0.0456$	*(i)* $P(|Z| < 1.96) = 0.95$

(a) $c = -0.67$	*(b)* $c = 0.67$
(c) $c = 0.67$	*(d)* $c = 1.04$
(e) $c = 1.96$	*(f)* $c = 1.96$

Problem 4.31

Problem 4.33
(a) Let η denote the minimum score to get an A; it satisfies the equation $P(X \geq \eta) = 0.90$. Therefore, $\eta = 65 + 1.28 \times 15 = 84.2$.
(b) Let η denote the minimum score to get a B; it satisfies the equation $P(X \geq \eta) = 0.80$. Therefore, $\eta = 65 + 0.7881 \times 15 = 76.8$.

Problem 4.35

$$P(|X - 0.15| > 0.005) = P(X > 0.155) + P(X < 0.145)$$
$$= P\left(\frac{X - 0.151}{0.003} > \frac{0.155 - 0.151}{0.003}\right) + P\left(\frac{X - 0.151}{0.003} < \frac{0.145 - 0.151}{0.003}\right)$$
$$= P(Z > 1.33) + P(Z < -2.00) = 0.0918 + 0.0228 = 0.1146$$

Problem 4.37 *(a) $P(X < 50) = 0.0228$ (b) $P(X > 65) = 0.1587$*
(c) $c = 9.8$ (d) $c = 48.35$

Problem 4.39
(a) $P(X > 650) = 0.0668$ (b) $P(X < 550) = 0.6915$
(c) The number of students who score 650 or less is a binomial random variable X with parameters $n = 10$, $p = P(X < 650) = 0.9332$. The probability that at least one out of the ten scored at least 650 or better equals $1 - (0.9332)^{10} = 0.50$.
(d) The probability that all ten scored at most 550 equals $P(X < 550)^{10} = (0.6915)^{10} = 0.025$.

Problem 4.41 *Insert the values $\mu = 0.10, \sigma = 0.20, t = 0.5, S_0 = 50$ into Equations 4.47, 4.48, and 4.49 as needed, and use the fact that $\ln(b/a) = \ln(b) - \ln(a)$. In detail:*

(a) $E(\ln(S(0.5)) = \ln(50) + 0.10 \times 0.5 = 3.1920 + 0.05 = 3.9620$;
$\sigma(\ln(S(0.5)) = \sqrt{0.5} \times 0.20 = 0.1414$;
$E(S(0.5)) = 50 \exp(0.5(0.10 + 0.20^2/2) = 53.09$;
$\sigma^2(S(0.5)/50) = \exp(0.5(2 \times 0.10 + 0.20^2)(\exp(0.5 \times 0.2^2)) - 1) = 0.0228$; *therefore*
$\sigma(S(0.5) = \sqrt{\sigma^2(S(0.5) = 50^2 \times 0.0228} = \sqrt{56.9424} = 7.55$.

(b) $P(S(0.5)/50 > 1.3) = P(\ln(S(0.5)/50) > \ln(1.3))$
$$= P\left(\frac{\ln(S(0.5)/50) - 0.05}{0.1414} > \frac{0.2624 - 0.05}{0.1414}\right)$$
$$= P(Z > 1.502) = 0.0668$$

(c) $P(S(0.5)/50 < 0.8) = P(\ln(S(0.5)/50) < \ln(0.8))$
$$= P\left(\frac{\ln(S(0.5)/50) - 0.05}{0.1414} > \frac{-0.2231 - 0.05}{0.1414}\right)$$
$$= P(Z < -1.93) = 0.0268$$

Problem 4.43 *Insert the values $\mu = 0.20, \sigma = 0.30, t = 0.25, S_0 = 60$ into Equations 4.47, 4.48, and 4.49 as needed. In detail:*

(a) $E(S(0.25)/60) = \exp(0.25(0.20 + (0.30)^2/2)) = 1.0632;$ *consequently*
$$E(S(0.5)) = 60 \times 1.0632 = 63.97.$$
(b) $\sigma^2(S(0.25)) = 60^2 \times \exp(0.5(2 \times 0.20 + 0.30^2)) \times (\exp(0.25 \times 0.30^2) - 1)$
$$= 3600 \times 0.0257 = 92.52; \ \textit{consequently}$$
$$\sigma(S(0.5)) = \sqrt{92.52} = 9.62.$$

(c) $P\left(S(0.25)/60 < 1\right) = P\left(\ln\left(S(0.5)/60\right) < 0\right)$
$$= P\left(\frac{\ln\left(S(0.5)/60\right) - 0.25 \times 0.20}{0.25 \times 0.30} < \frac{-0.05}{0.075}\right)$$
$$= P(Z < -0.67) = 0.2743$$

(d) $P\left(S(0.25)/60 > 1.15\right) = P\left(\ln\left(S(0.5)/60\right) > \ln(1.15)\right)$
$$= P\left(\frac{\ln\left(S(0.5)/60\right) - 0.05}{.075} > \frac{0.1398 - 0.05}{0.075}\right)$$
$$= P(Z > 1.12) = 0.1314$$

(e) $P\left(S(0.25)/60 < 0.9\right) = P\left(\ln\left(S(0.5)/60\right) < \ln(0.9)\right)$
$$= P\left(\frac{\ln\left(S(0.5)/60\right) - 0.05}{.075} > \frac{-0.1054 - 0.05}{0.075}\right)$$
$$= P(Z < 2.07) = 0.0192$$

Problem 4.45 *The following answers were obtained without the continuity correction.*

(a) $P(X \geq 160) = 0.124$	*(b)* $P(X \leq 140) = 0.124$
(c) $P(|X - 150| \geq 20) = 0.0208$	*(d)* $P(136 \leq X \leq 1.62) = 0.8948$

Problem 4.47 *The random variable Y is binomial with parameters $n = 100$, $p = 0.7$, so $npq = 21$, $\sqrt{npq} = 4.58$.*
We use the normal approximation without the continuity correction.
(a) $P(Y = y) = b(y; 100; 0.7), \ (y = 0, 1, \ldots, 100)$
(b) $P(Y \geq 75) \approx 0.1379$
(c) $P(Y \geq 80) \approx 0.0146$
(d) $P(Y \leq 65) \approx 0.1379$

Problem 4.49 *(a)* Y *is binomial with* $n = 500$, $p = e^{-4} = 0.0183$.
(b) $E(Y) = 500 \times 0.0183 = 9.1578; V(Y) = 500 \times 0.0183 \times 0.9817 = 8.9901$

$$\text{(c)} \ P\left(Y \geq 15\right) \approx P\left(\frac{Y - 9.1578}{\sqrt{8.9901}} \geq \frac{15 - 9.1578}{\sqrt{8.9901}}\right) \approx P(Z \geq 1.95) = 0.0256$$

Problem 4.51 *Let n denote the number of reservations accepted and let X denote the number of guests who actually show up. Then X is a binomial random variable with parameters $n, p = 0.9$. We seek the largest value of n such that $P(X \geq 1001) \leq 0.01$.*

$$P(X \geq 1001) = P\left(\frac{X - 0.9n}{\sqrt{0.09n}} \geq \frac{1001 - 0.9n}{\sqrt{0.09n}}\right) \leq 0.01; \ \ so$$

$$\frac{1001 - 0.9n}{\sqrt{0.09n}} \geq 2.33. \ \ Therefore,$$

$$0.9n + 2.33\sqrt{0.09}\sqrt{n} - 1001 = 0. \ \ Set \ x^2 = n:$$

$$0.9x^2 + 0.699x - 1001 = 0; \ \ solving \ for \ x \ yields$$

$$x = 32.95, \ \ so \ choose \ n = 1086 \geq x^2 = 1085.80.$$

Problem 4.53 *(a) The change of variable $x = u^2/2$ (so $dx = u\,du$) transforms the integral expression for $\Gamma(1/2)$ as follows:*

$$\Gamma(1/2) = \int_0^\infty x^{-1/2} e^{-x}\,dx$$

$$= \sqrt{2} \int_0^\infty e^{-u^2/2}\,du$$

$$= \sqrt{2} \times \frac{1}{2} \int_{-\infty}^\infty e^{-u^2/2}\,du$$

$$= \frac{1}{\sqrt{2}}\sqrt{2\pi} = \sqrt{\pi}.$$

(b) $\Gamma(3/2) = \Gamma\left(\frac{1}{2} + 1\right) = \frac{1}{2}\Gamma\left(\frac{1}{2}\right) = \frac{\sqrt{\pi}}{2}$

$\Gamma(5/2) = \Gamma\left(\frac{3}{2} + 1\right) = \frac{3}{2}\Gamma\left(\frac{3}{2}\right) = \frac{3}{2} \times \frac{\sqrt{\pi}}{2} = \frac{3\sqrt{\pi}}{4}$

Problem 4.55 *A direct calculation yields the following formula for $E(X^k)$:*

$$E(X^k) = \int_0^\infty x^k \left(\frac{1}{\Gamma(\alpha)\beta^\alpha}\right) x^{\alpha-1} e^{-x/\beta}\,dx$$

$$= \int_0^\infty \left(\frac{1}{\Gamma(\alpha)\beta^\alpha}\right) x^{(k+\alpha)-1} e^{-x/\beta}\,dx$$

$$= \frac{\Gamma(\alpha + k)\beta^{(\alpha+k)}}{\Gamma(\alpha)\beta^\alpha} = \alpha(\alpha + 1)\cdots(\alpha + k - 1)\beta^k$$

Setting $k = 1$ yields $E(X) = \alpha\beta$ and setting $k = 2$ yields $E(X^2) = \alpha(\alpha + 1)\beta^2$. Therefore, $V(X) = \alpha(\alpha + 1)\beta^2 - (\alpha\beta)^2 = \alpha\beta^2$.

Problem 4.57 *(a) We find the median by solving (for t) the equation $1 - \exp-(t/15)^{1/2} = 0.5$. The solution is $t = 15 \times \ln(2)^2 = 7.2068$, or $7,206.8$ miles.*
(b) $E(T) = 15\Gamma(1 + (1/0.5)) = 15\Gamma(3) = 30$, or $30,000$ miles
$P(T > 10) = \exp\left(-(10/15)^{1/2}\right) = 0.5134$

Problem 4.59 *(a) We obtain the median by solving the equation*

$$P(T < t) = 1 - \exp((-t/26710)^{1.053}) = 0.5$$

for t. *This leads to the following sequence of calculations:*

$$\exp((-t/26710)^{1.053}) = 0.5; \ \textit{taking logarithms of both sides yields}$$
$$(-t/26710)^{1.053} = \ln 0.5 = -0.6931, \ \textit{so}$$
$$1.053 \ln\left(\frac{t}{26710}\right) = \ln 0.6931. \ \textit{Therefore}$$
$$\ln\left(\frac{t}{26710}\right) = \frac{\ln 0.6931}{1.053} = -0.3481$$

Thus, the median $= 26710 \times \exp(-0.3481) = 18858.67 \ hrs.$

(b) The proportion of engine fans that fail on an 8000 hour warranty is given by
$1 - \exp(-(8000/26710)^{1.053}) = 0.245.$

Problem 4.61 *(a)* $B(1/2, 1) = 2$ *(b)* $B(3, 2) = 1/12$ *(c)* $B(3/2, 2) = 4/15$

Problem 4.63 $\alpha = 3$, $\beta = 4$, *and* $B(3, 4) = 1/60$; *therefore,* $C = 60$.

Problem 4.67

$$E(X) = \int_0^1 \frac{1}{B(\alpha, \beta)} \left(x \times x^{\alpha-1}(1-x)^{\beta-1}\right) dx = \int_0^1 \frac{1}{B(\alpha, \beta)} \times x^{\alpha}(1-x)^{\beta-1} dx$$
$$= \frac{B(\alpha+1, \beta)}{B(\alpha, \beta)} = \frac{\alpha}{\alpha + \beta}$$
$$E(X^2) = \int_0^1 \frac{1}{B(\alpha, \beta)} \left(x^2 \times x^{\alpha-1}(1-x)^{\beta-1}\right) dx = \int_0^1 \frac{1}{B(\alpha, \beta)} \times x^{\alpha+1}(1-x)^{\beta-1} dx$$
$$= \frac{B(\alpha+2, \beta)}{B(\alpha, \beta)} = \frac{\alpha\beta}{(\alpha + \beta + 1)(\alpha + \beta)^2}$$

Problem 4.69 *The distribution of Z is symmetric about 0; consequently*

$$\Phi(x) = P(Z \le x) = P(Z \ge -x) = P(-Z \le x).$$

Problem 4.71 *Let $Y = 1 - X$ with cdf and pdf $G(y)$ and $g(y)$, respectively.*

$$G(y) = P(Y \le y) = P(1 - X \le y) = P(X \ge 1 - y) = 1 - F(1 - y)$$
$$g(y) = G'(y) = F'(1 - y) = f(1 - y) = \frac{1}{B(\alpha, \beta)} \left((1-y)^{\alpha-1} y^{\beta-1}\right)$$
$$= \frac{1}{B(\beta, \alpha)} \left(y^{\beta-1}(1-y)^{\alpha-1}\right),$$

where we used the fact that $B(\alpha, \beta) = B(\beta, \alpha)$.

Problem 4.73 *Let U be a uniformly distributed random variable on the interval $[-1, 1]$. Let $Y = 4 - U^2$.*

$$\textit{(a)} \ G(y) = 0, \ y \le 3$$
$$G(y) = 1 - \sqrt{4 - y}, \ 3 \le y \le 4$$
$$G(y) = 1, \ y > 4$$
$$\textit{(b)} \ g(y) = \frac{1}{2\sqrt{4 - y}}, \ 3 < y < 4;$$
$$g(y) = 0, \ \textit{elsewhere}$$

Problem 4.75

$$F_R(r) = P(R \le r) = P(\sqrt{X} \le r) = P(X \le r^2)$$
$$= 1 - \exp(-r^2/2),\ 0 < r < \infty,\ so$$
$$f_R(r) = F'_R(r) = r \exp(-r^2/2),\ 0 < r < \infty$$
$$f_R(r) = F'_R(r) = 0,\ elsewhere.$$

Problem 4.77 *Let* $X = \tan U$, *where is uniformly distributed on the interval* $[-\pi/2,\ \pi/2]$.

$$(a)\quad F_X(x) = P(\tan U \le x) = P(\frac{-\pi}{2} \le U \le \tan^{-1} x)$$
$$= \frac{1}{\pi}\left(\tan^{-1} x - \left(-\frac{\pi}{2}\right)\right)$$
$$= \frac{1}{\pi}\left(\left(-\frac{\pi}{2}\right) + \tan^{-1} x\right)$$
$$= \frac{1}{2} + \frac{1}{\pi}\arctan x$$

$$(b)\quad f_X(x) = F'_X(x) = \frac{1}{\pi}\frac{1}{1+x^2}$$

Problem 4.79 *The df of* X *is symmetric about the origin, so* $P(-a \le X \le a) = 2P(0 \le X \le a)$.

$$F(y) = 0,\ y \le 0$$
$$F(y) = P(0 \le Y \le y) = P(0 \le X^2 \le y) = P(-\sqrt{y} \le X \le \sqrt{y})$$
$$= 2P(0 \le X \le \sqrt{y}) = 2\int_0^{\sqrt{y}} (1-x)\,dx$$
$$= 1 - (1 - \sqrt{y})^2,\ 0 \le y \le 1$$
$$F(y) = 1,\ y > 1$$
$$f(y) = F'(y) = 2(1 - \sqrt{y})\frac{1}{2\sqrt{y}} = \frac{1}{\sqrt{y}} - 1,\ 0 < y < 1$$
$$f(y) = 0,\ elsewhere$$

Problem 4.81

$$M'_X(s) = \frac{d}{ds}(\theta(\theta - s)^{-1}) = \theta(\theta - s)^{-2};\ M'_X(0) = \frac{1}{\theta};$$
$$M''_X(s) = \frac{d^2}{ds^2}(\theta(\theta - s)^{-1}) = 2\theta(\theta - s)^{-3};\ M''_X(0) = \frac{2}{\theta^2}$$

Problem 4.83 *All three statements are false; the counter examples follow.*
(a) If X *is* $N(0,1)$ *then* $E(X)^2 = 0^2 = 0 \ne E(X^2) = 1$.
(b) If X *is uniform on the interval* $[1,2]$ *then* $E(X) = 1.5$ *and*

$$E(1/X) = \int_1^2 x^{-1}dx = \ln 2 = 0.6931 \ne 1/1.5 = 0.6667 = 1/E(X).$$

(c) If X *is* $N(0,1)$ *then* $E(X) = 0$ *but* $P(X = 0) = 0$.

Chapter 5

Multivariate Probability Distributions

Problem 5.1

(a)

X\ Y	2	3	4	5	$f_X(x)$
0	1/24	3/24	1/24	1/24	6/24
1	1/12	1/12	3/12	1/12	6/12
2	1/12	1/24	1/12	1/24	6/24
$f_Y(y)$	5/24	6/24	9/24	4/24	

(b) X and Y are not independent. Note that $f(0,2) = 1/24 \neq f_X(0)f_Y(2) = (6/24) \times (5/24)$.
(c) $\mu_X = 2.5; \sigma(X) = 1$ (d) $\mu_Y = 1; \sigma(Y) = (0.5)^{1/2} = 0.7071$

Problem 5.3
(a) $P(Y \leq 1 | X = x) = 12/15, 1, 1$ for $x = 0, 1, 2$, respectively.
(b) $P(Y = y | X = 0) = 1/5, 3/5, 1/5$ for $y = 0, 1, 2$, respectively.

Problem 5.5
(a) $E(X) = 120/54 = 2.2222, E(X^2) = 300/54 = 5.5556, V(X) = 0.6173$
(b) $E(Y) = 114/54 = 2.1111, E(Y^2) = 276/54 = 5.1111, V(Y) = 0.6543$
(c) $E(XY) = 256/54 = 4.407$
(d) $Cov(X,Y) = 0.0494$ and $\rho(X,Y) = 0.0494/(\sqrt{0.6173 \times 0.6543}) = 0.0778$

Problem 5.7 (a) $E(X) = 7/3 = 2.3333, E(X^2) = 6, V(X) = 5/9$
(b) $E(Y) = 5/3, E(Y^2) = 3, V(Y) = 2/9$
(c) $E(XY) = 35/9$
(d) $Cov(X,Y) = 0$ and $\rho(X,Y) = 0$, because X, Y are independent.

Problem 5.9

(a)

Joint Probability Function of (Y, Z)				
Y\Z	0	1	2	$f_Y(y)$
0	1/28	6/28	3/28	12/28
1	6/28	9/28	0	15/28
2	3/28	0	0	3/28
$f_Z(z)$	10/28	15/28	3/28	

(b) $E(X) = 1/2, E(Y) = E(Z) = 3/4$, so $E(W) = E(5X + 10Y + 25Z) = 115/4 = 28.75$ cents.

Problem 5.11

$$(a) \quad \frac{\binom{13}{x}\binom{39}{13-x}}{\binom{52}{13}}, \quad x = 0, 1, \ldots, 13$$

$$(b) \quad \frac{\binom{13}{y}\binom{39}{13-y}}{\binom{52}{13}}, \quad y = 0, 1, \ldots, 13$$

$$(c) \quad \frac{\binom{13}{x}\binom{13}{y}\binom{26}{13-x-y}}{\binom{52}{13}}, \quad 0 \leq x + y \leq 13$$

Problem 5.13 *The probability functions of the weird dice and the standard dice are the same, as shown in the horizontal bar chart below.*

x	$f_X(x)$	$F_X(x)$	$\{s : X(s) = x\}$
2	1/36	1/36	$(1,1)$
3	2/36	3/36	$(2,1), (2,1)$
4	3/36	6/36	$(1,3), (3,1), (3,1)$
5	4/36	10/36	$(1,4), (2,3), (2,3), (4,1)$
6	5/36	15/36	$(1,5), (2,4), (2,4)(3,3)(3,3)$
7	6/36	21/36	$(1,6), (2,5), (2,5), (3,4), (3,4), (4,3)$
8	5/36	26/36	$(2,6), (2,6), (3,5), (3,5), (4,4)$
9	4/36	30/36	$(1,8), (3,6), (3,6), (4,5)$
10	3/36	33/36	$(2,8), (2,8), (4,6)$
11	2/36	35/36	$(3,8), (3,8)$
12	1/36	1	$(8,4)$

Problem 5.15

$x + y$	0	1	2	3
f_{X+Y}	1/8	3/8	3/8	1/8

Problem 5.17 *A die is thrown 6 times. Let $Y_i =$ the number of times that i appears.*

$$(a) \quad P(Y_1 = 1, \ldots, Y_6 = 1) = \binom{6}{1, 1, 1, 1, 1, 1}\left(\frac{1}{6}\right)^6 = 0.0154$$

$$(b) \quad P(Y_1 = 2, Y_3 = 2, Y_5 = 2) = \binom{6}{2, 0, 2, 0, 2, 0}\left(\frac{1}{6}\right)^2\left(\frac{1}{6}\right)^2\left(\frac{1}{6}\right)^2$$
$$= 0.0019$$

$$(c) \quad E(Y_i) = 6 \times \frac{1}{6} = 1$$

Problem 5.19 *(a)* $E(Y_1) = 10 \times 0.5 = 5$, $E(Y_2) = 10 \times = 0.3 = 3 E(Y_3) = 10 \times 0.2 = 2$

$$(b) \quad P(Y_1 = 5, Y_2 = 3, Y_3 = 2) = \binom{10}{5, 3, 2}(0.5)^5(0.3)^3(0.2)^2 = 0.0851$$

Problem 5.21
(a) $E(3X - 2Y) = 12$ (b) $V(3X) = 81$ (c) $V(-2Y) = 64$ (d) $V(3X - 2Y) = 145$

Problem 5.23 *(a) $E(X_i) = 0.4$; $V(X_i) = 0.64$; so $E(T) = 40 \times 0.4 = 16$, $V(T) = 40 \times 0.64 = 25.6$.*
(b) $-40 \le T \le 40$
(c) The distribution of T is approximately normal with $\mu_T = 16$, $\sigma(T) = \sqrt{25.6} = 5.0596$. Therefore, $P(T > 25) = P(Z > (25 - 16)/5.0596) \approx P(Z > 1.78) = 0.0375$, and $P(T < 0) \approx P(Z < -16/5.0596) = P(Z < -3.16) = 0.0008$.

Problem 5.25 *The proof is a consequence of the fact (established in the previous problem) that $Cov(aX + b, cY + d) = acCov(X,Y)$ and the fact that $\sigma_{aX}\sigma_{cY} = |ac|\sigma_X\sigma_Y$. So,*

$$\rho(aX + b, cY + d) = \frac{acCov(X,Y)}{|ac|\sigma_X\sigma_Y}$$
$$= \frac{ac}{|ac|}\rho(X,Y).$$

Now $ac/|ac| = \pm 1$ accordingly as $ac > 0$ or $ac < 0$.

Problem 5.27 *(a) $E(X_i) = 4$, $V(X_i) = 16$, so $E(\overline{X}) = 4$, $V(\overline{X}) = 1$.*

$$(b)\ P(2.4 < \overline{X} > 5.6) = P((2.4 - 4)/1 < (\overline{X} - 4)/1 < (5.6 - 4)/1)$$
$$\approx P(-1.6 < Z < 1.6) = 0.8994$$

Problem 5.29

$$M_{X_1+X_2}(s) = M_{X_1}(s)M_{X_2}(s) = \exp(\lambda_1(e^s - 1))\exp(\lambda_2(e^s - 1))$$
$$= \exp(\lambda_1 + \lambda_2)(e^s - 1)),$$

which is the mgf of a Poisson distribution with parameter $\lambda = \lambda_1 + \lambda_2$.

Problem 5.31
(a) E(stock price after 1 year) = \$47.85
(b) σ(stock price after 1 year) = 12.1769
(c) The probability of a loss after 1 year = 0.2743.
(d) The probability that the stock's price will be up at least 25% after 1 year = 0.3849.
(e) The probability that the stock's price will be down at least 15% after 1 year = 0.1665.

Problem 5.33
$$P\left(\ln\left(\frac{S_n(j/n)}{S_n((j-1)/n)}\right) = \frac{\mu}{n} \pm \frac{\sigma}{\sqrt{n}}\right) = \frac{1}{2}$$

Therefore,

$$E\left(\ln\left(\frac{S_n(j/n)}{S_n((j-1)/n)}\right)\right) = \frac{1}{2}\left((\frac{\mu}{n} + \frac{\sigma}{\sqrt{n}}) + (\frac{\mu}{n} - \frac{\sigma}{\sqrt{n}})\right) = \mu/n$$

$$E\left(\left(\ln\left(\frac{S_n(j/n)}{S_n((j-1)/n)}\right)\right)^2\right) = \frac{1}{2}\left((\frac{\mu}{n} + \frac{\sigma}{\sqrt{n}})^2 + (\frac{\mu}{n} - \frac{\sigma}{\sqrt{n}})^2\right)$$

$$= \left(\frac{\mu}{n}\right)^2 + .\frac{\sigma^2}{/}n.\ Therefore,$$

$$V\left(\ln\left(\frac{S_n(j/n)}{S_n((j-1)/n)}\right)\right) = \left(\frac{\mu}{n}\right)^2 + \frac{\sigma^2}{n} - \left(\frac{\mu}{n}\right)^2 = \frac{\sigma^2}{n}.$$

Problem 5.35 *Let* $R = 0.4 \times R_A + 0.6 \times R_B$. *Then* $E(R) = 0.13$, $\sigma(R) = 0.2066$.

Problem 5.37 *(a) Let* $R_1 = 0.5 \times R_B + 0.5 \times R_C$; $\sigma(R_1) = 0.1442$
(b) Let $R_2 = 0.4 \times R_A + 0.52 \times R_B + 0.4 \times R_C$; $\sigma(R_2) = 0.2043$

Problem 5.39
(a)

year	2001	2002	2003	2004	2005
value of portfolio	$9096	$8346	$10,526	$12,941	$14,436

(b) Annualized rate of return = 7.62% (c) Average annual rate of return = 8.67%

Problem 5.41 *(a) The monthly interest rate is* $r = 0.10/12 = 0.0833$. *The present value of dealer A's financing scheme is:*

$$1,000 + \sum_{1 \leq k \leq 30} \frac{300}{(1+0.0833)^k} = 1,000 + 7934.11 = \$8934.11.$$

The present value of dealer B's financing scheme is \$9,000. So, the customer should purchase the car from dealer A.

Problem 5.43
(a) $u_4 = \exp(0.16/4 + 0.35/2) = \exp(0.215 = 1.2399; d_4 = \exp(0.16/4 - 0.35/2) = \exp(-0.135) = 0.8737$

(c)

s	22.14	31.42	44.59	63.28	89.80
$P(S = s)$	1/16	4/16	6/16	4/16	1/16

Problem 5.45 *(a) \$14.58 (Use Equation 5.108 with* $q = 0.90, R = 1.08, u = 1.1, d = 0.9, f(2,2) = 21, f(2,0), f(2,1) = 0.)*
(b) \$0.32 = 32 cents (Here, $f(2,0) = 19$, $f(2,1) = 1$, $f(2,2) = 0.)*

Problem 5.47 *(a) \$0.69 (b) \$1.38*

Problem 5.49

$$\text{(a) } P(X(2) \leq 3) = P(2;3) = 0.857$$
$$\text{(b) } P(X(2) = 2, X(4) = 6) = P(X(2) = 2)P(X(4) - X(2) = 4)$$
$$= p(2;3) \times p(4;3) = 0.224 \times 0.1680 = 0.0376$$

Problem 5.51

$$F'_{W_n}(t) = = \left(1 - \sum_{k=0}^{n-1} e^{-\lambda t} \frac{(\lambda t)^k}{k!}\right)' = -\left(\sum_{k=0}^{n-1} e^{-\lambda t} \frac{(\lambda t)^k}{k!}\right)'$$

$$= -\left(\lambda e^{-\lambda t} + \sum_{k=1}^{n-1} \lambda e^{-\lambda t} \left(\frac{(\lambda t)^{k-1}}{(k-1)!} - \frac{(\lambda t)^k}{k!}\right)\right)$$

$$= \frac{\lambda(\lambda t)^{n-1} e^{-\lambda t}}{(n-1)!}$$

Problem 5.53 *Let $X(t)$ be a Poissont process with intensity λ. Then,*

$$(a)\ \lim_{t\to\infty} E\left(\left|\frac{X(t)}{t}-\lambda\right|^2\right) = \lim_{t\to\infty} V\left(\frac{X(t)}{t}\right) = \lim_{t\to\infty}\frac{\lambda t}{t^2} = 0$$

$$(b)\ \lim_{t\to\infty} P\left(\left|\frac{X(t)}{t}-\lambda\right|>d\right) \le \lim_{t\to\infty}\frac{V(x(t)/t)}{t} = \lim_{t\to\infty}\frac{\lambda}{td^2} = 0$$

Problem 5.55 *A two-out-of-three system is working if (i) all three components are working or (ii) two components are working and one is not. These possibilities are mutually exclusive. In case (i) $X_1 = X_2 = X_3 = 1$, so $X_1X_2X_3$ is the only non-zero term, since $1-X_i = 0$, $(i = 1,2,3)$. In case (ii) we have, say, $X_1 = 0$, $X_2 = 1$, $X_3 = 1$. In this case the only non-zero term is $(1-X_1)X_2X_3$. The other two possibilities are $X_1 = 1$, $X_2 = 0$, $X_3 = 1$, and $X_1 = 1$, $X_2 = 1$, $X_3 = 0$. For each of these cases we see that $X = 1$, so the two functions are equal.*

Problem 5.57 *(i) The system function for the second system is given by $X = X_1(1-(1-X_2)(1-X_3))$, and (ii) its reliability is given by*

$$E(X) = E(X_1(1-(1-X_2)(1-X_3)))$$
$$= p(1-(1-p)^2).$$

Problem 5.59
(a) $P(X < 0.4, Y < 0.6) = 0.56$ (b) $P(X < 0.4) = 0.7$; $P(Y < 0.6) = 0.8$
(c) $P(X < Y) = 0.5$ (d) $P(Y > 2X) = 0.5$
(e) $P(Y = X) = 0$ (f) $P(X + Y < 0.5) = 0.7188$; $P(X + Y < 1.5) = 0.9688$

Problem 5.61

$(a)\ \ f_X(x) = e^{-x}, 0 \le x \le \infty,\ f_X(x) = 0,\ elsewhere$

$\mu_X = 1, \sigma_X = 1$

$(b)\ \ f_Y(y) = (y+1)^{-2}, 0 \le y \le \infty,\ f_Y(y) = 0,\ elsewhere$

$\mu_Y = \int_0^\infty \frac{y}{(y+1)^2} = \infty;\ \sigma_Y = \infty$

$(c)\ No,\ f_x(x)f_Y(y) = \frac{e^{-x}}{(y+1)^2} \ne f(x,y)$

Problem 5.63

$(a)\ \ \int_0^1\int_0^1 (x+xy)dxdy = \int_0^1\int_0^1 x(1+y)dxdy$

$\int_0^1 xdx \int_0^1 (1+y)dy = \frac{3}{4},\ so\ c = \frac{4}{3}$

$(b)\ \ f_X(x) = 2x,\ 0 \le x \le 1,\ f_X(x) = 0,\ elsewhere$

$f_Y(y) = \frac{2}{3}(1+y),\ 0 \le y \le 1,\ f_Y(y) = 0,\ elsewhere$

$(c)\ \ Yes,\ f_X(x)f_Y(y) = \frac{4}{3}x(1+y) = f(x,y)$

$(d)\ \ P(X+Y \le 1) = \frac{4}{3}\int_0^1\int_0^{1-x} x(1+y)dydx = 0.6296$

Problem 5.65

$$(a)\ f_X(x) = \int_{-\sqrt{1-x^2}}^{\sqrt{1-x^2}} \frac{1}{\pi} dy$$

$$= \frac{2}{\pi}\sqrt{1-x^2},\ |x| \le 1$$

$$= 0,\ \ elsewhere$$

$$f_Y(y) = \int_{-\sqrt{1-y^2}}^{\sqrt{1-y^2}} \frac{1}{\pi} dx$$

$$= \frac{2}{\pi}\sqrt{1-y^2},\ |y| \le 1$$

$$= 0,\ \ elsewhere$$

(b)   The symmetry of the marginal probability functions about the origin, that is $f_X(x) = f_X(-x)$, *etc., implies that* $E(X) = 0$ *and* $E(Y) = 0$. *Consequently,*

$$Cov(X,Y) = E(XY) = \int_{-1}^{1}\left(\int_{-\sqrt{1-y^2}}^{\sqrt{1-y^2}} \frac{1}{\pi} dy\right) dx = 0.$$

On the other hand, X *and* Y *are not independent since the product of their pfs is given by*

$$f_X(x)f_Y(y) = \frac{4}{\pi^2}\sqrt{1-x^2}\sqrt{1-y^2} \ne \frac{1}{\pi} = f(x,y).$$

Problem 5.67

$$(a)\ \ f(y|x) = \frac{xe^{-xy-x}}{e^{-x}} = xe^{-xy}$$

$$(b)\ \ \mu_{Y|x} = \int_{0}^{\infty} yf(y|x)dy = \int_{0}^{\infty} yxe^{-xy}dy = \frac{1}{x}$$

Problem 5.69 *(a)* Y *is* $N(1,16)$, *so* $P(Y < 3) = P((Y-1)/4 < (3-1)/4) = P(Z < 0.5) = 0.6915$.
(b) $f(y|x=2)$ *is* $N(1,7)$, *so* $P(Y < 3|X = 2) = P(Y - 1/\sqrt{7} < 2/\sqrt{7}) = P(Z < 0.76) = 0.7764$.
(c) $E(Y|X=2) = 1$
(d) $Cov(X,Y) = -9$; *so* $V(X+Y) = 9 + 16 + 2Cov(X,Y) = 25 - 18 = 7$.

Chapter 6

Sampling Distribution Theory

Problem 6.1 *(a)* $E(X) = 50 - 100 + 100 = 50$ *(b)* $V(X) = 20 + 4 \times 20 + 4 \times 20 = 180$
(c) $P(\mid X - 50 \mid \leq 25) = P(\mid (X - 50)/\sqrt{180} \mid \leq 25/\sqrt{180}) = P(|Z| \leq 1.86) = 0.9372$
(d) $\eta_{(0.9)} = 50 + 1.28 \times \sqrt{180} = 67.173$

Problem 6.3 *(a)* $P(X < 60) = P((X - 65)/10 < (60 - 65)/10) = P(Z < -0.5) = 0.3085$
(b) $T = X_1 + \ldots X_8 \sim N(8 \times 65; 8 \times 100) = N(520; 800)$. *Eight hours equals 480 minutes,*
so $P(T < 480) = P((T - 520)/\sqrt{800} < (480 - 520)/\sqrt{800}) = P(Z < -1.414) = 0.0793$.

Problem 6.5 *(a)* $\overline{X} - \overline{Y} \sim N(1; 25/36)$

$$(b) \ P(\overline{X} > \overline{Y}) = P(\overline{X} - \overline{Y} > 0)$$
$$= P\left(\frac{\overline{X} - \overline{Y} - 1}{5/6} > \frac{-1}{5/6}\right)$$
$$= P(Z > -1.2) = 0.8849$$

Problem 6.7
(a) $\sigma(\overline{X} - \overline{Y}) = \sqrt{\sigma^2(\overline{X}) + \sigma^2(\overline{Y})} = 1347.22$. *Therefore,* $\overline{X} - \overline{Y} \sim N(0; (1347.22)^2)$.

$$(b) \ P(|\overline{X} - \overline{Y}| > 2500) = P\left(\frac{\overline{X} - \overline{Y}}{1347.22} > \frac{2500}{1347.22}\right)$$
$$= P(|Z| > 1.86) = 0.0628$$
$$(c) \ P(\overline{X} - \overline{Y} < -2500) = P\left(\frac{\overline{X} - \overline{Y}}{1347.22} < \frac{-2500}{1347.22}\right)$$
$$= P(Z < -1.86) = 0.0314$$

Problem 6.9 *(a)* $E(W_n) = n \times 175$ *and* $V(W_n) = n \times 400$
(b) $W_n \sim N(n \times 175; n \times 400)$
(c) $P(W_{18} > 3000) = 0.9616$
(d) $N = 16$

Problem 6.11 *(a)* $\chi_{10}^2(0.1) = 15.987$ *(b)* $\chi_{10}^2(0.05) = 18.307$
(c) $\chi_{15}^2(0.1) = 22.307$ *(d)* $\chi_{20}^2(0.9) = 12.443$

Problem 6.13 *(a)* $P(7.22\chi_2^2 < 10) = P(\chi_2^2 < 1.39) = 0.50$.
(b) The probability that n rounds fail to destroy the target equals 0.5^n. *The first n satisfying*
the condition $0.5^n \leq 0.1$ *satisfies the inequality* $n \geq \ln 0.1/\ln 0.5 = 3.32$. *Therefore* $n = 4$.

Problem 6.15

$$P(1.5 \leq s \leq 2.9) = P\left(\frac{15 \times (1.5)^2}{5} \leq \frac{15s^2}{5} \leq \frac{15 \times (2.9)^2}{5}\right)$$
$$= P(6.75 \leq \chi_{15}^2 \leq 25.23) \approx 0.95 - 0.05 = 0.90$$

Problem 6.17
(a) $t_9(0.05) = 1.833$ *(b)* $t_9(0.01) = 2.821$
(c) $t_{18}(0.025) = 2.101$ *(d)* $t_{18}(0.01) = 2.552$

Problem 6.19

(a) $P(14 < \overline{X} < 16) = P(|t_8| < 1.86) = 0.90$
(b) $P(13.76 < \overline{X} < 16.24) = P(|t_8| < 2.31) = 0.95$
(c) $P(13.20 < \overline{X} < 16.80) = P(|t_8| < 3.25) = 0.99$

Chapter 7

Point and Interval Estimation

Problem 7.1 *(a)* $75 \pm 1.47 = [73.53, 76.47]$ *(b)* $[73.86, \infty)$
(c) $75 \pm 1.75 = [73.25, 76.75]$ *(d)* $[73.53, \infty)$

Problem 7.3
(a) 15.88 ± 1.34 *(We use the value $t_{32}(0.05) \approx 1.69$.)*
(b) 15.88 ± 2.17 *(We use the value $t_{32}(0.005) \approx 2.74$.)*

Problem 7.5
(a) Equation 3.64 tells us that $E(X_i) = \lambda$. Therefore $E(\overline{X}) = \lambda$. Our unbiased estimator is therefore $\hat{\lambda} = \overline{X}$.
(b) Equation 3.64 tells us that $V(X_i) = \lambda$. Therefore $\sigma(\overline{X}) = \sqrt{\lambda/n}$. Since λ is unknown we replace it with $\hat{\lambda} = \overline{X}$. Consequently, $s(\overline{X}) = \sqrt{\hat{\lambda}/n}$.
(c) The total number of defects divided by the total number of items is given by $\hat{\lambda} = 43/35 = 1.2286$.
(d) $\sqrt{\hat{\lambda}/n} = \sqrt{1.2286/35} = 0.1874$

Problem 7.7
$$n - 1 = 7, \ \alpha/2 = 0.05, \ t_7(0.05) = 1.895, \ s = 0.8,$$

therefore,

$$\overline{X} \pm t_{n-1}(\alpha/2) \times s/\sqrt{n}) = 10.23 \pm 1.895 \times \frac{0.8}{\sqrt{8}} = 10.23 \pm 0.54$$

Problem 7.9

$$\overline{x} = 22.5067, \ s^2 = 76.7586, \ s = 8.7612$$
$$t_{29}(0.025) = 2.045$$
$$(CI \text{ for } \mu) \ 22.5067 \pm 3.271 = [19.2356 \ 25.778]$$
$$(CI \text{ for } \sigma^2) \ = [48.6855, 138.7175]$$

Problem 7.11
(a) A 95% lower bound is given by $L = 29.3 - 1.761 \times 3.2/\sqrt{15} = 27.845$ mpg.
(b) A lower bound for the cruising range is $12 \times 27.845 = 334.14$ miles.

Problem 7.13

$$\overline{x} = 0.359, \ s = 0.2897, \ n = 20, \ t_{19}(0.025) = 2.093$$
$$t_{19}(0.025) \times \frac{s}{\sqrt{20}} = 2.093 \times 0.2897 = 0.1356$$
$$0.359 \pm 0.1356 = [0.2234, 0.4946]$$

The confidence interval for μ is $[0.2234, 0.4946]$.

Problem 7.15

$$(a)\ \ 0.50 \pm 2.021 \times \frac{0.0086}{\sqrt{40}} = [0.4973, 0.5027]$$

$$(b)\ \ 0.50 \pm 2.971 \times \frac{0.0086}{\sqrt{40}} = [0.496, 0.5040]$$

Problem 7.17

$$(a)\ \sqrt{n} \geq \frac{1.96 \times 0.2}{0.1} \ \ so\ n = 16$$

$$(b)\ \sqrt{n} \geq \frac{2.58 \times 0.2}{0.1} \ \ so\ n = 27$$

$$(c)\ \sqrt{n} \geq \frac{1.96 \times 0.2}{0.05} \ \ so\ n = 62$$

Problem 7.19　$(n-1)s^2 = 39 \times 0.000075 = 0.002925$
$\chi^2_{39}(0.025) = 58.12,\ \chi^2_{39}(0.975) = 23.654$
Confidence interval for σ^2: [0.00005, 0.000124]

Problem 7.21

$$\chi^2_{47}(0.025) \approx 67.821,\ \chi^2_{32}(0.975) \approx 29.956;\ (n-1)s^2 = 47 \times 25.69 = 1207.43$$

The confidence interval for σ^2 is [17.80, 40.31].

Problem 7.23　*(a)* $\hat{\mu} = 52 \times 0.0091 = 0.473;\ \hat{\sigma} = \sqrt{52} \times 0.0399 = 0.288$
(b) $(1/12) \times 0.4732 = 0.039;\ \sqrt{1/12} \times 0.2877 = 0.083$
(c) $(1/4) \times 0.4732 = 0.118;\ \sqrt{1/4} \times 0.2877 = 0.144$

Problem 7.25　*(a)* $\hat{\mu} = 52 \times (-0.0011) = -0.057;\ \hat{\sigma} = \sqrt{52} \times 0.0058 = 0.042$
(b) $-0.057 \times (1/4) = -0.014;\ \sqrt{1/4} \times 0.0418 = 0.021$
(c) $[0.035, 0.052]$

Problem 7.27　*(a)* $-\$14,280$ *(b)* $-\$20,371$

Problem 7.29　*(a)* $x = -\$1,242$ *(b)* $x = -\$1,691$

Problem 7.31　*Solving* $-(\mu + x/W_0)/\sigma = z(\alpha)$ *for x yields the solution*
$x = -W_0(\mu + \sigma z(\alpha)).$

Problem 7.33
(a) $s_p^2 = (0.556776)^2 = 0.3100$ *(b)* $s(\overline{X}_1 - \overline{X}_2) = 0.2281$ *(c)* $[-1.073, -0.1270]$

Problem 7.35　*(a)* $s_p^2 = (0.556776)^2 = 0.3100$
(b) $s(\overline{X}_1 - \overline{X}_2) = 0.2281$
(c) $[-1.073, -0.1270]$

Problem 7.37 *A straightforward calculation yields the following results (rounded to two decimal places):*

$$\bar{x}_1 = 5.86, \bar{x}_2 = 10.79, \bar{x}_1 - \bar{x}_2 = -4.93;$$
$$s_1^2 = 11.85, s_2^2 = 10.67;$$
$$s_p^2 = 11.26, s_p = 3.36; t_{18}(0.025)s_p\sqrt{\frac{1}{10} + \frac{1}{10}} = 3.15;$$
$$L = -4.93 - 3.15 = -8.08;$$
$$U = -4.93 + 3.15 = -1.78.$$

The confidence interval is $[-8.08, -1.78]$, which does not include 0. The data imply that diet 2 is somewhat more effective in reducing weight.

Problem 7.39

$$\bar{x}_1 - \bar{x}_2 = 93.78 - 87.4 = 6.38, s_1 = 4.206, s_2 = 7.956$$
$$t_{12}(0.025) = 2.179, \quad s(\overline{X}_1 - \overline{X}_2) = 3.199$$
$$t_{12}(0.025) \times s(\overline{X}_1 - \overline{X}_2) = 2.179 \times 3.199 = 6.971$$

The confidence interval is 6.38 ± 6.971, equivalently, $[-0.5910, 13.351]$.

Problem 7.41 *The data are paired, so we use Equations 7.37 and 7.38 to construct the confidence interval.*

$$(a) \; \bar{d} = 15.97, s_d = 15.86, t_{32}(0.05) \approx 1.69, \frac{s_d}{\sqrt{33}} = 2.76;$$
$$15.97 \pm t_{32}(0.05) \times 2.76 = [11.30, 20.64]$$
$$(b) \; 15.97 \pm t_{32}(0.025) \times 2.76 = [10.34, 21.60]$$
$$(c) \; 15.97 \pm t_{32}(0.005) \times 2.76 = [8.41, 23.53]$$

Problem 7.43

$$(a) \; \bar{d} = 2.607, s_d = 4.713, t_{14}(0.025) = 2.145$$
$$2.145 \times \frac{4.173}{\sqrt{15}} = 2.6102$$
$$2.607 \pm 2.6102 = [-0.0032, 5.2172]$$
$$(b) \; \bar{d} = 2.607, s_d = 4.713, t_{14}(0.005) = 2.997$$
$$2.997 \times \frac{4.173}{\sqrt{15}} = 3.6230$$
$$2.607 \pm 2.6102 = [-1.016, 6.230]$$

Problem 7.45 $V(\hat{p}) = p(1-p)/n$ *implies that* $E(\hat{p}^2) = p(1-p)/n + p^2$. *Therefore,*

$$E(\hat{p}(1-\hat{p})) = E(\hat{p}) - E(\hat{p}^2) = p - (p(1-p)/n + p^2)$$
$$= p - p^2 - p(1-p)/n = p(1-p) - p(1-p)/n$$
$$= p(1-p)(1 - (1/n)) = \frac{n-1}{n} \times p(1-p) < p(1-p).$$

Problem 7.47

$$(a)\ \hat{p} = \frac{675}{1300} = 0.519;\ s(\hat{p}) = 0.0272;$$

$$[0.491, 0.546]\ (confidence\ interval)$$

$$(b)\ \hat{p} = \frac{625}{1300} = 0.481;\ s(\hat{p}) = 0.0272;$$

$$[0.454, 0.508]\ (confidence\ interval)$$

Problem 7.49

(a) The margin of error (when p is known) is

$$1.96 \times \sqrt{\frac{p(1-p)}{600}} \leq \frac{1.96}{2\sqrt{600}} = 0.04.$$

Thus, it cannot exceed 0.04 *no matter what the value of p is.*

(b) Similarly, the margin of error when the sample size is 1200 *cannot exceed*

$$\frac{1.96}{2\sqrt{1200}} = 0.028.$$

Problem 7.51

(a) $n \geq 1692$ *(b)* $n \geq 609$

Problem 7.53

(a)$[-0.382, 0.048]$ *(b)* $[-0.3472, 0.0138]$.
Conclusions: *Both confidence intervals contain* 0. *So the differences between the two teaching methods can be explained by random variation.*

Problem 7.55   *(a)* $E(X_i) = 1/\lambda$ *so* $E(\overline{X}) = 1/\lambda$

$$(b)\quad P(\min(X_1, \ldots, X_n) > x) = P(X_1 > x, \ldots X_n > x)$$
$$= P(X_1 > x) \cdots P(X_n > x) = \Pi_{1 \leq i \leq n} \exp(-\lambda x)$$
$$= \exp(-n\lambda x).\ Consequently,$$
$$(c)\quad E(U) = 1/n\lambda\ and\ V(U) = 1/(n\lambda)^2.\ Therefore,$$
$$E(nU) = 1/\lambda\ and\ V(nU) = 1/(\lambda)^2.$$
$$(d)\quad Since\ V(\overline{X}) = 1/n(\lambda)^2 \leq V(nU) = 1/(\lambda)^2,$$

we conclude that $\overline{X}$ *has the smaller variance, so it is the better estimator.*

Problem 7.57

Let $Q(\alpha)$ *denote the* 100α *percentile of a (non-standard) normal distribution.*

$$Q(\alpha) = x(1-\alpha) = \mu + \sigma z(1-\alpha)$$
$$Q(0.10) = \mu - 1.28\sigma$$
$$\hat{Q}(0.10) = \hat{\mu} - 1.28\hat{\sigma}$$
$$\hat{\mu} = 9830,\ \hat{\sigma} = \sqrt{\frac{35}{36}} \times 400 = 394.41$$
$$\hat{Q}(0.10) = 9830 - 504.84 = 9325.16$$

Problem 7.59 *(a)*

$$\mu_1 = \int_0^1 x f(x|\theta) = \int_0^1 x \theta x^{\theta-1} dx$$

$$= \theta \int_0^1 x^\theta dx = \frac{\theta}{\theta+1}$$

(b) Solving the equation $\mu_1 = \theta/(1+\theta)$ for θ yields: $\theta = \mu_1/(1-\mu_1)$. The MME for θ is therefore

$$\hat{\theta} = \frac{\overline{X}}{1-\overline{X}}.$$

Problem 7.61

$$E\left(\sum_{1\leq i\leq n} X_i^k\right) = \sum_{1\leq i\leq n} \mu_k = n\mu_k, \text{ thus,}$$

$$E\left(\frac{1}{n}\sum_{1\leq i\leq n} X_i^k\right) = \frac{n\mu_k}{n} = \mu_k.$$

Chapter 8

Hypothesis Testing

Problem 8.1 *(a) The null hypothesis is that the town's water supply is unsafe, so the null and alternative hyotheses are: $H_0 : \mu \geq 15$ against $H_1 : \mu < 15$. The null value is $\mu_0 = 15$.*
(b) Correctly rejecting H_0 means the water supply is safe.
(c) If the null hypothesis is erroneously rejected, then the water supply is unsafe, but the consumers are told that it is.

Problem 8.3 *(a) A type I error here means the water is declared unsafe to drink when it is, in fact, safe. A type II error means the water is declared safe to drink when it is, in fact, unsafe.*
(b) A type I error here means the water is declared safe to drink when it is, in fact, unsafe. A type II error means the water is declared unsafe to drink when it is, in fact, safe.
(c) Use (b) since it is a public health hazard to declare a water supply as safe when, in fact, it is not.

Problem 8.5 *If $Q(0.90) \leq 15$ then $P(X > 15) = 1 - P(X \leq 15) \leq 1 - 0.90 = 0.10$. That is the probability that the amount of lead in a water sample exceeds the EPA standard is less than 0.10. On the other hand, if $Q(0.90) > 15$ then $P(X > 15) > P(X > Q(0.90)) = 1 - P(X \leq Q(0.90)) = 0.10$. That is, the probability that the amount of lead in a water sample is greater than 0.10.*

Problem 8.7 *(a) A type I error (rejecting H_0 when true) means production is unnecessarily shut down while engineers waste time and money looking for trouble that is nowhere to be found. A type II error means that defective silicon wafers are being produced even though the sample means indicate otherwise.*
(b) In this case $\sigma/\sqrt{n} = 0.5$, so $P(|\overline{X} - 10| > 1.0) = P(|Z| > 2) = 0.0456$.
(c) Choose c so that $P(|\overline{X} - 10| > c) = P(|Z| > 2c) = 0.01$. This implies that $2c = 2.575$, so $c = 1.2875$.

Problem 8.9
(a) Because the engineer is more concerned about possibly accepting defective cable he puts the burden of proof on the supplier, that is, he assumes that the cables do not meet the specifications. Consequently, the null hypothesis is $H_0 : \mu = 10,000$ and $H_1 : \mu > 10,000$. A type I error means that one has accepted cable that is non-conforming. A type II error means that one has rejected cable that is conforming, that is, he has rejected cable that meets the specifications.
(b) We reject H_0 if

$$\frac{\overline{X} - 10,000}{400/\sqrt{16}} > 1.645.$$

When $\overline{x} = 10,100$ we get that

$$\frac{10,100 - 10,000}{400/\sqrt{16}} = 1 < 1.645,$$

so we do not reject H_0.
(c) The 95% lower confidence interval is $[9935.5, \infty)$.

Problem 8.11 *(a) The 95% confidence interval is $[\overline{X} - 3.92, \overline{X} + 3.92]$. We reject H_0 when 50 lies outside this interval.*
(b) We reject H_0 when $|\overline{X} - 50| > 3.92$.
(c) When $\overline{x} = 53$ the P-value equals $P(|Z| > 1.5) = 0.134$. When $\overline{x} = 46$ the P-value equals $P(|Z| > 2.0) = 0.046$.

Problem 8.13 *(a) $\pi(10) = 1 - \Phi(2(10.98 - 10)) = 1 - \Phi(1.96) = 0.025$*
(b) $1 - \pi(11.5) = \Phi(2(10.98 - 11.5)) = \Phi(-1.04) = 0.1492$

Problem 8.15

(a) If $\mu_1 < \mu_2$ then
$$\frac{\sqrt{n}(c - \mu_1)}{\sigma} > \frac{\sqrt{n}(c - \mu_2)}{\sigma}; \text{ consequently,}$$
$$\Phi\left(\frac{\sqrt{n}(c - \mu_1)}{\sigma}\right) > \Phi\left(\frac{\sqrt{n}(c - \mu_2)}{\sigma}\right) \text{ therefore,}$$
$$\pi(\mu_1) = 1 - \Phi\left(\frac{\sqrt{n}(c - \mu_2)}{\sigma}\right) < 1 - \Phi\left(\frac{\sqrt{n}(c - \mu_2)}{\sigma}\right) = \pi(\mu_2).$$
(b) If $\mu_1 < \mu_2$ then
$$\frac{\sqrt{n}(c - \mu_1)}{\sigma} > \frac{\sqrt{n}(c - \mu_2)}{\sigma}; \text{ consequently,}$$
$$\pi(\mu_1) = \Phi\left(\frac{\sqrt{n}(c - \mu_1)}{\sigma}\right) > \Phi\left(\frac{\sqrt{n}(c - \mu_2)}{\sigma}\right) = \pi(\mu_2).$$

Problem 8.17
$$\text{(a) Reject if } \overline{X} < c = 0.2 - t_6(\alpha)\frac{s}{\sqrt{7}}.$$

(b) The cutoff value $c = 0.2 - 1.943\frac{0.03}{\sqrt{7}} = 0.1780$. Since $\overline{x} = 0.17 < 0.178$ we reject H_0.
(c) P-value $= P(t_6 < -\sqrt{7}) = P(t_6 < -2.65)$
(d) Upper confidence interval $= (-\infty, 0.192]$, which does not contain 0.20. We reject H_0. The concentration of arsenic could be as high as 0.192%.

Problem 8.19 *(a) Because of the large sample size ($n = 100$) the distribution of $\overline{X}$ is approximately normal.*
(b) Reject H_0 if $\sqrt{n}(\overline{X} - 40,000)/3200 < -z(\alpha)$.
(c) P-value $= P(\overline{X} < 39,360|\mu = 40000) = 0.0228$, which is significant at the 5% level.

Problem 8.21
$$\Phi\left(\frac{c - \mu_0}{\sigma/\sqrt{n}}\right) = \alpha \text{ implies}$$
$$\frac{c - \mu_0}{\sigma/\sqrt{n}} = z(1 - \alpha) = -z(\alpha) \text{ and}$$
$$1 - \Phi\left(\frac{c - \mu}{\sigma/\sqrt{n}}\right) = \beta \text{ implies}$$
$$\frac{c - \mu_0}{\sigma/\sqrt{n}} = z(\beta).$$

Solving for the two unknowns n, c yields

$$\sqrt{n} = \left(\frac{-\sigma(z(\alpha) + z(\beta))}{\mu - \mu_0} \right) \text{ so}$$

$$n = \left(\frac{\sigma(z(\alpha) + z(\beta)}{\mu - \mu_0} \right)^2,$$

$$c = \frac{\mu_0 z(\beta) + \mu z(\alpha)}{z(\alpha) + z(\beta)}.$$

Problem 8.23 *Putting $\mu = \mu_0 + d$ and then $\mu = \mu_0 - d$ in Equation 8.16 yields*

$$\pi(\mu_0 + d) = 1 - \Phi\left(\frac{\sqrt{n}(c - d)}{\sigma} \right) + \Phi\left(\frac{\sqrt{n}(-c - d)}{\sigma} \right) \text{ and}$$

$$\pi(\mu_0 - d) = 1 - \Phi\left(\frac{\sqrt{n}(c + d)}{\sigma} \right) + \Phi\left(\frac{\sqrt{n}(d - c)}{\sigma} \right).$$

The proof of symmetry is completed by using the equation $\Phi(-x) = 1 - \Phi(x)$ with

$$x = \frac{\sqrt{n}(-c - d)}{\sigma} \text{ and then } x = \frac{\sqrt{n}(d - c)}{\sigma} \text{ implies that}$$

$$\Phi\left(\frac{\sqrt{n}(-c - d)}{\sigma} \right) = 1 - \Phi\left(\frac{\sqrt{n}(c + d)}{\sigma} \right) \text{ and}$$

$$1 - \Phi\left(\frac{\sqrt{n}(c - d)}{\sigma} \right) = \Phi\left(\frac{\sqrt{n}(d - c)}{\sigma} \right).$$

Problem 8.25
(a) $P(t_{11} < -1.155)$ (b) $P(t_{11} < -1.732)$ (c) $P(t_{11} < -2.079)$
(d) (a) and (b) are not significant at the 5% level while (c) is significant.

Problem 8.27 $\pi(\mu) = \Phi(21 - 0.75\mu)$

Problem 8.29 $n = 21$ $c = 53.30$

Problem 8.31
(a) The 95% confidence interval $[-0.0013, 0.0057]$ contains 0, so the null hypothesis of no difference between the means is not rejected.
(b) The P-value is $P(|t_{10}| > 1.4113) > P(|t_{10}| > 1.812) = 0.10$. Consequently, the P-value is greater than 0.10.

Problem 8.33 $W_1 = 108.5$, $n_1 = 11$, $n_2 = 15$, $E(W_1) = 148.5$, $\sigma(W_1) = 19.27$
$(W_1 - E(W_1) + 0.5)/\sigma(W_1) = (108.5 - 148.5 + 0.5)/19.27 = -2.05$
The P-value is $P(Z < -2.05) = 0.0202$. So we reject H_0 at the 5% level.

Problem 8.35 *We use the paired sample t test with $\overline{d} = -62.2$, $s_D = 109.5317$, $n = 5$, $t_4 = -1.2698$. We do not reject $H_0 : \Delta = 0$ because $P(|t_4| > 1.2698) > P(|t_4| > 2.776) = 0.05$.*

Problem 8.37 *We use the paired sample t test because each car receives two treatments: standard and premium (this is an example of self pairing). A straightforward computation yields $\overline{d} = -0.66$, $s_D = 0.563915$, $n = 5$, so $t = -2.617$ and the P-value is $P(|t_4| > 2.617) = 0.059$. So the difference between the two blends is not statistically significant.*

Problem 8.39 *(a) Let μ_A and μ_B denote the means from the diet A and diet B populations, respectively. The null hypothesis is $H_0 : \mu_A = \mu_B$ against $H_1 : \mu_A \neq \mu_B$. The value of the t statistic is*

$$t = \frac{9.09 - 5.86}{1.59} = 2.0296.$$

The P-value is $P(|t_{18}| > 2.0296) = 0.0574$. Conclusion: Diet A appears to be more effective in reducing weight but the evidence is not strong enough, at least at the 5% level.
  (b)

$$\frac{|W_1 - E(W_1)| - 0.5}{\sigma(W_1)} = \frac{|130 - 105| - 0.5}{13.2288} = 1.852$$

The P-value is $Prob(|Z| > 1.852) = 0.0639$.

Problem 8.41 *(a) The null hypothesis is $H_0 : \mu_A = \mu_A$ and the alternative is $H_1 : \mu_A \neq \mu_A$. We apply the two-sample t test for independent samples to A and B. We note that $t_{14} = -1.7762$, so the P-value is $P(|t_{14}| > 1.7762) = 0.0974$.*

These results suggest that typists using the word processing program A are the most efficient. A formal analysis using the t test leads to the conclusion that the differences between these programs are not statistically significant.

(b) Notation: Word processing programs A and B correspond to populations 1 and 2 respectively. $n_1 = n_2 = 8$, $W_1 = 53$, $E(W_1) = 68$, $\sigma(W_1) = 9.5219$

$$\frac{W_1 - E(W_1) - 0.5}{\sigma(W_1)} = \frac{|53 - 68| - 0.5}{9.5219} = 1.52$$

The P-value is $P(|Z| > 1.52) = 2 \times 0.0643 = 0.1286$. Consequently, the differences between these programs are not statistically significant. This is consistent with the results of the t test in part (a).

Problem 8.43 *The null hypothesis is that the mass of nitrogen gas obtained from air is equal to that obtained by chemical decomposition. The value of the t statistic is $t_{13} = 20.2137$ and the P-value is $Prob(|t_{13}| > 20.2137) < 0.0001$. This is powerful evidence against the null hypothesis and is consistent with Lord Rayleigh's claim that nitrogen gas obtained from air is heavier than nitrogen gas obtained from chemical decomposition.*

Problem 8.45 $s_1^2/s_2^2 \sim F_{5,9}$. *Accept H_0 because*

$$0.0984 = \frac{1}{F_{9,5}(0.01)} \leq \frac{s_1^2}{s_2^2} = 0.3235 \leq F_{5,9}(0.01) = 6.06.$$

Since we do not reject the null hypothesis that $\sigma_1^2 = \sigma_2^2$, one is justified in using the two sample t test.

Problem 8.51 *(ii)* $W = 0.9818$; $Prob < W = 0.6541$

Problem 8.53 *(ii)* $W = 0.9856$; $Prob < W = 0.7864$

Problem 8.55 $W = 0.9750$; $Prob < W = 0.3386$

Problem 8.57
(a) $\alpha = 0.078$, $\beta(0.30) = 0.850$, $\beta(0.5) = 0.377$
(b) $\alpha = 0.02$, $\beta(0.3) = 0.953$, $\beta(0.5) = 0.623$

Problem 8.59 $c = 11$, $\beta(0.6) = 0.2173$

Problem 8.61
(a) $\pi(p) = 1 - B(8; 15, p)$ *(b)* $\alpha = 0.095$ *(c)* $\beta(0.6) = 0.39$

Problem 8.63 $H_0 : p = 0.50$ *against* $H_1 : p < 0.50$. *Reject* H_0 *if* $\hat{p} < 0.4772$. *However,* $\hat{p} = 625/1300 = 0.4808 > 0.4772$; *we do not reject. P-value* $= 0.0823$

Problem 8.65 $n = 87$.

Problem 8.67 *The two sided* 95% *confidence interval for* $p_A - p_B$ *is* $[-0.0562, 0.0162]$; *it contains* 0, *so the difference is not statistically significant.*

Problem 8.69
(a) X_i *is binomial with parameters* n_i, p_i.
(b) $H_0 : p_1 \leq p_2$, $H_1 : p_1 > p_2$
(c) *The one sided lower confidence interval is* $[0.1830, \infty)$.

Chapter 9

Statistical Analysis of Categorical Data

Problem 9.1

Seeds	Observed frequency	Cell probability	Expected frequency
ry	273	9/16	$480 \times 9/16 = 270$
wy	94	3/16	$480 \times 3/16 = 90$
rg	88	3/16	$480 \times 3/16 = 90$
wg	25	1/16	$480 \times 1/16 = 30$

$\chi^2 = 294/270 = 1.0889$. *Since* $k - 1 = 3$ *it follows that* $\chi^2 \approx \chi_3^2$. *From the chi-square table (Table A.5) we see that* $\chi^2 = 1.0889 < \chi_3^2(0.05) = 7.815$; *consequently, we do not reject the null hypothesis. The data are consistent with Mendel's theory.*

Problem 9.3 *(a)*

Number of heads	Observed frequency y_i	Expected frequency np_{i0}	$y_i - np_{i0}$
0	1	1.5625	-0.5625
1	17	7.8125	9.1875
2	15	15.6250	-0.6250
3	10	15.6250	-5.6250
4	6	7.8125	-1.8125
5	1	1.5625	-0.5625

(b) Combining cells 0 and 1 and cells 4 and 5 yields the following table.

Number of heads	Observed frequency y_i	Expected frequency np_{i0}	$y_i - np_{i0}$
{0, 1}	18	9.375	8.8250
2	15	15.6250	-0.6250
3	10	15.6250	-5.6250
{4, 5}	7	9.375	-2.375

$\chi_3^2 = 10.5867$. *Since* $P(\chi_3^2 > 10.5867) < P(\chi_3^2 > 9.348) = 0.025$ *we reject the null hypothesis that* $p = 0.5$.

Problem 9.5 *After combining classes 6 and 7 into one class* $= (\geq 6)$ *we obtain*

Number of electron-positron pairs on photograph i	Number of photos with k electron-positron pairs f_i
0	47
1	69
2	84
3	76
4	49
5	16
≥ 6	14

$\lambda = 828/355 = 2.3324$; *Pearson* χ^2 *statistic* $= 13.7577$, *degrees of freedom=5, P-value=0.0185. The data are consistent with the theory.*

Problem 9.7 *Set* $k = 2, y_1 = k, y_2 = n - k, p_1 = p, p_2 = 1 - p$ *in Equation 5.17. We can do this because* $y_1 + y_2 = n$ *and* $p_1 + p_2 = 1$*. We obtain*

$$P(Y_1 = y_1, Y_2 = y_2) = \frac{n!}{y_1!y_2!}p_1^{y_1} \cdots p_2^{y_2}$$

$$= \frac{n!}{k!(n-k)!}p^k(1-p)^{n-k}$$

$$= \binom{n}{k}p^k(1-p)^{n-k}.$$

Problem 9.9 $\chi^2 = 34.076 > 12.592 = \chi_6^2(0.05)$. *We conclude that the three classes of voters differ with respect to the importance of these four public policy issues.*

Problem 9.11   *Expected counts are printed below observed counts.*

	Brittle	Non-Brittle	Total
Length of heating cycle 77	323		400
30	127.00	273.00	
90	177	223	400
Total	127.00	273.00	
Total	254	546	800

$\chi_1^2 = 19.685 + 9.158 + 19.685 + 9.158 = 57.685$. *Since* $P(\chi_1^2 > 57.685) < P(\chi_1^2 > 7.879) = 0.005$ *we reject the null hypothesis that the length of the heating cycle has no effect on the brittlenes of nylon bars.*

Problem 9.13   *Expected counts are printed below observed counts.*

	Porous	Non-porous	Total
defective	142	331	473
	95.57	377.43	
non defective	1233	5099	6332
	1279.43	5052.57	
Total	1375	5430	6805

$\chi_1^2 = 22.553 + 5.711 + 1.685 + 0.427 = 30.375$. *Since* $P(\chi_1^2 > 30.375) < P(\chi_1^2 > 7.879) = 0.005$ *we reject the null hypothesis that the porosity and dimension of the molded vulcanite are independent.*

Problem 9.15 *(a) Each father-daughter, considered as a single unit, is classified according to their blood pressures. Are these attributes independent or associated? We use the χ^2 test of independence, with $DF = 4$ and null hypothesis:*

$$H_0 : P(A_i \cap B_j) = P(A_i)P(B_j) \text{ for } i = 1, 2, 3 \text{ and } j = 1, 2, 3.$$

(b) Reject H_0 if $\chi^2 > 9.488 = \chi_4^2(0.05)$. Now $\chi^2 = 3.814$, with P-value $P(\chi^2 > 3.814) = 0.4317$. We do not reject H_0; the blood pressures of fathers and daughters do not appear to be associated.

Problem 9.17 *Expected counts are printed below observed counts.*

	Helped	Harmed	No Effect	Total
Treatment	50	10	22	82
	46.00	11.00	25.00	
Control	42	12	28	82
	46.00	11.00	25.00	
Total	92	22	50	164

$\chi_2^2 = 0.348 + 0.091 + 0.360 + 0.348 + 0.091 + 0.360 = 1.597$. *Since* $P(\chi_2^2 > 1.597) > P(\chi_2^2 > 4.605) = 0.1$ *We conclude that there is no difference between the treatment and control groups.*

Chapter 10

Linear Regression and Correlation

Problem 10.1 *(c)* $\hat{\beta} = 0.667$ *and* $\hat{\beta}_1 = 1.000$

Problem 10.3 *(b)* $\hat{\beta}_0 = 1.5975$ *and* $\hat{\beta}_1 = 0.01645$
(c) $\rho = 0.9728$, $R^2 = 0.9464$, $SSE = 0.09660$
(d) $\hat{y}_1 = \hat{y}(30) = 2.0911$ *and* $\hat{e}_1 = 0.0189$
$\hat{y}_4 = \hat{y}(75) = 2.88$ *and* $\hat{e}_4 = 0.0486$.

	Dep Var	Predict	
Obs	Y	Value	Residual
1	2.1100	2.0911	0.0189
2	2.2700	2.3379	-0.0679
3	2.5000	2.5846	-0.0846
4	2.8800	2.8314	0.0486
5	3.2100	3.0782	0.1318
6	3.4800	3.3250	0.1550
7	3.3700	3.5718	-0.2018

Problem 10.5 *(b)* $\hat{\beta}_0 = 31.446$, $\hat{\beta}_1 = -1.186$
(c) $r = -0.6998$, $R^2 = 0.4897$, $SSE = 25.64$
(d) $\hat{y}(1.6) = 29.5487$, $y - \hat{y} = -0.5487$
$\hat{y}(4.6) = 25.9909$, $y - \hat{y} = -0.9909$
(e) $R^2(weight) = 0.6919 > 0.4897 = R^2(displacement)$

Problem 10.7 *(b)* $\hat{\beta}_0 = 3.133$, $\hat{\beta}_1 = 0.157$
(c) $r = 0.7160$, $R^2 = 0.5127$, $SSE = 0.411$
(d) $\hat{y}(1.6) = 3.3848$, $y - \hat{y} = 0.0522$ $\hat{y}(4.6) = 3.8566$, $y - \hat{y} = 0.1434$
(e) $R^2(weight) = 0.7052 > R^2(displacement) = 0.5127$

Problem 10.9 *(b) Calculate the least squares estimates* $\hat{\beta}_0 = 0.183$ *and* $\hat{\beta}_1 = 1.074$.
(c) $r = 0.946$, $R^2 = 0.895$, $SSE = 0.891$
(d) Compute the predicted values and residuals for $x = 2.78$, $x = 3.64$, $x = 5.59$
(e) Model(Y = gpm |X = weight) is a better fit because its $R^2 = 0.895 > 0.831$ *is greater.*

Problem 10.11 *(b)* $\hat{\beta}_0 = 1.735$ *and* $\hat{\beta}_1 = 0.694$
(c) $r = 0.785$, $R^2 = 0.617$, $SSE = 3.265$
(d) Compute the predicted value and residual for $x = 2.0$, $x = 3.2$, $x = 5.3$.
(e) Weight is a much better predictor of fuel efficiency (gpm) because its $R^2 = 0.895 > 0.617$.

Problem 10.15 *If* $y_i = \beta_0 + \beta x + \epsilon_i$ *and* $y_j = \beta_0 + \beta x + \epsilon_j$, *then* $y_i - y_j = \epsilon_i - \epsilon_j \neq 0$ *because* $\epsilon_i - \epsilon_j$ *is normally distributed.*

Problem 10.17
(a)

Analysis of Variance

Source	DF	Sum of Squares	Mean Square	F Value	Prob>F
Model	1	250708.69157	250708.69157	25.121	0.0010
Error	8	79841.30843	9980.16355		
C Total	9	330550.00000			

Root MSE	99.90077	R-square	0.7585

(b) $R^2 = 0.7585$ *(c) 95% confidence interval for* $\hat{\beta}_0$ *is* $[-1283.79, 144.85]$; *95% confidence interval for* $\hat{\beta}_1$ *is* $[3.72, 10.07]$.
(d) $\hat{y}(220) = 948.1$; *95% confidence interval is* $[874.2, 1022.0]$.

Problem 10.19
(a) 95% confidence interval for $\hat{\beta}_0$ *is* $[33.54, 44.03]$; *95% confidence interval for* $\hat{\beta}_1$ *is* $[-5.05, -1.82]$.
(b) $[28.73, 31.46]$
(c) $[27.01, 33.19]$

Problem 10.21 *(b)* $SSR = 816.44 = MSR$; $SSE = 190.87$, $MSE = 17.35$, $F = 47.05$
(c) $R^2 = 0.8105$
(d) 95% confidence interval for $\hat{\beta}_0$ *is* $[22.485, 31.311]$;
95% confidence interval for $\hat{\beta}_1$ *is* $[0.0114, 0.0226]$.

Problem 10.23 *(c)* $SSR = 22426.67 = MSR$; $SSE = 952.5$, $MSE = 95.25$; $F = 235.51$
(d) $R^2 = 0.9593$
(e) 95% confidence interval for $\hat{\beta}_0$ *is* $[135.944, 171.890]$
95% confidence interval for $\hat{\beta}_1$ *is* $[2.066, 2.768]$. *(f)* 269.9

Problem 10.25 *(a)* $SSR = 7.625 = MSR$; $SSE = 0.891$, $MSE = 0.056$, $F = 136.94$
(b) $[-0.576, 0.942]$; $[40.858, 1.236]$
(c) $[4.036, 4.292]$ *(d)* $[3.650, 4.684]$; *predicted value* $= 4.010$

Problem 10.27 *(a)* $SSR = 5.252 = MSR$; $SSE = 3.265$; $MSE = 0.204$; $F = 25.74$
(b) $[0.631, 2.840]$; $0.404, 0.984]$
(c) $[3.940, 4.394]$ *(d)* $[3.183, 5.151]$; *predicted value* $= 4.370$

Problem 10.29 *The least squares estimates for Model* $(Y = \ln P | x = \ln D)$ *are*
$\hat{\beta}_0 = 0.000016833$, $\hat{\beta}_1 = 1.4997$. *Consequently,* $a = \exp(0.000016833) = 1.000$ *and* $b = 1.4997015 \approx 1.5$. *So these results are consistent with Kepler's predicted values.*

Problem 10.31 *(a) Looking at the scatter plot it is obvious that the simple linear regression model is inappropriate.*
(d) The equation $pv = k$ implies that $p = kv^{-1}$. Thus Boyle's law implies that $b = -1$, which is consistent with the least squares estimate $\hat{\beta}_1 = -1.002195$.

Problem 10.33 *(c) The points of the scatter plot appear to lie close to a straight line; the points of the residual plot appear to be randomly scattered about the line $y = 0$, and the normal probability of the residuals appears to confirm that they are normally distributed.*

Problem 10.35 *The residuals plot suggest that the variances are not constant; they appear to decrease. The Q–Q plot does not indicate significant departures from normality.*

Problem 10.37 $r = 0.9165$

Problem 10.39 *(b)* $r = 0.7673$

Problem 10.41 *(b)* $r(X_2, EX_3) = 0.695$ *(d)* $r(X_3, X_4) = -0.6025$

Problem 10.43 $r = 04591$; *observation 6 (Hungary) departs from the pattern.*

Problem 10.45

$$E(\overline{Y}) = E(\hat{\beta}_0 + \hat{\beta}\overline{x}) = E(\hat{\beta}_0) + (\hat{\beta}\overline{x}) = \beta_0 + \beta_1\overline{x}$$

Problem 10.47

$$\left(\frac{1}{n} - \overline{x}\left(\frac{x - \overline{x}}{S_{xx}}\right)\right)^2 = \frac{1}{n^2} - \frac{2\overline{x}}{n}\left(\frac{x - \overline{x}}{S_{xx}}\right) + \left(\overline{x}\left(\frac{x - \overline{x}}{S_{xx}}\right)\right)^2 ; \ therefore,$$

$$\sum_i \left(\frac{1}{n} - \overline{x}\left(\frac{x - \overline{x}}{S_{xx}}\right)\right)^2 = \sum_i \left(\frac{1}{n^2} - \frac{2\overline{x}}{n}\left(\frac{x - \overline{x}}{S_{xx}}\right) + \left(\overline{x}\left(\frac{x - \overline{x}}{S_{xx}}\right)\right)^2\right)$$

$$= \frac{1}{n} + \overline{x}^2\sum_i \left(\frac{x - \overline{x}}{S_{xx}}\right)^2 = \frac{1}{n} + \overline{x}^2\left(\frac{S_{xx}}{S_{xx}^2}\right)$$

$$= \frac{1}{n} + \frac{\overline{x}^2}{S_{xx}}.$$

Note: We used the fact that $\sum_i(x_i - \overline{x}) = 0$.

Chapter 11

Multiple Linear Regression

Problem 11.3 *(a)*

$$\begin{pmatrix} Y_1 \\ Y_2 \\ Y_3 \\ Y_4 \\ Y_5 \\ Y_6 \\ Y_7 \end{pmatrix} = \begin{pmatrix} 1 \; 14 \\ 1 \; 16 \\ 1 \; 27 \\ 1 \; 42 \\ 1 \; 39 \\ 1 \; 50 \\ 1 \; 83 \end{pmatrix} \begin{pmatrix} \beta_0 \\ \beta_1 \end{pmatrix} + \begin{pmatrix} \epsilon_1 \\ \epsilon_2 \\ \epsilon_3 \\ \epsilon_4 \\ \epsilon_5 \\ \epsilon_6 \\ \epsilon_7 \end{pmatrix}$$

(b)

$$\begin{pmatrix} 7 & 271 \\ 271 & 13855 \end{pmatrix} \begin{pmatrix} \hat{\beta}_0 \\ \hat{\beta}_1 \end{pmatrix} = \begin{pmatrix} 66 \\ 3375 \end{pmatrix}$$

(d)

$$s^2(\boldsymbol{X'X})^{-1} = \begin{pmatrix} 0.6907 & -0.0135 \\ -0.0135 & 0.0003 \end{pmatrix}$$

Problem 11.5

$$det(\boldsymbol{X'X}) = n \sum x_i^2 - \left(\sum x_i \right)^2 = n \sum x_i^2 - (n\bar{x})^2$$
$$= n \left(\sum x_i^2 - n(\bar{x})^2 \right) = n S_{xx}$$

Problem 11.7 *Using the result that $((\boldsymbol{X'X})^{-1})' = (\boldsymbol{X'X})^{-1}$ it follows that*

$$\boldsymbol{H'} = (\boldsymbol{X}(\boldsymbol{X'X})^{-1}\boldsymbol{X'})' = ((\boldsymbol{X'})'((\boldsymbol{X'X})^{-1})'\boldsymbol{X'}$$
$$= \boldsymbol{X}(\boldsymbol{X'X})^{-1}\boldsymbol{X'};$$
$$\boldsymbol{H}^2 = \boldsymbol{HH} = \boldsymbol{X}(\boldsymbol{X'X})^{-1}\boldsymbol{X'X}(\boldsymbol{X'X})^{-1}\boldsymbol{X'}$$
$$= \boldsymbol{X}(\boldsymbol{X'X})^{-1}(\boldsymbol{X'X})(\boldsymbol{X'X})^{-1}\boldsymbol{X'}$$
$$= \boldsymbol{XI}(\boldsymbol{X'X})^{-1}\boldsymbol{X'} = \boldsymbol{X}(\boldsymbol{X'X})^{-1}\boldsymbol{X'} = \boldsymbol{H}.$$

Now $\boldsymbol{H}^2 = \boldsymbol{H}$ implies that

$$0 = \boldsymbol{H} - \boldsymbol{H}^2 = \boldsymbol{H}(\boldsymbol{I} - \boldsymbol{H}), \text{ and finally,}$$
$$(\boldsymbol{I} - \boldsymbol{H})^2 = \boldsymbol{I} - 2\boldsymbol{H} + \boldsymbol{H}^2 = \boldsymbol{I} - \boldsymbol{H} + (\boldsymbol{H}^2 - \boldsymbol{H}) = \boldsymbol{I} - \boldsymbol{H}.$$

Problem 11.9 *(a)* $y = -28.3654 + 0.6013x_1 + 1.0279x_2$
(b) 65.2695 *(c)* $\hat{y}(841, 48) = 526.7$, *residual* $= 8.3028$

(d)

Analysis of Variance

Source	DF	Sum of Squares	Mean Square	F Value	Prob>F
Model	2	171328.66023	85664.33012	1304.326	0.0001
Error	7	459.73977	65.67711		
C Total	9	171788.40000			

Root MSE	8.10414	R-square	0.9973

$R^2 = 0.9973$ *and adjusted* $R^2 = 0.9966$

(e) $SSE(r) = 1166.75898$, $DF(r) = 8$ *and* $SSE(f) = 459.73977$, $DF(f) = 7$. *The F ratio* $F = 10.7651 > F_{1,7}(0.05) = 5.59$. *On the other hand* $F = 10.7651 < F_{1,7}(0.01) = 12.25$. *The P-value is* 0.0135.

Problem 11.11   (b) $y = -26219 + 189.204551x - 0.331194x^2$
(d) $SSE(r) = 22508.10547$, $DF(r) = 6$ *and* $SSE(f) = 10213.02766$, $DF(f) = 5$. *The F ratio is* $F = 6.0172 > F1, 5(0.05) = 6.61$. *The P-value is* 0.0577.

Problem 11.13   *Looking at the ANOVA table we see that the F ratio is* 1.774 *with a P-value of* 0.1925. *The coefficient of multiple determination is* 0.2496, *which is quite low. The model is not a good one.*

Analysis of Variance

Source	DF	Sum of Squares	Mean Square	F Value	Prob>F
Model	3	0.02685	0.00895	1.774	0.1925
Error	16	0.08072	0.00505		
C Total	19	0.10757			

Root MSE	0.07103	R-square	0.2496

Chapter 12

Single Factor Experiments: Analysis of Variance

Problem 12.1 $IQR = Q_3 - Q_1 = 29 - 27 = 2$ *and* $1.5 \times IQR = 3$. *There is only one observation in group 3 that lies outside the interval*

$$(Q_1 - 1.5 \times IQR, Q_3 + 1.5 \times IQR] = [24, 32]$$

and that is 23.

Problem 12.3 *(a)*

$$Y_{31} = 28 = \mu_3 + \epsilon_{31}, \ Y_{32} = 30 = \mu_3 + \epsilon_{32}$$

Consequently, the difference between the two values $Y_{31} - Y_{32} = -2 = \epsilon_{31} - \epsilon_{32}$ *is accounted for by the difference in the error terms (which are random variables).*
(b) The arithmetic test grades for each group are modeled by a normal distribution with (possibly) different means. There is a positive probability that an individual taught by method 1 will have a test grade higher than someone taught by method 3, even though the latter has a higher sample mean.

Problem 12.5
(a) $\bar{y}_{1.} = 221.67$, $\bar{y}_{2.} = 202.67$, $\bar{y}_{3.} = 177.00$, $\bar{y}_{4.} = 249.00$, $\bar{y}_{..} = 212.58$
(b) $SSTr = 8319.58$, $SSE = 3523.33$, $SST = 11842.92$

(c) Dependent Variable: Y   Burst Strength (psi)

Source	DF	Sum of Squares	Mean Square	F Value	Pr > F
Model	3	8319.5833333	2773.1944444	6.30	0.0168
Error	8	3523.3333333	440.4166667		
Corrected Total	11	11842.9166667			

$F = 6.30 > 4.07 = F_{3,8}(0.05)$. *Reject* H_0.

Problem 12.7

Source	DF	Sum of Squares	Mean Square	F Value	Pr > F
Model	2	76.85405000	38.42702500	74.21	0.0001
Error	9	4.66045000	0.51782778		
Corrected Total	11	81.51450000			

$F = 74.21 > 4.26 = F_{2,9}(0.05)$. *Reject* H_0.

Problem 12.9
(a) The factor is the type of diet, there are three levels corresponding to the three different diets, and the response variable is the weight loss.
(b) The null hypothesis is the mean weight loss is the same for the three diets.

(c) Dependent Variable: POUNDS

Source	DF	Sum of Squares	Mean Square	F Value	Pr > F
Model	2	155.1450286	77.5725143	6.01	0.0083
Error	22	284.1725714	12.9169351		
Corrected Total	24	439.3176000			

Problem 12.11   *(a) $E(y_{ij}) = \mu + \alpha_i$; therefore*

$$E\left(\sum_{1 \le j \le J} y_{ij}\right) = J(\mu + \alpha_i), \; so$$

$$E(\bar{y}_{i.}) = E\left(\frac{1}{J}\sum_{1 \le j \le J} y_{ij}\right) = (\mu + \alpha_i).$$

(b) Using the fact that $E(\bar{y}_{i.} - \bar{y}_{..}) = E(\bar{y}_{i.}) - E(\bar{y}_{..})$, the result follows by noting that $E(\bar{y}_{i.}) = (\mu + \alpha_i)$ and $E(\bar{y}_{..}) = \mu$ as will now be shown.

$$E(y_{..}) = \left(\sum_i \sum_j E(y_{ij})\right) = \left(\sum_i \sum_j (\mu + \alpha_i)\right) = IJ\mu, \; so$$

$$E(\bar{y}_{..}) = E\left(\frac{y_{..}}{IJ}\right) = \mu,$$

where we used the result that $\sum_i \alpha_i = 0$.

Problem 12.13
(a) $t_{n-I}(0.025)\sqrt{MSE/J} = t_{12}(0.025)\sqrt{12584.14583/4} = 122.2191$
The four 95% confidence intervals are: [2755.7809, 3000.2191], [3032.2809, 3276.7191], [2815.0309, 3059.4691], [2551.7809, 2796.2191].

(b)   T tests (LSD) for variable: Y
        NOTE: This test controls the type I comparisonwise error rate not the
                experimentwise error rate.
            Alpha= 0.05 Confidence= 0.95 df= 12 MSE= 12584.15
                    Critical Value of T= 2.17881
                Least Significant Difference= 172.83
        Comparisons significant at the 0.05 level are indicated by '***'.

LEVEL Comparison	Lower Confidence Limit	Difference Between Means	Upper Confidence Limit	
2 - 3	44.42	217.25	390.08	***
2 - 1	103.67	276.50	449.33	***

2	- 4	307.67	480.50	653.33	***
3	- 2	-390.08	-217.25	-44.42	***
3	- 1	-113.58	59.25	232.08	
3	- 4	90.42	263.25	436.08	***
1	- 2	-449.33	-276.50	-103.67	***
1	- 3	-232.08	-59.25	113.58	
1	- 4	31.17	204.00	376.83	***
4	- 2	-653.33	-480.50	-307.67	***
4	- 3	-436.08	-263.25	-90.42	***
4	- 1	-376.83	-204.00	-31.17	***

(c)   Tukey's Studentized Range (HSD) Test for variable: Y
NOTE: This test controls the type I experimentwise error rate.
Alpha= 0.05 Confidence= 0.95 df= 12 MSE= 12584.15
Critical Value of Studentized Range= 4.199
Minimum Significant Difference= 235.5
Comparisons significant at the 0.05 level are indicated by '***'.

LEVEL Comparison		Simultaneous Lower Confidence Limit	Difference Between Means	Simultaneous Upper Confidence Limit	
2	- 3	-18.25	217.25	452.75	
2	- 1	41.00	276.50	512.00	***
2	- 4	245.00	480.50	716.00	***
3	- 2	-452.75	-217.25	18.25	
3	- 1	-176.25	59.25	294.75	
3	- 4	27.75	263.25	498.75	***
1	- 2	-512.00	-276.50	-41.00	***
1	- 3	-294.75	-59.25	176.25	
1	- 4	-31.50	204.00	439.50	
4	- 2	-716.00	-480.50	-245.00	***
4	- 3	-498.75	-263.25	-27.75	***
4	- 1	-439.50	-204.00	31.50	

Problem 12.15
(a) In this problem $n = 24$, $I = 3$, $n - I = 21$, $J = 8$, $MSE = 10.7262$. *The confidence intervals are:*

$$\overline{Y}_{1.} \pm t_{n-I}(\alpha/2)\sqrt{\frac{MSE}{J}} = 15.75 \pm 2.4085 = [13.3415, 18.1585]$$

$$\overline{Y}_{2.} \pm t_{n-I}(\alpha/2)\sqrt{\frac{MSE}{J}} = 14.625 \pm 2.4085 = [12.2165, 17.0335]$$

$$\overline{Y}_{3.} \pm t_{n-I}(\alpha/2)\sqrt{\frac{MSE}{J}} = 16.625 \pm 2.4085 = [15.2165, 20.0335]$$

(b) *The three means arranged in increasing order are:*

$$\overline{y}_{2.} < \overline{y}_{1.} < \overline{y}_{3.}$$
$$14.625 < 15.75 < 17.625$$

$HSD = 4.1275$. *There are no significant differences among the means.*

Problem 12.17 (b)   Tukey's Studentized Range (HSD) Test for variable: POU
        NOTE: This test controls the type I experimentwise error rate.
        Alpha= 0.05 Confidence= 0.95 df= 22 MSE= 12.91694
                Critical Value of Studentized Range= 3.553
        Comparisons significant at the 0.05 level are indicated by '***'.

		Simultaneous Lower Confidence	Difference Between	Simultaneous Upper Confidence	
L Comparison		Limit	Means	Limit	
3	- 1	-2.248	2.035	6.318	
3	- 2	1.667	6.339	11.012	***
1	- 3	-6.318	-2.035	2.248	
1	- 2	-0.145	4.304	8.754	
2	- 3	-11.012	-6.339	-1.667	***
2	- 1	-8.754	-4.304	0.145	

Problem 12.19

$$E(\hat{\theta}) = \sum c_i E(\overline{Y}_{i.}) = \sum c_i \mu_i = \theta$$
$$V(\hat{\theta}) = \sum c_i^2 V(Y_{i.}) = \sum c_i^2 \sigma^2 / J$$
$$= \sigma^2 \left(\sum \frac{c_i^2}{J} \right)$$

Problem 12.21 Dependent Variable: Y Resistivities

Source	DF	Sum of Squares	Mean Square	F Value
Model	5	261.50648	52.3013	69.0108
Error	12	9.09446	0.7579	
Corrected Total	17	270.60094		

$Pr(F_{5,12} > 69.0108) < 0.0001$. *We reject* H_0.
Components of variance: $\hat{\sigma}_A^2 = (52.3013 - 0.7579)/3 = 17.811$, *and* $\hat{\sigma}^2 = 0.7579$.

Problem 12.23

(a)   Analysis of Variance Procedure
Dependent Variable: X

Source	DF	Sum of Squares	Mean Square	F Value	Pr > F
Model	3	33.45833333	11.15277778	2.41	0.0969
Error	20	92.50000000	4.62500000		
Corrected Total	23	125.95833333			

Looking at the ANOVA table we see that the F ratio is 2.41 and the P-value is $Prob(F_{3,20} >$

$2.41) = 0.0969 > 0.05.$ *Consequently we do not reject* H_0.
(b) The components of variance are $\hat{\sigma}_A^2 = 1.0880$ *and* $\hat{\sigma}^2 = 4.625$.

Problem 12.25
(a) $E(Y_{ij}) = \mu + \alpha_i$, $j = 1, \ldots J$; *therefore*, $E(\overline{Y}_{i.}) = \mu + \alpha_i$.
(b) Since $E(\overline{Y}_{..}) = \mu$ *it follows from part (a) that*

$$E(\overline{Y}_{i.} - \overline{Y}_{..}) = E(\overline{Y}_{i.}) - E(\overline{Y}_{..}) = \mu + \alpha_i - \mu = \alpha_i.$$

Chapter 13

Design and Analysis of Multi-Factor Experiments

Problem 13.1

Source	DF	Sum of Squares	Mean Square	F Value	Pr > F
Model	2	219.5000000	109.7500000	0.49	0.6299
Error	9	2029.5000000	225.5000000		
Corrected Total	11	2249.0000000			

Conclusions: *Ignoring the genetic strain sharply increases the unexplained variability from $SSE = 11.83$ to $SSE = 2029.50$. Consequently, we do not reject the null hypothesis that the treatment means are equal.*

Problem 13.3 *(a) Grand mean is $\hat{\mu} = 102.56$. The treatment means are: $\hat{\mu}_1 = 97.67$, $\hat{\mu}_2 = 103.33$, $\hat{\mu}_3 = 106.67$. Block means are: $100.67, 100.33, 101.00, 104.00, 103.00, 106.33$.*

(b)

Source	DF	Sum of Squares	Mean Square	F Value	Pr > F
A	2	248.4444444	124.2222222	48.61	0.0001
B	5	82.4444444	16.4888889	6.45	0.0063
Error	10	25.5555556	2.5555556		
Corrected Total	17	356.4444444			

Problem 13.5 *(a) There are three treatments corresponding to the three configurations. There are five blocks corresponding to the five workloads.*

(b)

Source	DF	Sum of Squares	Mean Square	F Value	Pr > F
A	2	12857.20000	6428.60000	217.18	0.0001
B	4	308.40000	77.10000	2.60	0.1161
Error	8	236.80000	29.60000		
Corrected Total	14	13402.40000			

(c) The three estimated treatment means are 51.0, 52.0, 113.6. $HSD = 9.8322$ and $MSE = 29.6$. Consequently, the mean execution time for the no cache memory is significantly different from the first two mean execution times. The difference in execution times between the two cache memory and the one cache memory is not significant.

Problem 13.7 *(a) In this case the blocking variable is the soil type (B) and the furnace slags (A) are the treatments. Looking at the ANOVA table we see that the P-value of the F ratio is 0.2457, so the differences among the treatments do not appear to be significant.*

Source	DF	Sum of Squares	Mean Square	F Value	Pr > F
SLAGS	6	731.0580952	121.8430159	1.54	0.2457
SOIL	2	5696.3400000	2848.1700000	36.07	0.0001
Error	12	947.433333	78.952778		
Corrected Total	20	7374.831429			

$W = 0.940208$, $Prob < W = 0.22$. *The hypothesis of normality is not rejected.*
Conclusions: *The graph indicates that there is a treatment-block interaction.*

Problem 13.9

(a)

Source	DF	SS	Mean Square	F Value	Pr > F
FUNGCIDE	1	1.29323105	1.29323105	18.41	0.0001
PESTCIDE	2	0.64873201	0.32436600	4.62	0.0154
FUNGCIDE*PESTCIDE	2	0.71247627	0.35623813	5.07	0.0106
Error	42	2.94960161	0.07022861		
Corrected Total	47	5.60404094			

Conclusions: *Each of the main effects and interactions is significant. No improvement is discernable.*

Problem 13.11   *We only give the proof that $\overline{Y}_{i..}$ is an unbiased estimator for $\overline{\mu}_i$; the proof for $\overline{Y}_{.j.}$ is similar. Since $E(Y_{ijk}) = \mu_{ij}$, $(k = 1, \ldots, n)$ it follows that*

$$E(\overline{Y}_{i..}) = E\frac{1}{bn}\left(\sum_j \sum_k E(Y_{ijk})\right)$$

$$= \frac{1}{bn}\sum_j n\mu_{ij} = \frac{1}{b}\left(\sum_j \mu_{ij}\right) = \overline{\mu}_i.$$

Problem 13.13   *Using the facts that $\sum_i \alpha_i = 0$ and $\overline{\mu}_{.j} = \mu + \beta_j$ we see that*

$$\sum_{i=1}^a (\alpha\beta)_{ij} = \sum_{i=1}^a (\mu_{ij} - \mu - \alpha_i - \beta_j) = a(\overline{\mu}_{.j} - \mu - \beta_j) = 0.$$

The proof that $\sum_j (\alpha\beta)_{ij} = 0$ is similar and is therefore omitted.

Problem 13.15

	Estimated Cell Means			
	Factor (B)			
Factor (A)	*1*	*2*	*3*	*Row means*
1	*5340.00*	*4990.00*	*4815.00*	*5048.33*
2	*5045.00*	*5345.00*	*4515.00*	*4968.33*
3	*5675.00*	*5415.00*	*4917.50*	*5335.83*
Column means	*5353.33*	*5250.00*	*4749.17*	$\overline{y}_{...} = 5117.50$

Source	DF	Type I SS	Mean Square	F Value	Pr > F
GAUGER	2	896450.000	448225.000	1.63	0.2142
BREAKER	2	2506116.667	1253058.333	4.56	0.0196
GAUGER*BREAKER	4	663833.333	165958.333	0.60	0.6629
Error	27	7415475.000	274647.222		
Corrected Total	35	11481875.000			

Problem 13.17 *(a)*

	Estimated Cell Means				
	Factor (B)				
Factor (A)	40	60	80	100	Row means
50	18.5	18.5	23.0	27.5	21.875
75	10.5	15.5	14.5	29.0	17.375
100	14.0	19.5	24.0	26.5	21.0
125	19.0	22.0	22.5	30.0	21.
Column means	15.5	18.875	21.0	28.25	$\bar{y}_{...} = 20.9065$

(c)

Dependent Variable: WARPING

Source	DF	Sum of Squares	Mean Square	F Value	Pr > F
TEMP	3	156.0937500	52.0312500	7.67	0.0021
COPPER	3	698.3437500	232.7812500	34.33	0.0001
TEMP*COPPER	9	113.7812500	12.6423611	1.86	0.1327
Error	16	108.5000000	6.7812500		
Corrected Total	31	1076.7187500			

Problem 13.19

$$(a) \ \frac{L_{X_1}}{8} = 0.352, \ \frac{L_{X_8}}{8} = 0.5658, \ \frac{L_{X_1 X_8}}{8} = 0.125$$

$$(b) \ SSX1 = \frac{2.817^2}{16} = 0.496, \ SSX1 = \frac{(-4.526)^2}{16} = 1.280, \ SSX1X8 = \frac{1^2}{16} = 0.0625$$

(c) The sample mean of $Y2$ at $X8 = 0$ is -0.3656; the sample mean of $Y2$ at $X8 = 1$ is -0.9313. Setting the nozzle position at $X8 = 1$ produces a smaller variance, since $\exp(-0.9313) < \exp(-0.3656)$.

Problem 13.21 *(a)*

$$X4 = 0, X8 = 0 \quad 14.821, 14.757, 14.415, 14.932$$
$$X4 = 1, X8 = 0 \quad 13.880, 13.860, 13.972, 13.907$$
$$X4 = 0, X8 = 1 \quad 14.888, 14.921, 14.843, 14.878$$
$$X4 = 1, X8 = 1 \quad 14.037, 14.165, 14.032, 13.914$$

(b)

Dependent Variable: Y1   Epitaxial thickness

Source	DF	Sum of Squares	Mean Square	F Value	Pr > F
X8	1	0.08037225	0.08037225	5.05	0.0441
X4	1	2.79558400	2.79558400	175.82	0.0001
X8*X4	1	0.00036100	0.00036100	0.02	0.8827
Error	12	0.19080650	0.01590054		
Corrected Total	15	3.06712375			

(c)

$$X4 = 0, X8 = 0 \quad -0.4425, -0.3267, -0.3131, -0.2292$$
$$X4 = 1, X8 = 0 \quad -0.6505, -0.4969, -0.3467, -0.1190$$
$$X4 = 0, X8 = 1 \quad -1.1989, -0.6270, -0.4369, -0.6154$$
$$X4 = 1, X8 = 1 \quad -1.4307, -1.4230, -0.8663, -0.8625$$

(d)

Dependent Variable: Y2   Log of s-square

Source	DF	Sum of Squares	Mean Square	F Value	Pr > F
X8	1	1.28034883	1.28034883	18.55	0.0010
X4	1	0.24897605	0.24897605	3.61	0.0818
X8*X4	1	0.12122583	0.12122583	1.76	0.2097
Error	12	0.82812271	0.06901023		
Corrected Total	15	2.47867342			

(e) Yes. Deposition time (X4) is significant for Y1 (P-value of F ratio is < 0.0001) but not for Y2 (P-value of F ratio is 0.0818).

Problem 13.23

contrasts	SS_{effect}	SS_{effect}/SST	SS_{effect}/SST (complete factorial)
$l_A = 339$	9576.75	0.4588	0.4463
$l_B = 313$	8164.083	0.3911	0.4463
$l_C = 183$	2790.750	0.1337	0.0735

Source	DF	Sum of Squares	Mean Square	F Value	Pr > F
A	1	9576.750000	9576.750000	224.46	0.0001
B	1	8164.083333	8164.083333	191.35	0.0001
C	1	2790.750000	2790.750000	65.41	0.0001
Error	8	341.33333	42.66667		
Corrected Total	11	20872.91667			

These results are consistent with the full factorial model which indicate that the factors A, B, and C are significant.

Problem 13.25

contrasts	SS_{effect}	SS_{effect}/SST	SS_{effect}/SST (complete factorial)
$l_A = 37.5$	175.78	0.8953	0.5302
$l_B = 10.3$	13.26	0.0675	0.0054
$l_C = 3.1$	1.20	0.0061	0.0023

Source	DF	Sum of Squares	Mean Square	F Value	Pr > F
A	1	175.7812500	175.7812500	115.55	0.0004
B	1	13.2612500	13.2612500	8.72	0.0419
C	1	1.2012500	1.2012500	0.79	0.4244
Error	4	6.0850000	1.5212500		
Corrected Total	7	196.3287500			

In the full factorial model A, A*C, and B*C are significant while B is not. In the fractional replicate, B is significant because it is aliased with AC, which is significant. C is not significant in the fractional model because it is aliased with AB, which is not significant in the full factorial. So the results of the fractional factorial model are not consistent with the results of the full factorial.

Problem 13.27   *The pattern of plus and minus signs for the A,B,C, and D factors are*

$$A \; -,+,-,+,-,+,-,+,-,+,-,+,-,+,-,+$$
$$B \; -,-,+,+,-,-,+,+,-,-,+,+,-,-,+,+$$
$$C \; -,-,-,-,+,+,+,+,-,-,-,-,+,+,+,+$$
$$D \; -,-,-,-,-,-,-,-,+,+,+,+,+,+,+,+$$

The pattern of $\pm$ signs for the ABCD interaction is the product of these rows. Thus, block 1 contains $\{1, ab, ac, bc, ad, bd, cd, abcd\}$ and block 2 contains $\{a, b, c, abc, d, abd, acd, bcd\}$.

Chapter 14

Statistical Quality Control

Problem 14.1 *(a)* $LCL_{\overline{X}} = 13.9$, $UCL_{\overline{X}} = 15.1$ *(b)* $LCL_R = 0$, $UCL_R = 1.88$
(c) $P(|\overline{X} - 14.5| > 0.5) = 0.0155$

Problem 14.3 *Assume $\mu = \mu_0 + \delta\sigma$.*

$$
\begin{aligned}
P(|\overline{X} - \mu_0| > 3\sigma/\sqrt{n}) &= P(\overline{X} - \mu_0 > 3\sigma/\sqrt{n}) + P(\overline{X} - \mu_0 < -3\sigma/\sqrt{n}) \\
&= P\left(\frac{\overline{X} - \mu_0 - \delta\sigma}{\sigma/\sqrt{n}} > \frac{3\sigma/\sqrt{n} - \delta\sigma}{\sigma/\sqrt{n}}\right) \\
&\quad + P\left(\frac{\overline{X} - \mu_0 - \delta\sigma}{\sigma/\sqrt{n}} < \frac{-3\sigma/\sqrt{n} - \delta\sigma}{\sigma/\sqrt{n}}\right) \\
&= P(Z > 3 - \delta\sqrt{n}) + P(Z < -3 - \delta\sqrt{n}) \\
&= \Phi(-3 - \delta\sqrt{n}) + (1 - \Phi(3 - \delta\sqrt{n}))
\end{aligned}
$$

Problem 14.5 *(a) Forty-two rheostats are non-conforming, that is, approximately 31% failed to meet the specifications.*
(b) $LCL_{\overline{X}} = 135.67$, $UCL_{\overline{X}} = 145.62$, $\overline{\overline{x}} = 140.64$; $LCL_R = 0$, $UCL_R = 18.2$, $\overline{R} = 8.6$. The $\overline{x}$ and R control charts do not show a lack of control.

Problem 14.7 *(a) $LCL_{\overline{p}} = 0$, $UCL_{\overline{p}} = 0.0794$*
(b) $P(\hat{p} > 0.0794 | p = 0.04) = 0.0778$. T has a geometric distribution with parameter $p = 0.0778$. $E(T) = 12.85$.

Problem 14.9 $LCL_{\overline{c}} = 0$, $UCL_{\overline{c}} = 14.702$. *The maximum number of observed defects is 12, so the control chart does not detect a lack of control.*

Index

Acceptance number, 132
Acceptance sampling, 132
Adjusted R-square, R_a^2, 460
Aliases, of fractional contrasts, 536
Alternative hypothesis, 326
Amortization, 235
ANOVA
    analysis of variance, 469
    one-way, 470
    two-way, 497
       additive effects, 500
    two-way table, 517

Bayes' theorem, 94, 97
Bayesian estimation, 324
Bear market, 10
Bernoulli trials process, 211
    definition of, 212
Beta
    interpretation of, 419
    of a security, 42
    of a stock, 417
Beta function, 189
Binary string, 86
Binomial
    approximation to the hypergeometric
       distribution, 138
    coefficient, 90
    experiment, 212, 355
    lattice model for stock prices, 140,
       142
    tree, 86
Block totals, 538
Blocking, 498
Blocking variables, 498
Blocks, 498
Bootstrap sample, 85
Box plots, 30
Brownian motion, 226
Bull market, 10

C total, as in corrected total, 408
Call option, 237
Capital asset pricing model (CAPM), 42

Cash flow, 235
Categorical variable, 8, 377
Categories, 212, 373
Cell, 373, 497
Cell means, 510
Cell probabilities, 212, 373
Cells, 212
Census, 36
Center line (CL)
    for the ranges (R chart), 556
    of R chart with known parameters,
       553
    of control chart, 553
    of three-sigma control chart for aver-
       ages, 556
Central limit theorem, 222
    sample mean, 223
central limit theorem, 172
Central value, 27, 32
Chebyshev's inequality, 130
Class
    boundaries, 23
    frequency, 23
    intervals, 23
    marks, 23
    widths, 23
Coefficient of determination, 396
Coefficient of multiple determination, 456
Column space of design matrix, 450
Combinations, $C_{n,k}$, 90
Complement of event A, 74
Complete $a \times b$ factorial experiment, 509
Complete trial, 85
Completely randomized design (CRD), 470
Components of variance, 488
Conditional
    distribution of Y given $X = x$, 209
    expectation
       continuous random variable, 257
       discrete random variable, 210
    probability, 94
    probability of the event $Y = y$ given
       that $X = x$, 209

Conditional density, 257
Confidence interval
    for mean of an arbitrary distribution,
        300
    lower one–sided, 297
    normal distribution, 296, 299
    population proportion, 307
    variance, 302
Confounding, in a 2^k factorial experiment,
    537
Contingency table, 377
Contingency table, $r \times c$, 378
Continuity correction, 182
Continuously compounded rate of return,
    234
Contrast, 483, 484
Contrasts, sum of, 536
Control chart, 551
    for averages, three-σ control chart,
        553
    for averages, unknown parameters, 556
    for sample ranges (R chart), 553, 556
Correlation
    analysis, 416
    coefficient, 38, 218
    negative, 39
    positive, 38
Covariance, 215
Cox-Ross-Rubinstein model for stock prices,
    142
Craps, dice game, 75
Critical value $x(\alpha)$, 176
Critical values, of the normal distribution,
    175
Cumulative frequency function, 18
Cumulative sum control chart (CUSUM),
    565
Cutoff value, 326

Data
    categorical, 5
    definition, 6
    multivariate, 5, 7
    time series, 5
    univariate, 5
De Morgan's laws, 76
Defining contrast, 536
Defining relation, 536
Degrees of freedom (DF)
    ANOVA, 475
    SSR, 401

SST, 406
Deming's red bead experiment, 4, 560
Design matrix, 442
DF as in degrees of freedom, 408
Dichotomous population, 131
Discount factor, 235
Distribution
    bell shaped, 172
    beta, 188
    binomial, 135
    bivariate normal, 257
    chi-square, 187, 280
    empirical distribution function, 19
    Erlang, 187
    exponential, 163
    F, 285
    frequency, 18
    function (df), 114
    gamma, 185
    gaussian, 172
    geometric, 119
    hypergeometric, 130
    lognormal, 178
    marginal, 208
    multinomial, 213, 374
    non-central t, 341
    normal, 172
    of a variable defined on a population,
        17
    parent, 220
    Poisson, 146
    service, 243
    standard normal, 174
    uniform, 165
        discrete, 126
    Weibull, 188
Distributive law, 76
Diversification, 34, 230
Dot plot, 21
Dot subscript notation, 378
Doubling the bet, a gambling strategy, 124
Drift parameter, in geometric Brownian
    motion, 227

Effect, importance of, 527
Effect, of factor level, 510
Effective annual interest rate, 234
Efficient market hypothesis, 140, 224
Empirical distribution function, 20
Empirical probability function $\hat{f}(x)$, 80
Empirical probability measure, 80

Environmental Protection Agency (EPA), 361
Error sum of squares, 474
Estimate, 293
Estimated
    regression function, 395
    standard error ith residual, $s(e_i)$, 460
    standard error $s(\hat{\theta})$, 294
    standard error $s(\overline{X})$, 294
Estimator
    bias of, 294
    biased, 294
    consistent, 295
    definition, 293
    maximum likelihood (MLE), 295
    method of moments, 310
    minimum variance unbiased (MVUE), 295
    MLE for θ, 312
    point, 292
    standard error of, 294
    unbiased, 294
Event A, 73
Expected frequency, 213, 374
Expected value, 121
Experiment, 72
Experimental design, 470, 498
    balanced, 472
    unbalanced, 480
Experimentwise error rate, 487
Explanatory variable, 389
Exploratory data analysis, 18
Extra sum of squares, 458

Factor, 470
Factor levels, 470
Factorial, 87
Fair game, 123
Favorable game, 123
Finite population correction factor, 133
Fitted value, 392
Fractional contrast, 536
Fractional replication, 535
Full model, 458
Function of the random variable X, 120
Fundamental theorem of mathematical statistics, 222, 352
Funnel shape, of residual plot, 519
Future value, 232

Gamma function $\Gamma(\alpha)$, 186

Generator, of design, 536
Geometric Brownian motion, 227
Geometric series, 119
Grand mean, 472
Greatest integer function, 228

Half fraction of the 2^k factorial, 535
Histogram, 22
Honest significant difference (HSD), 486
Horizontal bar chart, 18
House odds, 124
House percentage, 123

Incomplete blocks, 537
Independent
    events, 99
    samples, 344
Independent and identically distributed random variables (iid), 220
Input process, to a queue, 243
Inter-arrival time, 246
Interactions, 513
    between factors, 511
    between treatments and blocks, 503
    completely confounded with blocks, 538
    effect, 513
Intercept, of regression line, 399
Interest
    compound, 233
    simple, 232
Interquartile range (IQR), 30
Intersection of A and B, 75
Intersection of events, 76
Intrinsically linear model, 410

Joint distribution function, 207
Joint probability mass function (jpmf), 206

K-out-of-n system, 248

Law of large numbers, 220, 222
Least significant difference (LSD), 484
Least squares method, 393
Least squares problem, matrix formulation, 450
Limits of variation attributable to the system, 183
Linear functions, 37
Logarithmic return, 143
logistic distribution, 193

Lower bound on a population parameter, 297
Lower control limit ($LCL_{\overline{p}}$), 559
Lower control limit ($LCL_{\overline{X}}$), 552

Main effect, 524
Main effects model, 511
Margin of error, 298
Marginal
    column frequencies, 378
    distribution function (discrete case), 207
    probability density functions, 252
    row frequencies, 378
Market model, 417
Market risk, 70
Matrix
    estimated variance–covariance, $s^2(\hat{\boldsymbol{\beta}})$, 447
    hat $\boldsymbol{H}$, 444
    random, 446
    transpose of, 443
    variance–covariance $Cov(\boldsymbol{Y})$, 446
Maturity
    of option, 237
Maximum contaminant level (MCL), 361
Mean square
    due to regression (MSR), 455
    error (MSE), 407, 455, 476
    treatment (MSTr), 476
Mean time before failure (MTBF), 32
Mean vector, $E(\boldsymbol{Y})$, 446
Mean-standard deviation diagram, 36
Median, 26
    of distribution, 167
Missing values, 54
Mode, 18
Model checking, 411
Model statement, 390, 451
Moment generating function (mgf)
    continuous random variable, 171
    discrete random variable, 146
Moments of a random variable, 125
MSR, mean square due to regression, 407
Multicollinearity, 458
Multinomial coefficient, 93, 94
Multinomial experiment, 212
Multiple linear regression, 441, 450
Multiplication principle, in counting, 84
Mutually disjoint events, 75

Natural, throwing a, 74
No interaction model, 511
Nodes, of a tree, 86
Non-parametric statistics, 372
Normal approximation to the binomial distribution, 179
Normal equations, 443
Normal probability density function, 172
Normal probability plot, 351
Null distribution, 326
Null hypothesis, 326
Null value, 326

Observed frequency, 213
One price axiom, 239
One sided alternatives, 326
Option, 237
    American, 237
    call, 237
    European, 237
    expiration date of, 237
    payoff, 238
    put, 237
Order statistics, 18
Outcome, 73
Outlier, 30
Overall F ratio, 529
Overall mean, 472
Overbooking, 184

P-value, 336
Paired samples, 344
Parallel system, 248
Parameter space Θ, 293
Pareto distribution, 169
Partial F test, 458
Partition, of sample space, 75
Pascal triangle, 92
Path, of a tree, 86
Payoff function, 238
Percentiles
    of distribution, 167
    of empirical distribution function, 26
Permutations, $P_{n,k}$, 87
Piecewise defined function, 20
Point, throwing a, 74
Poisson approximation, 145
Poisson process, 245
Polynomial regression, 451
Population, 5
Population mean, 278

Population parameters, 36
Population proportion p, 293
Population variance, 278
Portfolio, 39, 230
   definition, 10
   diversification, 40
   efficient
      example of, 36
   inefficient, 35, 231
      example of, 36
   market, 14, 42, 70
      example of, 40
Power function, 329
Power of a test, 328
Predicted value, 392
Prediction interval, 405
Predictor variable, 389
Present value, 235
Present value analysis, 232
Principal, 232
Principal block, 538
Probability density function (pdf), 162
Probability function (pf), 114, 116
Probability histogram, 116
Probability mass function (pmf), 114
Probability measure, 78
   definition, 81
   discrete, 79
   equally likely, 78
Product of two experiments, 84
Product rule, for computing a probability, 95
Proportion of the total variability explained by the model, R^2, 397, 456
Put option, 237

Q–Q plot, 352
Quantile, of empirical distribution function, 26
Quartiles, 28
Queueing system, 243
Quintiles, 27

Random number generator, 190
Random sample, 84
   mathematical model of, 220
   taken from a df $F(x)$, 221
Random sampling, 71
Random telegraph signal, 245
Random variable
   Bernoulli, 119

   continuous, 162
   definition, 114
   discrete, 114
   expected value(continuous), 168
   functions of, 189
   payoff, 117
   range of, 114
Random variables
   iid sequence of, 212
   uncorrelated, 218
Random vector, 206, 446
   jpmf, 206
   uniformly distributed, 253
Random walk model, 224
Randomization, 470
Randomized complete block design (RCBD), 498
Recurrence relation, 91
Reduced model, 458
Regression
   coefficients, 394
   function (true), 399
   line, 41
   model, 390
   parameters, 394
Regression of Y on X, 257
Regression parameters, 441
Regressor variable, 389
Rejection region, 326
Relative class frequency, 23
Relative frequency histogram, 24
Reliability, 77
Replicates, 470
Replicating portfolio, 239
Residual, 392
   plots, 412
   studentized, 412
      normal probability plot of, 462
Response variable, 389
Return
   negative, 13
   of an investment, 12
   positive, 13
   rate of, 13
   risk free, 14
   total, 13
Risk
   averse, 233
   free return, 232
   neutral, 233
   neutral probability measure, 233, 239

of an investment, 33
seeking, 233
tolerance, 233
total, of a stock, 419
unique, of a stock, 419
Root, of a tree, 86

Safe life, 26
Sample, 5
correlation, 38
covariance, 38
distribution function, $\hat{F}_n(x)$, 20
mean, 32, 220, 223
normal distribution, 278
point, 73
proportion, 220
range, 18, 553
size, 6
space, 72
space (discrete), 73
standard deviation, 33, 34, 282
total, 220, 222
normal distribution, 278
unordered, 90
variance, 34, 282
Sample proportion $\hat{p}$, 185
Sampling, biased, 71
Sampling, from a normal population, 277
Scatter plot, 38, 390
Securities, 10
Series system, 248
Shapiro–Wilk test, 352
Sharp symbol $\sharp(A)$, 19
Shooter, craps, 73
Significance level of the test, 327
Simple linear regression model, 398
Single-factor experiment, 470
Size of the rejection region, 327
Skewed positive, 21
Slope, of regression line, 399
Sorting, data, 18
Source of variation, 408
Special notation, 523
SSE, error sum of squares, 392, 397
SSR, regression sum of squares, 397
SST, total sum of squares, 397
Standard assumptions, of linear regression, 399
Standard deviation, of random variable, 128
Standard error of prediction, 448, 463

Standard error of sample mean, 294
Standard error of the mean $\sigma(\overline{X})$, 279
Standard order, of factor-level combinations, 525
Standardized normal scores, 353
Statistic, 292, 293
Statistical control, definition, 552
Statistical hypothesis, 325
Statistical inference, 71
Statistically significant, 337
Statistics, 1
Statistics, inferential, 4
Stem and leaf plots, 21
Stochastic process, 226, 245
Stock prices
lognormal model for, 177
Stocks
aggressive, 420
defensive, 420
Strike price, 237
String, of symbols, 86
System function, 248

Tail
of distribution, 21
Tail probability
of binomial distribution, 188
of normal distribution, 175
Test
chi-square test of homogeneity, 381
chi-square test of homogeneity, 380
chi-square test of independence, 378
lower tailed, 329
of a statistical hypothesis, 326
one sided, 329
paired sample t, 346
upper tailed, 329
Wilcoxon rank sum, 347
Ties, in Wilcoxon rank sum test, 349
Time series, 9
Time series, plot, 9
Time value of money, 235
Tolerance limits, 554
Treatment, 470
Treatment effect, 472
Treatment sum of squares, 474
Tree diagrams, 86
True odds, 124
Tukey's multiple comparison procedure, 486, 507
Two sample t test, 346

Two way table, 377
type I error, 326
type II error, 326
    $\beta(\mu)$, 328

Unfavorable game, 123
Union of events, 74, 76
Upper control limit ($UCL_{\overline{p}}$), 559
Upper control limit ($UCL_{\overline{X}}$), 552
Urn model, 131

Value at risk (VaR), 303
Variable defined on a population, 6
Variance
    continuous random variable, 168
    discrete random variable, 128
    pooled estimator of, 304
    shortcut formula, 128
Variation
    common cause, 552
    special cause, 552
Vector notation, 312
Vectors
    distance between two of them, 449
    length of, 449
    orthogonal, 449
    scalar product, $\boldsymbol{x'y}$, 449
Venn diagrams, 76
Volatility, 34, 178, 258, 291
    in geometric Brownian motion, 227

Waiting time until the first success, 119